Advances in Neurobiology

Volume 9

Series Editor

Arne Schousboe

For further volumes:
http://www.springer.com/series/8787

Robert K. Yu • Cara-Lynne Schengrund
Editors

Glycobiology of the Nervous System

Springer

Editors
Robert K. Yu
Department of Neuroscience
and Regenerative Medicine
Medical College of Georgia
Georgia Regents University
Augusta, GA, USA

Cara-Lynne Schengrund
Department of Biochemistry
and Molecular Biology
The Pennsylvania State University
College of Medicine
Hershey, PA, USA

ISSN 2190-5215 ISSN 2190-5223 (electronic)
ISBN 978-1-4939-4637-2 ISBN 978-1-4939-1154-7 (eBook)
DOI 10.1007/978-1-4939-1154-7
Springer New York Heidelberg Dordrecht London

Printed on acid-free paper

Springer is part of Springer Science+Business Media (www.springer.com)

Preface

The growing recognition of the structural complexity and functional importance of glycoconjugates, including glycoproteins, glycolipids, and glycosaminoglycans, in biological systems has prompted the need for intensive research in recent years. These molecules are known to not only serve as cell surface biomarkers at various stages of cellular differentiation and proliferation but also play important functional roles in cell–cell recognition, adhesion, and signal transduction. Despite the critical roles of these molecules in biological systems, their investigation remains an under-explored and under-appreciated area for neurobiological exploration. Some of the reasons for the lack of attention could be due to the structural diversity of these molecules and their temporal and spatial expression in tissues, especially in the nervous system. To promote more attention to research on glycoconjugates in the nervous system, the Editors organized a Colloquium entitled "Glycobiology of the Nervous System" at the 43rd meeting of the American Society for Neurochemistry, Baltimore, MD. This prompted Springer Publishing Company to endorse the publication of this monograph with the objective of introducing students and researchers to this fascinating and important facet of biological sciences. We are particularly encouraged by the recent publication of a National Academy of Sciences-sponsored Committee Report "Transforming Glycoscience, a Roadmap for the Future" that reiterates the importance and impact of glycomics and glycosciences (The National Academies Press 2012). In the present volume, we have assembled a broad range of topics, contributed by experts knowledgeable in the specific areas to provide a framework on glycobiology as related to neurobiology in health and disease. It is our sincere hope that the monograph with 25 topics on the chemistry, structure, and biological functions of glycoconjugates will provide a stimulus for further research on this important topic.

Glycomics, youngest of the "omics" and perhaps the most difficult to study, deals with identification of sugar moieties added to proteins or lipids and their functional roles. Unlike proteins that are synthesized using a template, addition of sugar moieties is dependent upon the presence of specific glycosyltransferases, sugar nucleotides, and the environment within the cell. Glycoconjugates formed upon addition of one sugar to another on either a protein or lipid can have more variability than compounds formed by addition of one amino acid to another (Chap. 1). The importance of an appropriately functioning nervous system is a given, but that functioning can be severely disrupted by errors in glycosylation. An introduction to the cellular complexity of the central nervous system (CNS) is provided in Chap. 2. Understanding the interplay between various cell types, such as glia and neurons, is essential if we are to understand neurodegenerative diseases and develop approaches for their treatment. Specifics regarding the pathways used by cells to synthesize N- and O-linked glycoconjugates including glycosaminoglycans are provided in Chaps. 3–5.

Over time a number of methods have been developed to enhance our ability to characterize glycoconjugates. These range from the use of glycan-targeted antibodies and lectins to indicate what types of glycans are present (Chap. 6) to the use of mass spectrometry to identify new glyco-epitopes and to confirm structural details (Chap. 7) and nuclear magnetic resonance spectroscopy which enables one to define the stereochemistry of carbohydrates comprising the glycan portion of a molecule (Chap. 8). The latter is also capable of providing solution conformation of a glycoconjugate, which is important for deciphering its biological function.

Accumulating evidence indicates that glycoconjugates play a significant role in neural development. Expression of both glycoproteins and glycolipids changes during development where they function in cell proliferation, differentiation, interactions, migration, and signal transduction (Chap. 9). Because glycosylation is affected by the activity of both glycosyltransferases and glycosidases (glycohydrolases) (Chap. 10), understanding the mechanism by which their activities are controlled in order to maintain expression of specific glycans during development is essential. Defining the roles of sialic acid-binding lectins, siglecs, such as MAG (myelin-associated glycoprotein), SMP (Schwann cell myelin protein), and neural cell adhesion molecules (NCAM), has enhanced our understanding of myelination and of the problems such interactions can cause in axon regeneration (Chap. 11). More detailed information about the interaction of oligodendrocytes and myelin is provided by identification of interactions between the carbohydrate components of galactosylceramide and sulfatide that provide transmembrane signaling. Interestingly, both GSLs are also needed for normal neuronal function (Chap. 12).

The observations that carbohydrate–carbohydrate interactions occur and that glycosphingolipids (GSLs) are enriched in lipid rafts, sites enriched in molecules needed for signal transduction, indicate the need to be aware of the roles played by glycans (Chap. 13). Research into the possible roles of lipid raft-associated GSLs has led to the hypothesis that in some cases their function is nonspecific and can be carried out by multiple GSLs while others are specific (Chap. 14). At a more basic level, possible roles for GSLs and N-glycosylated glycoproteins in the regulation of ion transport are being investigated (Chap. 15). The exciting observation that failure to appropriately modify specific proteins by addition of single O-GlcNAcs can affect learning and memory (too few) or the neurodegeneration seen in disease (too many or too few) points out just how important characterization of the glycome is to our understanding of the nervous system (Chap. 16). Specific roles of N-glycans in the regulation of neural transmission is a current field of investigation.

Because results obtained from in vitro studies can vary with cell type and the metabolic state of the cell the study of in vivo model organisms such as Drosophila can be advantageous for looking at conserved functions of N-glycosylation (Chap. 17). As our understanding of specific pathways affected by glycans develops it may provide clues that can be applied for developing treatments to ameliorate clinical problems that arise from defects in those pathways.

The ability of oligo-/poly-saccharides to function as ligands for proteins and/or other saccharides has been co-opted by various pathogens. Many viruses, bacteria, and bacterial toxins use either a carbohydrate-binding protein on their surface or a saccharide recognized by the target cell, as a first step in the infection process (Chap. 18). Development of inhibitors of this binding is an expanding area of research. While pathogens can use carbohydrate–carbohydrate and carbohydrate–protein interactions to facilitate infection, accruing evidence indicates that glycoconjugate changes during aging may contribute to age-related diseases (Chap. 19). In this regard, the ability of ganglioside GM1 to ameliorate damage to the nervous system has been extensively studied. Of particular importance was the observation that it might be useful in treating patients with Parkinson's disease (PD) (Chap. 20). While exogenous GM1 was used in the treatment of PD, inborn errors affecting GSL catabolism, resulting in their accumulation in CNS neurons and glia, can have severe consequences on affected individuals (Chap. 21). Current investigations are studying the possibility that caloric restriction coupled with use of inhibitors of GSL synthesis might provide relief in those instances where enzyme replacement therapy is not yet a viable option (Chap. 22). Transformation of cells may be accompanied by changes in their glycan composition. An example of this is the expression of ganglioside GD2 by neuroblastomas, a GSL that can be detected in the circulation prior to detection of recurrent tumors (Chap. 23). Evidence indicates that the ganglioside patterns of these tumors may provide an insight into their lethality and has been hypothesized to correlate with the degree of differentiation of the cells.

The question of whether the CNS is "immune privileged" has been re-evaluated with the result that it is now thought to interact with the immune system and that these interactions are needed to maintain neuronal function and aid in control of CNS infection and injury. While galactose-binding lectins (galectins) participate in a variety of functions in the CNS, they also have important immune modulatory functions and may be of use in the treatment of inflammatory diseases (Chap. 24). Neuroimmunological diseases can affect the nervous system and a number of peripheral neuropathies have been identified in which the auto-antibodies recognize specific glycan structures. Knowledge of the biology of the glycan recognized has enhanced our ability to understand resultant clinical symptoms and in some cases to develop successful treatments (Chap. 25).

From this brief introduction to the material covered it can be seen that understanding the role(s) of specific glycans is essential for understanding cell function. As one reads the chapters, one will note that despite the progress made the determination of glycan structures is not as easy as determining the amino acid sequence of a protein. The fact that glycan synthesis is not encoded by a template but depends on the expression of glycosyltransferases and glycosidases and the environment in which they exist further complicates the issue as it can result in molecules of a specific protein being glycosylated differently. If nothing else it should be obvious that errors in glycosylation and degradation can have a severe impact on development and function of the CNS and that it is only when we understand the biological functions of specific glycans that we will be able to develop treatments to correct those errors.

Finally, we wish to thank all the authors for their contributions, which made this monograph possible. We are also deeply indebted to the staff of Springer whose invaluable logistical support facilitated its publication. It is our sincere hope that readers find this a useful source of information about the glycan-decorated molecules that comprise the fourth set of "omics," glycomics, methods for their study, and their functions in the highly heterogeneous nervous system.

Hershey, PA, USA Cara-Lynne Schengrund
Augusta, GA, USA Robert K. Yu

Contents

Contributors

Susumu Ando Tokyo Metropolitan Institute of Gerontology, Tokyo, Japan

Soka Royal Care Center, Saitama, Japan

Hisashi Ashida Faculty of Biology-Oriented Science and Technology, Kinki University, Kinokawa-shi, Wakayama, Japan

Massimo Aureli Department of Medical Biotechnology and Translational Medicine, University of Milano, Segrate, Milan, Italy

Rosaria Bassi Department of Medical Biotechnology and Translational Medicine, University of Milano, Segrate, Milan, Italy

Veer P. Bhavanandan Department of Biochemistry and Molecular Biology, The Pennsylvania State University College of Medicine, Hershey, PA, USA

Simpsonville, SC, USA

Erhard Bieberich Department of Neuroscience and Regenerative Medicine, Medical College of Georgia, Georgia Regents University, Augusta, GA, USA

Joan M. Boggs Molecular Structure and Function Program, Research Institute, Hospital for Sick Children, Toronto, ON, Canada

Department of Laboratory Medicine and Pathobiology, University of Toronto, Toronto, ON, Canada

Hung-Lin Chen Institute of Biomedical Sciences, Academia Sinica, Taipei, Taiwan

James R. Connor Department of Neurosurgery, Penn State Milton S. Hershey Medical Center, Hershey, PA, USA

Glyn Dawson Department of Pediatrics, University of Chicago, Chicago, IL, USA

Miriam S. Domowicz Department of Pediatrics, University of Chicago, Chicago, IL, USA

Keiko Furukawa Department of Biochemistry II, Nagoya University Graduate School of Medicine, Nagoya, Japan

Department of Biomedical Sciences, College of Life and Health Sciences, Chubu University, Kasugai, Japan

Koichi Furukawa Department of Biochemistry II, Nagoya University Graduate School of Medicine, Nagoya, Japan

Shinji Go Division of Glycopathology, Institute of Molecular Biomembranes and Glycobiology, Tohoku Pharmaceutical University, Sendai, Japan

D. Channe Gowda Department of Biochemistry and Molecular Biology, The Pennsylvania State University College of Medicine, Hershey, PA, USA

Gerald W. Hart Department of Biological Chemistry, Johns Hopkins University School of Medicine, Baltimore, MD, USA

Karie A. Heinecke Biology Department, Boston College, Chestnut Hill, MA, USA

Yoshio Hirabayashi Laboratory for Molecular Membrane Neuroscience, Riken Brain Science Institute, Saitama, Japan

Jin-Ichi Inokuchi Division of Glycopathology, Institute of Molecular Biomembranes and Glycobiology, Tohoku Pharmaceutical University, Sendai, Japan

Yutaka Itokazu Department of Neuroscience and Regenerative Medicine, Medical College of Georgia, Georgia Regents University, Augusta, GA, USA

Charlie Norwood VA Medical Center, Augusta, GA, USA

Kenji Kanekiyo Disease Glycomics Team, Systems Glycobiology Research Group, RIKEN-Max Planck Joint Research Center, Global Research Cluster, RIKEN, Hirosawa, Wako, Saitama, Japan

Koichi Kato Institute for Molecular Science and Okazaki Institute for Integrative Bioscience, Okazaki, Japan

Graduate School of Pharmaceutical Sciences, Nagoya City University, Nagoya, Japan

The Glycoscience Institute, Ochanomizu University, Tokyo, Japan

Kay-Hooi Khoo Institute of Biological Chemistry, Academia Sinica, Taipei, Taiwan

Shinobu Kitazume Disease Glycomics Team, Systems Glycobiology Research Group, RIKEN-Max Planck Joint Research Center, Global Research Cluster, RIKEN, Hirosawa, Wako, Saitama, Japan

Yasuhiko Kizuka Disease Glycomics Team, Systems Glycobiology Research Group, RIKEN-Max Planck Joint Research Center, Global Research Cluster, RIKEN, Hirosawa, Wako, Saitama, Japan

Stephan Ladisch Center for Cancer and Immunology Research, Children's National Medical Center, Washington, DC, USA

George Washington University School of Medicine, Washington, DC, USA

Olof Lagerlöf Department of Biological Chemistry, Johns Hopkins University School of Medicine, Baltimore, MD, USA

Robert W. Ledeen Department of Neurology and Neurosciences of Rutgers, RBHS, New Jersey Medical School, The State University of New Jersey, Newark, NJ, USA

Department of Pharmacology and Physiology, RBHS, New Jersey Medical School, The State University of New Jersey, Newark, NJ, USA

Yu-Teh Li Department of Biochemistry and Molecular Biology, Tulane University School of Medicine, New Orleans, LA, USA

Fang Liao Institute of Biomedical Sciences, Academia Sinica, Taipei, Taiwan

Teng-Nan Lin Institute of Biomedical Sciences, Academia Sinica, Taipei, Taiwan

Fu-Tong Liu Institute of Biomedical Sciences, Academia Sinica, Taipei, Taiwan

Yihui Liu Center for Cancer and Immunology Research, Children's National Medical Center, Washington, DC, USA

Nicoletta Loberto Department of Medical Biotechnology and Translational Medicine, University of Milano, Segrate, Milan, Italy

Pablo H.H. Lopez Laboratorio de Neurobiología, Instituto de Investigación Médica Mercedes y Martín Ferreyra INIMEC-CONICET, Universidad Nacional de Córdoba, Cordoba, Argentina

Facultad de Psicología, Universidad Nacional de Córdoba, Córdoba, Argentina

Douglas R. Martin Scott-Ritchey Research Center and Department of Anatomy, Physiology and Pharmacology, Auburn University College of Veterinary Medicine, Auburn, AL, USA

Laura Mauri Department of Medical Biotechnology and Translational Medicine, University of Milano, Segrate, Milan, Italy

Valentina Murdica Department of Medical Biotechnology and Translational Medicine, University of Milano, Segrate, Milan, Italy

Martha C. Nowycky Department of Pharmacology and Physiology, RBHS, New Jersey Medical School, The State University of New Jersey, Newark, NJ, USA

Yuki Ohkawa Department of Biochemistry II, Nagoya University Graduate School of Medicine, Nagoya, Japan

Yusuke Ohmi Department of Biochemistry II, Nagoya University Graduate School of Medicine, Nagoya, Japan

Vladislav M. Panin Department of Biochemistry and Biophysics, Texas A&M University, College Park, TX, USA

Douglas G. Peters Department of Neurosurgery, Penn State Milton S. Hershey Medical Center, Hershey, PA, USA

Alessandro Prinetti Department of Medical Biotechnology and Translational Medicine, University of Milano, Segrate, Milan, Italy

Hannah E. Rockwell Biology Department, Boston College, Chestnut Hill, MA, USA

Maura Samarani Department of Medical Biotechnology and Translational Medicine, University of Milano, Segrate, Milan, Italy

Cara-Lynne Schengrund Department of Biochemistry and Molecular Biology, The Pennsylvania State University College of Medicine, Hershey, PA, USA

J.S. Schneider Department of Pathology, Anatomy and Cell Biology, Thomas Jefferson University, Philadelphia, PA, USA

Nancy B. Schwartz Department of Pediatrics, University of Chicago, Chicago, IL, USA

Department of Biochemistry and Molecular Biology, University of Chicago, Chicago, IL, USA

Hilary Scott Department of Biochemistry and Biophysics, Texas A&M University, College Park, TX, USA

Miguel Sena-Esteves Department of Neurology and Gene Therapy Center, University of Massachusetts Medical School, Worcester, MA, USA

Thomas N. Seyfried Biology Department, Boston College, Chestnut Hill, MA, USA

Sandro Sonnino Department of Medical Biotechnology and Translational Medicine, University of Milano, Segrate, Milan, Italy

Orie Tajima Department of Biomedical Sciences, College of Life and Health Sciences, Chubu University, Kasugai, Japan

Naoyuki Taniguchi Disease Glycomics Team, Systems Glycobiology Research Group, RIKEN-Max Planck Joint Research Center, Global Research Cluster, RIKEN, Hirosawa, Wako, Saitama, Japan

Hugh J. Willison Institute of Infection, Immunity and Inflammation, College of Medical, Veterinary and Life Sciences, University of Glasgow, Glasgow, UK

Glasgow Biomedical Research Centre, University of Glasgow, Glasgow, UK

Gusheng Wu Department of Neurology and Neurosciences of Rutgers, The State University of New Jersey, Newark, NJ, USA

Takumi Yamaguchi Institute for Molecular Science and Okazaki Institute for Integrative Bioscience, Okazaki, Japan

Graduate School of Pharmaceutical Sciences, Nagoya City University, Nagoya, Japan

Yoshiki Yamaguchi Structural Glycobiology Team, Systems Glycobiology Research Group, RIKEN-Max Planck Joint Research Center, RIKEN Global Research Cluster, Wako-City, Japan

Robert K. Yu Department of Neuroscience and Regenerative Medicine, Medical College of Georgia, Georgia Regents University, Augusta, GA, USA

Charlie Norwood VA Medical Center, Augusta, GA, USA

Chapter 1
Introduction to the Complexity of Cell Surface and Tissue Matrix Glycoconjugates

Veer P. Bhavanandan and D. Channe Gowda

Abstract This chapter provides an overview of structures and functions of complex carbohydrates (commonly called glycans) that are covalently linked to proteins or lipids to form glycoconjugates known as glycoproteins, glycolipids, and proteoglycans. To understand the complexity of the glycan structures, the nature of their monosaccharide building blocks, how the monomeric units are covalently linked to each other, and how the resulting glycans are attached to proteins or lipids are discussed. Then, the classification, nomenclature, structural features, and functions of the glycan moieties of animal glycoconjugates are briefly described. All three classes of glycoconjugates are constituents of plasma membranes of all animal cells, including those of the nervous system. Glycoproteins and, particularly, proteoglycans are also found abundantly as constituents of tissue matrices. Additionally, glycan-rich mucin glycoproteins are the major constituents of mucus secretions of epithelia of various organs. Furthermore, the chapter draws attention to the incredible structural complexity and diversity of the glycan moieties of cell surface and extracellular glycoconjugates. Finally, the involvement of the glycans as informational molecules in a wide range of essential functions in almost all known biological processes, which are crucial for development, differentiation, and normal functioning of animals, is discussed.

Keywords Complex carbohydrates • *N*-Glycans • *O*-Glycans • Glycosaminoglycans • Glycoconjugates • Glycoproteins • Glycolipids • GPI anchors • Proteoglycans • Structure and functions of glycoconjugates

---

V.P. Bhavanandan (✉)
Department of Biochemistry and Molecular Biology, H171, The Pennsylvania State University College of Medicine, Hershey, PA 17033, USA

43 Waters Reach Lane, Simpsonville, SC 29681, USA
e-mail: veerbhavan@yahoo.com

D.C. Gowda (✉)
Department of Biochemistry and Molecular Biology, H171, The Pennsylvania State University College of Medicine, Hershey, PA 17033, USA
e-mail: gowda@psu.edu

R.K. Yu and C.-L. Schengrund (eds.), *Glycobiology of the Nervous System*, Advances in Neurobiology 9, DOI 10.1007/978-1-4939-1154-7_1,

1.1 Introduction

Carbohydrates are the most abundant, structurally complex, and functionally diverse organic compounds found on earth. They comprise monosaccharides (e.g., glucose), oligosaccharides (e.g., sucrose, lactose, components of glycoproteins, and glycolipids; see below), and polysaccharides (e.g., starch, cellulose, plant and microbial cell wall polysaccharides, arthropod chitin, and animal glycosaminoglycans). Carbohydrates are important dietary components for animals, including humans, and play crucial roles in energy metabolism as exemplified by glucose homeostasis. In addition to their occurrence as free molecules, when conjugated to proteins and lipids, they form glycoconjugates (glycoproteins, proteoglycans, and glycolipids) (Brooks et al. 2002; Gabius 2009; Taylor and Drickamer 2011; Varki et al. 2008). The oligo- and polysaccharides covalently conjugated to proteins and lipids are referred to as complex carbohydrates or glycans. The glycan moieties of glycoconjugates are structurally complex and are involved in a myriad of functions, which are crucial for differentiation, development, and all aspects of normal functioning of animals. For instance, they constitute the major blood group and other antigens, serve as informational molecules in cell–cell and cell–molecule interactions, function as receptors for biological processes, and assist in protein folding, targeting, and secretion. The glycan moieties of glycoconjugates are also involved in tissue organization, trafficking of lymphocytes, cell signaling, and immune regulation. In many cases, the true functions of glycan moieties are still not fully understood. Research during the past several decades and recent technological developments have deciphered the structures of thousands of glycans and substantially increased our knowledge of their biological roles. However, it is not exaggerating to state that what we know about the functions of these molecules represents only the tip of an iceberg.

Glycoconjugates occur extensively in all cells and tissues of animals, including those of the nervous system (Brooks et al. 2002; Gabius 2009; Taylor and Drickamer 2011; Varki et al. 2008). They are found in tissue matrices and extracellular fluids such as serum, spinal fluid, saliva, and are especially highly abundant in respiratory, gastrointestinal, and urogenital mucus. However, the main reason for the increased interest in these molecules is that they occur ubiquitously on cell surfaces, where they play important roles in many biological processes. The intrinsic plasma membrane glycoconjugates of cells have their glycan moieties projecting outward far and wide and thus are disposed for functional interactions. The anionic sugar residues of glycans, which impart hydrophilicity and bestow negative charges to the cell surface, are important determinants in the social behavior of cells.

In this introductory chapter, we will briefly discuss (1) the nature of the monomeric building blocks of glycans, (2) how they are linked to one another to form glycans, (3) how glycans are conjugated to proteins and lipids to form different glycoconjugates, and (4) provide an overview of the classification, key structural features, and functions of glycans. Since it is impossible to cover all aspects of glycan biology in this brief overview, readers are referred to more detailed books, monographs, and reviews for additional information (Allen and Kisailus 1992; Bertozzi and Rabuka 2008; Brooks et al. 2002; Gabius 2009; Miljkovic 2010; Taylor and Drickamer 2011; Varki et al. 2008; Voet and Voet 2010; Yuriev and Ramsland 2012).

1.2 Monosaccharides: Building Blocks of Glycans

The structural complexity of glycans arises from the fact that monosaccharides, the monomeric units or building blocks of these molecules, have multiple functional groups and exhibit stereoisomerism (Allen and Kisailus 1992; Bertozzi and Rabuka 2008; Brooks et al. 2002; Gabius 2009; Miljkovic 2010; Taylor and Drickamer 2011; Varki et al. 2008; Voet and Voet 2010; Yuriev and Ramsland 2012). Monosaccharides (generally called sugars) are polyhydroxy aldehydes or ketones with the general formula $(CH_2O)_n$ and are referred to as aldoses or ketoses, respectively. Both aldoses and ketoses are classified into different groups based on the number of carbon atoms they contain; for example, tetroses, pentoses, and hexoses are aldoses containing 4, 5, and 6 carbon atoms, respectively. Each of these groups is further classified into D- and L-series. All D-sugars have the same stereochemistry (i.e., identical absolute configuration) as D-glyceraldehyde at the asymmetric carbon atom that is most remote (e.g., C-5 in hexoses; Fig. 1.1) from the carbonyl group. As mentioned above, a detailed discussion of the structures and stereochemistry of various sugars is beyond the scope of this chapter as it is available in the introductory chapters of biochemistry textbooks and in specialty glycobiology books (Bertozzi and Rabuka 2008; Brooks et al. 2002; Gabius 2009; Miljkovic 2010; Taylor and Drickamer 2011; Varki et al. 2008; Voet and Voet 2010). The predominant constituents of the glycan moieties of animal glycoconjugates are hexoaldoses and their derivatives. The structures and stereochemistry of the eight possible

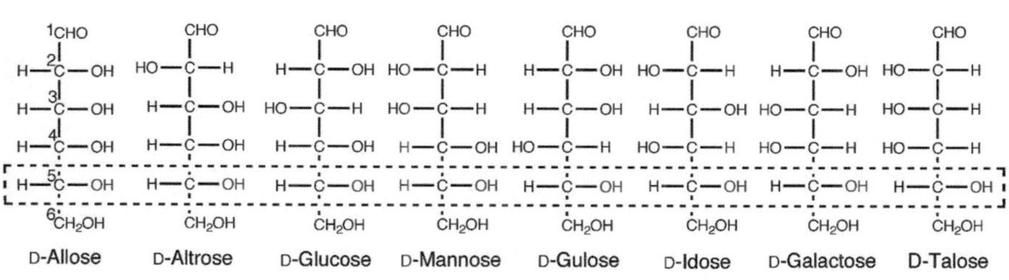

Fig. 1.1 Structures of eight possible D-aldohexoses are shown as open-chain Fischer projection formulas. They all have the same stereochemistry as that of D-glyceraldehyde at C-5, the asymmetric carbon that is most remote from the carbonyl group. The eight L-aldohexoses are the mirror images of these D-sugars. Please see refs. (Brooks et al. 2002; Taylor and Drickamer 2011; Varki et al. 2008; Voet and Voet 2010) for the structures of aldopentoses and other monosaccharides. Ketopentoses and ketohexoses are not present as constituents of the glycan moieties of animal glycoconjugates

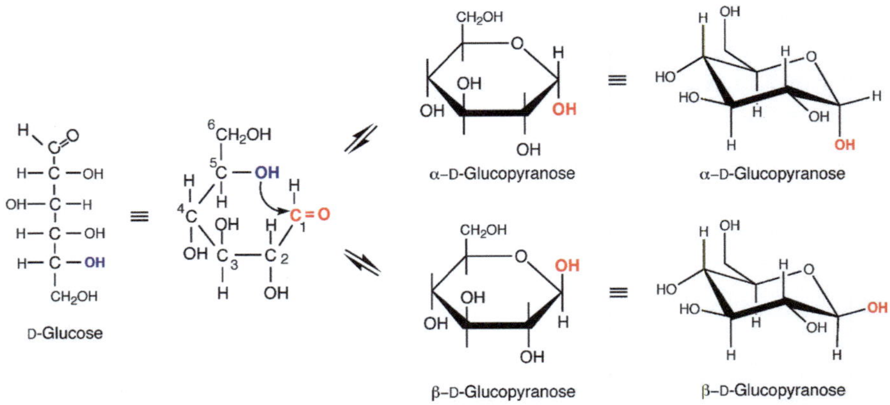

Fig. 1.2 The folding of open-chain monosaccharides (e.g., D-glucose shown here) to relieve carbon bond angle strain facilitates intramolecular nucleophilic condensation of the C-5 hydroxyl group to the carbonyl group. This intramolecular reaction results in the formation of cyclic hemiacetal structures and the creation of a chiral center at carbon atom 1 (C-1). Therefore, the majority of sugars exist primarily as two thermodynamically stable six-membered cyclic structures called α- and β-anomers. Haworth projection formulas and the chair conformational structures of α-D-glucopyranose and β-D-glucopyranose are also shown. Some pentoses and hexoses can exist as five-membered cyclic structures see refs. (Brooks et al. 2002; Taylor and Drickamer 2011; Varki et al. 2008; Voet and Voet 2010)

D-series hexoses in open-chain Fischer projection formula are depicted in Fig. 1.1; the L-series sugars are the mirror images of D-sugars. Of the eight hexoses shown in Fig. 1.1, D-galactose and D-mannose are found widely in glycoproteins. D-Glucose is present in glycolipids but is rarely found in glycoproteins. Glucose is also the biosynthetic precursor for all other sugars, which occur in nature. Similarly, there are four each of the D-series and L-series pentoses, namely, ribose, arabinose, xylose, and lyxose. Of these, xylose occurs in animals as a constituent of proteoglycans.

Because of the tetrahedral bond angle requirement for carbon, sugars are not present as highly strained open-chain linear structures. They are folded in such a way that either C-4 or C-5 hydroxyl groups come in close proximity to the electrophilic carbonyl group and react to form intramolecular hemiacetal bonds; Fig. 1.2 shows the cyclic structure of D-glucose. Thus, five or more carbon atom-containing monosaccharides exist as thermodynamically stable six-membered or five-membered cyclic structures, designated as pyranose or furanose, respectively. The formation of ring structures, which are depicted by Haworth projection formulas, results in the creation of an additional chiral center at C-1, and thus, each sugar exists as two isomers. These isomers are referred to as α- and β-anomers, and the carbonyl carbon of sugar in the ring form is the anomeric carbon. By convention, hexopyranoses in which the –OH group at C-1 and the $-CH_2OH$ group at C-5 have a *trans* or *cis* configuration when depicted by Haworth formulas are named α- and β-anomers, respectively (Fig. 1.2). In aqueous solutions the cyclic anomers are in equilibrium with open chain forms having either a free aldehyde or a keto group.

This allows sugars to display the characteristic properties of aldehydes and ketones by shifting the equilibrium toward the open chain forms. For example, all aldoses exhibit reducing property. If the anomeric hydroxyl group of a sugar in cyclic structure is locked in covalent bond formation, for example, with alcohols forming glycosides, the reducing property of the sugar is lost.

Two sugars differing in configuration at a specific carbon, other than C-1, are referred to as epimers. Thus, glucose and mannose, which differ in configuration only at C-2, are C-2 epimers. Similarly, glucose and galactose, which differ in configuration only at C-4, are C-4 epimers, and D-glucuronic acid and L-iduronic acid (see below) are C-5 epimers. Because mannose and galactose differ in configuration at more than one position, they are not epimers; they are isomers.

The ring carbon and oxygen atoms of six-membered pyranose structures of aldohexoses and their derivatives, including those that constitute glycoconjugate glycans, do not adopt coplanar structures as depicted in Haworth projection formulas (Fig. 1.2). To relieve bond angle strain and steric interactions between bulky functional groups, in most cases, sugars assume "chair" conformations with a large number of different arrangements of their various functional groups in space. The conformations in which bulky substituents occupy equatorial positions (lying in parallel to plane of the ring) are the more stable ones. The stable chair conformations of α- and β-anomers of D-glucose are illustrated in Fig. 1.2.

As mentioned above, D-galactose (D-Gal), D-mannose (D-Man), D-glucose (D-GLC), and D-xylose (D-Xyl) are the unmodified sugars that occur as constituents of glycans in animal glycoconjugates (Brooks et al. 2002; Gabius 2009; Taylor and Drickamer 2011; Varki et al. 2008). In addition, several modified forms of sugars are also common constituents of glycans. These include: 6-deoxy-L-galactose known as L-fucose (L-Fuc); the C-6 carboxyl derivative of D-Glc and L-idose (L-Ido), called D-glucuronic acid (D-GlcA) and L-iduronic acid (L-IdoA), respectively; and the *N*-acetylated forms of 2-amino-2-deoxy-D-glucose (D-GlcN) and 2-amino-2-deoxy-galactose (D-GalN), called *N*-acetyl-D-glucosamine (D-GlcNAc) and *N*-acetyl-D-galactosamine (D-GalNAc), respectively. Other types of modified sugars found in animal glycans have certain hydroxyl groups acetylated, sulfated, or phosphorylated. An unusual 9-carbon monosaccharide acid widely distributed in animals is D-neuraminic acid (5-amino-3,5-dideoxy-D-nonulosonic acid) (Yu and Ledeen 1969), which has –COOH, –C=O, and $–NH_2$ functions (Schauer 2004; Varki 1992; Varki and Schauer 2008). The amino group is either acetylated ($–NHCOCH_3$) or glycolylated ($–NHCOCH_2OH$) resulting in *N*-acetylneuraminic acid (NANA, NeuAc) or *N*-glycolylneuraminic acid (NeuGc); humans synthesize only NeuAc, but traces of NeuGc are found as a result of eating meat (Bardor et al. 2002). Some of the hydroxyl groups of these neuraminic acids are either acetylated, methylated, or sulfated forming a family of more than 20 derivatives, which are collectively called sialic acid (SA) (Schauer 2004; Varki 1992). Sialic acid occurs abundantly in brain as a constituent of glycolipids (Nagai and Iwamori 1995; Wang et al. 1998).

All these modifications of monosaccharides contribute to the incredible structural diversity of glycans. The monosaccharides and their derivatives commonly found as constituents of glycans of animal glycoconjugate are shown in Fig. 1.3; note that, of these sugars, L-iduronic acid occurs only in proteoglycans.

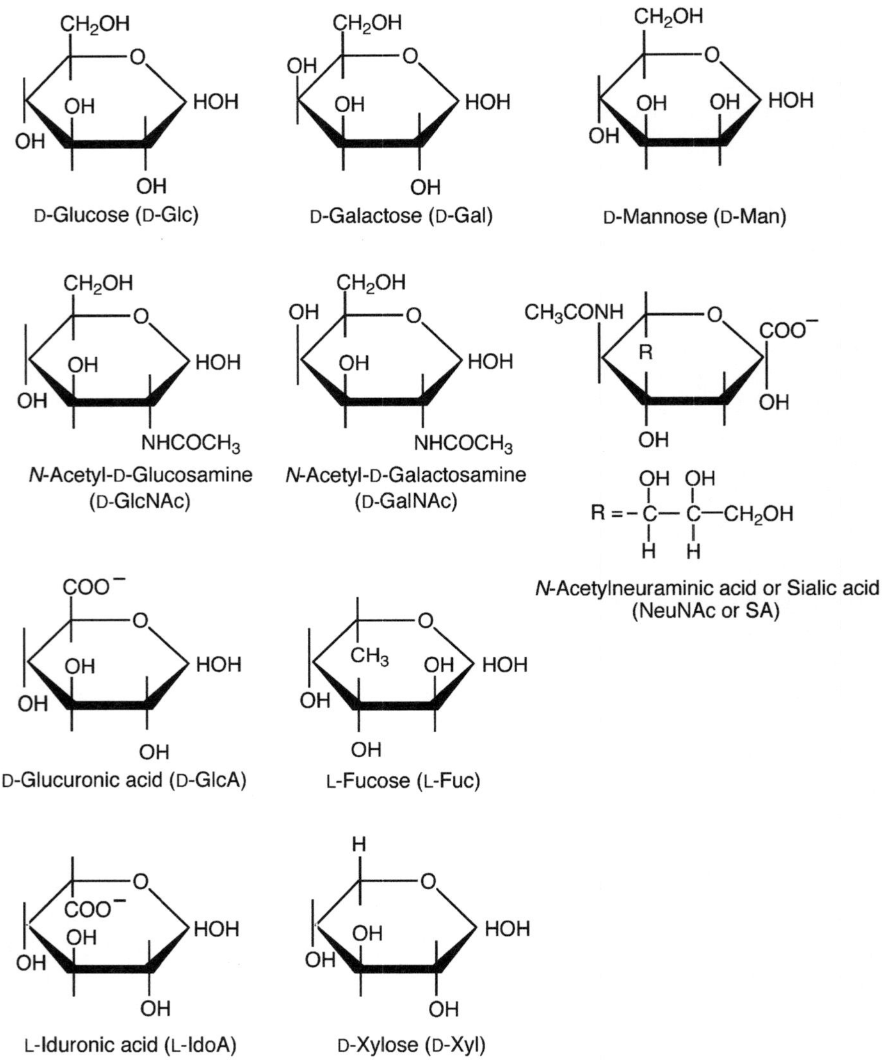

Fig. 1.3 Haworth projection formulas of monosaccharides and their derivatives found in glycans of animal glycoconjugates. Note that of these ten sugars, eight are D-sugars and two are L-sugars

1.3 Formation of Oligosaccharides

The anomeric hydroxyl group of monosaccharides is more reactive than the other hydroxyl groups and undergoes protonation, becoming a relatively good leaving group (Brooks et al. 2002; Miljkovic 2010). This makes C-1 relatively electrophilic and susceptible to attack by negatively charged or electron-rich atoms such as the oxygen of hydroxyl groups. Thus, alcohols and other compounds containing hydroxyl groups can condense with the carbonyl group of monosaccharides, and the

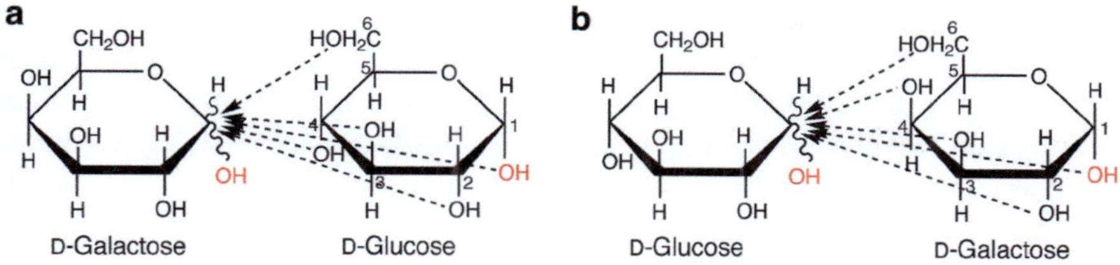

Fig. 1.4 Glycosidic bond formation between the α- or β-anomeric hydroxyl group of D-galactopyranose and the hydroxyl groups on carbons 2, 3, 4, and 6 of D-glucopyranose results in eight different disaccharides with glucose at the reducing end (**a**). When the positions of the sugars are reversed as in (**b**), eight disaccharides with galactose at the reducing end are formed. Four additional nonreducing disaccharides are formed when the anomeric hydroxyls of both sugars are involved in the interaction

bond formed by this condensation is called a glycosidic bond or a glycosidic linkage (Brooks et al. 2002; Gabius 2009; Taylor and Drickamer 2011; Varki et al. 2008). If the attacking nucleophile is a hydroxyl group of another monosaccharide, then the two monosaccharides are linked to each other, forming a disaccharide (Fig. 1.4). When a glycosidic bond is formed by involving the anomeric hydroxyl of one sugar with a non-anomeric hydroxyl of another sugar, the anomeric hydroxyl of the latter sugar is free. Therefore, this sugar retains its reducing property and is called the reducing end of the resulting disaccharide. The other monosaccharide moiety whose anomeric group is involved in the stable glycosidic bond has no reducing power and is called the nonreducing end sugar. The nonreducing and reducing ends of linear oligo- and polysaccharide chains are analogous to the N- and C-terminal amino acids of polypeptides. The α- or β-anomers of one sugar can form two different glycosidic bonds with multiple hydroxyl groups (five in the case of hexopyranoses) of a second sugar to form several different disaccharides. Thus, two molecules of the same sugar, for example, glucose, can form a total of 11 disaccharides, and two different monosaccharides can form 20 disaccharides (Fig. 1.4).

An additional complexity is involved when more than two sugars are attached to form higher oligosaccharides (Bertozzi and Rabuka 2008; Brooks et al. 2002). For example, in the case of trisaccharides that are formed from three molecules of the same sugar, D-Glc, the third molecule can be attached to either the nonreducing glucose of disaccharides, Glc–Glc, to form linear trisaccharides [Glc–Glc–Glc] or the reducing end glucose to form branched trisaccharides [Glc–(Glc)–Glc]. Thus, oligosaccharides have three ways of generating structural diversity: one is by using different hydroxyl groups, the second is by the formation of α- and β-anomeric linkages, and the third is by branching. Thus, three molecules of the same sugar can form 176 trisaccharides, and three different sugars can form 1,056 trisaccharides. Four or more different sugars can form several thousands of tetra- or higher oligosaccharides. Note that linear oligosaccharides will have one each of reducing and nonreducing sugar ends. Whereas, branched oligosaccharides will have one reducing and multiple nonreducing sugar ends and the number of nonreducing ends indicate the degree of branching.

Since several sugars (see Fig. 1.3) and their acetylated, methylated, and sulfated derivatives constitute the glycan moieties of glycoconjugates, the combinatorial diversity of glycan structures is incredibly large and literally thousands of glycans occur in glycoconjugates (Brooks et al. 2002; Gabius 2009; Taylor and Drickamer 2011; Varki et al. 2008). Since a single glycoprotein molecule may contain hundreds of different glycans at different glycosylation sites and even one glycosylation site may contain different glycan chains, determining the structures and understanding functional roles of glycans is challenging. However, recent technological advancements, particularly mass spectrometry (Leymarie and Zaia 2012; Li et al. 2009; Nishimura 2011; North et al. 2009; Orlando 2013; Schiel 2012; Taylor and Drickamer 2011; Zaia 2010), have greatly eased structural determination efforts, and the structures of tens of thousands of glycans have been determined (Taylor and Drickamer 2011; Consortium for Functional Glycomics: http://www.functionalglycomics.org/glycomics/common/jsp/firstpage.jsp). However, as mentioned above, our understanding of the functional roles of glycans is still limited.

1.4 Classification of Glycans

The glycans of glycoconjugates (glycoproteins and proteoglycans) are classified into two major groups: (1) *N*-glycans and (2) *O*-glycans (Bhavanandan and Furukawa 1995; Brockhausen et al. 2008; Brooks et al. 2002; Fukuda 2000; Patsos and Corfield 2009; Stanley and Cummings 2008; Stanley et al. 2008; Taylor and Drickamer 2011; Zuber and Roth 2009). *N*-Glycans are linked via an *N*-glycosidic bond formed between the reducing terminal of GlcNAc and the amide nitrogen atom of asparagine (Asn) residues of proteins that are found in the sequon Asn-Xaa-Ser/Thr, where Xaa is any amino acid other than proline. Although serum glycoproteins such as alpha-1-acid glycoprotein contain exclusively *N*-linked glycans, the vast majority of glycoproteins contain both *N*- and *O*-glycans. Some examples of these are fetuin, immunoglobulin A, human chorionic gonadotropin, and many cell surface glycoproteins. *O*-Glycans are linked via an *O*-glycosidic bond formed between anomeric hydroxyl groups of sugars and the hydroxyl group of serine (Ser) or threonine (Thr) residues of proteins. The sugars involved in *O*-glycosidic linkages are predominantly GalNAc and to lesser extents Man, Fuc, Glc, and Gal. In the case of mucins and the mucin-type glycoproteins of the plasma membrane and secreted glycoproteins, the glycan substituents are linked via GalNAc. In mannan-type yeast glycoproteins, and in certain neuronal glycoproteins and proteoglycans and in a few animal muscle glycoproteins, *O*-glycans are linked through α-Man to Ser/Thr (Brooks et al. 2002; Kleene and Schachner 2004; Krusius et al. 1986; Nakamura et al. 2010; Patsos and Corfield 2009; Stanley and Cummings 2008). In proteoglycans, the glycosaminoglycan polysaccharide chains are *O*-linked via β-D-Xyl to Ser residues of proteins (see below). In proteins such as Notch, coagulation factors, and urokinase-type and tissue-type plasminogen activator, *O*-glycans are linked through α-L-Fuc to Ser/Thr (Brooks et al. 2002; Freeze and Haltiwanger 2008; Gebauer et al. 2008; Luther and Haltiwanger 2009; Patsos and Corfield 2009; Stanley and Cummings 2008). In collagens, Galβ1- monosaccharides and Glcα1-2Galβ1- disaccharides are *O*-linked to the hydroxyl group of hydroxylysines.

In various nuclear and cytoplasmic proteins, a special type of glycosylation involving single β-GlcNAc *O*-linked to Ser/Thr is found in almost all eukaryotes and believed to be involved in the regulation of signaling, transcription, and various other biological processes. This special type of glycan modification is discussed by Lagerlof and Hart in Chap. 16.

1.5 Structural Features of *N*-Glycans

All *N*-glycans have a common pentasaccharide "inner core," consisting of three Man and two GlcNAc (residues shown in blue/bold in Fig. 1.5). The occurrence of this conserved core structure is due to the involvement of common biosynthetic

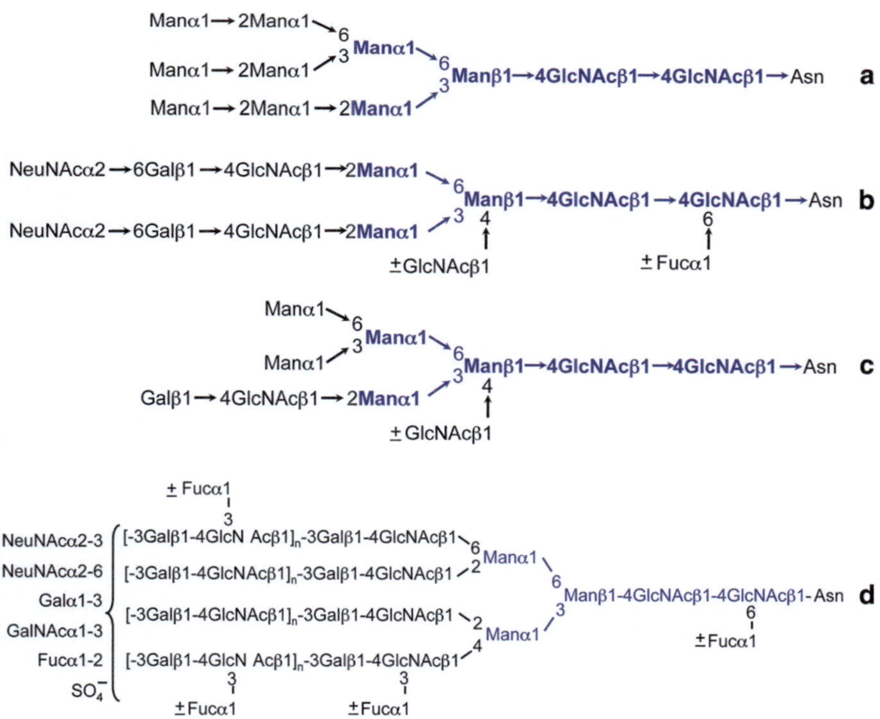

Fig. 1.5 The structural features of *N*-glycans of animal glycoproteins. Because of the involvement of a common biosynthetic lipid-linked oligosaccharide precursor, all *N*-glycans contain a common inner core structure, consisting of three D-mannose (Man) and two *N*-acetyl-D-glucosamine (GlcNAc) residues (shown in *blue/bold*). After transferring to proteins, the common oligosaccharide precursor is modified differently, but retaining the pentasaccharide core, to yield three subclasses of *N*-glycans referred to as high-mannose- or oligomannose-type (**a**), complex- or *N*-acetyllactosamine-type (**b**), and hybrid-type (**c**) *N*-glycans. The complex di-antennary *N*-glycans (**b**) may be further modified to form tri-antennary, tetra-antennary (e.g., **d**), and penta-antennary structures; $n = 2$ to ~50

pathway for all types of *N*-glycan chains found in glycoproteins (Brooks et al. 2002; Fukuda 2000; Stanley et al. 2008; Stanley and Cummings 2008; Taylor and Drickamer 2011; Zuber and Roth 2009). The *N*-glycans are classified into three subgroups based on the types of modifications peripheral to the inner core structures and are referred to as the (1) oligomannose or high mannose type, (2) complex type, and (3) hybrid type (Fig. 1.5). All these subgroups show enormous structural variations due to different types of peripheral modifications, including variation in chain lengths, attachment of different terminal sugars, incomplete addition of terminal sugars, and incomplete chain formation.

Oligomannose-type *N*-glycans contain only Man and GlcNAc residues (Fig. 1.5a), and the heterogeneity in this type of *N*-glycans is due to variations in the numbers and locations of outer Man residues linked to the two α-Man of the inner core. Complex-type *N*-glycans (Fig. 1.5b) exhibit a far wider spectrum of structural variation compared to high-mannose-type *N*-glycans (Brooks et al. 2002; Fukuda 2000; Stanley et al. 2008; Stanley and Cummings 2008; Taylor and Drickamer 2011; Zuber and Roth 2009). Typically, each of the two α-Man residues of the inner core is substituted with one or more *N*-acetyllactosamine (Galβ1-4GlcNAcβ1-) moieties, which form outer chains. Those *N*-glycans that carry two *N*-acetyllactosamine substituents—one chain on each of the α-linked Man are called complex-type di-antennary oligosaccharides (Fig. 1.5b). The glycans that contain more than one *N*-acetyllactosamine substituent on either one or both α-Man are called multi-antennary structures and are referred to as tri-, tetra-, penta-antennary based on the total number of antennas present; an example of a complex tetra-antennary structure is shown in Fig. 1.5d. In *N*-glycans of many animal glycoproteins, the *N*-acetyllactosamine moieties are elongated with repetitive sequential additions of β-GlcNAc and β-Gal, resulting in poly-*N*-acetyllactosamine chains containing two to as many as fifty or more of the repeating disaccharide, -3Galβ1-4GlcNAcβ1- (type 2 *N*-acetyllactosamine), or -3Galβ1-3GlcNAcβ1- (type 1 *N*-acetyllactosamine) units; for example, Fig. 1.5d in which $n=2$ to ~50. The sugar chains are terminated by the substitution of β-Gal with α(2-3)- and/or α(2-6)-linked SA, α(1-2)-linked Fuc, α(1-3)-linked Gal, α(1-3)-linked GalNAc or sulfate groups. The terminal residues within one *N*-glycan may be the same sugar or two or more different α-linked sugars (SA, Fuc, Gal or GalNAc). In some *N*-glycans, terminal β-Gal residues are unsubstituted or even absent, exposing β-GlcNAc as the terminal sugar. Additionally, a wide range of different types of substitutions on the inner and subterminal type 2*N*-acetyllactosamine moieties exist, including substitution of inner GlcNAc with α(1-3)-linked Fuc and that of β-Gal with β(1-6)-linked GlcNAc on which *N*-acetyllactosamine chains can be formed and elongated (Stanley and Cummings 2008). Type 1 *N*-acetyllactosamine (-3Galβ1-3GlcNAcβ1-) structures carrying α(1-4)-linked Fuc substitution on GlcNAc also occur (Brockhausen et al. 2008; Brooks et al. 2002; Fukuda 2000; Stanley and Cummings 2008). Moreover, β-Man of the core structure is substituted with a single GlcNAc, forming bisecting structures (Fig. 1.5). The *N*-glycans of secretory glycoproteins and erythrocyte surface proteins exhibit blood A, B, H, Lewis^a, Lewis^b, and other blood group antigenic structures (Brooks et al. 2002; Schachter and Brockhausen 1992; Cummings 1992;

Stanley and Cummings 2008). Many *N*-glycans of animal cells carry sialyl Lewis^x and related antigens, and their expression is regulated during development and differentiation; highly expressed in fetuses but rarely in adults. Sialyl Lewis^x and related structures are also highly expressed by cancer cells and therefore, are referred to as onco-fetal antigens. The hybrid-type *N*-glycans contain structural features of both oligomannose-type and complex-type oligosaccharides (see Fig. 1.5c). These glycans also show variations in the number of Man substitutions and length and substitutions of the complex-type chain, resulting in multiple structures. Because of all the different modifications mentioned above, literally tens of thousands of *N*-glycans occur as constituents of various glycoproteins.

In the central nervous system, in addition to many of the *N*-glycans described above, some unique structures are present (Bruses and Rutishauser 2000; Fukuda 2000; Gascon et al. 2007; Hildebrandt and Dityatev 2013; Kleene and Schachner 2004; Ledeen and Wu 2009; Stanley and Cummings 2008). For example, the neuronal cell adhesion molecules (N-CAMs) of developing brain contains α(2-8)-linked polysialic acid chains made up of as many as 50 or more sialic acid residues that are attached to β-Gal of one or more outer chains of *N*-glycans. This unique modification is implicated in cell migration, neurite outgrowth, and the development of nervous system. Another example of an uncommon glycan present in neural cells is the HNK-1 antigen, sulfate-3GlcAβ1-3Galβ1-4GlcNAc-, found as a terminal structure in glycan chains (Fukuda 2000; Freeze and Haltiwanger 2008; Gebauer et al. 2008; Kleene and Schachner 2004; Luther and Haltiwanger 2009; Schachter and Brockhausen 1992; Stanley and Cummings 2008). HNK-1 epitope specificity is provided by the terminal sulfate-3GlcAβ1-3Galβ1- moiety (Tokuda et al. 1998). The HNK1 antigen was originally identified in human natural killer cells and subsequently found as an antigen involved in the autoimmune disease, peripheral demyelinative neuropathy (Ariga et al. 1987). The antigen is regulated both temporally and spatially in the developing nervous system (Schwarting et al. 1987) and is found in several neuronal cell adhesion molecules, including N-CAM, myelin-associated protein, contactin, L1, and P0. HNK-1 mediates cell–cell and cell–substrate interactions (Ariga 2011; Fukuda 2000; Kizuka and Oka 2012; Kleene and Schachner 2004; Morita et al. 2008; Stanley and Cummings 2008).

1.6 Structural Features of *O*-Glycans

As discussed above, mucin-type glycans that are linked via α-GalNAc to Ser/Thr of proteins are the most abundant *O*-linked glycans in eukaryotic cells, including cells of the central nervous system (Bhavanandan and Furukawa 1995; Brockhausen et al. 2008; Brooks et al. 2002; Patsos and Corfield 2009; Schachter and Brockhausen 1992; Taylor and Drickamer 2011). The α-GalNAc *O*-linked to Ser/Thr can be substituted with β-Gal and/or β-GlcNAc or with an additional α-GalNAc residue, forming eight distinct core structures (Fig. 1.6). The wide variety of *O*-glycans found in animal cells is formed by the addition of different sugars to the core structures.

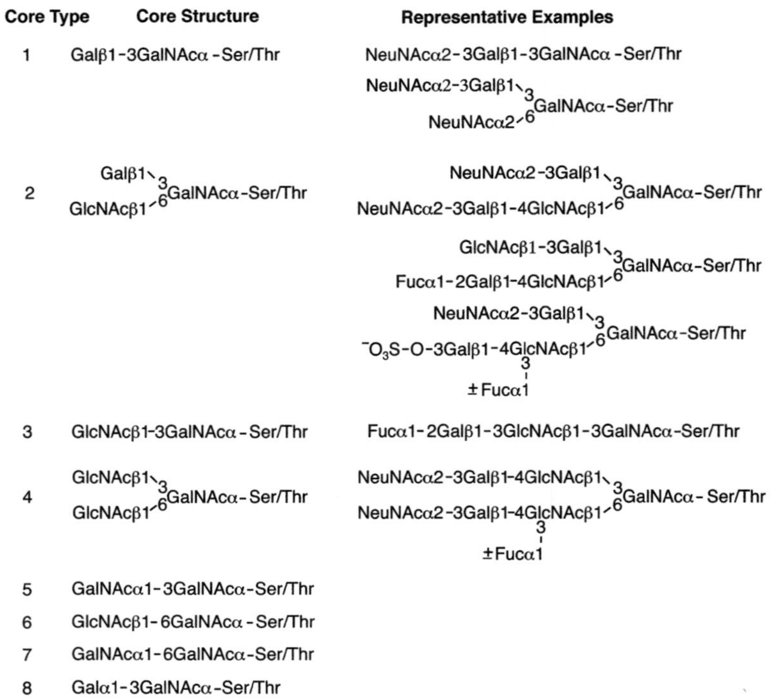

Fig. 1.6 The structures of core sugar moieties of *O*-glycans and structures of representative *O*-glycans found in animal cells and tissues

O-Glycans having core structures 1–4 are widely distributed in animal cells and tissues, and those with core structures 5–8 are rarely found (Brockhausen et al. 2008; Brooks et al. 2002; Fukuda 2000; Schachter and Brockhausen 1992; Taylor and Drickamer 2011). The unsubstituted GalNAcα1 linked to Ser/Thr constitutes the Tn antigen (Brooks et al. 2002). The substitution of α-GalNAcα1 linked to Ser/Thr with sialic acid forms NeuAcα2-6GalNAcα1, the simplest sialylated disaccharide called the sialyl Tn antigen. The core 1 structure, Galβ-1-3GalNAcα1, is known as the Thomsen–Friedenreich antigen or T antigen. The Tn and T antigens are found at low levels in normal tissues, but are highly expressed in certain tumors (Brooks et al. 2002; Varki et al. 2008). The β-Gal residue of core 1 structure Galβ-1-3GalNAcα1-Ser/Thr can be substituted with α(2-3)-linked NeuAc, and the α-GalNAc can be substituted with α(2-6)-linked NeuAc. Further, the β-Gal residue of the core 2 structure can be substituted with α(2-3)-linked NeuAc, and β-GlcNAc can be substituted with β(1-4)-linked Gal to form a type 2 *N*-acetyllactosamine structure (Fig. 1.6). Similarly, both β-GlcNAc of core 4 can be substituted with β(1,3)-linked Gal to form two branches having type 1 *N*-acetyllactosamine structure. As in the case of *N*-glycans, in both core 2 and core 4 structures, the *N*-acetyllactosamine is either terminated by the substitution of α-linked NeuAc, Fuc, Gal, and GalNAc or elongated to form polymeric *N*-acetyllactosamine chains, which are then terminated with α-linked sugars. The *O*-linked glycans of some animal mucins are also sulfated, typically at the terminal ends and/or on internal Gal and GlcNAc. Many *O*-glycans of secretary proteins also contain the repeating units of type 1 *N*-acetyllactosamine -3Galβ1-3GlcNAcβ1- or type 2 *N*-acetyllactosamine structure, -3Galβ1-4GlcNAcβ1- structures (Brockhausen et al. 2008; Brooks et al. 2002; Patsos and Corfield 2009; Schachter and Brockhausen 1992; Taylor and Drickamer 2011). As in *N*-glycans, the inner β-Gal can be substituted with β(1-6)-linked GlcNAc to form a branched structure, which can also be elongated with *N*-acetyllactosamine moieties. In addition, the β-GlcNAc residues of type 1 and type 2 *N*-acetyllactosamine can be substituted with, respectively, α(1-4)-linked and α(1-3)-linked Fuc. These and the substitution of subterminal β-Gal with α-linked sugars results in the formation of *O*-glycans carrying the blood group A, B, H, and Lewis antigens, and development- and differentiation-specific sialyl Lewis antigens (Brockhausen et al. 2008; Brooks et al. 2002; Patsos and Corfield 2009; Schachter and Brockhausen 1992; Taylor and Drickamer 2011). All the above mentioned and various other type of modifications give rise to numerous distinct *O*-glycans found in animal glycoproteins.

Although *O*-glycans linked via α-GalNAcα1-Ser/Thr are found in most glycoproteins, they are abundantly present in two groups of glycoproteins: (1) mucins produced by epithelia of salivary glands, and the respiratory, gastrointestinal and urogenital tracts, and (2) membrane-associated mucin-like glycoproteins (Bhavanandan and Furukawa 1995; Brockhausen et al. 2008; Brooks et al. 2002; Patsos and Corfield 2009; Schachter and Brockhausen 1992; Taylor and Drickamer 2011). Epithelial mucin glycoproteins contain hundreds of α-GalNAc-linked *O*-glycans, which account for as much as 50–80 % of the mass of mucin molecules. In these mucin molecules, the *O*-glycans are found clustered at certain regions of the protein backbones. Because of the high levels of *O*-glycan substitution, mucin molecules can hold large amount of water and assume extended structures. In addition, because of the high net negative charge imparted by sialic acid and sulfate groups, mucin molecules in solution are randomly oriented, exhibiting high viscosity. This property allows mucin to serve as a protective barrier for epithelia against physical abrasion and inhibit infection by functioning as decoys for the adherence of pathogens. The latter is a double-edged sword since pathogens such as influenza virus and *Helicobacter pylori* exploit these molecules for their invasion/attachments and infect the respiratory system and stomach, respectively. Abnormality in mucin structure/function is implicated in the pathobiology of several human diseases (Brooks et al. 2002; Hennet et al. 2009; Taylor and Drickamer 2011; Varki and Freeze 2008). These include cystic fibrosis, chronic bronchitis, Crohn's disease, duodenal ulceration, colonic adenocarcinomas, infertility problems, and inflammatory ulcerative colitis.

Compared to secretory epithelial mucins, the carbohydrate content in many membrane-associated cell surface glycoproteins is relatively less because they have fewer and shorter glycan chains. Examples of such glycoproteins are the erythrocyte membrane glycophorins, and most of the cell membrane and tissue matrix glycoproteins of animal tissues, including the central nervous system. Other examples of membrane-associated, mucin-like glycoproteins are human white blood cell-associated leukosialin (CD43, sialophorin), red blood cell membrane decay-accelerating factor, low-density lipid receptor, platelet membrane-associated CD42b (glycocalicin), and human milk fat granule membrane glycoproteins. These glycoproteins contain hundreds of shorter *O*-glycans and are present as highly extended structures, projecting far beyond the cell surface (Bhavanandan and Furukawa 1995; Brooks et al. 2002; Fukuda 2000; Taylor and Drickamer 2011). As in the case of *N*-glycans, expression of the membrane-associated cell surface *O*-glycans is markedly increased or altered in malignantly transformed cells. Because of their important role in the pathobiology of cancer there has been extensive research on this subclass of *O*-glycans. These glycans are also widely implicated as ligands for selectins, which are involved in thrombosis, inflammation, allergy, auto immunity, and cancer metastasis. For example, the endothelial cell-associated mucin-type glycoproteins such as GlyCAM-1, CD34, and MAdCAM-1 function as ligands for L-selectin, and the leukocyte membrane mucin-type glycoprotein PSGL-1 has been shown to be a ligand for P-selectin and E-selectin. Thus, these mucin-like glycoproteins are involved in recognition, binding and recruitment of immune cells to the site of injury, and in homing of the circulating lymphocytes to lymph nodes (Homeister and Lowe 2000).

It has been shown that Fuc *O*-linked to the Ser/Thr residues of EGF-like repeat-containing proteins is substituted to form GlcNAcβ1-3Fucα1-, Galβ1-4GlcNAcβ1-3Fucα1-, Galβ1-4GlcNAcβ1-3Fucα1, NeuAcα2-6 Galβ1-4GlcNAcβ1-3Fucα1-oligosaccharides, which play important roles in intracellular signaling (Brooks et al. 2002; Gebauer et al 2008; Freeze and Haltiwanger 2008; Kleene and Schachner 2004; Luther and Haltiwanger 2009; Stanley and Cummings 2008; Taylor and Drickamer 2011). Some glycoproteins and proteoglycans of the nervous system, and neuronal and muscle α-dystroglycan, in addition to having the GalNAc-linked *O*-glycans, are modified with significant levels of *O*-glycans linked via α-Man to Ser/Thr (Chai et al. 1999; Krusius et al. 1986; Nakamura et al. 2010). The α-Man that is *O*-linked to Ser/Thr is substituted with either β1-2-linked GlcNAc residues or β1-2- and β1-6-linked GlcNAc residues. These GlcNAc residues are further substituted by sequential addition of β(1-4)-linked Gal and terminated with NeuAc or 3-sulfated glucuronic acid; the latter forming the HNK-1 epitope (Ariga 2011; Fukuda 2000; Kleene and Schachner 2004; Kizuka and Oka 2012; Morita et al. 2008; Stanley and Cummings 2008). Deficiency in the *O*-Man-linked oligosaccharide modification is associated with several types of severe brain and eye abnormalities, mental retardation, severely impaired mobility, muscle weakness, reduced muscle bulk, and dystrophic muscle (Kleene and Schachner 2004; Nakamura et al. 2010).

1.7 Biosynthesis of *N*- and *O*-Glycans

One important difference in the biosynthesis of glycans compared to nucleic acids and proteins is that the synthesis of glycans is not template dependent. Instead the genetic control is exerted by the expression of enzymes called glycosyltransferases, which catalyze the biosynthesis of glycans (Brockhausen et al. 2008; Brooks et al. 2002; Cummings 1992; Stanley et al. 2008; Schachter and Brockhausen 1992; Taylor and Drickamer 2011; Zuber and Roth 2009). The glycosidic bond formation involves the transfer of sugars from activated donors (nucleotide sugars) to acceptor monosaccharide or oligosaccharide substrates, which exist either as free molecules or as moieties linked to proteins/lipids. The activated sugars are derivatives of either uridine or guanidine diphosphate and of cytidine monophosphate in the case of sialic acids. The biosynthesis occurs at several locations, cytoplasm, lumen of the endoplasmic reticulum (ER), and lumen of the Golgi. The nucleotide sugars are formed in the cytoplasm and, for glycan biosynthesis, are transported to the Golgi by membrane transporters. In contrast, the ER membrane has no nucleotide-sugar transporters; sugar donors for glycan synthesis in the ER lumen are dolichol phosphate sugars formed by the transfer of sugars from nucleotide sugars in the cytoplasm to dolichol phosphate in the membrane. The dolichol phosphate sugars in which the sugar residues face the cytoplasmic side are then flipped to the luminal side and thus can donate sugar residues to acceptor molecules. *N*-Glycans are initially synthesized in the ER as dolichol diphosphate-linked high-mannose-type oligosaccharides containing three terminal Glc, eight Man, and two internal GlcNAc residues, and the oligosaccharide is transferred *en bloc* to the Asn residues of Asn-X-Ser/Thr motifs of polypeptide chains while the latter are still being synthesized. The oligosaccharides on proteins are then processed beginning in the ER but mostly in the Golgi to complex-type or hybrid oligosaccharides (Fig. 1.5). The *O*-glycans and glycosaminoglycans are synthesized in the Golgi by the addition of sugars sequentially after the first sugar is added to the Ser/Thr residues of proteins.

The general reaction for glycan biosynthesis is represented as the following: nucleotide-sugar donors + acceptor substrates → products + nucleotide diphosphate. The reaction is driven by the energy released by the hydrolysis of nucleoside diphosphate into nucleoside monophosphate and phosphate. Sugars are transferred from donor nucleotide sugars to different hydroxyl groups of acceptor monosaccharides, oligosaccharides, proteins, or lipids. The glycosidic linkages thus formed vary with respect to their positions and animeric configarations. The glycosyltransferases determine the types of glycosidic linkages formed by exhibiting specificity toward the donor sugars and acceptor mono- or oligosaccharide substrates. Therefore, it is the specificity of glycosyltransferases that primarily directs and controls the formation of a particular glycan. In general, one glycosyltransferase is required for the formation of each type of glycosidic linkage between two sugars in glycans, although a few exceptions exist. The glycosyltransferases involved in *N*-and *O*-glycan biosynthesis are membrane-bound enzymes and are involved in an assembly line-like process in the synthesis of specific glycans. The biosyntheses of *O*-and *N*-glycans are discussed in detail in Chap. 3 and 4, respectively.

1.8 Glycosphingolipids

Glycosphingolipids (GSLs) are a class of glycolipids that are conjugates of glycans and ceramide-containing lipids and are distributed widely in all organisms (Brooks et al. 2002; Kopitz 2009; Kundu 1992; Leeden and Wu 2009; Schnaar et al. 2008; Taylor and Drickamer 2011). They are ubiquitous plasma membrane components and are primarily localized in the outer leaflet with their hydrophilic sugar moieties projecting outside of the cell. GSLs are abundantly found in the brain. The glycan moieties on the cell surface function as specific receptors for certain pituitary hormones, growth factors, viruses, and cholera, tetanus, and botulinum bacterial toxins. They also serve as specific determinants of cell–cell interactions and as tumor-associated antigens. Disorders of GSL catabolism are responsible for several genetic diseases, which are referred to as glycosphingolipid storage diseases. These diseases are caused by defects in specific lysosomal glycosidases, involved in the degradation of the glycan chains or in sphingolipid activator proteins. For example, Tay–Sachs and Gaucher's diseases are caused by defects in *N*-acetyl-β-hexosaminidase and β-glucosidase, respectively. The inborn errors of GSL catabolism are discussed in detail in Chap. 21.

1.9 Structural Features of the Glycan Moieties of Glycosphingolipids

In the majority of glycolipids, glycans are conjugated to the terminal (C-1) primary hydroxyl group of ceramides (Cer), which are *N*-acyl fatty acid derivatives of sphingosine. Both the fatty acid (acyl chain) and sphingosine moieties of ceramides are heterogeneous (Brooks et al. 2002; Hakomori 2003; Kopitz 2009; Kundu 1992; Leeden and Wu 2009; Schnaar et al. 2008; Taylor and Drickamer 2011). The acyl chain can be 14–26 carbons in length, saturated, unsaturated, or 2-hydroxylated. The sphingosine can be 14–20 carbons in length. The carbohydrate moieties of the GSLs consist of the following sugars: D-Glc, D-Gal, L-Fuc, D-GlcNAc, D-GalNAc, and NeuAc. The simplest GSLs are the monohexosylcerebrosides, glucocerebroside (Glcβ1-1Cer), and galactocerebroside (Galβ1-1Cer). While the Glc residue of Glcβ1-1Cer is the link sugar in complex GSLs having a wide range of short to large glycan chains, the Gal residue of Galβ1-1Cer is elaborated only to a limited degree to form GSLs with only short glycan chains and the sulfatide, 3-*O*-sulfo-Galβ1-ceramide (Kundu 1992). Thus, the glycan chains of different compositions, linkage types, and chain lengths are formed by the sequential addition of other sugars to the core Glc residue linked to Cer. Further, like in the case of glycoprotein glycans, the glycan chain lengths of glycolipids in certain cells such as erythrocytes and granulocytes are highly extended with repeating *N*-acetyllactosamine units, and inner GlcNAc are substituted with Fuc to form Lewis antigens (Fukuda et al. 1986; Fukuda and Hakomori 1982; Müthing 1996; Stanley and Cummings 2008). In addition, these chains may be terminated by one or more of α-linked NeuAc, Fuc, Gal, or GalNAc to yield various blood group and sialyl Lewis antigens (Stanley and Cummings 2008). Based on the core structures of glycan moieties that are attached to the Cer moiety, GSLs are classified into different subclasses: lacto-, lactoneo-, globo-, isoglobo-, ganglio-, muco-, and gal-series (Table 1.1). Lactosylceramide (Galβ1-4Glcβ1-1Cer) serves as the precursor for the synthesis of five families of GSLs. The GSLs that contain sialic acid are

Table 1.1 The core structures of major glycolipids found in animals

Type	Structure
Lacto	Galβ1-3GlcNAcβ1-3**Galβ1-4Glcβ1-1-ceramide**
Lactoneo	Galβ1-4GlcNAcβ1-3**Galβ1-4Glcβ1-1-ceramide**
Globo	GalNAcβ1-3Galα1-4**Galβ1-4Glcβ1-1-ceramide**
Globoneo	GalNAcβ1-3Galα1-3**Galβ1-4Glcβ1-1-ceramide**
Isoglobo	GalNAcβ1-3Galα1-3**Galb1-4Glcβ1-1-ceramide**
Ganglio	Galβ1-3GalNAcβ1-4**Galβ1-4Glcβ1-1-ceramide**
Muco	Galβ1-3Galβ1-3**Galβ1-4Glcβ1-1-ceramide**
Galacto	Galα1-4Galβ1-1-ceramide
Sulfatides	3-*O*-Sulfo-Galβ1-ceramide

The sugars highlighted in bold represent common core residues

called gangliosides. As in glycoproteins, the structures of GSLs are defined by the specificity of glycosyltransferase and their relative levels and distribution in the cells.

GSLs are present ubiquitously in cells, and the subclasses are differentially distributed in different tissues (Kundu 1992). For example, gangliosides are present at high levels in neural tissues. Among the non-sialylated GSLs, only Gal-Cer and sulfatide, SO_3-3Gal-Cer, but not Glc-Cer are present in brain and other neural tissues. However, fetal and neonatal brains contain almost equal amounts of Gal-Cer and Glc-Cer. Erythrocytes contain mainly Glc-Cer and complex GSLs having large glycan chains, whereas the kidney and intestine have substantial levels of both Glc-Cer and Gal-Cer (Kundu 1992).

Sialic acid-containing GSLs were first isolated in high yield from the brain and were thought to be exclusively present in ganglia and hence named gangliosides. However, gangliosides were subsequently found to be present throughout the body, albeit at relatively lower levels. NeuAc is the almost exclusive sialic acid in gangliosides of humans, whereas animal gangliosides contain both NeuAc and NeuGc and their acetylated and/or methylated derivatives (Brooks et al. 2002; Schauer 2004; Varki 1992; Varki et al. 2008). The gangliosides of various animal brains contain one to as many as five sialic acid residues. Gangliosides are named according to the rules proposed by Dr. Lars Svennerholm (1994). G_M, G_D, G_T, G_Q, and G_P, where G stands for ganglio, and the subscript letters define the total number of sialyl residues, indicating mono-, di- tri-, tetra- and penta-sialylated gangliosides, respectively. The numerical numbers 1, 2, 3, and 4 following these letters define 5 minus the number of neutral sugars in the molecule. The lowercase letters, *a, b,* or *c* after the numerical number define, respectively, one, two, or three SA residues on the inner Gal. For example, G_{T1b} refers to trisialyl globotetraosyl ganglioside having SAα2-8SA linked through α(1-3) glycosidic bond to the inner Gal residue. The SA residues on Gal are α(2-3)-linked, while the SA–SA glycosidic bond is α(2-8)-linked. One or more SA residues can be similarly linked to the terminal Gal. The structures of the lacto-ceramide G_M gangliosides are shown in Fig. 1.7 and examples of other gangliosides are available elsewhere (Hakomori 2003; Kopitz 2009; Kundu 1992; Schnaar et al. 2008).

An important difference in GSLs compared to glycoproteins and proteoglycans is the absence of heterogeneity at specific glycosylation sites. Glycan microheterogeneity is an inherent factor in glycoproteins and proteoglycans. Not only does the number

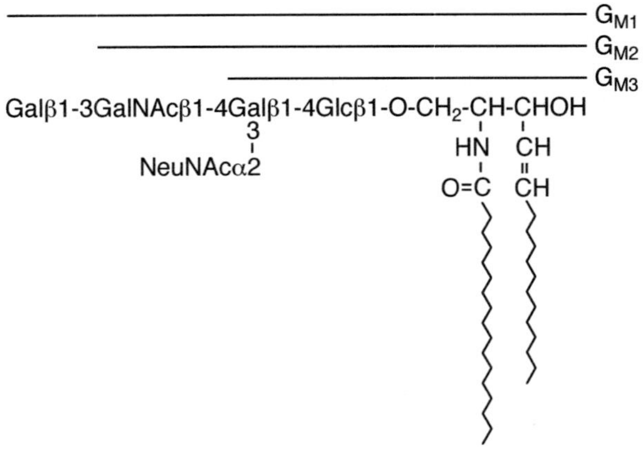

Fig. 1.7 The structures of monosialogangliosides, G_{M1}, G_{M2}, and G_{M3}, containing four, three, and two neutral monosaccharides, respectively. G_{M4} is NeuAcα2-3Galβ1-1Cer

of glycan chains attached to core proteins vary widely, there can also be heterogeneity in the glycan at a single site. For example, the single *N*-linked oligomannose chain attached to a specific Asn residue in ribonuclease varies widely in its number of mannose residues. In contrast, in GSLs such as G_{D1a} ganglioside (NeuAcα2-3Galβ1-3GalNAcb1-4(NeuAc2α-3)Galβ1-4Glcβ1-ceramide) and G_{M3} ganglioside (NeuAcα2-3Galβ1-4Glcβ1-ceramide), the glycan chains are well defined and homogenous. While there is heterogeneity in the length of fatty acyl chain and sphingosine moiety of ceramide as explained earlier, there is no glycan chain heterogeneity. In fact, if one monosaccharide is missing in the carbohydrate chain, the molecule becomes a different glycolipid.

1.10 Synthesis and Functions of the Glycan Moieties of Glycolipids

As in the case of glycoprotein *O*-glycans, glycolipids are synthesized in the Golgi by the sequential addition of sugars to the C-1 hydroxyl group of ceramide. The newly synthesized glycolipids are then transported to plasma membranes (Brooks et al. 2002; Kopitz 2009; Kundu 1992; Schnaar et al. 2008).

Like *N*- and *O*-glycans of glycoproteins, GSLs perform numerous physiological functions, including cell–cell and cell–molecule interactions that are critical for biological processes such as development, differentiation, defining cell–cell communications, cell social behavior, antigenicity, modulation of immune responses, and cell signaling. As in the case of glycoproteins, the glycan moieties of GSLs can also function as various blood group- and development- and differentiation-specific antigens such as sialyl Lewis antigen (Stanley and Cummings 2008). In the central nervous system, GSLs are involved in many tissue-specific functions, including neuritogenesis, nerve repair, inhibition of neurite outgrowth, neuromuscular formation, cell social behavior, and various other functions (Hakomori 2003; Kopitz 2009; Kundu 1992; Schnaar et al. 2008).

Alterations in GSL composition, levels of expression, and distribution appear to contribute to tumor growth and spreading. For example, high levels of G_{M2} and G_{D2} gangliosides are found in melanoma and neuroectodermal tumors, and G_{D3} ganglioside containing 9-*O*-acetyl neuraminic acid is highly expressed by melanoma tumors (Kundu 1992). Various sialyl Lewis antigen-containing GSLs are also highly expressed by teratocarcinoma, colorectal adenocarcinoma, pancreatic cancer, ovarian cancer, and other cancers. A number of monoclonal antibodies such as CA50, N-19-9, OFA-1, and OFA-2 that specifically recognize GSL glycan antigens are being used for diagnosis of cancers, including breast cancer, brain tumors, and colon cancer (Kundu 1992).

1.11 Glycosylphosphatidylinositol Anchors: A Special Group of Glycolipids

Glycosylphosphatidylinositol (GPI) anchors represent a distinct class of glycolipids and consist of glycan moieties attached to the C-6 position of the *myo*-inositol residue of phosphatidylinositol (PI) (Brooks et al. 2002; Ferguson et al. 2008; Paulick and Bertozzi 2008; Shams-Eldin et al. 2009; Taylor and Drickamer 2011). The glycan moieties of GPIs consist of a conserved trimannosyl core substituted with ethanolamine at the C-6 hydroxyl group of the terminal mannose, ethanolamine-phosphate-6Manα1-2Manα1-6Manα1-4GlcNα1-. The trimannosyl core moiety in GPIs from many sources is modified with additional sugars and one or more ethanolamine phosphate residues (Fig. 1.8). The lipid portion contains saturated or

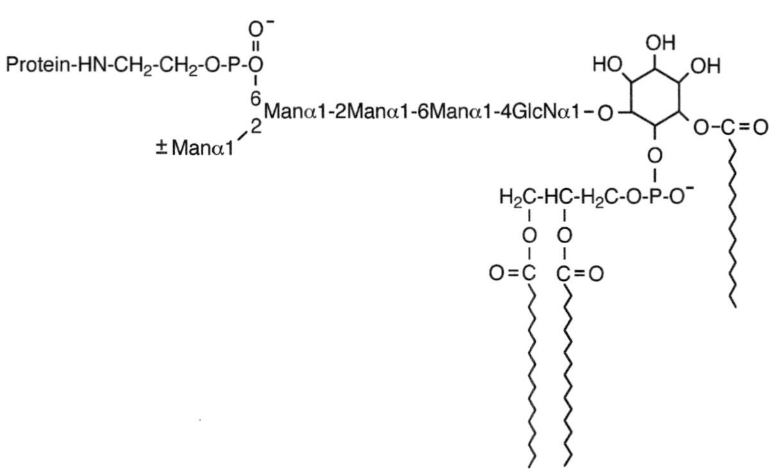

Fig. 1.8 Representative structure of a GPI anchor. The inner trimannosyl moiety is usually modified with additional sugars and ethanolamine phosphate residue(s). The lipid chains on glycerol are either acyl/acyl or acyl/alkyl chains. The chains lengths of both acyl and alkyl residues are variable and may contain unsaturated bonds (see ref. (Ferguson et al. 2008) for GPIs from many sources). GPIs from some sources carry acyl substituent, usually C16:0 at C-2 of *myo*-inositol

unsaturated fatty acyl/alkyl residues of 16–24 carbon atoms in length at the C-1 and C-2 positions of glycerol. In addition, the C-2 position of the *myo*-inositol residue is either substituted with fatty acid or unsubstituted. GPIs from different cells and species exhibit broad structural diversity due to heterogeneity in the structures of the glycan and lipid moieties (Ferguson et al. 2008).

GPIs are expressed ubiquitously by all eukaryotic cells and are found in all animal tissues. GPIs are particularly abundant in parasites, including species of the genus *Trypanosoma*, *Leishmania*, and *Plasmodium* (Ferguson et al. 2008; Shams-Eldin et al. 2009). The primary role of GPIs is to anchor certain proteins, glycoproteins, and proteoglycans of cells, including those in the brain, to the plasma membrane via formation of an amide bond between the amino group of the ethanolamine residue and the C-terminal carboxyl group of the protein. In addition to anchoring proteins to the plasma membrane, GPIs seem to be involved in many biological functions (Ferguson et al. 2008; Paulick and Bertozzi 2008; Shams-Eldin et al. 2009). GPI modification is not limited to a specific class of proteins/glycoproteins. Proteins including enzymes, receptors, cell surface antigens, cell adhesion molecules, transporters, and other functional proteins are anchored to plasma membranes via GPIs. Specific examples of GPI-anchored proteins include acetylcholine esterase and decay-accelerating factor present on the erythrocyte membrane, placental membrane alkaline phosphatase, the Thy-1 antigen of lymphocytes, the neural cell adhesion molecule and brevican proteoglycan.

Biosynthesis of GPIs occurs exclusively in the ER, although some modifications to the glycan moieties may occur after the GPI-anchored proteins move to the Golgi. In the ER, first GlcNAc is added to the membrane PI moiety on the cytoplasmic side, and after *N*-deacetylation of GlcNAc and acylation of *myo*-inositol residue at C-2, the GlcN-PI intermediate flips over to the luminal side of the ER, where Man residues donated by dolichol phosphate-Man are sequentially added. The fully assembled GPI is then transferred to the carboxyl group of the C-terminal amino acid of the acceptor protein to form an amide bond by the action of a transamidase, thereby anchoring the protein to the membrane.

1.12 Glycosaminoglycans and Proteoglycans

Glycosaminoglycans (GAGs) are linear, anionic polysaccharides consisting of alternating residues of an uronic acid or galactose and an *N*-acetylhexosamine (Bhavanandan and Davidson 1992; Buddecke 2009; Esko et al. 2008; Hascall and Esko 2008; Sasisekharan et al. 2008; Volpi 2006). An additional structural feature is that all GAGs except hyaluronic acid (see below) contain sulfate. GAGs are made up of repeating disaccharide units with or without sulfate; two such repeats of each GAG are illustrated in Fig. 1.9. Several types of GAGs occur in animals and are classified into two broad groups based on the type of *N*-acetylhexosamine present: (1) Glucosaminoglycans, containing either GlcNAc or *N*-sulfated glucosamine ($GlcNSO_3^-$), include hyaluronic acid (HA), heparan sulfate (HS), heparin, and

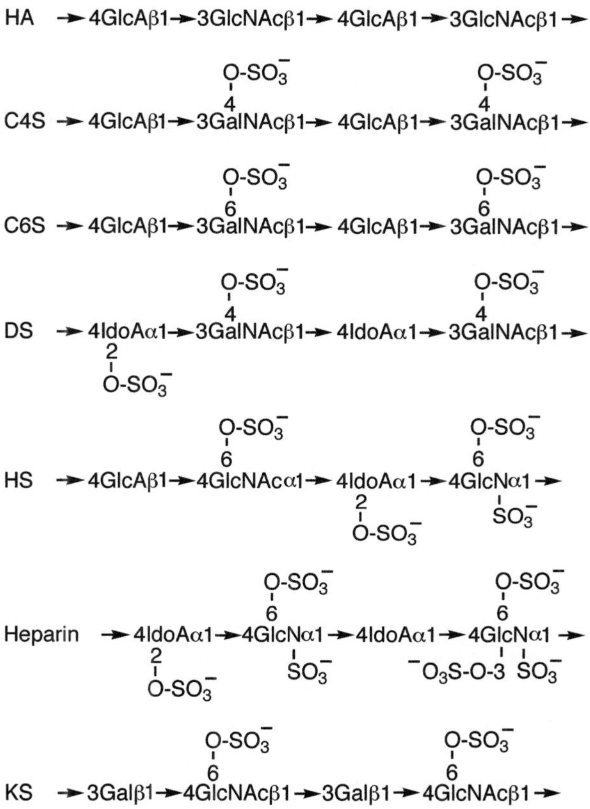

Fig. 1.9 The structural features of glycosaminoglycans (GAGs) that are commonly found in animal cells and tissues. Two repeating disaccharide units each consisting of alternating residues of an uronic acid or galactose and an *N*-acetylated or *N*-sulfated hexosamine are shown. Note that all GAGs except hyaluronic acid (HA) are sulfated

keratan sulfate (KS). (2) Galactosaminoglycans, containing GalNAc, are chondroitin sulfates (CS). In early studies, three types of galactosaminoglycans were discovered and initially named CSA, CSB, and CSC. After the structures were elucidated, these GAGs were renamed chondroitin-4-sulfate (C4S), dermatan sulfate (DS), and chondroitin-6-sulfate (C6S), respectively. DS differs from C4S and C6S in having predominantly IdoA instead of GlcA. There are several other chondroitin sulfates that can be distinguished based on the position and number of sulfate groups present, for example, chondroitin 2,6-sulfate (C2,6diS; CSD) and chondroitin 4,6-sulfate (C4,6diS or CSE) (Malavaki et al. 2008; Nandini and Sugahara 2006). In CS and DS, the sulfate groups are mainly present on GalNAc residues. However, in DS and in C2,6diS, additional sulfate groups are present at C-2 of certain IdoA and GlcA residues, respectively.

The nonsulfated GAG, HA, consists exclusively of uniformly defined repeating disaccharide units (see Fig. 1.9). In contrast, the sulfated GAGs from various tissues of different animals are highly heterogeneous with respect to uronic acid composition and sulfate content. In addition, the disaccharide moieties with different compositions are variously distributed in the polymer chains. Thus, not all repeating disaccharide units of C4S and C6S from different animal tissues are exclusively and uniformly sulfated at C-4 and C-6, respectively. C4S from many sources contains significant amounts of nonsulfated and 6-sulfated disaccharide units. For example, bovine cartilage C4S consists of ~53 % 4-sulfated, ~39 % 6-sulfated, and ~8 % nonsulfated disaccharide units. CS from some sources, including those of the brain, have, in addition to significant levels of both C4S and C6S repeating disaccharide units, one or more elements of 2,4-di-, 3,4-di-, 3,6-di-, and 3,4,6-trisulfated disaccharide moieties (Malavaki et al. 2008; Nandini and Sugahara 2006). Some tissues such as the cornea and placenta contain uniquely low sulfated CS with 80–95 % of the disaccharides not sulfated. In placental CS, only 5–10 % of the disaccharides are sulfated exclusively at C-4, and in corneal CS, the sulfation is mainly at C-4 and to certain extent at C-6 (Achur et al. 2000, 2004). DS from different tissues, including those from nervous tissues, while containing mainly IdoA, have low to moderate levels of GlcA. As in the case of oversulfated CS, 2,4-di-, 3,6-di, 4,6-di, and 2,4,6-trisulfated disaccharide units are also found in some DS (Nandini and Sugahara 2006). The variations in uronic acid composition and the position and degree of sulfation lead to incredible structural diversity and enormous microheterogeneity in the structures of GAGs. Interestingly, as in the case of *N*- and *O*-glycans, it is this structural variation and microheterogeneity that confer the ability for GAGs to perform a wide range of biological functions. The structure and functions of GAGs of the nervous system are discussed in detail in Chap. 5.

Heparan sulfate (HS) and heparin vary substantially in the proportions of GlcA and IdoA as well as sulfate content (Bhavanandan and Davidson 1992; Buddecke 2009; Esko et al. 2008). Usually, HS contains nearly equal levels of GlcA and IdoA, whereas in heparin, >80 % of the uronic acid is IdoA and the remainder is GlcA. HS has 0.5–1.5 sulfate groups per disaccharide unit, but the average sulfate content of heparin varies from 1.5 to 2.5 sulfate groups per disaccharide moiety. In heparin, the majority of glucosamine is, in addition to having sulfate at C-6, *N*-sulfated, and some *N*-sulfated glucosamine residues are sulfated at both C-3 and C-6. Because of variations in GlcA and IdoA contents, the level of sulfate, and sulfation at different positions, including *N*-sulfation and variations in modifications at different regions of the polymer chains, both HS and heparin exhibit enormous microheterogeneity. The microheterogeneity in HS and heparin is far higher than that of CS and DS.

GAGs are ubiquitous constituents of all animal tissues and are expressed by almost all cell types of eukaryotes (Achur et al. 2000, 2004; Bernfield et al. 1999; Bhavanandan and Davidson 1992; Bishop et al. 2007; Brooks et al. 2002; Buddecke 2009; Esko et al. 2008; Hascall and Esko 2008; Malavaki et al. 2008; Nandini and Sugahara 2006; Sasisekharan et al 2008; Taylor and Drickamer 2011; Volpi 2006). CS, DS, and HA occur abundantly as gel-like ground substances in the extracellular matrix of connective tissues such as cartilage, tendon, skin, cornea, blood vessels, and umbilical cord. They also occur at substantial levels in matrices of almost all other tissues, including those of the central nervous system, and at moderate levels in the form of proteoglycans in plasma membrane. Although present at significant levels in the matrices of connective and skeletal tissues and brain (Funderburgh 2000; Krusius et al. 1986; Zhang et al. 2006), keratan sulfate (KS) is particularly abundant in the cornea (Funderburgh et al. 1987). HS is ubiquitously found as a cell surface component and also occurs as an extracellular component in blood vessels, in the brain, and in basement membranes, particularly in the kidney, where it is involved in filtration (Bhavanandan and Davidson 1992). Heparin, on the other hand, is exclusively found in intracellular granules of mast cells that line the arterial walls. It is secreted in response to injury and functions as an anticoagulant to regulate the blood-clotting cascade (Bhavanandan and Davidson 1992).

Because of the presence of numerous hydrophilic hydroxyl groups and the high net negative charge imparted by carboxyl and sulfate groups, GAGs have the capacity to hold large amounts of water and stay as extended molecules in solution. They form highly viscous and slimy mucus-like solutions and hence their older name, mucopolysaccharides, which is still used occasionally. The ability to form gel-like substance enables GAGs to function as shock absorbers and lubricants in joints and umbilical cord and to impart resilience to tissues. GAGs also provide nutrients to cartilage and other connective tissues lacking blood vessels by their property to absorb and release extracellular fluid in the absence and presence of mechanical shear force and to regulate tissue calcification (Bhavanandan and Davidson 1992). More importantly, GAGs are involved in cell–cell and cell–molecule interactions, cell–pathogen binding, cell signaling, binding and mobilizing growth factors, chemokines and cytokines, promoting growth, and regulating immune responses (Achur et al. 2000, 2004; Bernfield et al. 1999; Bishop et al. 2007; Bhavanandan and Davidson 1992; Brooks et al. 2002; Buddecke 2009; Esko et al. 2008; Funderburgh 2000; Funderburgh et al. 1987; Hascall and Esko 2008; Iozzo and Schaefer 2010; Malavaki et al. 2008; Nandini and Sugahara 2006; Sasisekharan et al 2008; Taylor and Drickamer 2011; Varki et al. 2008; Volpi 2006; Zhang et al. 2006). Through these interactions, GAGs play important roles in biological processes such as development, differentiation, cell migration, tissue organization, cartilage and bone formation, wound healing, and in disease processes, including cancer and atherosclerosis. Furthermore, deficiencies in GAG catabolism due to lysosomal enzyme deficiencies lead to many diseases called mucopolysaccharidoses, which are discussed in detail in Chap. 5.

All GAGs, except HA, regardless of whether they are the components of various extracellular matrices or cell membranes, occur as moieties conjugated to proteins to form proteoglycans (PGs). A common tetrasaccharide core covalently attaches GAG (CS, DS, HS, and heparin) chains via xylose to the hydroxyl groups of Ser residues: **[HexNR-HexA]$_n$-GlcAβ1-4Galβ1-3Galβ1-4Xylβ1-Ser**, where HexNR is variously sulfated GlcNAc or GalNAc, HexA is nonsulfated or sulfated GlcA or IdoA, and n is number of repeating disaccharide units.

In proteoglycan biosynthesis, the addition of GlcNAc to the protein-linked tetrasaccharide core commits to the formation of heparan sulfate and heparin chains, whereas the addition of GalNAc leads to the formation of CS or DS chains. Once commitment to synthesize either CS or HS is made, presumably by the specific recognition of core proteins by the glycosyltransferase, the GAG chains are synthesized by the sequential addition of HexNAc and GlcA. Then, the sulfate residues are added, and in the case of DS, HS, and heparin, IdoA residues are formed by the epimerization of D-GlcA to L-IdoA residues. The Ser–Gly motifs of proteins are sites for the attachment of GAG chains. If several Ser–Gly are present, the proteins are modified with multiple GAG chains. For example, the core proteins of serglycin and aggrecan contain 49 and >100 tandem repeats of Ser–Gly motifs, respectively, and all or most of these motifs are substituted with CS chains. The majority of PGs contain one type of GAG chain, CS, DS, or HS, and are designated as CSPGs, DSPGs, and HSPGs, respectively. However, some PGs contain two types of GAG chains such as CS plus HS on the same core protein molecule and thus are hybrid PGs. For example, syndecan, a cell surface PG expressed by a number of cell types, contains both CS and HS chains. KS also is found in PGs, and examples of KS containing PGs include lumican, keratocan, fibromodulin, mimecan, and aggrecan. In the case of KS, the disaccharide-containing polymer chains are built on the core residues of either *N*-linked or *O*-linked glycans in a manner similar to those of poly-*N*-acetyllactosamine chains. The KS are classified into three types, KS-I, KS-II, and KS-III (Funderburgh 2000). KS-I chains are mainly found in the cornea and are linked to proteins via the core glycan moiety of *N*-glycans to Asn residues. KS-II chains are present mainly in skeletal tissues and are linked through α-GalNAc to Ser/Thr of proteins. KS-III is present in the brain and is linked via *O*-linked β-Man residues to Ser/Thr of proteins (Krusius et al. 1986).

In many PGs, including those of the central nervous system, the GAG chains define their functions, and hence the functions of PGs are the same as those mentioned above for GAGs. In addition, the core proteins of many PGs contain functional domains, including the C-type lectin domain that interacts with HA, FGF, fibronectin III, laminin-G, and fibronectin. As such, PGs interact with cell surface and extracellular matrix proteins and these interactions define their function.

1.13 Glycans of Cell Surface Glycoconjugates Perform a Variety of Functions

The majority of animal cell surface proteins, including those on cells of the central nervous system, are modified with variable amounts of *N*- and *O*-linked glycans and/or glycosaminoglycan chains (Brooks et al. 2002; Collins and Paulson 2005; Fukuda 2000; Hattrup and Gendler 2008; Iozzo and Schaefer 2010; Schauer 2009; Springer and Gagneux 2013; Taylor and Drickamer 2011; Varki and Lowe 2008). In some instances, such as in mucin-type glycoproteins, the mass of glycans exceeds that of proteins to which they are conjugated. In addition, a significant portion of the outer head group of cell surface ceramide is modified with glycans. Thus, the *N*- and *O*-glycans of glycoproteins, the glycan moieties of glycolipids, and the chondroitin sulfate and heparan sulfate chains of proteoglycans are prominently displayed on the outer leaflet of the plasma membrane (Fig. 1.10). As mentioned earlier, because of high hydrophilicity, glycans have a large capacity to hold water and exist as bulky and highly extended molecules. The eukaryotic cell surface is thus covered with a glycan-rich zone, referred to as the glycocalyx (Weinbaum et al. 2007; Salmon and

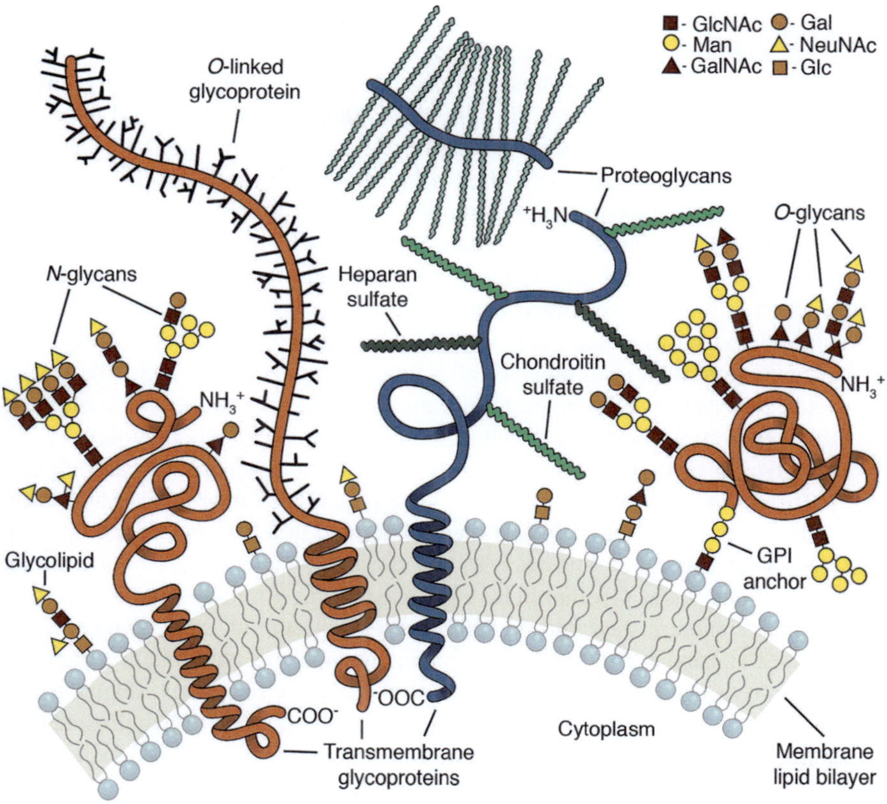

Fig. 1.10 Schematic illustration of a section of animal cell plasma membrane showing the placement of glycan moieties of glycoconjugates in the outer leaflet of the lipid bilayer. The *N*- and *O*-glycans of glycoproteins, the glycan moieties of glycolipids, and the GAG chains of integral membrane and secreted proteoglycans that together form the glycocalyx are shown. Also shown is a highly glycosylated mucin-type glycoprotein present as an extended molecule on the cell surface

Satchell 2012). Moreover, many glycan chains of glycoproteins and glycolipids carry terminal sialic acid residues and proteoglycans carry anionic chondroitin and heparan sulfate chains. Therefore, the glycocalyx is negatively charged and can be seen by electron microscopy after staining cells with a dye such as ruthenium red. The thickness of the glycocalyx is typically greater than that of the plasma membrane itself.

Because of their strategic locations, the cell surface glycans that constitute the glycocalyx play several nonspecific yet crucial physical roles. By holding large amounts of water, they help to maintain cell shape, assist tissues in their function (e.g., keeping the lungs and airways moist and open), protect epithelial barrier against mechanical damage and proteolysis, and provide an aqueous environment for biochemical interactions. They also function to stabilize protein conformation and provide the aqueous environment essential for cell–cell and cell–matrix interactions and communication. More importantly, cell surface glycans perform numerous specific, biological functions through interactions with proteins and glycans of other cells and tissue matrices (Brooks et al. 2002; Gabius 2009; Taylor and Drickamer 2011; Varki et al. 2008). These functions are dependent on specific structural features of glycans and also on their clustering patterns and densities. The wide range of structural diversity, microheterogeneity, and distributions along protein chains enable glycans to serve as specific recognition and information molecules. Thus, they play essential and specific roles in almost all biological processes involved in fertilization and embryogenesis through the sperm–egg interactions, differentiation and development, and survival. Other biological processes in which glycans play important roles include cell migration and recruitment of cells to specific sites; turnover of cells and proteins; removal of hormones, receptors, and aged erythrocytes from the circulation; cell signaling; and immune modulation.

The glycans also play important roles in the pathogenesis of various diseases (Brooks et al. 2002; Gabius 2009; Taylor and Drickamer 2011; Varki et al. 2008). For example, when cells become transformed to a malignant state, the cell surface glycan structures and profiles are drastically altered, promoting tumor growth and metastasis. Glycoproteins expressed by most tumors have altered glycan structures and are shed into the circulation (Brooks et al. 2002; Gabius 2009; Taylor and Drickamer 2011; Varki et al. 2008a). Monoclonal antibodies recognizing the cancer-associated glycan antigens are used clinically for cancer diagnosis and management. For example, neuroblastomas express high levels of ganglioside G_{D2}, and antibodies to the glycan portion of the G_{D2} are currently being used experimentally to target drugs to these tumors (Tivnan et al. 2012). Furthermore, glycans are involved in bacterial, viral, and parasitic infections. Infectious diseases such as influenza, stomach ulcer caused by *H. pylori*, and several airway and urinary tract infections arise by the glycan-dependent recognition and binding of viruses and bacteria to host target cells and tissues (Brooks et al. 2002; Gabius 2009; Taylor and Drickamer 2011; Varki et al. 2008). For example, influenza infection involves the viral hemagglutinin-mediated attachment to cell surface sialic acid residues. The surface of the causative agent of AIDS, the human immunodeficiency virus (HIV-1), is studded with a major envelope glycoprotein (gp120). About 50 % of the molecular mass of gp120 is carbohydrate, and it is involved in the viral invasion of lymphocytes.

1.14 Summary

Carbohydrates in the form of oligosaccharides and polysaccharides (called glycans) occur widely in almost all animal cells as moieties linked to proteins or lipids to form glycoproteins, glycolipids, and proteoglycans, which are collectively called glycoconjugates. The glycan moieties are structurally highly complex, and their complexity is attributed to the multiple ways the constituent sugar residues are linked to one another. More importantly, the glycan structural complexity arises from the variations in their sugar composition and size. The biosynthesis of these complex glycan structures is not template dependent as in the case of proteins, but instead is determined by the specificity of glycosyltransferases that catalyze their formation and the orderly manner these enzymes are organized in the biosynthetic compartments. The glycans function as informational molecules through varied and specific interactions with partner proteins and complementary glycans. As such, the wide range of glycan structural complexity is translated to a myriad of biological roles. The glycoconjugates are abundantly found in extracellular matrix and on cell surfaces, where the glycan moieties participate in numerous biological interactions, which are crucial for differentiation and development and life of the organism. Therefore, abnormal glycan metabolism leads to disease pathogenesis. Ironically, many pathogens and toxins produced by them exploit the cell surface glycans to attach and invade cells, causing debilitating and fatal disease.

Acknowledgments We thank Dr. Rajeshwara Achur for the preparation of Figs. 1.1–1.9 and Ms. Jillian Dunbar, Devon Medical Art, Hershey, for the artwork in Fig. 1.10. DCG is partly supported by the grant AI41139 from National Institute of Allergy and Infectious Diseases, National Institutes of Health, USA.

Conflicts of Interest The authors have no conflict of interest.

References

Achur RN, Valiyaveettil M, Alkhalil A, Ockenhouse CF, Gowda DC. Characterization of proteoglycans of human placenta and identification of unique chondroitin sulfate proteoglycans in the intervillous spaces that mediate the adherence of *Plasmodium falciparum*-infected erythrocytes to the placenta. J Biol Chem. 2000;275(51):40344–56.

Achur RN, Muthusamy A, Madhunapantula SV, Bhavanandan V, Seudieu C, Gowda DC. Chondroitin sulfate proteoglycans of bovine cornea: Structural characterization and assessment for the adherence of *Plasmodium falciparum*-infected erythrocytes. Biochim Biophys Acta. 2004;1701(1–2):109–19.

Allen HJ, Kisailus EC, editors. Glycoconjugates: composition, structure, and function. 1st ed. New York: Marcel-Dekker; 1992.

Ariga T. The role of sulfoglucuronosyl glycosphingolipids in the pathogenesis of monoclonal IgM paraproteinemia and peripheral neuropathy. Proc Jpn Acad Ser B Phys Biol Sci. 2011;87(7): 386–404.

Ariga T, Kohriyama T, Freddo L, Latov N, Saito M, Kon K, Ando S, et al. Characterization of sulfated glucuronic acid containing glycolipids reacting with IgM M-proteins in patients with neuropathy. J Biol Chem. 1987;262:848–53.

Bardor M, Nguyen DH, Diaz S, Varki A. Mechanism of uptake and incorporation of the non-human sialic acid *N*-glycolylneuraminic acid into human cells. J Biol Chem. 2002;280(6): 4228–37.

Bernfield M, Götte M, Park PW, Reizes O, Fitzgerald ML, Lincecum J, Zako M. Functions of cell surface heparan sulfate proteoglycans. Annu Rev Biochem. 1999;68:729–77.

Bertozzi CR, Rabuka D. Structural basis of glycan diversity. In: Varki A, Cummings RD, Esko JD, Freeze HH, Stanley P, Bertozzi CR, Hart GW, Etzler ME, editors. Essentials of glycobiology. New York: Cold Spring Harbor Laboratory Press; 2008. p. 23–36.

Bhavanandan VP, Davidson EA. Proteoglycans: structure, synthesis, function. In: Allen HJ, Kisailus EC, editors. Glycoconjugates: composition, structure, and function. New York: Marcel-Dekker; 1992. p. 167–202.

Bhavanandan VP, Furukawa K. Biochemistry and oncology of sialoglycoproteins. In: Rosenberg A, editor. Biology of the sialic acids. New York: Plenum Press; 1995. p. 145–96.

Bishop JR, Schuksz M, Esko JD. Heparan sulfate proteoglycans fine-tune mammalian physiology. Nature. 2007;446(7139):1030–7.

Brockhausen I, Schachter H, Stanley P. *O*-Glycans. In: Varki A, Cummings RD, Esko JD, Freeze HH, Stanley P, Bertozzi CR, Hart GW, Etzler ME, editors. Essentials of glycobiology. New York: Cold Spring Harbor Laboratory Press; 2008. p. 115–27.

Brooks SA, Dwek MV, Schumacher U. Functional and Molecular Glycobiology. 1st ed. Oxford: BIOS Scientific; 2002.

Bruses JL, Rutishauser U. Polysialic acid in neural cell development: roles, regulation and mechanism. In: Fukuda M, Hindsgaul O, editors. Molecular and Cellular Glycobiology. Oxford: Oxford University Press; 2000. p. 116–32.

Buddecke E. Proteoglycans. In: Gabius H-J, editor. The sugar code: fundamentals of glycosciences. Weinheim: Wiley-VCH; 2009. p. 199–216.

Chai W, Yuen CT, Kogelberg H, Carruthers RA, Margolis RU, Feizi T, Lawson AM. High prevalence of 2-mono- and 2,6-di-substituted manol-terminating sequences among *O*-glycans released from brain glycopeptides by reductive alkaline hydrolysis. Eur J Biochem.1999; 263(3):879–88.

Collins BE, Paulson JC. Cell surface biology mediated by low affinity multivalent protein-glycan interactions. Curr Opin Chem Biol. 2005;8(6):617–25.

Cummings RD. Synthesis of asparagine-linked oligosaccharides: pathways, genetics, and metabolic regulation. In: Allen HJ, Kisailus EC, editors. Glycoconjugates. New York: Marcel-Dekker; 1992. p. 333–60.

Esko JD, Kimata K, Lindahl U. Proteoglycans and sulfated glycosaminoglycans. In: Varki A, Cummings RD, Esko JD, Freeze HH, Stanley P, Bertozzi CR, Hart GW, Etzler ME, editors. Essentials of glycobiology. New York: Cold Spring Harbor Laboratory Press; 2008. p. 229–48.

Ferguson MAJ, Kinoshita T, Hart GW. Glycosylphosphatidylinositol anchors. In: Varki A, Cummings RD, Esko JD, Freeze HH, Stanley P, Bertozzi CR, Hart GW, Etzler ME, editors. Essentials of glycobiology. New York: Cold Spring Harbor Laboratory Press; 2008. p. 143–61.

Freeze H, Haltiwanger RS. Other classes of ER/Golgi-derived glycans. In: Varki A, Cummings RD, Esko JD, Freeze HH, Stanley P, Bertozzi CR, Hart GW, Etzler ME, editors. Essentials of glycobiology. New York: Cold Spring Harbor Laboratory Press; 2008. p. 163–73.

Fukuda M. Cell surface carbohydrates: cell type-specific expression. In: Fukuda M, Hindsgaul O, editors. Molecular and Cellular Glycobiology. Oxford: Oxford University Press; 2000. p. 1–61.

Fukuda MN, Hakomori S. Structures of branched blood group A-active glycosphingolipids in human erythrocytes and polymorphism of A- and H-glycolipids in A1 and A2 subgroups. J Biol Chem. 1982;257(1):446–55.

Fukuda MN, Bothner B, Scartezzini P, Dell A. Isolation and characterization of poly-*N*-acetyllactosaminylceramides accumulated in the erythrocytes of congenital dyserythropoietic anemia type II patients. Chem Phys Lipids. 1986;42(1–3):185–97.

Funderburgh JL. Keratan sulfate: structure, biosynthesis, and function. Glycobiology. 2000;10(10): 951–8.

Funderburgh JL, Caterson B, Conrad GW. Distribution of proteoglycans antigenically related to corneal keratan sulfate proteoglycan. J Biol Chem. 1987;262(24):11634–40.

Gabius H-J, editor. The sugar code: fundamentals of glycosciences. 1st ed. Weinheim: Wiley-VCH; 2009.

Gascon E, Vutskits L, Kiss JZ. Polysialic acid-neural cell adhesion molecule in brain plasticity: from synapses to integration of new neurons. Brain Res Rev. 2007;56(1):101–18.

Gebauer JM, Müller S, Hanisch FG, Paulsson M, Wagener R. *O*-Glucosylation and *O*-fucosylation occur together in close proximity on the first epidermal growth factor repeat of AMACO (VWA2 protein). J Biol Chem. 2008;283(26):17846–54.

Hakomori S. Structure, organization, and functions of glycosphingolipids in membrane. Curr Opin Hematol. 2003;10(1):16–24.
Hascall V, Esko JD. Hyaluronan. In: Varki A, Cummings RD, Esko JD, Freeze HH, Stanley P, Bertozzi CR, Hart GW, Etzler ME, editors. Essentials of glycobiology. New York: Cold Spring Harbor Laboratory Press; 2008. p. 219–27.
Hattrup CL, Gendler SJ. Structure and function of the cell surface (tethered) mucins. Ann Rev Physiol. 2008;70:431–57.
Hennet T. Diseases of glycosylation. In: Gabius H-J, editor. The sugar code: fundamentals of glycosciences. Weinheim: Wiley-VCH; 2009. p. 365–83.
Hildebrandt H, Dityatev A. Polysialic acid in brain development and synaptic plasticity. Top Curr Chem. (2013) 10.1007/128_2013_446, Springer-Verlag Berlin Heidelberg 2013.
Homeister J, Lowe JB. Carbohydrate recognition in leukocyte-endothelial cell interactions. In: Hindsgaul O, Fukuda M, editors. Molecular and cellular glycobiology. Oxford: Oxford University Press; 2000. p. 62–115.
Iozzo RV, Schaefer L. Proteoglycans in health and disease: novel regulatory signaling mechanisms evoked by the small leucine-rich proteoglycans. FEBS J. 2010;277(19):3864–75.
Kizuka Y, Oka S. Regulated expression and neural functions of human natural killer-1 (HNK-1) carbohydrate. Cell Mol Life Sci. 2012;69(24):4135–47.
Kleene R, Schachner M. Glycans and neural cell interactions. Nat Rev Neurosci. 2004;5(3): 195–208.
Kopitz J. Glycolipids. In: Gabius H-J, editor. The sugar code: fundamentals of glycosciences. Weinheim: Wiley-VCH; 2009. p. 177–98.
Krusius T, Finne J, Margolis RK, Margolis RU. Identification of an *O*-glycosidic mannose-linked sialylated tetrasaccharide and keratan sulfate oligosaccharides in the chondroitin sulfate proteoglycan of brain. J Biol Chem. 1986;261(18):8237–42.
Kundu SK. Glycolipids: structure, synthesis, functions. In: Allen HJ, Kisailus EC, editors. Glycoconjugates: composition, structure, and function. New York: Marcel-Dekker; 1992. p. 203–62.
Leeden RW, Wu G. Neurobiology meets glycosciences. In: Gabius H-J, editor. The sugar code: fundamentals of glycosciences. Weinheim: Wiley-VCH; 2009. p. 495–516.
Leymarie N, Zaia J. Effective use of mass spectrometry for glycan and glycopeptide structural analysis. Anal Chem. 2012;84(7):3040–8.
Li B, An HJ, Hedrick JL, Lebrilla CB. Collision-induced dissociation tandem mass spectrometry for structural elucidation of glycans. Methods Mol Biol. 2009;534:133–45.
Luther KB, Haltiwanger RS. Role of unusual *O*-glycans in intercellular signaling. Int J Biochem Cell Biol. 2009;41(5):1011–24.
Malavaki C, Mizumoto S, Karamanos N, Sugahara K. Recent advances in the structural study of functional chondroitin and dermatan sulfate in health and disease. Connect Tissue Res. 2008; 49(3):133–9.
Miljkovic M. Carbohydrates: synthesis, mechanisms, and stereoelectronic effects. 1st ed. New York: Springer; 2010.
Morita I, Kizuka Y, Kakuda S, Oka S. Expression and function of the HNK-1 carbohydrate. J Biochem. 2008;143(6):719–24.
Müthing J. Influenza A, and Sendai viruses preferentially bind to fucosylated gangliosides with linear poly-N-acetyllactosaminyl chains from human granulocytes. Carbohydr Res. 1996;290(2):217–24.
Nagai Y, Iwamori M. Cellular biology of gangliosides. In: Rosenberg A, editor. Biology of the sialic acids. New York: Plenum Press; 1995. p. 197–241.
Nakamura N, Lyalin D, Panin VM. Protein *O*-mannosylation in animal development and physiology: from human disorders to *Drosophila* phenotypes. Semin Cell Dev Biol. 2010;21(6):622–30.
Nandini CD, Sugahara K. Role of the sulfation pattern of chondroitin sulfate in its biological activities and in the binding of growth factors. Adv Pharmacol. 2006;53:253–79.
Nishimura S. Toward automated glycan analysis. Adv Carbohydr Chem Biochem. 2011;65: 219–71.

North SJ, Hitchen PG, Haslam SM, Dell A. Mass spectrometry in the analysis of *N*-linked and *O*-linked glycans. Curr Opin Struct Biol. 2009;19(5):498–506.
Orlando R. Quantitative analysis of glycoprotein glycans. Methods Mol Biol. 2013;951:197–215.
Patsos G, Corfield A. *O*-Glycosylation. In: Gabius H-J, editor. The sugar code: fundamentals of glycosciences. Weinheim: Wiley-VCH; 2009. p. 111–37.
Paulick MG, Bertozzi CR. The glycosylphosphatidylinositol anchor: a complex membrane-anchoring structure for proteins. Biochem. 2008;47(27):6991–7000.
Salmon AH, Satchell SC. Endothelial glycocalyx dysfunction in disease: albuminuria and increased microvascular permeability. J Pathol. 2012;226(4):562–74.
Sasisekharan R, Raman R, Prabhakar V. Glycomics approach to structure-function relationships of glycosaminoglycans. Ann Rev Biomed Eng. 2008;8:181–231.
Schachter H, Brockhausen I. The biosynthesis of serine(threonine)-*N*-acetylgalactosamine-linked carbohydrate moieties. In: Allen HJ, Kisailus EC, editors. Glycoconjugates. New York: Marcel-Dekker; 1992. p. 263–332.
Schauer R. Sialic acids: fascinating sugars in higher animals and man. Zoology (Jena). 2004;107(1): 49–64.
Schauer R. Sialic acids as regulators of molecular and cellular interactions. Curr Opin Struct Biol. 2009;19(5):507–14.
Schiel JE. Glycoprotein analysis using mass spectrometry: unraveling the layers of complexity. Anal Bioanal Chem. 2012;404(4):1141–9.
Schnaar R, Suzuki A, Stanley P. Glycospingolipids. In: Varki A, Cummings RD, Esko JD, Freeze HH, Stanley P, Bertozzi CR, Hart GW, Etzler ME, editors. Essentials of glycobiology. New York: Cold Spring Harbor Laboratory Press; 2008. p. 129–41.
Schwarting GA, Jungalwala FB, Chou DK, Boyer AM, Yamamoto M. Sulfated glucuronic acid-containing glycoconjugates are temporally and spatially regulated antigens in the developing mammalian nervous system. Dev Biol. 1987;120:65–76.
Shams-Eldin H, Debierre-Grockiego F, Schwarz RT. Glycosylphosphatidylinositol anchors: structure, biosynthesis and functions. In: Gabius H-J, editor. The sugar code: fundamentals of glycosciences. Weinheim: Wiley-VCH; 2009. p. 155–73.
Springer SA, Gagneux P. Glycan evolution in response to collaboration, conflict, and constraint. J Biol Chem. 2013;288(10):6904–11.
Stanley P, Cummings RD. Structures common to different glycans. In: Varki A, Cummings RD, Esko JD, Freeze HH, Stanley P, Bertozzi CR, Hart GW, Etzler ME, editors. Essentials of glycobiology. New York: Cold Spring Harbor Laboratory Press; 2008. p. 175–98.
Stanley P, Schachter H, Taniguchi N. *N*-Glycans. In: Varki A, Cummings RD, Esko JD, Freeze HH, Stanley P, Bertozzi CR, Hart GW, Etzler ME, editors. Essentials of glycobiology. New York: Cold Spring Harbor Laboratory Press; 2008. p. 101–14.
Svennerholm L. Designation and schematic structure of gangliosides and allied glycosphingolipids. Prog Brain Res. 1994;101:XI–XIV.
Taylor ME, Drickamer K. Introduction to Glycobiology. 3rd ed. Oxford: Oxford University Press; 2011.
Tivnan A, Shannon W, Gubala V, Nooney R, Williams DE, McDonagh C, et al. Inhibition of neuroblastoma tumor growth by targeted delivery of microRNA-34a using anti-disialoganglioside G_{D2} coated nanoparticles. PLoS One. 2012;7(5):e38129.
Tokuda A, Ariga T, Isogai Y, Komba S, Kiso M, Hasegawa A, et al. On the specificity of anti-sulfoglucuronosyl glycolipid antibodies. J Carbohydrate Chem. 1998;17:535–46.
Varki A. Diversity of the sialic acids. Glycobiology. 1992;2(1):25–40.
Varki A, Freeze HH. Glycans in acquired human diseases. In: Varki A, Cummings RD, Esko JD, Freeze HH, Stanley P, Bertozzi CR, Hart GW, Etzler ME, editors. Essentials of glycobiology. New York: Cold Spring Harbor Laboratory Press; 2008. p. 601–15.
Varki A, Lowe JB. Biological roles of glycans. In: Varki A, Cummings RD, Esko JD, Freeze HH, Stanley P, Bertozzi CR, Hart GW, Etzler ME, editors. Essentials of glycobiology. New York: Cold Spring Harbor Laboratory Press; 2008. p. 75–88.

Varki A, Schauer R. Sialic acids. In: Varki A, Cummings RD, Esko JD, Freeze HH, Stanley P, Bertozzi CR, Hart GW, Etzler ME, editors. Essentials of glycobiology. New York: Cold Spring Harbor Laboratory Press; 2008. p. 199–217.

Varki A, Cummings RD, Esko JD, Freeze HH, Stanley P, Bertozzi CR, Hart GW, Etzler ME, editors. Essentials of glycobiology. 2nd ed. New York: Cold Spring Harbor Laboratory Press; 2008.

Voet D, Voet JG. Biochemistry. 4th ed. New York: John Wiley & Sons; 2010.

Volpi N. Chondroitin sulfate; structure, role and pharmacological activity. Ad Pharmacol. 2006;53: 1–58.

Wang B, Miller JB, McNeil Y, McVeagh P. Sialic acid concentration of brain gangliosides: variation among eight mammalian species. Comp Biochem Physiol A Mol Integr Physiol. 1998; 119(1):435–9.

Weinbaum S, Tarbell JM, Damiano ER. The structure and function of the endothelial glycocalyx layer. Annu Rev Biomed Eng. 2007;9:121–67.

Yu RK, Ledeen RW. Configuration of the ketosidic bond of sialic acid. J Biol Chem. 1969; 24:1306–13.

Yuriev E, Ramsland PA, editors. Structural glycobiology. 1st ed. Boca Raton: CRC Press; 2012.

Zaia J. Mass spectrometry and glycomics. OMICS. 2010;14(4):401–18.

Zhang H, Uchimura K, Kadomatsu K. Brain keratan sulfate and glial scar formation. Ann NY Acad Sci. 2006;1086:81–90.

Zuber C, Roth J. *N*-Glycosylation. In: Gabius H-J, editor. The sugar code: fundamentals of glycosciences. Weinheim: Wiley-VCH; 2009. p. 87–110.

Chapter 2
Introduction to Cells Comprising the Nervous System

Douglas G. Peters and James R. Connor

Abstract The brain consists of neurons and glial cells. Neurons are responsible for integrating input and responding to stimuli from both the internal and the external environment. The integration occurs via electrical and chemical signals that impinge on the receptive area of neurons known as dendrites, and the response is via propagation of an axonal potential. Glial cells have three functionally distinct subtypes, astrocytes, oligodendrocytes, and microglia. Astrocytes perform a variety of functions responsible for maintaining homeostasis in the brain through functions such as formation of the blood–brain barrier, preserving osmolarity, and the uptake, degradation, and secretion of neurotransmitters. Oligodendrocytes are responsible for the production of myelin, a lipid-rich substance that encapsulates neuronal axons. Microglia are responsible for immune surveillance and remodeling of the CNS during both normal development and injury. Together the cells of the brain form a highly metabolic and dynamic unit with robust requirements for oxygen and nutrients.

Keywords Neuron • Axon • Glia • Oligodendrocyte • Myelin • Astrocyte • Microglia

Abbreviations

ATP	Adenosine triphosphate
CNS	Central nervous system
GFAP	Glial fibrillary acidic protein
MAG	Myelin-associated glycoprotein

D.G. Peters • J.R. Connor (✉)
Department of Neurosurgery, Penn State Milton S. Hershey Medical Center,
500 University Drive, PO Box 850, Hershey, PA 17033, USA
e-mail: jconnor@hmc.psu.edu

R.K. Yu and C.-L. Schengrund (eds.), *Glycobiology of the Nervous System*,
Advances in Neurobiology 9, DOI 10.1007/978-1-4939-1154-7_2,

MAP	Microglia-associated protein
MBP	Myelin basic protein
OPC	Oligodendrocyte precursor cell
RER	Rough endoplasmic reticulum

2.1 Introduction

The intricate network of cells in the central nervous system (CNS) consists of neurons and glial cells. For the most part, these cells develop from the neuroectoderm and neural crest during embryogenesis in a preprogrammed activity-independent manner. Neurons and glial cells have special protein expressions that subserve their function within the CNS. For example, neurons are primarily responsible for propagation of action potential so they contain evenly distributed voltage-gated ion channels to propagate these electrical signals. There are multiple types of glial cells (astrocytes, microglia, and oligodendrocytes), and each of these subtypes subserves a different function that will be discussed later. In addition to their specific responsibilities within the brain, the various cell types within the CNS contain many similar components present in eukaryotes found in cells outside the brain.

The cell body of all cells in the brain contains a nucleus and cytoplasmic organelles. The nucleus is surrounded by a nuclear envelope, which is continuous with the rough endoplasmic reticulum (RER). The RER is studded with ribosomes, which help translate mRNAs transported from the nucleus into protein. Vesicles that bud from the endoplasmic reticulum fuse with the Golgi complex, which further modifies proteins for cell sorting to specific organelles or secretion outside of the cell. Within neurons, areas with an increased need for ATP, such as the presynaptic and postsynaptic terminals, active growth cones or branch points of axons and dendrites, and the nodes of Ranvier, have a greater density of mitochondria than other parts of the neuron (Sheng and Cai 2012). Other organelles found in cells of the brain are proteasomes which are large multi-enzymatic organelles that identify and degrade proteins tagged with ubiquitin and lysosomes which are smaller organelles that utilize vesicular transportation to collect molecules to degrade.

The cells of the brain share similar cytoskeletal components to other cell types, but these components support unique functions and enable cells of the brain, especially neurons, to have complex and unique morphological appearances. There are three main components of the cytoskeleton classified by diameter: microfilaments (7 nm), intermediate filaments (10 nm), and microtubules (25 nm). Actin is the predominant microfilament in neurons where it clusters at presynaptic terminals, dendritic spines, and growth cones. The breakdown and elongation of actin fibers are highly dynamic and responsive to both extracellular and intracellular factors. Intermediate filaments convey structural rigidity and are not found in dynamic structures. There are six classes of intermediate filaments with distinctive patterns of expression specific to cell type and developmental stage (see Table 2.1). The unique biochemical composition of the intermediate filaments among cell types has

Table 2.1 Types of intermediate filaments

Class	Example	Cell types/location	Cell function
I/II (acidic/basic)	Cytokeratin	Epithelial cells/cytoplasm	Cell–cell adhesion, tensile strength, and intracellular communication
III	Glial fibrillary acidic protein (GFAP)	Astrocytes/cytoplasm	Cell structure and intercellular communication
IV	α-Internexin, neurofilaments	Neurons/cytoplasm	Axon structure and growth
V	Lamins	Ubiquitous/nucleus	Help form the nuclear envelope and position nuclear pores
VI	Nestin	Neural precursor/extracellular surface	Axon guidance and growth cone anchor during development

been exploited by immunocytologists for identifying cell types. Microtubules are comprised of tubulin subunits that form polarized hollow tubes with plus and minus ends. Posttranslational modifications of tubulin subunits and microtubule-associated proteins (MAPs) increase microtubule diversity and confer unique microtubule functional regions (Wloga and Gaertig 2010). Tau is a MAP with 71 phosphorylation sites that are regulated by more than 20 protein kinases (Sergeant et al. 2008). *O*-glycosylation of tau occurs on phosphorylated serine and threonine residues adjacent to proline residues. *O*-glycosylation is involved in protein formation and degradation (Hart et al. 2007). It is possible that *O*-glycosylation is involved in the molecular pathology or neuronal dysfunction observed in neurological disorders, such as Alzheimer's disease, but further work is necessary to address this.

2.2 Neurons

Neurons in their cytological entirety were first visualized using the silver staining techniques of the Italian physician Camillo Golgi. The extensive dendritic patterns of neurons led Golgi to develop the Reticular Theory for the cell-based function of the brain, in which he postulated that the nervous system is a continuous structure of networked neurons. However, the Spanish histologist Ramón y Cajal countered with the Neuron Doctrine, which states that the brain is composed of individual neuron units that are functionally linked. Identification of the functional linkage of neurons, by units subsequently named "synapses" by Sherrington, supported the Neuron Doctrine which is the cellular basis of nervous system function used today (Grant 2007).

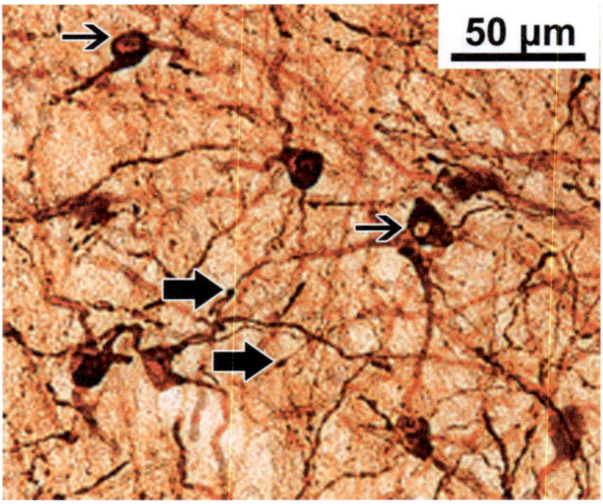

Fig. 2.1 Mouse tyrosine hydroxylase-positive (TH+) dopaminergic neurons in the substantia nigra pars compacta (SNc) (200×). In this image the neuronal cell bodies appear with a central *light-colored* nuclei (examples are indicated by *thin arrows*). The immunostained processes running throughout the image are dendrites that have emanated from the cell bodies (two examples are shown by the *thicker arrows*). These dendrites receive excitatory inputs from axons of glutamatergic and cholinergic cells (Image provided by Dr. Amanda Snyder)

Neurons are the specialized cells in the CNS that serve as units of information processing, integrating an organism's response to external stimuli. The human brain is comprised of about 100 billion neurons with roughly 100 trillion synapses between them. There are about 100 unique types of neurons within the CNS. They come in many different sizes and forms but have common components that subserve their specific functions. The basic form of a neuron is polarized, consisting of cell body or soma with a receptive bundle of dendrites on one side and a transmitting axon on the other side. A neuron's unique ability to transmit directional information is due to their asymmetry and polarization. An example of a tyrosine hydroxylase-positive (TH+) dopaminergic neuron in the substantia nigra pars compacta is shown in Fig. 2.1. These neurons receive information via synapses from other neurons and output information to the striatum, and they subserve voluntary movement as well as other broad behavioral processes.

Neurons communicate with one another via polarizing waves generated by action potentials. An action potential is a short-lived "all-or-none" depolarizing wave that begins at the neuronal body and ends at the axon terminal. It is initiated when the membrane potential shifts from a resting state of −65 mV to the threshold −50 mV due to the influx of Na^+. Voltage-gated Na^+ channels embedded within the axon help propagate the signal. Since these voltage-gated Na^+ channels automatically close upon a membrane potential of +50 mV, the electrical signal is propagated unidirectionally. Also at around +50 mV, voltage-gated K^+ channels open, and potassium flows out of the cell. This shifts the membrane potential toward the resting potential, and the membrane reaches steady-state basal levels due primarily to the constitutively opened potassium leak channels.

The neuronal cell body receives direct depolarizing (excitatory) and hyperpolarizing (inhibitory) impulses from other neurons via synapses on its cell surface. The synapse is a region where the cell membrane of the signaling (presynaptic) neuron comes into close proximity to the target (postsynaptic) neuron by means of synaptic cell adhesion molecules (synCAMs). Certain synCAMs can increase or decrease their adhesion by means of site-specific *N*-glycosylation, thus altering synapse formation (Fogel et al. 2010). Membranous vesicles are formed and packed with neurotransmitter within the presynaptic neuron. Neurotransmitter vesicles fuse with the presynaptic neuron upon binding with calcium. The neurotransmitter enters the space between the pre- and postsynaptic neuron called the synaptic cleft. There, the neurotransmitter can bind a specific receptor on the postsynaptic neuron, leading to an excitatory or inhibitory signal depending on the type (positive or negative) and direction (inward or outward) of ion flux. The flow of polarizing waves travels from dendrites to the cell body. Summation is the combination of temporal and spatial polarizing waves. The closer the depolarizing waves occur in time and space, the more likely they are to overlap and reach the action potential threshold. Summation of synaptic inputs occurs at a region between the soma and axon called the axon hillock, which is characterized by microtubules packed into bundles that enter the axon as parallel fascicles. The axon hillock is where an action potential is generated. A region adjacent to the axon hillock termed the initial segment has a thickened specialized membrane, which contains many voltage-gated ion channels. Inhibitory synapses on this region may act to prevent an action potential from occurring (O'Rourke et al. 2012).

2.2.1 Dendrites

Multiple dendrites can radiate from the neuronal cell body. These dendrites branch locally, and the differences in the morphological appearance of the dendritic mass of different types of neurons are used to characterize them (i.e., unipolar, bipolar, and multipolar). Dendrites increase a neuron's receptive field by as much as 95 %. Dendrites are highly plastic, meaning that they can rapidly change size with development, activity, and aging. Experiments have shown that enriched environments are essential for facilitating proper dendritic arborization in a variety of animals at different ages (Uylings et al. 1978; Connor et al. 1981). Some dendrites contain small, actin- and tubulin-rich protrusions scattered across their surface, which are known as dendritic spines. Most of these spines contain the postsynaptic components of excitatory synapses. These tiny dendritic spines are also highly plastic. Findings of mRNA and ribosomes at the base of the spine suggest local synthesis of proteins and local regulation of the spine morphology.

2.2.2 *Axon*

All neurons have a single axon emanating from the soma. An axon is an elongated process by which a neuron makes contact with a target cell and facilitates site-to-site conduction. Axon diameter and length can vary; however, the diameter of a given axon remains the same so as not to alter the flow of the action potential. Axons in short pathways, for example, within the associative cortex, may be as short as 1 mm, whereas those within the spinal cord may be up to a meter long (Debanne et al. 2011). Longer axons are usually myelinated and project from neurons with large cell bodies, which facilitates propagation of the axon potential (*infra vide*).

Axons are dynamic structures composed of long expanses of polarized microtubules. The alignment of microtubules varies throughout a neuron. However, in axons, the "plus" (growing) end of parallel microtubules faces away from the soma and the "minus" (shrinking) end toward it. The alignment of microtubules allows for axonal transport. Because ribosomes are not present in the axon to generate protein and passive diffusion is too slow for shipping molecules over long expanses, essential proteins and growth signals need to be actively transported in both anterograde and retrograde fashion between the soma and axonal terminal. For example, membrane-bound organelles are transported in both the retrograde and anterograde systems at a rate faster than 400 mm per day via motor proteins by a process termed fast axonal transport. Whereas cytosolic and cytoskeletal proteins are transported in an anterograde direction at a slower rate, ~2.5 mm per day, by a process termed slow axonal transport. The transport method utilizes the sliding action of short preassembled microtubules along existing microtubules. The short microtubule segments are connected to their protein cargo with cross-linked protein bridges. The molecular motor proteins responsible for axonal transport are dynein and kinesin. In most cases, cytoplasmic dynein directs cargo toward the "minus end" of microtubules and soma via retrograde transport. Whereas kinesin moves its cargo toward the "plus end" of microtubules and the axonal terminal in a process termed anterograde transport. Early experiments showed that microtubules were essential for fast transport because addition of alkaloids known to disrupt microtubules blocked it.

2.3 Glia

Glial cells were discovered by Rudolf Virchow and were given their name because they were first believed to just hold neurons in place, like glue (Kettenmann and Verkhratsky 2008). But over time, these cells have emerged as vital to maintaining CNS homeostasis—increasing synaptic fidelity, providing trophic support, and contributing to the immune response within the brain. The three main types of glial cells found in the CNS are astrocytes, oligodendrocytes, and microglia.

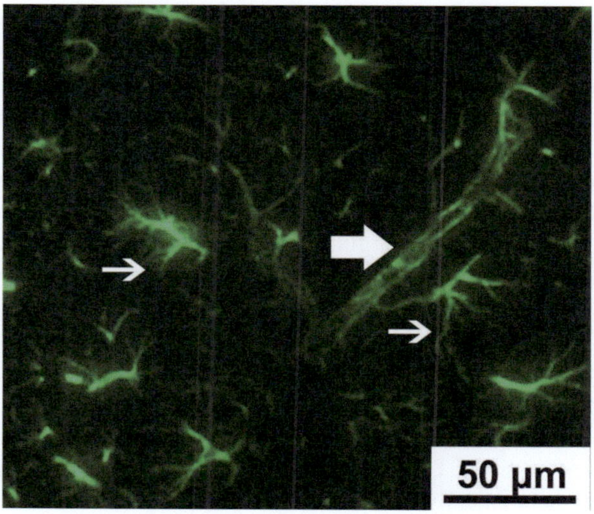

Fig. 2.2 Astrocytes in the mouse hippocampus following immunostaining for glial fibrillary acidic protein (GFAP), a cytoskeletal protein specific to astrocytes (Table 2.1). The *thinner arrows* indicate one of many short processes extending from a protoplasmic astrocyte. The *thicker arrow* indicates the outline of a small capillary that traverses the image form the *upper right* to the *middle* that is surrounded by the GFAP-positive glial end feet. An example of an astrocyte with a foot process leading to the blood vessel is indicated by the *thin arrow* on the *right* (Image provided by Dr. Wint Nandar)

2.3.1 Astrocytes

Astrocytes are the most numerous of the glial cells in the vertebrate brain, taking up 20–50 % of all cerebral volume and outnumbering neurons 5–1 (Sofroniew and Vinters 2010). Immunostaining for astrocyte-associated markers, such as GFAP, S-100, glutathione-S-transferase-μ, and glutamine synthetase, has helped to characterize astrocytes as highly branched and stellated cells and to reveal their unique metabolic roles (see Fig. 2.2). They are organized in a tile-like web throughout the brain and can be divided into two main morphological types: fibrous and protoplasmic. Fibrous astrocytes are found predominantly in white matter and display many long thin filaments projecting from oval-shaped nuclei, while protoplasmic astrocytes found in gray matter have short thick processes around spherical nuclei. The morphological differences between the two subtypes are most likely artificial based on their ability to adapt to location. In development, astrocytes divide from radial glial cells in the ventricular zone and progenitors in the subventricular zone (Ge et al. 2012). After development, astrocytes are retained to radial glial domains, perhaps important for establishing and sustaining an astrocyte network (Merkle et al. 2007). Astrocytes also play a role in forming the blood–brain barrier (BBB) by secreting angiogenic factors and making bidirectional connections with blood vessels through glial end feet.

Astrocytes form a syncytium, whereby gap junctions between neighboring astrocytes act to relay nutrients and ions between the blood and neurons. They are responsible for preserving osmolarity in the brain by means of ion intake, spatial buffering, and controlled water diffusion through expression of aquaporins. Since the brain is contained inside a rigid skull, minor changes in water metabolism may lead to compressive cerebral edema, hydrocephalus, or alterations in neuronal activity (Rash et al. 1998). Astrocytes can also take up, degrade, and secrete neurotransmitters and their precursors (i.e., glutamate) thereby helping to regulate synaptic activity. Thus, astrocytes are key in prevention of excitotoxicity. Astrocytes are the only storage site in the brain for glycogen and are therefore an important regulator of brain glucose levels. They also secrete neurotrophins, such as BDNF, GDNF, and HGF, which are important for neuron growth, survival, and synaptogenesis (Sofroniew and Vinters 2010).

Astrocytes also migrate and change shape as they react to an acute trauma in a process called astrogliosis. During this time, astrocytes become bigger, more plastic, mobile, and metabolically active. Astrogliosis has a positive and negative impact on brain recovery. Immediately after injury, astrocytes scavenge for debris and control proinflammatory events dependent on the type of stimulation. Eventually, astrocytes mesh together with microglia and other astrocytes to form a dense glial scar. The scar compartmentalizes the injury and inflammation, which helps modulate a local immune response, prevents abhorrent axonal growth, and helps revascularize the damaged tissue (Rolls et al. 2009). Initially, astrogliosis helps to preserve neuronal survival and homeostasis in the brain; however, the scars produced can restrict axonal growth and thus attempts to reestablish functional connections.

2.3.2 Oligodendrocytes and Myelin

Oligodendrocytes are unique glial cells responsible for providing support and insulation of axons within the CNS. Unlike astrocytes, oligodendrocytes do not contain a large number of integral transport proteins to maintain metabolic and ion distribution within the CNS. Oligodendrocytes arise from a large population of oligodendrocyte precursor cells (OPC), often called NG2 cells because they express the proteoglycan protein NG2 (Nave 2010). NG2 cells have been shown to have a variety of complex cellular processes that are preserved even after cell division (Ge et al. 2009). Oligodendrocytes are similar to Schwann cells found in the peripheral nervous system because both are responsible for enveloping axons with their plasma membranes. There are some mechanistic differences as Schwann cells will envelope one axon segment of one neuron, while oligodendrocytes can envelope axonal segments of up to 30 different neurons (Butt 2012). Mature oligodendrocytes wrap their cell membranes several times around an axonal segment in a spiral pattern, leaving a multilayered lipid sheath termed myelin. Mature oligodendrocytes align in rows in white matter tracts but can also be present individually (Fig. 2.3).

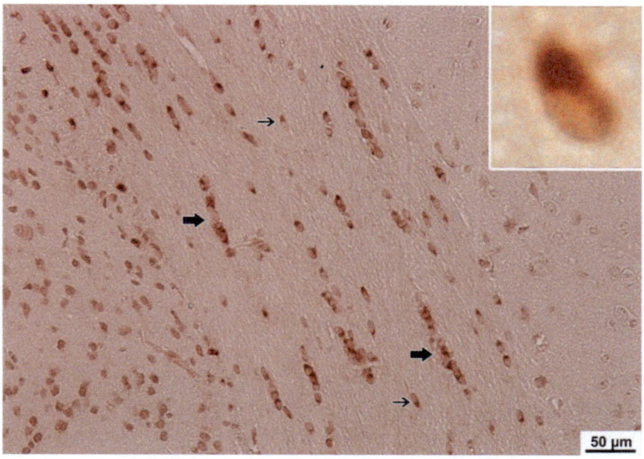

Fig. 2.3 Oligodendrocytes in the white matter from a mouse brain. The oligodendrocytes in this micrograph are expressing transferrin, an iron mobilization protein, which has been detected using immunocytochemistry. The oligodendrocytes appear in *rows* (*thick arrows*) but also individual cells can be observed (*thin arrows*). A single immunostained oligodendrocyte is magnified in the *inset* (*top right*). In this cell the eccentric nucleus (*unstained area*) is clearly demarcated from the cytoplasm (appears as a *cap* in the image) that contains the immunoreaction product for transferrin (This image was provided by Dr. Wint Nandar)

Myelin is an extension of an oligodendrocyte's plasma membrane. It contains bilipid membrane layers interspersed between layers of protein. The proteins including myelin basic protein (MBP), myelin-associated glycoprotein (MAG), and proteolipid protein help to stabilize the myelin. MBP is a small positively charged protein located on the cytoplasmic side of the myelin. MAG is an adhesive cell surface protein with a structure similar to that of members of the Ig superfamily (see Chap. 11). MAG contains a carbohydrate epitope necessary for the adhesion and enzymatic interactions between neurons and oligodendrocytes, inhibiting neurite outgrowth. The carbohydrate moieties linked to MAG have also been implicated as possible antigens for autoimmune neuropathies (Quarles 2007). Proteolipid protein is a large hydrophobic integral membrane protein.

There are several ways in which neurons benefit from myelination. Firstly, it reduces the number of Na+/K+ ATPase channels on the axonal surface, greatly decreasing the energy needed to create a resting membrane potential. The areas of axon between myelin segments termed nodes of Ranvier contain high concentrations of Na^+ channels responsible for propagating a depolarizing action potential wave. This process is called saltatory conduction. Secondly, myelin increases the resistance across the membrane surface, reducing the ions that leak across and increasing the conductive velocity of action potentials. Thirdly, myelin is important for axonal longevity and health. Some recent work claims that the lactate transporter, monocarboxylate transporter 1 (MCT1), found in the extracellular membrane of oligodendrocytes provides metabolic support to axons and neurons (Lee et al. 2012). MCT1 also transports pyruvate, ketone bodies and protons, needed for neuronal metabolism (Morrison et al. 2013). However, it is unclear whether oligodendrocytes provide a direct energy link to neurons or if astrocytes are a necessary intermediary.

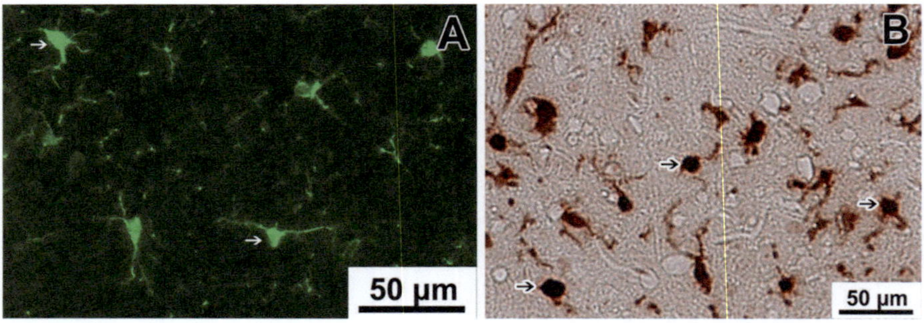

Fig. 2.4 Ionized calcium-binding adapter molecule 1 (IBA-1) was used to visualize rat microglial cells (200×) in the spinal cord. Microglia have two major appearances. In *Panel A*, we show resting microglia. These cells have small cell bodies (*arrows*) from which short branching processes are learly visible. When microglia encounter inflammatory cytokines or cellular debris, they become activated (*Panel B*; examples indicated by the *arrows*). In this state their cell bodies become enlarged, more rounded, and processes are short or no longer present (Image provided by Dr. Wint Nandar)

2.3.3 Microglia

Microglia are the smallest glial cells with highly branched processes and dense, heterochromatic nuclei when in the resting state. They are the only resident immune cells of the CNS, and they belong to a monocyte–macrophage lineage (Prinz and Mildner 2011). Microglia are identified histologically using antibodies against F4/80, Mac-1, and Iba-1 (macrophage antigens), lectins (plant proteins binding carbohydrates), CD11, and CD45 (see Fig. 2.4). Microglia are not formed from the neuroectoderm like other cells in the CNS but are derived from erythromyeloid precursors in the yolk sack. The erythromyeloid precursors migrate to early brain structures beginning at embryonic day 9 (E9) in a process termed early embryonic hematopoiesis (EHp) (Neumann and Wekerle 2013). This process is short lived during embryogenesis. Beyond embryogenesis, circulatory monocytes derived from bone marrow have also been shown to enter the brain in response to microglial loss in the brain and inflammation. This process branches from “definitive,” lifelong hematopoiesis (DHp), which is also responsible for generating neutrophils, peripheral phagocytes, and perivascular macrophages important for forming the BBB (Neumann and Wekerle 2013). Upon entering the brain, the monocytes differentiate into active microglia (Soulet and Rivest 2008). Microglia, with their small circumscribed dendritic mass, act as an immune-like surveillance cell, patrolling the healthy brain in a quiescent state and making temporary contacts with pre- and postsynapses, perisynaptic astrocytes, and the synaptic cleft (Nimmerjahn et al. 2005; Miyamoto et al 2013). Microglia can quickly sample multiple factors in their environment and respond accordingly to alter multiple elements in the brain’s milieu. The surfaces of microglia are covered with receptors indicating their involvement in innate, bacterial, and viral immunity. They also express receptors for neurotransmitters and trophic factors. Activation of microglia can be induced by the presence of a foreign substance (i.e., viral particle) or the absence of a normal signal (i.e., trophic factor). When activated, microglia retract their processes and become rounded and move in response to a gradient of chemokines (Nimmerjahn et al. 2005). Once at the end of their guided course, microglia have been shown to phagocytose neuronal debris, sequester heavy metals, shape synaptic connections, and alter synaptic transmission by removing synapses (Graeber and Streit 2010; Pascual et al. 2012). Thus, activation of microglia can be related to normal neuronal development, normal neuronal function, as well as response to infection or trauma.

Microglia are functionally distinguished by classification as M1 or M2. M1 microglia are commonly associated with pro-inflammation, and M2 microglial are associated with anti-inflammation and phagocytosis. Microglia can express the M1 or M2 phenotype upon activation. Active microglia have been shown to secrete a variety of factors that may be either neuroprotective or neurotoxic depending on the response and cytokines released. Recent research suggests that this variation in responsive profile may allow microglia to respond differently at different times, perhaps contributing to age-associated brain diseases (Crain et al. 2013). For example, M2 microglia induce oligodendrocyte differentiation during remyelination, and thus they may be a critical target in multiple sclerosis therapies (Miron et al. 2013). In addition, a study of Rett syndrome, an autistic spectral disorder, showed that microglial engraftment via bone marrow transplant to the mouse model brain arrested many facets of the pathology (Derecki et al. 2012). It is assumed that the therapeutic microglia in this experiment are M2 because preventing phagocytosis with annexin V reduced their benefit. In contrast, as an example of the M1 phenotype, in a neuron–microglia coculture model of Alzheimer's activation of microglia increases release of interleukin-1 (IL-1), leading to a loss of synaptophysin through tau phosphorylation (Li et al. 2003). Moreover, selectively inhibiting M1 expression reduced the onset of progressive amyotrophic lateral sclerosis in mice (Kobayashi et al. 2013).

2.4 Summary

Throughout life, the cells of the central nervous system are dynamic. They form direct, indirect, and complementary networks to enhance the metabolism and homeostasis of the brain. Neurons and glial cells are able to modify their shape and normal physiological function when there is a significant change in their environmental milieu. The stochastic interplay of countless enzymatic pathways affecting transcriptional, translational, and posttranslational alterations allows all the cells of the CNS to adapt to changes in input and alterations in their milieu. A failure of adaptation results, particularly in terminally differentiated neurons, in cell death. Understanding the interplay between glia and neurons, including oligodendrocytes and myelin surrounding neuronal axons, is key to understanding neurodegenerative diseases. How the adaptive mechanisms can be utilized to enhance neuronal survival and how stress responses may be minimized are critical areas requiring additional investigation if we are going to optimize cognitive performance following disease or injury.

Conflict of Interest The authors declare no conflicts of interest.

References

Butt AM. Structure and function of oligodendrocytes. In: Kettenmann H, Ransom BR, editors. Neuroglia. 3rd ed. New York: Oxford University Press; 2012. p. 62–73.

Connor JR, Diamond MC, Connor JA, Johnson RE. A Golgi study of the dendritic morphology of socially reared aged rats. Exp Neurol. 1981;73:525–33.

Crain JM, Nikodemova M, Watters JJ. Microglia express distinct M1 and M2 phenotypic markers in the postnatal and adult central nervous system in male and female mice. J Neurosci Res. 2013;91(9):1143–51.

Debanne D, Campanac E, Bialowas A, Carlier E, Alcaraz G. Axon physiology. Physiol Rev. 2011;91:555–602.

Derecki NC, Cronk JC, Lu Z, Xu E, Abbott SB, Guyenet PG, Kipnis J. Wild-type microglia arrest pathology in a mouse model of Rett syndrome. Nature. 2012;484(7392):105–9.

Fogel AI, Li Y, Giza J, Wang Q, Lam TT, Modis Y, Biederer T. N-glycosylation at the SynCAM (synaptic cell adhesion molecule) immunoglobulin interface modulates synaptic adhesion. J Biol Chem. 2010;285(45):34864–74.

Ge WP, Zhou W, Lou Q, Jan LY, Jan YN. Dividing glia cells maintain differentiated properties including complex morphology and functional synapses. Proc Natl Acad Sci U S A. 2009;106(1):328–33.

Ge WP, Miyawaki A, Gage FH, Jan YN, Jan LY. Local generation of glia is a major astrocyte source in postnatal cortex. Nature. 2012;484(7394):376–80.

Graeber MB, Streit WJ. Microglia: active sensor and versatile effector cells in the pathologic brain. Nat Neurosci. 2010;10:1387–94.

Grant G. How the 1906 Nobel Prize in physiology or medicine was shared between Golgi and Cajal. Brain Res Rev. 2007;55(2):490–8.

Hart GW, Housley MP, Slawson C. Cycling of O-linked β-N-acetylglucosamine on nucleocytoplasmic proteins. Nature. 2007;446(7139):1017–22.

Kettenmann H, Verkhratsky A. Neuroglia: 150 years later. Trends Neurosci. 2008;31(12):653–9.

Kobayashi K, Imagama S, Ohgomori T, Hirano K, Uchimura K, Sakamoto K, Hirakawa A, Takeuchi H, Suzumura A, Ishiguro N, Kadomatsu K. Minocycline selectively inhibits M1 polarization of microglia. Cell Death Dis. 2013;4:e525.

Lee Y, Morrison BM, Li Y, Lengacher S, Farah MH, Hoffman PN, et al. Oligodendroglia metabolically support axons and contribute to neurodegeneration. Nature. 2012;487(7408):442–8.

Li Y, Liu L, Barger SW, Griffith WS. Interleukin-1 mediates pathological effects of microglia on tau phosphorylation and on synaptophysin synthesis in cortical neurons through a p38-mapk pathway. J Neurosci. 2003;23:1605–11.

Merkle FT, Mirzadeh Z, Alvarez-Buylla A. Mosaic organization of neural stem cells in the adult brain. Science. 2007;317(5836):381–4.

Miron VE, Boyd A, Zhao JW, Yuen TJ, Ruckh JM, Shadrach JL, van Wijngaarden P, Wagers AJ, Williams A, Franklin RJ, ffrench-Constant C. M2 microglia and macrophages drive oligodendrocyte differentiation during CNS remyelination. Nat Neurosci. 2013;16(9):1211–8.

Miyamoto A, Wake H, Moorhouse AJ, Nabekura J. Microglia and synapse interactions: fine tuning neural circuits and candidate molecules. Front Cell Neurosci. 2013;7:70.

Morrison BM, Lee Y, Rothstein JD. Oligodendroglia: metabolic supporters of axons. Trends Cell Biol. 2013;23(12):644–51.

Nave KA. Myelination and support of axonal integrity by glia. Nature. 2010;468(7321):244–52.

Neumann H, Wekerle H. Brain microglia: watchdogs with pedigree. Nat Neurosci. 2013;16(3): 253–5.

Nimmerjahn A, Kirchoff F, Helmchen F. Resting microglial cells are highly dynamic surveillants of brain parenchyma in vivo. Science. 2005;308:1314–8.

O'Rourke NA, Weilelr NC, Micheva KD, Smith SJ. Deep molecular diversity of mammalian synapses: why it matters and how to measure it. Nat Rev Neurosci. 2012;13(6):365–79.

Pascual O, Ben Achour S, Rostaing P, Triller A, Bessis A. Microglia activation triggers astrocyte-mediated modulation of excitatory neurotransmission. Proc Natl Acad Sci U S A. 2012;109(4):E197–205.

Prinz M, Mildner A. Microglia in the CNS: Immigrants from another world. Glia. 2011;59: 177–87.

Quarles RH. Myelin-associated glycoprotein (MAG): past, present, and beyond. J Neurochem. 2007;100(6):1431–48.

Rash JE, Yasumura T, Hudson CS, Agre P, Nielsen S. Direct immunogold labeling of aquaporin-4 in square arrays of astrocytes and ependymocyte plasma membranes in rat brain and spinal cord. Proc Natl Acad Sci U S A. 1998;95(20):11981–6.

Rolls A, Shechter R, Schwartz M. The bright side of the glial scar in CNS repair. Nat Rev Neurosci. 2009;10(3):235–41.

Sergeant N, Bretteville A, Hamdane M, Caillet-Boudin ML, Grognet P, Bombois S, et al. Biochemistry of Tau in Alzheimer's disease and related neurological disorders. Expert Rev Proteomics. 2008;5(2):207–24.

Sheng Z, Cai Q. Mitochondrial transport in neurons: impact on synaptic homeostasis and neurodegeneration. Nat Rev Neurosci. 2012;13(2):77–93.

Sofroniew M, Vinters H. Astrocytes: biology and pathology. Acta Neuropathol. 2010;119(1): 7–35.

Soulet D, Rivest S. Bone-marrow-derived microglia: myth or reality? Curr Opin Pharmacol. 2008;8(4):508–18.

Uylings HBM, Kuypers K, Diamond MC, Veltman WAM. Effects of differential environments on plasticity of dendrites of cortical pyramidal neurons in adult rats. Exp Neurol. 1978;62(3): 658–77.

Wloga D, Gaertig J. Post-translational modifications of microtubules. J Cell Sci. 2010;123: 3447–55.

Chapter 3
Synthesis, Processing, and Function of N-glycans in N-glycoproteins

Erhard Bieberich

Abstract Many membrane-resident and secrected proteins, including growth factors and their receptors, are N-glycosylated. The initial N-glycan structure is synthesized in the endoplasmic reticulum (ER) as a branched structure on a lipid anchor (dolichol pyrophosphate) and then co-translationally, "en bloc" transferred and linked via *N*-acetylglucosamine to asparagine within a specific N-glycosylation acceptor sequence of the nascent recipient protein. In the ER and then the Golgi apparatus, the N-linked glycan structure is modified by hydrolytic removal of sugar residues ("trimming") followed by re-glycosylation with additional sugar residues ("processing") such as galactose, fucose, or sialic acid to form complex N-glycoproteins. While the sequence of the reactions leading to biosynthesis, "en bloc" transfer and processing of N-glycans is well investigated, it is still not completely understood how N-glycans affect the biological fate and function of N-glycoproteins. This review discusses the biology of N-glycoprotein synthesis, processing, and function with specific reference to the physiology and pathophysiology of the nervous system.

Keywords Chaperones • Congenital disorders of glycosylation • ERAD • Glycolipids • N-glycoproteins • N-glycans • Processing • Trimming • Gangliosides • Glycosyltransferases

E. Bieberich (✉)
Department of Neuroscience and Regenerative Medicine, Medical College of Georgia, Georgia Regents University, 1120 15th Street Room CA4012, Augusta, GA 30912, USA
e-mail: ebieberich@gru.edu

R.K. Yu and C.-L. Schengrund (eds.), *Glycobiology of the Nervous System*, Advances in Neurobiology 9, DOI 10.1007/978-1-4939-1154-7_3,

Abbreviations

CDG	Congenital disorders of glycosylation
COG	Conserved oligomeric Golgi
dNM	Deoxynojirimycin (dNJ)
Dol-PP	Dolichol pyrophosphate
ER	Endoplasmic reticulum
ERAD	ER-assisted degradation
Gal	Galactose
GalNAc	*N*-acetylgalactosamine
GD2	Ganglioside GD2
GD3	Ganglioside GD3
Glc	Glucose
GlcNAc	*N*-acetylglucosamine
GM1	Ganglioside GM1
GM3	Ganglioside GM3
M6P	Mannose-6-phosphate
Man	Mannose
OST	Oligosaccharyl transferase

3.1 Introduction

Throughout the life cycle of neural cells undergoing a "metamorphosis" from neural stem cells to mature neurons, astrocytes, and oligodendrocytes (or Schwann cells in the peripheral nervous system), the differentiation and function of these cells relies upon their response to extracellular and intracellular signaling cues. This response depends on the specificity and sensitivity of receptor proteins. It becomes increasingly clear that the sensitivity of receptors is regulated by specific N-glycan residues that affect: (1) secretion, stability, and clearance of the receptor ligands; (2) surface expression, internalization, and recycling or turnover of the receptors; (3) adhesion of neurons and other cells via cell surface receptors and extracellular matrix proteins; and (4) signal induction and transduction by growth factor and neurotransmitter receptors and ion channels. In many of these cases, the N-glycan enhances (1) proper folding of ligand or receptor; (2) solubility or polarity of the ligand or receptor; and (3) binding to extracellular or intracellular factors that induce cell signaling pathways or mediate further processing of the N-glycoprotein. In particular, the latter has gained recent attention since specific N-glycans can regulate protein association in receptor/ligand complexes or sugar-specific binding proteins in the plasma membrane (e.g., galectins) that mediate endocytosis or exocytosis, transport or sorting, and recycling or turnover of the receptor (Boscher et al. 2011, 2012; Dennis et al. 2009a, b; Lajoie et al. 2009; Lajoie and Nabi 2010).

While these mechanisms are critical for the proportion of receptors expressed on the cell surface or the retrograde transport of signalosomes (protein complexes between ligand and receptor), it can also modulate the exocytotic transport and secretion of proteins with pathological effects such as amyloid or prion protein (Browning et al. 2011; McFarlane et al. 1999). It is not surprising that the particular structure of N-glycans and therefore, the sequence of enzymatic processing steps leading to their structure is the focus of intensive research, in particular for the identification of new drug targets. For more than three decades, specific inhibitors of glycosidases involved in N-glycoprotein processing have been tested for their application in antiviral and tumor therapy (Wohlfarth and Efferth 2009; Nash et al. 2011; Robina et al. 2004; Wrodnigg et al. 2008). On the other hand, mutations in proteins that mediate N-glycosylation and N-glycan processing can lead to severe diseases, including those of the nervous system (Goreta et al. 2012; Jaeken 2010, 2011; Freeze 2002). This is not limited to mutations in trimming or processing glycosidases, but encompasses proteins mediating the transport of N-glycoproteins for their processing in the ER or Golgi, as found in human congenital disorders of glycosylation (CDG) (Fung et al. 2012; Reynders et al. 2011; Kelleher and Gilmore 2006). To define the function of N-glycans in normal physiology and disease one needs to know their precise structure and the enzymatic steps generating this structure. Currently, the rapid progress in high throughput mass spectrometric analysis has opened a growing field of comprehensive glycomics studies on N-glycans and other proteinogenic glycoconjugates such as O-glycans and proteoglycans (Moremen et al. 2012). Mutation analysis and the genome projects have provided us with the information and tools needed to study proteins involved in N-glycoprotein biosynthesis and processing. And yet, the dynamics of biochemistry on the cellular level requires knowledge beyond the statics of structure and sequence, an insight into the flux of biological reactions.

To understand this flux, one may envision the cell as a gigantic factory with sequential assembly lines for the generation and functional editing of N-glycans. This processing works like a flow chart with individual yes/no decision points for: (1) glycosylation leading to non-glycosylated or glycosylated proteins; (2) trimming and reglucosylation leading to "high mannose" and "reglucosylated" intermediates; and (3) further processing and re-glycosylation leading to "mannose-6-phosphate," "hybrid," or "complex" N-glycoproteins as end products. As the result of this editing process, N-glycans act as specific addresses or tags that regulate the processing or function of their attached proteins by a simple rule: N-glycan-dependent association kinetics between enzymes, receptor proteins, and other factors keeping the N-glycoprotein in a particular compartment or moving it to the next. As a first step of understanding processing of N-glycans, we will discuss the initial assembly of the N-glycan and its transfer to the protein.

3.2 N-Glycans Are First Born on a Lipid and then Transferred "En Bloc" onto the Nascent N-Glycoprotein in the ER

N-glycans are oligosaccharides by their chemical nature: branched chains of sugar residues attached to each other by α- and β-glycosidic linkages (Fig. 3.1). However, the N-glycan is not made on the protein, but pre-manufactured on an ER-resident

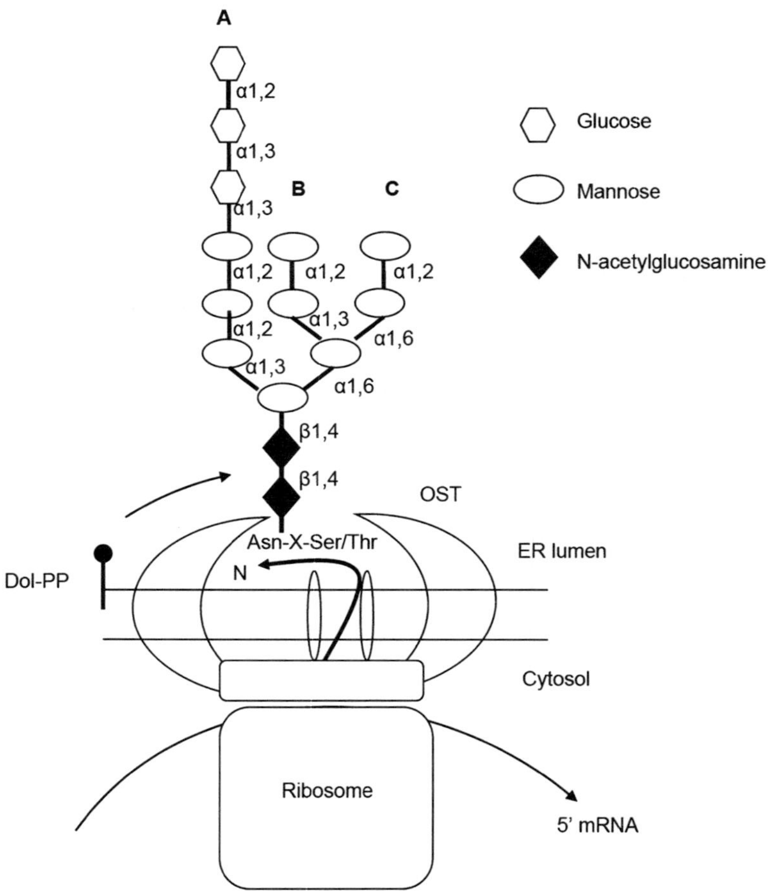

Fig. 3.1 Co-translational, "en bloc" transfer of the Glc3Man9GlcNAc2 precursor oligosaccharide from dolichol pyrophosphate to asparagine, catalyzed by oligosaccharyl transferase in the ER lumen

lipid, the polyprenol dolichol pyrophosphate. Further, while after attachment to the protein, the N-glycan points to the lumen of the ER, the assembly of the initial 7 of the 14 sugar residues in the dolichol pyrophosphate (Dol-PP)-linked precursor oligosaccharide is accomplished at the cytosolic side. Hence, the N-glycosylation reaction relies on two critical steps: transport of the partial precursor oligosaccharide (Dol-PP-GlcNAc2Man5) across the ER membrane from the cytosolic to the luminal side and then after further glycosylation reactions, "en bloc" transfer of the mature precursor to the protein. In eukaryotes, the assembly of this precursor oligosaccharide is achieved by a set of ER-resident, transmembrane protein glycosyltransferases of the types (a), *N*-aceytylglucosaminyltransferases that attach two GlcNAc residues to Dol-PP in a β1-N and then β1-4 linkage, (b) mannosyltransferases catalyzing four different glycosidic linkages: β-1,4 (first mannose attached to second GlcNAc), α-1,3 and α-1,6 (mannose at the two branching points of the biantennary oligosaccharide), and α-1,2 (mannose elongation of the middle (B) and two outer (A and C) branches, see Fig. 3.1) that attach a total of nine mannose residues to Dol-PP-GlcNAc2 to form a branched Dol-PP-GlcNAc2Man9 structure, and (c) glucosyltransferases that attach the terminal three glucose residues in two α1,3- and one terminal α1,2-glycosidic linkage onto the outer α1,3-mannosidic branch (A branch) of the precursor oligosaccharide (Fig. 3.1). These reactions depend on activated sugars that are provided on the cytosolic and luminal side of the ER in the form of UDP-GlcNAc, GDP-mannose (GDP-man), and UDP-glucose (UDP-glc).

Mutations in glycosyltransferases and the flippase transporting the Dol-PP-GlcNAc2Man5 partial precursor from the cytosol into the ER lumen lead to a spectrum of diseases known as congenital disorders of glycosylation type I (CDGs type I) (Leroy 2006; Mohorko et al. 2011). The symptoms of CDGs type I often involve the nervous system. For example, mutations of the RFT1 (requiring fifty three 1, CDG-In) yeast homolog, the mammalian flippase, cause sensorineural deafness (Jaeken 2010; Leroy 2006; Goreta et al. 2012). CDG-In is an extremely rare disease, only six patients are known so far. Other CDGs type I resulting from mutations in precursor glycosyltransferases often show defects in multiple tissues that can lead to multiorgan failure as observed with glucosyltransferase II deficiency (CDG-Ih). It should be noted that while CDGs type I can be very severe it is not clear which N-glycoprotein dysfunction due to hypoglycosylation accounts for a specific disease phenotype or symptom. The diagnosis of CDG type I is commonly based on testing for hypoglycosylation of transferrin in patient plasma, although this aberrant glycosylation does not account for the entire spectrum of symptoms observed with this disease. As discussed later, the consequences of hypoglycosylation can be severe for a variety of proteins, the proper folding of which relies on the intact N-glycan residue. Therefore, it may not be surprising that CDGs type I can lead to multiple tissue and organ failures.

Following assembly and the flipping reaction of the precursor oligosaccharide, co-translational "en bloc" transfer of the N-glycan from Dol-PP onto the nascent polypeptide is the next critical step in N-glycoprotein biosynthesis. This step is catalyzed by oligosaccharyl transferase (OST), a multimeric enzyme complex composed of nine subunits in yeast and four subunits in higher eukaryotes (Moremen et al. 2012; Roth et al. 2010; Ruiz-Canada et al. 2009; Parodi 2000; Kelleher et al. 1992; Kelleher and Gilmore 2006; Bause et al. 1995; Hardt et al. 2000; Sharma et al. 1981; Kaplan et al. 1987; Pless and Lennarz 1977). The precise function of these subunits is still a subject of ongoing research; however, it is clear that OST manages association with the ribosome and signal recognition particle receptor for the nascent membrane glycoprotein, the recognition of the acceptor sequence Asn-X-Ser/Thr (X cannot be proline), and the catalytic transfer of the oligosaccharide from Dol-PP onto an asparagine (Fig. 3.1).

As with mutations in glycosyltransferases involved in the precursor oligosaccharide assembly, deficiencies of OST subunits lead to several CDGs of type I. The catalytic subunit of OST, Stt3 in yeast or Stt3A and B in mammals, is a conserved subunit that is already expressed in archaebacteria. It has been speculated that because of the redundancy conferred by the two mammalian Stt3 homologs, CDGs have not yet been discovered (Kelleher and Gilmore 2006; Vleugels et al. 2009; Mohorko et al. 2011; Ruiz-Canada et al. 2009). However, mutations in other OST subunits have been associated with various CDGs of type Ix. For example, deficiencies of the ribophorin I subunit RPN2 have been found by screening for hypoglycosylation defects without impairment of precursor oligosaccharide assembly (Vleugels et al. 2009). At present, the pathology of these CDGs is unclear, but it is likely to involve nervous system defects. Once OST has transferred the precursor oligosaccharide onto the nascent polypeptide, a process starts that entails an important decision for the newly born N-glycoprotein: *fold* or *fail*. In particular, the glucose residues on the outer mannosidic (α1,3- or A) branch are instrumental in assisting the protein folding proofreading and refolding process, which will be discussed in the next section.

3.3 Trimming, Reglycosylation, and Remodeling: There Are Many Ways of N-Glycoprotein Processing in the ER and Golgi

There are two pathways by which the N-linked Glc3Man9GlcNac2 oligosaccharides are processed ("trimmed"): the glucosidase-dependent and independent pathway (Fig. 3.2). The glucosidase-dependent pathway occurs in the ER and is mediated by a sequential hydrolytic cleavage of the terminal glucose residues by glucosidases I and II (Fig. 3.2). This sequence is embedded into chaperone-assisted proofreading of protein folding: the calnexin–calreticulin cycle. The glucosidase-independent pathway is catalyzed by an endomannosidase which cleaves off a Glc1-3Man1 residue in the Golgi (Hamilton et al. 2005; Lubas and Spiro 1987; Alonzi et al. 2013). The function and regulation of this reaction is not known. Therefore, we will focus on the glucosidase-dependent pathway and its interaction with the calnexin–calreticulin cycle for protein folding.

To date, the calnexin–calreticulin cycle has been discussed as one of the most important processes underlying the function of N-glycosylation (Deprez et al. 2005; Ellgaard et al. 1999; Hammond et al. 1994; Hammond and Helenius 1993, 1994; Molinari and Helenius 2000). It was first described by Ari Helenius in 1994 based on two important observations: (a) calnexin and calreticulin are chaperones in the ER that bind to monoglucosylated (GlcMan9GlcNAc2) N-glycoproteins and (b) the ER harbors a glucosyltransferase that reattaches glucose to N-linked Man9GlcNAc2 (Hammond and Helenius 1993, 1994; Hammond et al. 1994; Trombetta et al. 1989, 1991; Sousa et al. 1992; Trombetta and Parodi 1992). Trimming or processing of the terminal glucose residues is thus crucial for the chaperone function of calnexin and calreticulin. The function of a chaperone is to recognize and bind misfolded proteins and then mediate refolding until the proper conformation of the protein is accomplished. If the misfolded protein cannot adopt its proper conformation it will be degraded in the ER by the ERAD (ER-assisted degradation) system, a rather

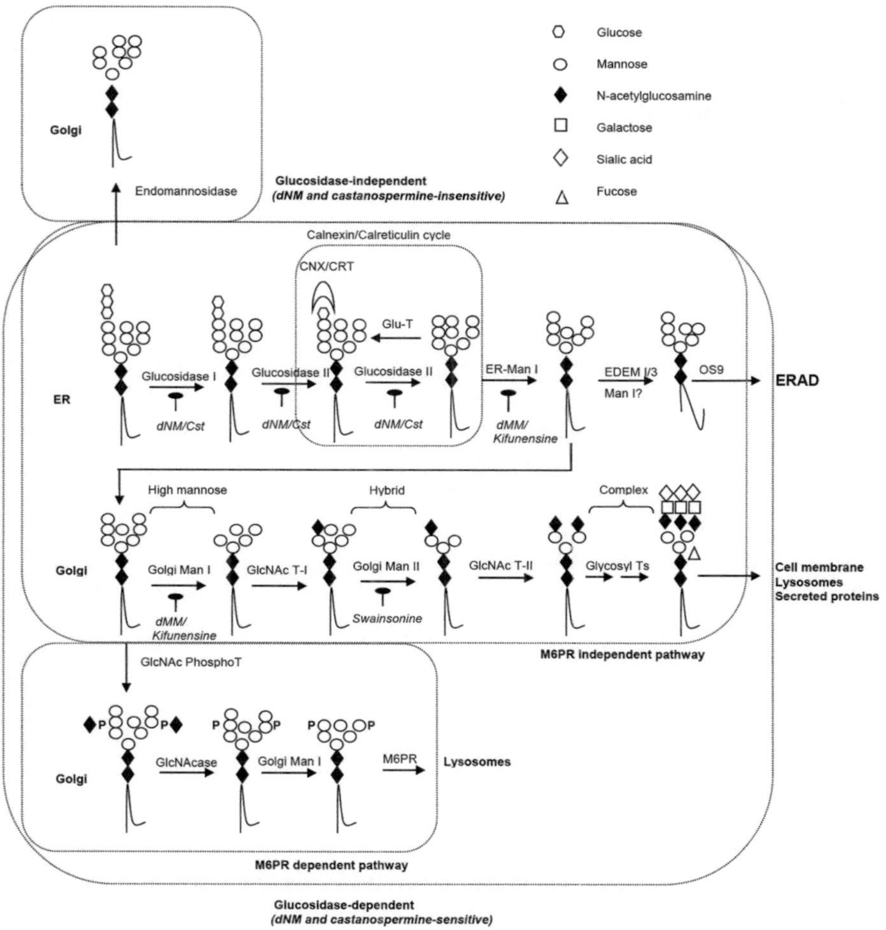

Fig. 3.2 N-glycoprotein processing: sequential removal (trimming) of glucose and mannose residues from the N-linked glycan in the ER and Golgi, followed by re-glycosylation. N-glycan processing generates signals for chaperone-assisted refolding, mannose-6-phosphate receptor-mediated transport of lysosomal enzymes, and other functions in protein trafficking, enzyme complex formation, and cell adhesion

complicated organized sequential process of "unfolded protein response" that first shuttles the misfolded protein into the cytosol, where it is ubiquitinated and then removed by proteasomal degradation (Banerjee et al. 2007). Another ERAD-associated system transports the misfolded proteins to the Golgi and then the lysosome for proteolysis. Failure of ERAD can result in severe diseases due to the accumulation and aggregation of misfolded proteins. A prominent example related to neurobiology is Parkinson's disease that can result from ERAD malfunction involving the E3-ubiquitinase parkin (Ron et al. 2010). Mutations in glucocerebrosidase (GCase), a lysosomal β-glucosidase deficient in Gaucher's disease, can lead to the accumulation of aberrant GCase attached to parkin, which is currently discussed as one of the causes of Parkinson's disease (Ron et al. 2010). Since GCase cleaves off glucose form glucosylceramide to generate ceramide, it is an example for the interdependence of N-glycoprotein and glycosphingolipid metabolism, which will be discussed later in this chapter. In addition to the aggregation and accumulation of aberrant proteins, ERAD malfunction may also lead to premature degradation of proteins that are still in the process of refolding, which has been discussed as one of the causes of cystic fibrosis (Farinha and Amaral 2005). From these examples, it is evident that the N-glycan assisted calnexin–calreticulin cycle for protein folding and thus the function of ERAD critically relies on proper N-glycan processing.

To generate the monoglucosylated N-glycan, glucosidases I and II remove the first two glucose residues from the N-linked Glc3Man9GlNAc2 oligosaccharide (Peyrieras et al. 1983; Kalz-Fuller et al. 1995; Parodi 2000; Volker et al. 2002). The resulting GlcMan9GlcNAc2 is bound to calreticulin or calnexin and the protein (partially) refolded, which is followed by removal of the innermost glucose residue by glucosidase II. After this, the N-glycan (Man9GlcNAc2) is transiently reglucosylated to GlcMan9GlcNAc2 by UDP-glucose:glycoprotein glucosyltransferase to prevent further N-glycan trimming of a glycoprotein that has not yet adopted its proper folding state. After glucosidase II has cleaved off the newly added glucose residue, the Man9GlcNAc2 oligosaccharide reenters the reglucosylation–refolding–trimming cycle until the native conformation of the N-glycoprotein is accomplished. Once the protein is correctly folded, the chaperones do not bind to the protein anymore and the Man9GlcNAc2 oligosaccharide is processed by a series of mannosidases (Fig. 3.2).

The removal of mannose residues from Man9GlcNAc2 is initiated in the ER and intimately associated with the ERAD response to misfolded proteins (Banerjee et al. 2007; Tokunaga et al. 2000; Wang and White 2000; Ruddock and Molinari 2006; Alonzi et al. 2013). If correct folding cannot be achieved the N-glycoprotein is not reglucosylated and ER mannosidase I cleaves off the terminal alpha1,2-mannose residue of the middle (B) branch in the N-glycan (Fig. 3.1) generating a Man8GlcNAc2 oligosaccharide (Fig. 3.2). How exactly the ERAD machinery distinguishes between properly folded and misfolded Man8GlcNAc2 bearing glycoproteins is not yet fully understood. However, it has been shown that overexpression of ER mannosidase I yields to the acceleration of ERAD-mediated protein degradation. In vivo, this function is thought to be mediated by EDEM (ER degradation enhancing alpha mannosidase like) proteins that accelerate the ERAD response to misfolded N-glycoproteins instead of allowing further exit to the Golgi (Helenius and Aebi 2004; Ruddock and Molinari 2006; Hosokawa et al. 2010). It is expected that mutations in trimming enzymes, in particular glucosidase I, II, and ER mannosidase I would lead to another type of CDGs, called CDG type II since these mutations will affect the N-glycan structure after assembly and transfer to the protein.

Indeed, the first CDG type II known to be caused by trimming glycosidase deficiency is CDG type IIb, which results from mutations in ER glucosidase I (Volker et al. 2002). CDG type IIb shows multiorgan deficiencies leading to a variety of symptoms such as hepatomegaly, hypoventilation, feeding problems, seizures, and in one patient a fatal outcome at 74 days after birth. Interestingly, other CDG type II diseases are not directly associated with mutations of trimming enzymes (or have not yet been found), but with proteins of the intercisternal Golgi transport machinery such as the "conserved oligomeric Golgi" (COG) complex proteins. Since trimming glycosidases and N-glycan-associated glycosyltransferases are located in distinct Golgi compartments it is expected that mutations in these transport proteins would also lead to disorders of N-glycosylation (Kelleher and Gilmore 2006; Reynders et al. 2011; Fung et al. 2012). The COG proteins form a complex of eight proteins critical for retrograde vesicle transport within the Golgi. It is known to regulate Golgi distribution of mannosidases and glycosyltransferases important for N- and O-glycoprotein processing (Moremen et al. 2012). Mutations in COG proteins are known to cause CDGs of type II, many of those with presentation of nervous system disorders.

The majority of CDGs caused by COG proteins are associated with abnormal reglycosylation of processed N-glycans due to mislocalization of the respective glycosyltransferases. This process of reshaping the N-linked oligosaccharide by reglycosylation is initiated by three distinct Golgi-resident mannosidases and follows (at least) two different routes (Fig. 3.2). The transport of lysosomal hydrolases requires the attachment of a Mannose-6-phosphate tag (mannose-6-phosphate or M6P-dependent pathway), while other N-glycoproteins of the "hybrid" and "complex" type are further trimmed by removal of additional mannose residues. We will discuss the pathway of further trimming first.

Golgi mannosidase I is distinct from ER mannosidase I in that it cleaves off 3 α1,2-residues from the Man8GlcNAc2 precursor to yield Man5GlcNAc2 (Helenius and Aebi 2004; Dunphy et al. 1981; Cummings et al. 1983; Moremen et al. 2012). This α1,2 exomannosidase is also different from Man9 mannosidase, which has been cloned and characterized by the author's former group and cleaves three mannose residues from Man9GlcNAc2 to yield Man6GlNAc2, which can then be substrate of Golgi mannosidase I that removes one additional mannose residue (Schweden et al. 1986; Bause et al. 1993; Bieberich and Bause 1995; Bieberich et al. 1997). Regardless of how the Man5GlcNAc2 oligosaccharide is generated, further trimming proceeds after attachment of one GlcNAc residue to the outer mannose (A) branch by UDP-GlcNAc transferase I (Fig. 3.2). The resulting GlcNAcMan5GlcNAc2 oligosaccharide is then either elongated by the addition of sugar derivatives such as GalNAc or sialic acid (hybrid N-glycans), or it is subjected to removal of an additional two mannose residues by Golgi mannosidase II, which generates N-linked GlcNAcMan3GlcNAc2, the "core" glycan structure that is the initial building block for all complex N-glycoproteins (Arumugham and Tanzer 1983; Cummings et al. 1983; Moremen and Touster 1985; Moremen and Robbins 1991; Herscovics et al. 1994; Moremen 2002). As with hybrid N-glycans, complex N-linked oligosaccharides are reglycosylated with additional sugar derivatives such as GalNAc or sialic acid, or fucose (Fig. 3.2). These complex N-glycans come in a large variety of highly branched structures, which can exceed the initial biantennary (two branching points) structure of high mannose oligosaccharides by far.

Biochemically, high mannose N-glycans can be distinguished from hybrid or complex ones by the use of two endoglycosidases that either cleave off the complete N-glycan (glycopeptidase F) or hydrolyze the β-glycosidic linkage between the two GlcNAc residues (chitobiose) of high mannose N-glycoproteins (endoglycosidase H). These enzymes and other glycosidases were discovered early in the history of glycobiology, which in combination with metabolic labeling using radioactive sugars tremendously facilitated the structural analysis of N-glycans (Tarentino et al. 1974).

Several glycosylation deficiencies are known to result from mutations of enzymes in the M6P-dependent pathway or from aberrant reglycosylation of hybrid or complex N-glycoproteins. In the M6P-dependent pathway for lysosomal enzymes, the Man8GlcNAc glycan is first endowed with two GlcNAc phosphate residues that are attached by a Golgi transferase to the subterminal mannose residues of the two outer mannose branches (Coutinho et al. 2012; Lemansky et al. 1984; Ghosh et al. 2003; Gary-Bobo et al. 2007; Waheed et al. 1981; Chao et al. 1990; Lubke et al. 1999). Next, the mannose-bound phosphate residues are uncovered by *N*-acetylglucosaminidase and the terminal mannose residues of the outer (A and C) branches are removed by Golgi mannosidase I (Fig. 3.2). The resulting P2Man6GlcNAc2 oligosaccharide is now recognized by the M6P receptor in the trans-Golgi, which binds to the N-glycoprotein and initiates its transport to the late endosome. While the late endosome matures to lysosomes, the pH value drops and the M6P receptor releases the lysosomal enzyme. The M6P receptor is then recycled to the trans-Golgi for further transport of lysosomal enzymes. Deficiencies in the M6P-dependent pathway, in particular caused by mutations of GlcNAc phosphotransferase, the enzyme attaching the GlcNAc phosphate residues to the N-glycan, can lead to severe disorders of glycosylation. Well-known examples are I-cell disease or mucolipidosis type II, and pseudo-Hurler polydystrophy or mucolipidosis type III (Coutinho et al. 2012). These diseases are usually not classified as CDG type II but as oligosaccharidosis or mucolipidosis-type lysosomal storage diseases because failure of transporting enzymes to lysosomes will lead to the accumulation and lysosomal storage of the enzyme substrates, in particular glycosaminoglycans. Mucolipidosis type II and III lead to severe abnormalities in multiple organs (hepatomegaly, splenomegaly) and delay in cognitive and motor skills development.

As mentioned earlier, the majority of CDGs type II are caused by mutations in glycosyltransferases that generate the complex N-glycan, e.g., CDG IId, a deficiency of β1,4 galactosyltransferase I, which leads to psychomotor delay and macrocephaly (Leroy 2006; Jaeken 2010). Also, as discussed earlier, many CDGs type II of complex N-glycoprotein processing are caused by aberrant COG proteins, e.g., CDG-IIe or COG7 deficiency, which leads to hyposialylation of complex N-glycans. In contrast to lysosomal storage diseases related to aberrant M6P-dependent transport of lysosomal enzymes, the molecular cause of the deficiency aka the malfunction of the N-glycoprotein in various CDGs is not clearly defined. The reason is twofold: (a) glycosylation defects affect not only one but a variety of complex N-glycoproteins; and (b) for many N-glycoproteins it is not well understood what the physiological function of the N-glycan is, which makes it difficult to understand the malfunction as well. In the next two sections, we will discuss some of these functions, in particular with respect to the significance of N-glycoproteins for brain development and physiology. However, before we move on to these sections, it is necessary to briefly discuss important tools that have helped to elucidate the sequence of N-glycoprotein processing and the function of N-glycoproteins: trimming enzyme-specific inhibitors.

3.4 The Essential Toolbox of a Glycobiologist: A Brief History of the Discovery of N-Glycoprotein Biosynthesis Inhibitors and Their Impact on Our Understanding of N-Glycan Processing

The discovery and development of N-glycosylation and processing inhibitors is intimately linked to the history and progress in N-glycoprotein research—and the professional careers of many leading glycobiologists. The first inhibitor of N-glycosylation, tunicamycin was found more than 40 years ago in an attempt to screen for antiviral drugs made by bacteria, in particular strains of *Streptomyces* (Takatsuki et al. 1971; Takatsuki and Tamura 1971a, b, c; Schwarz et al. 1976; Ericson et al. 1977; Leavitt et al. 1977; Hart and Lennarz 1978). This finding is not as surprising as it may seem today since the screening procedures at that time were often based on virus hemagglutination and in vitro proliferation assays, which were critically affected by the glycoprotein nature of serum proteins and the virus envelope, a well-known fact even decades ago (Hewitt 1937). Therefore, one of the tests routinely performed was a competition assay determining whether addition of sugars, in particular N-acetylamino sugars would reverse the effect of the antiviral antibiotic, as seen with tunicamycin (Takatsuki and Tamura 1971b). Later, Alan Elbein discovered that tunicamycin inhibits the transfer of *N*-acetylglucosamine to dolichol phosphate, the first step in the synthesis of the lipid-linked oligosaccharide that serves as the precursor for all N-glycoproteins (Ericson et al. 1977; Chambers and Elbein 1975). It was also found early on that tunicamycin induces the degradation of viral glycoproteins and it was hypothesized that proteolysis was due to the lack of glycosylation, the first inkling of what would later be known as chaperone-assisted N-glycoprotein proofreading or the calreticulin–calnexin cycle in the ERAD response to unfolded proteins (Schwarz et al. 1976). Nowadays, tunicamycin is commonly used to induce the "unfolded protein response" or ER stress, unfortunately often without paying attention to its effect on N-glycoprotein biosynthesis. Because of its toxicity tunicamycin has never made its way into use as a virus therapy, although recent studies suggest that it may have antiviral effects against Hepatitis C virus at subtoxic doses (Reszka et al. 2010).

Another antibiotic acting on N-glycoprotein biosynthesis and isolated from *Streptomyces*, the imino sugar deoxynojirimycin (dNM or dNJ) was also discovered more than 40 years ago. It was identified as a member of the validamycin family in a screening assay for antifungal agents that would inhibit the breakdown of trehalose, the storage disaccharide in insects and fungi (Nash et al. 2011). Deoxynojirimycin is an inhibitor of α-glucosidases and was first designed for treatment of type II diabetes because of its ability to prevent the breakdown of amylase (starch) and mobilization and uptake of glucose in the intestine. Nowadays, acarbose, a natural tetrasaccharide containing a structural isomer of deoxynojirimycin is widely used as an oral medication for diabetes type II. Deoxynojirimycin itself has done more to shape the history of research on N-glycoprotein processing than it has as a therapeutic for diabetes.

At the time when acarbose and dNM were discovered, it was well known that neuraminidase-sensitive, glycoprotein-bound sialic acid is important in virus agglutination. Gilbert Ashwell and Anatol Morell reported for the first time a galactose-specific receptor for asialoglycoproteins (Van Den Hamer et al. 1970; Morell et al. 1971). Yet, besides knowing that many glycoprotein-linked oligosaccharides contained terminal sialic acid, units of galactose-β1,4-*N*-acetylglucosamide, and other disaccharides, trisaccharides, and tetrasaccharides, the actual (branched) structure of the N-linked glycan remained unclear until 1978, when Ellen Li and Stuart Kornfeld proposed a structure with branched mannose chains linked to asparagine via *N*-acetylglucosamine (Li and Kornfeld 1978). The 1970s and early 1980s became a wellspring of new discoveries in glycobiology. Willam Lennarz worked out the biosynthesis pathway of the dolichol-linked precursor oligosaccharide and Amando Parodi reported the "en bloc" transfer of this precursor onto the nascent glycoprotein (Parodi et al. 1972; Waechter et al. 1973; Lucas et al. 1975).

Deoxynojirimycin and other inhibitors such as castanospermine played essential roles in this discovery since they inhibited not only broad spectrum α-glucosidases but also glucosidase I and II, which are the other components of the N-glycan-driven proofreading machinery. Similar to tunicamycin, dNM and castanospermine prevented correct folding and secretory exit of viral N-glycoproteins and induced ER-resident protein degradation instead. However, inhibitors of Golgi mannosidases such as swainsonine or deoxymannojirimycin did not. Based on this observation and the concurrent characterization of GlcMan9GlcNAc2-binding ER chaperones, Ari Helenius then proposed the calreticulin–calnexin cycle, which is probably the most prominent example for the general function of N-glycans (Hammond et al. 1994). Besides their value in understanding the unfolded protein response to N-glycoproteins, glucosidase I and II inhibitors were also instrumental in the discovery of the M6P-dependent pathway by Kurt von Figura et al. in 1984 (Lemansky et al. 1984).

I came into contact with dNM through my graduate student mentor Dr. Gunter Legler, who contributed much to our understanding of the catalytic mechanism of α- and β-glucosidases. He and his mentee, who then became my first postdoctoral mentor, Dr. Ernst Bause, were among the first to use dNM and its derivatives to characterize and purify glycosidases, including those involved in trimming of N-glycoproteins (Hettkamp et al. 1982, 1984; Peyrieras et al. 1983; Schweden et al. 1986). At this time, the pre-human genome era, one could not just "blast search" for cDNA sequences, but actually had to purify a protein to homogeneity and then identify the amino acid sequence in order to synthesize oligonucleotide primers that could be used for RT-PCR to generate a probe for screening of a lambda gt11 library, and eventually, isolate the protein-specific cDNA; a long-forgotten and extremely tedious technique. Using alkylated dNM derivatives for affinity chromatography of trimming enzymes proved to be extremely helpful in this endeavor. Despite this progress, it took several years and thousands of plasmid minipreps to generate specific probes that then led to the first cloned cDNAs for glucosidase I and Man9 mannosidase (Bause et al. 1993; Bieberich and Bause 1995; Kalz-Fuller et al. 1995).

Currently, alkylated dNM derivatives such as N-butyl dNM are being tested as antibiotics for treatment of several viral diseases, including HIV (Robina et al. 2004; Ratner and Vander 1993). Grabowski's and Legler's groups have reported that alkylated dNM derivatives inhibit the two enzymes that regulate the metabolic conversion of ceramide and glucosylceramide into each other, lysosomal β-glucosidase (the "Gaucher enzyme") and glucosyltransferase, with longer alkyl chain length being more specific for β-glucosidase and shorter chain length for glucosyltransferase (Greenberg et al. 1990; Osiecki-Newman et al. 1986, 1987; Legler and Liedtke 1985; Legler and Bieberich 1988). Therefore, N-butyl dNM (miglustat) has been discussed as a possible treatment for several lysosomal storage diseases involving accumulation of sphingolipids, including Gaucher's disease (glucosylceramide accumulation), Niemann–Pick disease (sphingomyelin accumulation), and Tay–Sachs and Sandhoff disease (GM2 accumulation) (Nash et al. 2011; Platt et al. 1994; Chavany and Jendoubi 1998; Jeyakumar et al. 1999; Baek et al. 2008; Mistry 2000; Venier and Igdoura 2012; Abian et al. 2011; Patterson et al. 2007; Cox et al. 2000). While the use of trimming enzyme inhibitors has not (yet) led to breakthroughs in virus therapy, dNM and swainsonine have certainly been invaluable in elucidating the sequence of trimming enzymes and the function of N-glycans. One of these more recently discovered functions intertwines N-glycoprotein processing with glycolipid biosynthesis: the role of N-glycosylation in the subcellular localization and enzyme complex formation of glycosyltransferases in ganglioside biosynthesis.

3.5 Sweet Encounters of Proteins and Lipids: N-Glycans Affect the Subcellular Distribution and Complex Formation of Enzymes in Glycolipid Biosynthesis

Glycosyltransferases transfer sugar residues not only onto protein linked-glycans but also onto lipids, in particular sphingolipids. Glycosphingolipids are synthesized from ceramide, a sphingolipid consisting of sphingosine attached to various fatty acids, by sequential glycosylation reactions catalyzed by a series of ER- or Golgi-resident glycosyltransferases (Gault et al. 2010; Sandhoff and Kolter 2003; Maccioni et al. 2011a, b; Yu et al. 2010; Ngamukote et al. 2007). As expected, the substrate specificity of these enzymes is different from that of glycosyltransferases in N-glycoprotein biosynthesis and processing. After attachment of glucose and then galactose to ceramide, which generates lactosylceramide, the most basic ganglioside, GM3, is synthesized by attachment of sialic acid. This reaction, catalyzed by GM3 synthase, is followed by enzymatic reactions that split ganglioside biosynthesis into three distinct pathways: a-, b-, and c-series gangliosides (Yu et al. 1988, 2004, 2012; Svennerholm 1956; Yu and Ledeen 1972; Yu 1984, 1994; Ando and Yu 1977; Yu and Ando 1980). If *N*-acetylgalactosamine is the next sugar residue added to GM3, a reaction catalyzed by GM2/GD2 synthase, ganglioside biosynthesis will exclusively follow the a-series pathway. However, if another sialic acid residue is added first, a reaction catalyzed by GD3 synthase, then ganglioside biosynthesis follows the b- or c-series pathway. Note that GD3 synthase can only act on GM3 (thereby generating GD3), while GM2/GD2 synthase can use GM3 (thereby generating GM2) or GD3 (thereby generating GD2) as the substrate. Therefore, the relative location of GD3 and GM2/GD2 synthase determines which pathway of ganglioside biosynthesis is taken. If GM2/GD2 synthase acts first, only a-series gangliosides are made, whereas a sequential reaction of first GD3 synthase and then GM2/GD2 synthase channels ganglioside biosynthesis towards the b-series pathway. Likewise, c-series gangliosides are made if GT3 synthase acts on GD3 before GM2/GD2 synthase does.

Regulation of biosynthetic pathways by the relative location of enzymes in a reaction sequence is not just of academic curiosity, but may actually determine the composition of gangliosides in an organism, tissue, or cell. Robert K. Yu, whose laboratory I joined after my postdoctoral work on N-glycoproteins, found more than 25 years ago that the ganglioside pathways undergo a rapid switch from a- to b-series during embryonic brain development (Yu et al. 1988), just at the time point when neuroprogenitor cells start to divide asymmetrically and many intermediate neurons are born (Yu 1994; Yu et al. 2012; Bieberich and Yu 1999). Bob Yu gave me the opportunity to pursue my own ideas about the regulation of ganglioside biosynthesis by the interdependence of glycosylation reactions in glycolipid and glycoprotein biology.

To understand the meaning of this interdependence one has to know that most glycosyltransferases are type II transmembrane proteins with 3–4 N-glycosylation sites. Hence, bearing in mind what we have discussed before—N-glycans are critical for protein folding and transport—N-glycoprotein processing may regulate the subcellular localization, and therefore relative location, of glycosyltransferases in ganglioside biosynthesis. I quickly realized that I was not the only one who pursued this idea. Hugo Maccioni's and our group published in 1998 that inhibition of trimming glucosidases I and II with dNM and castanospermine, but not inhibition of Golgi mannosidase I and II with deoxymannojirimycin and swainsonine prevented transport of GD3 synthase from the ER to the Golgi (Martina et al. 1998; Bieberich et al. 2000). There was some discrepancy between both studies with respect to the effect of glucosidase inhibition on enzyme activity. While in Maccioni's study, dNM and castanospermine still preserved the activity of GD3 synthase, castanospermine increased proteolytic turnover of this enzyme in our study. Since it was known from the Helenius model that N-glycans are required to achieve chaperone-assisted protein folding via the calreticulin–calnexin cycle, both groups concluded that N-glycosylation was necessary to attain and maintain catalytic activity, while

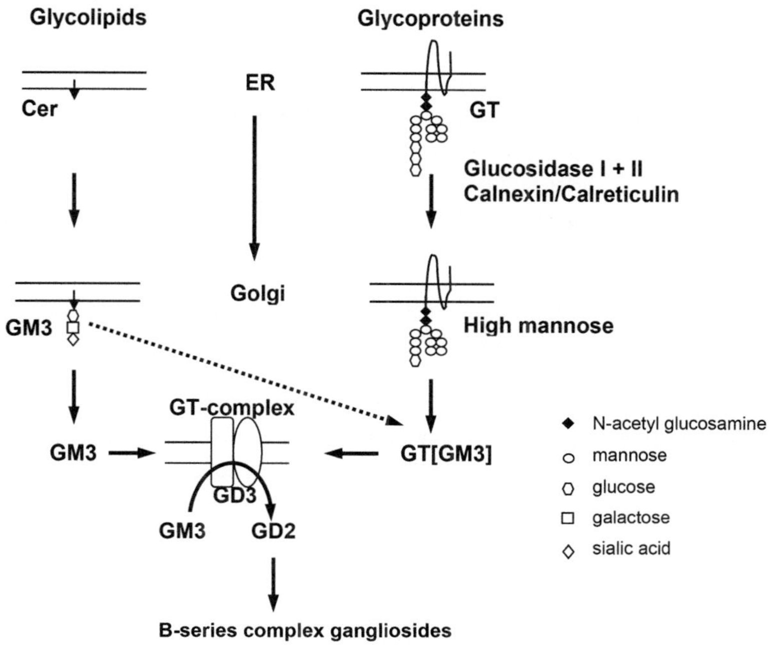

Fig. 3.3 Regulation of ganglioside biosynthetic pathways by N-glycan-dependent glycosyltransferase distribution and complex formation. Inhibition of trimming by glucosidases I and II increases proteolytic turnover of GD3 synthase and prevents enzyme complex formation with GM2/GD2 synthase in the Golgi, suggesting that N-glycoprotein processing of glycosyltransferases is critical for ganglioside metabolism. The GD3 synthase-GM2/GD2 synthase complex is hypothesized to promote b-series complex ganglioside biosynthesis. Moreover, our group has proposed that binding of GD3 synthase to GM3 may facilitate enzyme complex formation ("lipid co-chaperone" hypothesis)

glucose trimming was required for the ER-to-Golgi exit of the enzyme (Martina et al. 1998; Bieberich et al. 2000) .

In the following years, Maccioni's group made great strides toward understanding the regulation of glycosyltransferase transport and enzyme complex formation, and its significance for glycolipid biosynthesis (Maccioni et al. 1999, 2002, 2011a, b; Martina et al. 2000; Giraudo et al. 2001; Giraudo and Maccioni 2003; Maccioni 2007; Ferrari et al. 2012; Spessott et al. 2012). However, it still remained unclear how this would switch a- to b-series gangliosides and whether N-glycans are actually critical for this pathway switch. I was working in Bob Yu's group on this problem and discovered that GD3 synthase forms a disulfide bridge-mediated homodimer that turns into a heterodimer with GM2/GD2 synthase (Bieberich et al. 1998). Interestingly, binding to GM3 as well as inhibition of trimming by glucosidases I and II retained the GD3 synthase homodimer in the ER, suggesting that the enzyme-substrate complex may participate in protein folding or transport (Fig. 3.3). The observation of enzyme (–substrate) complexes in ganglioside metabolism was in line with Saul Roseman's original idea that the biosynthesis of specific gangliosides is best achieved by forming multi-enzyme complexes of glycosyltransferases ("cooperative sequential specificity" model) (Roseman 1970). Moreover, by forming the complex between GD3 synthase and GM2/GD2 synthase, we calculated that ganglioside biosynthesis would be efficiently channeled into the b-series pathway, an enzymatic switch that provided an explanation for the rapid change in the composition of gangliosides during embryonic brain development (Bieberich et al. 1998; Bieberich and Yu 1999).

In addition to the GD3 synthase homodimer and GD3 synthase-GM2/GD2 synthase heterodimer complex identified by Bob Yu's group, Hugo Maccioni's group discovered that enzyme complexes were also formed between LacCer-, GM3-, and GD3-synthase, and GM2/GD2- and GM1/GD1a synthase (Spessott et al. 2012; Ferrari et al. 2012; Maccioni et al. 2011b; Giraudo and Maccioni 2003; Giraudo et al. 2001; Martina et al. 2000). Common to two of these glycosyltransferase complexes described so far is that the stability and/or subcellular localization of at least two of their subunits, GD3 synthase and GM1/GD1a synthase, are critically dependent on N-glycosylation and trimming by glucosidases I and II. Moreover, inhibition of this trimming (by castanospermine) dramatically changes the ganglioside composition by preventing synthesis of higher sialylated, complex b-series gangliosides such as GT1b, the most prominent ganglioside after the "a-to-b series switch" in mouse (and human) embryonic brain at a time point of intense neural progenitor (and intermediate neuron) proliferation and migration (Bieberich et al. 1998). It should be noted that Bob Yu's group not only was the first to discover this ganglioside pathway switch but also demonstrated that the simplest b-series ganglioside, GD3, is a robust cell surface marker for mouse (and human) neural progenitor cells (Nakatani et al. 2010; Ngamukote et al. 2007; Yanagisawa et al. 2004). Yet it remains to be determined which functional role the pathway switch plays for brain development and how it is integrated with the regulation for N-glycan processing of glycosyltransferases in ganglioside metabolism.

3.6 Conclusions and Epilogue: The Tale of the Tail That Wags the Dog

At the end of this chapter on N-glycosylation and N-glycoprotein processing, one may miss a discussion of the important functions that cell surface N-glycans play in cell-to-cell recognition and adhesion, in particular in the brain. For example, galectins, cell surface lectins that bind to *N*-acetylgalactosamine in N-linked glycans, have been found to regulate growth factor receptor endocytosis/recycling, which may contribute to glioma metastasis (Le Mercier et al. 2010). Another non-discussed example for lectin-like cell surface binding is the interaction of myelin-associated glycoprotein (MAG) with specific gangliosides, which has been suggested to be critical for myelination of axons (Yang et al. 1996). Loss of b-series complex gangliosides as well as abnormal N-glycoprotein processing of MAG leads to severe nervous system symptoms such as Wallerian degeneration and demyelination (Konat et al. 1987; Sheikh et al. 1999). The tail (protein or lipid-linked glycan) wags the dog (neuron or glia), so to speak.

One may also criticize that other types of protein glycosylation such as O-glycosylation or protein-associated glycans in general, such as proteoglycans are not mentioned (see Moremen et al. (2012) for a comprehensive review on these). The reason for this is twofold: for one, these glycoconjugates are reviewed in Chaps. 4 and 5. More importantly, I wanted to focus on biological processes that are dynamically regulated by the morphing and reshaping of protein-linked N-glycans. As we have seen, the glycosylation and trimming machinery is intimately connected with proofreading and editing of N-glycoproteins. In this regard, it should be noted that the stability, subcellular distribution, and complex formation of glycosyltransferases in ganglioside biosynthesis is the first example of enzymes in a metabolic pathway that may actually be regulated by N-glycoprotein processing. Moreover, as we have discussed, the function of N-glycoprotein processing is by far not completely understood and may involve substrates such as the ganglioside GM3 as "lipid co-chaperones." And finally, the interdependence between protein-linked N-glycan processing and glycolipid metabolism, a theme that brought me as a researcher trained in glycoprotein biology to pursue studies on glycolipids, holds promise to make future discoveries in uncharted territories with impact on systems biology; a fascinating area of research that tears down the boundaries between over-specialized disciplines in biology and goes back to a more classical, Humboldtian view on life as emerging from the self-organized biology of interacting metabolic systems, such as glycoprotein processing and glycolipid biosynthesis. Therefore, this chapter is not only meant to give an account of what is known about N-glycosylation (certainly not in an exhaustive manner) but also to ignite interest in young scientists to pursue this area of research in their careers.

Acknowledgements This work was supported by the National Institutes of Health R01AG034389 and the National Science Foundation (NSF1121579). We also thank the Department of Neuroscience and Regenerative Medicine (Chair Dr. Lin Mei), Georgia Regents University, Augusta, GA, for institutional support.

Compliance with Ethics Requirements The author, Dr. Erhard Bieberich, declares that he does not have any conflict of interest. This chapter does not contain any studies with human or animal subjects.

References

Abian O, Alfonso P, Velazquez-Campoy A, Giraldo P, Pocovi M, Sancho J. Therapeutic strategies for Gaucher disease: miglustat (NB-DNJ) as a pharmacological chaperone for glucocerebrosidase and the different thermostability of velaglucerase alfa and imiglucerase. Mol Pharm. 2011;8(6):2390–7. doi:10.1021/mp200313e.

Alonzi DS, Kukushkin NV, Allman SA, Hakki Z, Williams SJ, Pierce L, et al. Glycoprotein misfolding in the endoplasmic reticulum: identification of released oligosaccharides reveals a

second ER-associated degradation pathway for Golgi-retrieved proteins. Cell Mol Life Sci. 2013. doi:10.1007/s00018-013-1304-6.
Ando S, Yu RK. Isolation and characterization of a novel trisialoganglioside, GT1a, from human brain. J Biol Chem. 1977;252(18):6247–50.
Arumugham RG, Tanzer ML. Abnormal glycosylation of human cellular fibronectin in the presence of swainsonine. J Biol Chem. 1983;258(19):11883–9.
Baek RC, Kasperzyk JL, Platt FM, Seyfried TN. N-butyldeoxygalactonojirimycin reduces brain ganglioside and GM2 content in neonatal Sandhoff disease mice. Neurochem Int. 2008;52(6):1125–33. doi:10.1016/j.neuint.2007.12.001. S0197-0186(07)00329-4. [pii].
Banerjee S, Vishwanath P, Cui J, Kelleher DJ, Gilmore R, Robbins PW, et al. The evolution of N-glycan-dependent endoplasmic reticulum quality control factors for glycoprotein folding and degradation. Proc Natl Acad Sci U S A. 2007;104(28):11676–81. doi:10.1073/pnas.0704862104. 0704862104 [pii].
Bause E, Bieberich E, Rolfs A, Volker C, Schmidt B. Molecular cloning and primary structure of Man9-mannosidase from human kidney. Eur J Biochem. 1993;217(2):535–40.
Bause E, Breuer W, Peters S. Investigation of the active site of oligosaccharyltransferase from pig liver using synthetic tripeptides as tools. Biochem J. 1995;312(Pt 3):979–85.
Bieberich E, Bause E. Man9-mannosidase from human kidney is expressed in COS cells as a Golgi-resident type II transmembrane N-glycoprotein. Eur J Biochem. 1995;233(2):644–9.
Bieberich E, Yu RK. Multi-enzyme kinetic analysis of glycolipid biosynthesis. Biochim Biophys Acta. 1999;1432(1):113–24. S0167-4838(99)00085-0 [pii].
Bieberich E, Treml K, Volker C, Rolfs A, Kalz-Fuller B, Bause E. Man9-mannosidase from pig liver is a type-II membrane protein that resides in the endoplasmic reticulum. cDNA cloning and expression of the enzyme in COS 1 cells. Eur J Biochem. 1997;246(3):681–9.
Bieberich E, Freischutz B, Liour SS, Yu RK. Regulation of ganglioside metabolism by phosphorylation and dephosphorylation. J Neurochem. 1998;71(3):972–9.
Bieberich E, Tencomnao T, Kapitonov D, Yu RK. Effect of N-glycosylation on turnover and subcellular distribution of N-acetylgalactosaminyltransferase I and sialyltransferase II in neuroblastoma cells. J Neurochem. 2000;74(6):2359–64.
Boscher C, Dennis JW, Nabi IR. Glycosylation, galectins and cellular signaling. Curr Opin Cell Biol. 2011;23(4):383–92. doi:10.1016/j.ceb.2011.05.001. S0955-0674(11)00062-7 [pii].
Boscher C, Zheng YZ, Lakshminarayan R, Johannes L, Dennis JW, Foster LJ, et al. Galectin-3 protein regulates mobility of N-cadherin and GM1 ganglioside at cell-cell junctions of mammary carcinoma cells. J Biol Chem. 2012;287(39):32940–52. doi:10.1074/jbc.M112.353334. M112.353334 [pii].
Browning S, Baker CA, Smith E, Mahal SP, Herva ME, Demczyk CA, et al. Abrogation of complex glycosylation by swainsonine results in strain- and cell-specific inhibition of prion replication. J Biol Chem. 2011;286(47):40962–73. doi:10.1074/jbc.M111.283978. M111.283978 [pii].
Chambers J, Elbein AD. Biosynthesis and characterization of lipid-linked sugars and glycoproteins in aorta. J Biol Chem. 1975;250(17):6904–15.
Chao HH, Waheed A, Pohlmann R, Hille A, von Figura K. Mannose 6-phosphate receptor dependent secretion of lysosomal enzymes. EMBO J. 1990;9(11):3507–13.
Chavany C, Jendoubi M. Biology and potential strategies for the treatment of GM2 gangliosidoses. Mol Med Today. 1998;4(4):158–65. doi:10.1016/S1357-4310(98)01227-1.
Coutinho MF, Prata MJ, Alves S. Mannose-6-phosphate pathway: a review on its role in lysosomal function and dysfunction. Mol Genet Metab. 2012;105(4):542–50. doi:10.1016/j.ymgme.2011.12.012. S1096-7192(11)00652-4 [pii].
Cox T, Lachmann R, Hollak C, Aerts J, van Weely S, Hrebicek M, et al. Novel oral treatment of Gaucher's disease with N-butyldeoxynojirimycin (OGT 918) to decrease substrate biosynthesis. Lancet. 2000;355(9214):1481–5. doi:10.1016/S0140-6736(00)02161-9. S0140-6736(00)02161-9 [pii].
Cummings RD, Kornfeld S, Schneider WJ, Hobgood KK, Tolleshaug H, Brown MS, et al. Biosynthesis of N- and O-linked oligosaccharides of the low density lipoprotein receptor. J Biol Chem. 1983;258(24):15261–73.

Dennis JW, Lau KS, Demetriou M, Nabi IR. Adaptive regulation at the cell surface by N-glycosylation. Traffic. 2009a;10(11):1569–78. doi:10.1111/j.1600-0854.2009.00981.x. TRA981 [pii].
Dennis JW, Nabi IR, Demetriou M. Metabolism, cell surface organization, and disease. Cell. 2009b;139(7):1229–41. doi:10.1016/j.cell.2009.12.008. S0092-8674(09)01555-4 [pii].
Deprez P, Gautschi M, Helenius A. More than one glycan is needed for ER glucosidase II to allow entry of glycoproteins into the calnexin/calreticulin cycle. Mol Cell. 2005;19(2):183–95. doi:10.1016/j.molcel.2005.05.029. S1097-2765(05)01384-5 [pii].
Dunphy WG, Fries E, Urbani LJ, Rothman JE. Early and late functions associated with the Golgi apparatus reside in distinct compartments. Proc Natl Acad Sci U S A. 1981;78(12):7453–7.
Ellgaard L, Molinari M, Helenius A. Setting the standards: quality control in the secretory pathway. Science. 1999;286(5446):1882–8. 8063 [pii].
Ericson MC, Gafford JT, Elbein AD. Tunicamycin inhibits GlcNAc-lipid formation in plants. J Biol Chem. 1977;252(21):7431–3.
Farinha CM, Amaral MD. Most F508del-CFTR is targeted to degradation at an early folding checkpoint and independently of calnexin. Mol Cell Biol. 2005;25(12):5242–52. doi:10.1128/MCB.25.12.5242-5252.2005. 25/12/5242 [pii].
Ferrari ML, Gomez GA, Maccioni HJ. Spatial organization and stoichiometry of N-terminal domain-mediated glycosyltransferase complexes in Golgi membranes determined by fret microscopy. Neurochem Res. 2012;37(6):1325–34. doi:10.1007/s11064-012-0741-1.
Freeze HH. Human disorders in N-glycosylation and animal models. Biochim Biophys Acta. 2002;1573(3):388–93. S0304416502004087 [pii].
Fung CW, Matthijs G, Sturiale L, Garozzo D, Wong KY, Wong R, et al. COG5-CDG with a mild neurohepatic presentation. JIMD Rep. 2012;3:67–70. doi:10.1007/8904_2011_61.
Gary-Bobo M, Nirde P, Jeanjean A, Morere A, Garcia M. Mannose 6-phosphate receptor targeting and its applications in human diseases. Curr Med Chem. 2007;14(28):2945–53.
Gault CR, Obeid LM, Hannun YA. An overview of sphingolipid metabolism: from synthesis to breakdown. Adv Exp Med Biol. 2010;688:1–23.
Ghosh P, Dahms NM, Kornfeld S. Mannose 6-phosphate receptors: new twists in the tale. Nat Rev Mol Cell Biol. 2003;4(3):202–12. doi:10.1038/nrm1050. nrm1050 [pii].
Giraudo CG, Maccioni HJ. Ganglioside glycosyltransferases organize in distinct multienzyme complexes in CHO-K1 cells. J Biol Chem. 2003;278(41):40262–71. doi:10.1074/jbc.M305455200. M305455200 [pii].
Giraudo CG, Daniotti JL, Maccioni HJ. Physical and functional association of glycolipid N-acetyl-galactosaminyl and galactosyl transferases in the Golgi apparatus. Proc Natl Acad Sci U S A. 2001;98(4):1625–30. doi:10.1073/pnas.031458398. 031458398 [pii].
Goreta SS, Dabelic S, Dumic J. Insights into complexity of congenital disorders of glycosylation. Biochem Med (Zagreb). 2012;22(2):156–70.
Greenberg P, Merrill AH, Liotta DC, Grabowski GA. Human acid beta-glucosidase: use of sphingosyl and N-alkyl-glucosylamine inhibitors to investigate the properties of the active site. Biochim Biophys Acta. 1990;1039(1):12–20. doi:10.1016/0167-4838(90)90220-A.
Hamilton SR, Li H, Wischnewski H, Prasad A, Kerley-Hamilton JS, Mitchell T, et al. Intact {alpha}-1,2-endomannosidase is a typical type II membrane protein. Glycobiology. 2005;15(6):615–24. doi:10.1093/glycob/cwi045. cwi045 [pii].
Hammond C, Helenius A. A chaperone with a sweet tooth. Curr Biol. 1993;3(12):884–6. doi:10.1016/0960-9822(93)90226-E.
Hammond C, Helenius A. Quality control in the secretory pathway: retention of a misfolded viral membrane glycoprotein involves cycling between the ER, intermediate compartment, and Golgi apparatus. J Cell Biol. 1994;126(1):41–52.
Hammond C, Braakman I, Helenius A. Role of N-linked oligosaccharide recognition, glucose trimming, and calnexin in glycoprotein folding and quality control. Proc Natl Acad Sci U S A. 1994;91(3):913–7.
Hardt B, Aparicio R, Bause E. The oligosaccharyltransferase complex from pig liver: cDNA cloning, expression and functional characterisation. Glycoconj J. 2000;17(11):767–79.

Hart GW, Lennarz WJ. Effects of tunicamycin on the biosynthesis of glycosaminoglycans by embryonic chick cornea. J Biol Chem. 1978;253(16):5795–801.

Helenius A, Aebi M. Roles of N-linked glycans in the endoplasmic reticulum. Annu Rev Biochem. 2004;73:1019–49. doi:10.1146/annurev.biochem.73.011303.073752.

Herscovics A, Schneikert J, Athanassiadis A, Moremen KW. Isolation of a mouse Golgi mannosidase cDNA, a member of a gene family conserved from yeast to mammals. J Biol Chem. 1994;269(13):9864–71.

Hettkamp H, Bause E, Legler G. Inhibition by nojirimycin and 1-deoxynojirimycin of microsomal glucosidases from calf liver acting on the glycoprotein oligosaccharides Glc1-3Man9GlcNAc2. Biosci Rep. 1982;2(11):899–906.

Hettkamp H, Legler G, Bause E. Purification by affinity chromatography of glucosidase I, an endoplasmic reticulum hydrolase involved in the processing of asparagine-linked oligosaccharides. Eur J Biochem. 1984;142(1):85–90.

Hewitt LF. Separation of serum albumin into two fractions: observations on the nature of the glycoprotein fraction. Biochem J. 1937;31(3):360–6.

Hosokawa N, Tremblay LO, Sleno B, Kamiya Y, Wada I, Nagata K, et al. EDEM1 accelerates the trimming of alpha1,2-linked mannose on the C branch of N-glycans. Glycobiology. 2010;20(5):567–75. doi:10.1093/glycob/cwq001. cwq001 [pii].

Jaeken J. Congenital disorders of glycosylation. Ann N Y Acad Sci. 2010;1214:190–8. doi:10.1111/j.1749-6632.2010.05840.x.

Jaeken J. Congenital disorders of glycosylation (CDG): it's (nearly) all in it! J Inherit Metab Dis. 2011;34(4):853–8. doi:10.1007/s10545-011-9299-3.

Jeyakumar M, Butters TD, Cortina-Borja M, Hunnam V, Proia RL, Perry VH, et al. Delayed symptom onset and increased life expectancy in Sandhoff disease mice treated with N-butyldeoxynojirimycin. Proc Natl Acad Sci U S A. 1999;96(11):6388–93.

Kalz-Fuller B, Bieberich E, Bause E. Cloning and expression of glucosidase I from human hippocampus. Eur J Biochem. 1995;231(2):344–51.

Kaplan HA, Welply JK, Lennarz WJ. Oligosaccharyl transferase: the central enzyme in the pathway of glycoprotein assembly. Biochim Biophys Acta. 1987;906(2):161–73. doi:10.1016/0304-4157(87)90010-4.

Kelleher DJ, Gilmore R. An evolving view of the eukaryotic oligosaccharyltransferase. Glycobiology. 2006;16(4):47R–62R. doi:10.1093/glycob/cwj066. cwj066 [pii].

Kelleher DJ, Kreibich G, Gilmore R. Oligosaccharyltransferase activity is associated with a protein complex composed of ribophorins I and II and a 48 kd protein. Cell. 1992;69(1):55–65. doi:10.1016/0092-8674(92)90118-V.

Konat G, Hogan EL, Leskawa KC, Gantt G, Singh I. Abnormal glycosylation of myelin-associated glycoprotein in quaking mouse brain. Neurochem Int. 1987;10(4):555–8. doi:10.1016/0197-0186(87)90084-2.

Lajoie P, Nabi IR. Lipid rafts, caveolae, and their endocytosis. Int Rev Cell Mol Biol. 2010;282:135–63. doi:10.1016/S1937-6448(10)82003-9. S1937-6448(10)82003-9 [pii].

Lajoie P, Goetz JG, Dennis JW, Nabi IR. Lattices, rafts, and scaffolds: domain regulation of receptor signaling at the plasma membrane. J Cell Biol. 2009;185(3):381–5. doi:10.1083/jcb.200811059. jcb.200811059 [pii].

Le Mercier M, Fortin S, Mathieu V, Kiss R, Lefranc F. Galectins and gliomas. Brain Pathol. 2010;20(1):17–27. doi:10.1111/j.1750-3639.2009.00270.x. BPA270 [pii].

Leavitt R, Schlesinger S, Kornfeld S. Tunicamycin inhibits glycosylation and multiplication of Sindbis and vesicular stomatitis viruses. J Virol. 1977;21(1):375–85.

Legler G, Bieberich E. Isolation of a cytosolic beta-glucosidase from calf liver and characterization of its active site with alkyl glucosides and basic glycosyl derivatives. Arch Biochem Biophys. 1988;260(1):427–36. doi:10.1016/0003-9861(88)90466-3.

Legler G, Liedtke H. Glucosylceramidase from calf spleen. Characterization of its active site with 4-n-alkylumbelliferyl beta-glucosides and N-alkyl derivatives of 1-deoxynojirimycin. Biol Chem Hoppe Seyler. 1985;366(12):1113–22.

Lemansky P, Gieselmann V, Hasilik A, von Figura K. Cathepsin D and beta-hexosaminidase synthesized in the presence of 1-deoxynojirimycin accumulate in the endoplasmic reticulum. J Biol Chem. 1984;259(16):10129–35.

Leroy JG. Congenital disorders of N-glycosylation including diseases associated with O- as well as N-glycosylation defects. Pediatr Res. 2006;60(6):643–56. doi:10.1203/01.pdr.0000246802.57692.ea. 01.pdr.0000246802.57692.ea [pii].

Li E, Kornfeld S. Structure of the altered oligosaccharide present in glycoproteins from a clone of Chinese hamster ovary cells deficient in N-acetylglucosaminyltransferase activity. J Biol Chem. 1978;253(18):6426–31.

Lubas WA, Spiro RG. Golgi endo-alpha-D-mannosidase from rat liver, a novel N-linked carbohydrate unit processing enzyme. J Biol Chem. 1987;262(8):3775–81.

Lubke T, Marquardt T, von Figura K, Korner C. A new type of carbohydrate-deficient glycoprotein syndrome due to a decreased import of GDP-fucose into the golgi. J Biol Chem. 1999;274(37):25986–9.

Lucas JJ, Waechter J, Lennarz WJ. The participation of lipid-linked oligosaccharide in synthesis of membrane glycoproteins. J Biol Chem. 1975;250(6):1992–2002.

Maccioni HJ. Glycosylation of glycolipids in the Golgi complex. J Neurochem. 2007;103 Suppl 1:81–90. doi:10.1111/j.1471-4159.2007.04717.x. JNC4717 [pii].

Maccioni HJ, Daniotti JL, Martina JA. Organization of ganglioside synthesis in the Golgi apparatus. Biochim Biophys Acta. 1999;1437(2):101–18. doi:10.1016/S1388-1981(99)00002-5.

Maccioni HJ, Giraudo CG, Daniotti JL. Understanding the stepwise synthesis of glycolipids. Neurochem Res. 2002;27(7–8):629–36.

Maccioni HJ, Quiroga R, Ferrari ML. Cellular and molecular biology of glycosphingolipid glycosylation. J Neurochem. 2011a;117(4):589–602. doi:10.1111/j.1471-4159.2011.07232.x.

Maccioni HJ, Quiroga R, Spessott W. Organization of the synthesis of glycolipid oligosaccharides in the Golgi complex. FEBS Lett. 2011b;585(11):1691–8. doi:10.1016/j.febslet.2011.03.030. S0014-5793(11)00183-9 [pii].

Martina JA, Daniotti JL, Maccioni HJ. Influence of N-glycosylation and N-glycan trimming on the activity and intracellular traffic of GD3 synthase. J Biol Chem. 1998;273(6):3725–31.

Martina JA, Daniotti JL, Maccioni HJ. GM1 synthase depends on N-glycosylation for enzyme activity and trafficking to the Golgi complex. Neurochem Res. 2000;25(5):725–31.

McFarlane I, Georgopoulou N, Coughlan CM, Gillian AM, Breen KC. The role of the protein glycosylation state in the control of cellular transport of the amyloid beta precursor protein. Neuroscience. 1999;90(1):15–25. doi:10.1016/S0306-4522(98)00361-3.

Mistry PK. Treatment of Gaucher's disease with OGT 918. Lancet. 2000;356(9230):676–7. doi:10.1016/S0140673600026180.

Mohorko E, Glockshuber R, Aebi M. Oligosaccharyltransferase: the central enzyme of N-linked protein glycosylation. J Inherit Metab Dis. 2011;34(4):869–78. doi:10.1007/s10545-011-9337-1.

Molinari M, Helenius A. Chaperone selection during glycoprotein translocation into the endoplasmic reticulum. Science. 2000;288(5464):331–3. 8432 [pii].

Morell AG, Gregoriadis G, Scheinberg IH, Hickman J, Ashwell G. The role of sialic acid in determining the survival of glycoproteins in the circulation. J Biol Chem. 1971;246(5):1461–7.

Moremen KW. Golgi alpha-mannosidase II deficiency in vertebrate systems: implications for asparagine-linked oligosaccharide processing in mammals. Biochim Biophys Acta. 2002;1573(3):225–35. S0304416502003884 [pii].

Moremen KW, Robbins PW. Isolation, characterization, and expression of cDNAs encoding murine alpha-mannosidase II, a Golgi enzyme that controls conversion of high mannose to complex N-glycans. J Cell Biol. 1991;115(6):1521–34.

Moremen KW, Touster O. Biosynthesis and modification of Golgi mannosidase II in HeLa and 3T3 cells. J Biol Chem. 1985;260(11):6654–62.

Moremen KW, Tiemeyer M, Nairn AV. Vertebrate protein glycosylation: diversity, synthesis and function. Nat Rev Mol Cell Biol. 2012;13(7):448–62. doi:10.1038/nrm3383. nrm3383 [pii].

Nakatani Y, Yanagisawa M, Suzuki Y, Yu RK. Characterization of GD3 ganglioside as a novel biomarker of mouse neural stem cells. Glycobiology. 2010;20(1):78–86. doi:10.1093/glycob/cwp149. cwp149 [pii].

Nash RJ, Kato A, Yu CY, Fleet GW. Iminosugars as therapeutic agents: recent advances and promising trends. Future Med Chem. 2011;3(12):1513–21. doi:10.4155/fmc.11.117.

Ngamukote S, Yanagisawa M, Ariga T, Ando S, Yu RK. Developmental changes of glycosphingolipids and expression of glycogenes in mouse brains. J Neurochem. 2007;103(6):2327–41. doi:10.1111/j.1471-4159.2007.04910.x. JNC4910 [pii].

Osiecki-Newman KM, Fabbro D, Dinur T, Boas S, Gatt S, Legler G, et al. Human acid beta-glucosidase: affinity purification of the normal placental and Gaucher disease splenic enzymes on N-alkyl-deoxynojirimycin-sepharose. Enzyme. 1986;35(3):147–53.

Osiecki-Newman K, Fabbro D, Legler G, Desnick RJ, Grabowski GA. Human acid beta-glucosidase: use of inhibitors, alternative substrates and amphiphiles to investigate the properties of the normal and Gaucher disease active sites. Biochim Biophys Acta. 1987;915(1):87–100. doi:10.1016/0167-4838(87)90128-2.

Parodi AJ. Role of N-oligosaccharide endoplasmic reticulum processing reactions in glycoprotein folding and degradation. Biochem J. 2000;348(Pt 1):1–13.

Parodi AJ, Behrens NH, Leloir LF, Carminatti H. The role of polyprenol-bound saccharides as intermediates in glycoprotein synthesis in liver. Proc Natl Acad Sci U S A. 1972;69(11):3268–72.

Patterson MC, Vecchio D, Prady H, Abel L, Wraith JE. Miglustat for treatment of Niemann-Pick C disease: a randomised controlled study. Lancet Neurol. 2007;6(9):765–72. doi:10.1016/S1474-4422(07)70194-1. S1474-4422(07)70194-1 [pii].

Peyrieras N, Bause E, Legler G, Vasilov R, Claesson L, Peterson P, et al. Effects of the glucosidase inhibitors nojirimycin and deoxynojirimycin on the biosynthesis of membrane and secretory glycoproteins. EMBO J. 1983;2(6):823–32.

Platt FM, Neises GR, Dwek RA, Butters TD. N-butyldeoxynojirimycin is a novel inhibitor of glycolipid biosynthesis. J Biol Chem. 1994;269(11):8362–5.

Pless DD, Lennarz WJ. Enzymatic conversion of proteins to glycoproteins. Proc Natl Acad Sci U S A. 1977;74(1):134–8.

Ratner L, Vander HN. Mechanism of action of N-butyl deoxynojirimycin in inhibiting HIV-1 infection and activity in combination with nucleoside analogs. AIDS Res Hum Retroviruses. 1993;9(4):291–7.

Reszka N, Krol E, Patel AH, Szewczyk B. Effect of tunicamycin on the biogenesis of hepatitis C virus glycoproteins. Acta Biochim Pol. 2010;57(4):541–6. 20101994 [pii].

Reynders E, Foulquier F, Annaert W, Matthijs G. How Golgi glycosylation meets and needs trafficking: the case of the COG complex. Glycobiology. 2011;21(7):853–63. doi:10.1093/glycob/cwq179. cwq179 [pii].

Robina I, Moreno-Vargas AJ, Carmona AT, Vogel P. Glycosidase inhibitors as potential HIV entry inhibitors? Curr Drug Metab. 2004;5(4):329–61.

Ron I, Rapaport D, Horowitz M. Interaction between parkin and mutant glucocerebrosidase variants: a possible link between Parkinson disease and Gaucher disease. Hum Mol Genet. 2010;19(19):3771–81. doi:10.1093/hmg/ddq292. ddq292 [pii].

Roseman S. The synthesis of complex carbohydrates by multiglycosyltransferase systems and their potential function in intercellular adhesion. Chem Phys Lipids. 1970;5(1):270–97.

Roth J, Zuber C, Park S, Jang I, Lee Y, Kysela KG, et al. Protein N-glycosylation, protein folding, and protein quality control. Mol Cells. 2010;30(6):497–506. doi:10.1007/s10059-010-0159-z.

Ruddock LW, Molinari M. N-glycan processing in ER quality control. J Cell Sci. 2006;119(Pt 21):4373–80. doi:10.1242/jcs.03225. 119/21/4373 [pii].

Ruiz-Canada C, Kelleher DJ, Gilmore R. Cotranslational and posttranslational N-glycosylation of polypeptides by distinct mammalian OST isoforms. Cell. 2009;136(2):272–83. doi:10.1016/j.cell.2008.11.047. S0092-8674(08)01562-6 [pii].

Sandhoff K, Kolter T. Biosynthesis and degradation of mammalian glycosphingolipids. Philos Trans R Soc Lond B Biol Sci. 2003;358(1433):847–61.

Schwarz RT, Rohrschneider JM, Schmidt MF. Suppression of glycoprotein formation of Semliki Forest, influenza, and avian sarcoma virus by tunicamycin. J Virol. 1976;19(3):782–91.

Schweden J, Legler G, Bause E. Purification and characterization of a neutral processing mannosidase from calf liver acting on (Man)9(GlcNAc)2 oligosaccharides. Eur J Biochem. 1986; 157(3):563–70.
Sharma CB, Lehle L, Tanner W. N-Glycosylation of yeast proteins. Characterization of the solubilized oligosaccharyl transferase. Eur J Biochem. 1981;116(1):101–8.
Sheikh KA, Sun J, Liu Y, Kawai H, Crawford TO, Proia RL, et al. Mice lacking complex gangliosides develop Wallerian degeneration and myelination defects. Proc Natl Acad Sci U S A. 1999;96(13):7532–7.
Sousa MC, Ferrero-Garcia MA, Parodi AJ. Recognition of the oligosaccharide and protein moieties of glycoproteins by the UDP-Glc:glycoprotein glucosyltransferase. Biochemistry. 1992;31(1):97–105.
Spessott W, Crespo PM, Daniotti JL, Maccioni HJ. Glycosyltransferase complexes improve glycolipid synthesis. FEBS Lett. 2012;586(16):2346–50. doi:10.1016/j.febslet.2012.05.041. S0014-5793(12)00427-9 [pii].
Svennerholm L. Composition of gangliosides from human brain. Nature. 1956;177(4507):524–5.
Takatsuki A, Tamura G. Effect of tunicamycin on the synthesis of macromolecules in cultures of chick embryo fibroblasts infected with Newcastle disease virus. J Antibiot (Tokyo). 1971a; 24(11):785–94.
Takatsuki A, Tamura G. Tunicamycin, a new antibiotic. 3. Reversal of the antiviral activity of tunicamycin by aminosugars and their derivatives. J Antibiot (Tokyo). 1971b;24(4):232–8.
Takatsuki A, Tamura G. Tunicamycin, a new antibiotic. II. Some biological properties of the antiviral activity of tunicamycin. J Antibiot (Tokyo). 1971c;24(4):224–31.
Takatsuki A, Arima K, Tamura G. Tunicamycin, a new antibiotic. I. Isolation and characterization of tunicamycin. J Antibiot (Tokyo). 1971;24(4):215–23.
Tarentino AL, Plummer Jr TH, Maley F. The release of intact oligosaccharides from specific glycoproteins by endo-beta-N-acetylglucosaminidase H. J Biol Chem. 1974;249(3):818–24.
Tokunaga F, Brostrom C, Koide T, Arvan P. Endoplasmic reticulum (ER)-associated degradation of misfolded N-linked glycoproteins is suppressed upon inhibition of ER mannosidase I. J Biol Chem. 2000;275(52):40757–64. doi:10.1074/jbc.M001073200. . M001073200 [pii].
Trombetta SE, Parodi AJ. Purification to apparent homogeneity and partial characterization of rat liver UDP-glucose:glycoprotein glucosyltransferase. J Biol Chem. 1992;267(13):9236–40.
Trombetta SE, Bosch M, Parodi AJ. Glucosylation of glycoproteins by mammalian, plant, fungal, and trypanosomatid protozoa microsomal membranes. Biochemistry. 1989;28(20):8108–16.
Trombetta SE, Ganan SA, Parodi AJ. The UDP-Glc:glycoprotein glucosyltransferase is a soluble protein of the endoplasmic reticulum. Glycobiology. 1991;1(2):155–61.
Van Den Hamer CJ, Morell AG, Scheinberg IH, Hickman J, Ashwell G. Physical and chemical studies on ceruloplasmin. IX. The role of galactosyl residues in the clearance of ceruloplasmin from the circulation. J Biol Chem. 1970;245(17):4397–402.
Venier RE, Igdoura SA. Miglustat as a therapeutic agent: prospects and caveats. J Med Genet. 2012;49(9):591–7. doi:10.1136/jmedgenet-2012-101070. jmedgenet-2012-101070 [pii].
Vleugels W, Schollen E, Foulquier F, Matthijs G. Screening for OST deficiencies in unsolved CDG-I patients. Biochem Biophys Res Commun. 2009;390(3):769–74. doi:10.1016/j.bbrc.2009.10.047. S0006-291X(09)02025-7 [pii].
Volker C, De Praeter CM, Hardt B, Breuer W, Kalz-Fuller B, Van Coster RN, et al. Processing of N-linked carbohydrate chains in a patient with glucosidase I deficiency (CDG type IIb). Glycobiology. 2002;12(8):473–83.
Waechter CJ, Lucas JJ, Lennarz WJ. Membrane glycoproteins. I. Enzymatic synthesis of mannosyl phosphoryl polyisoprenol and its role as a mannosyl donor in glycoprotein synthesis. J Biol Chem. 1973;248(21):7570–9.
Waheed A, Hasilik A, von Figura K. Processing of the phosphorylated recognition marker in lysosomal enzymes. Characterization and partial purification of a microsomal alpha-N-acetylglucosaminyl phosphodiesterase. J Biol Chem. 1981;256(11):5717–21.
Wang J, White AL. Role of calnexin, calreticulin, and endoplasmic reticulum mannosidase I in apolipoprotein(a) intracellular targeting. Biochemistry. 2000;39(30):8993–9000. bi000027v [pii].

Wohlfarth C, Efferth T. Natural products as promising drug candidates for the treatment of hepatitis B and C. Acta Pharmacol Sin. 2009;30(1):25–30. doi:10.1038/aps.2008.5. aps20085 [pii].
Wrodnigg TM, Steiner AJ, Ueberbacher BJ. Natural and synthetic iminosugars as carbohydrate processing enzyme inhibitors for cancer therapy. Anticancer Agents Med Chem. 2008;8(1): 77–85.
Yanagisawa M, Liour SS, Yu RK. Involvement of gangliosides in proliferation of immortalized neural progenitor cells. J Neurochem. 2004;91(4):804–12. doi:10.1111/j.1471-4159.2004.02750.x. JNC2750 [pii].
Yang LJ, Zeller CB, Shaper NL, Kiso M, Hasegawa A, Shapiro RE, et al. Gangliosides are neuronal ligands for myelin-associated glycoprotein. Proc Natl Acad Sci U S A. 1996;93(2):814–8.
Yu RK. Gangliosides: structure and analysis. Adv Exp Med Biol. 1984;174:39–53.
Yu RK. Development regulation of ganglioside metabolism. Prog Brain Res. 1994;101:31–44.
Yu RK, Ando S. Structures of some new complex gangliosides of fish brain. Adv Exp Med Biol. 1980;125:33–45.
Yu RK, Ledeen RW. Gangliosides of human, bovine, and rabbit plasma. J Lipid Res. 1972; 13(5):680–6.
Yu RK, Macala LJ, Taki T, Weinfield HM, Yu FS. Developmental changes in ganglioside composition and synthesis in embryonic rat brain. J Neurochem. 1988;50(6):1825–9.
Yu RK, Bieberich E, Xia T, Zeng G. Regulation of ganglioside biosynthesis in the nervous system. J Lipid Res. 2004;45(5):783–93. doi:10.1194/jlr.R300020-JLR200. R300020-JLR200 [pii].
Yu RK, Suzuki Y, Yanagisawa M. Membrane glycolipids in stem cells. FEBS Lett. 2010;584(9):1694–9. doi:10.1016/j.febslet.2009.08.026. S0014-5793(09)00656-5 [pii].
Yu RK, Tsai YT, Ariga T. Functional roles of gangliosides in neurodevelopment: an overview of recent advances. Neurochem Res. 2012; 37(6):1230–44.

Chapter 4
Synthesis of O-Linked Glycoconjugates in the Nervous System

Jin-Ichi Inokuchi, Shinji Go, and Yoshio Hirabayashi

Abstract Glycoproteins carrying O-linked N-acetylgalactosamine, N-acetylglucosamine, mannose, fucose, glucose, and xylose are found in the nervous system. Lipids can be glycosylated as well. Membrane lipid, ceramide, is modified by the addition of either glucose or galactose to form glycosphingolipid, galactosylceramide, or glucosylceramide. Recent analyses have identified glucosylated lipids of cholesterol and phosphatidic acid. These O-linked carbohydrate residues are found primarily on the outer surface of the plasma membrane or in the extracellular space. Their expression is cell or tissue specific and developmentally regulated. Due to their structural diversity, they play important roles in a variety of biological processes such as membrane transport and cell–cell interactions.

Keywords Ceramide • Glycosphingolipid • Glucosylceramide • Ganglioside • Heparan sulfate • Chondroitin sulfate • Keratan sulfate • Mucin • Glycosyltransferase • Sulfation

Abbreviations

CGT	UDP-Gal:ceramide galactosyltransferase
ChlGlc	Cholesterylglucoside
CST	Cerebroside sulfotransferase
GalCer	Galactosylceramide or cerebroside

J.-I. Inokuchi • S. Go
Division of Glycopathology, Institute of Molecular Biomembranes and Glycobiology, Tohoku Pharmaceutical University, 4-4-1 Komatsushima, Aoba-ku, Sendai, Miyagi 981-8558, Japan

Y. Hirabayashi (✉)
Laboratory for Molecular Membrane Neuroscience, Riken Brain Science Institute, 2-1 Hirosawa, Wako-shi, Saitama 351-0198, Japan
e-mail: hirabaya@riken.jp

R.K. Yu and C.-L. Schengrund (eds.), *Glycobiology of the Nervous System*, Advances in Neurobiology 9, DOI 10.1007/978-1-4939-1154-7_4,

GalNAc-T	UDP-GalNAc: polypeptide N-acetylgalactosaminyltransferase
GCS	UDP-Glc:ceramide glucosyltransferase, UGCG or GlcT-1
GlcCer	Glucosylceramide
GSL	Glycosphingolipid
LacCer	Lactosylceramide
LARGE	The like-acetylglucosaminyltransferase
NCAM	Neural cell adhesion molecule
POFUT1	Protein O-fucosyltransferase 1
POMGNT1	Protein O-linked mannose N-acetylglucosaminyltransferase 1 (beta 1,2-)

4.1 Introduction

O-linked glycans are found in proteins and lipids, both of which can carry a great variety of glycan chains. Most typical protein/peptide O-glycans in nonneural tissues possess an αGalNAc residue linked to a serine or threonine hydroxyl group to which other sugars can be added. These protein/peptide O-glycans are classified as mucin-type glycans and they also exist in brain. Glycoproteins carrying O-linked mannose, fucose, glucose, and xylose are also present in the nervous system. Lipids can be glycosylated as well. Ceramide, a membrane sphingolipid, can be modified by the addition of either glucose or galactose through a β-glycosidic linkage to form the glycosphingolipid identified as a cerebroside (ceramide-O-glc/gal). Recent analyses have identified glucosylated lipids of cholesterol and phosphatidic acid. These O-linked carbohydrate residues are found primarily on the outer surface of the plasma membrane or in the extracellular space. Their expression is cell or tissue specific and developmentally regulated. Due to their structural diversity, they play important roles in a variety of biological processes such as membrane transport and cell–cell interactions. The glycosylation reactions occur primarily in the Golgi apparatus and/or in ER membranes. Protein O-β-GlcNAc modification can also occur in the cytoplasm. This review describes the basic structures of O-glycans and the synthetic enzymes involved with focus on the initial step of glycosylation.

4.2 Biosynthesis of O-Linked Proteins

Cell surface glycan chains in the nervous system have critical roles during differentiation, development, regeneration, synaptic plasticity, and aging (Kleene and Schachner 2004). Their tremendous structural diversity contributes to cell–cell communication, including neuron–glia and cell–matrix interactions. There are at least six different O-glycan-protein linkages that have been identified in the mammalian nervous system (Fig. 4.1).

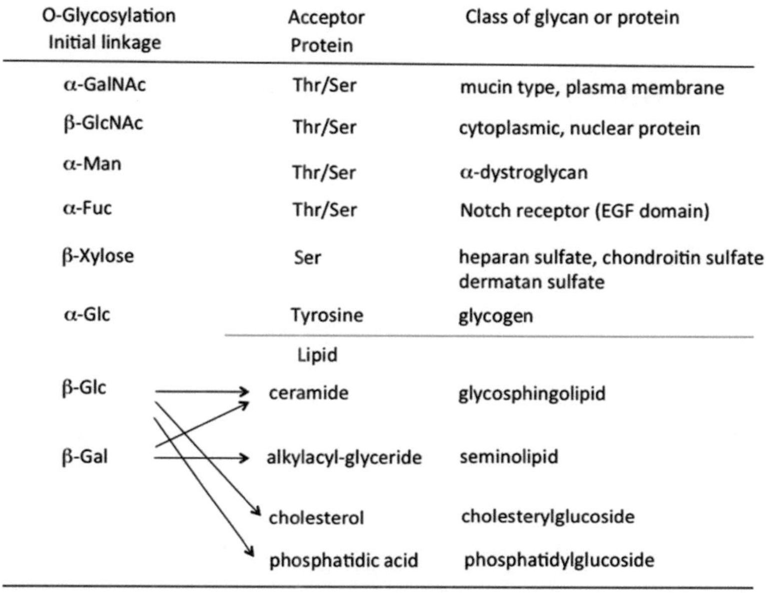

Fig. 4.1 Molecular diversity of vertebrate brain O-glycans

4.2.1 O-GalNAcylation

O-GalNAc modification of proteins occurs most frequently and is abundant in mucin-type glycoproteins. The glycan chains are very heterogeneous, and eight different core structures are known (Fig. 4.2). However, little is understood about their detailed structure, distribution, and function in the nervous system. To understand precisely how and where O-GalNAc is attached to each protein, a novel method of O-GalNAc glycoproteome was developed (Steentoft et al. 2013). This type of approach is absolutely necessary for advancing our knowledge of O-GalNAc glycans in the brain.

UDP-GalNAc:polypeptide N-acetylgalactosaminyltransferase (GalNAc-T) is the key Golgi enzyme catalyzing the initial step in the biosynthesis of O-GalNAc glycans. Among the 20 or 21 GalNAc-T genes (*GALNT*) that have been identified to date (Bennett et al. 2012), only a few genes, such as *GalNAc-T9*, *T-13*, and *T-17* (also called *T-16*), have been isolated and found in brain tissues. The latter gene is also known as WBSCR17 and is associated with Williams–Beuren syndrome (WBS) (Nakayama et al. 2012). Syndecan-3 is suggested to carry this type of glycan (Zhang et al. 2003).

Neural cell adhesion molecule (NCAM) and neuropilin-2 contain O-glycans, which are polysialylated with α2-8 sialosyl residues. Although a detailed functional study has not yet been carried out, modification of O-GalNAc chains is most likely to be involved with important physiological functions in the central nervous system.

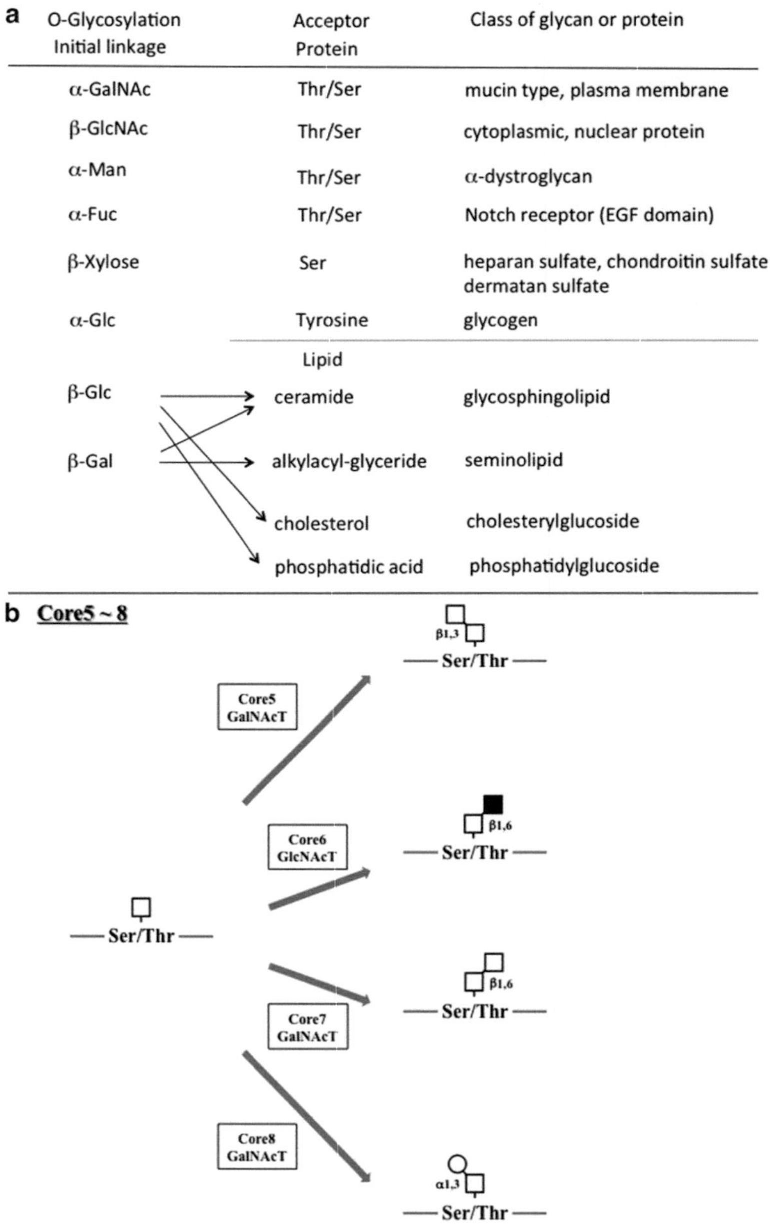

Fig. 4.2 (**a**) Synthesis of O-GalNAc glycan (Cores 1–4). Sugar species are depicted using the symbols given in "Essentials of Glycobiology" (Varki et al. 2009). (**b**) Synthesis of O-GalNAc glycans (Cores 5–8)

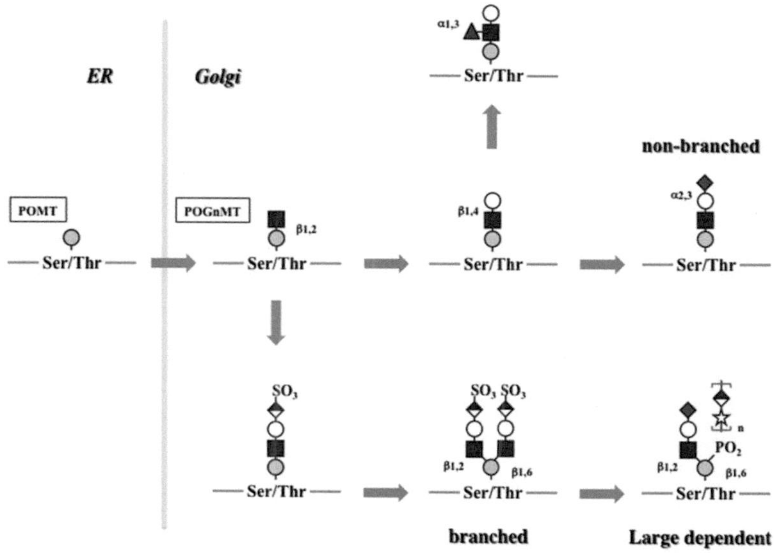

Fig. 4.3 Synthesis of O-Man glycans

4.2.2 O-Mannosylation

Krusius and co-workers first reported that more than half of the carbohydrate–peptide linkages in brain proteoglycans are of the mannosyl-O-serine/threonine type and contain keratan sulfate (Krusius et al. 1986). The same group identified the sialylated tetrasaccharide NeuAcα2-3Galβ1-4GlcNAcβ1-3Man-Ser/Thr and its related glycans (Krusius et al 1986).

O-Mannosylated glycans, isolated from peripheral nerve tissues, have slight structural differences compared to sialylated tetrasaccharides present in central nervous system tissues (Endo 1999). This glycan has the following structure: NeuAcα2-3Galβ1-4GlcNAcβ1-2Man-Ser/Thr. The linkage between the GlcNAc and the Man residue is 3-substituted in the case of the brain tetrasaccharide. The O-glycan in peripheral nerve tissue is a major sialylated O-glycan in α-dystroglycan. The O-glycans isolated from α-dystroglycan possess an evolutionarily conserved structure, being observed in animals from *Drosophila* to mammals. They have been found in only a limited number of proteins in the brain, such as α-dystroglycan, chondroitin sulfate, and PTPRZ1/RPTPβ (Krusius et al. 1986; Nakamura et al. 2010). Because O-mannosylated glycans in α-dystroglycan have essential biological roles as well as pathobiological roles in muscle and the nervous system, their structures have been extensively studied (Fig. 4.3) (Endo 1999).

O-Mannosylation is catalyzed by O-mannosyltransferase, which is encoded by the *POMT1* and *POMT2* genes. POMT1 and POMT2 complex formation is essential for POMT activity. Unlike other peptide O-glycosyltransferases, POMT is a member of the dolichyl-phosphate-mannose-dependent mannosyltransferase located in ER membranes. Mutations in either *POMT1* or *POMT2* cause Walker–Warburg syndrome, a congenital muscular dystrophy with abnormal neuronal migration (Reeuwijk et al. 2005).

Mannose residues are further glycosylated by protein O-linked mannose N-acetylglucosaminyltransferase 1 (POMGnT1) to form disaccharides that are further elongated by glycosyltransferases to make three major forms: non-branched, branched, and LARGE-dependent glycan chains (Fig. 4.3). *LARGE* encodes a bifunctional glycosyltransferase that has both xylosyltransferase and glucuronyltransferase activities. It is of interest to note that LARGE-dependent glycan structures have a phosphate residue at the 6-position of the core mannose. Recent studies indicated that SGK196 is a typical kinase that phosphorylates the 6-hydroxy position of mannose and is required for dystroglycan receptor function (Yoshida-Moriguchi et al. 2013).

4.2.3 O-GlcNAcylation

O-GlcNAc is covalently attached to serine or threonine residues of intracellular proteins through a β-glycosidic linkage. This modification is evolutionarily conserved and is found in nuclear and cytoplasmic proteins, such as nuclear pore proteins, chromatic histone proteins, transcriptional factors, and P53 (Wells et al. 2001). The GlcNAc residue is not modified by further glycosylation. Importantly, the modification site can also be phosphorylated by a serine/threonine kinase. O-GlcNAc modification is suggested to be involved in the homeostatic mechanism of energy metabolism, since UDP-GlcNAc—the donor for O-GlcNAc transferase (OGT)—is metabolically derived from all metabolites, including glucose, nucleotides, fatty acids, and nitrogen (Ruan et al. 2013). Further detailed information on the physiological significance of O-GlcNAc modification is described in Chap. 16 by Lagerlof and Hart.

4.2.4 O-Xylosylation

O-Xylosylated proteins are present in proteoglycans including chondroitin sulfate/dermatan sulfate and heparan sulfate (Fig. 4.4), important components for brain development and physiology. The most characteristic feature of these two groups is that they have highly complicated, heterogeneous glycan chains. These proteoglycans are found as constituents of the extracellular matrix and function as cell adhesion molecules in the brain. Proteoglycans interact with extracellular or cell surface molecules expressed in neighboring cells. Both chondroitin sulfate/dermatan sulfate and heparan sulfate have common basic tetrasaccharide structures: GlcUAβ1-3Galβ1-3Galβ1-4Xyl(±2-O-phosphate) β-Ser (Fig. 4.4). Xylosyltransferase encoded

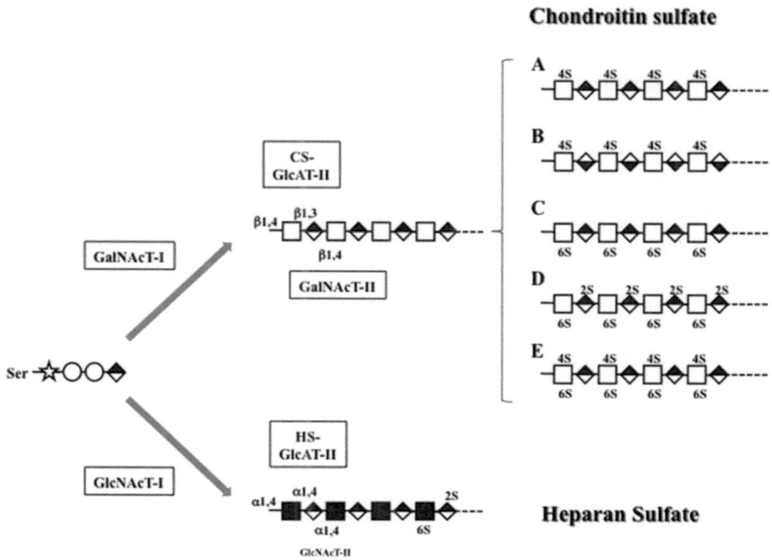

Fig. 4.4 Synthesis of both heparan sulfate and chondroitin sulfate

by the *XYLT1* gene catalyzes the transfer of xylose from UDP-xylose to *serine* residues within the transferase recognition sequences of protein substrates. The enzyme is localized in the early *cis*-Golgi apparatus (Sarrazin et al. 2011), and its activity is upregulated during myofibroblast differentiation in skin fibrosis (Faust et al. 2013).

Heparan Sulfate

The basic tetrasaccharide is further modified by an α4GlcNAc-transferase, termed Extl3. Then, an enzyme complex composed of Ext1 and Ext2 alternately adds GlcA and GlcNAc to the nascent chain (Fig. 4.4). *Ext* genes are evolutionarily conserved and are characterized as tumor suppressors. The oligosaccharide chains simultaneously undergo a series of processing reactions, including N-deacetylation and sulfation, to yield mature glycans (for details, see Ruan et al. 2013). Variation in the number and length of the chains produces enormous chemical diversity. Brain heparan sulfate glycan structures are typically found in neuropilin-1, syndecan-3, and glypican-1. Recent studies on proteoglycan syntheses and core proteins in genetically engineered animals have demonstrated that neural proteoglycans play critical roles not only in brain development and neuronal network formation but also in neuronal regeneration in injured nervous tissues, in formation and deposition of Aβ-amyloid peptide in Alzheimer brains, and in autism (Irie et al. 2012).

Chondroitin Sulfate

In the case of chondroitin sulfate, the core tetrasaccharide is first modified by the addition of GalNAc catalyzed by a β4GalNAc-transferase and then by the addition of GlcA catalyzed by a β4GlcA-transferase. The combination of the two glycosyltransferases is responsible for forming the repeating disaccharide unit (GlcA-GalNAc). Sulfation on the disaccharide unit generates the diverse structures of chondroitin sulfate proteoglycans (Fig. 4.4; structures A to E) (Silbert and Sugumaran 2002). The chondroitin-sulfated proteoglycans, NG2 and neurocan, are synthesized by both neurons and glia. Accumulating evidence has shown that chondroitin sulfate proteoglycans play key roles in neural development, axon guidance cues, neural plasticity, and neural repair after injury (Kwok et al. 2012).

4.2.5 O-Fucosylation

O-Fucosylation is a conserved posttranslational modification of proteins that is catalyzed by two glycosyltransferases, protein O-fucosyltransferase 1 (POFUT1) and protein O-fucosyltransferase 2 (POFUT2). Molecular genetic studies have proven that POFUT1 is essential for normal development in both flies and mice. Analysis of *POFUT1* null embryos demonstrated that it is implicated in the Notch signaling pathway (Okajima and Irvine 2002).

4.2.6 O-Glucosylation

O-Glc is present in glycogenin (GYG1 and 2). Transfer of glucose to glycogenin is catalyzed by the glycogen initiator synthases (UDP-glucose-glycogen α-glucosyltransferases; *GYS1, GYS2*), which enable the glucose residue to act as an acceptor for subsequent glucosylation. The glucose molecules are attached through a hydroxyl group on a specific tyrosine side chain of the glycogenin. Elongation of the chain is accomplished by formation of an α1,4-glycosidic linkage that is catalyzed by glycogen synthase. Recent studies have shown that the brain contains glycogen (Brown and Ransom 2007). Brain glycogen functions as an energy source when the ambient glucose concentration is unable to meet immediate energy demands (Brown and Ransom 2007).

4.3 Biosynthesis of O-Linked Lipids

There are three major classes of membrane lipids: cholesterol, glycerolipids, and sphingolipids. All of these lipids can be modified by the addition of sugars. Among the three, sphingolipids are the most abundant glycolipids in the brain. They have been extensively studied in efforts to elucidate their biological functions in the nervous system.

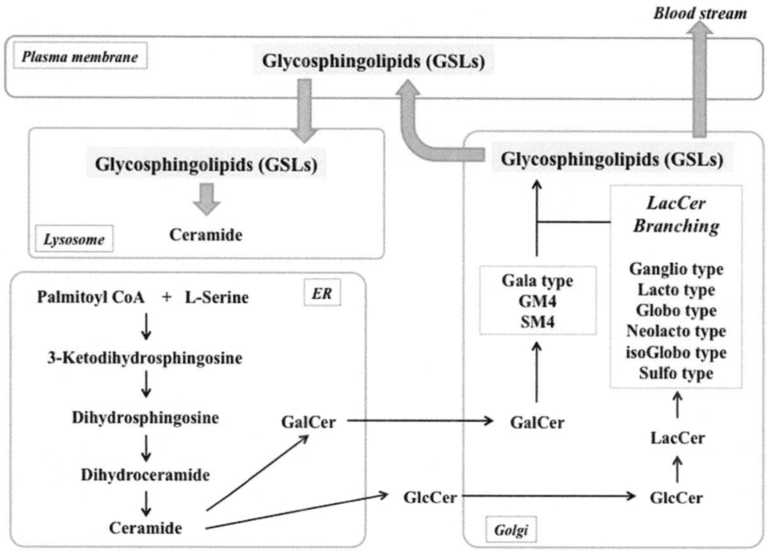

Fig. 4.5 The subcellular biosynthesis of GSLs

4.3.1 Glucosylceramide and Glycosphingolipids

Glycosphingolipids (GSLs) are a large and heterogeneous family of sphingolipids that form complex patterns on eukaryotic cell surfaces. Their diverse structures result from various combinations of their long-chain (sphingoid) bases, amide-linked fatty acids, and hundreds of head group variants. The subcellular biosynthetic machinery of GSLs is summarized in Fig. 4.5. Most GSLs are generated from glucosylceramide (GlcCer), which is formed when glucose is attached to the primary alcohol group (C1-OH) of a ceramide molecule of the endoplasmic reticulum (ER)/Golgi compartment. This glucosylation is catalyzed by UDP-Glc:ceramide glucosyltransferase, UGCG/GlcT-1/GCS (Ichikawa and Hirabayashi 1998).

Glycan chains of GSLs are mainly synthesized in the lumen of the Golgi apparatus. Especially notable is that GlcCer is biosynthesized on the cytoplasmic leaflet of the Golgi membrane. To execute further elongation, GlcCer has to be translocated from the cytosolic leaflet to the inner leaflet of the Golgi membrane by a specific translocase, which has yet to be identified. This topological orientation of the catalytic sites of glycosyltransferases is supported by their membrane protein type III structures. GlcCer synthase is a type III membrane protein with an N-terminal signal-anchor sequence, whereas other glycosyltransferases are type II membrane proteins with an N-terminal membrane-spanning domain and catalytic domains in the C-terminal region.

GlcCer is then modified by a β1-4galactosyltransferase to generate lactosylceramide (LacCer). LacCer plays a pivotal role as a precursor for the synthesis of

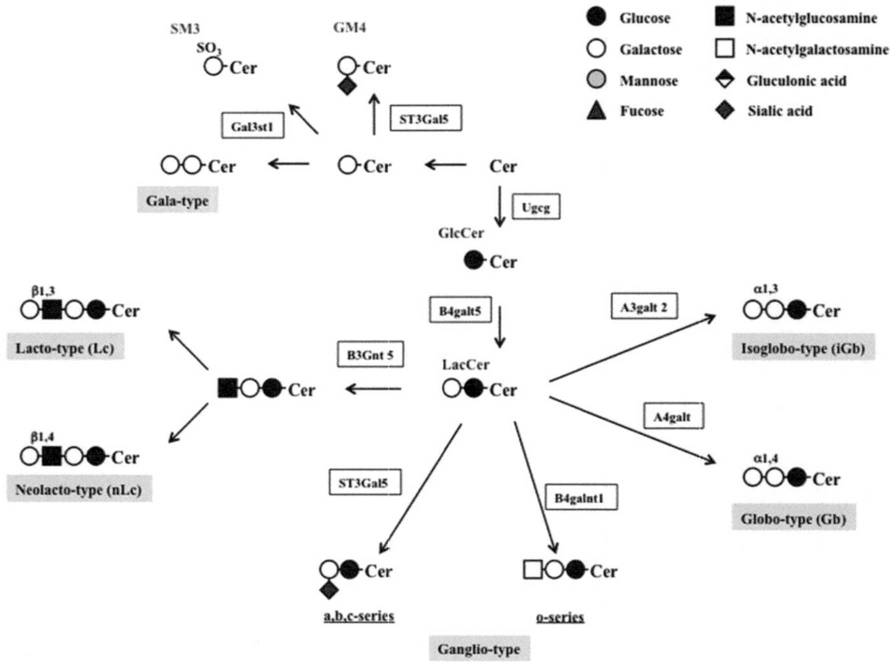

Fig. 4.6 Overview of the biosynthetic pathways for the different classes of GSLs

complex GSLs. The common LacCer structure is then elongated by different glycosyltransferases, to form the six classes of GSLs bearing the core structures known as gala-, globo- (Gb), isoglobo- (iGb), ganglio (Gg), lacto (Lc), and neolacto- (nLc). Detailed maps of the biosynthetic pathways for each class of GSL in mammals are depicted schematically in Figs. 4.5, 4.6, 4.7, 4.8, 4.9, 4.10, and 4.11. Gangliosides are typical membrane GSLs and are found in both neurons and glia. Some gangliosides are markers for particular sets of neuronal cells. For example, α-series gangliosides such as GQ1bα are specific to cholinergic neurons (see also Chap. 19).

Every tissue or cell has a unique set of glycosyltransferases, which can be regulated via tissue- or cell-specific transcriptional control and by posttranslational modifications. Additionally, the possibility of an epigenetic control mechanism for glycosyltransferases has been demonstrated, which could result from environmental factors (Tsai and Yu 2013). Many genetically modified animal models, including tissue-specific (conditional) glycosyltransferase knockout mice, have been generated. These models have contributed greatly to our understanding of the evolutional, physiological, and pathological significance of GSL synthesis. In particular, studies of mice in which the GlcCer synthase gene (*Ugcg*) was knocked out in specific tissues indicated that GSLs have tissue-specific biological functions (Ishibashi et al. 2013; Jennemann and Gröne 2013). Importantly, GlcCer itself is thought to play an important role in the homeostatic regulation of energy metabolism (Fig. 4.12) (Ishibashi et al. 2013).

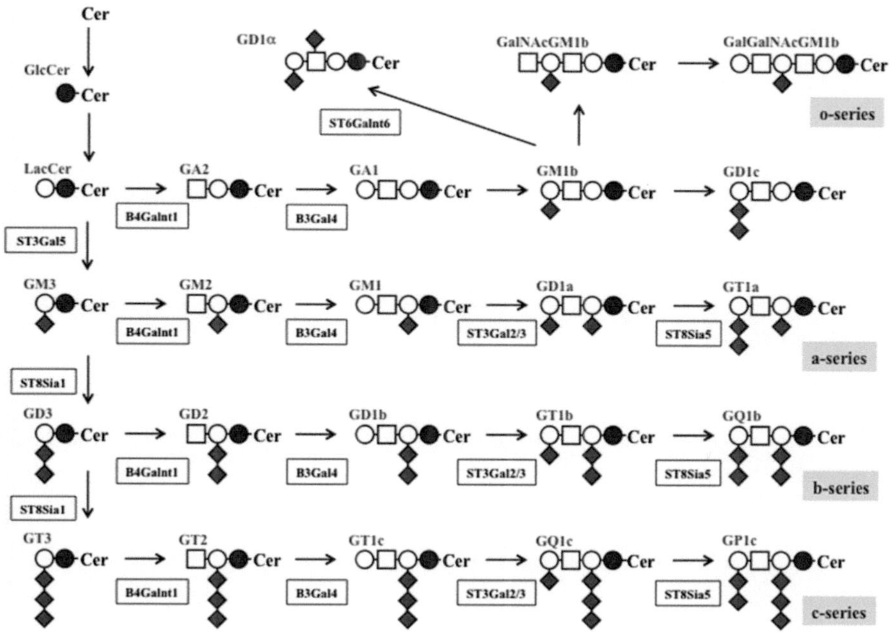

Fig. 4.7 Ganglio-type GSLs (I)

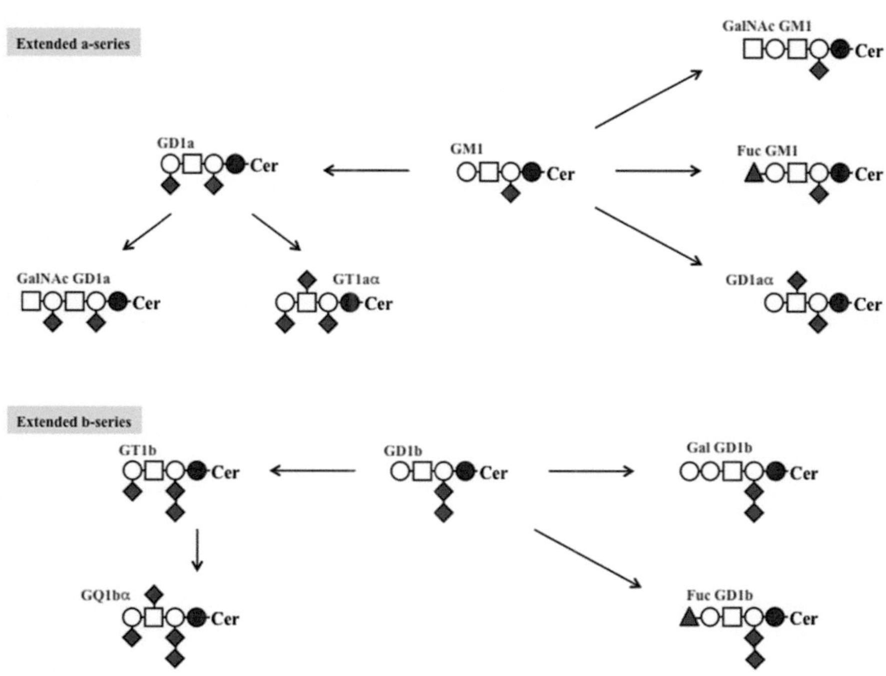

Fig. 4.8 Ganglio-type GSLs (2)

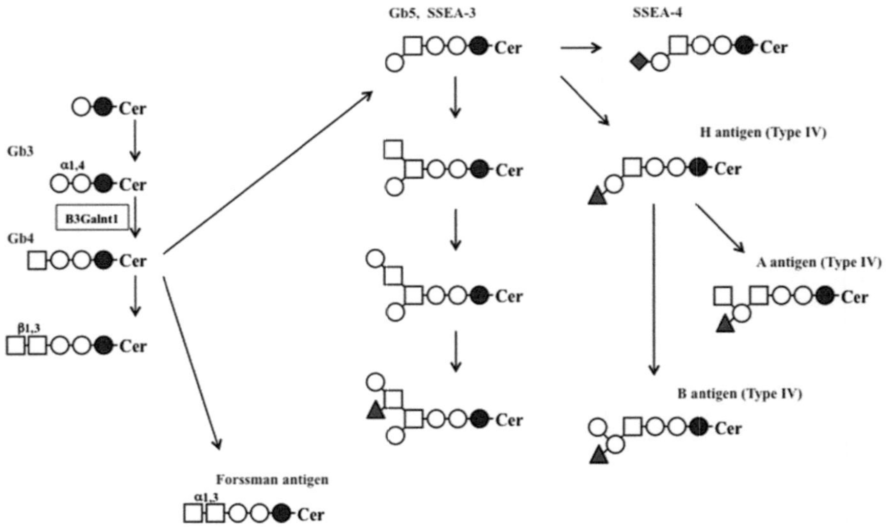

Fig. 4.9 Globo-type GSLs

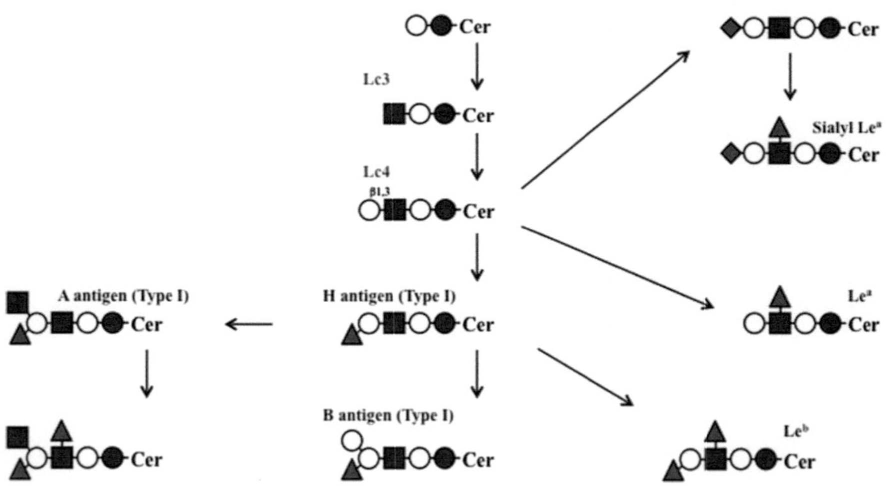

Fig. 4.10 Lacto-type GSLs

4.3.2 *Galactosylceramide and Glycosphingolipids*

GalCer is characteristically abundant in the myelin membranes of oligodendrocytes. GalCer formation occurs on the luminal side of ER membranes in a reaction catalyzed by UDP-Gal:ceramide galactosyltransferase (CGT). GalCer is translocated from the inner leaflet of the ER to the lumen of the Golgi apparatus prior to

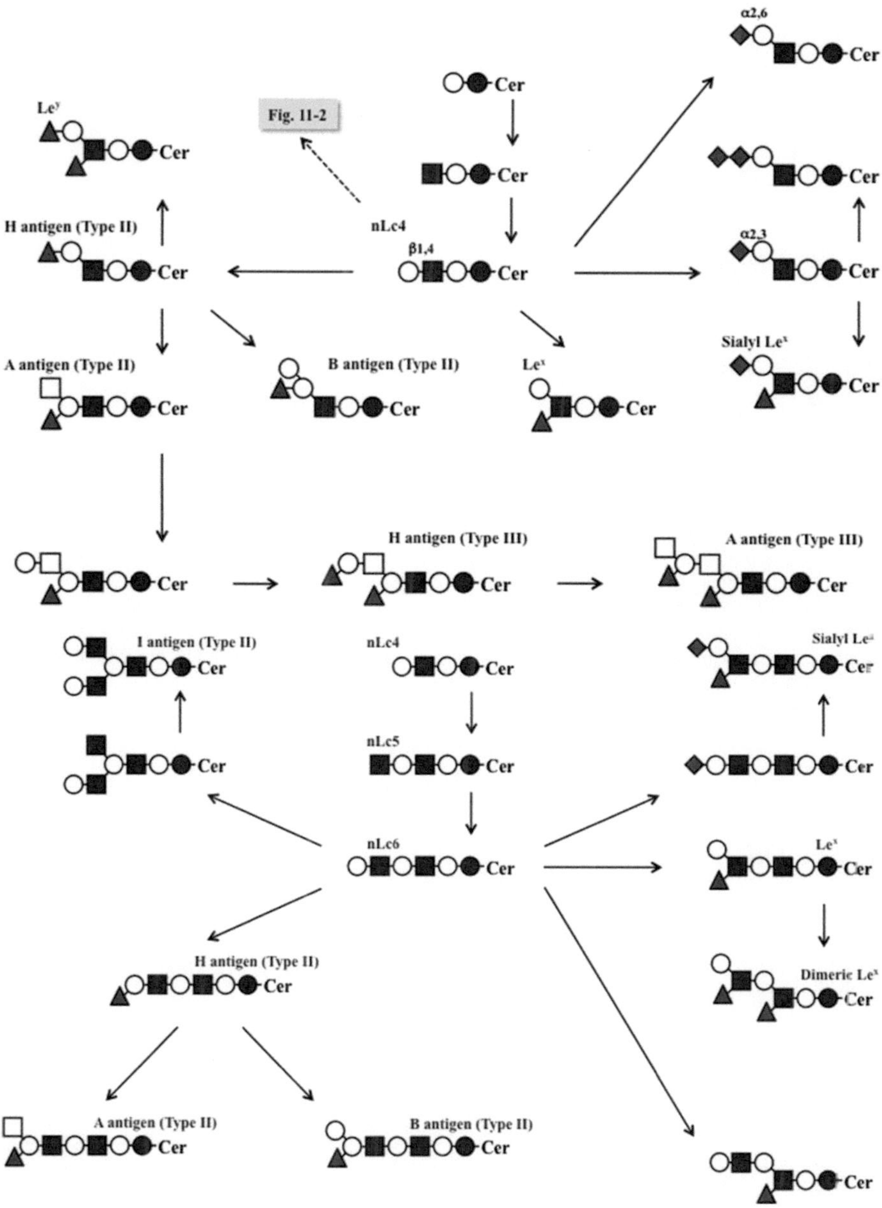

Fig. 4.11 Neolacto-type GSLs (1). Neolacto-type GSLs (2)

synthesis of NeuAcα2-3GalCer (GM4) and sulfated GalCer (SM4), both of which are also abundant in myelin membranes. The sialylation and sulfation of GalCer to form GM4 and sulfatide are catalyzed by GM3 synthase (ST3Gal5) and sulfotransferase (CST), respectively. These glycolipids play an essential role in myelin functions.

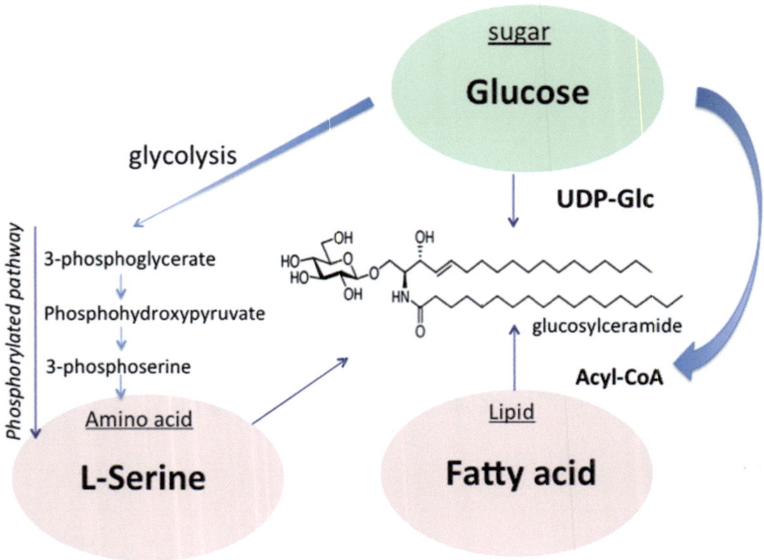

Fig. 4.12 Glucosylceramide synthesis links to cellular energy metabolism. Animal model studies show that glycolipid synthesis regulates the accumulation and release of stored lipids, such as triacylglycerides in adipose tissue. This regulatory role of glucosylceramide and its downstream GSLs on energy homeostasis is not surprising, since the basic building blocks of glucosylceramide synthesis, namely, UDP-glucose, palmitoyl-CoA, and serine (derived from glucose), are directly related to energy metabolism

4.3.3 Other O-Linked lipids

Recent studies show that cholesterol and phosphatidic acid are also modified with glucose (Fig. 4.13) to form cholesterylglucoside and phosphatidylglucoside, respectively.

Cholesterylglucoside

Mammalian glucosylated cholesterol (1-O-cholesteryl-β-D-glucopyranoside, ChlGlc) is found in mouse brain (Nakajima and Akiyama, unpublished observations). Mammalian ChlGlc synthesis is not dependent on UDP-Glc. Interestingly, GlcCer is a glucose donor for ChlGlc synthesis. UGCG-deficient GM-95 cells are incapable of synthesizing ChlGlc without the addition of exogenous GlcCer. The physiological function of brain ChlGlc is still unknown. Glucocerebrosidase has been shown to transfer glucose residues from GlcCer to cholesterol in vitro (Akiyama et al. 2013).

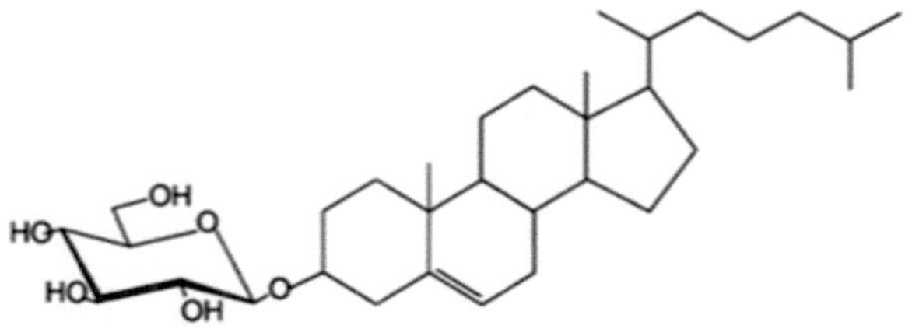

(1-O-cholesteryl-β-D-glucopyranoside, ChlGlc)

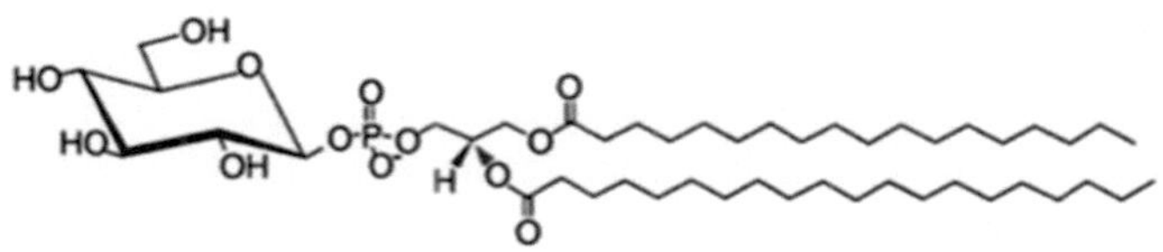

1-stearyl-2-arachidoyl-*sn*-glycerol-3-phosphoryl-β-D-glucopyranoside

Fig. 4.13 Novel glucosylated lipids

Phosphatidylglucoside

PtdGlc was first isolated from fetal rodent brains and purified to homogeneity in 2006 (Nagatsuka et al. 2006). Its complete structure is 1-stearyl-2-arachidoyl-sn-glycerol-3-phosphoryl-β-D-glucopyranoside (Fig. 4.13). Acetylated PtdGlc (1-stearyl-2-arachidoyl-*sn*-glycerol-3-phosphoryl β-D-(6-O-acetyl)glucopyranoside) also exists in fetal rat brain. Very interestingly and importantly, PtdGlc isolated from fetal rat brain has only one fatty acid combination: its *sn*-1 and *sn*-2 chains are exclusively stearic acid (C18:0) and arachidic acid (C20:0), respectively. A single molecular species rarely occurs in natural phospholipids. It is not fully understood how this glycolipid is biosynthesized in cells. Glycosylation is dependent on UDP-glucose synthesized by a β-glucosyltransferase in the ER. The glycolipid is localized in lipid rafts and is involved in astroglial differentiation (Ishibashi et al. 2013). PtdGlc is degraded by phospholipase A2 to form a water-soluble Lyso-PtdGlc. This lyso-glycolipid is thought to play an important role in neuron–glia communication.

4.4 Conclusions

Brain contains a large number of carbohydrate chains with different core structures on both proteins and lipids. To understand their physiological and pathophysiological functions, it is crucial to identify and characterize the glycosyltransferases involved in the initial step in the synthesis of each type of glycan. While many of the glycosyltransferases have been identified, endogenous structures of O-GalNAc chains and the corresponding GalNAc-Ts have not been completely identified.

Many questions remain to be addressed including how glycosyltransferase expression is regulated during development and how their glycosylation activities are related to neural activities such as memory and learning, formation, and maintenance.

Recent developments in MS-technology (see Chap. 7) have resulted in the identification of cell-type-specific and disease-associated glycan chains present in mammalian cell surface membranes. Examples of this can be seen in the identification of an O-mannose modified with phosphate in α-dystroglycan and of new lipids modified with glucose in the nervous system. Since glucose is a conserved essential compound for life, it is anticipated that all glucosylated lipids will have conserved, basic biological functions. The sialic acid containing GlcCer-derived lipid, GM3, is dominantly expressed in insulin responsive organs such as human skeletal muscle, liver and adipose tissue, and brain. GM3-dependent membrane microdomains (lipid rafts) involved in regulation of energy metabolism. In order to understand the precise roles of GlCer- and/or GM3-depedent lipid rafts, generation of tissue-specific knockout mice for each synthase is essential since metabolic homeostasis is maintained by communication between tissues.

Although not discussed in detail in this chapter, characterization of the structural diversity of the sphingoid base and the N-acyl chain of ceramide is essential for understanding the dynamics of GSLs in living cell membranes as well as their interactions with lipid raft-associated proteins. "Sphingolipidomics" should enable investigators to determine the precise structure of sugar chains as well as ceramide components of GSLs. This will aid in elucidation of the functional supra-biomolecular complex consisting of GSLs (gangliosides) and functional proteins in lipid microdomains. Here again, the advancements made in MS will accelerate our ability to understand the physiological role(s) of each glycosylated lipid.

Conflict of Interest The authors declare that they have no conflict of interest.

References

Akiyama H, Kobayashi S, Hirabayashi Y, Murakami-Murofushi K. Cholesterol glucosylation is catalyzed by transglucosylation reaction of β-glucosidase 1. Biochem Biophys Res Commun. 2013;441(4):838–43.

Bennett EP, Mandel U, Clausen H, Gerken TA, Fritz TA, Tabak LA. Control of mucin-type O-glycosylation: a classification of the polypeptide GalNAc-transferase gene family. Glycobiology. 2012;22(6):736–56.

Brown AM, Ransom BR. Astrocyte glycogen and brain energy metabolism. Glia. 2007;55(12): 1263–71.

Endo T. O-mannosyl glycans in mammals. Biochim Biophys Acta. 1999;1473(1):237–46.

Faust I, Roch C, Kuhn J, Prante C, Knabbe C, Hendig D. Human xylosyltransferase-1—a new marker for myofibroblast differentiation in skin fibrosis. Biochem Biophys Res Commun. 2013;436(3):449–54.

Ichikawa S, Hirabayashi Y. Glucosylceramide synthase and glycosphingolipid synthesis. Trends Cell Biol. 1998;8(5):198–202.

Irie F, Badie-Mahdavi H, Yamaguchi Y. Autism-like socio-communicative deficits and stereotypes in mice lacking heparin sulfate. Proc Natl Acad Sci U S A. 2012;109(13):5052–6.

Ishibashi Y, Kohyama A, Hirabayashi Y. New insights on glycosylated lipids: metabolism and functions. Biochim Biophys Acta. 2013;1831(9):1475–85.

Jennemann R, Gröne HJ. Cell-specific in vivo functions of glycosphingolipids: lessons from genetic deletions of enzymes involved in glycosphingolipid synthesis. Prog Lipid Res. 2013;52(2):231–48.

Kleene R, Schachner M. Glycans and neural cell interactions. Nat Rev Neurosci. 2004;5(3): 195–208.

Krusius T, Finne J, Margolis RK, Margolis RU. Identification of an O-glycosidic mannose-linked sialylated tetrasaccharide and keratan sulfate oligosaccharides in the chondroitin sulfate proteoglycan of brain. J Biol Chem. 1986;261(18):8237–42.

Kwok JC, Warren P, Fawcett JW. Chondroitin sulfate: a key molecule in the brain matrix. Int J Biochem Cell Biol. 2012;44(4):582–6.

Nagatsuka Y, Horibata Y, Yamazaki Y, Kinoshita M, Shinoda Y, Hashikawa T, et al. Phosphatidylglucoside exists as a single molecular species with saturated fatty acyl chains in developing astroglial membranes. Biochemistry. 2006;45(29):8742–50.

Nakamura N, Lyalin D, Panin VM. Protein O-mannosylation in animal development and physiology: from human disorders to Drosophila phenotypes. Semin Cell Dev Biol. 2010;21(6): 622–30.

Nakayama Y, Nakamura N, Oki S, Wakabayashi M, Ishihama Y, Miyake A, et al. A putative polypeptide N-acetylgalactosaminyltransferase/Williams-Beuren syndrome chromosome region 17 (WBSCR17) regulates lamellipodium formation and macropinocytosis. J Biol Chem. 2012; 287(38):32222–35.

Okajima T, Irvine KD. Regulation of notch signaling by O-linked fucose. Cell. 2002;111: 893–904.

Reeuwijk J, Janssen M, Elzen C, Beltran-Valero D, Sabatelli P, Merlini L, et al. *POMT2* mutations cause α-dystroglycan hypoglycosylation and Walker-Warburg syndrome. J Med Genet. 2005; 42(12):907–12.

Ruan HB, Singh JP, Li MD, Wu J, Yang X. Cracking the O-GlcNAc code in metabolism. Trends Endocrinol Metab. 2013;24(6):301–9.

Sarrazin S, Lamanna WC, Esko JD. Heparan sulfate proteoglycan. Cold Spring Harb Perspect Biol. 2011;3:a004952.

Silbert JE, Sugumaran G. Biosynthesis of chondroitin/dermatan sulfate. IUBMB Life. 2002; 54(4):177–86.

Steentoft C, Vakhrushev S, Joshi HJ, Kong Y, Vester-Christensen MB, Schjoldager KT, et al. Precision mapping of the human O-GalNAc glycoproteome through Simple Cell technology. EMBO J. 2013;32(10):1478–889.

Tsai Y-T, Yu RK. Epigenetic activation of mouse ganglioside synthase genes: implications for neurogenesis. J Neurochem. 2013. doi:10.1111/jnc.12456.

Varki A, Cummings RD, Esko JD, Freeze HH, Stanly P, Bertozzi CR, Hart GW, Etzler ME. Essentials of glycobiology. Cold Spring Harbor: Cold Spring Harbor Laboratory Press; 2009.

Wells L, Vosseller K, Hart GW. Glycosylation of nucleocytoplasmic proteins: signal transduction and O-GlcNAc. Science. 2001;291:2376–8.

Yoshida-Moriguchi T, Willer T, Anderson ME, Venzke D, Whyte T, Muntoni F, et al. SGK196 is a glycosylation-specific O-mannose kinase required for dystroglycan function. Science. 2013;341(6148):896–9.

Zhang Y, Iwasaki H, Wang H, Kudo T, Kalka TB, Hennet T, et al. Cloning and characterization of a new human UDP-N-acetyl-alpha-D-galactosamine:polypeptide N-acetylgalactosaminyltransferase, designated pp-GalNAc-T13, that is specifically expressed in neurons and synthesizes GalNAc alpha-serine/threonine antigen. J Biol Chem. 2003;278(1):573–84.

Chapter 5
Chemistry and Function of Glycosaminoglycans in the Nervous System

Nancy B. Schwartz and Miriam S. Domowicz

Abstract The glycosaminoglycan (GAG) family is characterized by covalently linked repeating disaccharides forming long unbranched polysaccharide chains. Thus far in higher eukaryotes, the family consists of chondroitin sulfate (CS), heparin/heparan sulfate (HS), dermatan sulfate (DS), and hyaluronan (HA). All GAG chains (except HA) are characteristically modified by varying amounts of esterified sulfate. One or more GAG chains are usually found in nature bound to polypeptide backbones in the form of proteoglycans; HA is the exception and is not synthesized covalently bound to a protein. Proteoglycans, and especially their GAG components, participate in numerous biologically significant interactions with growth factors, chemokines, morphogens, guidance molecules, survival factors, and other extracellular and cell-surface components. These interactions are often critical to the basic developmental processes of cellular proliferation and differentiation, as well as to both the onset of disease sequelae and the prevention of disease progression. In the nervous system, GAG/proteoglycan-mediated interactions participate in proliferation and synaptogenesis, neural plasticity, and regeneration. This review focuses on the structure, chemistry, and function of GAGs in nervous system development, disease, and injury response.

Keywords Proteoglycan • Glycosaminoglycan • Chondroitin sulfate • Heparan sulfate • Glycosyltranferase • Sulfotransferase • Brain injury response • Axon guidance molecule • Growth factor interaction

---

N.B. Schwartz (✉)
Department of Pediatrics, University of Chicago, Chicago, IL 60637, USA

Department of Biochemistry and Molecular Biology, University of Chicago, Chicago, IL 60637, USA
e-mail: n-schwartz@uchicago.edu

M.S. Domowicz
Department of Pediatrics, University of Chicago, Chicago, IL 60637, USA
e-mail: mdxx@uchicago.edu

R.K. Yu and C.-L. Schengrund (eds.), *Glycobiology of the Nervous System*,
Advances in Neurobiology 9, DOI 10.1007/978-1-4939-1154-7_5,

Abbreviations

APP	β-Amyloid precursor protein
CS	Chondroitin sulfate
DS	Dermatan sulfate
Ext1	Exostosin glycosyltransferase 1
FGF	Fibroblast growth factor
GAG	Glycosaminoglycan
Gal	Galactose
GalNAc	*N*-Acetylgalactosamine
GalNAc4S6ST	*N*-Acetylgalactosamine 4-sulfate 6-*O*-sulfotransferase
Gdf5	Growth differentiation factor 5
GlcA	Glucuronic acid
GlcAT-1	Glucuronosyltransferase-I
GlcN	Glucosamine
GlcNAc	*N*-Acetylglucosamine
HA	Hyaluronan
HS	Heparan sulfate
IdoA	Iduronic acid
LAR	Leukocyte common antigen-related phosphatase
NDST	N-Deacetylase/N-sulfotransferases
NSC	Neural stem cell
PAPS	3′-Phosphoadenosyl 5′-phosphosulfate
PNN	Perineuronal net
VZ	Ventricular zone

5.1 Introduction

Glycosaminoglycan (GAG) chains are predominantly found in nature covalently attached to multidomain core proteins, together denoted as proteoglycans (Schwartz 2000). Proteoglycans are chemically complex and structurally diverse due to variation in: (1) primary sequence, modular arrangements and repetition of domains; (2) abundance, distribution and composition of the GAG chains; and (3) position and distribution of GAG modifications including sulfation, phosphorylation, and epimerization. From their early discovery, proteoglycans were characterized based on their GAG constituents, of which there are six distinct classes: heparan sulfate (HS), heparin, chondroitin sulfate (CS), dermatan sulfate (DS), keratan sulfate (KS), and hyaluronan (HA) (only HA is not synthesized covalently bound to protein). Increasingly, this nomenclature of proteoglycans is complicated by the identification of molecules having more than one type of GAG chain attached to the same core protein. Both proteoglycan core proteins and their GAG constituents contribute significantly to the structural and functional roles proteoglycans play in numerous

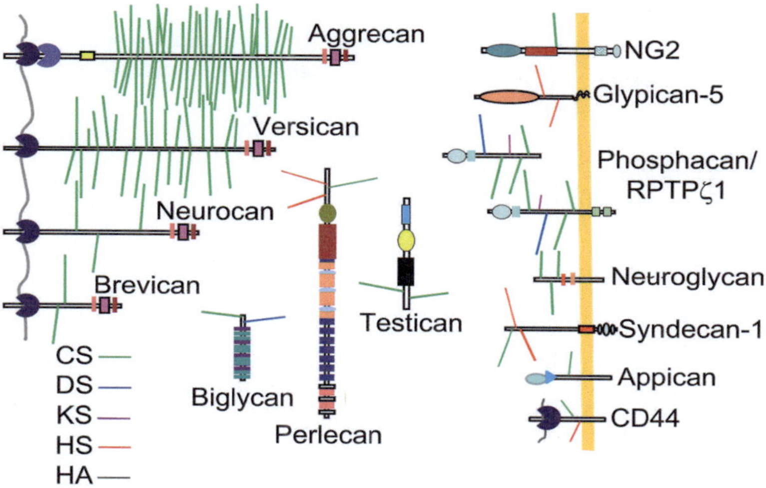

Fig. 5.1 Representation of the extracellular and membrane-bound proteoglycan families in the central nervous system. Potential GAG modifications are color-coded according to the type. *Yellow bar* signifies the plasma membrane. All the members of the lectican family of CSPGs are represented, while only representative member of the syndecan, glypican, and small leucine-rich repeat PG (biglycan) families are shown. Many of the PGs have also alternative splice forms

biological interactions critical to development and disease progression. In particular, the brain contains significant amounts and types of proteoglycans and HA that change with stage of development and state of cellular differentiation, migration, and regional specificity. In this review, we focus on the chemistry and function of GAGs as components of proteoglycans in the nervous system. Some of the major types of proteoglycans in brain are represented in Fig. 5.1.

5.2 Glycosaminoglycan Structure and Chemistry

Although six classes of GAGs exist, certain features are common across classes: (1) the long, unbranched heteropolysaccharide chains consist largely of repeating disaccharide units composed of a hexosamine and uronic acid which is characteristic of each type of GAG; (2) the most common substituents are sulfate groups, linked either by ester bonds to certain monosaccharides or by amide bonds to the amino group of the hexosamine; the exception again is HA which is not modified. The abundance of sulfate groups on the GAG chains, as well as the carboxyl groups of the uronic acids, contributes to the high net negative charges on GAG chains. Recent findings that correlate structural and chemical properties of the various GAG families with biological activity are summarized in this section.

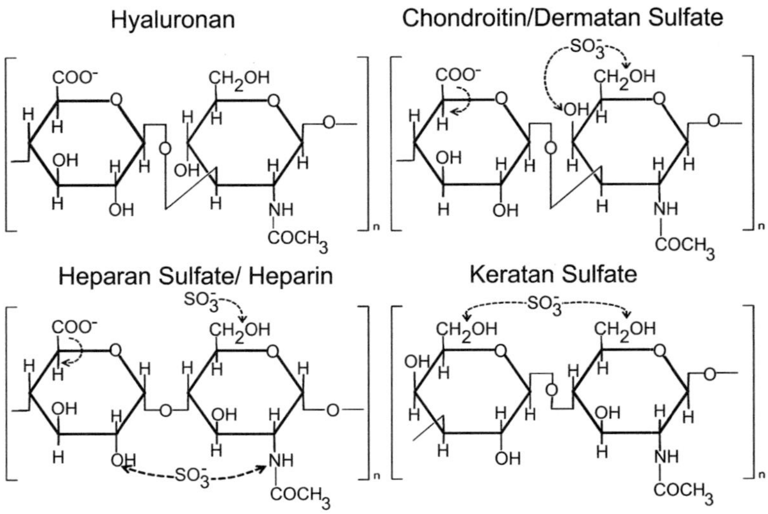

Fig. 5.2 Chemical structure of GAGs. The repeating disaccharide backbone structure of the six classes of GAGs is shown. *Arrows* indicate possible modifications

5.2.1 Chondroitin Sulfate

Throughout nature, chondroitin sulfate (CS) is the most abundant type of GAG.

Structure and Chemistry

CS chains are characterized by a repeating disaccharide of *N*-acetylgalactosamine (GalNAc) linked by a β1,3-glycosidic bond to a glucuronic acid (GlcA), which is then linked via a β1-4 bond to another disaccharide unit (Fig. 5.2). The disaccharide units may be sulfated on the C-4 (designated CS-A) or C-6 (CS-C) position of GalNAc. A functionally important disulfated CS disaccharide has been identified that is sulfated on both the 4 and 6 position of GalNAc (CS-E), is expressed in brain, and may modulate neurite out-growth (Mikami et al. 2009). CS chains typically contain between 30 and 50 disaccharide units (15–25,000 Da) and are linked covalently to the various protein cores of different chondroitin sulfate proteoglycans (CSPG) via a specific tetrasaccharide linkage region, consisting of xylose-galactose-galactose-glucuronic acid, which is O-glycosyl linked to serines within the various core proteins, i.e., GlcAβ1-3Galβ1-3 Galβ1-4Xylβ1-*O*-Ser (Schwartz 2009). The number, length, distribution, and degree of sulfation of the CS chains are highly variable among the various CSPGs.

Synthesis and Modification of CS Chains

In general, all GAG chains are synthesized by the sequential action of glycosyltransferases which transfer a monosaccharide from a nucleotide-linked derivative to an appropriate acceptor, either the non-reducing end of another sugar or a polypeptide, rather than en bloc as in glycoprotein or bacterial polysaccharide synthesis. This mechanism imposes strict substrate specificity for donor, acceptor and linkage type. For CS chains, formation of the unique tetrasaccharide linkage region is the first step, sequentially catalyzed by xylosyltransferase, β1-4-galactosyltransferase, β1-3-galactosyltransferase, and β1-3-glucuronosyltransferase activities, which exist in multiple isoforms for each activity (Schwartz 2000). Polymerization then proceeds via alternating addition of GlcA and GalNAc catalyzed by repeated glucuronosyltransferase and *N*-acetylgalactosaminyltransferase activities, respectively, to form the characteristic CS disaccharide units (Schwartz 2010). To date, six homologous glycosyltransferases responsible for synthesis of the repeating disaccharide regions have been cloned. A recent review by Mikami and Kitagawa provides a comprehensive summary of CS biosynthetic enzymes (Mikami and Kitagawa 2013). Sulfation of GalNAc on the C-4 or C-6 positions occurs concomitantly with, or shortly after, chain polymerization, and is catalyzed by two types of sulfotransferases, 4-O-sulfotransferases forming CS-A and 6-O-sulfotransferases forming CS-C; multiple isoforms have been cloned and characterized for each. These may be functionally redundant, except for one of the 4-O-sulfotransferases (C4ST-1), which may play a distinct regulatory role in CS synthesis (Izumikawa et al. 2011). To form the disulfated CS-E species, a GalNAc 4-sulfate 6-O-sulfotransferase catalyzes the transfer of sulfate to the C-6 position of a preexisting 4-*O*-sulfated GalNAc residue (Ohtake et al. 2001). An uronyl 2-*O*-sulfotransferase catalyzes 2-O sulfation of a GlcA residue in CS, resulting in formation of a second disulfated disaccharide designated CS-D. For sulfation of all GAGs, the sulfate donor is 3′-phosphoadenosyl 5′-phosphosulfate (PAPS), which is formed from ATP and sulfate in two steps catalyzed by the bifunctional PAPS synthetase (Schwartz 2005, 2010).

Lastly, it has been reported that the tetrasaccharide linkage region can be modified by a phosphate group on the C-2 position of xylose and by sulfate groups on the C-6 position of the first galactose (Gal) and the C-4 and/or C-6 positions of the second Gal (Sugahara and Kitagawa 2000; Uyama et al. 2007). These modifications may be involved in the processing of the growing linkage region, although the biological significance of these modifications is still under investigation (Koike et al. 2009; Tone et al. 2008). Formation of the tetrasaccharide linkage region starts in the endoplasmic reticulum with addition of xylose, continues with addition of the two galactoses in the *cis*/medial Golgi and addition of the first GlcA in the medial/trans Golgi (Kearns et al. 1993; Vertel et al. 1993). Most ensuing sulfation/polymerization reactions occur in the trans Golgi (Silbert and Sugumaran 2002).

CS Proteoglycans in Brain

The nervous system is rich in proteoglycans, especially CS proteoglycans (CSPGs) (5.2.1), which function in neural cell growth and plasticity (Yamaguchi 2000) and are upregulated in scar formation thereby inhibiting axonal regeneration after injury. Two types of CSPGs are found in the brain: (1) secreted extracellular CSPGs called lecticans that bind to HA and link proteins and include aggrecan, neurocan, versican, and brevican; and (2) membrane-bound CSPGs such as neuroglycan C and PTPRζ. The transcript encoding the latter CSPG also can be alternatively spliced to generate an extracellular variant, phosphacan (Sakurai et al. 1996; Snyder et al. 1996). The expression of these CSPGs is developmentally and spatially regulated; moreover, the CS composition changes during development: CS-C is expressed in developing brain, while CS-A is more highly expressed in adult brain (Sakurai et al. 1996; Snyder et al. 1996). Presumably these changes reflect the observed interactions of CSPGs with various growth factors, chemokines, and guidance molecules during brain development. Various fibroblast growth factors (FGFs) bind highly sulfated CS/DS chains and guidance molecules such as slit2, netrin1, and ephrin A1 and A5. Semaphorin 3A, 5A, and 5B bind CS in a sulfate-dependent manner (Maeda 2010) (see Sect. II.1. for further discussion of functions of CSPGs in the nervous system).

5.2.2 Dermatan Sulfate

Dermatan sulfate (DS) is a stereoisomer of CS, differing from CS in that some of its GlcA residues are converted to iduronic acid (IdoA) by dermatan sulfate epimerases, although the glycosidic linkages have the same positions and configurations as in CS (Fig. 5.2). Two DS epimerases have been identified, and one (DS-epi2) is expressed in developing brain and appears to regulate the proportion of IdoA-containing units in cerebellar DS during postnatal development. The iduronated disaccharides are sulfated in the C-4 position of GalNAc by a dermatan sulfate 4-O-sulfotransferase. IdoA may also be sulfated at the C-2 position by the same uronyl 2-O-sulfotransferase that sulfates GlcA in CS, yielding another type of di-sulfated disaccharide (IdoA (2-SO_4)-GalNAc (4-SO_4)). Interestingly, over-sulfated DS has been implicated in neuritogenesis in the brain (Hikino et al. 2003).

5.2.3 Heparin

Heparin (and HS, see Sect. 5.2.4) differ from other GAGs by containing α-glycosidic linkages and having glucosamine (GlcN) as the hexosamine joined with either GlcA or IdoA to form the disaccharide repeat. Because of their highly varied modifications, heparin and HS are considered the most complex GAGs. In heparin, almost all GlcN residues contain sulfamide linkages, and some GlcN residues are N-acetylated. Heparin may be N-sulfated or O-sulfated on C-6 of GlcN, and O-sulfated on C-3 of the hexosamine and C-2 of GlcA, with the average disaccharide unit having 2.5–2.7 sulfate groups. Heparin functions mainly as an anticoagulant and lipid-clearing agent (Cui et al. 2012; Hughes 2012; Oyagi and Hara 2012), although heparin binds strongly to numerous brain-specific chemokines and growth factors (see Section II for further discussion).

5.2.4 Heparan Sulfate

As a proteoglycan, heparan sulfate (HS) may be extracellular or an integral membrane component of many cell types, including those that constitute blood vessel walls and brain.

Structure and Chemistry

HS contains disaccharide repeat units similar to those of heparin, i.e., GlcA/IdoA β1-4GlcN, but has more N-acetyl groups and fewer N-sulfate and O-sulfate groups in general (Fig. 5.2). However, structural heterogeneity arises due to variability in epimerization at the C-5 of uronic acid, in having either acetyl or sulfo groups on the N-position of GlcN, and in the sulfation at the C-2 of the uronic acid or at C-6 and C-3 of GlcN. In addition to this microheterogeneity, an HS chain can be differentially modified in specific regions, leading to domain-specific patterns along the GAG chain, many of which may be biologically important (Kreuger and Kjellen 2012; Shriver et al. 2012).

Synthesis and Modification of HS Chains

The structural diversity of HS is biosynthetically generated in a non-template-dependent manner by specific glyco- and sulfo-transferases and sugar-modifying enzymes acting in a highly sequential pattern. Modification of a growing HS chain, consisting of the same tetrasaccharide linkage region found in CS and repeating disaccharide units of GlcU and GlcNAc, is initiated by N-deacetylation of GlcNAc residues and replacement of the acetyl groups with sulfate groups on the resulting GlcN residues by N-deacetylase/N-sulfotransferases (NDST). Some of the GlcA residues are epimerized to IdoA by heparan sulfate C-5-epimerase, followed by O-sulfation at either the C-2 of uronic acid by heparan sulfate 2-O-sulfotransferase or on the C-6 of GlcN by heparan sulfate 6-O-sulfotransferase. The final modification step is O-sulfation of the C-3 of GlcN by heparan sulfate 3-O-sulfotransferases. Each subsequent reaction is dependent to some extent on the preceding modification, i.e., the product of one reaction is the substrate for the next. While reactions often do not proceed to completion, yielding chains with variable modifications in different regions, the ultimate structure is not entirely random and seems to be regulated during development in a tissue-specific manner (Maeda et al. 2011). Although HS contains the same modifications found in heparin, the modifications are not as complete leading to greater structural heterogeneity; an average sulfate content of only one sulfate group per disaccharide is common in HS chains. Interestingly, due to the presence of IdoA in their linear sequences, heparin and HS exhibit conformational flexibility, allowing better binding to some proteins (Shriver et al. 2012).

HS Proteoglycans in Brain

HS-containing proteoglycans are usually associated with cell surfaces, either as integral membrane proteins such as syndecans, as glycosylphosphatidylinositol (GPI)-anchored proteins like glypican, or as secreted extracellular HSPGs, e.g., like perlecan (Fig. 5.1). Each heparan sulfate proteoglycan (*HSPG*) type has multiple family members that may be differentially expressed temporally or spatially. The HS GAG constituents of these HSPGs are the preferred binding partners for target proteins due to their specific sulfated and modified motifs within the HS chains. The structure/chemistry-dependent binding of HS has been demonstrated for several biologically critical interactions: e.g., FGF family members fall into five distinct groups based on specific HS structures required for efficient binding (Ashikari-Hada et al. 2004); axon guidance molecules like slit2 prefer HS with 6-O- and N-sulfate, netrin1 prefers 2-O-, 6-O- and N-sulfate, and semaphorin 5B, ephrin A1 and ephrin A5 prefer 2-O- and N-sulfate modifications (Shipp and Hsieh-Wilson 2007). Also notable is that many HSPGs are found to be hybrids containing CS and HS chains (Fig. 5.1).

5.2.5 Keratan Sulfate

Keratan sulfate (KS) is composed predominantly of a repeating disaccharide in which Gal is linked β1-4 to *N*-acetylglucosamine (GlcNAc), and it differs from the other GAGs by containing no uronic acid but may contain other monosaccharides such as mannose, fucose, sialic acid, or GalNAc (Fig. 5.2); it is the only GAG that may contain these sugars in branched positions. Sulfate content is variable, with sulfation on C-6 of either or both of the Gal and hexosamine residues. KS also exists in two types that differ in their linkage to protein, carbohydrate content and tissue distribution. KS-I is linked to protein by a GlcNAc asparaginyl bond, typical of glycoproteins, and KS-II is linked through GalNAc to serine or threonine and is often found on the same core proteins as CS. KS expression, as well as the previously mentioned CS, is upregulated in glial scar formation in response to injury; this has been shown both by detection of KS with a KS-specific monoclonal antibody and in a mouse model deficient in *N*-acetylglucosamine 6-O-sulfotransferase-1 (Zhang et al. 2006a, b).

5.2.6 Hyaluronan

Hyaluronan (HA) is a copolymer of GlcNAC and GlcA (Fig. 5.2). HA differs from other GAGs in several respects: (1) it contains no modifications (no sulfation); (2) it is not covalently linked to a core protein (not a proteoglycan component); and (3) it is produced by bacteria as well as by eukaryotic cells. Although often considered to be the least complex GAG structurally and chemically, HA chains can reach molecular masses of 10^5–10^7 Da, which contributes to its roles as a lubricant and shock absorber in many tissues. In brain it helps, along with CSPGs, to structure the extracellular milieu and is especially abundant in developing brain (Frischknecht and Gundelfinger 2012; Preston and Sherman 2011).

5.3 Function of GAGs in the Brain

GAGs, as components of proteoglycans, have been implicated in an extensive list of brain-development, aging, and disease processes. The structural differences between GAG chain types may dictate distinct binding partners, thus associating them with unique functionalities. The types of core proteins to which they are covalently bound determine distinct subcellular localizations, association with unique receptors, and participation in intracellular-signaling cascades. Furthermore, GAG localization can be controlled by transcriptional regulation of core protein expression, availability of biosynthetic enzymes, intracellular degradation, and extracellular modification by proteases, sulfatases, and deacetylases. All these aspects have added to the complexity of studying these molecules and continue to limit our understanding of GAG function; nonetheless, the problems posed maintain continuing and enthusiastic interest in this field. This section covers recent advances in understanding the functional properties of GAG chains with respect to brain development and disease.

5.3.1 GAG Interactions and Binding Partners

As mentioned, HS chains interact specifically and in a structure-dependent manner with growth factors (FGFs, midkine, pleiotrophin), axon guidance molecules (slit2, netrin1, semaphorin 5B, ephrins), chemokines (chemokine ligand 2), and morphogens (hedgehogs, Wnts) (Bornemann et al. 2008; Haerry et al. 1997; Lin 2004; Mizumoto et al. 2013a; Shipp and Hsieh-Wilson 2007; Sweeney et al. 2006; Zou et al. 2003). Consequently, they play critical roles in cell proliferation, survival, differentiation, and migration during brain development. One of the best-described systems is the tripartite molecular complex of FGF2, FGFR1, and the HS chain of a proteoglycan. This interaction is required to induce receptor dimerization and ensuing intracellular signaling. The functional dependency on particular sulfation and acetylation patterns has been highlighted by results of studies using carbohydrate microarrays (Shipp and Hsieh-Wilson 2007) which indicated that even for modifications resulting in equal charge distribution, binding is dependent on the specific position of the charge. Thus, the expression of specific sulfotransferases in a tissue or cell type may dictate the type of composition and/or concentration gradient that a particular HS-binding protein can form.

CS chains also have been implicated in interactions with many of the same molecules reported to interact with HS and heparin (Deepa et al. 2002; Kawashima et al. 2002; Kuschert et al. 1999; Nandini et al. 2005) but often do so with reduced affinity. In general, higher binding affinities are found with CS-E polysaccharides. For example, netrin1, slit2, and semaphorin 5B bind to CS-E with higher affinity than to heparin and HS, while they bind weakly to CS-A, CS-B, CS-C, and CS-D (Shipp and Hsieh-Wilson 2007) while heparin-binding proteins such as midkine and pleiotrophin bind highly sulfated CS-E with an affinity similar to that of heparin (Deepa et al. 2002; Maeda et al. 2006) and with lower affinity to CS-A, CS-C, and CS-D (Mizumoto et al. 2013a). Usually, HSPGs are on the cell surface while CSPGs for the most part are secreted; in general, HSPGs are found in lower concentrations than CSPGs. Together, all of these differing properties can lead to complex biological consequences.

Structural influence also has been observed in the axon-guidance action of semaphorin 5A in the diencephalon, where HS exerts an attractive signal and CS a repulsive one (Kantor et al. 2004). Furthermore, functional complexity may be based in variants of one PG class, as occurs in neural stem niches during brain development, where significant changes in types and quantity of HSPGs occur during neurogenesis (Hagihara et al. 2000; Litwack et al. 1998). Likewise, later in development CSPG composition changes also occur during the critical switch from neurogenesis to gliogenesis (Domowicz et al. 2008; Ishii and Maeda 2008; Shimazaki et al. 2005).

5.3.2 GAG Functions

The above-mentioned interactions are critical to the function of GAGs in neural development, disease and injury response.

Neural System Development

Knockout animal models for the individual proteoglycan core proteins exhibit relatively mild phenotypes (Brakebusch et al. 2002; Rauch et al. 2005), indicating possible redundant functions for HSPGs and CSPGs; while knockout animal models for GAG-synthesizing enzymes suggest that GAGs are essential for early embryogenesis (reviewed in Izumikawa and Kitagawa 2010 and Maeda et al. 2011). In fact, because of embryonic lethality due to loss of some enzymes, such as GlcAT1$^{-/-}$

(Izumikawa et al. 2013) which dies at E2.5 and $Ext1^{-/-}$ which dies at E8.5 (Lin et al. 2000), studying GAG function during brain development has required the use of conditional knockout mice. Furthermore, the multiple isoforms of some biosynthetic enzymes have made it difficult to ascertain many essential functions. On the other hand, the use of GAG-degrading enzymes like chondroitinase and hyaluronidase has become a common strategy for assessing the functions of CS chains under various experimental conditions (Maeda 2010).

Brain Patterning

Based on the important interactions of HS in FGF-signaling paradigms, it is not surprising that brain patterning defects are associated with HS-biosynthetic-enzyme mutations. Nestin-Cre;Ext1 conditional-knockout mice (Inatani et al. 2003) lack a cerebellum due to defects in establishing the midbrain/hindbrain boundary, which is known to be controlled by FGF8 and Wnt signaling, rather than to specification defects (reviewed in Yamaguchi et al. 2010). Another patterning defect in this model is agenesis of the olfactory bulb; however, the molecular basis for this phenotype is less clear, although both FGF8 hypomorphic mutants (Meyers et al. 1998) and FGFR1-null mice (Hebert et al. 2003) lack olfactory bulb formation. Also, patterning defects with variable penetrance have been observed in heparan-sulfate-$Ndst1^{-/-}$ mutants, leading to lack of olfactory bulbs, absence of the hippocampal and anterior commissures, and microcephaly (Grobe et al. 2005). However, it is difficult to interpret these brain defects due to the constrains imposed by the craniofacial phenotypes. No other mutations affecting HS or CS synthesis have associated patterning defects, but this could be due to the existence of multiple isoforms for the various enzymes involved.

Neurite Outgrowth and Migration

Several lines of evidence have strongly supported the participation of GAGs in axon guidance and pathfinding in both the peripheral and central nervous systems. Axonal-guidance defects have been described in many of the HS-synthesizing enzyme mutants. Most notably, Nestin-cre;Ext1 knockouts show severe axon guidance defects in three major commissural fiber tracts of the forebrain (corpus callosum, hippocampal commissure, and anterior commissure) (Inatani et al. 2003), and even though these defects, as well as defects in guidance through the optic chiasm, are similar to those of slit1 and 2 double mutants, it is possible that some of these defects are due to defective FGF signaling (Yamaguchi et al. 2010). Studies in a Wnt1-Cre-mediated ablation of Ext1, in which HS was eliminated only in the dorsal part of the spinal cord, also found commissural axon-pathfinding defects, demonstrating that commissural neurons need to express HS for their axons to respond to netrin1 signals (Matsumoto et al. 2007). Brain phenotypes of $Ndst1^{-/-}$ mice include axon guidance defects similar to those of the Nestin-cre;Ext1 knockout, implying that sulfation of HS chains is essential to their axonal-guidance function (Grobe et al. 2005). The subtle phenotypes in axon-guidance observed in 2-O-sulfotransferase-(HS2ST), 6-O-sulfotransferase-(HS6ST1), and C5-epimerase-knockout mice further support this notion (Li et al. 2003; McLaughlin et al. 2003; Pratt et al. 2006). Recently, the proteoglycan perlecan has been identified as a requirement for sema-1a-PlexA-mediated repulsive guidance of motor axons in Drosophila (Cho et al. 2012).

Despite strong evidence for the influence of CS chains on dendritic and axonal growth in culture, similar evidence for their function in axonal guidance in vivo is limited, in part due to the lack of effective models for neural knockdown of CS-chain expression. Most studies rely on de-glycosylation experiments with chondroitinase or tissue culture studies in artificial matrices. In culture, CS and purified CSPG have been shown to inhibit neuritogenesis and axonal growth by a number of neuronal types on different substrates (Bandtlow and Zimmermann 2000; Bao et al. 2005; Schmalfeldt et al. 2000; Ughrin et al. 2003), but there have also been reports of CS and DS promoting neurite outgrowth (Fernaud-Espinosa et al. 1994; Hikino et al. 2003; Lafont et al. 1992). In primary hippocampal neuronal cultures, CS is localized to focal contacts between neurons, and its removal by chondroitinase treatment or by knockdown of sulfotransferases involved in CS-E synthesis was followed by destabilization of focal contacts and induced formation of multiple axons (Nishimura et al. 2010). Thus, in culture the effects of CS chains and their core proteins on neurogenesis strongly depend on the type of neuron, plating substrate, and developmental stage.

In vivo, CSPGs also participate in axonal pathfinding in different areas of the brain and at different developmental stages. Trigeminal-ganglion neurons show abnormal axon growth into mesenchymal regions after treatment with chondroitinase, and this effect is mediated by semaphorin 5A (Kantor et al. 2004). CS also modulates retinal axonal growth towards the optic nerve (Brittis et al. 1992) and across the chiasm (Chung et al. 2000a, b; Ichijo and Kawabata 2001), and lastly, prevents the axons from crossing the midline when they reach the optic tectum (Carulli et al. 2005; Hoffman-Kim et al. 1998). Surprisingly, no biosynthetic knockout model of a single CSPG has exhibited major axonal-guidance problems, perhaps highlighting the functional redundancy of individual CSPGs (Rauch and Kappler 2006), e.g., defects in axonal pathfinding have not been described in a mouse knockout of chondroitin-6-sulfotransferase (Uchimura et al. 2002). Nevertheless, a zebrafish model of morpholino-knockdown of chondroitin-4-sulfate exhibited aberrant projections from spinal motor axons, implying a function in axonal guidance (Mizumoto et al. 2009).

Differentiation and Stem-Cell Niche

During development the brain is formed in a complex sequence of events from a simple neuroepithelium that lines the cerebral ventricles and spinal canal. Differentiation is characterized by a progressive wave of neurogenesis from radial glial progenitors, and later by gliogenesis which predominantly depletes the radial glial progenitors from the brain ventricular zone (VZ), leaving only two adult neurogenic niches (Rowitch and Kriegstein 2010). Complex changes in proteoglycans and associated GAG-chains have been described in the areas where neural stem cells reside during brain development, supporting the idea that proteoglycans could regulate neural stem cell survival, proliferation, and/or differentiation (Akita et al. 2008; Sirko et al. 2010a). Even though we are in the early stages of understanding how proteoglycans influence neural stem cell niches, several lines of evidence in vivo and in culture indicate interesting connections in this area (see also Chap. 9).

Members of the glypican family of HSPGs have been found to be expressed in a developmentally regulated manner in the VZ during neurogenesis. In rodents, glypican-4 expression in neural stem cells is down-regulated after neuronal commitment (Hagihara et al. 2000), while glypican-1 continues to be expressed in postmitotic neurons (Litwack et al. 1998). Glypicans-2 and -5 are only expressed in committed neurons and not in their precursors (Saunders et al. 1997). Interestingly, knockout of glypican-1 is characterized by a reduction in brain size, accompanied by impairment in FGF signaling and premature differentiation of post-mitotic neurons (Jen et al. 2009). The HSPG perlecan is a component of the basal lamina of the neuroepithelium during development, and null mutants also display microcephaly with variable penetrance due to reduction of early mitotic precursors and impaired cell-cycle progression (Giros et al. 2007). Other HSPGs, like syndecan-1 and -4, also are localized in the VZ where neural precursors reside (Ford-Perriss et al. 2003). Knock-down of syndecan-1 in vivo by in utero electroporation reduces neural stem cell proliferation and induces premature neuronal differentiation, possibly acting through the Wnt signaling pathway (Wang et al. 2012). Furthermore, in models of altered HSPG biosynthesis like the Nestin-cre;Ext1 knockout and Ndst1$^{-/-}$ mutants, thinning of the frontal cortex was also observed (Grobe et al. 2005; Inatani et al. 2003).

CSPGs have also been linked to neural stem/progenitor cell proliferation, survival and differentiation in culture. Several CSPGs are enriched in the VZ of the telencephalon during neurogenesis, and their expression is maintained in neurosphere cultures derived from VZ cells (Ida et al. 2006). Degradation of CS with chondroitinase in neural stem cell (NSC) cultures results in reduced cell proliferation and impaired neurosphere formation (Sirko et al. 2007), while addition of exogenous proteoglycans to these cultures increase survival of the precursors (Tham et al. 2010), and alter their differentiation potential resulting in increased numbers of astrocytic precursors and decreases in neurogenesis (Sirko et al. 2007). Intrauterine injection of chondroitinase into the lateral ventricle during mid-neurogenesis provided results consistent with the in culture observations (Sirko et al. 2007), indicating a function for CSPG in precursor self-renewal, proliferation, and differentiation during the neuron-glial differentiation switch. Also, studies in neurosphere cultures established that CS mediates proliferation and maintenance through FGF-2, and that CS restrains maturation and gliogenesis through an EGF-dependent pathway (Sirko et al. 2010b). CSPG regulation of radial glial cell differentiation and maturation was also linked to the integrin signaling pathways (Gu et al. 2009).

Some members of the lectican family are more firmly associated with the gliogenesis process. Brevican in rodents and aggrecan in birds starts to be expressed in radial glial cells of the VZ during the *switch* from *neuronal* to *glial* precursor production (Domowicz et al. 2008; Jaworski et al. 1995); in particular, aggrecan expression is developmentally regulated and observed only in glial precursors and not in mature cells (Domowicz et al. 2008). Furthermore, in mutants in which expression of aggrecan is knocked out, increased differentiation to the astrocytic pathway is observed in vivo and in culture; addition of purified aggrecan into the culture system rescues this phenotype (Domowicz et al. 2008).

Interestingly, there is increasing evidence that HA, the main binding partner of lecticans, is also an important component of many stem cell niches (Preston and Sherman 2011), and in particular, HA in brain may play a role in neural stem cell precursor proliferation and differentiation during development and possibly in adult brain-stem-cell niches as well (Preston and Sherman 2011). HA degradation induces proliferation of quiescent astrocytes in adult spinal cord (Struve et al. 2005), indicating that HA might act to slow proliferation. On the other hand, an HA-rich matrix has been hypothesized to inhibit oligodendrocyte differentiation (Back et al. 2005).

Synaptic Plasticity

GAGs are important components of the perineuronal nets (PNNs). PNNs are specialized extracellular matrices surrounding many neurons and their dendrites in the CNS, and PNNs are established as the final, mature synaptic circuitry is stabilized. Importantly, PNN components have been linked to regulating brain plasticity during synapse stabilization and maturation. Neurocan, versican, and phosphacan/PTPRζ are components of some PNNs, while aggrecan and link protein expression is upregulated postnatally during PNN formation, suggesting they also play a role in PNN formation (Galtrey et al. 2008; Wang and Fawcett 2012).

The mammalian visual cortex has the ability to undergo plastic changes controlled by visual experiences during a critical period of postnatal development. Monocular deprivation causes an ocular dominance shift toward the non-deprived eye after enzymatic degradation of CS chains, indicating that GAGs are inhibitory for experience-dependent plasticity and their removal reactivates ocular plasticity, further suggesting GAGs play an essential role in the age-dependent decrease in ocular dominance plasticity (Pizzorusso et al. 2002, 2006). Furthermore, enhancement of long-term recognition memory has been observed after in vivo treatment with chondroitinase, and these effects are lost over time as PNNs are rebuilt (Carulli et al. 2010; Romberg et al. 2013). Similar memory effects also have been described in link protein (Hapln1) knockout mice, which retain juvenile levels of ocular dominance plasticity (Pizzorusso et al. 2006), suggesting that regulation of memory and experience-driven synaptic plasticity may involve lectican-rich matrices (Carulli et al. 2010; Romberg et al. 2013). Other examples of synaptic plasticity implicating action by the proteoglycan components of PNNs have been described in rodent barrel cortex after sensory deprivation (McRae et al. 2007), in hippocampal slices after inducing spine remodeling (Orlando et al. 2012) and by interfering with induction of long-term potentiation (Bukalo et al. 2001), in neuronal hippocampal cultures by regulating AMPA receptor motility (Frischknecht et al. 2009), in the lateral vestibular nucleus of the rat after unilateral labyrinthectomy (Deak et al. 2012), and in neostriatum during the emergence of behavior (Lee et al. 2008). The proteoglycan-interacting axonal-guidance molecule, semaphorin 3 (Vo et al. 2013), and the recently described CSPG receptors (Ye and Miao 2013), LAR (leukocyte common antigen-related phosphatase) (Fisher et al. 2011) and Nogo-receptor (Dickendesher et al. 2012), have also been localized to PNNs. The mechanisms through which CSPGs might regulate synaptic plasticity have been elusive, but it has been suggested that PNN components could inhibit axonal growth and sprouting (Crespo et al. 2007; Fitch and Silver 1997; Grumet et al. 1996; Oohira et al. 1991; Wang and Fawcett 2012). Further support comes from the area of axonal injury and regeneration, where clear evidence of inhibition of axonal growth by proteoglycans continues to be accumulated (See next section).

In contrast with CSPGs, association of HSPGs with late synaptic maturation has been only minimally documented, in particular for syndecan-2 (Ethell and Yamaguchi 1999) and -3 (Reizes et al. 2001), but as mentioned earlier HSPGs are important to the synapse-formation process (Irie and Yamaguchi 2004; Johnson et al. 2006; Raulo et al. 2005). Recently, mice with a conditional Ext1-knockout targeted to postnatal neurons developed autistic-like socio-communicative deficits with attenuation of excitatory synaptic transmission in glutaminergic amygdala pyramidal neurons (Irie et al. 2012), strongly linking HS to synaptic plasticity.

5.3.3 Injury Response

CSPGs are major components of the glial scar after CNS injury and are largely responsible for the inhibition of axonal growth that hampers functional recovery after injury (Bradbury and Carter 2011; Galtrey et al. 2007; Galtrey and Fawcett 2007; Silver and Miller 2004). The efficacy of removing GAGs with chondroitinase treatment in improving axonal growth and/or functional recovery has been demonstrated in numerous adult-injury models (Bradbury and Carter 2011; Bradbury et al. 2002; Cafferty et al. 2007; Carter et al. 2011; Fawcett 2009; Garcia-Alias et al. 2011; Moon et al. 2001; Tom et al. 2009; Zhao et al. 2013; Zuo et al. 2002). Even though functional connections below the injury level have been confirmed in some cases (Bradbury et al. 2002), activation of alternative sprouting patterns cannot be discarded as a mechanism of repair (Bradbury and Carter 2011; Massey et al. 2006), nor can neuroprotective effects due to release of growth factors from the digested areas (Carter et al. 2011; Dudas and Semeniken 2012). The type of CS sulfation that mediates these effects is still not totally clear, but C6ST-1 knockout in mice has a positive influence on axonal regeneration in the CNS (Lin et al. 2011), while down-regulation of CS-E by siRNA targeting GalNAc4S6ST (*N*-acetylgalactosamine 4-sulfate 6-O-sulfotransferase) produced inhibition of neuronal attachment and neurite extensions in culture (Karumbaiah et al. 2011). This type of dichotomy could explain the lack of functional recovery often observed after chondroitinase treatment (Tom et al. 2009). However, other possibilities have been suggested such as an inhibitory effect on axonal growth by the remainder core protein, which provides potential therapeutic targets that could be exploited like extracellular matrix proteases (Cua et al. 2013) and transcription factors that regulate proteoglycan synthesis in the glia scar (Iseki et al. 2012). Toward this end, conditional ablation of Sox9, a transcription factor known to regulate matrix synthesis, reduced levels of CSPG biosynthetic enzymes and increased hind-limb function and locomotor recovery after spinal cord injury (McKillop et al. 2013).

This area of research has also unveiled a number of CSPG receptors that may regulate CSPG inhibition of axonal growth through the glia scar. PTPRζ binds to GAG chains, and RPTPσ knockout mice show enhanced axonal growth after spinal cord injury (Shen et al. 2009). Similarly, LAR binds CSPGs with high affinity, and treatment with LAR-targeting peptides after thoracic spinal cord transection injuries in mice also promotes axonal growth (Fisher et al. 2011). More recently, Nogo-receptors have been shown to bind CSPGs, and triple-null mutants for Nogo-receptors 1, 2, and 3 exhibited enhanced axonal regeneration after retro-orbital optic nerve crush injury (Dickendesher et al. 2012). All these receptors could represent new therapeutic targets for novel traumatic-injury treatments.

Fewer examples exist for potential roles of HSPGs in regulation of axonal growth after injury although increased levels of HS and HS2ST mRNA levels in glial scars after injury have been reported (Properzi et al. 2008). Glypican-1 is expressed by reactive astrocytes after brain injury in concert with its binding partner slit2, which is known to be a negative regulator of axonal growth (Hagino et al. 2003a, b; Lau and Margolis 2010). Testican-1, a heparan/chondroitin sulfate hybrid proteoglycan, is also expressed around necrotic areas in similar experimental paradigms (Iseki et al. 2011). Glypican infusion into the brain infarct cavity after cerebral ischemia produced the same level of improvement in functional recovery after injury as chondroitinase treatment (Hill et al. 2012), indicating that HS can also exert a positive influence on axonal regeneration.

5.3.4 GAGs and Human Neural Diseases

A number of genetic diseases are rooted in mutations of GAG biosynthetic and degrading enzymes and some of them exhibit neural phenotypes. In addition, GAG changes have been described in several neuro-pathological conditions such as Alzheimer's disease, schizophrenia, Parkinson's disease and epilepsy; how these changes affect the disease pathologies is currently an area of active investigation.

Accumulation of GAGs is the biochemical hallmark of human mucopolysaccharidoses. These are genetic metabolic diseases caused by mutations in genes encoding GAG-degrading (hydrolyzing) lysosomal enzymes. Some mucopolysaccharidoses have no neurological involvement, while some are characterized by intellectual disabilities and/or a variety of CNS deficits such as hydrocephalus, spasticity, seizures, cerebral infarction, ataxia, sleep apnea, optic atrophy, hearing impairment, and hyperactivity/aggressive behavior. Extensive examination of this family of diseases can be found in the reviews from Zafeiriou and Batzios 2013; Banecka-Majkutewicz et al. 2012; Lehman et al. 2011. Although there are mucopolysaccharidoses associated with failure to degrade CSPGs, the most severe neurological symptoms are linked with accumulation of HS in neural cells (Wegrzyn et al. 2010). The mucopolysaccharidoses with associated neurological impact illustrate that normal brain function requires dynamic turnover of GAGs and appropriate homeostasis of proteoglycans.

Genetic diseases associated with defects in GAG biosynthetic enzymes have been recently reviewed by Mizumoto et al. (2013b). Some of these are embryonic lethal, many exhibit skeletal and muscle defects with no neurological involvements, and some manifest clear neurological pathologies. Two missense mutations in chondroitin β1,4-*N*-acetylgalactosaminyltransferase-1 have been found in patients with different types of motor and sensory neuropathy (Saigoh et al. 2011); in both cases, total loss of enzymatic activity was associated with recombinant enzymes bearing the mutated amino acids. In studies of two large cohorts of European ancestry genetic association was found between autism and the HS3ST5 gene encoding one of the heparan sulfate 3-O-sulfotransferases (Wang et al. 2009). Chondroitin synthase-1 mutations were reported in patients with Temtamy preaxial brachydactyly who exhibit numerous skeletal defects, sensorineural hearing loss, and varied degrees of learning disabilities (Li et al. 2010; Tian et al. 2010). Evidence of disrupted Gdf5 (Li et al. 2010) and Notch (Tian et al. 2010) signaling was found in two animal models created with chondroitin synthase-1 mutations, and a patient with neuropathy harbored a missense mutation of the chondroitin synthase-1 gene which reduced the enzymatic activity by half (Izumikawa et al. 2013). In familial and sporadic cases of two connective tissue diseases, Ehlers–Danlos syndrome and Adducted thumb-clubfoot syndrome, patients harbored mutations in dermatan 4-O-sulfotransferase, and exhibited mild cranial ventricular enlargement and psychomotor retardation as part of their pathological manifestations (Dundar et al. 2009; Malfait et al. 2010; Miyake et al. 2010).

A genome-wide association study carried out on subjects with recurrent early-onset major depressive disorder revealed a chromosome 18q22.1 site as having the strongest evidence for association (Shi et al. 2011). This site encompasses the promoter regulatory area for the DS epimerase-2 gene, which has also been linked to other bipolar disorders (Goossens et al. 2003). In light of the recent findings indicating that DS epimerase-2 is the predominant epimerase expressed in postnatal developing mouse brain (Akatsu et al. 2011), and that over-sulfated dermatan structures could actively promote neurite growth, the association study results make phosphacan (Hikino et al. 2003), the major DS proteoglycan in PNNs (Deepa et al. 2006), an attractive target for future studies in the schizophrenia field. Interestingly, over-expression in mice of PTPRζ1, of which phosphacan is an alternative splicing form, produces schizophrenia-like behavior (Takahashi et al. 2011), and removal of PNNs with chondroitinase produced deficits in dopamine system function and enhanced response to psychostimulants, as is observed in schizophrenia patients (Shah and Lodge 2013).

A number of degenerative diseases exhibit abnormal GAG distribution, including Alzheimer's and Parkinson's diseases (van Horssen et al. 2003). The accumulation of Aβ peptide in Alzheimer's disease is produced by sequential cleavage of β-amyloid precursor protein (APP) by the β-site APP cleaving enzyme, BACE1, and γ-secretase. Aβ peptide in senile plaques and neurofibrillary tangles colocalized with HS- and CSPGs (reviewed in Ariga et al. 2010; Cui et al. 2013). Sulfated GAGs regulate aggregation and/or stabilization of Aβ amyloid in a structure dependent manner (Cui et al. 2012), and heparin treatment reduces Aβ production by disrupting BACE-1 processing of APP. Treatment with low-molecular-weight heparin also reduces plaques and Aβ accumulation in a mouse model of Alzheimer's disease (Beckman et al. 2006; Klaver et al. 2010; Leveugle et al. 1997). Thus, the design of GAG derivatives which act specifically to inhibit BACE-1 cleavage of APP and efficiently cross the blood–brain barrier is being actively pursued.

5.4 Concluding Remarks

This brief synopsis of the structure-function relationship of GAGs and GAG containing proteoglycans highlights the importance of these cell-surface and extracellular constituents in numerous biologically significant interactions critical to the basic developmental processes of proliferation, differentiation, patterning, axonal pathfinding and synapsis formation which underlie the formation of a functional nervous system. Increasingly, GAGs are implicated in regeneration, injury response and disease pathology of the brain, functional revelations that have been made concomitantly with elucidation of the complex chemistry and structure of these intricate molecules.

Acknowledgment This work was supported by grants from the National Institute of Child Health and Human Disorders (P01-HD 09402, P30-HD0054275, R03 HD00354235).

Conflict of Interest The authors declare that they have no conflict of interest.

References

Akatsu C, Mizumoto S, Kaneiwa T, Maccarana M, Malmstrom A, Yamada S, et al. Dermatan sulfate epimerase 2 is the predominant isozyme in the formation of the chondroitin sulfate/dermatan sulfate hybrid structure in postnatal developing mouse brain. Glycobiology. 2011;21(5):565–74.

Akita K, von Holst A, Furukawa Y, Mikami T, Sugahara K, Faissner A. Expression of multiple chondroitin/dermatan sulfotransferases in the neurogenic regions of the embryonic and adult central nervous system implies that complex chondroitin sulfates have a role in neural stem cell maintenance. Stem Cells. 2008;26(3):798–809.

Ariga T, Miyatake T, Yu RK. Role of proteoglycans and glycosaminoglycans in the pathogenesis of Alzheimer's disease and related disorders: amyloidogenesis and therapeutic strategies–a review. J Neurosci Res. 2010;88(11):2303–15.

Ashikari-Hada S, Habuchi H, Kariya Y, Itoh N, Reddi AH, Kimata K. Characterization of growth factor-binding structures in heparin/heparan sulfate using an octasaccharide library. J Biol Chem. 2004;279(13):12346–54.

Back SA, Tuohy TM, Chen H, Wallingford N, Craig A, Struve J, et al. Hyaluronan accumulates in demyelinated lesions and inhibits oligodendrocyte progenitor maturation. Nat Med. 2005; 11(9):966–72.

Bandtlow CE, Zimmermann DR. Proteoglycans in the developing brain: new conceptual insights for old proteins. Physiol Rev. 2000;80(4):1267–90.

Banecka-Majkutewicz Z, Jakobkiewicz-Banecka J, Gabig-Ciminska M, Wegrzyn A, Wegrzyn G. Putative biological mechanisms of efficiency of substrate reduction therapies for mucopolysaccharidoses. Arch Immunol Ther Exp (Warsz). 2012;60(6):461–8.

Bao X, Mikami T, Yamada S, Faissner A, Muramatsu T, Sugahara K. Heparin-binding growth factor, pleiotrophin, mediates neuritogenic activity of embryonic pig brain-derived chondroitin sulfate/dermatan sulfate hybrid chains. J Biol Chem. 2005;280(10):9180–91.

Beckman M, Holsinger RM, Small DH. Heparin activates beta-secretase (BACE1) of Alzheimer's disease and increases autocatalysis of the enzyme. Biochemistry. 2006;45(21):6703–14.

Bornemann DJ, Park S, Phin S, Warrior R. A translational block to HSPG synthesis permits BMP signaling in the early Drosophila embryo. Development. 2008;135(6):1039–47.

Bradbury EJ, Carter LM. Manipulating the glial scar: chondroitinase ABC as a therapy for spinal cord injury. Brain Res Bull. 2011;84(4–5):306–16.

Bradbury EJ, Moon LD, Popat RJ, King VR, Bennett GS, Patel PN, et al. Chondroitinase ABC promotes functional recovery after spinal cord injury. Nature. 2002;416(6881):636–40.

Brakebusch C, Seidenbecher CI, Asztely F, Rauch U, Matthies H, Meyer H, et al. Brevican-deficient mice display impaired hippocampal CA1 long-term potentiation but show no obvious deficits in learning and memory. Mol Cell Biol. 2002;22(21):7417–27.

Brittis PA, Canning DR, Silver J. Chondroitin sulfate as a regulator of neuronal patterning in the retina. Science. 1992;255:733–6.

Bukalo O, Schachner M, Dityatev A. Modification of extracellular matrix by enzymatic removal of chondroitin sulfate and by lack of tenascin-R differentially affects several forms of synaptic plasticity in the hippocampus. Neuroscience. 2001;104(2):359–69.

Cafferty WB, Yang SH, Duffy PJ, Li S, Strittmatter SM. Functional axonal regeneration through astrocytic scar genetically modified to digest chondroitin sulfate proteoglycans. J Neurosci. 2007;27(9):2176–85.

Carter LM, McMahon SB, Bradbury EJ. Delayed treatment with chondroitinase ABC reverses chronic atrophy of rubrospinal neurons following spinal cord injury. Exp Neurol. 2011;228(1): 149–56.

Carulli D, Laabs T, Geller HM, Fawcett JW. Chondroitin sulfate proteoglycans in neural development and regeneration. Curr Opin Neurobiol. 2005;15(1):116–20.

Carulli D, Pizzorusso T, Kwok JC, Putignano E, Poli A, Forostyak S, et al. Animals lacking link protein have attenuated perineuronal nets and persistent plasticity. Brain. 2010;133(Pt 8): 2331–47.

Cho JY, Chak K, Andreone BJ, Wooley JR, Kolodkin AL. The extracellular matrix proteoglycan perlecan facilitates transmembrane semaphorin-mediated repulsive guidance. Genes Dev. 2012;26(19):2222–35.

Chung KY, Shum DK, Chan SO. Expression of chondroitin sulfate proteoglycans in the chiasm of mouse embryos. J Comp Neurol. 2000a;417(2):153–63.

Chung KY, Taylor JS, Shum DK, Chan SO. Axon routing at the optic chiasm after enzymatic removal of chondroitin sulfate in mouse embryos. Development. 2000b;127(12):2673–83.

Crespo D, Asher RA, Lin R, Rhodes KE, Fawcett JW. How does chondroitinase promote functional recovery in the damaged CNS? Exp Neurol. 2007;206(2):159–71.

Cua RC, Lau LW, Keough MB, Midha R, Apte SS, Yong VW. Overcoming neurite-inhibitory chondroitin sulfate proteoglycans in the astrocyte matrix. Glia. 2013;61(6):972–84.

Cui H, Hung AC, Freeman C, Narkowicz C, Jacobson GA, Small DH. Size and sulfation are critical for the effect of heparin on APP processing and Abeta production. J Neurochem. 2012; 123(3):447–57.

Cui H, Freeman C, Jacobson GA, Small DH. Proteoglycans in the central nervous system: role in development, neural repair, and Alzheimer's disease. IUBMB Life. 2013;65(2):108–20.

Deak A, Bacskai T, Gaal B, Racz E, Matesz K. Effect of unilateral labyrinthectomy on the molecular composition of perineuronal nets in the lateral vestibular nucleus of the rat. Neurosci Lett. 2012;513(1):1–5.

Deepa SS, Umehara Y, Higashiyama S, Itoh N, Sugahara K. Specific molecular interactions of oversulfated chondroitin sulfate E with various heparin-binding growth factors. Implications as a physiological binding partner in the brain and other tissues. J Biol Chem. 2002; 277(46):43707–16.

Deepa SS, Carulli D, Galtrey C, Rhodes K, Fukuda J, Mikami T, et al. Composition of perineuronal net extracellular matrix in rat brain: a different disaccharide composition for the net-associated proteoglycans. J Biol Chem. 2006;281(26):17789–800.

Dickendesher TL, Baldwin KT, Mironova YA, Koriyama Y, Raiker SJ, Askew KL, et al. NgR1 and NgR3 are receptors for chondroitin sulfate proteoglycans. Nat Neurosci. 2012;15(5):703–12.

Domowicz MS, Sanders TA, Ragsdale CW, Schwartz NB. Aggrecan is expressed by embryonic brain glia and regulates astrocyte development. Dev Biol. 2008;315(1):114–24.

Dudas B, Semeniken K. Glycosaminoglycans and neuroprotection. Handb Exp Pharmacol. 2012;207:325–43.

Dundar M, Muller T, Zhang Q, Pan J, Steinmann B, Vodopiutz J, et al. Loss of dermatan-4-sulfotransferase 1 function results in adducted thumb-clubfoot syndrome. Am J Hum Genet. 2009;85(6):873–82.

Ethell IM, Yamaguchi Y. Cell surface heparan sulfate proteoglycan syndecan-2 induces the maturation of dendritic spines in rat hippocampal neurons. J Cell Biol. 1999;144(3):575–86.

Fawcett J. Molecular control of brain plasticity and repair. Prog Brain Res. 2009;175:501–9.

Fernaud-Espinosa I, Nieto-Sampedro M, Bovolenta P. Differential effects of glycosaminoglycans on neurite outgrowth from hippocampal and thalamic neurones. J Cell Sci. 1994;107 (Pt 6):1437–48.

Fisher D, Xing B, Dill J, Li H, Hoang HH, Zhao Z, et al. Leukocyte common antigen-related phosphatase is a functional receptor for chondroitin sulfate proteoglycan axon growth inhibitors. J Neurosci. 2011;31(40):14051–66.

Fitch MT, Silver J. Glial cell extracellular matrix: boundaries for axon growth in development and regeneration. Cell Tissue Res. 1997;290(2):379–84.

Ford-Perriss M, Turner K, Guimond S, Apedaile A, Haubeck HD, Turnbull J, et al. Localisation of specific heparan sulfate proteoglycans during the proliferative phase of brain development. Dev Dyn. 2003;227(2):170–84.

Frischknecht R, Gundelfinger ED. The brain's extracellular matrix and its role in synaptic plasticity. Adv Exp Med Biol. 2012;970:153–71.

Frischknecht R, Heine M, Perrais D, Seidenbecher CI, Choquet D, Gundelfinger ED. Brain extracellular matrix affects AMPA receptor lateral mobility and short-term synaptic plasticity. Nat Neurosci. 2009;12(7):897–904.

Galtrey CM, Fawcett JW. The role of chondroitin sulfate proteoglycans in regeneration and plasticity in the central nervous system. Brain Res Rev. 2007;54(1):1–18.

Galtrey CM, Asher RA, Nothias F, Fawcett JW. Promoting plasticity in the spinal cord with chondroitinase improves functional recovery after peripheral nerve repair. Brain. 2007;130 (Pt 4):926–39.

Galtrey CM, Kwok JC, Carulli D, Rhodes KE, Fawcett JW. Distribution and synthesis of extracellular matrix proteoglycans, hyaluronan, link proteins and tenascin-R in the rat spinal cord. Eur J Neurosci. 2008;27(6):1373–90.

Garcia-Alias G, Petrosyan HA, Schnell L, Horner PJ, Bowers WJ, Mendell LM, et al. Chondroitinase ABC combined with neurotrophin NT-3 secretion and NR2D expression promotes axonal plasticity and functional recovery in rats with lateral hemisection of the spinal cord. J Neurosci. 2011;31(49):17788–99.

Giros A, Morante J, Gil-Sanz C, Fairen A, Costell M. Perlecan controls neurogenesis in the developing telencephalon. BMC Dev Biol. 2007;7:29.

Goossens D, Van Gestel S, Claes S, De Rijk P, Souery D, Massat I, et al. A novel CpG-associated brain-expressed candidate gene for chromosome 18q-linked bipolar disorder. Mol Psychiatry. 2003;8(1):83–9.

Grobe K, Inatani M, Pallerla SR, Castagnola J, Yamaguchi Y, Esko JD. Cerebral hypoplasia and craniofacial defects in mice lacking heparan sulfate Ndst1 gene function. Development. 2005; 132(16):3777–86.

Grumet M, Friedlander DR, Sakurai T. Functions of brain chondroitin sulfate proteoglycans during developments: interactions with adhesion molecules. Perspect Dev Neurobiol. 1996;3(4): 319–30.

Gu W-L, Fu S-L, Wang Y-X, Li Y, Lu H-Z, Xu X-M, et al. Chondroitin sulfate proteoglycans regulate the growth, differentiation and migration of multipotent neural precursor cells through the integrin signaling pathway. BMC Neurosci. 2009;10:128.

Haerry TE, Heslip TR, Marsh JL, O'Connor MB. Defects in glucuronate biosynthesis disrupt Wingless signaling in Drosophila. Development. 1997;124(16):3055–64.

Hagihara K, Watanabe K, Chun J, Yamaguchi Y. Glypican-4 is an FGF2-binding heparan sulfate proteoglycan expressed in neural precursor cells. Dev Dyn. 2000;219(3):353–67.

Hagino S, Iseki K, Mori T, Zhang Y, Hikake T, Yokoya S, et al. Slit and glypican-1 mRNAs are coexpressed in the reactive astrocytes of the injured adult brain. Glia. 2003a;42(2):130–8.

Hagino S, Iseki K, Mori T, Zhang Y, Sakai N, Yokoya S, et al. Expression pattern of glypican-1 mRNA after brain injury in mice. Neurosci Lett. 2003b;349(1):29–32.

Hebert JM, Lin M, Partanen J, Rossant J, McConnell SK. FGF signaling through FGFR1 is required for olfactory bulb morphogenesis. Development. 2003;130(6):1101–11.

Hikino M, Mikami T, Faissner A, Vilela-Silva AC, Pavao MS, Sugahara K. Oversulfated dermatan sulfate exhibits neurite outgrowth-promoting activity toward embryonic mouse hippocampal neurons: implications of dermatan sulfate in neuritogenesis in the brain. J Biol Chem. 2003;278(44):43744–54.

Hill JJ, Jin K, Mao XO, Xie L, Greenberg DA. Intracerebral chondroitinase ABC and heparan sulfate proteoglycan glypican improve outcome from chronic stroke in rats. Proc Natl Acad Sci U S A. 2012;109(23):9155–60.

Hoffman-Kim D, Lander AD, Jhaveri S. Patterns of chondroitin sulfate immunoreactivity in the developing tectum reflect regional differences in glycosaminoglycan biosynthesis. J Neurosci. 1998;18(15):5881–90.

Hughes GR. Heparin, antiphospholipid antibodies and the brain. Lupus. 2012;21(10):1039–40.

Ichijo H, Kawabata I. Roles of the telencephalic cells and their chondroitin sulfate proteoglycans in delimiting an anterior border of the retinal pathway. J Neurosci. 2001;21(23):9304–14.

Ida M, Shuo T, Hirano K, Tokita Y, Nakanishi K, Matsui F, et al. Identification and functions of chondroitin sulfate in the milieu of neural stem cells. J Biol Chem. 2006;281(9):5982–91.

Inatani M, Irie F, Plump AS, Tessier-Lavigne M, Yamaguchi Y. Mammalian brain morphogenesis and midline axon guidance require heparan sulfate. Science. 2003;302(5647):1044–6.

Irie F, Yamaguchi Y. EPHB receptor signaling in dendritic spine development. Front Biosci. 2004;9:1365–73.

Irie F, Badie-Mahdavi H, Yamaguchi Y. Autism-like socio-communicative deficits and stereotypies in mice lacking heparan sulfate. Proc Natl Acad Sci U S A. 2012;109(13):5052–6.

Iseki K, Hagino S, Zhang Y, Mori T, Sato N, Yokoya S, et al. Altered expression pattern of testican-1 mRNA after brain injury. Biomed Res. 2011;32(6):373–8.

Iseki K, Hagino S, Nikaido T, Zhang Y, Mori T, Yokoya S, et al. Gliosis-specific transcription factor OASIS coincides with proteoglycan core protein genes in the glial scar and inhibits neurite outgrowth. Biomed Res. 2012;33(6):345–53.

Ishii M. Maeda N. Glycobiology: Spatiotemporal expression of chondroitin sulfate sulfotransferases in the postnatal developing mouse cerebellum; 2008.
Izumikawa T, Kitagawa H. Mice deficient in glucuronyltransferase-I. Prog Mol Biol Transl Sci. 2010;93:19–34.
Izumikawa T, Okuura Y, Koike T, Sakoda N, Kitagawa H. Chondroitin 4-O-sulfotransferase-1 regulates the chain length of chondroitin sulfate in co-operation with chondroitin N-acetylgalactosaminyltransferase-2. Biochem J. 2011;434(2):321–31.
Izumikawa T, Saigoh K, Shimizu J, Tsuji S, Kusunoki S, Kitagawa H. A chondroitin synthase-1 (ChSy-1) missense mutation in a patient with neuropathy impairs the elongation of chondroitin sulfate chains initiated by chondroitin N-acetylgalactosaminyltransferase-1. Biochim Biophys Acta. 2013;1830(10):4806–12.
Jaworski DM, Kelly GM, Hockfield S. The CNS-specific hyaluronan-binding protein BEHAB is expressed in ventricular zones coincident with gliogenesis. J Neurosci. 1995;15(2):1352–62.
Jen Y-HL, Musacchio M, Lander AD. Glypican-1 controls brain size through regulation of fibroblast growth factor signaling in early neurogenesis. Neural Dev. 2009;4:33.
Johnson KG, Tenney AP, Ghose A, Duckworth AM, Higashi ME, Parfitt K, et al. The HSPGs Syndecan and Dallylike bind the receptor phosphatase LAR and exert distinct effects on synaptic development. Neuron. 2006;49(4):517–31.
Kantor DB, Chivatakarn O, Peer KL, Oster SF, Inatani M, Hansen MJ, et al. Semaphorin 5A is a bifunctional axon guidance cue regulated by heparan and chondroitin sulfate proteoglycans. Neuron. 2004;44(6):961–75.
Karumbaiah L, Anand S, Thazhath R, Zhong Y, McKeon RJ, Bellamkonda RV. Targeted downregulation of N-acetylgalactosamine 4-sulfate 6-O-sulfotransferase significantly mitigates chondroitin sulfate proteoglycan-mediated inhibition. Glia. 2011;59(6):981–96.
Kawashima H, Atarashi K, Hirose M, Hirose J, Yamada S, Sugahara K, et al. Oversulfated chondroitin/dermatan sulfates containing GlcAbeta1/IdoAalpha1-3GalNAc(4,6-O-disulfate) interact with L- and P-selectin and chemokines. J Biol Chem. 2002;277(15):12921–30.
Kearns AE, Vertel BM, Schwartz NB. Topography of glycosylation and UDP-xylose production. J Biol Chem. 1993;268(15):11097–104.
Klaver DW, Wilce MC, Gasperini R, Freeman C, Juliano JP, Parish C, et al. Glycosaminoglycan-induced activation of the beta-secretase (BACE1) of Alzheimer's disease. J Neurochem. 2010;112(6):1552–61.
Koike T, Izumikawa T, Tamura J, Kitagawa H. FAM20B is a kinase that phosphorylates xylose in the glycosaminoglycan-protein linkage region. Biochem J. 2009;421(2):157–62.
Kreuger J, Kjellen L. Heparan sulfate biosynthesis: regulation and variability. J Histochem Cytochem. 2012;60(12):898–907.
Kuschert GS, Coulin F, Power CA, Proudfoot AE, Hubbard RE, Hoogewerf AJ, et al. Glycosaminoglycans interact selectively with chemokines and modulate receptor binding and cellular responses. Biochemistry. 1999;38(39):12959–68.
Lafont F, Rouget M, Triller A, Prochiantz A, Rousselet A. In vitro control of neuronal polarity by glycosaminoglycans. Development. 1992;114(1):17–29.
Lau E, Margolis RU. Inhibitors of slit protein interactions with the heparan sulphate proteoglycan glypican-1: potential agents for the treatment of spinal cord injury. Clin Exp Pharmacol Physiol. 2010;37(4):417–21.
Lee H, Leamey CA, Sawatari A. Rapid reversal of chondroitin sulfate proteoglycan associated staining in subcompartments of mouse neostriatum during the emergence of behaviour. PLoS One. 2008;3(8):e3020.
Lehman TJ, Miller N, Norquist B, Underhill L, Keutzer J. Diagnosis of the mucopolysaccharidoses. Rheumatology. 2011;50 Suppl 5:v41–8.
Leveugle B, Ding W, Durkin JT, Mistretta S, Eisle J, Matic M, et al. Heparin promotes beta-secretase cleavage of the Alzheimer's amyloid precursor protein. Neurochem Int. 1997;30(6): 543–8.
Li JP, Gong F, Hagner-McWhirter A, Forsberg E, Abrink M, Kisilevsky R, et al. Targeted disruption of a murine glucuronyl C5-epimerase gene results in heparan sulfate lacking L-iduronic acid and in neonatal lethality. J Biol Chem. 2003;278(31):28363–6.

Li Y, Laue K, Temtamy S, Aglan M, Kotan LD, Yigit G, et al. Temtamy preaxial brachydactyly syndrome is caused by loss-of-function mutations in chondroitin synthase 1, a potential target of BMP signaling. Am J Hum Genet. 2010;87(6):757–67.

Lin X. Functions of heparan sulfate proteoglycans in cell signaling during development. Development. 2004;131(24):6009–21.

Lin X, Wei G, Shi Z, Dryer L, Esko JD, Wells DE, et al. Disruption of gastrulation and heparan sulfate biosynthesis in EXT1- deficient mice. Dev Biol. 2000;224(2):299–311.

Lin R, Rosahl TW, Whiting PJ, Fawcett JW, Kwok JC. 6-Sulphated chondroitins have a positive influence on axonal regeneration. PLoS One. 2011;6(7):e21499.

Litwack ED, Ivins JK, Kumbasar A, Paine-Saunders S, Stipp CS, Lander AD. Expression of the heparan sulfate proteoglycan glypican-1 in the developing rodent. Dev Dyn. 1998;211(1): 72–87.

Maeda N. Structural variation of chondroitin sulfate and its roles in the central nervous system. Cent Nerv Syst Agents Med Chem. 2010;10(1):22–31.

Maeda N, Fukazawa N, Hata T. The binding of chondroitin sulfate to pleiotrophin/heparin-binding growth-associated molecule is regulated by chain length and oversulfated structures. J Biol Chem. 2006;281(8):4894–902.

Maeda N, Ishii M, Nishimura K, Kamimura K. Functions of chondroitin sulfate and heparan sulfate in the developing brain. Neurochem Res. 2011;36(7):1228–40.

Malfait F, Syx D, Vlummens P, Symoens S, Nampoothiri S, Hermanns-Le T, et al. Musculocontractural Ehlers-Danlos Syndrome (former EDS type VIB) and adducted thumb clubfoot syndrome (ATCS) represent a single clinical entity caused by mutations in the dermatan-4-sulfotransferase 1 encoding CHST14 gene. Hum Mutat. 2010;31(11):1233–9.

Massey JM, Hubscher CH, Wagoner MR, Decker JA, Amps J, Silver J, et al. Chondroitinase ABC digestion of the perineuronal net promotes functional collateral sprouting in the cuneate nucleus after cervical spinal cord injury. J Neurosci. 2006;26(16):4406–14.

Matsumoto Y, Irie F, Inatani M, Tessier-Lavigne M, Yamaguchi Y. Netrin-1/DCC signaling in commissural axon guidance requires cell-autonomous expression of heparan sulfate. J Neurosci. 2007;27(16):4342–50.

McKillop WM, Dragan M, Schedl A, Brown A. Conditional Sox9 ablation reduces chondroitin sulfate proteoglycan levels and improves motor function following spinal cord injury. Glia. 2013;61(2):164–77.

McLaughlin D, Karlsson F, Tian N, Pratt T, Bullock SL, Wilson VA, et al. Specific modification of heparan sulphate is required for normal cerebral cortical development. Mech Dev. 2003;120(12):1481–8.

McRae PA, Rocco MM, Kelly G, Brumberg JC, Matthews RT. Sensory deprivation alters aggrecan and perineuronal net expression in the mouse barrel cortex. J Neurosci. 2007;27(20): 5405–13.

Meyers EN, Lewandoski M, Martin GR. An Fgf8 mutant allelic series generated by Cre- and Flp-mediated recombination. Nat Genet. 1998;18(2):136–41.

Mikami T, Kitagawa H. Biosynthesis and function of chondroitin sulfate. Biochim Biophys Acta 2013;1830(10):4719–33.

Mikami T, Yasunaga D, Kitagawa H. Contactin-1 is a functional receptor for neuroregulatory chondroitin sulfate-E. J Biol Chem. 2009;284(7):4494–9.

Miyake N, Kosho T, Mizumoto S, Furuichi T, Hatamochi A, Nagashima Y, et al. Loss-of-function mutations of CHST14 in a new type of Ehlers-Danlos syndrome. Hum Mutat. 2010;31(8): 966–74.

Mizumoto S, Mikami T, Yasunaga D, Kobayashi N, Yamauchi H, Miyake A, et al. Chondroitin 4-O-sulfotransferase-1 is required for somitic muscle development and motor axon guidance in zebrafish. Biochem J. 2009;419(2):387–99.

Mizumoto S, Fongmoon D, Sugahara K. Interaction of chondroitin sulfate and dermatan sulfate from various biological sources with heparin-binding growth factors and cytokines. Glycoconj J. 2013a;30(6):619–32.

Mizumoto S, Ikegawa S, Sugahara K. Human genetic disorders caused by mutations in genes encoding biosynthetic enzymes for sulfated glycosaminoglycans. J Biol Chem. 2013b;288(16):10953–61.

Moon LD, Asher RA, Rhodes KE, Fawcett JW. Regeneration of CNS axons back to their target following treatment of adult rat brain with chondroitinase ABC. Nat Neurosci. 2001;4(5):465–6.

Nandini CD, Itoh N, Sugahara K. Novel 70-kDa chondroitin sulfate/dermatan sulfate hybrid chains with a unique heterogeneous sulfation pattern from shark skin, which exhibit neuritogenic activity and binding activities for growth factors and neurotrophic factors. J Biol Chem. 2005;280(6):4058–69.

Nishimura K, Ishii M, Kuraoka M, Kamimura K, Maeda N. Opposing functions of chondroitin sulfate and heparan sulfate during early neuronal polarization. Neuroscience. 2010;169(4):1535–47.

Ohtake S, Ito Y, Fukuta M, Habuchi O. Human N-acetylgalactosamine 4-sulfate 6-O-sulfotransferase cDNA is related to human B cell recombination activating gene-associated gene. J Biol Chem. 2001;276(47):43894–900.

Oohira A, Matsui F, Katoh-Semba R. Inhibitory effect of brain chondroitin sulphate proteoglycans on neurite outgrowth from PC12D cells. J Neurosci. 1991;11:822–7.

Orlando C, Ster J, Gerber U, Fawcett JW, Raineteau O. Perisynaptic chondroitin sulfate proteoglycans restrict structural plasticity in an integrin-dependent manner. J Neurosci. 2012;32(50):18009-17, 17a.

Oyagi A, Hara H. Essential roles of heparin-binding epidermal growth factor-like growth factor in the brain. CNS Neurosci Ther. 2012;18(10):803–10.

Pizzorusso T, Medini P, Berardi N, Chierzi S, Fawcett JW, Maffei L. Reactivation of ocular dominance plasticity in the adult visual cortex. Science. 2002;298(5596):1248–51.

Pizzorusso T, Medini P, Landi S, Baldini S, Berardi N, Maffei L. Structural and functional recovery from early monocular deprivation in adult rats. Proc Natl Acad Sci U S A. 2006;103(22):8517–22.

Pratt T, Conway CD, Tian NM, Price DJ, Mason JO. Heparan sulphation patterns generated by specific heparan sulfotransferase enzymes direct distinct aspects of retinal axon guidance at the optic chiasm. J Neurosci. 2006;26(26):6911–23.

Preston M, Sherman LS. Neural stem cell niches: roles for the hyaluronan-based extracellular matrix. Front Biosci. 2011;3:1165–79.

Properzi F, Lin R, Kwok J, Naidu M, van Kuppevelt TH, Ten Dam GB, et al. Heparan sulphate proteoglycans in glia and in the normal and injured CNS: expression of sulphotransferases and changes in sulphation. Eur J Neurosci. 2008;27(3):593–604.

Rauch U, Kappler J. Chondroitin/Dermatan sulfates in the central nervous system: their structures and functions in health and disease. Adv Pharmacol. 2006;53:337–56.

Rauch U, Zhou XH, Roos G. Extracellular matrix alterations in brains lacking four of its components. Biochem Biophys Res Commun. 2005;328(2):608–17.

Raulo E, Tumova S, Pavlov I, Pekkanen M, Hienola A, Klankki E, et al. The two thrombospondin type I repeat domains of the heparin-binding growth-associated molecule bind to heparin/heparan sulfate and regulate neurite extension and plasticity in hippocampal neurons. J Biol Chem. 2005;280(50):41576–83.

Reizes O, Lincecum J, Wang Z, Goldberger O, Huang L, Kaksonen M, et al. Transgenic expression of syndecan-1 uncovers a physiological control of feeding behavior by syndecan-3. Cell. 2001;106(1):105–16.

Romberg C, Yang S, Melani R, Andrews MR, Horner AE, Spillantini MG, et al. Depletion of perineuronal nets enhances recognition memory and long-term depression in the perirhinal cortex. J Neurosci. 2013;33(16):7057–65.

Rowitch DH, Kriegstein AR. Developmental genetics of vertebrate glial-cell specification. Nature. 2010;468(7321):214–22.

Saigoh K, Izumikawa T, Koike T, Shimizu J, Kitagawa H, Kusunoki S. Chondroitin beta-1,4-N-ace tylgalactosaminyltransferase-1 missense mutations are associated with neuropathies. J Hum Genet. 2011;56(2):143–6.

Sakurai T, Friedlander DR, Grumet M. Expression of polypeptide variants of receptor-type protein tyrosine phosphatase beta: the secreted form, phosphacan, increases dramatically during embryonic development and modulates glial cell behavior in vitro. J Neurosci Res. 1996;43(6):694–706.

Saunders S, Paine-Saunders S, Lander AD. Expression of the cell surface proteoglycan glypican-5 is developmentally regulated in kidney, limb, and brain. Dev Biol. 1997;190(1):78–93.

Schmalfeldt M, Bandtlow CE, Dours-Zimmermann MT, Winterhalter KH, Zimmermann DR. Brain derived versican V2 is a potent inhibitor of axonal growth. J Cell Sci. 2000;113(Pt 5):807–16.

Schwartz NB. Biosynthesis and regulation of expression of proteoglycans. Front Biosci. 2000;5:D649–55.

Schwartz NB. PAPS and sulfoconjugation. In: Coughtrie MW, Pacifici GM, editors. Human cytosolic sulfotransferases. London: Taylor & Francis; 2005. p. 43–60.

Schwartz NB. Special pathways and glycoconjugates. In: Devlin TM, editor. Textbook of biochemistry. 7th ed. New York: Wiley Liss; 2010. p. 647–73.

Schwartz, Nancy B. Proteoglycans. In: Encyclopedia of Life Sciences. John Wiley & Sons Ltd, Chichester. http://www.els.net [doi: 10.1002/9780470015902.a0000623.pub2]; 2009.

Shah A, Lodge DJ. A loss of hippocampal perineuronal nets produces deficits in dopamine system function: relevance to the positive symptoms of schizophrenia. Transl Psychiatry. 2013;3:e215.

Shen Y, Tenney AP, Busch SA, Horn KP, Cuascut FX, Liu K, et al. PTPsigma is a receptor for chondroitin sulfate proteoglycan, an inhibitor of neural regeneration. Science. 2009;326(5952): 592–6.

Shi J, Potash JB, Knowles JA, Weissman MM, Coryell W, Scheftner WA, et al. Genome-wide association study of recurrent early-onset major depressive disorder. Mol Psychiatry. 2011;16(2):193–201.

Shimazaki Y, Nagata I, Ishii M, Tanaka M, Marunouchi T, Hata T, et al. Developmental change and function of chondroitin sulfate deposited around cerebellar Purkinje cells. J Neurosci Res. 2005;82(2):172–83.

Shipp EL, Hsieh-Wilson LC. Profiling the sulfation specificities of glycosaminoglycan interactions with growth factors and chemotactic proteins using microarrays. Chem Biol. 2007; 14(2):195–208.

Shriver Z, Capila I, Venkataraman G, Sasisekharan R. Heparin and heparan sulfate: analyzing structure and microheterogeneity. Handb Exp Pharmacol. 2012;207:159–76.

Silbert JE, Sugumaran G. Biosynthesis of chondroitin/dermatan sulfate. IUBMB Life. 2002;54(4): 177–86.

Silver J, Miller JH. Regeneration beyond the glial scar. Nat Rev Neurosci. 2004;5(2):146–56.

Sirko S, von Holst A, Wizenmann A, Gotz M, Faissner A. Chondroitin sulfate glycosaminoglycans control proliferation, radial glia cell differentiation and neurogenesis in neural stem/progenitor cells. Development. 2007;134(15):2727–38.

Sirko S, Akita K, Von Holst A, Faissner A. Structural and functional analysis of chondroitin sulfate proteoglycans in the neural stem cell niche. Methods Enzymol. 2010a;479:37–71.

Sirko S, von Holst A, Weber A, Wizenmann A, Theocharidis U, Gotz M, et al. Chondroitin sulfates are required for fibroblast growth factor-2-dependent proliferation and maintenance in neural stem cells and for epidermal growth factor-dependent migration of their progeny. Stem Cells. 2010b;28(4):775–87.

Snyder SE, Li J, Schauwecker PE, McNeill TH, Salton SR. Comparison of RPTP zeta/beta, phosphacan, and trkB mRNA expression in the developing and adult rat nervous system and induction of RPTP zeta/beta and phosphacan mRNA following brain injury. Brain Res Mol Brain Res. 1996;40(1):79–96.

Struve J, Maher PC, Li YQ, Kinney S, Fehlings MG, Kuntz CT, et al. Disruption of the hyaluronan-based extracellular matrix in spinal cord promotes astrocyte proliferation. Glia. 2005;52(1): 16–24.

Sugahara K, Kitagawa H. Recent advances in the study of the biosynthesis and functions of sulfated glycosaminoglycans. Curr Opin Struct Biol. 2000;10(5):518–27.

Sweeney MD, Yu Y, Leary JA. Effects of sulfate position on heparin octasaccharide binding to CCL2 examined by tandem mass spectrometry. J Am Soc Mass Spectrom. 2006;17(8): 1114–9.

Takahashi N, Sakurai T, Bozdagi-Gunal O, Dorr NP, Moy J, Krug L, et al. Increased expression of receptor phosphotyrosine phosphatase-beta/zeta is associated with molecular, cellular, behavioral and cognitive schizophrenia phenotypes. Transl Psychiatry. 2011;1:e8.

Tham M, Ramasamy S, Gan HT, Ramachandran A, Poonepalli A, Yu YH, et al. CSPG is a secreted factor that stimulates neural stem cell survival possibly by enhanced EGFR signaling. PLoS One. 2010;5(12):e15341.

Tian J, Ling L, Shboul M, Lee H, O'Connor B, Merriman B, et al. Loss of CHSY1, a secreted FRINGE enzyme, causes syndromic brachydactyly in humans via increased NOTCH signaling. Am J Hum Genet. 2010;87(6):768–78.

Tom VJ, Kadakia R, Santi L, Houle JD. Administration of chondroitinase ABC rostral or caudal to a spinal cord injury site promotes anatomical but not functional plasticity. J Neurotrauma. 2009;26(12):2323–33.

Tone Y, Pedersen LC, Yamamoto T, Izumikawa T, Kitagawa H, Nishihara J, et al. 2-o-phosphorylation of xylose and 6-o-sulfation of galactose in the protein linkage region of glycosaminoglycans influence the glucuronyltransferase-I activity involved in the linkage region synthesis. J Biol Chem. 2008;283(24):16801–7.

Uchimura K, Kadomatsu K, Nishimura H, Muramatsu H, Nakamura E, Kurosawa N, et al. Functional analysis of the chondroitin 6-sulfotransferase gene in relation to lymphocyte subpopulations, brain development, and oversulfated chondroitin sulfates. J Biol Chem. 2002; 277(2):1443–50.

Ughrin YM, Chen ZJ, Levine JM. Multiple regions of the NG2 proteoglycan inhibit neurite growth and induce growth cone collapse. J Neurosci. 2003;23(1):175–86.

Uyama T, Kitagawa K, Sugahara H. Biosynthesis of glycosaminoglycans and proteoglycans. In: Kamerling JP, editor. Comprehensive glycoscience, vol. 3. Amsterdam: Elsevier; 2007. p. 79–104.

van Horssen J, Wesseling P, van den Heuvel LP, de Waal RM, Verbeek MM. Heparan sulphate proteoglycans in Alzheimer's disease and amyloid-related disorders. Lancet Neurol. 2003;2(8): 482–92.

Vertel BM, Walters LM, Flay N, Kearns AE, Schwartz NB. Xylosylation is an endoplasmic reticulum to Golgi event. J Biol Chem. 1993;268(15):11105–12.

Vo T, Carulli D, Ehlert EM, Kwok JC, Dick G, Mecollari V, et al. The chemorepulsive axon guidance protein semaphorin3A is a constituent of perineuronal nets in the adult rodent brain. Mol Cell Neurosci. 2013;56C:186–200.

Wang D, Fawcett J. The perineuronal net and the control of CNS plasticity. Cell Tissue Res. 2012;349(1):147–60.

Wang K, Zhang H, Ma D, Bucan M, Glessner JT, Abrahams BS, et al. Common genetic variants on 5p14.1 associate with autism spectrum disorders. Nature. 2009;459(7246):528–33.

Wang Q, Yang L, Alexander C, Temple S. The niche factor syndecan-1 regulates the maintenance and proliferation of neural progenitor cells during mammalian cortical development. PLoS One. 2012;7(8):e42883.

Wegrzyn G, Jakobkiewicz-Banecka J, Narajczyk M, Wisniewski A, Piotrowska E, Gabig-Ciminska M, et al. Why are behaviors of children suffering from various neuronopathic types of mucopolysaccharidoses different? Med Hypotheses. 2010;75(6):605–9.

Yamaguchi Y. Lecticans: organizers of the brain extracellular matrix. Cell Mol Life Sci. 2000;57(2):276–89.

Yamaguchi Y, Inatani M, Matsumoto Y, Ogawa J, Irie F. Roles of heparan sulfate in mammalian brain development current views based on the findings from Ext1 conditional knockout studies. Prog Mol Biol Transl Sci. 2010;93:133–52.

Ye Q, Miao QL. Experience-dependent development of perineuronal nets and chondroitin sulfate proteoglycan receptors in mouse visual cortex. Matrix Biol. 2013;32(6):352–63.

Zafeiriou DI, Batzios SP. Brain and spinal MR imaging findings in mucopolysaccharidoses: a review. AJNR Am J Neuroradiol. 2013;34(1):5–13.

Zhang H, Muramatsu T, Murase A, Yuasa S, Uchimura K, Kadomatsu K. N-Acetylglucosamine 6-O-sulfotransferase-1 is required for brain keratan sulfate biosynthesis and glial scar formation after brain injury. Glycobiology. 2006a;16(8):702–10.

Zhang H, Uchimura K, Kadomatsu K. Brain keratan sulfate and glial scar formation. Ann N Y Acad Sci. 2006b;1086:81–90.

Zhao RR, Andrews MR, Wang D, Warren P, Gullo M, Schnell L, et al. Combination treatment with anti-Nogo-A and chondroitinase ABC is more effective than single treatments at enhancing functional recovery after spinal cord injury. The European journal of neuroscience. 2013;38(6): 2946–61.

Zou P, Zou K, Muramatsu H, Ichihara-Tanaka K, Habuchi O, Ohtake S, et al. Glycosaminoglycan structures required for strong binding to midkine, a heparin-binding growth factor. Glycobiology. 2003;13(1):35–42.

Zuo J, Neubauer D, Graham J, Krekoski CA, Ferguson TA, Muir D. Regeneration of axons after nerve transection repair is enhanced by degradation of chondroitin sulfate proteoglycan. Exp Neurol. 2002;176(1):221–8.

Chapter 6
Use of Glycan-Targeted Antibodies/Lectins to Study the Expression/Function of Glycosyltransferases in the Nervous System

Yasuhiko Kizuka, Kenji Kanekiyo, Shinobu Kitazume, and Naoyuki Taniguchi

Abstract In the nervous system, various unique glycans not found in other tissues are expressed on glycoproteins, and their expression/functions have been studied using specific antibodies/lectins. Among brain-specific glycans in mammals, we focus on human natural killer-1 (HNK-1) and related Cat-315 epitopes, which can be detected using specific antibodies. It is known that the HNK-1 epitope is expressed on N- and O-mannosylated glycans and that Cat-315 mAb preferentially recognizes the HNK-1 epitope on brain-specific "branched O-mannose glycan." The β1,6-branched O-mannose structure is synthesized by a brain-specific glycosyltransferase, N-acetylglucosaminyltransferase-IX (GnT-IX, also designated as GnT-Vb). Using GnT-IX gene-deficient mice and specific antibodies/lectins, the function of GnT-IX was found to be quite different from that of its ubiquitous homologue, GnT-V. Using Cat-315 mAb, the receptor protein tyrosine phosphatase-beta (RPTPβ) was identified as an in vivo target glycoprotein for GnT-IX. Analysis of the function of branched O-mannose glycan on RPTPβ indicated that its loss promoted the recovery process after myelin injury (called remyelination) in brain and that this phenomenon is probably caused in vivo by reduced activation of astrocytes in GnT-IX-deficient brain.

Keywords HNK-1 • GnT-V • GnT-IX(Vb) • Cat-315 • Demyelination • O-mannose glycan • RPTPβ

Y. Kizuka • K. Kanekiyo • S. Kitazume • N. Taniguchi (✉)
Disease Glycomics Team, Systems Glycobiology Research Group, RIKEN-Max Planck Joint Research Center, Global Research Cluster, RIKEN, Hirosawa, Wako, Saitama, Japan
e-mail: tani52@wd5.so-net.ne.jp

R.K. Yu and C.-L. Schengrund (eds.), *Glycobiology of the Nervous System*, Advances in Neurobiology 9, DOI 10.1007/978-1-4939-1154-7_6,

6.1 Introduction

The mammalian nervous system is composed of various cell types, including neurons, astrocytes, myelinating cells (oligodendrocytes and Schwann cells), and microglia. Each type of cell communicates with the others to elaborate a complicated neural network. In the nervous system, unique glycan structures are expressed which are quite different from those in other tissues and are exemplified by the human natural killer-1 (HNK-1) epitope and polysialic acid (PSA) (Kleene and Schachner 2004). Accumulating evidence has shown that brain-specific glycans are essential for maintaining brain functions, such as learning/memory, cognition, and behavior (Weinhold et al. 2005; Yamamoto et al. 2002). Specific lectins and antibodies are useful for both carbohydrate detection and expression/functional studies. For instance, good monoclonal antibodies for detecting HNK-1 and PSA, anti-HNK-1 and 12E3 mAbs, respectively, are now widely used (Yamamoto et al. 2000; 2002). In addition to the use of these antibodies for the identification of target glycoproteins (e.g., neural cell adhesion molecule), they are used for elucidation of the biosynthetic pathways of glycosylation.

Lectins, proteins which recognize specific glycan structures, are now widely used to detect glycans. Plant lectins such as conA (concanavalin A), WGA (wheat germ agglutinin), PHA (phytohemagglutinin), etc. (Sharon and Lis 2004), are often used to characterize glycans in cells or tissues. Meanwhile, in mammals, the functions of animal lectins have been well studied in the immune system. For instance, some Siglecs (sialic acid-binding immunoglobulin type lectins) or C-type lectins are specifically expressed in subsets of immune cells and involved in regulation of immune cell activity or recognition of pathogens (Crocker et al. 2007; Hardison and Brown 2012). Recent advances in lectin biology allowed us to perform cellular glycomic analyses using lectin microarrays (Tateno et al. 2010). Although lectins are a useful tool, their affinity and specificity for their ligand glycans are often lower than that of anti-carbohydrate antibodies, which sometimes makes it difficult to use them to determine glycan structure.

This chapter focuses on a brain-specific glycan, the HNK-1 epitope. We describe its expression on branched O-mannose glycans and the function of a brain-specific glycosyltransferase, N-acetylglucosaminyltransferase-IX (GnT-IX) (Inamori et al. 2006), in its synthesis. Using specific antibodies/lectins, an in vivo target glycoprotein expressing this modification was identified. Using a mouse model, we also found that branched O-mannose glycans play key roles in the remyelination process after myelin injury by regulating astrocyte activation (Kanekiyo et al. 2013).

6.2 HNK-1 Epitope

The HNK-1 epitope was first found as an antigen on the surface of human natural killer cells in 1981 (Abo and Balch 1981). However, subsequent studies revealed that the HNK-1 glycan is almost exclusively found in the nervous system (Kruse et al. 1984).

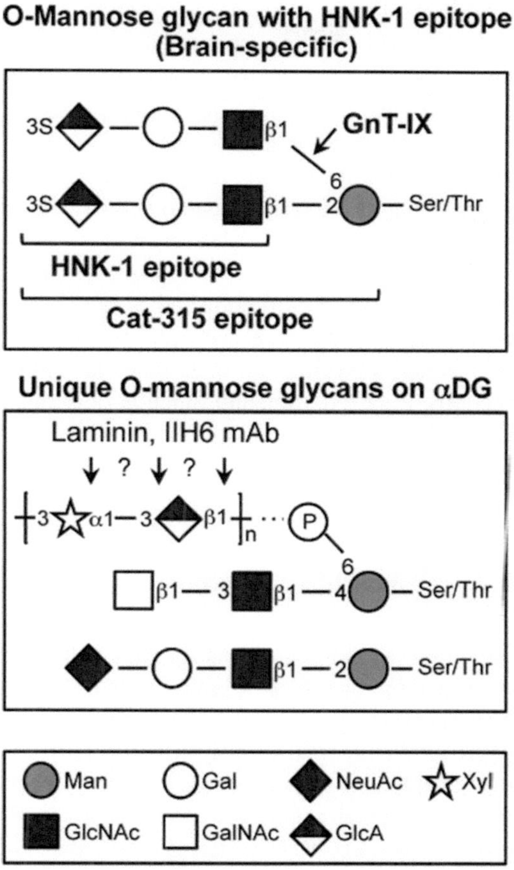

Fig. 6.1 Putative structure of the Cat-315 epitope and unique O-mannose glycan structure on αDG. Cat-315-reactive glycan, which possesses terminal HNK-1 epitope(s), is specifically expressed in brain. The Cat-315-reactive glycan is expressed on RPTPβ and other molecules expressed in brain. In contrast, unique phosphorylated O-mannose glycans that have never been identified on other glycoproteins are expressed on αDG. Phosphorylated O-mannose glycan is essential for binding to ligands or IIH6 mAb

The glycan structure of HNK-1 was identified as HSO_3-3GlcAβ1-3Galβ1-4GlcNAc- (Fig. 6.1) (Ariga et al. 1987; Chou et al. 1986; Tokuda et al. 1998). The terminal sulfated glucuronic acid residue is a rare structure in mammalian N- and O-glycans other than glycosaminoglycans. A key step in its biosynthesis is the addition of GlcA, and the major enzyme responsible for this step, glucuronyltransferase-P (GlcAT-P), is specifically expressed in the nervous system (Terayama et al. 1997). Later, by expression cloning and use of an HNK-1 mAb, the sulfotransferase designated HNK-1ST was cloned (Bakker et al. 1997).

Function of the HNK-1 glycan was examined using GlcAT-P-deficient mice in which HNK-1 mAb reactivity almost disappears. GlcAT-P-deficient mice have reduced synaptic plasticity and impaired learning/memory functions (Yamamoto et al. 2002). Spine (postsynapse) structure in primary neurons from GlcAT-P-deficient brains was consistently abnormally immature (Morita et al. 2009b). To explore the molecular mechanism by which the HNK-1 glycan regulates spine maturation and learning, HNK-1 mAb was used to identify glycoproteins expressing the HNK-1 glycan. This resulted in identification of the glutamate receptor subunit GluA2 (Morita et al. 2009a). Compared with other HNK-1-carrying glycoproteins, GluA2 is highly enriched in the postsynaptic density fraction, and among glutamate receptor subunits was the one selectively modified with HNK-1. GluA2 is also known to be an essential molecule for spine maturation and synaptic plasticity (Isaac et al. 2007), and the loss of HNK-1 in GlcAT-P-deficient neurons causes instability of GluA2 at the synaptic membrane. These findings strongly suggest that the HNK-1 glycan on GluA2 plays a key role in learning and memory function in vivo.

HNK-1 glycan is expressed on both N-glycan and O-mannose glycans in brain. Although the glycan structure and carrier glycoprotein for N-linked HNK-1 are well characterized (Kruse et al. 1984; Morita et al. 2009a; Voshol et al. 1996), the expression pattern and function of O-mannose-linked HNK-1 are not. In the following sections, the expression and function of O-mannose glycans in brain are described prior to focusing on brain-specific "branched" O-mannose glycans carrying a terminal HNK-1 modification.

6.3 O-Mannose Glycans in the Brain

Although O-mannose glycans are abundantly expressed in mammalian brain, with evidence that one third of brain O-glycans are O-mannosylated (Chai et al. 1999), their functions remain to be clarified. The most well-studied O-mannosylated glycoprotein in mammals is α-dystroglycan (αDG), and a number of studies have demonstrated that O-mannosylation of αDG is essential for skeletal muscle function in mice and humans (Godfrey et al. 2011). Previous reports have shown that αDG binds to ligands such as laminin or pikachurin via its unusual phosphorylated O-mannose glycan, which was detected using the IIH6 mAb (Fig. 6.1) (Chiba et al. 1997; Inamori et al. 2012; Sato et al. 2008; Yoshida-Moriguchi et al. 2010). However, in brain, the content of O-mannose glycan on αDG is almost negligible (Stalnaker et al. 2011), indicating that the O-mannose glycan in brain modulates functions of glycoproteins other than αDG. To date, several neural glycoproteins such as neurocan, neurofascin 186, and CD24, enriched and purified using specific lectins or antibodies, have been shown to be modified by O-mannose glycans (Bleckmann et al. 2009; Pacharra et al. 2012; 2013); however, the functions of their O-mannose glycans are still unclear. Meanwhile, a neural phosphatase, receptor protein tyrosine phosphatase-beta (RPTPβ), was found to be modified by branched O-mannose glycans with terminal HNK-1 epitope(s) (Abbott et al. 2008; Kanekiyo et al. 2013). These glycans are detected by the HNK-1 mAb as well as the Cat-315 mAb. Although the epitope for the Cat-315 mAb has not been accurately determined, it binds preferentially to O-mannosylated HNK-1 (6.1) (Dwyer et al. 2012). Moreover, the major glycoprotein carrying the Cat-315 epitope changes from RPTPβ to aggrecan during mouse brain development (Dino et al. 2006). Using these specific mAbs, the biological function of O-mannose modifications in brain was studied as described below.

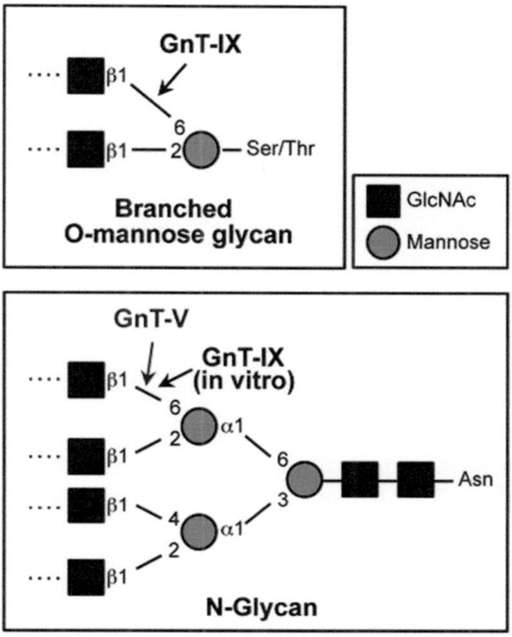

Fig. 6.2 Structures of a branched O-mannose glycan and an N-glycan. GnT-IX catalyzes formation of the β1,6-branch of an O-mannose glycan. GnT-V catalyzes β1,6-branch formation on the α6-mannose arm in an N-glycan. GnT-IX also shows modest activity toward N-glycans in vitro

Expression in brain of a "branched" O-mannose glycan not found in O-mannosylated glycans from other tissues (Figs. 6.1 and 6.2) (Chai et al. 1999; Yuen et al. 1997) indicated the unique activity of a branching enzyme present specifically in brain. We and others identified the brain-specific enzyme as GnT-IX (Vb).

6.4 GnT-IX Is a Brain-Specific O-Mannose Branching Enzyme

GnT-IX was originally identified in silico as a homologue of GnT-V by both our laboratory and that of Dr. Pierce (Inamori et al. 2003; Kaneko et al. 2003), and for that reason, GnT-IX is also designated as GnT-Vb. GnT-V is ubiquitously expressed and is responsible for N-glycan branching at the α6-mannose arm (Fig. 6.2). Many studies have shown that GnT-V is involved in cancer progression and growth factor receptor signaling (Granovsky et al. 2000; Lau et al. 2007). In brain, the GnT-V product, which can be detected with the L4-PHA lectin, is involved in depression-like behavior (Soleimani et al. 2008). Although GnT-IX shows high sequence identity to GnT-V (42 % in the case of humans), its expression and function are different from those of GnT-V. GnT-IX is highly specific to brain (Kizuka et al. 2011) and exhibits weak activity toward N-glycans in vitro. We and others have reported that GnT-IX has preferential branching activity toward O-mannose glycans (Fig. 6.2) (Alvarez-Manilla et al. 2010; Inamori et al. 2004). As described above, branched

O-mannosyl glycans terminally modified with sialic acid, Lewis epitope, or HNK-1 are found only in the brain (Chai et al. 1999; Yuen et al. 1997). To date, no lectin/antibody has been developed that can specifically detect all O-mannose glycans or distinguish linear O-mannose glycans from branched ones. However, using some specific detection probes and mouse models, the functions and expression patterns of O-mannose glycans in brain have been gradually identified.

6.5 In Vivo Enzymatic Functions of GnT-IX and GnT-V

Our group and Dr. Pierce's group independently generated GnT-IX (Vb) gene-deficient mice (Kanekiyo et al. 2013; Lee et al. 2012). These mutant mice showed no overt phenotype in terms of brain morphology and fertility. However, they showed reduced reactivity to both Cat-315 and HNK-1 mAbs, supporting the idea that Cat-315 recognizes O-mannosylated glycans with terminal HNK-1 modification(s) in brain. Using immunoprecipitation, we identified RPTPβ as one of the major target proteins for GnT-IX in brain, consistent with previous findings in cultured cells (Abbott et al. 2008). Dr. Pierce's group analyzed glycan structures from GnT-V- or GnT-IX-deficient brain and found that GnT-IX acts on O-mannose glycans but not on N-glycans and that GnT-IX cannot compensate for GnT-V's function in the biosynthesis of N-glycans in vivo (Lee et al. 2012). In addition, lectin blot analysis using L4-PHA showed that loss of lectin reactivity is solely caused by knockout of GnT-V but not by knockout of GnT-IX, indicating that GnT-V acts on N-glycans, while GnT-IX acts on O-mannose glycans in vivo (Kanekiyo et al. 2013). These findings indicate that the in vivo function of GnT-IX is different from that of its homologue GnT-V. Moreover, Dr. Pierce's group observed no apparent change in αDG binding to IIH6 mAb or laminin in GnT-IX-deficient mouse brain tissue, indicating that the target glycoprotein for GnT-IX function is probably not αDG.

6.6 GnT-IX-Deficient Mice Show Enhanced Recovery from Demyelinating Damage

Based on the fact that a target of GnT-IX, RPTPβ, is known to be critical for recovery from demyelinating damage in vivo (Harroch et al. 2002), we hypothesized that O-mannose glycans on RPTPβ produced by GnT-IX are also involved in the demyelination/remyelination process. Demyelination is found in many pathological conditions including multiple sclerosis, and promoting remyelination is a rational strategy for treating demyelinating diseases (Fancy et al. 2011). We induced demyelination in wild-type and GnT-IX-deficient mice using the copper chelator "cuprizone," which induces oligodendrocyte damage through oxidative stress (Torkildsen et al. 2008). In wild-type mouse brain, myelin is progressively damaged by

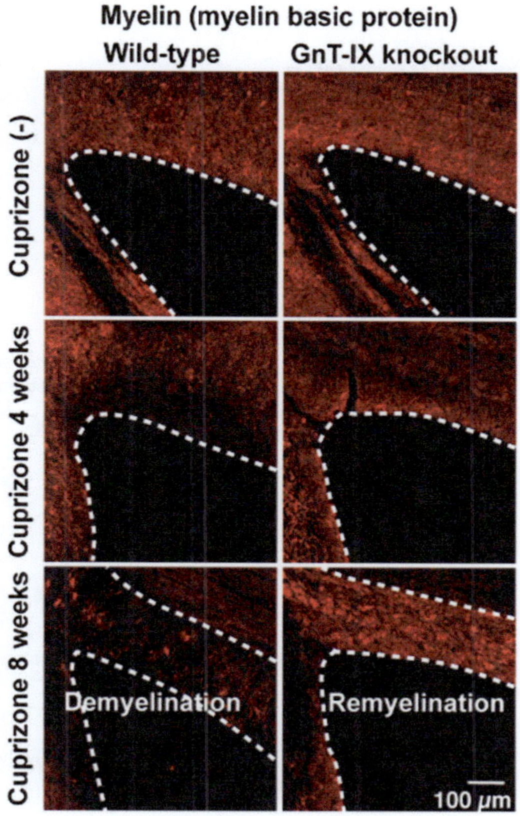

Fig. 6.3 Recovery from myelin injury is promoted in GnT-IX-deficient mice. Myelin in the corpus callosum was stained with anti-myelin basic protein antibody. Cuprizone feeding for 4 weeks causes demyelination in both wild-type and GnT-IX-deficient mice. Further treatment with cuprizone causes more severe demyelination in wild-type mice, while GnT-IX-deficient mice show recovery from myelin damage (remyelination). Figure is modified from that published previously by Kanekiyo et al. (2013)

cuprizone. In contrast, in GnT-IX-deficient mouse brain, once early-phase demyelination has occurred as in the case of wild-type mice, significant recovery of myelin (remyelination) was observed (Fig. 6.3) (Kanekiyo et al. 2013).

As a potential mechanism for this effect, we found that astrocyte activation was reduced in the damaged region in GnT-IX-deficient brain (Kanekiyo et al. 2013). In wild-type brain, Cat-315-positive astrocytes accumulated in demyelinated corpus callosum, and astrocyte activation is one of the mechanisms underlying suppression of the remyelination process by inhibiting oligodendrocyte differentiation from oligodendrocyte precursor cells (OPCs) (Wang et al. 2011). Concomitantly, oligodendrocyte lineage analysis revealed that oligodendrocyte differentiation is actually enhanced in GnT-IX-deficient mice (Kanekiyo et al. 2013). Based on these data, we suggest that the branched O-mannose glycans on RPTPβ are involved in astrocyte activation, which probably suppresses oligodendrocyte differentiation and remyelination. A schematic model for the role of branched O-mannosyl glycans on astrocytic RPTPβ in remyelination is shown in Fig. 6.4. Cultured primary astrocytes isolated from nascent pups also showed decreased activation in vitro (Kanekiyo et al. 2013), strongly suggesting that GnT-IX can regulate the astrocyte activation process with or without cuprizone treatment. Further analysis of the role of GnT-IX

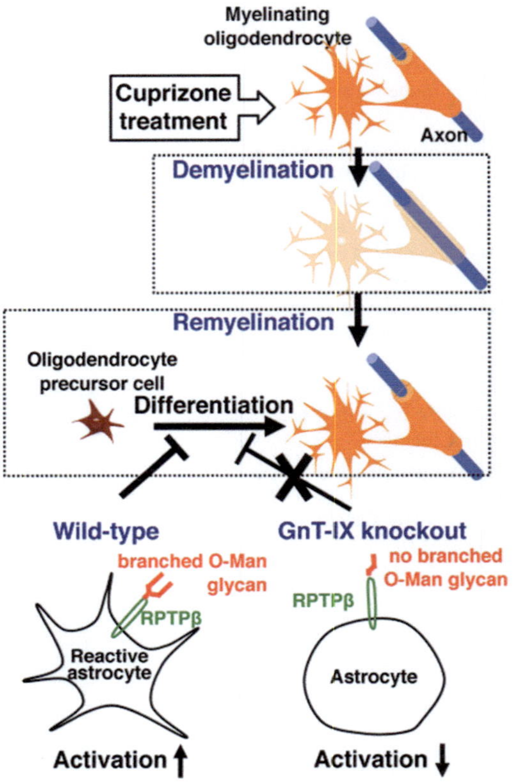

Fig. 6.4 Putative model for the regulatory role of branched O-mannosyl glycans on astrocytic RPTPβ in remyelination after myelin injury. Cuprizone administration causes reversible demyelination. Activated astrocytes are known to inhibit differentiation of oligodendrocyte precursor cells, leading to impairment of remyelination. In the absence of branched O-mannosyl glycans (GnT-IX knockout), astrocyte activation is suppressed, and remyelination is enhanced. Figure modified from the paper by Kanekiyo et al. (2013)

is needed to clarify how branched O-mannose glycans are involved in astrocyte activation. On the basis of these findings, we suggest that targeting of protein glycosylation may be a novel therapeutic strategy for demyelinating disorders.

Conflict of Interest No conflict of interest is declared.

References

Abbott KL, Matthews RT, Pierce M. Receptor tyrosine phosphatase beta (RPTPbeta) activity and signaling are attenuated by glycosylation and subsequent cell surface galectin-1 binding. J Biol Chem. 2008;283:33026–35.

Abo T, Balch CM. A differentiation antigen of human NK and K cells identified by a monoclonal antibody (HNK-1). J Immunol. 1981;127:1024–9.

Alvarez-Manilla G, Troupe K, Fleming M, Martinez-Uribe E, Pierce M. Comparison of the substrate specificities and catalytic properties of the sister N-acetylglucosaminyltransferases, GnT-V and GnT-Vb (IX). Glycobiology. 2010;20:166–74.

Ariga T, et al. Characterization of sulfated glucuronic acid containing glycolipids reacting with IgM M-proteins in patients with neuropathy. J Biol Chem. 1987;262:848–53.

Bakker H, Friedmann I, Oka S, Kawasaki T, Nifant'ev N, Schachner M, Mantei N. Expression cloning of a cDNA encoding a sulfotransferase involved in the biosynthesis of the HNK-1 carbohydrate epitope. J Biol Chem. 1997;272:29942–6.

Bleckmann C, et al. O-glycosylation pattern of CD24 from mouse brain. Biol Chem. 2009; 390:627–45.

Chai W, Yuen CT, Kogelberg H, Carruthers RA, Margolis RU, Feizi T, Lawson AM. High prevalence of 2-mono- and 2,6-di-substituted manol-terminating sequences among O-glycans released from brain glycopeptides by reductive alkaline hydrolysis. Eur J Biochem. 1999;263:879–88.

Chiba A, Matsumura K, Yamada H, Inazu T, Shimizu T, Kusunoki S, Kanazawa I, Kobata A, Endo T. Structures of sialylated O-linked oligosaccharides of bovine peripheral nerve alpha-dystroglycan. The role of a novel O-mannosyl-type oligosaccharide in the binding of alpha-dystroglycan with laminin. J Biol Chem. 1997;272:2156–62.

Chou DK, Ilyas AA, Evans JE, Costello C, Quarles RH, Jungalwala FB. Structure of sulfated glucuronyl glycolipids in the nervous system reacting with HNK-1 antibody and some IgM paraproteins in neuropathy. J Biol Chem. 1986;261:11717–25.

Crocker PR, Paulson JC, Varki A. Siglecs and their roles in the immune system. Nat Rev Immunol. 2007;7:255–66.

Dino MR, Harroch S, Hockfield S, Matthews RT. Monoclonal antibody Cat-315 detects a glycoform of receptor protein tyrosine phosphatase beta/phosphacan early in CNS development that localizes to extrasynaptic sites prior to synapse formation. Neuroscience. 2006;142:1055–69.

Dwyer CA, Baker E, Hu H, Matthews RT. RPTPzeta/phosphacan is abnormally glycosylated in a model of muscle-eye-brain disease lacking functional POMGnT1. Neuroscience. 2012;220: 47–61.

Fancy SP, Chan JR, Baranzini SE, Franklin RJ, Rowitch DH. Myelin regeneration: a recapitulation of development? Annu Rev Neurosci. 2011;34:21–43.

Godfrey C, Foley AR, Clement E, Muntoni F. Dystroglycanopathies: coming into focus. Curr Opin Genet Dev. 2011;21:278–85.

Granovsky M, Fata J, Pawling J, Muller WJ, Khokha R, Dennis JW. Suppression of tumor growth and metastasis in Mgat5-deficient mice. Nat Med. 2000;6:306–12.

Hardison SE, Brown GD. C-type lectin receptors orchestrate antifungal immunity. Nat Immunol. 2012;13:817–22.

Harroch S, Furtado GC, Brueck W, Rosenbluth J, Lafaille J, Chao M, Buxbaum JD, Schlessinger J. A critical role for the protein tyrosine phosphatase receptor type Z in functional recovery from demyelinating lesions. Nat Genet. 2002;32:411–4.

Inamori K, Endo T, Ide Y, Fujii S, Gu J, Honke K, Taniguchi N. Molecular cloning and characterization of human GnT-IX, a novel beta1,6-N-acetylglucosaminyltransferase that is specifically expressed in the brain. J Biol Chem. 2003;278:43102–9.

Inamori K, et al. N-Acetylglucosaminyltransferase IX acts on the GlcNAc beta 1,2-Man alpha 1-Ser/Thr moiety, forming a 2,6-branched structure in brain O-mannosyl glycan. J Biol Chem. 2004;279:2337–40.

Inamori K, Mita S, Gu J, Mizuno-Horikawa Y, Miyoshi E, Dennis JW, Taniguchi N. Demonstration of the expression and the enzymatic activity of N-acetylglucosaminyltransferase IX in the mouse brain. Biochim Biophys Acta. 2006;1760:678–84.

Inamori K, Yoshida-Moriguchi T, Hara Y, Anderson ME, Yu L, Campbell KP. Dystroglycan function requires xylosyl- and glucuronyltransferase activities of LARGE. Science. 2012;335:93–6.

Isaac JT, Ashby MC, McBain CJ. The role of the GluR2 subunit in AMPA receptor function and synaptic plasticity. Neuron. 2007;54:859–71.

Kanekiyo K, et al. Loss of branched o-mannosyl glycans in astrocytes accelerates remyelination. J Neurosci. 2013;33:10037–47.

Kaneko M, Alvarez-Manilla G, Kamar M, Lee I, Lee JK, Troupe K, Zhang W, Osawa M, Pierce M. A novel beta(1,6)-N-acetylglucosaminyltransferase V (GnT-VB)(1). FEBS Lett. 2003;554: 515–9.

Kizuka Y, Kitazume S, Yoshida M, Taniguchi N. Brain-specific expression of N-acetylglucosaminyltransferase IX (GnT-IX) is regulated by epigenetic histone modifications. J Biol Chem. 2011;286:31875–84.

Kleene R, Schachner M. Glycans and neural cell interactions. Nat Rev Neurosci. 2004;5: 195–208.

Kruse J, Mailhammer R, Wernecke H, Faissner A, Sommer I, Goridis C, Schachner M. Neural cell adhesion molecules and myelin-associated glycoprotein share a common carbohydrate moiety recognized by monoclonal antibodies L2 and HNK-1. Nature. 1984;311:153–5.

Lau KS, Partridge EA, Grigorian A, Silvescu CI, Reinhold VN, Demetriou M, Dennis JW. Complex N-glycan number and degree of branching cooperate to regulate cell proliferation and differentiation. Cell. 2007;129:123–34.

Lee JK, Matthews RT, Lim JM, Swanier K, Wells L, Pierce JM. Developmental expression of the neuron-specific N-acetylglucosaminyltransferase Vb (GnT-Vb/IX) and identification of its in vivo glycan products in comparison with those of its paralog, GnT-V. J Biol Chem. 2012;287:28526–36.

Morita I, Kakuda S, Takeuchi Y, Itoh S, Kawasaki N, Kizuka Y, Kawasaki T, Oka S. HNK-1 glycoepitope regulates the stability of the glutamate receptor subunit GluR2 on the neuronal cell surface. J Biol Chem. 2009a;284:30209–17.

Morita I, Kakuda S, Takeuchi Y, Kawasaki T, Oka S. HNK-1 (human natural killer-1) glycoepitope is essential for normal spine morphogenesis in developing hippocampal neurons. Neuroscience. 2009b;164:1685–94.

Pacharra S, Hanisch FG, Breloy I. Neurofascin 186 is O-mannosylated within and outside of the mucin domain. J Proteome Res. 2012;11:3955–64.

Pacharra S, Hanisch FG, Muhlenhoff M, Faissner A, Rauch U, Breloy I (2013) The lecticans of mammalian brain perineural net are O-mannosylated. J Proteome Res. 12(4):1764-71

Sato S, et al. Pikachurin, a dystroglycan ligand, is essential for photoreceptor ribbon synapse formation. Nat Neurosci. 2008;11:923–31.

Sharon N, Lis H. History of lectins: from hemagglutinins to biological recognition molecules. Glycobiology. 2004;14:53R–62.

Soleimani L, Roder JC, Dennis JW, Lipina T. Beta N-acetylglucosaminyltransferase V (Mgat5) deficiency reduces the depression-like phenotype in mice. Genes Brain Behav. 2008;7: 334–43.

Stalnaker SH, et al. Glycomic analyses of mouse models of congenital muscular dystrophy. J Biol Chem. 2011;286:21180–90.

Tateno H, Kuno A, Itakura Y, Hirabayashi J. A versatile technology for cellular glycomics using lectin microarray. Methods Enzymol. 2010;478:181–95.

Terayama K, Oka S, Seiki T, Miki Y, Nakamura A, Kozutsumi Y, Takio K, Kawasaki T. Cloning and functional expression of a novel glucuronyltransferase involved in the biosynthesis of the carbohydrate epitope HNK-1. Proc Natl Acad Sci U S A. 1997;94:6093–8.

Tokuda A, Ariga T, Isogai Y, Komba S, Kiso M, Hasegawa A, Tai T, Yu RK. On the specificity of anti-sulfoglucuronosyl glycolipid antibodies. J Carbohydr Chem. 1998;17:535–46.

Torkildsen O, Brunborg LA, Myhr KM, Bo L. The cuprizone model for demyelination. Acta Neurol Scand Suppl. 2008;188:72–6.

Voshol H, van Zuylen CW, Orberger G, Vliegenthart JF, Schachner M. Structure of the HNK-1 carbohydrate epitope on bovine peripheral myelin glycoprotein P0. J Biol Chem. 1996;271: 22957–60.

Wang Y, Cheng X, He Q, Zheng Y, Kim DH, Whittemore SR, Cao QL. Astrocytes from the contused spinal cord inhibit oligodendrocyte differentiation of adult oligodendrocyte precursor cells by increasing the expression of bone morphogenetic proteins. J Neurosci. 2011;31: 6053–8.

Weinhold B, Seidenfaden R, Rockle I, Muhlenhoff M, Schertzinger F, Conzelmann S, Marth JD, Gerardy-Schahn R, Hildebrandt H. Genetic ablation of polysialic acid causes severe neurodevelopmental defects rescued by deletion of the neural cell adhesion molecule. J Biol Chem. 2005;280:42971–7.

Yamamoto N, Inui K, Matsuyama Y, Harada A, Hanamura K, Murakami F, Ruthazer ES, Rutishauser U, Seki T. Inhibitory mechanism by polysialic acid for lamina-specific branch formation of thalamocortical axons. J Neurosci. 2000;20:9145–51.
Yamamoto S, et al. Mice deficient in nervous system-specific carbohydrate epitope HNK-1 exhibit impaired synaptic plasticity and spatial learning. J Biol Chem. 2002;277:27227–31.
Yoshida-Moriguchi T, et al. O-mannosyl phosphorylation of alpha-dystroglycan is required for laminin binding. Science. 2010;327:88–92.
Yuen CT, Chai W, Loveless RW, Lawson AM, Margolis RU, Feizi T. Brain contains HNK-1 immunoreactive O-glycans of the sulfoglucuronyl lactosamine series that terminate in 2-linked or 2,6-linked hexose (mannose). J Biol Chem. 1997;272:8924–31.

Chapter 7
From Mass Spectrometry-Based Glycosylation Analysis to Glycomics and Glycoproteomics

Kay-Hooi Khoo

Abstract Mass spectrometry (MS) is the pivotal technique driving most current day protein glycosylation analysis. It is unrivaled in combining high sensitivity, resolution, and precision for de novo identification and therefore most conducive to discovery mapping of new or biologically implicated glyco-epitopes. This chapter attempts to provide an account of current advances and limitations in MS-based glycosylation analysis as it rapidly evolves into all-encompassing glycomics and glycoproteomics. Given the omic-scale complexity, an urgent need is to orchestrate advances in chromatographic separation, chemical derivatization, and innovative online LC-MS/MS scan functions for more penetrative and purposeful data acquisition, along with empowering computational tools to overcome the bottleneck in automated data analysis, in order to increase the breadth and depth of glycomic coverage. This entails not only MS^n-level resolution and mapping of isomeric variations but also addressing often overlooked structural features such as poly-N-acetyllactosamine extension and the widespread occurrence of sulfation in relation to immuno-activation and malignant transformation. Only then can functional glycotopes of relevance be uncovered, and further localized to particular glycan and protein carriers, the latter by means of target glycoproteomics. Mapping of site occupancy without addressing the full range of occupying glycans by direct sequencing of glycopeptides is essentially inadequate for any glycobiology driven venture. This chapter aims to conceptualize the required experimental workflows from glycomics to glycoproteomics, with MS analysis of permethylated glycans occupying the central node.

Keywords Mass spectrometry • Glycosylation • Glycomics • Glycoproteomics • MS/MS • Glycan sequencing

K.-H. Khoo (✉)
Institute of Biological Chemistry, Academia Sinica,
128, Academia Road Sec 2 Nankang, Taipei 115, Taiwan
e-mail: kkhoo@gate.sinica.edu.tw

R.K. Yu and C.-L. Schengrund (eds.), *Glycobiology of the Nervous System*, Advances in Neurobiology 9, DOI 10.1007/978-1-4939-1154-7_7,

Abbreviations

CID	Collision induced dissociation
DMB	1,2-Diamino-4,5-methylenedioxybenzene
DMSO	Dimethyl sulfoxide
ECD	Electron capture dissociation
EDD	Electron detachment dissociation
EI	Electron impact
ESI	Electrospray
ETD	Electron transfer dissociation
FAB	Fast atom bombardment
GAGs	Glycosaminoglycans
GC	Gas chromatography
GSL	Glycosphingolipid
HCD	Higher energy (C-trap) collision dissociation
Hex	Hexose
HexNAc	*N*-acetylhexosamine
HILIC	Hydrophilic interaction
HPAEC	High performance anion exchange chromatography
HPLC	High performance liquid chromatography
IRMPD	Infrared multiple photon dissociation
LacdiNAc	*N,N′*-diacetyllactosamine
LacNAc	*N*-acetyllactosamine
LC	Liquid chromatography
MALDI	Matrix-assisted laser desorption ionization
MS	Mass spectrometry
MS/MS	Tandem mass spectrometry
MS^n	Multistage fragmentation
PAD	Pulsed amperometric detection
PGC	Poros graphitized carbon
polyLacNAc	Poly-*N*-acetyllactosamine
Q/TOF	Quadrupole/time-of-flight
RP	Reverse phase
SPE	Solid phase extraction
TMT	Tandem mass tags
TOF/TOF	Time-of-flight/time-of-flight

7.1 Overview and Scope

Over the course of the last decade, mass spectrometry (MS)-based protein glycosylation analysis has evolved into glycomics and glycoproteomics, which largely signifies the increasingly holistic scale and systems view being taken by the scientific community rather than representing fundamental changes in how we determine the structural details of glycoconjugates. This chapter aims to provide a concise account of the current status of mammalian glycosylation analysis without aiming to be comprehensive. First, it presupposes the readers have some basic knowledge of the diverse range of mammalian glycoconjugates that exist in nature. Although there may be some nervous system-specific glycoconjugates, the underlying structural features are universal and the occurrence of N- and O-glycans, glycosphingolipids, and proteoglycans is ubiquitous. So are the analytical methods employed to define their precise structures or to simply map their occurrences and changes in development and disease states, in response to activation or malignant transformation. This chapter does not attempt to tailor the discussion specifically to the nervous system but takes a broader perspective. Readers are referred to Chaps. 1, 3, 4, 5 of this monograph for an introduction to distinct classes of glycoconjugates found in the nervous system.

This chapter does not aim to provide an extensive review of published work on methodology developments and applications, nor does it provide experimental protocols and technical details (see North et al. 2010b; Bleckmann et al. 2011; Doherty et al. 2012; Hayes et al. 2012; Jensen et al. 2012; Kolarich et al. 2012; Ruhaak et al. 2012; Patrie et al. 2013; Kenny et al. 2013; Lin and Lubman 2013; Orlando 2013; Zauner et al. 2013; Zhao et al. 2013) because many such articles including excellent introductory overviews and perspectives can be found in the recent literature (Marino et al. 2010; Zaia 2010; Pan et al. 2011; Leymarie and Zaia 2012; Zauner et al. 2012; Alley et al. 2013; Han and Costello 2013; Novotny et al. 2013; Wuhrer 2013; Mechref et al. 2013; Reinhold et al. 2013). It is more a summary of what is currently feasible and where we are headed to. A few take home simple and unifying conceptual understandings are presented. The discussion is restricted to MS-based approaches most often used to study the prevalent forms of protein N- and O-glycosylation. Finally, while analysis of naturally occurring free glycans or those released from the glycosphingolipids (GSLs) may be appropriately considered as part of the glycomics venture, the term lipidomics or glycosphingolipidomics (Levery 2005; Meisen et al. 2011) has been additionally introduced when not only the glycans but also the heterogeneity of the sphingolipid anchors to which they attached is considered and mapped in total. The polymeric sulfated glycosaminoglycans (GAGs) carried on proteoglycans should likewise be included in the glycomic constituency (Zaia 2013) but often are treated separately, as the requisite techniques and experimental approaches are distinct. Neither the GSLs nor the GAGs are discussed individually although some of the general principles elaborated are equally applicable.

MS analysis distinguishes itself from other analytical techniques for glycosylation analysis by virtue of being most conducive to discovery of new glyco-epitopes (glycotopes) or features and to validate structural details, such as identification of its underlying glycan and protein carriers. In that respect, it is unrivaled in resolution, sensitivity and precision, capable of handling complex mixtures as necessitated by omics ventures, and not precluded by the need of reference standards. NMR analysis, as covered in Chap. 8, is the only complementary biophysical technique capable of discovery and de novo sequencing. Its major limitations are sensitivity and throughput while its most obvious strength is its ability to define stereochemistry, which is the Achilles' heel of MS analysis. On the other hand, the highest sensitivity and multiplexing capabilities is afforded not by MS analysis but by use of a range of biological probes such as antibodies and lectins, which also permits analysis of live cells and real time imaging. The most serious drawback, however, is that these probes only detect predefined or anticipated glycotopes, sometimes with severe cross-reactivities and are not always available for the target of interest. It also selectively "sees" only the target it is meant to "see" without giving a full picture of all glycomic constituents acting in concert as would be expected in most biological processes. Overall, MS-based mapping offers the best prospect for an unbiased view and this chapter aims to convey a personal view on what is currently limiting MS-based glycomic analysis from completely meeting expectations of glycobiologists.

7.2 Mass Spectrometry for Glycosylation Analysis

7.2.1 A Primer for MALDI-MS and LC-ESI-MS

For readers who are not familiar with MS, the first thing to know is that MS measures the molecular weight of molecules in the form of *m/z* values, which stipulates that the analytes need to be ionized in the first instance. Although many ionization techniques exist, electrospray ionization (ESI) and matrix-assisted laser desorption ionization (MALDI) are currently dominant in biological MS. The former ionization technique is most often coupled with online liquid chromatography (LC)-MS applications although direct infusion of analytes in an appropriate solvent without LC separation is also possible. In contrast, the latter is essentially an offline technique requiring the analytes to be co-crystallized with matrix on a target plate before being subjected to laser induced desorption. Consequently, MALDI-MS takes an overall picture, or mass profile, of whatever analyte is placed onto the target spot, whereas an online LC-MS experiment takes hundreds or thousands of snapshots across the entire LC period, each corresponding to a mass profile of what is coming out of the capillary LC column at that time-point.

A common practice for achieving the highest sensitivity is to take advantage of a nanoflow system delivering an LC flow rate of about a few hundred nL/min using a 75-μm capillary column of variable length, dictated by the separation needed, and coupled to the MS system via a nanospray (nanoESI) ionization source. However, in applications where sample amount is not limiting it may be advantageous to increase sample loading capacity by using a micro flow rate and spray interface, which is more robust than the nanoLC system preferred by most laboratories. For the purpose of this chapter, LC-MS is used as a generic term covering all flow rates and respective ESI sources. More importantly, it implies an online ESI coupling distinct from an offline LC-MALDI-MS. For the latter, the LC eluates are sequentially spotted onto a MALDI target plate along with added matrix in predefined steps of volume or time span and then analyzed offline by MALDI-MS at a later stage instead of being directly sprayed into the MS system in real time.

By far, online LC-MS and offline MALDI-MS are the most common complementary set ups used by most glyco-analytical laboratories today. MALDI-MS is used for a rapid first screen of released glycans before subjecting them to further bulk fractionation if needed, while LC-MS provides the additional dimension of online separation immediately preceding MS analysis. LC-MS is also the prevalent mode of glycoproteomic analysis of peptides and glycopeptides.

7.2.2 LC Separation

Due to mutual ion suppression or suppression by analytes of better ionization efficiency or higher abundance, any MS profiling covering a specified mass range can afford only a limited number of signals spanning several orders of ion intensity as defined by the inherent dynamic range of the detector. Thus, even without interference from matrix peaks, adducts, salts, and detergent signals, there is a finite number of glycan or glycopeptide peaks that can be detected at any one time. The main advantage afforded by LC preceding MS is to reduce crowding of ion signals either among themselves, or from other contaminants while allowing species of lower abundance or poorer ionization efficiency to be better detected or resolved from background noise. This is particularly true if as many molecular ion signals as possible are to be selected for MS/MS analysis. A preliminary screen by MALDI-MS can inform about the complexity of the sample being analyzed, indicating whether it is contaminated, and, in the case of chemically derivatized glycan samples, the completeness of the chemistry. Except for relatively clean samples of low complexity, further offline and/or online fractionation is always beneficial.

In the case of peptides and glycopeptides, from either a single protein or subproteomic digests, a reverse phase (RP) C18 LC preceding MS is the "standard" practice. In this respect, the glycopeptides of interest would need to be retained efficiently during sample loading and wash while being readily eluted off upon increasing the concentration of organic solvent, typically acetonitrile, in acidic conditions. Different brands of C18 packing materials offer slightly different retention and separation but, in general, retention is determined by the hydrophobicity of the peptide backbone. In contrast, different glycoforms are poorly separated and usually elute within a few minutes in an average 1–2 h LC run. Less commonly used but a potentially complementary approach is hydrophilic interaction (HILIC) phase chromatography (Wuhrer et al. 2009a; Zauner et al. 2011). It is particularly effective for glycopeptides not retained on a RP column. Neither RP nor HILIC phase chromatography adequately resolves glycoforms carrying isomeric glycans. At present, only a porous graphitized carbon (PGC) column may allow a reasonable degree of isomeric separation, but the peptide portion of the glycopeptides needs to be trimmed down to only a few amino acids. This requirement means that this approach is mostly used for analysis of nonspecific proteolytic digests such as those provided by pronase or proteinase K digestions (Hua et al. 2013a; Zauner et al. 2013). The gain in isomeric glycan separation is thus at the expense of being able to confidently identify specific glycosylated peptides unless one is dealing with glycosylation analysis of single or several known proteins with only a limited total number of potential N-glycosylation sites.

The ability of PGC chromatography to separate glycan isomers is actually more commonly utilized in glycan (Ruhaak et al. 2009; Jensen et al. 2012) than glycopeptide analysis. One of the most successful applications is LC-MS analysis of reduced, native O-glycans from mucins (Hayes et al. 2012; Kenny et al. 2013), in which isomers of different glycosidic linkages can be resolved. Even so, it should be noted that no single dimensional LC is capable of resolving the full complexity of the glycan repertoire and therefore two or multiple dimensional online separations have been attempted. These were initially developed not for LC-MS but for conventional HPLC mapping. To allow high sensitivity detection, a fluorescent tag most commonly in the form of 2-aminopyridine (Tomiya et al. 1988; Takahashi et al. 1995), 2-aminobenzamide (Bigge et al. 1995), or 2-aminobenzoic acid (Bigge et al. 1995; Anumula and Dhume 1998), is introduced via reductive amination at the reducing end of each released glycan (Anumula 2006; Ruhaak et al. 2010). Since the detector response is solely dependent on the single fluorescent tag per glycan chain, this approach allows a relatively straightforward quantification based on peak area, without the need to derive a full complement of standard glycans. By virtue of carrying a sufficiently hydrophobic tag, it allows native glycans to be retained on a RP C18 column allowing for a more efficient separation of isomers than that which can be accomplished by amide or amine-based normal phase size separation. These two complementary approaches are also commonly preceded by the use of anion ion exchange chromatography to bulk separate the glycan pool according to charge as contributed by degree of sialylation and sulfation.

An inherent problem with the HPLC mapping approach is that glycan identification relies solely on ability to match the multidimensional LC coordinates against information in a pre-derived database defined by standards (Takahashi and Kato 2003; Royle et al. 2008) and is therefore less likely to identify novel structures. Often, sequential exoglycosidase digestions are needed to stepwise convert the peak of interest into expected smaller core structures, that can then be used to infer the original structure or for confirmation of structural assignment. To increase the confidence of assignment, one can couple LC with MS detection, either offline for each collected peak (Yagi et al. 2005) or, more efficiently, via online LC-MS concomitant with miniaturization to micro or nanoscale (Wuhrer et al. 2009b; Doherty et al. 2013). To date, practical LC-MS applications of reducing end tagged glycans have not been as widely adopted as PGC-based LC-MS analysis of native glycans. Fluorescent detection is still the norm for quantitative 2D or 3D HPLC-mapping analysis with supplementary offline MS analysis for select peaks, but such an approach is ultimately too laborious to cope with the demands of current glycomics. The highest added value of LC-MS is separation by isomeric structural features and not by size since MS alone provides the highest resolution possible in that respect. Thus, a PGC-based LC-MS appears to offer the best orthogonal coupling permutation, which can be further preceded by bulk charge separation. However, its wider application to resolving a full range of larger complex N-glycans (Hua et al. 2013b) as opposed to the smaller O-glycans remain to be demonstrated.

Newer column matrices are being developed and introduced commercially, which capitalize mostly on size separation by HILIC but also offering certain degree of resolution for important isomeric features such as differences in sialylation positions and antennary extensions (Ahn et al. 2010). These may not be sufficient for full-blown glycomics but are more than adequate for a variety of biopharmaceutical applications including high resolution mapping of the precise glycosylation profile of recombinant glycoprotein therapeutics such as IgG (Doherty et al. 2013). Another commonly used analytical LC mapping method capable of resolving structural isomers uses the Dionex high performance anion exchange chromatography (HPAEC) LC system using high pH and hydroxide eluent conditions and pulsed amperometric detection (PAD) (Townsend et al. 1989). Its popularity has waned recently probably because it needs a dedicated LC system and is less readily adapted to online LC-MS analysis due to the high salt content. Finally, RP C18 remains the only viable option for LC-MS analysis of permethylated glycans (Alley et al. 2010; Hu and Mechref 2012; Ritamo et al. 2013) despite poor chromatographic resolution. This reflects its inability to resolve isomeric components adequately, which leads to many tailing and split peaks. Currently this is a major limitation for analysis of permethylated glycans, which to be addressed will require strategically different thinking and implementation of the LC-MS/MS data acquisition functions (see later Sect. 7.3.4).

7.2.3 Chemical Derivatization

For any structural analysis, it is important to remember that MS only provides molecular weight and therefore is blind to isomeric structures unless the isomeric differences can be translated into mass differences or chromatographic retention time differences. An obvious approach is to rely on the stereospecificity of exoglycosidases to resolve the epimeric and anomeric configuration of the non-reducing terminal glycosyl residue, and possibly its linkage position. A combination of exoglycosidases can be judiciously employed in a sequential manner or as distinct sets of cocktails and the digestion products determined by LC and/or MS mapping. A major limitation of this approach is uncertainty in tracing the products back to specific precursor glycans when applied to glycomic mixtures. It is less of a problem when applied to purified or non-overlapping components in conjunction with 2D LC mapping techniques. Other possible issues include structure-dependent incomplete digestion due to steric hindrance, inactivity of stored enzymes, incorrect or suboptimal buffer conditions, which can introduce ambiguity especially when dealing with negative results. Another problem is the nonavailability of enzymes able to afford both the required digestion efficiency and specificity, a particular problem when handling novel glycotopes.

A number of chemical derivatizations have been used to introduce structure-specific mass differences, e.g., the use of chromium trioxide to distinguish anomeric configuration (Khoo and Dell 1990), periodate oxidation to determine linkage position and branching pattern (Khoo et al. 1995; Reinhold et al. 1995), and sialic acid derivatization to differentiate 2–3 versus 2–6 sialylation (Wheeler et al. 2009; Tousi et al. 2013). Most have not been used routinely possibly because the chemistry is not sufficiently clean, complete, and/or specific. This can translate into messy MS data with many unassigned by-products. In practice, the single most widely used chemical derivatization for glycan analysis other than the aforementioned reducing end tagging is permethylation (Wada et al. 2007, 2010). This converts every exposed OH group to an O-methyl (OMe), which is stable under most conditions including acid hydrolysis. In fact, the first use of permethylation in the history of glycosylation analysis was for linkage analysis by gas chromatography (GC)-electron impact (EI)-MS, also referred to as methylation analysis (Bjorndal et al. 1970). The permethylated glycans are acid hydrolyzed, reduced and then further peracetylated to yield partially methylated alditol acetates, which can be resolved by GC and give distinctive EI-fragmentation patterns depending on the relative positions of OMe versus O-acetyl. Larger permethylated oligosylalditols can also be analyzed by EI directly, but the field of MS analysis of intact glycans really took off with the introduction of fast-atom-bombardment (FAB)-MS of permethylated glycans in the 1980s (Dell 1987; Egge and Peter-Katalinic 1987).

The disadvantages of permethylation boil down largely to two issues. First, it was often argued that it is not amenable to non-specialists or chemists but, in fact, the NaOH/DMSO slurry method (Ciucanu and Kerek 1984) as opposed to the original methylsulfinyl carbanion method introduced by Hakomori (1964) is quite robust and can be performed by any initiated laboratory. Efforts are being made not only to offer practical courses but also to introduce commercial kits for high throughput solid phase permethylation (Kang et al. 2008). High sensitivity and high yield for routine microscale derivatization is now possible although admittedly it does introduce additional steps of chemistry and cleanup, which lead to loss that may not be compensated for by the overall MS detection sensitivity gained when only picomolar amounts of starting material are available. Second, alkaline-labile non-glycosyl substituents are lost during the reaction and this is particularly relevant for the naturally occurring and biologically important O-acetyl groups. However, phosphates and sulfates can be fully retained. In fact, the phosphodiester can be readily methylated, the monoester can acquire up to two O-Me groups, whereas the sulfate remains unmethylated and therefore offers a simple mass difference to distinguish between sulfate and phosphate (Yu et al. 2009).

Countering the perceived disadvantages are the many positive gains obtained by permethylation. First, permethylated glycans ionize better than native glycans thereby affording a lower detection limit. However, the most important and practical advantage gained by permethylation is a more reliable sequence and linkage informative MS/MS fragmentation than that afforded by native glycans (see Sect. 7.2.4). Any extra effort incurred and possible shortcomings are more than compensated for by better MS/MS sequencing prospects, which provide overall more structural information. Moreover, permethylation is the necessary first step for methylation analysis by GC–MS, which remains the only definitive technique available for establishing the linkage positions of the glycosyl constituents of a glycan or glycome. Finally, permethylation neutralizes the sialic acids (and uronic acids if present) and results in non-charged glycans except for those carrying sulfates. It thus provides a better overall quantitative profiling of the sialylated complex glycans alongside those non-sialylated ones in positive ion mode (Wada et al. 2007), while allowing selective detection of the low abundant sulfated glycans in negative ion mode (Yu et al. 2009). This has in fact formed the technical basis of the recently developed sulfoglycomics platform (Khoo and Yu 2010) (see Sect. 7.3.3), allowing one to explore the hitherto underappreciated realm of the glycome.

7.2.4 MS/MS Sequencing

Molecular mass measurement at the MS level provides at most the overall glycosyl composition of a detected glycan species. An experienced analyst can, however, proceed to deduce structures in terms of how many LacNAcs ($Hex_1HexNAc_1$), terminal NeuAc/NeuGc, Fuc, extra Hex (usually due to α3-Gal cap in nonhuman structures), extra HexNAcs (due to incomplete β-galactosylation, bisecting β-GlcNAc, GalNAcβ1-4GlcNAc), etc. that can be sensibly fitted onto a trimannosyl core ($Man_3GlcNAc_2$) structure for the N-glycans or simple core 1 (Galβ1-3GalNAc) structure for mucin type O-glycans, within the presumptive constraints imposed by known mammalian glycan biosynthetic pathways (Moremen et al. 2012). Preceding or subsequent compositional analysis may provide the missing stereochemistry and linkages of individual monosaccharide constituents, while sequential exoglycosidase digestion and/or selective chemical derivatization followed by MS analysis of the products may allow piecing together of tentatively assigned structures represented by the *m/z* values of the detected molecular ion signals.

It is clear that the structures thus defined would be "ambiguous" since most structural information except the molecular mass itself is inferred indirectly from analysis of the total glycan pool. On the other hand, detailed analysis of purified individual glycans is incompatible with the sensitivity and throughput required for current day glycomic complexity. This is compounded further by the inherent problems associated with chemo-enzymatic manipulations and chromatographic resolution (see preceding Sects. 7.2.2 and 7.2.3). Only direct tandem MS analysis, namely MS/MS, or MS^n if more than one stage of MS/MS is needed, on isolated precursor ions can define the precise terminal epitopes, core type, overall sequence and branching pattern of a glycan at sufficiently high sensitivity and speed within the online chromatographic time scale.

By and large, current MS/MS analysis of glycans relies predominantly on low energy collision induced dissociation (CID) usually introduced by resonance excitation and/or collision gas although more recently, other MS/MS modes such as infrared multiple photon dissociation (IRMPD), electron capture dissociation

(ECD), electron transfer dissociation (ETD), electron detachment dissociation (EDD), etc. have been developed and applied in select applications (reviewed in Han and Costello 2013). CID, as implemented on a quadruple/time-of-flight (Q/TOF) MS instrument or ion trap, and the higher energy collision dissociation (HCD) available on the hybrid Orbitrap series (Olsen et al. 2007) are all considered to be low energy cleavages and the most favored pathways are similar irrespective of instrument type (Fig. 7.1a).

A key to obtaining sequence and linkage information is to fragment glycans selectively and reproducibly through well-defined pathways to yield the so-called diagnostic fragment ions. This is the case when CID or HCD is applied to permethylated glycans. By virtue of having an OMe tag on each non-substituted OH position, product ions generated by single cleavage are readily distinguishable from those arising via multiple cleavages (Fig. 7.2a). This useful feature is not afforded by analysis of native glycans in which loss of one or more glycosyl residues attached to different sites would give rise to products having the same mass as one losing the same glycosyl residues consecutively from a single site. When in source fragmentation occurs, fragment ions arising through loss of one or more glycosyl residues cannot be differentiated from genuinely existing glycosylation heterogeneity (Fig. 7.2b). Moreover for native glycans subjected to CID in positive ion mode, neutral losses of the more labile terminal glycosyl residues such as sialic acids and fucoses occur far more readily than cleavages along the glycan chains constituted by LacNAc repeats and therefore often informs only their presence without being able to localize them to a particular site along the chain. For all these plus the shortcomings described in the preceding section, MS and MS/MS analysis of permethylated glycans is preferable to analysis of native glycans as it gives more definitive sequencing, based on well conserved and characterized sets of fragment ions (Dell 1987; Lemoine et al. 1991, 1993; Reinhold et al. 1995; Viseux et al. 1998; Morelle et al. 2004; Yu et al. 2006).

In brief, protonated molecular ions of permethylated glycans yield abundant oxonium ions due primarily to glycosidic cleavages at HexNAc, often followed by secondary cleavage amounting to elimination of the glycosyl substituent from the C3-position (Dell 1987) (Fig. 7.1b). These two features alone are often sufficient to define the terminal glycotopes in the form of R-(R′)HexNAc-, which includes all the commonly found epitopes such as ±NeuAc-Hex-(±Fuc)HexNAc- (sialyl and non-sialyl Lewis a/x or LacNAc), Fuc-Hex-(Fuc)HexNAc- (Lewis b/y), Hex-Hex-HexNAc- (αGal-capped LacNAc), HexNAc-HexNAc- (LacdiNAc), ±Hex/HexNAc-(Fuc)Hex-HexNAc- (blood group ABH). Type 1 and 2 chains, along with the corresponding Lewis a/b or x/y epitopes, can be inferred from elimination of the 3-substituents although ambiguity will arise if different terminal epitopes coexist on different antennae. Unfortunately this is usually the norm rather than exception. In such cases, it is beneficial to select the respective oxonium ions for a second stage of fragmentation, i.e., MS^3, so that the secondary loss from the 3-position can be directly attributed to a particular epitope in question.

Despite its simplicity, it is often inadequate to rely solely on the oxonium ions to pin down detailed glycan structures. Moreover, unless special cleanup or reducing

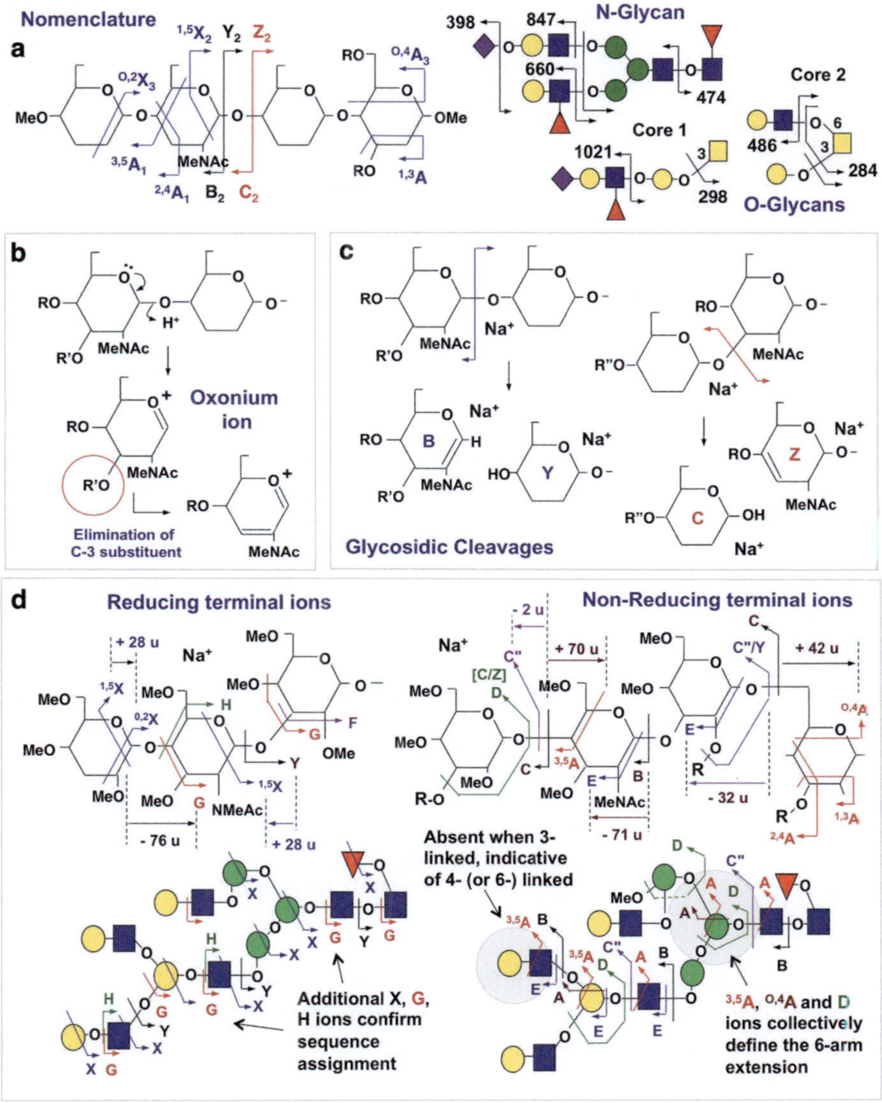

Fig. 7.1 CID MS/MS fragmentation pathways for permethylated glycans. (**a**) The commonly adopted Domon and Costello nomenclature (Domon and Costello 1988) for glycosidic and ring cleavage ions. Also depicted on the right is how the preferred low energy cleavages at HexNAc result in major diagnostic fragment ions that allow determination of the non-reducing terminal glycotopes and core types for the N- and O-glycans. With low energy CID, the protonated molecular ions tend to give mostly oxonium ions corresponding to the b ions by the nomenclature (**b**), whereas sodiated precursor ions would afford the sodiated b and y ions via glycosidic cleavage at HexNAc and less frequently c and z ions (**c**). In contrast, high energy CID by MALDI-TOF/TOF gives additional series of D, E, F, G, and H ions at each residue, allowing a more complete linkage specific sequencing (**d**). Also shown are the mass differences between some of the major ion series and how the high energy cleavages can be utilized to determine branching of polyLacNAc, as well as 3-arm versus 6-arm extension from the trimannosyl cores. Practical applications can be found in many published works (Wu et al. 2007; Fan et al. 2008; Yu et al. 2008; Lin et al. 2011; Wang et al. 2011)

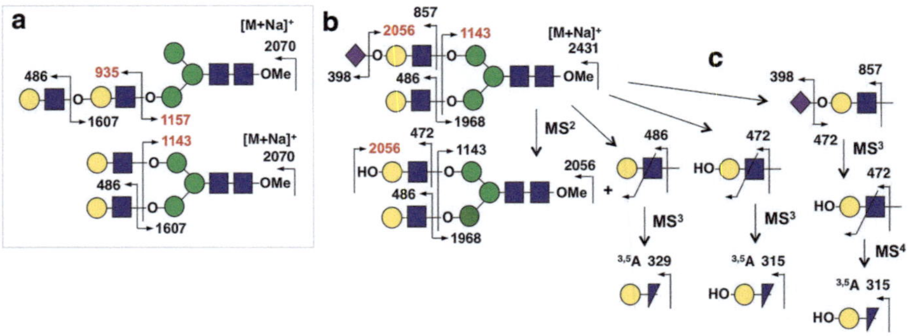

Fig. 7.2 General principles for defining branching pattern and to elicit linkage specific fragmentation via low energy CID MS/MS. (**a**) Permethylation allows better discrimination between losing glycosyl substituents from a single site or two separate sites. In this example, the two isomers can be distinguished by virtue of the trimannosyl core ion at *m/z* 1,157 versus 1,143, the latter carrying an extra free OH group. The presence of diLacNAc is further indicated by detecting the B ion at *m/z* 935 in only the monoantennary and not the biantennary structure. (**b**) The single sialyl moiety is prone to loss upon ionization of native glycans, and therefore would not be differentiated from non-sialylated molecular ions. In contrast, for permethylated glycans, any in source loss of the sialic acids is similar to CID MS/MS loss, creating a free OH group. In this case, the desialylated fragment ion at *m/z* 2,056 would be readily distinguished from genuine non-sialylated molecular ions at *m/z* 2,070. (**c**) The readily formed B and Y ions by MS^2 will not allow defining the linkage between the Gal and GlcNAc. Either it is assumed to be 4-linked as would be expected for the most common LacNAc unit, or MS^n can be performed. If one selects the MS^2 ion at *m/z* 857 for MS^3, it is likely to lose sialic acid and give *m/z* 472 as the major MS^3 ion without yielding any linkage information. Only when this MS^3 ion at *m/z* 472 is subjected to further MS^4 would a ^{3,5}A ion be produced to confirm its 4-linkage. Thus, one needs to be able to select *m/z* 472 and 486 and not 857 among the MS^2 ion for MS^3 to avoid a non-informative MS^3 stage. Alternatively, the same information can also be obtained by direct high energy CID MS/MS (Fig. 7.1d)

end tagging is applied, direct infusion nanoESI or MALDI-MS analysis of permethylated glycans most often produces sodiated and not protonated molecular ions. Even in LC-MS with extensive wash applied to the pre-column, followed by elution with an acidic solvent, one still gets sodiated molecular ions along with protonated and/or ammoniated ones. This propensity to form sodiated molecular ions may be taken advantage of and further promoted by addition of low (mM) levels of sodium hydroxide to the LC eluent (Aoki et al. 2007; Ritamo et al. 2013), so that all glycans ionize primarily as sodiated species and thus eliminating ion intensity spread caused by multiple distinct cation adducts.

Instead of producing mainly the oxonium ions, the sodiated glycan species tend to yield the complementary pairs of sodiated B and Y ions due to glycosidic cleavages at HexNAc, and occasionally also sodiated C and Z ions at other glycosyl residues particularly if they are 3-substituted (Fig. 7.1c). Even with these additional ions, which allow one to readily identify the R-(R′)HexNAc- termini along with the core types, information is still lacking with respect to linkage. The preferred B/Y glycosidic cleavages directed to the HexNAc residue facilitates mapping of the non-reducing terminal epitopes extending from the last HexNAc of the glycan chain but produce virtually no sequence or linkage information with respect to the specific moiety. It is also not possible to identify from which antenna of a multiantennary glycan the terminal epitope extends, whether it is from the 3- or 6-arm of the trimannosyl core or core 2 O-glycans, or a branch point along the polyLacNAc chains. To allow complete sequencing, linkage specific cleavages in the form of the various ring cleavages or concerted cleavages around the ring at each HexNAc and Hex site (Fig. 7.1d), particularly at branch points or multiply substituted residues, are needed.

At present, there are only two practical solutions to this problem of lacking linkage information. The first is to resort to multi-stage MS^n, based on the premise that the unfavorable ring cleavage type of fragmentation at less preferred sites would only occur when the isolated precursor ion is already devoid of the more labile glycosidic bonds. Often, many successive stages of isolation and fragmentation are needed to achieve the cleavage needed to obtain the necessary linkage information (Ashline et al. 2005; Reinhold et al. 2013) (Fig. 7.2c). Previously, Q/TOF or triple quadrupole types of instruments only allowed up to MS^2 unless they were coupled with in source fragmentation to give pseudo-MS^3. This applies also to the HCD now available on the Orbitrap series. However, a Q/TOF platform configured with two collision cells sandwiching an ion mobility cell has recently become available (Pringle et al. 2007). Intriguing and potentially very useful applications are emerging as investigators explore ion mobility separation of MS^2 products coupled with further MS^3 analysis (Olivova et al. 2008). It is also now possible using the new Orbitrap Fusion Tribrid system (Hebert et al. 2014) to subject the HCD MS^2 products to further MS^3 although its practical applications and effectiveness await further investigation. Other than these developments, ion trap remains the only genuine mass analyzer capable of MS^n. The obvious advantage is being able to successively go through deeper levels of fragmentation. However, this requires more sample and an increase level of expertise in manual selection. In fact, a major problem is that an intelligent decision tree is required in order to automate on the fly which product ions of each MS^n stage should be selected for the next MS^{n+1} fragmentation in order to obtain productive information (Lapadula et al. 2005).

An alternative and more straightforward approach is to rely on high energy CID cleavages which, at present, is offered only by a handful of MALDI–TOF/TOF instruments. At sufficiently high collision energy, usually a few kV, it has been noted that $^{1,5}X$ ions formed at each glycosyl residue are preferred over Y ions at HexNAc (Harvey et al. 1997; Mechref et al. 2003; Spina et al. 2004; Stephens et al. 2004; Yu et al. 2006). This allows precise sequencing along the entire glycan chain without gaps. Importantly, the plethora of reducing end fragment ions particularly of the A, D, and E ion series coupled with the non-reducing end G ion series are characteristic enough to allow systematic but painstaking de novo assignment of the linkages and branching pattern (Yu et al. 2006) (Fig. 7.1d). In this context, a complementary use of both low and high energy CID MS/MS is most effective since the low energy CID spectrum is easier to interpret and allows one to rapidly sketch out the overall sequence of the glycan, if not already known, whereas the additional cleavages afforded by high energy CID allow one to fill in the missing information (Yu et al. 2006). Only a single MS^2 spectrum and no decision making for higher order of MS^n is needed resulting in a far more efficient data acquisition process. Unfortunately, sensitivity is equally limiting. The truly useful high energy cleavages should also be distinguished from essentially just a lifting of energy fragment ions afforded by post-source decay mode, or in source prompt fragmentation (Wuhrer and Deelder 2006), or when energy is not sufficiently high and only the fragmentation pattern akin to low energy CID mode is attained despite the use of MALDI MS instruments capable of high energy CID.

7.2.5 Summary and Perspectives

Considering everything, the most productive MS platform for glycosylation analysis should include both MALDI-MS /MS and LC-MS/MS as a two-tier setup, a MALDI-TOF/TOF instrument for rapid mass profiling followed by low and high energy MALDI CID MS/MS on select peaks to confirm tentatively assigned structures. Such a setup is unrivaled in its simplicity, robustness for walk-in analysis by nonexpert users, rapid turnover, ease of handling MS and MS/MS data, as well as ease of comparative analysis across several samples to see which glycan peaks may change in relative intensity. Assignment of signals relies primarily on fitting observed *m*/*z* values against probable glycosyl composition from which tentative structures including the types of terminal epitopes and core can be sensibly deduced based on well founded glycobiology knowledge. Representative peaks carrying such inferred structural features, as well as those that do not appear to fit the usual suspects, should then be selected for MS/MS analysis, to either confirm assignment or for de novo sequencing. If needed, reiterative rounds of analysis can be performed after sequential exoglycosidase treatments, the choice of which is informed by the tentatively assigned structures based on preceding MS and MS/MS data.

This first-screen by MALDI-MS will suffice for most applications. Importantly, it will inform if the sample quality and complexity merit a follow-up, more comprehensive but also more time-consuming and labor-consuming offline nanoESI-MS/MS and/or online LC-MS/MS analysis and, if so, which structural aspects or target epitopes should be focused on. This is needed to specifically tailor the desirable MS/MS data acquisition functions and to facilitate ensuing data analysis that otherwise would be too intimidating and manual intensive, since prototype informatics solutions are still being developed (Ceroni et al. 2008; Maxwell et al. 2012; Peltoniemi et al. 2013; Tsai et al. 2013; Yu et al. 2013a, b) and most are not up and running for routine general use. With current advances in LC-MS/MS systems, it is now possible to auto-select up to 20 precursors for MS/MS analysis per second across the entire LC duration and still mass measure each of the product ion spectra at high mass accuracy and sensitivity (Michalski et al. 2012; Jones et al. 2013; Hebert et al. 2014). Alternatively, less MS^2 can be performed but each can be further accompanied by one or more MS^3 events if certain predefined criteria are met including the production of desirable neutral losses or diagnostic fragment ions. Thus, in theory at least one can attempt a comprehensive global glycomic analysis in a manner similar to global proteomics, complemented further by more targeted analysis. In practice, the glycomic field is limited not by brute force data acquisition but by data analysis and ability to intelligently reach down to the least abundant isomeric components while covering as broadly as possible diverse classes of glycans under a single optimized LC-MS/MS condition. The questions are: can we achieve sufficient depth and breadth for a biologically meaningful glycomic coverage, including discovery and characterization in fine structural detail of any novel glyco-feature or epitope? How can one sensibly utilize the extra dimensions afforded by LC-MS/MS analysis to obtain these goals?

7.3 Increasing the Breadth and Depth of MS-Based Glycomic Coverage

A glycobiology driven glycomic venture attempts to identify or confirm the occurrence of specific glycosylation events having possible biological relevance. It aims to describe either the key glycomic features of a cell or tissue at a particular pathophysiological state or the corresponding glycosylation profile of target glycoproteins. The latter is ultimately limited by the need to first identify the glycoproteins of interest and then to isolate them in sufficient purity and amount for MS-based analysis. This may not always be feasible especially when one targets the endogenous membrane glycoproteins and not the secreted and/or over-expressed ones. Without a priori reason to presuppose protein-specific glycosylation changes, it is arguably more sensible to first identify glycosylation anomalies and onco-developmental changes by a global high throughput approach with sufficient specificity and sensitivity at the glycomic level. Under this premise, one can initiate the discovery by either MS-based glycomic analysis leading eventually to the use of specific monoclonal antibody detection coupled with genetic manipulations for biological validation or, in reverse order, to prescreen the biological samples with lectin array and glycogene-targeted transcriptomics before using MS to home in on validating the observed changes, along with identification of the underlying glycan chains and protein carriers.

Using MS-based glycomics as a discovery mapping tool will ultimately test its ability to detect subtle changes that could be carried on minor components that appear after immuno-activation, malignant transformation or genetic manipulation. Often these appear as minor quantitative changes, and may not be apparent at the MS level but only when subjected to MS^2 or higher levels of MS^n. If it is a protein specific glycosylation change, the glycans implicated will constitute only a very minor proportion of the glycomic repertoire analyzed as a total pool. Similarly, using MS-based glycomics in reverse for target validation also often fails to meet expectations. It is possible that a change in expression of glycosyltransferases or other known glycogenes does not directly translate into distinctive glycomic phenotypes due to higher order controls, redundancies of the implicated glycosyltransferases, and/or other less well defined contributing factors. It is also difficult to ascertain epitope specificity of the biological probes used, whether they cross-react and thus nonspecific and misleading, or the MS-based technique is of insufficient sensitivity or penetrative depth to reveal the glycomic change. Added to these shortcomings, there are several problematic structural features and/or glycan classes that are often rendered cryptic to current glycomic analysis.

7.3.1 Addressing Poly-N-acetyllactosaminoglycans

In MS terms, any N- or O-glycan assigned with a number of LacNAc units in excess of the implicated maximum number of antennae is assumed to have a poly-N-acetyllactosmine (-[3Galβ1-4GlcNAcβ1-]$_n$, polyLacNAc) extension. However, this simplistic picture does not reveal structural details of functional relevance, e.g., which of the few or all antennae are extended, the absolute and average length of the polyLacNAc, their length distribution, terminal capping, internal branching, and substituents. A fundamental problem is that as the number of LacNAc units carried increases, so does the size of the glycan, which can easily extend beyond *m/z* 5,000 thereby preventing not only high sensitivity detection but also productive and detailed sequencing by current MS/MS (Wang et al. 2011). Despite recent advances made (Sutton-Smith et al. 2007; North et al. 2010a; Bern et al. 2013) in extending the maximum number of LacNAc units carried on a permethylated glycan that can be detected by MALDI-MS, along with impressive MS/MS that can be directly applied, it is still not feasible to compare, for example, a glycomic content that carries at most 30 LacNAcs per glycan versus one carrying up to 50 LacNAcs or more, in order to "confirm" by MS that there is an increase in polyLacNAc extension. It does not distinguish between a change in the length but "amount" as implied by tomato lectin staining (Merkle and Cummings 1987), whether there is an increase in polyLacNAc extension from the 6-arm of the trimannosyl core as indicated by L-PHA lectin (Fernandes et al. 1991), and identifying an increase in branching or an alteration in terminal substituents, is almost impossible.

For mammalian protein glycosylation, polyLacNAc extension from the cores is the main contributing factor for increasing the size of N- and O-glycans beyond that which can be feasibly detected and sequenced by MS in detail. The current best bet for assaying polyLacNAc from the glycomic perspective is to incorporate a digestion step using endo-β-galactosidases, a method used 30 years ago (Fukuda et al. 1984a, b; Spooncer et al. 1984; Oates et al. 1985) but can now be coupled with MS and MS/MS analysis of greater sensitivity. The released GlcNAc-Gal disaccharide units can be mapped and quantified against the non-reducing terminal cap in the form of R-GlcNAc-Gal, where R is most commonly ±NeuAc-Gal-, with or without prior separation from the resistant glycomic pool (Fig. 7.3). A caveat though is that any α3-fucosylation of the internal GlcNAc or β6-branching of the internal Gal of the LacNAc units renders it resistant to digestion by a normal dose of endo-β-galactosidase (Wang et al. 2011). This results in the production of internal resistant

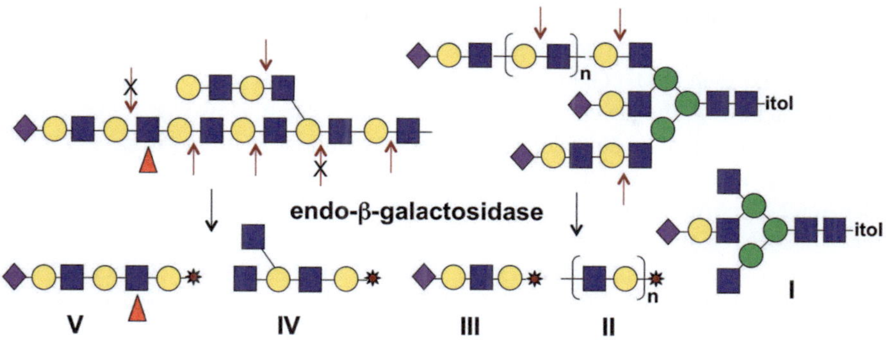

Fig. 7.3 Specificity of endo-β-galactosidase and expected digestion products. The *arrows* indicate positions that should be cleaved by a normal dose of endo-β-galactosidase but will be resistant if fucosylated at an adjacent GlcNAc or 6-branched at the Gal. Released products can be readily distinguished from the resistant core by virtue of mass differences at the reducing end if the glycans are pre-reduced prior to digestion. Reduction will not be an extraneous step since it would normally be performed for N-glycans or effected through the reductive elimination step for O-glycans in the first place in order to avoid chromatographic splitting due to different reducing end anomeric configurations. The free reducing ends of the released products can be further tagged with a fluorescent probe to facilitate their purification and quantification. The extent of polyLacNAc can be quantified based on the relative amount of product II compared to product III but one may need to consider other non-reducing end products such as product V. Product IV would inform the existence and degree of branching

units other than the expected disaccharides. This can complicate quantification for an estimation of length based on the internal units but, on the positive side, allows one to map internal structural details of polyLacNAc other than simple linear extension (Wang et al. 2011). By comparison with an untreated sample, it is possible to identify antennae that carry the polyLacNAc extension by virtue of their being trimmed to a non-extended GlcNAc. Considering many reports of glycomic changes attributed to increase in polyLacNAc extension from the 6-arm of N-glycans based primarily on tomato lectin and L-PHA lectin staining (Fernandes et al. 1991; Demetriou et al. 2001), a better developed quantitative MS-based analysis along the line described above is clearly needed. Likewise the functional relevance of the degree of branching and extension of polyLacNAc, along with the internal fucosylation, is still largely unknown and unexplored due to lack of definitive mapping tools.

7.3.2 *Addressing Terminal Disialyl Motif and Polysialylation*

Similar to polyLacNAc, the degree of polysialylation cannot be accurately and quantitatively mapped by MS analysis of intact glycans (Galuska et al. 2007, 2010, 2012). In fact the mere presence of polysialic acids may prevent the glycan carriers from ionization and being detected at all. The most prevailing mode of detection still relies on antibody staining, often complemented by 1,2-diamino-4,5-methylenedioxybenzene (DMB)-derivatization of the released polysialic acid chains for an estimation of the degree of polymerization (Sato et al. 1998, 1999). Thus far, the occurrence of polysialylation appears to be rather restricted to a couple of proteins (Liedtke et al. 2001; Galuska et al. 2010a), with the best known being N-CAM, which functions in the development of the nervous system. However, it may not be ubiquitous enough to be considered in a generic glycomic mapping strategy. On the other hand, it is increasingly found that both N- and O-glycans carry a terminal NeuAcα2-8NeuAc disialyl capping motif (Sato et al. 2000; Sato 2004). The occurrence of such disialyl units on simple core 1 and 2 O-glycan structures can be readily identified by MS analysis (Fukuda et al. 1987; Avril et al. 2006; Powlesland et al. 2009; Canis et al. 2010; Wang et al. 2013). However, those on N-glycans cannot (Lin et al. 2011; Wang et al. 2013) because the additional sialic acids may be due to internal 6-sialylation of GlcNAc or simply not apparent when the total number of sialic acids does not exceed that of the number of antennae, unless subjected to MS/MS analysis. Thus, glycomic mapping of terminal disialylation may require no specific treatment other than MS/MS analysis of each of the multi-sialylated glycans. Use of endo-β-galactosidase digestion as described above is also helpful in mapping the occurrence of sialic acid on the termini of polyLacNAc chains (Fukuda et al. 1985; Wang et al. 2013).

7.3.3 Addressing Sulfoglycomics

Other than its presence in GAGS, the occurrence of sulfate is best known and most commonly associated with the epithelial and secreted mucins. Together with sialylation, sulfates endow glycans with a high negative charge and contribute to the lubricant property of the mucosae. Moreover, the great diversity of glycotopes including the various combinations of sulfates, fucoses, and sialic acids on the O-glycans are a rich source of target ligands serving the interests of our microbiota and infectious pathogens (Robbe et al. 2004; Kufe 2009; McGuckin et al. 2011; Johansson et al. 2013). Specific sulfated epitopes on other N- or O-glycans can mediate a variety of biological processes (see Khoo and Yu 2010). For example, sulfated LacdiNAc on the pituitary hormones (Baenziger and Green 1988) serves as a recognition code for the hepatic receptor in mediating their clearance from blood circulation (Fiete et al. 1991), and sulfo-sialyl Lewis X on the high endothelial venule of peripheral lymph nodes mediates the rolling and recruitment of leukocytes to secondary lymphoid organs (Rosen 2004; McEver 2005). The HNK-1 epitope is a well-known sulfated glycotope of the nervous system (Morita et al. 2008) but its function is unclear. More recently, due to advances in MS and the associated development of optimized sample preparation techniques (Khoo and Yu 2010; Cheng et al. 2013; Kumagai et al. 2013), sulfation of N- and O-glycans is now known to have a wider and almost ubiquitous occurrence. In parallel, sulfated glycotopes in glycan arrays were often found to be better ligands for several C-type lectins, galectins, and Siglecs than unsulfated ones (Ideo et al. 2002; Galustian et al. 2004; Bochner et al. 2005; Tateno et al. 2005; Campanero-Rhodes et al. 2006; Kimura et al. 2007). This prompted a search for the naturally occurring sulfated glycotopes, a few of which have not been found including the 6′-sulfo 3′-sialyl LacNAc or Lewis X, thought to be the preferred ligands of Siglec F/8 (Tateno et al. 2005; Patnode et al. 2013a).

An inherent problem in detecting sulfated glycans is their extra negative charge and generally lower abundance relative to sialylated ones. Thus, other than the secreted and epithelial mucins and the pituitary hormones, sulfated versions of N- and O-glycans are generally not detected in conventional glycomic analysis, be it for the native or permethylated glycans. For native glycans, the main problem is due to their low abundance. They tend to be overshadowed by the sialylated glycans since the additionally charged sulfo-sialylated glycans are not expected to ionize as well. For the permethylated glycans, the main technical issues are (1) the often used chloroform/water partitioning after the NaOH/DMSO slurry permethylation, can result in sulfated glycans that retain a negative charge being partitioned into the discarded aqueous layer; and (2) when not particularly anticipated, the permethylated glycans were normally analyzed in positive ion mode, whereby ionization of sulfated permethylated glycans can be suppressed by the non-sulfated ones. Both issues can be solved by not cleaning up the permethylated glycans by simple organic solvent partitioning but by applying them directly to a reverse phase (RP) C18 column or by using another suitable solid phase fractionation after careful neutralization of the reaction mixtures (Yu et al. 2009). For fully methylated samples, the sulfate would remain the only substituent carrying a negative charge and hence can be selectively detected in negative ion mode, with or without being separated from non-sulfated ones. The field of sulfoglycomics was created using this simple approach (Khoo and Yu 2010; Yu et al. 2013c; Cheng et al. 2013).

In essence, what transpires is that sulfation is more ubiquitous than previously appreciated but often rendered cryptic because most analyses do not specifically look for it using the appropriate method. To be more precise, sulfoglycomics is simply a subbranch of glycomics just as phosphoproteomics is part of proteomics. Both share the common point that if no specific sample preparation treatment is carried out, one will not detect these ubiquitous modifications. A more comprehensive analytical scheme should incorporate these additional steps in order to obtain a more complete glycomic picture (Fig. 7.4). In most cases the function gained as a result of the extra sulfation may is unknown although examples exist in which sulfation was shown to increase affinity of the glycotope against its cognate lectins, such as sulfated sialyl LacNAc for Siglec 2 or CD22 on B cells and that the degree of sulfation decreased upon B cell activation (Kimura et al. 2007).

In general, if one analyzes permethylated sulfated glycans in positive ion mode, there is a tendency to observe loss of sulfite along with its counter-cation via in source fragmentation (Yu et al. 2009). For example, it is possible to detect disulfated glycans as $[M+3Na\text{–}2H]^+$ by MALDI-MS, but most likely they will be accompanied by species having lost one and two sodium sulfites, creating free OH groups in place of OSO_3Na in the MS spectrum. If full methylation is achieved, such species carrying additional OH groups may be considered as fragment ions that originated from parents initially carrying a sulfate at that position but this is less definitive than observing the species with both sulfates retained. The general trend, when one analyzes by MALDI-MS in positive ion mode, is that monosulfated species retain the

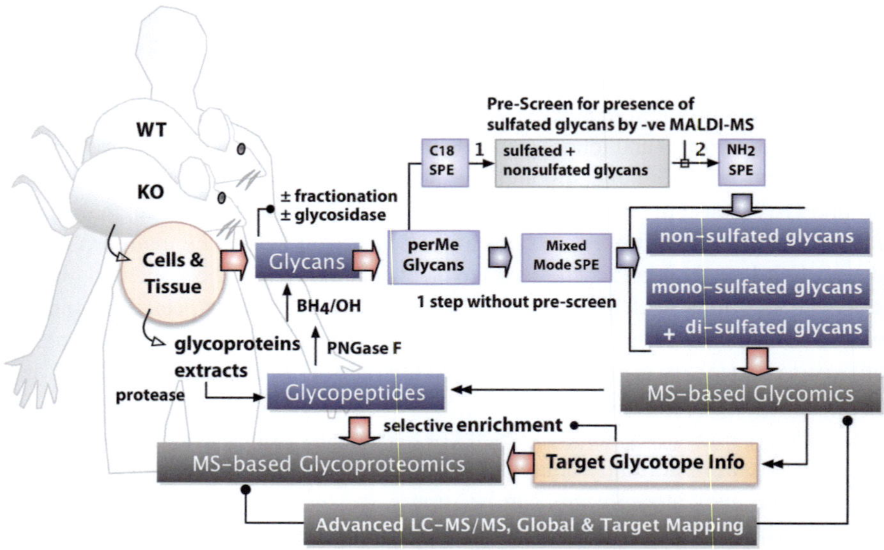

Fig. 7.4 Integrated workflows for MS-based glycomics and glycoproteomics. For a concerted workflow, a portion of the glycopeptides derived from cell extracts can be subjected to direct glycoproteomic analysis after the selective enrichment as dictated by the information gleaned from biological studies and/or preceding glycomic analysis. Another portion of the same glycopeptide pool, or from separate extractions, can be taken through sequential N- and O-glycan releases based on PNGase F and reductive elimination, respectively. The N-glycans can first be separated from the de-*N*-glycosylated peptides by C18 solid phase extraction (SPE) and reduced to oligoglycosyl alditols to facilitate subsequent LC-MS/MS analysis, or be kept in the same pool and subjected to reductive elimination. Endo-**β**-galactosidase digestion may best be performed at the glycopeptide level but is equally applicable at the glycan level. The separate or total glycan pools obtained will then be permethylated and directly cleaned up by an Oasis-Max SPE, which allows one step separation into non-sulfated flow-through, mono- and higher sulfated fractions for separate analysis (Cheng et al. 2013). Alternatively, a C18 SPE can be used and all permethylated glycans collected into one pool, and subjected to negative ion mode MALDI-MS screening, before further fractionation by amine-based SPE if sulfated glycan signals are detected (Yu et al. 2009). Glycomic analysis of permethylated glycans are further complemented in subsequent glycoproteomic analysis of native glycopeptides during which CID or HCD MS/MS produce abundant non-reducing end oxonium ions informing the presence of for example an O-acetyl group, if present. These oxonium ions also guide triggering of additional ETD MS/MS (Saba et al. 2012) and can be utilized in informatics solution for filtering out glycopeptides (Mayampurath et al. 2011), while peptide core sequencing is ultimately dependent upon the ability to produce sufficient b and y ions by HCD or *c* and *z* ions by ETD in multimode, decision tree-dependent MS/MS acquisition

single sulfate giving rise to disodiated molecular ions, whereas di- and higher sulfated species rarely retain more than one sulfate (Dell et al. 1991; Lei et al. 2009; Cheng et al. 2013). In addition, poor ionization efficiency often precludes their detection in positive ion mode even when purified away from the predominantly non-sulfated glycans. In contrast, when relatively free of suppressing non-sulfated glycans, either due to prior removal or online LC separation, it is possible to achieve reasonably good mapping of mono and disulfated glycans in positive ion modes by LC-MS/MS, where multiple positive charges tend to bring down the effective *m/z* for facile CID fragmentation. The number of sodium cations acquired often corresponds to those required to balance the sulfates plus additional ones imparting the multiply positive charge.

In negative ion mode, MALDI-MS analysis primarily yields single negatively charged $[M-H]^-$ molecular ions and is thus fairly robust for detecting monosulfated glycans even in the presence of non-sulfated glycans which would not be observed (Yu et al. 2013c) unless under-methylated at the sialic acid residue to give negative charge at the carboxylic group. However, the same problem exists for glycans carrying more than one sulfate in that the additional sulfate is prone to loss (Khoo and Yu 2010; Cheng et al. 2013). By negative ion mode LC-MS/MS, the charge state afforded by the sulfated glycans tends to equal the number of sulfates they carried in the absence of anion adducts and therefore disulfated glycans can be readily picked up as doubly charged, whereas monosulfated ones are seen as singly charged (Patnode et al. 2013a, b).

Despite the tendency to lose sulfite from multiply sulfated glycans during ionization, if one selects the sulfated glycan species for CID MS^n analysis, the sulfate surprisingly appears to be fairly stable compared to glycosidic cleavage. This trait allows one to readily sequence the sulfated glycans in a manner similar to the sequencing of non-sulfated glycans. It should be noted that in the positive ion mode, both sulfated and non-sulfated fragment ions can be detected along with other cleavages at the core and thus it is more informative in defining the overall glycan structure (Yu et al. 2009; Khoo and Yu 2010; Kobayashi et al. 2011). However, the location of sulfate is difficult to ascertain without resorting to successive stages of MS^n, which is often not feasible due to low sample amount. In contrast, CID or HCD in negative ion mode only allows fragment ions retaining the sulfate group and hence the negative charge to be detected. Importantly, diagnostic cleavages allow one to readily define the location of sulfate on the 3 or 6 position of Gal or GlcNAc of the LacNAc unit (Patnode et al. 2013a, b), thus generating very useful and specific information at the expense of defining the structural details of other parts of the glycans. A dual positive and negative ion mode analysis either in one LC-MS/MS run with successive polarity switch or as separate runs is therefore very desirable for sulfoglycomics.

7.3.4 Summary and Perspectives

In view of the not-too-difficult to adapt experimental approaches discussed above, all glycomic analyses should now incorporate the extra steps that would enable efficient mapping of the polyLacNAcs and sulfated glycans with or without the additional use of exoglycosidases. In essence, this broadens the glycomic coverage by tackling the larger size glycans and those endowed with extra negative charges due to sulfation but are often overwhelmed by the more abundant and ubiquitous sialylation. Although single dimensional PGC-based LC-MS/MS analysis of native O-glycans in negative ion mode is effective and sequence informative with linkage specific cleavages (Thomsson et al. 1999, 2000; Karlsson and McGuckin 2012), applications to more challenging samples, e.g., sulfated and multiply sialylated N-glycans, or glycomic samples with very low abundance of sulfated glycans, have yet to be demonstrated. For reasons already described at length, the permethyl derivatives offer the best sensitivity and versatility in an integrated glycomic workflow with multifaceted MS/MS analysis (Fig. 7.4).

First, permethylation allows facile fractionation of abundant non-sulfated sialylated glycans away from sulfated ones, a rather efficient enrichment step not readily accomplished with native glycans since negative charges are contributed by both sialic acids and sulfates. Second, it significantly improves sensitivity of detection by both MALDI and LC-MS/MS analysis. This is particularly relevant for species of lower abundance, larger size, and/or sulfated. Third, the permethylated sulfated glycans are amenable to both positive and negative ion mode analysis to take advantage of complementary fragmentation. Fourth, it directs more sequence-informative and linkage specific fragmentation and allows more definitive assignment of either the high energy CID MS/MS ions or the low energy sequential MS^n cleavages without the problem of glycosyl residue or sulfate migration often reported for native glycans (Kenny et al. 2011; Wuhrer et al. 2011). These last aspects are critical for increasing the depth of glycomic coverage, which is dictated by the ability to comprehensively or selectively perform at least MS^2 and preferably MS^n, as needed to at least resolve all coexisting structural isomers, if not all stereoisomers. The only drawback for permethylated derivatives is that individual isomeric species are not as well resolved by either offline or online chromatography. However, since sufficient chromatographic resolution is not currently available, a more promising approach may be to focus on developing innovative MS^n scan functions such as product ion dependent MS^3, which is a glycotope centric data acquisition mode to ensure obtaining the necessary linkage information out of all detectable glycofeatures as defined by a repertoire of diagnostic ions, followed by post-acquisition data filtering and mining.

Committing to extensive fractionation and comprehensive MS^n is not only sample and time-consuming, it requires an expertise not available in most non-specialist laboratories or general MS facilities. It makes no sense to prepare more than 10^8 cells from precious biological tissues or cell lines if the sole purpose is "discovery" in the absence of either a particular scientific rationale or supporting preliminary biological data that would make it hypothesis or target driven. It is a difficult problem, particularly for the field of glyco-biomarker discovery, since without investing much from the onset, one is unlikely to find any interesting differences in glycomic features among the most abundant species to make a biological, and hopefully clinically, important discovery! Conversely, it is equally frustrating if one cannot extract useful glycomic insights from the vast chunk of MS/MS data acquired. Added to that is the acute problem of quantification. Taking a cue from proteomics, efforts are being made to incorporate stable isotopes either metabolically or chemically onto glycans, most commonly at the reducing end, but so far success has been limited to proof of concept for relatively simple glycomic ventures (Orlando et al. 2009;

Orlando 2013; Mechref et al. 2013). The quantification problem at the glycomic level is essentially a chromatographic resolution problem and there is no easy solution on the horizon unless one glosses over the many co-eluting or near co-eluting isomeric forms and treats them as one composite peak. Often, these many slightly different isomers can only be resolved not by chromatography nor at MS^2 level but require MS^n analysis to demonstrate their presence, which renders quantification by extracted ion chromatograms unreliable if not impossible. Otherwise, a great many label free quantification software developed for proteomics (Cappadona et al. 2012; Nahnsen et al. 2013) can be readily imported to glycomics. Notwithstanding the quantitative impasse, an informatics solution is urgently needed in order to handle the increasingly large and rich dataset that an automated LC-MS/MS run can provide. Only when analysis of glycomic data is as simple and rapid as it is for proteomics allowing for identification and accurate quantification of thousands of glycans by a nonexpert will glycomics really take off, yielding the much-coveted results. Unfortunately we still have a long way to go to reach that point.

7.4 From Glycomics to Glycobiology-Driven Glycoproteomics

7.4.1 Raison d'etre and the Inadequacy of Only Defining Site Occupancy

It has been discussed from the start that a natural extension of glycomics is glycoproteomics. Based on the biological data which implicates a particular glycotope and MS data which not only confirms its structure but also the glycan carriers, the next key question is which proteins carry the glycotope and at which sites. Given that the biosynthesis of any glycotope is dependent on the concerted action of the requisite glycosyltransferases, one would expect all glycoproteins that pass through the ER/Golgi secretory pathway would be equally susceptible to similar modifications unless one or more of the glycosyltransferases are protein site-specific. There are currently a handful of such examples including the β4-GalNAcT which apparently recognizes specifically the pituitary hormones and synthesizes the LacdiNAc epitope that is subsequently 4-O-sulfated (Miller et al. 2008; Fiete et al. 2012a, b). Even so, this seemingly tight specificity is currently being challenged as LacdiNAc, with and without the sulfate, has been found elsewhere (see Yu et al. 2013c). Alternatively, a particular subset of proteins may share some common structural conformation, a certain unique physicochemical patch surrounding the glycosylation site, or some yet unknown traits that collectively allow them to be sought out from among hundreds or thousands of other proteins in order to be acted upon by the glycosylation machinery in a specific way. A simple question here is that as MS-based glycoproteomics becomes more powerful, will it be successfully employed to address the issue of protein specific glycosylation? The idea is that if one can identify most, if not all, proteins within a cell carrying a particular glycotope, one may then be able to define any common underlying recognition elements or sequence motifs. Identification of such elements will permit, investigation of how they function.

It may well turn out that no glycosylation is really site-specific or protein-specific. Each glycosylation site may carry any permutation of the glycomic constituents based simply on chance encounter with glycosyltransferases as permitted by local accessibility constraint. Such fundamental glycobiology puzzles cannot be solved if glycoproteomics is simply to define occupied glycosylation sites rather than the full repertoire of the occupying glycans. Unfortunately analysis of de-N-glycosylated peptides is still the most prevalent mode of glycoproteomic analysis at present (Pan et al. 2011; Kaji et al. 2003, 2012, 2013; Larsen et al. 2007; McDonald et al. 2009; Wollscheid et al. 2009; Zielinska et al. 2010, 2012). By virtue of the action of PNGase or endoF/endoH, the de-*N*-glycosylated peptides are mass-tagged by conversion of Asn to Asp within the consensus sequon or retention of a GlcNAc at the Asn, respectively. These can then be subjected to LC-MS/MS analysis and the MS/MS data searched against database for rapid identification of the glycosylation sites. Apart from not being able to inform the glycoforms, the most commonly criticized aspect of this approach is false positives introduced by spontaneous deamidation of Asn to Asp, which has nothing to do with the action of the PNGase (Palmisano et al. 2012). Even the use of heavy water to incorporate ^{18}O does not rule out this possibility, it just makes the site assignment more confident due to providing a mass difference of 3 instead of 1 (Kaji et al. 2003). By comparison, the action of endoF/endoH and any other recently identified endoglyosidases provides a more reliable identification of *N*-glycosylation sites (Hagglund et al. 2004, 2007; Segu et al. 2010; Zhang et al. 2011) but often suffers from incomplete digestion, and again does not inform the natural glycoforms. More recently, an innovative approach using the zinc finger nuclease gene targeting technique to impair O-glycosylation pathways by preventing extension of the GalNAc core has led to the so-called SimpleCell (Steentoft et al. 2011; Vakhrushev et al. 2013) in which all mucin type O-glycans synthesized carry only a single GalNAc or sialyl GalNAc. This greatly facilitated glycoproteomic analysis and led to a spectacular explosion of experimentally defined O-glycosylation sites (Steentoft et al. 2013). A caveat is that it actually only defined all the potential sites that could be O-glycosylated by the expressed ppGaN-Tases of a cell under the unnatural conditions seen when the GalNAc cannot be extended. The actual sites occupied under any particular physiological state remain unknown. As seen with the N-glycosylation sites one is still not defining the respective O-glycoforms for each known site.

7.4.2 MS/MS Sequencing and Identification of Glycopeptides

The obvious deficiency of current glycoproteomics as discussed above stems from the simple fact that direct LC-MS/MS analysis of intact glycopeptides cannot be completed in an automated fashion to allow simple database search of the generated MS/MS datasets for glycopeptide identification. For N-glycopeptides, any CID be it carried out in ion trap or the Q/TOF platform, or increasingly by HCD on LTQ-Orbitrap, would induce mostly glycosidic cleavage, giving rise to highly abundant glycan oxonium ions at low mass region, to be complemented by successive neutral losses of glycosyl residues from the precursors down to a single GlcNAc at the Asn (Wuhrer et al. 2007; Stavenhagen et al. 2013; Wu et al. 2013). In favorable cases, either manually or assisted by software, this so-called Y1 ion (peptide backbone + GlcNAc) can be identified and the *m*/*z* used to define the molecular mass of the peptide. This information alone does not allow unambiguous peptide identification in a shotgun glycoproteomic analysis of enriched glycopeptides derived from whole cell or the sub-proteome, instead of a single known, purified glycoprotein.

To allow confident identification of N-glycopeptides, there are currently two general approaches, both aimed at achieving or enhancing fragmentation along the peptide backbone. The first and perhaps most straightforward is to employ ETD instead of CID mode of MS^2 fragmentation (Yin et al. 2013). In principle, ETD leads mostly to c/z type of peptide bond cleavages along the peptide backbone without inducing much glycosidic cleavages. The practical problem though is that its efficiency is highly dependent on charge density, namely significant intensity of c and z series of ions will only be produced if the *z* is high while the overall *m*/*z* remain low. It does not work well with doubly and triply charged glycopeptides occurring at *m*/*z* over 1,400 or so (Alley et al. 2009), which one gets predominantly for a normal tryptic digests. Hence the trick is to increase the charge state by chemical derivatization such as using the tandem mass tag (TMT) (Yin et al. 2013) and/or using proteolytic enzyme such as LysC instead of trypsin.

The second approach is to increase the number of sequence informative b and y ions produced via CID and/or to mass measure them at high mass accuracy. Recent generations of Q/TOF and the Orbitrap instruments equipped with HCD are increasingly capable of meeting these challenges without compromising the required speed and thus the number of MS^2 per run. Alternatively the Y1 ion pre-identified in previous runs can be target isolated in the ion trap during a second run for MS^3 and the data obtained integrated. In this case, informatics is needed to facilitate Y1 identification in the first place. This is not trivial but computational tools are currently available (Goldberg et al. 2007; Irungu et al. 2007; Joenvaara et al. 2008; Ozohanics et al. 2008; Mayampurath et al. 2011; Wu et al. 2013). The handling of the second stage target MS^3 data is more straightforward since it requires only a direct database search using an existing search engine. Even so, it would benefit from a further optimized algorithm to discriminate among the few already narrowed down possibilities based on only a few b and y ions rather than implementing a de novo search. A direct database search with ETD or HCD MS^2 data is still problematic because one needs to take into consideration the fact that the glycan moiety is not a fixed mass but rather heterogeneous. A potential solution is to use either a manually input and thus customized glycan library or a full public library curated from an existing public glycan database, as implemented by the commercially available Byonics software (Bern et al. 2012; Zhu et al. 2013).

None of the methods described above provides an immediate foolproof solution for identification of N- and O-glycopeptides but we are getting there. MS instruments have gained in sensitivity, speed, and MS^2 versatility over the years while improved software is being developed and introduced by both the academic and commercial sectors. With time, it is conceivable that the technical platform will be established for glycopeptide MS/MS analysis leading to automated LC-MS/MS analysis. What remains to be developed is the critical step of target capture or selective enrichment of specific subsets of glycopeptides since no single LC-MS/MS run on any instrument will be sufficient to handle the full dynamic range of the entire glycopeptide pool derived from a glycoproteome. It would be prohibitively costly for instrument time, data mining and storage to undertake such an endeavor, which can probably be tackled as the human genome was, i.e., by several leading centers working together as a consortium. For each individual laboratory engaged in glycobiology, and for most small to middle scale central facilities for glyco-analysis, the most practically useful approach will always be selective or targeted, based on the biological problems in hand.

7.5 Closing and Future Prospects

The holy grail of MS-based glycomics and glycoproteomics is to progress from discovery mode glycomic mapping to target glycoproteomics, in order to not just identify specific glycosylation features associated with a particular process or disease, which would then be validated using a large human sample cohort, but also to localize the implicated glyco-features to a handful of specific glycoprotein carriers. It is a general belief, until experimentally proven otherwise, that a protein site-specific glycosylation or glycotope exists and the ability to sort these out among more universal protein glycosylation features is the driving force behind glycoproteomics. While glycomics aims to be as comprehensive as possible in both breadth and depth of coverage in order to "discover" and to define the full repertoire unique to each cell type at any particular pathophysiological state, glycoproteomics would more sensibly be focused. Whether it starts from global mapping or from target-captured subsets, the endpoint is the same. Applied MS/MS analysis is required to unambiguously define whether a predefined glycotope of interest is indeed carried on one or more glycosylation sites of the identified glycoproteins. To accomplish this, efficient MS/MS sequencing of glycopeptides and ensuing data analysis are indispensable.

Much remains to be done to harness the power of intelligent MS/MS data acquisition so that it is more directed and penetrative. Equally and perhaps even more urgently needed are informatics solutions to automate the process of data analysis via a combination of de novo sequencing and database search algorithms. The fundamentals governing MS-based protein glycosylation analysis are essentially the same as those guiding glycomics and glycoproteomics—only on a larger scale. Extra considerations needed are more fractionation versus target capture,

more MS/MS data acquisition versus target analysis, and the ability to handle them efficiently. We can confidently predict that MS instruments and MS/MS scan functions will keep improving in performance, notably in speed and sensitivity. Likewise informatics solutions along with the glycan database (Campbell et al. 2011; Hayes et al. 2011; Ranzinger et al. 2011; Artemenko et al. 2012) and the required Web interface and algorithms to interrogate them (Aoki-Kinoshita 2013) will be perfected. Eventually we will be back to square one in contemplating what biological questions we can ask and want to solve and, with it, the requisite sample preparation and enrichment steps at all levels—from subcellular fractionation, efficient membrane glycoprotein extractions, solubilization and digestions, down to enrichment of target glycopeptides and glycans, all of which require a separate treatise on their own as these are the true limiting factors, not mass spectrometry.
Conflict of Interest PolicyThe author declares that he has no conflict of interest.

Acknowledgment The author gratefully acknowledge Academia Sinica and Taiwan national Science Council (Grants 99-2311-B-001-021-MY3; 102-2311-B-001-026-MY3) for support on works carried out related to glycomics and glycoproteomics.

References

Ahn J, Bones J, Yu YQ, Rudd PM, Gilar M. Separation of 2-aminobenzamide labeled glycans using hydrophilic interaction chromatography columns packed with 1.7 microm sorbent. J Chromatogr B Analyt Technol Biomed Life Sci. 2010;878(3–4):403–8.

Alley Jr WR, Mechref Y, Novotny MV. Characterization of glycopeptides by combining collision-induced dissociation and electron-transfer dissociation mass spectrometry data. Rapid Commun Mass Spectrom. 2009;23(1):161–70.

Alley Jr WR, Madera M, Mechref Y, Novotny MV. Chip-based reversed-phase liquid chromatography-mass spectrometry of permethylated N-linked glycans: a potential methodology for cancer-biomarker discovery. Anal Chem. 2010;82(12):5095–106.

Alley Jr WR, Mann BF, Novotny MV. High-sensitivity analytical approaches for the structural characterization of glycoproteins. Chem Rev. 2013;113(4):2668–732.

Anumula KR. Advances in fluorescence derivatization methods for high-performance liquid chromatographic analysis of glycoprotein carbohydrates. Anal Biochem. 2006;350(1):1–23.

Anumula KR, Dhume ST. High resolution and high sensitivity methods for oligosaccharide mapping and characterization by normal phase high performance liquid chromatography following derivatization with highly fluorescent anthranilic acid. Glycobiology. 1998;8(7):685–94.

Aoki K, Perlman M, Lim JM, Cantu R, Wells L, Tiemeyer M. Dynamic developmental elaboration of N-linked glycan complexity in the Drosophila melanogaster embryo. J Biol Chem. 2007;282(12):9127–42.

Aoki-Kinoshita KF. Using databases and web resources for glycomics research. Mol Cell Proteomics. 2013;12(4):1036–45.

Artemenko NV, McDonald AG, Davey GP, Rudd PM. Databases and tools in glycobiology. Methods Mol Biol. 2012;899:325–50.

Ashline D, Singh S, Hanneman A, Reinhold V. Congruent strategies for carbohydrate sequencing. 1. Mining structural details by MSn. Anal Chem. 2005;77(19):6250–62.

Avril T, North SJ, Haslam SM, Willison HJ, Crocker PR. Probing the cis interactions of the inhibitory receptor Siglec-7 with alpha2,8-disialylated ligands on natural killer cells and other leukocytes using glycan-specific antibodies and by analysis of alpha2,8-sialyltransferase gene expression. J Leukoc Biol. 2006;80(4):787–96.

Baenziger JU, Green ED. Pituitary glycoprotein hormone oligosaccharides: structure, synthesis and function of the asparagine-linked oligosaccharides on lutropin, follitropin and thyrotropin. Biochim Biophys Acta. 1988;947(2):287–306.

Bern M, Kil YJ, Becker C. Byonic: advanced peptide and protein identification software. Curr Protoc Bioinformatics. 2012;Chapter 13:Unit13 20.

Bern M, Brito AE, Pang PC, Rekhi A, Dell A, Haslam SM. Polylactosaminoglycan glycomics: enhancing the detection of high-molecular-weight N-glycans in matrix-assisted laser desorption ionization time-of-flight profiles by matched filtering. Mol Cell Proteomics. 2013;12(4):996–1004.

Bigge JC, Patel TP, Bruce JA, Goulding PN, Charles SM, Parekh RB. Nonselective and efficient fluorescent labeling of glycans using 2-amino benzamide and anthranilic acid. Anal Biochem. 1995;230(2):229–38.

Bjorndal H, Hellerqv CG, Lindberg B, Svensson S. Gas-liquid chromatography and mass spectrometry in methylation analysis of polysaccharides. Angew Chem Int Ed Engl. 1970;9(8):610–9.

Bleckmann C, Geyer H, Geyer R. Nanoelectrospray-MS(n) of native and permethylated glycans. Methods Mol Biol. 2011;790:71–85.

Bochner BS, Alvarez RA, Mehta P, Bovin NV, Blixt O, White JR, Schnaar RL. Glycan array screening reveals a candidate ligand for Siglec-8. J Biol Chem. 2005;280(6):4307–12.

Campanero-Rhodes MA, Childs RA, Kiso M, Komba S, Le Narvor C, Warren J, Otto D, Crocker PR, Feizi T. Carbohydrate microarrays reveal sulphation as a modulator of siglec binding. Biochem Biophys Res Commun. 2006;344(4):1141–6.

Campbell MP, Hayes CA, Struwe WB, Wilkins MR, Aoki-Kinoshita KF, Harvey DJ, Rudd PM, Kolarich D, Lisacek F, Karlsson NG, Packer NH. UniCarbKB: putting the pieces together for glycomics research. Proteomics. 2011;11(21):4117–21.

Canis K, McKinnon TA, Nowak A, Panico M, Morris HR, Laffan M, Dell A. The plasma von Willebrand factor O-glycome comprises a surprising variety of structures including ABH antigens and disialosyl motifs. J Thromb Haemost. 2010;8(1):137–45.

Cappadona S, Baker PR, Cutillas PR, Heck AJ, van Breukelen B. Current challenges in software solutions for mass spectrometry-based quantitative proteomics. Amino Acids. 2012;43(3):1087–108.

Ceroni A, Maass K, Geyer H, Geyer R, Dell A, Haslam SM. GlycoWorkbench: a tool for the computer-assisted annotation of mass spectra of glycans. J Proteome Res. 2008;7(4):1650–9.

Cheng PF, Snovida S, Ho MY, Cheng CW, Wu AM, Khoo KH. Increasing the depth of mass spectrometry-based glycomic coverage by additional dimensions of sulfoglycomics and target analysis of permethylated glycans. Anal Bioanal Chem. 2013;405(21):6683–95.

Ciucanu I, Kerek F. A simple and rapid method for the permethylation of carbohydrates. Carbohydr Res. 1984;131(2):209–17.

Dell A. F.A.B.-mass spectrometry of carbohydrates. Adv Carbohydr Chem Biochem. 1987; 45:19–72.

Dell A, Morris HR, Greer F, Redfern JM, Rogers ME, Weisshaar G, Hiyama J, Renwick AG. Fast-atom-bombardment mass spectrometry of sulphated oligosaccharides from ovine lutropin. Carbohydr Res. 1991;209:33–50.

Demetriou M, Granovsky M, Quaggin S, Dennis JW. Negative regulation of T-cell activation and autoimmunity by Mgat5 N-glycosylation. Nature. 2001;409(6821):733–9.

Doherty M, McManus CA, Duke R, Rudd PM. High-throughput quantitative N-glycan analysis of glycoproteins. Methods Mol Biol. 2012;899:293–313.

Doherty M, Bones J, McLoughlin N, Telford JE, Harmon B, Defelippis MR, Rudd PM. An automated robotic platform for rapid profiling oligosaccharide analysis of monoclonal antibodies directly from cell culture. Anal Biochem. 2013;442(1):10–8.

Domon B, Costello CE. A systematic nomenclature for carbohydrate fragmentations in FAB-MS/MS spectra of glycoconjugates. Glycoconj J. 1988;5:397–409.

Egge H, Peter-Katalinic J. Fast atom bombardment mass spectrometry for structural elucidation of glycoconjugates. Mass Spectrom Rev. 1987;6(3):331–93.

Fan YY, Yu SY, Ito H, Kameyama A, Sato T, Lin CH, Yu LC, Narimatsu H, Khoo KH. Identification of further elongation and branching of dimeric type 1 chain on lactosylceramides from colonic adenocarcinoma by tandem mass spectrometry sequencing analyses. J Biol Chem. 2008;283(24):16455–68.

Fernandes B, Sagman U, Auger M, Demetrio M, Dennis JW. Beta 1-6 branched oligosaccharides as a marker of tumor progression in human breast and colon neoplasia. Cancer Res. 1991; 51(2):718–23.

Fiete D, Srivastava V, Hindsgaul O, Baenziger JU. A hepatic reticuloendothelial cell receptor specific for SO4-4GalNAc beta 1,4GlcNAc beta 1,2Man alpha that mediates rapid clearance of lutropin. Cell. 1991;67(6):1103–10.

Fiete D, Beranek M, Baenziger JU. Molecular basis for protein-specific transfer of N-acetylgalactosamine to N-linked glycans by the glycosyltransferases beta1,4-N-acetylgalactosaminyl transferase 3 (beta4GalNAc-T3) and beta4GalNAc-T4. J Biol Chem. 2012a;287(34):29194–203.

Fiete D, Beranek M, Baenziger JU. Peptide-specific transfer of N-acetylgalactosamine to O-linked glycans by the glycosyltransferases beta1,4-N-acetylgalactosaminyl transferase 3 (beta4GalNAc-T3) and beta4GalNAc-T4. J Biol Chem. 2012b;287(34):29204–12.

Fukuda M, Dell A, Fukuda MN. Structure of fetal lactosaminoglycan. The carbohydrate moiety of Band 3 isolated from human umbilical cord erythrocytes. J Biol Chem. 1984a;259(8):4782–91.

Fukuda M, Dell A, Oates JE, Fukuda MN. Structure of branched lactosaminoglycan, the carbohydrate moiety of band 3 isolated from adult human erythrocytes. J Biol Chem. 1984b; 259(13):8260–73.

Fukuda MN, Dell A, Oates JE, Fukuda M. Embryonal lactosaminoglycan. The structure of branched lactosaminoglycans with novel disialosyl (sialyl alpha 2-9 sialyl) terminals isolated from PA1 human embryonal carcinoma cells. J Biol Chem. 1985;260(11):6623–31.

Fukuda M, Lauffenburger M, Sasaki H, Rogers ME, Dell A. Structures of novel sialylated O-linked oligosaccharides isolated from human erythrocyte glycophorins. J Biol Chem. 1987; 262(25):11952–7.

Galuska SP, Geyer R, Muhlenhoff M, Geyer H. Characterization of oligo- and polysialic acids by MALDI-TOF-MS. Anal Chem. 2007;79(18):7161–9.

Galuska SP, Geyer H, Bleckmann C, Rohrich RC, Maass K, Bergfeld AK, Muhlenhoff M, Geyer R. Mass spectrometric fragmentation analysis of oligosialic and polysialic acids. Anal Chem. 2010;82(5):2059–66.

Galuska SP, Rollenhagen M, Kaup M, Eggers K, Oltmann-Norden I, Schiff M, Hartmann M, Weinhold B, Hildebrandt H, Geyer R, Muhlenhoff M and Geyer H. Synaptic cell adhesion molecule SynCAM 1 is a target for polysialylation in postnatal mouse brain. Proc Natl Acad Sci USA 2010a:107(22):10250–10255.

Galuska SP, Geyer H, Mink W, Kaese P, Kuhnhardt S, Schafer B, Muhlenhoff M, Freiberger F, Gerardy-Schahn R, Geyer R. Glycomic strategy for efficient linkage analysis of di-, oligo- and polysialic acids. J Proteomics. 2012;75(17):5266–78.

Galustian C, Park CG, Chai W, Kiso M, Bruening SA, Kang YS, Steinman RM, Feizi T. High and low affinity carbohydrate ligands revealed for murine SIGN-R1 by carbohydrate array and cell binding approaches, and differing specificities for SIGN-R3 and langerin. Int Immunol. 2004;16(6):853–66.

Goldberg D, Bern M, Parry S, Sutton-Smith M, Panico M, Morris HR, Dell A. Automated N-glycopeptide identification using a combination of single- and tandem-MS. J Proteome Res. 2007;6(10):3995–4005.

Hagglund P, Bunkenborg J, Elortza F, Jensen ON, Roepstorff P. A new strategy for identification of N-glycosylated proteins and unambiguous assignment of their glycosylation sites using HILIC enrichment and partial deglycosylation. J Proteome Res. 2004;3(3):556–66.

Hagglund P, Matthiesen R, Elortza F, Hojrup P, Roepstorff P, Jensen ON, Bunkenborg J. An enzymatic deglycosylation scheme enabling identification of core fucosylated N-glycans and O-glycosylation site mapping of human plasma proteins. J Proteome Res. 2007;6(8):3021–31.

Hakomori S. A rapid permethylation of glycolipid, and polysaccharide catalyzed by methylsulfinyl carbanion in dimethyl sulfoxide. J Biochem. 1964;55:205–8.

Han L, Costello CE. Mass spectrometry of glycans. Biochemistry (Mosc). 2013;78(7):710–20.

Harvey DJ, Bateman RH, Green MR. High-energy collision-induced fragmentation of complex oligosaccharides ionized by matrix-assisted laser desorption/ionization mass spectrometry. J Mass Spectrom. 1997;32(2):167–87.

Hayes CA, Karlsson NG, Struwe WB, Lisacek F, Rudd PM, Packer NH, Campbell MP. UniCarb-DB: a database resource for glycomic discovery. Bioinformatics. 2011;27(9):1343–4.

Hayes CA, Nemes S, Issa S, Jin C, Karlsson NG. Glycomic work-flow for analysis of mucin O-linked oligosaccharides. Methods Mol Biol. 2012;842:141–63.

Hebert AS, Richards AL, Bailey DJ, Ulbrich A, Coughlin EE, Westphall MS, Coon JJ. The One Hour Yeast Proteome. Mol Cell Proteomics. 2014;13(1):339–47.

Hu Y, Mechref Y. Comparing MALDI-MS, RP-LC-MALDI-MS and RP-LC-ESI-MS glycomic profiles of permethylated N-glycans derived from model glycoproteins and human blood serum. Electrophoresis. 2012;33(12):1768–77.

Hua S, Hu CY, Kim BJ, Totten SM, Oh MJ, Yun N, Nwosu CC, Yoo JS, Lebrilla CB, An HJ. Glyco-analytical multispecific proteolysis (Glyco-AMP): a simple method for detailed and quantitative glycoproteomic characterization. J Proteome Res. 2013a;12(10):4414–23.

Hua S, Jeong HN, Dimapasoc LM, Kang I, Han C, Choi JS, Lebrilla CB, An HJ. Isomer-specific LC/MS and LC/MS/MS profiling of the mouse serum N-glycome revealing a number of novel sialylated N-glycans. Anal Chem. 2013b;85(9):4636–43.

Ideo H, Seko A, Ohkura T, Matta KL, Yamashita K. High-affinity binding of recombinant human galectin-4 to SO(3)(−)→3Galbeta1→3GalNAc pyranoside. Glycobiology. 2002;12(3):199–208.

Irungu J, Go EP, Dalpathado DS, Desaire H. Simplification of mass spectral analysis of acidic glycopeptides using GlycoPep ID. Anal Chem. 2007;79(8):3065–74.

Jensen PH, Karlsson NG, Kolarich D, Packer NH. Structural analysis of N- and O-glycans released from glycoproteins. Nat Protoc. 2012;7(7):1299–310.

Joenvaara S, Ritamo I, Peltoniemi H, Renkonen R. N-glycoproteomics—an automated workflow approach. Glycobiology. 2008;18(4):339–49.

Johansson ME, Sjovall H, Hansson GC. The gastrointestinal mucus system in health and disease. Nat Rev Gastroenterol Hepatol. 2013;10(6):352–61.

Jones KA, Kim PD, Patel BB, Kelsen SG, Braverman A, Swinton DJ, Gafken PR, Jones LA, Lane WS, Neveu JM, Leung HC, Shaffer SA, Leszyk JD, Stanley BA, Fox TE, Stanley A, Hall MJ, Hampel H, South CD, de la Chapelle A, Burt RW, Jones DA, Kopelovich L, Yeung AT. Immunodepletion plasma proteomics by TripleTOF 5600 and Orbitrap Elite/LTQ-Orbitrap Velos/Q Exactive mass spectrometers. J Proteome Res. 2013;12(10):4351–65.

Kaji H, Saito H, Yamauchi Y, Shinkawa T, Taoka M, Hirabayashi J, Kasai K, Takahashi N, Isobe T. Lectin affinity capture, isotope-coded tagging and mass spectrometry to identify N-linked glycoproteins. Nat Biotechnol. 2003;21(6):667–72.

Kaji H, Shikanai T, Sasaki-Sawa A, Wen H, Fujita M, Suzuki Y, Sugahara D, Sawaki H, Yamauchi Y, Shinkawa T, Taoka M, Takahashi N, Isobe T, Narimatsu H. Large-scale identification of N-glycosylated proteins of mouse tissues and construction of a glycoprotein database, GlycoProtDB. J Proteome Res. 2012;11(9):4553–66.

Kaji H, Ocho M, Togayachi A, Kuno A, Sogabe M, Ohkura T, Nozaki H, Angata T, Chiba Y, Ozaki H, Hirabayashi J, Tanaka Y, Mizokami M, Ikehara Y, Narimatsu H. Glycoproteomic discovery of serological biomarker candidates for HCV/HBV infection-associated liver fibrosis and hepatocellular carcinoma. J Proteome Res. 2013;12(6):2630–40.

Kang P, Mechref Y, Novotny MV. High-throughput solid-phase permethylation of glycans prior to mass spectrometry. Rapid Commun Mass Spectrom. 2008;22(5):721–34.

Karlsson NG, McGuckin MA. O-Linked glycome and proteome of high-molecular-mass proteins in human ovarian cancer ascites: Identification of sulfation, disialic acid and O-linked fucose. Glycobiology. 2012;22(7):918–29.

Kenny DT, Issa SM, Karlsson NG. Sulfate migration in oligosaccharides induced by negative ion mode ion trap collision-induced dissociation. Rapid Commun Mass Spectrom. 2011; 25(18):2611–8.

Kenny DT, Gaunitz S, Hayes CA, Gustafsson A, Sjoblom M, Holgersson J, Karlsson NG. Mass spectrometric analysis of O-linked oligosaccharides from various recombinant expression systems. Methods Mol Biol. 2013;988:145–67.

Khoo KH, Dell A. Assignment of anomeric configurations of pyranose sugars in oligosaccharides using a sensitive FAB-MS strategy. Glycobiology. 1990;1(1):83–91.

Khoo KH, Yu SY. Mass spectrometric analysis of sulfated N- and O-glycans. Methods Enzymol. 2010;478:3–26.

Khoo KH, Sarda S, Xu X, Caulfield JP, McNeil MR, Homans SW, Morris HR, Dell A. A unique multifucosylated -3GalNAc beta 1–>4GlcNAc beta 1–>3Gal alpha 1- motif constitutes the repeating unit of the complex O-glycans derived from the cercarial glycocalyx of Schistosoma mansoni. J Biol Chem. 1995;270(29):17114–23.

Kimura N, Ohmori K, Miyazaki K, Izawa M, Matsuzaki Y, Yasuda Y, Takematsu H, Kozutsumi Y, Moriyama A, Kannagi R. Human B-lymphocytes express alpha2-6-sialylated 6-sulfo-N-acetyllactosamine serving as a preferred ligand for CD22/Siglec-2. J Biol Chem. 2007; 282(44):32200–7.

Kobayashi M, Mitoma J, Hoshino H, Yu SY, Shimojo Y, Suzawa K, Khoo KH, Fukuda M, Nakayama J. Prominent expression of sialyl Lewis X-capped core 2-branched O-glycans on high endothelial venule-like vessels in gastric MALT lymphoma. J Pathol. 2011;224(1):67–77.

Kolarich D, Jensen PH, Altmann F, Packer NH. Determination of site-specific glycan heterogeneity on glycoproteins. Nat Protoc. 2012;7(7):1285–98.

Kufe DW. Mucins in cancer: function, prognosis and therapy. Nat Rev Cancer. 2009;9(12):874–85.

Kumagai T, Katoh T, Nix DB, Tiemeyer M, Aoki K. In-Gel beta-Elimination and Aqueous-Organic Partition for Improved O- and Sulfoglycomics. Anal Chem. 2013;85(18):8692–9.

Lapadula AJ, Hatcher PJ, Hanneman AJ, Ashline DJ, Zhang H, Reinhold VN. Congruent strategies for carbohydrate sequencing. 3. OSCAR: an algorithm for assigning oligosaccharide topology from MSn data. Anal Chem. 2005;77(19):6271–9.

Larsen MR, Jensen SS, Jakobsen LA, Heegaard NH. Exploring the sialiome using titanium dioxide chromatography and mass spectrometry. Mol Cell Proteomics. 2007;6(10):1778–87.

Lei M, Mechref Y, Novotny MV. Structural analysis of sulfated glycans by sequential double-permethylation using methyl iodide and deuteromethyl iodide. J Am Soc Mass Spectrom. 2009;20(9):1660–71.

Lemoine J, Strecker G, Leroy Y, Fournet B, Ricart G. Collisional-activation tandem mass spectrometry of sodium adduct ions of methylated oligosaccharides: sequence analysis and discrimination between alpha-NeuAc-(2–3) and alpha-NeuAc-(2–6) linkages. Carbohydr Res. 1991;221:209–17.

Lemoine J, Fournet B, Despeyroux D, Jennings KR, Rosenberg R, de Hoffmann E. Collision-induced dissociation of alkali metal cationized and permethylated oligosaccharides: Influence of the collision energy and of the collision gas for the assignment of linkage position. J Am Soc Mass Spectrom. 1993;4(3):197–203.

Levery SB. Glycosphingolipid structural analysis and glycosphingolipidomics. Methods Enzymol. 2005;405:300–69.

Leymarie N, Zaia J. Effective use of mass spectrometry for glycan and glycopeptide structural analysis. Anal Chem. 2012;84(7):3040–8.

Liedtke S, Geyer H, Wuhrer M, Geyer R, Frank G, Gerardy-Schahn R, Zahringer U, Schachner M. Characterization of N-glycans from mouse brain neural cell adhesion molecule. Glycobiology. 2001;11(5):373–84.

Lin Z, Lubman DM. Permethylated N-glycan analysis with mass spectrometry. Methods Mol Biol. 2013;1007:289–300.

Lin CW, Chen JM, Wang YM, Wu SW, Tsai IH, Khoo KH. Terminal disialylated multiantennary complex-type N-glycans carried on acutobin define the glycosylation characteristics of the Deinagkistrodon acutus venom. Glycobiology. 2011;21(4):530–42.

Marino K, Bones J, Kattla JJ, Rudd PM. A systematic approach to protein glycosylation analysis: a path through the maze. Nat Chem Biol. 2010;6(10):713–23.
Maxwell E, Tan Y, Tan Y, Hu H, Benson G, Aizikov K, Conley S, Staples GO, Slysz GW, Smith RD, Zaia J. GlycReSoft: a software package for automated recognition of glycans from LC/MS data. PLoS One. 2012;7(9):e45474.
Mayampurath AM, Wu Y, Segu ZM, Mechref Y, Tang H. Improving confidence in detection and characterization of protein N-glycosylation sites and microheterogeneity. Rapid Commun Mass Spectrom. 2011;25(14):2007–19.
McDonald CA, Yang JY, Marathe V, Yen TY, Macher BA. Combining results from lectin affinity chromatography and glycocapture approaches substantially improves the coverage of the glycoproteome. Mol Cell Proteomics. 2009;8(2):287–301.
McEver RP. A sulfated address for lymphocyte homing. Nat Immunol. 2005;6(11):1067–9.
McGuckin MA, Linden SK, Sutton P, Florin TH. Mucin dynamics and enteric pathogens. Nat Rev Microbiol. 2011;9(4):265–78.
Mechref Y, Novotny MV, Krishnan C. Structural characterization of oligosaccharides using MALDI-TOF/TOF tandem mass spectrometry. Anal Chem. 2003;75(18):4895–903.
Mechref Y, Hu Y, Desantos-Garcia JL, Hussein A, Tang H. Quantitative glycomics strategies. Mol Cell Proteomics. 2013;12(4):874–84.
Meisen I, Mormann M, Muthing J. Thin-layer chromatography, overlay technique and mass spectrometry: a versatile triad advancing glycosphingolipidomics. Biochim Biophys Acta. 2011;1811(11):875–96.
Merkle RK, Cummings RD. Relationship of the terminal sequences to the length of poly-N-acetyllactosamine chains in asparagine-linked oligosaccharides from the mouse lymphoma cell line BW5147. Immobilized tomato lectin interacts with high affinity with glycopeptides containing long poly-N-acetyllactosamine chains. J Biol Chem. 1987;262(17):8179–89.
Michalski A, Damoc E, Lange O, Denisov E, Nolting D, Muller M, Viner R, Schwartz J, Remes P, Belford M, Dunyach JJ, Cox J, Horning S, Mann M, Makarov A. Ultra high resolution linear ion trap Orbitrap mass spectrometer (Orbitrap Elite) facilitates top down LC MS/MS and versatile peptide fragmentation modes. Mol Cell Proteomics. 2012;11(3):O111.013698.
Miller E, Fiete D, Blake NM, Beranek M, Oates EL, Mi Y, Roseman DS, Baenziger JU. A necessary and sufficient determinant for protein-selective glycosylation in vivo. J Biol Chem. 2008;283(4):1985–91.
Morelle W, Slomianny MC, Diemer H, Schaeffer C, van Dorsselaer A, Michalski JC. Fragmentation characteristics of permethylated oligosaccharides using a matrix-assisted laser desorption/ionization two-stage time-of-flight (TOF/TOF) tandem mass spectrometer. Rapid Commun Mass Spectrom. 2004;18(22):2637–49.
Moremen KW, Tiemeyer M, Nairn AV. Vertebrate protein glycosylation: diversity, synthesis and function. Nat Rev Mol Cell Biol. 2012;13(7):448–62.
Morita I, Kizuka Y, Kakuda S, Oka S. Expression and function of the HNK-1 carbohydrate. J Biochem. 2008;143(6):719–24.
Nahnsen S, Bielow C, Reinert K, Kohlbacher O. Tools for label-free peptide quantification. Mol Cell Proteomics. 2013;12(3):549–56.
North SJ, Huang HH, Sundaram S, Jang-Lee J, Etienne AT, Trollope A, Chalabi S, Dell A, Stanley P, Haslam SM. Glycomics profiling of Chinese hamster ovary cell glycosylation mutants reveals N-glycans of a novel size and complexity. J Biol Chem. 2010a;285(8):5759–75.
North SJ, Jang-Lee J, Harrison R, Canis K, Ismail MN, Trollope A, Antonopoulos A, Pang PC, Grassi P, Al-Chalabi S, Etienne AT, Dell A, Haslam SM. Mass spectrometric analysis of mutant mice. Methods Enzymol. 2010b;478:27–77.
Novotny MV, Alley Jr WR, Mann BF. Analytical glycobiology at high sensitivity: current approaches and directions. Glycoconj J. 2013;30(2):89–117.
Oates JE, Dell A, Fukuda M, Fukuda MN. A rapid mass-spectrometric procedure for probing the non-reducing structures of lactosaminoglycan-containing glycoconjugates. Carbohydr Res. 1985;141(1):149–52.

Olivova P, Chen W, Chakraborty AB, Gebler JC. Determination of N-glycosylation sites and site heterogeneity in a monoclonal antibody by electrospray quadrupole ion-mobility time-of-flight mass spectrometry. Rapid Commun Mass Spectrom. 2008;22(1):29–40.

Olsen JV, Macek B, Lange O, Makarov A, Horning S, Mann M. Higher-energy C-trap dissociation for peptide modification analysis. Nat Methods. 2007;4(9):709–12.

Orlando R. Quantitative analysis of glycoprotein glycans. Methods Mol Biol. 2013;951:197–215.

Orlando R, Lim JM, Atwood 3rd JA, Angel PM, Fang M, Aoki K, Alvarez-Manilla G, Moremen KW, York WS, Tiemeyer M, Pierce M, Dalton S, Wells L. IDAWG: Metabolic incorporation of stable isotope labels for quantitative glycomics of cultured cells. J Proteome Res. 2009;8(8):3816–23.

Ozohanics O, Krenyacz J, Ludanyi K, Pollreisz F, Vekey K, Drahos L. GlycoMiner: a new software tool to elucidate glycopeptide composition. Rapid Commun Mass Spectrom. 2008;22(20):3245–54.

Palmisano G, Melo-Braga MN, Engholm-Keller K, Parker BL, Larsen MR. Chemical deamidation: a common pitfall in large-scale N-linked glycoproteomic mass spectrometry-based analyses. J Proteome Res. 2012;11(3):1949–57.

Pan S, Chen R, Aebersold R, Brentnall TA. Mass spectrometry based glycoproteomics—from a proteomics perspective. Mol Cell Proteomics. 2011;10(1):R110.003251.

Patnode ML, Cheng CW, Chou CC, Singer MS, Elin MS, Uchimura K, Crocker PR, Khoo KH, Rosen SD. Galactose 6-o-sulfotransferases are not required for the generation of siglec-f ligands in leukocytes or lung tissue. J Biol Chem. 2013a;288(37):26533–45.

Patnode ML, Yu SY, Cheng CW, Ho MY, Tegesjo L, Sakuma K, Uchimura K, Khoo KH, Kannagi R, Rosen SD. KSGal6ST generates galactose-6-O-sulfate in high endothelial venules but does not contribute to L-selectin-dependent lymphocyte homing. Glycobiology. 2013b;23(3):381–94.

Patrie SM, Roth MJ, Kohler JJ. Introduction to glycosylation and mass spectrometry. Methods Mol Biol. 2013;951:1–17.

Peltoniemi H, Natunen S, Ritamo I, Valmu L, Rabina J. Novel data analysis tool for semiquantitative LC-MS-MS2 profiling of N-glycans. Glycoconj J. 2013;30(2):159–70.

Powlesland AS, Hitchen PG, Parry S, Graham SA, Barrio MM, Elola MT, Mordoh J, Dell A, Drickamer K, Taylor ME. Targeted glycoproteomic identification of cancer cell glycosylation. Glycobiology. 2009;19(8):899–909.

Pringle SD, Giles K, Wildgoose JL, Williams JP, Slade SE, Thalassinos K, Bateman RH, Bowers MT, Scrivens JH. An investigation of the mobility separation of some peptide and protein ions using a new hybrid quadrupole/travelling wave IMS/oa-ToF instrument. Int J Mass Spectrom. 2007;261(1):1–12.

Ranzinger R, Herget S, von der Lieth CW, Frank M. GlycomeDB—a unified database for carbohydrate structures. Nucleic Acids Res. 2011;39(Database issue):D373–6.

Reinhold VN, Reinhold BB, Costello CE. Carbohydrate molecular weight profiling, sequence, linkage, and branching data: ES-MS and CID. Anal Chem. 1995;67(11):1772–84.

Reinhold V, Zhang H, Hanneman A, Ashline D. Toward a platform for comprehensive glycan sequencing. Mol Cell Proteomics. 2013;12(4):866–73.

Ritamo I, Rabina J, Natunen S, Valmu L. Nanoscale reversed-phase liquid chromatography-mass spectrometry of permethylated N-glycans. Anal Bioanal Chem. 2013;405(8):2469–80.

Robbe C, Capon C, Coddeville B, Michalski JC. Structural diversity and specific distribution of O-glycans in normal human mucins along the intestinal tract. Biochem J. 2004;384(Pt 2):307–16.

Rosen SD. Ligands for L-selectin: homing, inflammation, and beyond. Annu Rev Immunol. 2004;22:129–56.

Royle L, Campbell MP, Radcliffe CM, White DM, Harvey DJ, Abrahams JL, Kim YG, Henry GW, Shadick NA, Weinblatt ME, Lee DM, Rudd PM, Dwek RA. HPLC-based analysis of serum N-glycans on a 96-well plate platform with dedicated database software. Anal Biochem. 2008;376(1):1–12.

Ruhaak LR, Deelder AM, Wuhrer M. Oligosaccharide analysis by graphitized carbon liquid chromatography-mass spectrometry. Anal Bioanal Chem. 2009;394(1):163–74.

Ruhaak LR, Zauner G, Huhn C, Bruggink C, Deelder AM, Wuhrer M. Glycan labeling strategies and their use in identification and quantification. Anal Bioanal Chem. 2010;397(8):3457–81.

Ruhaak LR, Huhn C, Koeleman CA, Deelder AM, Wuhrer M. Robust and high-throughput sample preparation for (semi-)quantitative analysis of N-glycosylation profiles from plasma samples. Methods Mol Biol. 2012;893:371–85.

Saba J, Dutta S, Hemenway E, Viner R. Increasing the productivity of glycopeptides analysis by using higher-energy collision dissociation-accurate mass-product-dependent electron transfer dissociation. Int J Proteomics. 2012;2012:560391.

Sato C. Chain length diversity of sialic acids and its biological significance. Trends Glycosci Glycotechnol. 2004;16(91):331–44.

Sato C, Inoue S, Matsuda T, Kitajima K. Development of a highly sensitive chemical method for detecting alpha2→8-linked oligo/polysialic acid residues in glycoproteins blotted on the membrane. Anal Biochem. 1998;261(2):191–7.

Sato C, Inoue S, Matsuda T, Kitajima K. Fluorescent-assisted detection of oligosialyl units in glycoconjugates. Anal Biochem. 1999;266(1):102–9.

Sato C, Fukuoka H, Ohta K, Matsuda T, Koshino R, Kobayashi K, Troy 2nd FA, Kitajima K. Frequent occurrence of pre-existing alpha 2→8-linked disialic and oligosialic acids with chain lengths up to 7 Sia residues in mammalian brain glycoproteins. Prevalence revealed by highly sensitive chemical methods and anti-di-, oligo-, and poly-Sia antibodies specific for defined chain lengths. J Biol Chem. 2000;275(20):15422–31.

Segu ZM, Hussein A, Novotny MV, Mechref Y. Assigning N-glycosylation sites of glycoproteins using LC/MSMS in conjunction with endo-M/exoglycosidase mixture. J Proteome Res. 2010; 9(7):3598–607.

Spina E, Sturiale L, Romeo D, Impallomeni G, Garozzo D, Waidelich D, Glueckmann M. New fragmentation mechanisms in matrix-assisted laser desorption/ionization time-of-flight/time-of-flight tandem mass spectrometry of carbohydrates. Rapid Commun Mass Spectrom. 2004;18(4):392–8.

Spooncer E, Fukuda M, Klock JC, Oates JE, Dell A. Isolation and characterization of polyfucosylated lactosaminoglycan from human granulocytes. J Biol Chem. 1984;259(8):4792–801.

Stavenhagen K, Hinneburg H, Thaysen-Andersen M, Hartmann L, Varon Silva D, Fuchser J, Kaspar S, Rapp E, Seeberger PH, Kolarich D. Quantitative mapping of glycoprotein micro-heterogeneity and macro-heterogeneity: an evaluation of mass spectrometry signal strengths using synthetic peptides and glycopeptides. J Mass Spectrom. 2013;48(6):627–39.

Steentoft C, Vakhrushev SY, Vester-Christensen MB, Schjoldager KT, Kong Y, Bennett EP, Mandel U, Wandall H, Levery SB, Clausen H. Mining the O-glycoproteome using zinc-finger nuclease-glycoengineered SimpleCell lines. Nat Methods. 2011;8(11):977–82.

Steentoft C, Vakhrushev SY, Joshi HJ, Kong Y, Vester-Christensen MB, Schjoldager KT, Lavrsen K, Dabelsteen S, Pedersen NB, Marcos-Silva L, Gupta R, Bennett EP, Mandel U, Brunak S, Wandall HH, Levery SB, Clausen H. Precision mapping of the human O-GalNAc glycoproteome through SimpleCell technology. EMBO J. 2013;32(10):1478–88.

Stephens E, Maslen SL, Green LG, Williams DH. Fragmentation characteristics of neutral N-linked glycans using a MALDI-TOF/TOF tandem mass spectrometer. Anal Chem. 2004;76(8):2343–54.

Sutton-Smith M, Wong NK, Khoo KH, Wu SW, Yu SY, Patankar MS, Easton R, Lattanzio FA, Morris HR, Dell A, Clark GF. Analysis of protein-linked glycosylation in a sperm-somatic cell adhesion system. Glycobiology. 2007;17(6):553–67.

Takahashi N, Kato K. GALXY (glycoanalysis by the three axes of ms and chromatography): a Web application that assists structural analyses of N-glycans. Trends Glycosci Glycotechnol. 2003;15(84):235–51.

Takahashi N, Nakagawa H, Fujikawa K, Kawamura Y, Tomiya N. Three-dimensional elution mapping of pyridylaminated N-linked neutral and sialyl oligosaccharides. Anal Biochem. 1995;226(1):139–46.

Tateno H, Crocker PR, Paulson JC. Mouse Siglec-F and human Siglec-8 are functionally convergent paralogs that are selectively expressed on eosinophils and recognize 6′-sulfo-sialyl Lewis X as a preferred glycan ligand. Glycobiology. 2005;15(11):1125–35.

Thomsson KA, Karlsson NG, Hansson GC. Liquid chromatography-electrospray mass spectrometry as a tool for the analysis of sulfated oligosaccharides from mucin glycoproteins. J Chromatogr A. 1999;854(1–2):131–9.

Thomsson KA, Karlsson H, Hansson GC. Sequencing of sulfated oligosaccharides from mucins by liquid chromatography and electrospray ionization tandem mass spectrometry. Anal Chem. 2000;72(19):4543–9.

Tomiya N, Awaya J, Kurono M, Endo S, Arata Y, Takahashi N. Analyses of N-linked oligosaccharides using a two-dimensional mapping technique. Anal Biochem. 1988;171(1):73–90.

Tousi F, Bones J, Hancock WS, Hincapie M. Differential chemical derivatization integrated with chromatographic separation for analysis of isomeric sialylated N-glycans: a nano-hydrophilic interaction liquid chromatography-MS platform. Anal Chem. 2013;85(17):8421–8.

Townsend RR, Hardy MR, Lee YC. Separation of oligosaccharides using high-performance anion-exchange chromatography with pulsed amperometric detection. Methods Enzymol. 1989;179: 65–76.

Tsai TH, Tadesse MG, Di Poto C, Pannell LK, Mechref Y, Wang Y, Ressom HW. Multi-profile Bayesian alignment model for LC-MS data analysis with integration of internal standards. Bioinformatics. 2013;29(21):2774–80.

Vakhrushev SY, Steentoft C, Vester-Christensen MB, Bennett EP, Clausen H, Levery SB. Enhanced mass spectrometric mapping of the human GalNAc-type O-glycoproteome with SimpleCells. Mol Cell Proteomics. 2013;12(4):932–44.

Viseux N, de Hoffmann E, Domon B. Structural assignment of permethylated oligosaccharide subunits using sequential tandem mass spectrometry. Anal Chem. 1998;70(23):4951–9.

Wada Y, Azadi P, Costello CE, Dell A, Dwek RA, Geyer H, Geyer R, Kakehi K, Karlsson NG, Kato K, Kawasaki N, Khoo KH, Kim S, Kondo A, Lattova E, Mechref Y, Miyoshi E, Nakamura K, Narimatsu H, Novotny MV, Packer NH, Perreault H, Peter-Katalinic J, Pohlentz G, Reinhold VN, Rudd PM, Suzuki A, Taniguchi N. Comparison of the methods for profiling glycoprotein glycans–HUPO Human Disease Glycomics/Proteome Initiative multi-institutional study. Glycobiology. 2007;17(4):411–22.

Wada Y, Dell A, Haslam SM, Tissot B, Canis K, Azadi P, Backstrom M, Costello CE, Hansson GC, Hiki Y, Ishihara M, Ito H, Kakehi K, Karlsson N, Hayes CE, Kato K, Kawasaki N, Khoo KH, Kobayashi K, Kolarich D, Kondo A, Lebrilla C, Nakano M, Narimatsu H, Novak J, Novotny MV, Ohno E, Packer NH, Palaima E, Renfrow MB, Tajiri M, Thomsson KA, Yagi H, Yu SY, Taniguchi N. Comparison of methods for profiling O-glycosylation: Human Proteome Organisation Human Disease Glycomics/Proteome Initiative Multi-Institutional Study of IgA1. Mol Cell Proteomics. 2010;9(4):719–27.

Wang SH, Wu SW, Khoo KH. MS-based glycomic strategies for probing the structural details of polylactosaminoglycan chain on N-glycans and glycoproteomic identification of its protein carriers. Proteomics. 2011;11(14):2812–29.

Wang SH, Tsai CM, Lin KI, Khoo KH. Advanced mass spectrometry and chemical analyses reveal the presence of terminal disialyl motif on mouse B-cell glycoproteins. Glycobiology. 2013;23(6):677–89.

Wheeler SF, Domann P, Harvey DJ. Derivatization of sialic acids for stabilization in matrix-assisted laser desorption/ionization mass spectrometry and concomitant differentiation of alpha(2→3)- and alpha(2→6)-isomers. Rapid Commun Mass Spectrom. 2009;23(2):303–12.

Wollscheid B, Bausch-Fluck D, Henderson C, O'Brien R, Bibel M, Schiess R, Aebersold R, Watts JD. Mass-spectrometric identification and relative quantification of N-linked cell surface glycoproteins. Nat Biotechnol. 2009;27(4):378–86.

Wu AM, Khoo KH, Yu SY, Yang Z, Kannagi R, Watkins WM. Glycomic mapping of pseudomucinous human ovarian cyst glycoproteins: identification of Lewis and sialyl Lewis glycotopes. Proteomics. 2007;7(20):3699–717.

Wu SW, Liang SY, Pu TH, Chang FY, Khoo KH. Sweet-Heart—an integrated suite of enabling computational tools for automated MS2/MS3 sequencing and identification of glycopeptides. J Proteomics. 2013;84:1–16.

Wuhrer M. Glycomics using mass spectrometry. Glycoconj J. 2013;30(1):11–22.

Wuhrer M, Deelder AM. Matrix-assisted laser desorption/ionization in-source decay combined with tandem time-of-flight mass spectrometry of permethylated oligosaccharides: targeted characterization of specific parts of the glycan structure. Rapid Commun Mass Spectrom. 2006;20(6):943–51.

Wuhrer M, Catalina MI, Deelder AM, Hokke CH. Glycoproteomics based on tandem mass spectrometry of glycopeptides. J Chromatogr B Analyt Technol Biomed Life Sci. 2007;849(1–2):115–28.
Wuhrer M, de Boer AR, Deelder AM. Structural glycomics using hydrophilic interaction chromatography (HILIC) with mass spectrometry. Mass Spectrom Rev. 2009a;28(2):192–206.
Wuhrer M, Koeleman CA, Deelder AM. Two-dimensional HPLC separation with reverse-phase-nano-LC-MS/MS for the characterization of glycan pools after labeling with 2-aminobenzamide. Methods Mol Biol. 2009b;534:79–91.
Wuhrer M, Deelder AM, van der Burgt YE. Mass spectrometric glycan rearrangements. Mass Spectrom Rev. 2011;30(4):664–80.
Yagi H, Takahashi N, Yamaguchi Y, Kimura N, Uchimura K, Kannagi R, Kato K. Development of structural analysis of sulfated N-glycans by multidimensional high performance liquid chromatography mapping methods. Glycobiology. 2005;15(10):1051–60.
Yin X, Bern M, Xing Q, Ho J, Viner R, Mayr M. Glycoproteomic analysis of the secretome of human endothelial cells. Mol Cell Proteomics. 2013;12(4):956–78.
Yu SY, Wu SW, Khoo KH. Distinctive characteristics of MALDI-Q/TOF and TOF/TOF tandem mass spectrometry for sequencing of permethylated complex type N-glycans. Glycoconj J. 2006;23(5–6):355–69.
Yu SY, Khoo KH, Yang Z, Herp A, Wu AM. Glycomic mapping of O- and N-linked glycans from major rat sublingual mucin. Glycoconj J. 2008;25(3):199–212.
Yu SY, Wu SW, Hsiao HH, Khoo KH. Enabling techniques and strategic workflow for sulfoglycomics based on mass spectrometry mapping and sequencing of permethylated sulfated glycans. Glycobiology. 2009;19(10):1136–49.
Yu CY, Mayampurath A, Hu Y, Zhou S, Mechref Y, Tang H. Automated annotation and quantification of glycans using liquid chromatography-mass spectrometry. Bioinformatics. 2013a;29(13):1706–7.
Yu CY, Mayampurath A, Tang H. Software tools for glycan profiling. Methods Mol Biol. 2013b;951:269–76.
Yu SY, Chang LY, Cheng CW, Chou CC, Fukuda MN, Khoo KH. Priming mass spectrometry-based sulfoglycomic mapping for identification of terminal sulfated lacdiNAc glycotope. Glycoconj J. 2013c;30(2):183–94.
Zaia J. Mass spectrometry and glycomics. OMICS. 2010;14(4):401–18.
Zaia J. Glycosaminoglycan glycomics using mass spectrometry. Mol Cell Proteomics. 2013;12(4):885–92.
Zauner G, Deelder AM, Wuhrer M. Recent advances in hydrophilic interaction liquid chromatography (HILIC) for structural glycomics. Electrophoresis. 2011;32(24):3456–66.
Zauner G, Kozak RP, Gardner RA, Fernandes DL, Deelder AM, Wuhrer M. Protein O-glycosylation analysis. Biol Chem. 2012;393(8):687–708.
Zauner G, Koeleman CA, Deelder AM, Wuhrer M. Nano-HPLC-MS of glycopeptides obtained after nonspecific proteolysis. Methods Mol Biol. 2013;951:113–27.
Zhang W, Wang H, Zhang L, Yao J, Yang P. Large-scale assignment of N-glycosylation sites using complementary enzymatic deglycosylation. Talanta. 2011;85(1):499–505.
Zhao P, Stalnaker SH, Wells L. Approaches for site mapping and quantification of o-linked glycopeptides. Methods Mol Biol. 2013;951:229–44.
Zhu Z, Hua D, Clark DF, Go EP, Desaire H. GlycoPep Detector: a tool for assigning mass spectrometry data of N-linked glycopeptides on the basis of their electron transfer dissociation spectra. Anal Chem. 2013;85(10):5023–32.
Zielinska DF, Gnad F, Wisniewski JR, Mann M. Precision mapping of an in vivo N-glycoproteome reveals rigid topological and sequence constraints. Cell. 2010;141(5):897–907.
Zielinska DF, Gnad F, Schropp K, Wisniewski JR, Mann M. Mapping N-glycosylation sites across seven evolutionarily distant species reveals a divergent substrate proteome despite a common core machinery. Mol Cell. 2012;46(4):542–8.

Chapter 8
Structural Analysis of Oligosaccharides and Glycoconjugates Using NMR

Yoshiki Yamaguchi, Takumi Yamaguchi, and Koichi Kato

Abstract Carbohydrate chains play critical roles in cellular recognition and subsequent signal transduction in the nervous system. Furthermore, gangliosides are targets for various amyloidogenic proteins associated with neurodegenerative disorders. To better understand the molecular mechanisms underlying these biological phenomena, atomic views are essential to delineate dynamic biomolecular interactions. Nuclear magnetic resonance (NMR) spectroscopy provides powerful tools for studying structures, dynamics, and interactions of biomolecules at the atomic level. This chapter describes the basics of solution NMR techniques and their applications to the analysis of 3D structures and interactions of glycoconjugates in the nervous system.

Keywords Nuclear magnetic resonance • Structure • Dynamics • Interaction • Oligosaccharide • Glycolipid

---

Y. Yamaguchi (✉)
Structural Glycobiology Team, Systems Glycobiology Research Group,
RIKEN-Max Planck Joint Research Center, RIKEN Global Research Cluster,
2-1 Hirosawa, Wako-City, Saitama 351-0198, Japan
e-mail: yyoshiki@riken.jp

T. Yamaguchi
Institute for Molecular Science and Okazaki Institute for Integrative Bioscience,
5-1 Higashiyama, Myodaiji, Okazaki 444-8787, Japan

Graduate School of Pharmaceutical Sciences, Nagoya City University,
3-1 Tababe-dori, Mizuho-ku, Nagoya 467-8603, Japan

K. Kato (✉)
Institute for Molecular Science and Okazaki Institute for Integrative Bioscience,
5-1 Higashiyama, Myodaiji, Okazaki 444-8787, Japan
e-mail: kkato@phar.nagoya-cu.ac.jp

Graduate School of Pharmaceutical Sciences, Nagoya City University,
3-1 Tababe-dori, Mizuho-ku, Nagoya 467-8603, Japan

The Glycoscience Institute, Ochanomizu University,
2-1-1 Ohtsuka, Bunkyo-ku, Tokyo 112-8610, Japan

R.K. Yu and C.-L. Schengrund (eds.), *Glycobiology of the Nervous System*,
Advances in Neurobiology 9, DOI 10.1007/978-1-4939-1154-7_8,

Abbreviations

Aβ	Amyloid β
DIS	Deuterium-induced isotope shift
FID	Free induction decay
HSQC	Heteronuclear single-quantum coherence
MD	Molecular dynamics
NMR	Nuclear magnetic resonance
NOE	Nuclear Overhauser effect
NOESY	NOE spectroscopy
PCS	Pseudocontact shift
PRE	Paramagnetic relaxation enhancement
REMD	Replica exchange MD
RF	Radio frequency
STD	Saturation transfer difference
TRNOE	Transferred NOE
TROSY	Transverse relaxation optimized spectroscopy

8.1 Introduction

In the nervous system, cellular processes, including division, development, migration, and morphological changes, are dynamically controlled through molecular recognition events on cell surfaces. In these bio-organization processes, carbohydrate chains that modify proteins and lipids play critical roles in recognition and adhesion during cell–cell communication. To better understand the molecular mechanisms underlying these neurophysiological functions, atomic views are desirable to describe dynamic interactions of biomolecules such as glycoconjugates. In addition, recent evidence has demonstrated that gangliosides on neuronal cell membranes are targets for various amyloidogenic proteins that are associated with neurodegenerative disorders, e.g., α-synuclein in Parkinson's disease, amyloid β (Aβ) in Alzheimer's disease, and prion proteins in Creutzfeldt–Jakob disease (Taylor and Hooper 2006; Ariga et al. 2008; Fantini and Yahi 2010; Matsuzaki et al. 2010). Atomic descriptions of these pathological processes provide a basis for designing novel therapeutic molecules.

Nuclear magnetic resonance (NMR) spectroscopy is one of the most widely used techniques for atomic visualizations of biomolecules. The unique feature of this method is its ability to determine atomic coordinates of biomacromolecules in solution and those embedded in membranes and to characterize their dynamic motion at the atomic level. NMR spectroscopy also serves as a powerful tool for detailed analyses of functional intermolecular interactions and is now routinely used by a wide range of researchers, including glycobiologists. This chapter provides the basics of biomolecular NMR spectroscopy needed for applying this useful technique to address neuroglycobiological issues.

8.2 Basic NMR Phenomena

NMR is a physical phenomenon that reflects quantum mechanical magnetic properties of atomic nuclei in orientation with a strong magnetic field (Abragam 1961; Levitt 2008). Although all isotopes containing odd numbers of protons and/or neutrons have intrinsic magnetic moments and are therefore NMR active, the most commonly studied nuclei are ^{1}H and ^{13}C, which have spin quantum numbers of 1/2 and therefore exhibit high-resolution NMR spectra. These nuclei have two spin states with energy differences that depend on intrinsic magnetic moments and a given magnetic field. In comparison with ^{1}H-NMR, ^{13}C-NMR measurements suffer from low sensitivity because of the lower natural abundance of this isotope (1.1%) and its smaller magnetic moment. Hence, NMR samples are often enriched with ^{13}C using metabolic labeling and chemical synthesis (Yamaguchi and Kato 2007b; Ohki and Kainosho 2008; Kato et al. 2010; Zhang et al. 2013).

Alignment of the nuclear magnetic moment in the magnetic field is perturbed by an electromagnetic field with a resonant radio frequency (RF) pulse that corresponds with the energy difference. After applying the RF pulse, nonequilibrium magnetization, as the sum of all the individual nuclear magnetic moments in the sample, precesses around the magnetic field with the resonant frequency and produces corresponding voltage oscillations in the detection coil. The duration of the oscillating signal is limited and decays exponentially. This time-domain NMR signal, known as free induction decay (FID), is Fourier transformed to produce a frequency-domain spectrum. Figure 8.1a shows FID and Fourier-transformed ^{1}H-NMR spectrum of the pentasaccharide of ganglioside GM1 dissolved in D_2O. The real part of the complex spectrum is typically displayed as the NMR spectrum, showing the absorptive Lorentzian line shape.

8.3 Chemical Shifts as Structural Probes

Individual protons in molecules are generally surrounded by differing electronic environments, which shield each proton against the magnetic field with various modes. As a result, resonant frequencies differ among protons depending on chemical environments, even under the same magnetic field. Hence, a ^{1}H-NMR spectrum exhibits a number of peaks at different positions, and these chemical shifts are each assigned to distinct protons in the molecule. The same is true for the other NMR-active nuclei.

In practice, chemical shift (δ) is measured in parts per million (ppm) relative to a reference resonance signal from a standard compound:

$$\delta = \left(\nu - \nu_{\text{ref}}\right) / \nu_{\text{ref}} \times 10^{6}$$

In the equation above, ν and ν_{ref} represent the resonance frequencies of sample and reference signals, respectively.

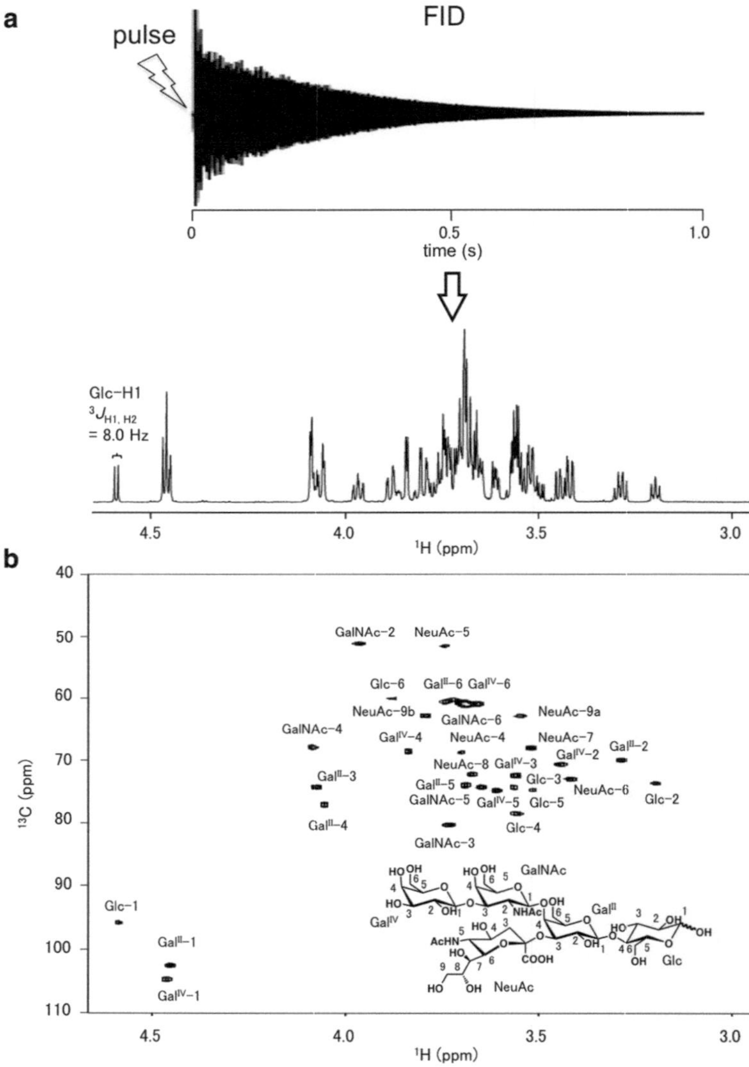

Fig. 8.1 NMR spectral examples of the GM1 pentasaccharide. (**a**) A free induction decay (FID) and Fourier-transformed ^{1}H-NMR spectrum and (**b**) ^{1}H–^{13}C HSQC spectrum of the GM1 pentasaccharide

In carbohydrate NMR spectroscopy, ^{1}H-NMR chemical shifts have traditionally been used as *structural reporters* for identifying chemical structures of oligosaccharides (Vliegenthart 1980). Chemical shifts also offer unique conformational probes for biomolecules. For example, secondary structures of proteins can be determined by inspecting chemical shifts of backbone ^{1}H and ^{13}C atoms (Wishart and Sykes 1994; Cornilescu et al. 1999).

In carbohydrate NMR spectra, signals are typically in very close proximity of one another (Fig. 8.1a). To resolve chemical shift degeneracy, applications of higher magnetic fields are certainly advantageous (Kato et al. 2008).

8.4 Through "BOND" and Through "SPACE" Interactions

Chemical shift values are influenced by surrounding chemical environments. In contrast, scalar coupling (or *J*-coupling) splits NMR peaks and reflects indirect interactions between NMR-active nuclei that are mediated by the electrons participating in chemical bonds between nuclei. Scalar coupling constants (*J*) are defined by magnitudes of peak splitting and are independent of molecular orientations with respect to the magnetic field but depend on molecular geometry. The vicinal scalar coupling constant ^3J, which pertains to atoms separated by three covalent bonds, is typically employed for conformational analyses of biomolecules. This constant is related to the dihedral angle θ as described by the Karplus equation as follows:

$$^3J(\theta) = \mathrm{A}\cos 2\theta + \mathrm{B}\cos\theta + \mathrm{C}$$

In this equation, A, B, and C are coupling coefficients, and θ is the dihedral angle. This relationship has been applied to determinations of sugar-ring stereochemistry and characterization of glycosidic linkage conformations of oligosaccharides (Zhao et al. 2007). In Fig. 8.1a, the anomeric proton (H1) of the β-glucose residue at the reducing terminus of the GM1 pentasaccharide exhibits peak spits that originate from vicinal scalar coupling between H1 and H2 protons ($^3J_{\mathrm{H1,H2}} = 8.0$ Hz).

In many two- and multidimensional NMR experiments, migration of magnetization among correlated nuclei is a crucial process. Magnetization transfer that proceeds from one spin to another is mostly achieved through scalar coupling. Due to increased numbers and line widths of resonances, 2D homonuclear ^{1}H-NMR methods are ineffective for biomolecules with molecular masses >10 kDa. In addition, larger line widths result in decreased sensitivity for ^{1}H correlation experiments that rely on small (<10 Hz) homonuclear ^3J scalar coupling for coherence transfer. The efficiency of magnetization transfer in heteronuclear NMR spectroscopy is improved by employing relatively large one-bond scalar coupling interactions of ^{1}H with ^{15}N or ^{13}C. To measure ^{1}H–^{13}C heteronuclear single-quantum coherence (HSQC) spectra (Bodenhausen and Ruben 1980) (Fig. 8.1b), one-bond scalar coupling $^1J_{\mathrm{C–H}}$ (approximately 145 Hz) is used to transfer magnetization from ^{1}H to ^{13}C and vice versa.

In addition to the through-bond scalar coupling interaction, through-space dipolar interactions are utilized to transfer magnetization between spins. Due to the nuclear Overhauser effect (NOE), perturbation of populations of stationary states within a spin system causes time-dependent changes in the intensity of dipolar-coupled resonance signals, which follows polarization transfers between spin populations via dipolar cross-relaxation. The efficiency of the NOE depends on the

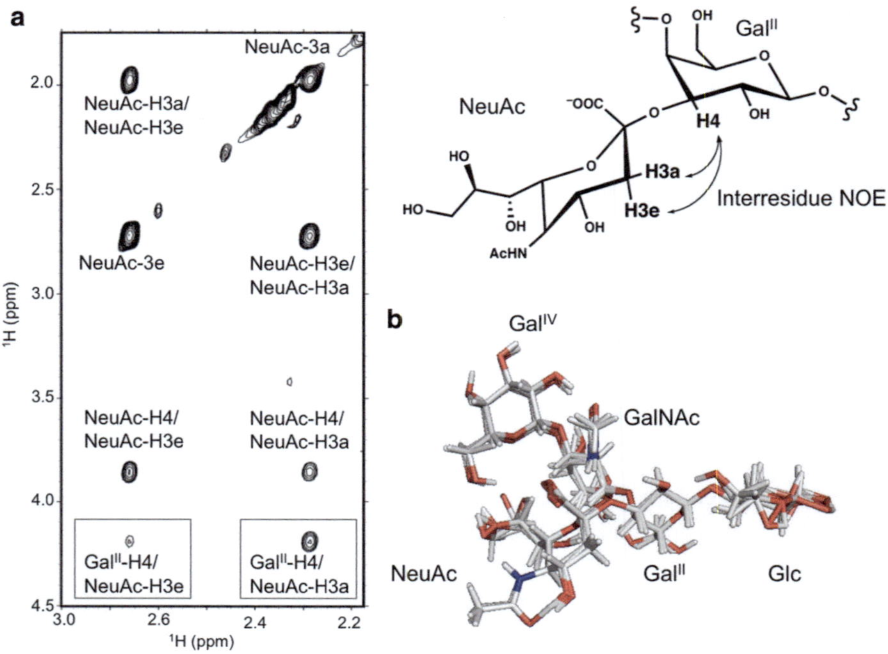

Fig. 8.2 Atomic distance information provided through NOEs. (**a**) A part of ^{1}H–^{1}H NOESY spectrum of the micellar lysoGM1 and (**b**) the lowest-penalty 3D models of the GM1 pentasaccharide calculated from interresidue NOE data. Reprinted from (Yagi-Utsumi et al. 2010) with permission from Elsevier

distance between interacting spins. Thus, through-space rather than through-bond magnetization transfer generates cross-peaks according to the NOE in NMR spectra. ^{1}H–^{1}H NOE spectroscopy (NOESY) provides a measure of interproton distances of up to 5 Å, enabling identification of atomic coordinates of biomacromolecules such as proteins (Wüthrich 1986). The intensity of the NOE (I) is related to the distance (r) between proton pairs, as in

$$I = f\left(\tau_c\right) \times r^{-6},$$

where $f(\tau_c)$ is a function of the rotational correlation time τ_c of the molecule. Figure 8.2a shows a part of the ^{1}H–^{1}H NOESY spectrum of micellar lyso-GM1. In addition to intraresidue NOEs, interresidue NOEs are observed between spatially proximal pairs of protons, as exemplified by the NeuAc H3 (axial)–Gal^{II} H4 proton pair. Interresidual NOE observations are used to identify glycosidic linkages of unknown compounds and estimate dihedral angles along glycosidic bonds (Homans et al. 1987; Cumming and Carver 1987; Voisin et al. 2005). Combined NMR analyses based on through-bond and through-space interactions enable sequence-specific resonance assignments of oligosaccharide NMR signals and subsequently provide 3D structural information (Acquotti et al. 1990; Brocca et al. 2000; Prestegard et al. 1982; Yagi-Utsumi et al. 2010; Yu et al. 1986) (Fig. 8.2b).

8.5 Relaxation and Molecular Motion

In principle, NMR experiments begin from the equilibrium state, in which all populations of energy levels of the system are described by the Boltzmann distribution. Although multiple pulses and multidimensional NMR techniques permit generation of nonequilibrium states, the equilibrium state is eventually restored. As in other spectroscopic techniques, recovery from the nonequilibrium state to equilibrium is called relaxation. Relaxation in NMR involves recovery of the nuclear spin magnetization component with an orientation that is parallel to the static magnetic field (called spin–lattice relaxation) and/or loss of phase coherence of individual nuclear spins (called spin–spin relaxation). Time constants of these two processes are termed T_1 and T_2, respectively. In solution NMR spectroscopy, relaxation is governed by the dynamic properties of molecules, including overall molecular tumbling and internal motions. For example, T_2 determines natural line widths of resonances detected during the acquisition period. In the comparison of ^{1}H-NMR spectra for micellar ganglioside GM1 and the free oligosaccharide derived from it shown in Fig. 8.3, molecules with slower tumbling rates exhibit broader signal line widths, which are inversely proportional to T_2. Thus, T_2 is a critical factor for detecting NMR peaks with higher signal-to-noise ratios. A sophisticated pulse sequence (transverse relaxation optimized spectroscopy (TROSY)) has been developed for

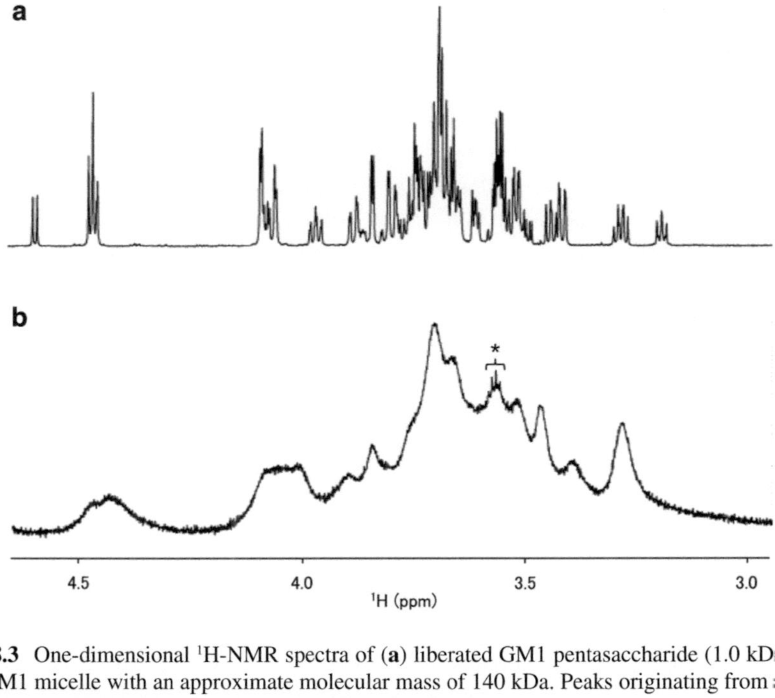

Fig. 8.3 One-dimensional ^{1}H-NMR spectra of (**a**) liberated GM1 pentasaccharide (1.0 kDa) and (**b**) GM1 micelle with an approximate molecular mass of 140 kDa. Peaks originating from a low-molecular-weight contaminant are indicated by an asterisk

NMR analyses of larger macromolecular complexes, in combination with optimal (generally higher) magnetic fields and sample deuteration for suppressing the magnetic dipole–dipole interactions caused by protons that are strong sources of relaxation (Pervushin et al. 1997).

8.6 Paramagnetic Effects as Sources of Long-Distance Information

Unpaired electrons can dramatically perturb NMR spectra, due to stronger dipole–dipole interactions with nuclei that have much larger magnetic moments than atomic nuclei. For example, through-space interactions between a paramagnetic center and neighboring protons cause increased relaxation rates with r^{-6} dependence between paramagnetic spin-proton distances (Solomon 1955). Such paramagnetic relaxation enhancement (PRE) offers long-distance information. Paramagnetic probes such as nitroxide radicals are used to characterize oligosaccharide conformations and lectin–carbohydrate interactions (Johnson et al. 1999; Jain et al. 2001; Yamaguchi et al. 2013a).

Chemical shifts can also be modulated in the presence of paramagnetic lanthanide ions (such as Er^{3+} and Tm^{3+}). This perturbation is known as pseudocontact shift (PCS), which occurs when the magnetic susceptibility of metal ion is anisotropic. PCS is exploited to determine geometrical arrangements of individual nuclei in relation to the position of the metal ion with r^{-3} dependence (McConnell and Robertson 1958; Kurland and McGarvey 1970). Therefore, the atomic long-distance information for determining biomacromolecular conformations can be obtained by observing PCS following introduction of lanthanide probes into specific target molecule sites. Several NMR studies have used PCS to restrain protein and oligosaccharide conformations (Bertini et al. 2005; Otting 2010; Zhang et al. 2013). Recently, PCS of the GM3 trisaccharide (NeuAcα2-3Galβ1-4Glc) and the GM2 tetrasaccharide (GalNAcβ1-4(NeuAcα2-3) Galβ1-4Glc) were analyzed by attaching a lanthanide-chelating tag to reducing ends (Yamamoto et al. 2012; Zhang et al. 2012). Using two-dimensional ^{1}H–^{13}C HSQC spectra, PCS values were measured as differences between ^{1}H and ^{13}C chemical shifts and those of diamagnetic compounds (Fig. 8.4). These analyses provide conformational information related to oligosaccharides. However, in general, oligosaccharide conformations dynamically fluctuate in solution. Therefore, observed PCS should be interpreted as averages of the dynamic conformational ensemble (*vide infra*).

8.7 Chemical Exchange: Dynamic Aspects in NMR

NMR spectroscopy provides unique information on the *exchange* of nuclei between different environments due to conformational transitions and/or intermolecular interactions (Lian and Roberts 1993). Suppose that a given nucleus exchanges with

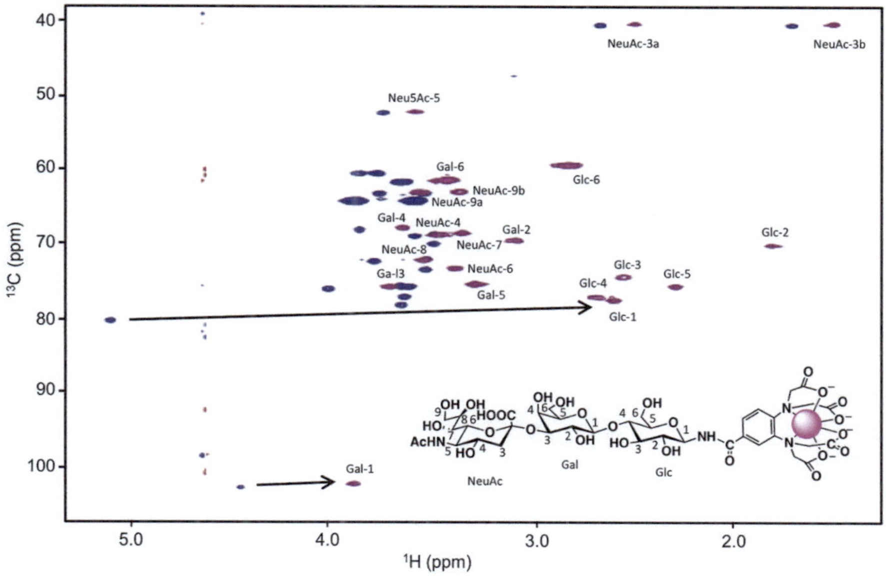

Fig. 8.4 Comparison of ^{1}H–^{13}C HSQC spectra of the GM3 trisaccharide with lanthanide-chelating tag complexed with paramagnetic Tm^{3+} (*red*) and diamagnetic La^{3+} (*blue*). Reprinted from (Yamamoto et al. 2012) with permission from The Royal Society of Chemistry

rate constant k between two sites with resonance frequencies that differ by $\Delta\omega$ (Fig. 8.5a). If k is slow in terms of the frequency of chemical shift differences ($k \ll \Delta\omega$), then two distinct signals corresponding to the nuclei of two sites are observed. In contrast, if the exchange rate is fast ($k \gg \Delta\omega$), then a single resonance is observed, which reflects the population-weighted average chemical shift. Because conformational transitions of free oligosaccharides occur in nanosecond time range (Yamamoto et al. 2012), the observed PCS shown in Fig. 8.4 are analyzed in the fast exchange regime. If the exchange rate is of the order of the chemical shift difference between two sites, the lines become considerably broad and coalesce at $k \sim \Delta\omega$. This is known as the intermediate exchange regime or coalescence, where k and $\Delta\omega$ as well as populations of individual states can be estimated using sophisticated relaxation dispersion experiments (Loria et al. 1999; Mittermaier and Kay 2006; Sugase et al. 2007).

Biomolecules possess exchangeable protons, such as those in hydroxyl and amide groups. Proton exchange rates of these groups provide useful probes for characterizing conformational fluctuations and interactions of biomolecules such as oligosaccharides, because slower exchange rates indicate protecting factors such as hydrogen-bonding interactions at corresponding sites (Englander and Mayne 1992; Englander et al. 2007). Exchanges between protein amide protons and water occur in the slow exchange regime. The rates of these processes can be measured in several ways depending on the rate of exchange. When the exchange rate is comparable to or faster than the spin–lattice relaxation rate (typically, $k_{ex} > 0.1\ s^{-1}$), the rate constant

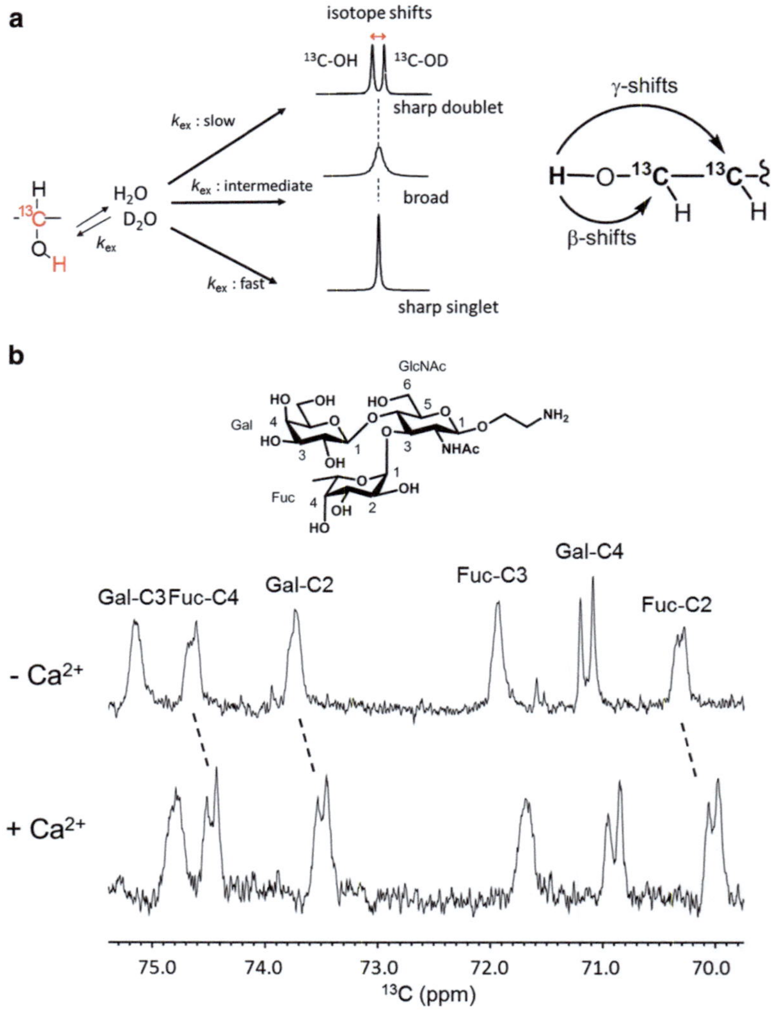

Fig. 8.5 ^{13}C-NMR isotope shifts for analyzing proton exchange rates of sugar hydroxyl groups. (**a**) Schematic of NMR experiment using isotope shifts. All hydroxyl protons on glycans rapidly exchange to deuterons in $H_2O/D_2O = 50{:}50$ solution. The ^{13}C-NMR signal of the geminal carbons at the H/D exchanging hydroxyl protons shows characteristic signal shape, which is dependent on the H/D exchanging rate. Especially under significantly slow exchange conditions, a set of sharp doublet is provided due to isotope shifts (β-shifts). The chemical shifts difference is ~0.15 ppm. k_{ex}; exchanging rates of protons. (**b**) Parts of ^{13}C-NMR spectra of 40 mM Lewis X trisaccharide without 1.0 M $CaCl_2$ (*upper*) and with 1.0 M $CaCl_2$ (*lower*) at 5 °C. Adapted with modifications from (Hanashima et al. 2011) with permission from The Royal Society of Chemistry

is most easily determined in saturation transfer experiments (Forsén and Hoffman 1963, 1964). In these experiments, saturation is performed by selectively irradiating the frequency of the water signal, and the exchange of amide protons with saturated water protons is quantitatively assessed according to NMR signal intensities (Spera et al. 1991). For slower rates (k_{ex} < 0.01 s^{-1}), exchanges are measured by observing progressive changes of NMR spectra, which exhibit time-dependent attenuation of peak intensities after rapidly transferring proteins from H_2O into D_2O (Jeng et al. 1990; Paterson et al. 1990).

Deuterium exchanges of rapidly exchanging protons, such as hydroxyl protons of oligosaccharides, cannot be quantitatively characterized using conventional H_2O/D_2O exchange monitoring. To overcome this difficulty in aqueous solution, an NMR strategy has been developed using deuterium-induced isotope shifts (DIS; Hanashima et al. 2011). This method provided detailed characterization of Ca^{2+}-dependent homophilic interactions of Lewis X trisaccharides (Fucα1-3(Galβ1-4) GlcNAc), which have been implicated as having a role in mediating compaction of the mouse embryo at the morula stage (Fenderson et al. 1984; Eggens et al. 1989). ^{13}C-NMR of this trisaccharide was measured in a 50:50 mixture of H_2O/D_2O (Fig. 8.5). DISs were observed to be dependent on Ca^{2+} concentrations. Sample conditions of 1.0 M Ca^{2+} provided doublets originating from Fuc-C2, Fuc-C4, and Gal-C2. In contrast, under Ca^{2+}-free conditions, no doublets originated from these carbon atoms, indicating that proton exchange became significantly slower upon Ca^{2+} coordination.

8.8 NMR Tools for Intermolecular Interaction Analysis

8.8.1 Oligosaccharide–Protein Interactions

Analysis of sugar–protein interactions is an important step for elucidating structure–function relationships of glycans and designing drugs that target carbohydrate recognition systems (Kamiya et al. 2011). NMR spectroscopy provides invaluable tools for this purpose because it allows detection of weak sugar–protein interactions (K_d ~ mM), identification of glycotopes that are recognized by proteins, characterization of protein-bound oligosaccharide conformations, and determination of the modes of atomic interaction between oligosaccharides and proteins in solution (Fig. 8.6) (Yamaguchi and Kato 2007a).

Saturation transfer difference (STD) NMR is now frequently used to analyze protein–ligand interactions (Mayer and Meyer 1999, 2001). One of the great advantages of this method is that it does not require expensive time-consuming stable-isotope labeling of either proteins or ligands. STD-NMR requires the alternate collection of an on-resonance spectrum for saturation of protein protons and an off-resonance spectrum, for reference. Upon irradiation of the protein with a saturation pulse, the saturation effect immediately spreads from irradiated points over the entire protein–ligand complex (Fig. 8.7). If ligand exchange between free and bound states is fast in terms of the time scale of spin–lattice relaxation, the saturation effect is readily transferred to the free fraction of the ligand, particularly to

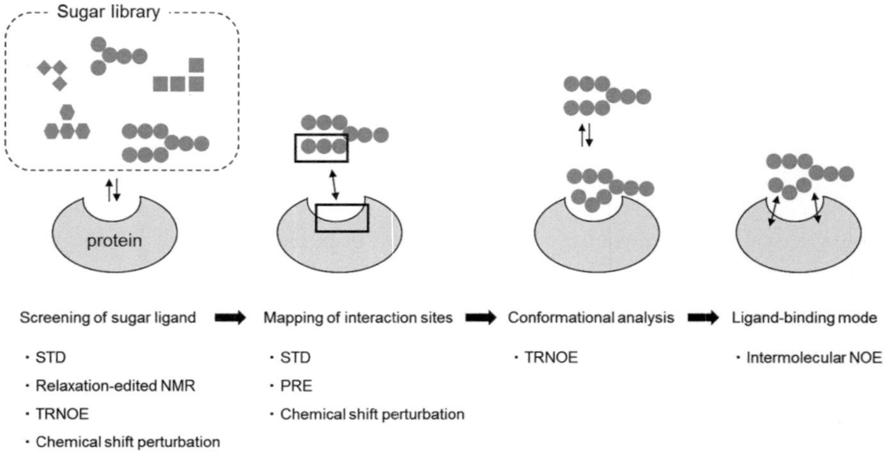

Fig. 8.6 A strategy for analyzing carbohydrate–protein interactions. Adapted with modifications from (Yamaguchi and Kato 2007a) with permission from Yodosha Co., Ltd.

ligand protons located at the interaction interface. An example of STD-NMR is shown in Fig. 8.7, in which the interaction between trisialic acid and a specific monoclonal antibody is analyzed (Hanashima et al. 2013). The nonreducing terminus residue (c) and the central residue (b) showed higher relative values of STD amplification factor compared with the reducing terminal residue (a). The protons at C4, C6, and C7 on residues b and c also had higher values than those at C3, C5, and C8, indicating that this antibody preferentially binds to the α-face of pyranose rings at residues b and c.

Atomic contacts can be identified by observing intermolecular NOE correlations between proteins and cognate ligands. Conformations of protein-bound ligands can also be determined by analyzing intramolecular NOE connectivities within ligands. The sign of the NOE signal depends on τ_c of the molecular tumbling motion and becomes opposite when a fast-tumbling low-molecular-weight ligand binds to a slow-tumbling protein. Under conditions of excess ligand, if the ligand undergoes chemical exchange between the free and bound state more rapidly than longitudinal relaxation, intramolecular NOE connectivities reflecting the protein-bound state can be observed even for peaks exhibiting chemical shifts of free ligand (Clore and Gronenborn 1982, 1983; Glaudemans et al. 1990; Ni and Scheraga 1994). This type of NOE is referred to as transferred NOE (TRNOE). Figure 8.8 displays TRNOE data that characterize conformations of a trimannosyl ligand bound to the carbohydrate recognition domain of VIP36, an animal lectin involved in vesicular transport of glycoproteins between the endoplasmic reticulum and the Golgi (Yamaguchi and Kato 2008). TRNOE can also determine atomic contacts between proteins and carbohydrate ligands (Satoh et al. 2010).

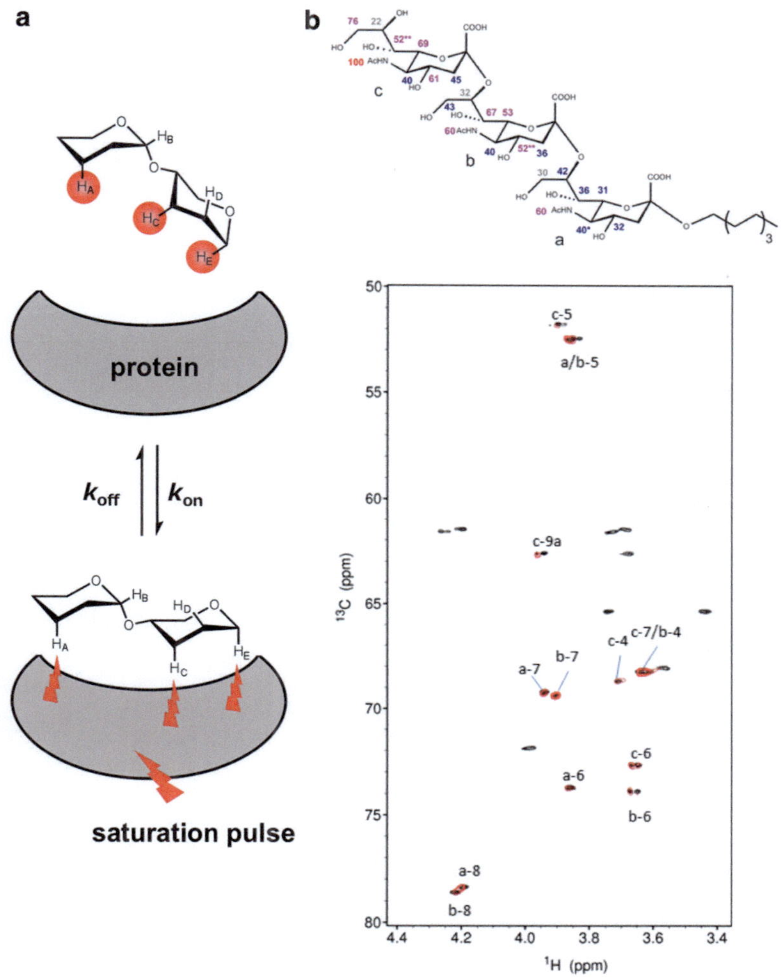

Fig. 8.7 STD experiment. (**a**) Schematic of saturation transfer used for characterizing a carbohydrate–protein interaction; in this experiment, saturation transfer from protein to ligand is observed, thereby identifying the protons involved in the interaction, i.e., H_A, H_C, and H_E; (**b**) Overlay of 2D ^1H–^{13}C STD-HSQC spectra of octyl-(NeuAc)$_3$ (50 equiv) with 20 μM anti-oligo/polysialic acid IgM antibody 12E3 (*red*) and 2D ^1H–^{13}C HSQC spectrum (*black*) in PBS with 99 % D_2O; Protein signal at 7 ppm was irradiated for saturation. Adapted with modifications from (Hanashima et al. 2013) with permission from Elsevier

8.8.2 *Protein Binding to Glycolipid Clusters*

In cell membranes, glycolipids such as gangliosides form clusters and play important roles in various biomolecular recognition events (Hakomori 2004). Because of their dynamic properties, crystallographic structural analyses of these glycolipid clusters

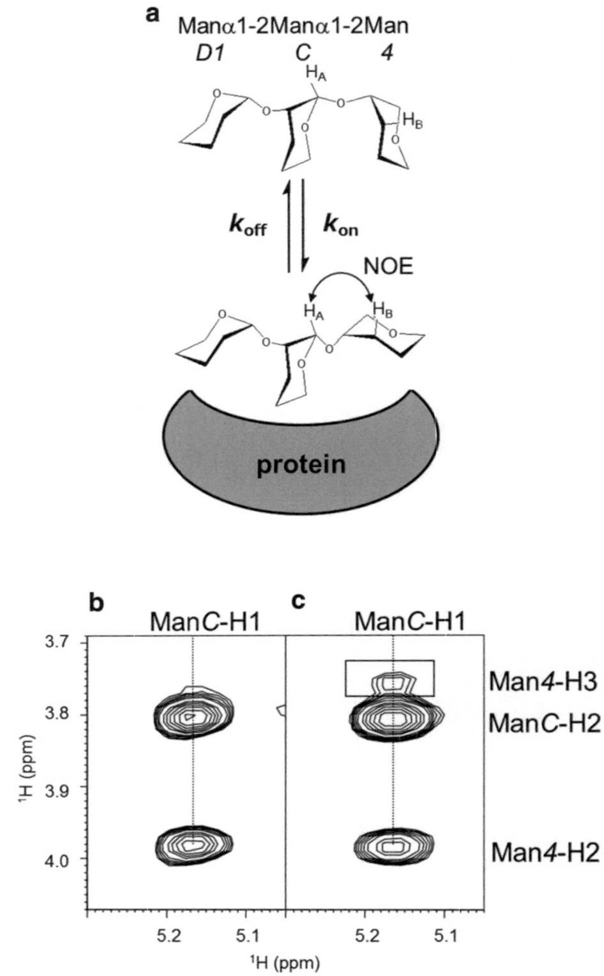

Fig. 8.8 TRNOE experiment. (**a**) Schematic of TRNOE used for characterizing carbohydrate–protein interactions; in this experiment, a spatially proximal proton pair, H_A and H_B, in the protein-bound state provides an NOE peak corresponding to chemical shifts of the unbound state under conditions of excess ligand. A part of NOESY spectrum of Manα1-3Manα1-3Man in (**b**) the absence and (**c**) the presence of the carbohydrate recognition domain of VIP36. An interglycosidic NOE between Man*C*-H1 and Man*4*-H3 (*boxed*) is observed only in the presence of lectin, indicating that these protons are spatially in close proximity of each other in the lectin-bound state of the trimannose. Reprinted from (Yamaguchi and Kato 2008) with permission from Springer

are difficult. In contrast, NMR techniques provide detailed structural analyses of such dynamic clusters of glycolipids and their specific interactions with proteins.

Aβ has been reported to interact with GM1 gangliosides in Alzheimer's disease patients, thereby undergoing conformational transitions that result in pathogenic assemblies (Matsuzaki et al. 2010). To determine the interaction modes of Aβ with ganglioside clusters, NMR experiments were conducted using deuterated ^{15}N-labeled Aβ(1–40) and aqueous gangliosidic micelles (Utsumi et al. 2009). Analyses of backbone chemical shift data of Aβ(1–40) indicated that this peptide forms discontinuous α-helices upon binding to GM1 micelles. The saturation transfer data demonstrate that Aβ(1–40) lies on the hydrophobic/hydrophilic interface of the ganglioside clusters, exhibiting an up-and-down topological mode in which the two α-helices (His14-Val24 and Ile31-Val36) and the C-terminal dipeptide are in contact with the hydrophobic interior (Fig. 8.9).

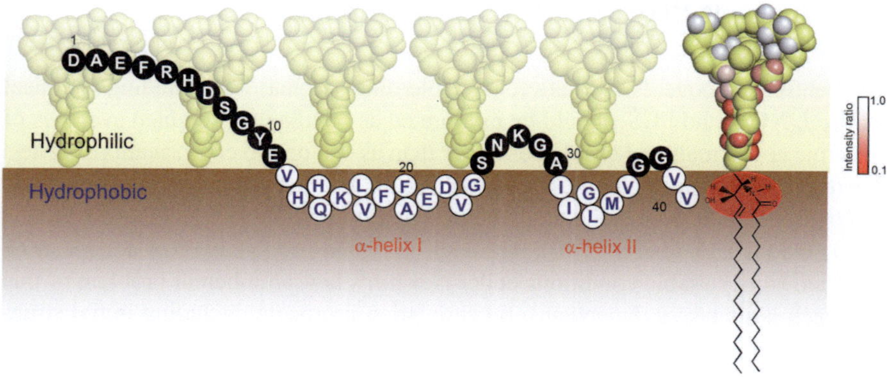

Fig. 8.9 A topological model of Aβ(1–40) bound to a ganglioside cluster, as deduced from NMR experiments; the regions of Aβ(1–40) buried inside the hydrophobic interior of lyso-GM1 micelles and those exposed to hydrophilic environments were identified using saturation transfer experiments. Moreover, the PRE effect was used to identify atomic groups of lyso-GM1 that are proximal to the spin-labeled Aβ peptide. The observed PRE effects are mapped on the 3D model of the carbohydrate moiety of lyso-GM1 with a color gradient from *red* to *white*. Adapted with modifications from (Utsumi et al. 2009; Yagi-Utsumi et al. 2010) with permission from Springer and Elsevier, respectively

The Aβ–glycolipid interaction was also characterized by PRE (Yagi-Utsumi et al. 2010). The Aβ(1–40) peptide with an extra C-terminal cysteine residue was recombinantly produced and conjugated through a disulfide bond with a spin-labeled probe. On addition of the spin-labeled Aβ peptide to the solution containing micellar lyso-GM1, the ^{1}H–^{13}C HSQC peaks originating from Glc and Gal^{II} and those originating from the head group of the lyso-GM1 lipid moiety exhibited significant attenuation of intensity due to PRE line broadening (Fig. 8.9). These results indicate that the sugar–lipid interface was primarily perturbed upon interactions of Aβ with the micelles.

Ganglioside micelle assemblies vary in size and curvature, depending on the size of the carbohydrate moiety. Such morphological variability can be a determining factor for ganglioside–protein interactions. Hence, for structural characterization of biomolecular interactions of glycolipid clusters, it is crucial to design appropriate membrane models that are suitable for sophisticated high-resolution spectral measurements. Small bicelles, in which a series of gangliosides were successfully embedded, have been developed as mimics of ganglioside-containing membranes for detailed NMR studies (Yamaguchi et al. 2013b). Using these standardized bicelles, chemical shift perturbation and relaxation data clearly indicated the ganglioside-specific involvement of N-terminal regions of α-synuclein in membrane interactions (Yamaguchi et al. 2013b).

8.9 Liaisons Between NMR and Computation

As mentioned above, NMR data for flexible oligosaccharides, including chemical shifts, *J*, NOE, and PCS, should be interpreted as population-weighted averages of dynamic conformational ensembles rather than as one or two conformational states. Therefore, quantitative interpretations of NMR data are supported by theoretical calculations such as molecular dynamics (MD) simulations (Fadda and Woods 2010; Woods and Tessier 2010). Although motional properties of systems can be obtained from suitable conditions under Newton's law, an inherent problem of this approach is its heavy dependence on simulation protocols, including initial structures, computational times, and force fields. It is therefore important to validate simulations by comparing with experimental observations. For example, the PCS-assisted NMR method has been successfully used to validate oligosaccharide conformational spaces sampled by MD simulations (Yamamoto et al. 2012; Zhang et al. 2012). Replica exchange MD simulations (REMD) can enhance sampling using a parallel tempering technique (Sugita and Okamoto 1999). This method overcomes the multiple-minima problem by exchanging noninteracting replicas of the system at several temperatures. REMD simulations were recently applied to biantennary *N*-glycan and were consistent with both experimental NMR data (Re et al. 2011; Nishima et al. 2012) and with the collisional cross sections determined using ion mobility spectrometry (Yamaguchi et al. 2012). Thus, the combination of NMR spectroscopy and theoretical approaches promises atomic descriptions of dynamic conformations and interactions of glycoconjugates of neurophysiological and neuropathological interest.

Acknowledgments This study was partly supported by JSPS/MEXT KAKENHI Grant-in-Aid for Scientific Research on Innovation Areas (20107004 and 25102008), Scientific Research (A) (24249002), Scientific Research (C) (25460054), Challenging Exploratory, Research (26560451), and Young Scientists (B) (24750170).

Compliance with Ethics Requirements The authors declare that they have no conflict of interest and that they have used no human subjects in work cited that was done in their laboratory.

References

Abragam A. The principles of nuclear magnetism. The international series of monographs on physics. Oxford: Clarendon; 1961.

Acquotti D, Poppe L, Dabrowski J, von der Lieth CW, Sonnino S, Tettamanti G. Three-dimensional structure of the oligosaccharide chain of GM1 ganglioside revealed by a distance-mapping procedure: a rotating and laboratory frame nuclear overhauser enhancement investigation of native glycolipid in dimethyl sulfoxide and in water- dodecylphosphocholine solutions. J Am Chem Soc. 1990;112(21):7772–8.

Ariga T, McDonald MP, Yu RK. Role of ganglioside metabolism in the pathogenesis of Alzheimer's disease – a review. J Lipid Res. 2008;49(6):1157–75.

Bertini I, Luchinat C, Parigi G, Pierattelli R. NMR spectroscopy of paramagnetic metalloproteins. ChemBioChem. 2005;6(9):1536–49.

Bodenhausen G, Ruben DJ. Natural abundance nitrogen-15 NMR by enhanced heteronuclear spectroscopy. Chem Phys Lett. 1980;69(1):185–9.

Brocca P, Bernardi A, Raimondi L, Sonnino S. Modeling ganglioside headgroups by conformational analysis and molecular dynamics. Glycoconj J. 2000;17(5):283–99.

Clore GM, Gronenborn AM. Theory and applications of the transferred nuclear Overhauser effect to the study of the conformations of small ligands bound to proteins. J Magn Reson. 1982;48(3): 402–17.

Clore GM, Gronenborn AM. Theory of the time-dependent transferred nuclear Overhauser effect – applications to structural-analysis of ligand protein complexes in solution. J Magn Reson. 1983;53(3):423–42.

Cornilescu G, Delaglio F, Bax A. Protein backbone angle restraints from searching a database for chemical shift and sequence homology. J Biomol NMR. 1999;13(3):289–302.

Cumming DA, Carver JP. Virtual and solution conformations of oligosaccharides. Biochemistry. 1987;26(21):6664–76.

Eggens I, Fenderson B, Toyokuni T, Dean B, Stroud M, Hakomori S. Specific interaction between Le^x and Le^x determinants. A possible basis for cell recognition in preimplantation embryos and in embryonal carcinoma cells. J Biol Chem. 1989;264(16):9476–84.

Englander SW, Mayne L. Protein folding studied using hydrogen-exchange labeling and two-dimensional NMR. Annu Rev Biophys Biomol Struct. 1992;21:243–65.

Englander SW, Mayne L, Krishna MM. Protein folding and misfolding: mechanism and principles. Q Rev Biophys. 2007;40(4):287–326.

Fadda E, Woods RJ. Molecular simulations of carbohydrates and protein–carbohydrate interactions: motivation, issues and prospects. Drug Discov Today. 2010;15(15–16):596–609.

Fantini J, Yahi N. Molecular insights into amyloid regulation by membrane cholesterol and sphingolipids: common mechanisms in neurodegenerative diseases. Expert Rev Mol Med. 2010;12:e27.

Fenderson BA, Zehavi U, Hakomori S. A multivalent lacto-*N*-fucopentaose III-lysyllysine conjugate decompacts preimplantation mouse embryos, while the free oligosaccharide is ineffective. J Exp Med. 1984;160(5):1591–6.

Forsén S, Hoffman RA. Study of moderately rapid chemical exchange reactions by means of nuclear magnetic double resonance. J Chem Phys. 1963;39(11):2892–901.

Forsén S, Hoffman RA. Exchange rates by nuclear magnetic multiple resonance. III. Exchange reactions in systems with several nonequivalent sites. J Chem Phys. 1964;40(5):1189–96.

Glaudemans CP, Lerner L, Daves Jr GD, Kováč P, Venable R, Bax A. Significant conformational changes in an antigenic carbohydrate epitope upon binding to a monoclonal antibody. Biochemistry. 1990;29(49):10906–11.

Hakomori S. Carbohydrate-to-carbohydrate interaction, through glycosynapse, as a basis of cell recognition and membrane organization. Glycoconj J. 2004;21(3–4):125–37.

Hanashima S, Kato K, Yamaguchi Y. ^{13}C-NMR quantification of proton exchange at LewisX hydroxyl groups in water. Chem Commun. 2011;47(38):10800–2.

Hanashima S, Sato C, Tanaka H, Takahashi T, Kitajima K, Yamaguchi Y. NMR study into the mechanism of recognition of the degree of polymerization by oligo/polysialic acid antibodies. Bioorg Med Chem. 2013;21(19):6069–76.

Homans SW, Dwek RA, Rademacher TW. Tertiary structure in N-linked oligosaccharides. Biochemistry. 1987;26(20):6553–60.

Jain NU, Venot A, Umemoto K, Leffler H, Prestegard JH. Distance mapping of protein-binding sites using spin-labeled oligosaccharide ligands. Protein Sci. 2001;10(11):2393–400.

Jeng MF, Englander SW, Elöve GA, Wand AJ, Roder H. Structural description of acid-denatured cytochrome *c* by hydrogen exchange and 2D NMR. Biochemistry. 1990;29(46):10433–7.

Johnson PE, Brun E, MacKenzie LF, Withers SG, McIntosh LP. The cellulose-binding domains from *Cellulomonas fimi* β-1,4-glucanase CenC bind nitroxide spin-labeled cellooligosaccharides in multiple orientations. J Mol Biol. 1999;287(3):609–25.

Kamiya Y, Yagi-Utsumi M, Yagi H, Kato K. Structural and molecular basis of carbohydrate–protein interaction systems as potential therapeutic targets. Curr Pharm Design. 2011;17(17):1672–84.
Kato K, Sasakawa H, Kamiya Y, Utsumi M, Nakano M, Takahashi N, et al. 920 MHz ultra-high field NMR approaches to structural glycobiology. Biochim Biophys Acta. 2008;1780(3):619–25.
Kato K, Yamaguchi Y, Arata Y. Stable-isotope-assisted NMR approaches to glycoproteins using immunoglobulin G as a model system. Prog Nucl Magn Reson Spectrosc. 2010;56:346–59.
Kurland RJ, McGarvey BR. Isotropic NMR shifts in transition metal complexes: the calculation of the fermi contact and pseudocontact terms. J Magn Reson. 1970;2(3):286–301.
Levitt MH. Spin dynamics: basics of nuclear magnetic resonance. 2nd ed. New York: Wiley; 2008.
Lian LY, Roberts GCK. Effects of chemical exchange on NMR spectra. In: Roberts GCK, editor. NMR of macromolecules. Oxford: Oxford University Press; 1993. p. 153–82.
Loria JP, Rance M, Palmer AGI. A relaxation-compensated Carr-Purcell-Meiboom-Gill sequence for characterizing chemical exchange by NMR spectroscopy. J Am Chem Soc. 1999;121(10):2331–2.
Matsuzaki K, Kato K, Yanagisawa K. Aβ polymerization through interaction with membrane gangliosides. Biochim Biophys Acta. 2010;1801(8):868–77.
Mayer M, Meyer B. Characterization of ligand binding by saturation transfer difference NMR spectroscopy. Angew Chem Int Ed. 1999;38(12):1784–8.
Mayer M, Meyer B. Group epitope mapping by saturation transfer difference NMR to identify segments of a ligand in direct contact with a protein receptor. J Am Chem Soc. 2001;123(25):6108–17.
McConnell HM, Robertson RE. Isotropic nuclear resonance shifts. J Chem Phys. 1958;29(6):1361–5.
Mittermaier A, Kay LE. New tools provide new insights in NMR studies of protein dynamics. Science. 2006;312(5771):224–8.
Ni F, Scheraga HA. Use of the transferred nuclear Overhauser effect to determine the conformations of ligands bound to proteins. Acc Chem Res. 1994;27(9):257–64.
Nishima W, Miyashita N, Yamaguchi Y, Sugita Y, Re S. Effect of bisecting GlcNAc and core fucosylation on conformational properties of biantennary complex-type N-Glycans in solution. J Phys Chem B. 2012;116(29):8504–12.
Ohki SY, Kainosho M. Stable isotope labeling methods for protein NMR spectroscopy. Prog Nucl Magn Reson Spectrosc. 2008;53(4):208–26.
Otting G. Protein NMR using paramagnetic ions. Annu Rev Biophys. 2010;39:387–405.
Paterson Y, Englander SW, Roder H. An antibody binding site on cytochrome c defined by hydrogen exchange and two-dimensional NMR. Science. 1990;249(4970):755–9.
Pervushin K, Riek R, Wider G, Wüthrich K. Attenuated T_2 relaxation by mutual cancellation of dipole-dipole coupling and chemical shift anisotropy indicates an avenue to NMR structures of very large biological macromolecules in solution. Proc Natl Acad Sci U S A. 1997;94(23):12366–71.
Prestegard JH, Koerner TAW, Demou PC, Yu RK. Complete analysis of oligosaccharide primary structure using two-dimensional high-field proton NMR. J Am Chem Soc. 1982;104(18):4993–5.
Re S, Miyashita N, Yamaguchi Y, Sugita Y. Structural diversity and changes in conformational equilibria of biantennary complex-type *N*-glycans in water revealed by replica-exchange molecular dynamics simulation. Biophys J. 2011;101(10):L44–6.
Satoh T, Chen Y, Hu D, Hanashima S, Yamamoto K, Yamaguchi Y. Structural basis for oligosaccharide recognition of misfolded glycoproteins by OS-9 in ER-associated degradation. Mol Cell. 2010;40(6):905–16.
Solomon I. Relaxation processes in a system of two spins. Phys Rev. 1955;99(2):559–65.

Spera S, Ikura M, Bax A. Measurement of the exchange rates of rapidly exchanging amide protons: application to the study of calmodulin and its complex with a myosin light chain kinase fragment. J Biomol NMR. 1991;1(2):155–65.

Sugase K, Dyson HJ, Wright PE. Mechanism of coupled folding and binding of an intrinsically disordered protein. Nature. 2007;447(7147):1021–5.

Sugita Y, Okamoto Y. Replica-exchange molecular dynamics method for protein folding. Chem Phys Lett. 1999;314(1–2):141–51.

Taylor DR, Hooper NM. The prion protein and lipid rafts. Mol Membr Biol. 2006;23(1):89–99.

Utsumi M, Yamaguchi Y, Sasakawa H, Yamamoto N, Yanagisawa K, Kato K. Up-and-down topological mode of amyloid β-peptide lying on hydrophilic/hydrophobic interface of ganglioside clusters. Glycoconj J. 2009;26(8):999–1006.

Vliegenthart JF. High resolution ^{1}H-NMR spectroscopy of carbohydrate structures. Adv Exp Med Biol. 1980;125:77–91.

Voisin S, Houliston RS, Kelly J, Brisson JR, Watson D, Bardy SL, et al. Identification and characterization of the unique *N*-linked glycan common to the flagellins and S-layer glycoprotein of *Methanococcus voltae*. J Biol Chem. 2005;280(17):16586–93.

Wüthrich K. NMR of proteins and nucleic acids. New York: Wiley; 1986.

Wishart DS, Sykes BD. The ^{13}C chemical-shift index: a simple method for the identification of protein secondary structure using ^{13}C chemical-shift data. J Biomol NMR. 1994;4(2):171–80.

Woods RJ, Tessier MB. Computational glycoscience: characterizing the spatial and temporal properties of glycans and glycan-protein complexes. Curr Opin Struct Biol. 2010;20(5):575–83.

Yagi-Utsumi M, Kameda T, Yamaguchi Y, Kato K. NMR characterization of the interactions between lyso-GM1 aqueous micelles and amyloid β. FEBS Lett. 2010;584(4):831–6.

Yamaguchi T, Kamiya Y, Choo YM, Yamamoto S, Kato K. Terminal spin labeling of a high-mannose-type oligosaccharide for quantitative NMR analysis of its dynamic conformation. Chem Lett. 2013a;42(5):544–6.

Yamaguchi T, Uno T, Uekusa Y, Yagi-Utsumi M, Kato K. Ganglioside-embedding small bicelles for probing membrane-landing processes of intrinsically disordered proteins. Chem Commun. 2013b;49(12):1235–7.

Yamaguchi Y, Kato K. NMR analyses of the carbohydrate–protein interactions. Exp Med. 2007a;25(7):231–8.

Yamaguchi Y, Kato K. Structural glycobiology by stable-isotope-assisted NMR spectroscopy. In: Webb GA, editor. Modern Magnetic Resonance. The Netherlands: Springer; 2007b. p. 219–25.

Yamaguchi Y, Kato K. Analysis of sugar–protein interactions by NMR, Experimental Glycoscience Glycochemistry. Berlin: Springer; 2008. p. 121–3.

Yamaguchi Y, Nishima W, Re SY, Sugita Y. Confident identification of isomeric *N*-glycan structures by combined ion mobility mass spectrometry and hydrophilic interaction liquid chromatography. Rapid Commun Mass Spect. 2012;26(24):2877–84.

Yamamoto S, Zhang Y, Yamaguchi T, Kameda T, Kato K. Lanthanide-assisted NMR evaluation of a dynamic ensemble of oligosaccharide conformations. Chem Commun. 2012;48(39):4752–4.

Yu RK, Koerner TA, Scarsdale JN, Prestegard JH. Elucidation of glycolipid structure by proton nuclear magnetic resonance spectroscopy. Chem Phys Lipids. 1986;42(1–3):27–48.

Zhang Y, Yamaguchi T, Kato K. New NMR tools for characterizing the dynamic conformations and interactions of oligosaccharides. Chem Lett. 2013;42(12):1455–62.

Zhang Y, Yamamoto S, Yamaguchi T, Kato K. Application of paramagnetic NMR-validated molecular dynamics simulation to the analysis of a conformational ensemble of a branched oligosaccharide. Molecules. 2012;17(6):6658–71.

Zhao H, Pan Q, Zhang W, Carmichael I, Serianni AS. DFT and NMR studies of $^2J_{COH}$, $^3J_{HCOH}$, and $^3J_{CCOH}$ spin-couplings in saccharides: C-O torsional bias and H-bonding in aqueous solution. J Org Chem. 2007;72(19):7071–82.

Chapter 9
Glycolipid and Glycoprotein Expression During Neural Development

Robert K. Yu and Yutaka Itokazu

Abstract In mammals, the central and peripheral nervous systems are developmentally derived from cells in the neural plate. Specific ectodermal cells in this area form the neural tube and neural crest during the early developmental stage. The neural tube is the origin of the central nervous system which consists of both the brain and spinal cord, whereas neural crest cells are precursors of the peripheral nervous system. During neural tube formation and neural crest development, carbohydrate-rich molecules, including glycolipids, glycoproteins, and proteoglycans, are expressed primarily on the outer surface of cell plasma membranes. The structural diversity of their carbohydrate moieties coupled with their expression at different stages of development makes these molecules excellent biomarkers for various cell types. In addition, these molecules play crucial functional roles in cell proliferation, differentiation, interaction, migration, and signal transduction. In this chapter, we discuss the expression profiles and potential functional roles of glycoconjugates during neural development.

Keywords Neural stem cell • Neural development • Neurogenesis • Gliogenesis • Glycolipid • Carbohydrate • Glycoconjugate • Glycosphingolipid • Ganglioside • Glycoprotein • Proteoglycan

---

R.K. Yu (✉) • Y. Itokazu
Department of Neuroscience and Regenerative Medicine, Medical College of Georgia, Georgia Regents University, Augusta, GA 30912, USA

Charlie Norwood VA Medical Center, Augusta, GA 30904, USA
e-mail: ryu@gru.edu

R.K. Yu and C.-L. Schengrund (eds.), *Glycobiology of the Nervous System*, Advances in Neurobiology 9, DOI 10.1007/978-1-4939-1154-7_9,

Abbreviations

BLBP	Brain lipid-binding protein
BMB	Bone morphogenetic protein
CD	Cluster of differentiation
Cer	Ceramide
CNS	Central nervous system
CNTF	Ciliary neurotrophic factor
CS	Chondroitin sulfate
CSPG	Chondroitin sulfate proteoglycan
CST	Cerebroside sulfotransferase
Dll1	Delta-like1
EGF	Epidermal growth factor
FABP7	Fatty acid-binding protein 7
FGF	Fibroblast growth factor
Fuc	Fucose
FUT	Fucosyltransferase
GAG	Glycosaminoglycan
GalCer	Galactosylceramide
GalNAcT	*N*-acetylgalactosaminyltransferase
GalT	Galactosyltransferase
GFAP	Glial fibrillary acidic protein
GlcAT-P	UDP-glucuronyltransferase-P
GlcCer	Glucosylceramide
GlcT	Glucosyltransferase
GRP	Glial-restricted precursor
GSL	Glycosphingolipid
HA	Hyaluronic acid
HNK-1	Human natural killer-1 antigen
HS	Heparin sulfate
HSPG	Heparin sulfate proteoglycan
IL-6	Interleukin 6
INP	Intermediate neuronal progenitor cell
IPC	Intermediate progenitor cell
JAK-STAT	Janus kinase (JAK)-signal transducer and activator of transcription 3 (STAT3)
LacCer	Lactosylceramide
Lfng	Lunatic fringe
MA	Cerebral mantle
mAb	Monoclonal antibody
MAPK	Mitogen-activated protein kinase
MZ	Marginal zone
NEC	Neuroepithelial cell
NG-2	Nerve/glial antigen 2

NRP	Neuronal restricted progenitor
NSC	Neural stem cell
OPC	Oligodendrocyte precursor cell
PDGF	Platelet-derived growth factor
PG	Proteoglycan
PHA-E4	*Phaseolus vulgaris* erythroagglutinating lectin
PNA	Peanut agglutinin
PNS	Peripheral nervous system
PSA-NCAM	Polysialic acid-neural cell adhesion molecule
PST	ST8SiaIV
RGC	Radial glial cell
SGZ	Subgranular zone
SSEA	Stage-specific embryonic antigen
ST	Sialyltransferase
STX	ST8SiaII
SVZ	Subventricular zone
VZ	Ventricular zone

9.1 Introduction

During neural development, dramatic and consistent changes in the composition of glycoconjugates, including glycolipids, glycoproteins, and proteoglycans (PGs), occur (Ngamukote et al. 2007; Yanagisawa and Yu 2007; Yu et al. 1988). It is known that changes in the expression of glycolipids, including gangliosides, in the nervous system correlate with neurodevelopmental events (Yu et al. 2009). For example, in fertilized eggs, the globo-series of glycolipids are robustly expressed. As cell division proceeds, the lacto-series glycosphingolipids (GSLs) are expressed, followed by the ganglio-series GSLs in the developing brain. The lipid portion of GSLs, including gangliosides, is the ceramide, which is synthesized in the endoplasmic reticulum (ER) from a sphingosine base and a fatty acid residue. Ceramide is transferred to the Golgi apparatus where it is modified by the sequential addition of carbohydrate moieties (Fig. 9.1) (Yu et al. 2012). Each step is catalyzed by a unique, specifically controlled glycosyltransferase. In early embryonic rodent brains, the pattern of ganglioside expression is characterized by the expression of a large amount of simple gangliosides, such as GM3 and GD3. In the later developmental stages, more complex gangliosides prevail, particularly GM1, GD1a, GD1b, and GT1b (Fig. 9.2). Correlations between ganglioside expression in the nervous system and neurodevelopmental events are summarized schematically in Fig. 9.3. This unique expression pattern suggests that the presence of specific gangliosides may reflect the functional roles they play at specific developmental stages. Abundant evidence supports the notion that GSLs, including gangliosides, serve regulatory roles in cellular events, including proliferation and neural differentiation, as exemplified by neuritogenesis, axonogenesis, and synaptogenesis (Bieberich et al. 2001;

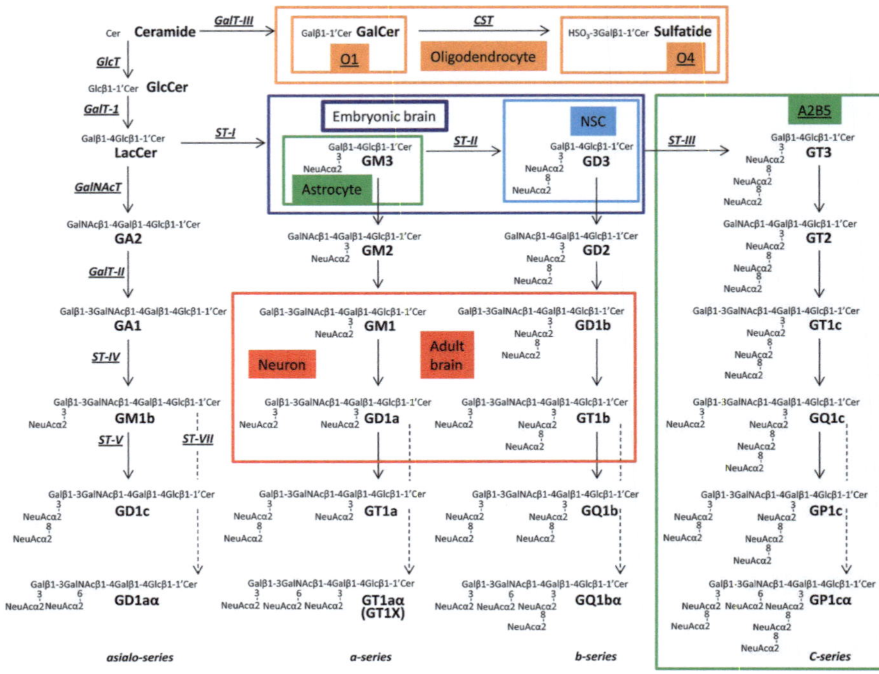

Fig. 9.1 Structures and biosynthetic pathways of glycosphingolipids (GSLs). The nomenclature for gangliosides and their components are based on that of Svennerholm (1963) and the IUPAC–IUBMB Joint Commission on Biochemical Nomenclature (1977). *Cer* ceramide, *CST* cerebroside sulfotransferase (*Gal3st1*, sulfatide synthase), *GalNAc-T N*-acetylgalactosaminyltransferase I (*B4galnt1*, GA2/GM2/GD2/GT2 synthase), *GalT-I* galactosyltransferase I (*B4galt6*, lactosylceramide synthase), *GalT-II* galactosyltransferase II (*B3galt4*, GA1/GM1/GD1b/GT1c synthase), *GalT-III* galactosyltransferase III (*Ugt8a*, galactosylceramide synthase), *GlcT* glucosyltransferase (Ugcg, glucosylceramide synthase), *ST-I* sialyltransferase I (*St3gal5*, GM3 synthase), *ST-II* sialyltransferase II (*St8Sia1*, GD3 synthase), *ST-III* sialyltransferase III (*St8Sia3*, GT3 synthase), *ST-IV* sialyltransferase IV (*St3gal2*, GM1b/GD1a/GT1b/GQ1c synthase), *ST-V* sialyltransferase V (*St8sia5*, GD1c/GT1a/GQ1b/GP1c synthase), *ST-VII* sialyltransferase VII (*St6galnac6*, GD1aα/GT1aα/GQ1bα/GP1cα-synthase). Official symbols of genes are represented in *italics* in this figure legend. GM3 and GD3 are abundant in embryonic brain (*blue*) and NSCs express GD3 (*light blue*). c-series gangliosides are A2B5 antigens (*green*) and astrocytes express GM3 (*green*). GM1, GD1a, GD1b, and GT1b are the most abundant ganglioside species in adult mammalian brain (*red*). Oligodendrocyte markers O1 and O4 are GalCer and sulfatide, respectively (*orange*)

Fang et al. 2000; Ngamukote et al. 2007; Wu et al. 1998, 2001; Yu et al. 2004, 2009). In recent years, with the advent of contemporary molecular genetics and biology, several lines of genetically modified mice have been established in which the expression of gangliosides and other GSLs has been altered or depleted, and this has greatly facilitated the unraveling of their biological functions. For example, GM2/GD2 synthase (GalNAcT) is one of the key enzymes needed for synthesis of the major "brain-type" gangliosides, including GM1, GD1a, GD1b, and GT1b. Mice lacking this enzyme do not express GalNAc-containing gangliosides. As a result

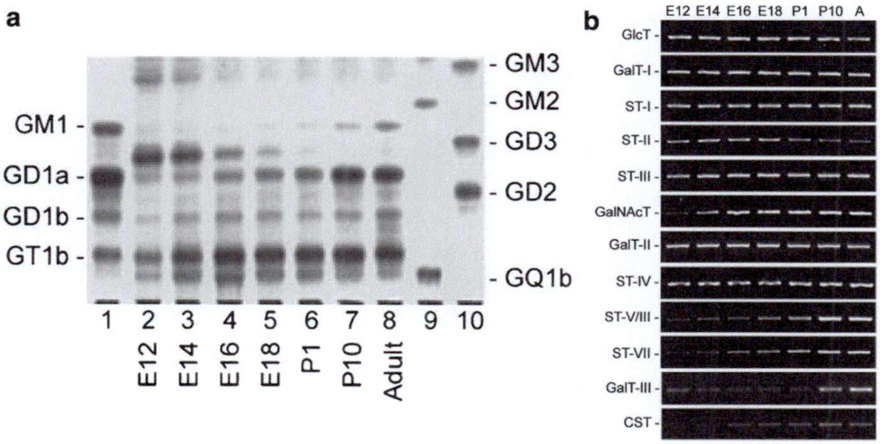

Fig. 9.2 Ganglioside and glycosyltransferase expression in the developing mouse brain. (**a**) Ganglioside expression patterns analyzed by thin-layer chromatography. Expression in mouse brain shift, with age, from simple gangliosides such as GM3 and GD3 to complex gangliosides such as GM1 and GD1a. (**b**) Glycosyltransferases expressed in developing mouse brains analyzed by RT-PCR. During early development, the message levels of GalNAcT (GA2/GM2/GD2/GT2 synthase) and ST-II (GD3 synthase) are developmentally regulated. "A" indicates adult mouse brain (reproduced from Ngamukote et al. 2007)

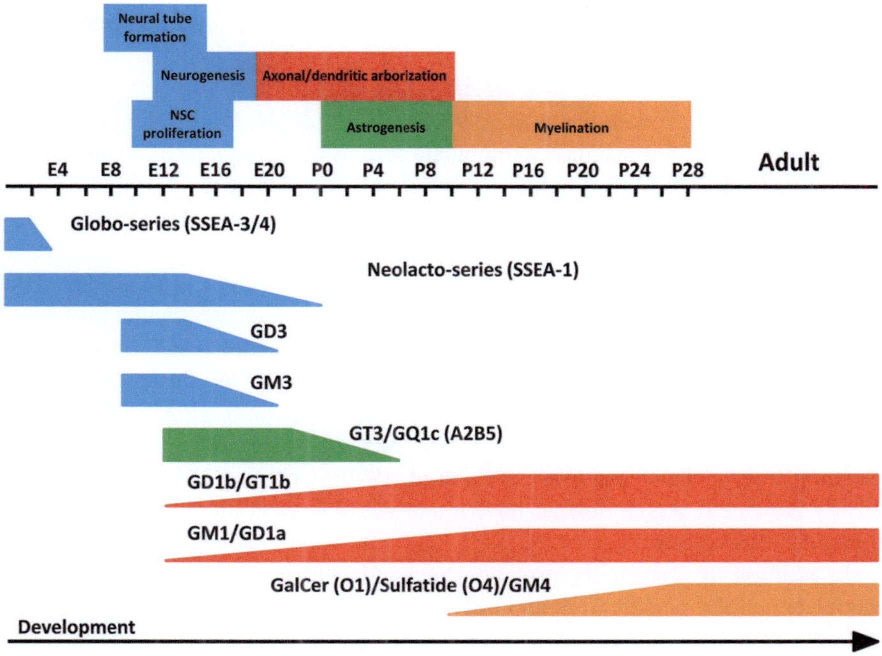

Fig. 9.3 Neurodevelopmental events and concurrent changes in GSL expression. "E" denotes embryonic day and "P" postnatal day

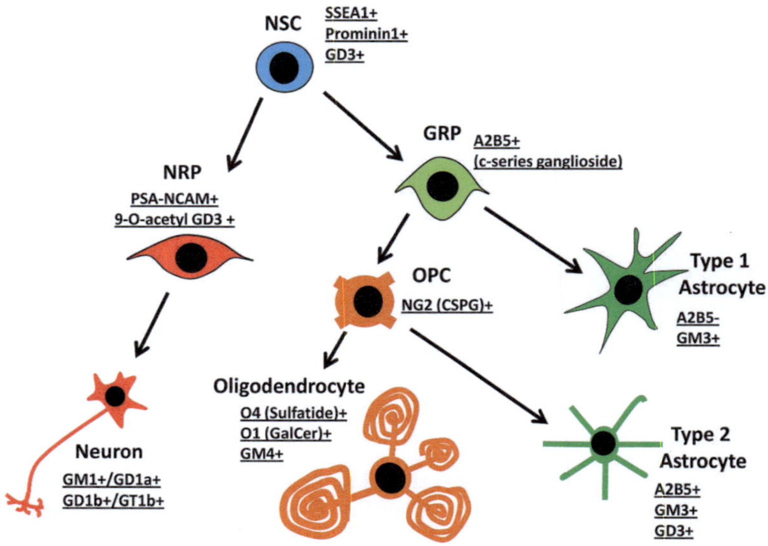

Fig. 9.4 A model for neural cell lineages derived from mouse neural stem cells (NSCs). The known glycoconjugate markers are *underlined. NSC* neural stem cell, *NRP* neuronal restricted precursor, *GRP* glial restricted precursor, *OPC* oligodendrocyte precursor cell

they are developmentally abnormal and appear to have neurological problems such as axonal degeneration; sensory, motor, and behavioral deficits; and other neurological dysfunctions (Furukawa et al. 2008; Sheikh et al. 1999; Sugiura et al. 2005; Susuki et al. 2007; Takamiya et al. 1996; Wu et al. 2011). During brain development, gangliosides are assumed to modulate ceramide (Cer)-induced apoptosis and to maintain cellular survival and differentiation (Bieberich et al. 2001). GM3 synthase (sialyltransferase I, ST-I) is a critical enzyme for the synthesis of all complex gangliosides. Mutation of GM3 synthase is associated with human autosomal recessive infantile-onset symptomatic epilepsy syndrome (Simpson et al. 2004). This study clearly demonstrated that deletion of complex gangliosides can be associated with human diseases. A lack of b- and c-series gangliosides results in clear and subtle developmental and behavioral deficits with mice lacking these gangliosides exhibiting sudden death from audiogenic seizures (Kawai et al. 2001). Both GalNAcT- and ST-I-deficient mice, which lack all gangliosides, die soon after weaning at 3 weeks of age (Yamashita et al. 1999). Combined these observations clearly indicate that gangliosides have important biological functions in the developing nervous system.

In addition to glycolipids, proteoglycans and glycoproteins are also known to modulate cellular proliferation and differentiation by participating in signal transduction in response to external stimuli and in mediating cell–cell interactions and adhesion. In this chapter, we will introduce these glycoconjugates expressed during neural development (Fig. 9.4).

9.2 Glycobiology During Early Embryogenesis

After fertilization, the fertilized egg undergoes cleavage to 2-, 4-, and 8-cell stages. From the 8-cell to 32-cell stage, the spherical cells undergo changes in morphology to a cubic shape. The cells bind tightly to each other forming compact spheres, and this stage is called the compaction stage (Purves and Lichtman 1985). At this stage, cell surface glycoconjugate markers, composed of fucose, *N*-acetyllactosamine, stage-specific embryonic antigen-1 (SSEA-1), and others, start to emerge (Fig. 9.5). SSEA-1 is also known as Lewis X antigen, belonging to cluster of differentiation (CD) 15. Strictly, "SSEA-1" is not equal to "Lewis X." Lewis X structure is defined by a minimal Lewis X motif consisting of the structural element Galβ1-4(Fucα1-3)GlcNAcβ-. The structure of SSEA-1 is shown in Fig. 9.5. However, in this chapter we will describe both as SSEA-1 because SSEA-1 and Lewis X have not been clearly distinguished in the literature. Since Lewis haptens have been reported to inhibit the cell compaction process in mouse embryos (Fenderson et al. 1984; Solter and Knowles 1978), it is believed that SSEA-1 may play an important role in early embryogenesis. Other stage-specific antigens, such as SSEA-3 and SSEA-4, are

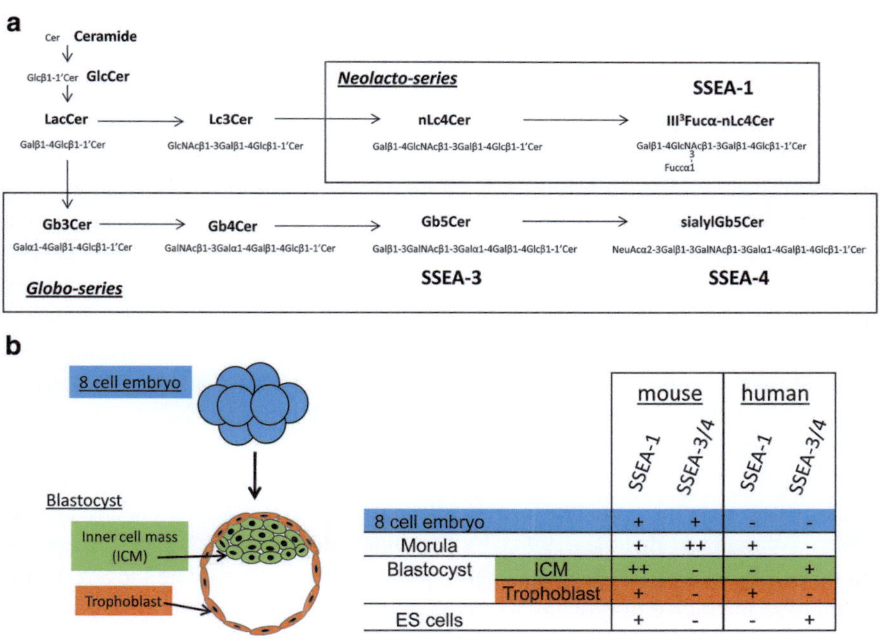

		mouse SSEA-1	mouse SSEA-3/4	human SSEA-1	human SSEA-3/4
8 cell embryo		+	+	-	-
Morula		+	++	+	-
Blastocyst	ICM	++	-	-	+
	Trophoblast	+	-	+	-
ES cells		+	-	-	+

Fig. 9.5 Expression of SSEAs and their biosynthetic pathways in early embryogenesis. (**a**) Structures and synthetic pathway of globo- and neolacto-series glycosphingolipids (GSLs). The abbreviations for GSLs follow the nomenclature systems of IUPAC–IUBMB Joint Commission on Biochemical Nomenclature (1977) and Svennerholm (1963). SSEA-1 is carried not only by neolacto-series GSLs but also by proteoglycans, glycoproteins, and lacto-, ganglio-, and globo-series GSLs. (**b**) A summary of the expression patterns of SSEAs in mouse early embryogenesis and embryonic stem (ES) cells

also expressed at the early stages of mouse embryogenesis. The expression of SSEA-3 usually peaks at the 4- to 8-cell stages, whereas SSEA-4 peaks at the morula and early blastocyst stages with some overlap with that of SSEA-3 (Fenderson et al. 1990). The expression patterns of these stage-specific antigens are different in human and mouse (Fig. 9.5). Thus, SSEA-1 has been utilized as a specific marker of mouse embryonic stem (ES) cells. SSEA-1 is not expressed in human ES cells. Instead, human ES cells express SSEA-3, SSEA-4, and keratan sulfate antigens (TRA)-1-60, TRA-181, GCTM2, and GCT343 (Adewumi et al. 2007; Muramatsu and Muramatsu 2004).

Analysis of mice deficient in SSEA-1 [fucosyltransferase 9 (FUT9)-deficient mice] revealed increased anxiety-like behavior, but no distinguishable morphological phenotypes in brain development (Kudo et al. 1998, 2007). While mice deficient in SSEA-3 and SSEA-4 expression (α1,4galactosyltransferase-deficient mice) were resistant to Shiga-like toxins, they showed no apparent abnormality in development (Okuda et al. 2006). These studies suggest that the functions of SSEAs may be compensated for by other carbohydrate molecules or are not essential for neural development.

9.3 Neural Tube Formation

Both the central nervous system (CNS) and peripheral nervous system (PNS) originate from ectodermal cells in the neural plate (Purves and Lichtman 1985). Cells at the neural plate undergo a series of divisions and morphological changes, and they form a neural groove that has neural folds on either side (Fig. 9.6). Cells in the neural folds constitute the precursors of neural crest cells. In mice, these neural folds approach each other in the median plane, become fused at embryonic day (E) 8.5, and eventually form the neural tube. All the cells for the CNS emanate from the neural tube. On the other hand, the PNS originates from the neural crest cells. With respect to neural tube formation, glycosaminoglycans (GAGs) and high molecular weight unbranched polysaccharides made up of repeating disaccharide subunits of an amino sugar and a uronic acid play an important role in its genesis. In addition, nonsulfated GAGs, hyaluronans, or hyaluronic acid (HA)-containing glycoconjugates support structural and tensile strengths during neural tube folding and closure (Morris-Wiman and Brinkley 1990a, b). Enzymatic degradation of the HA matrix in the neural plate with exogenous hyaluronidase leads to incomplete closure of the neural tube in chick embryos (Schoenwolf and Fisher 1983). Other GAGs are also important for early embryogenesis. For example, mice deficient in heparin sulfate (HS) die by E8.5, indicating that HS also has an important role(s) in early embryogenesis. In addition, mice deficient in glucuronyltransferase I (GlcAT-1), an enzyme required for the synthesis of the linkage tetrasaccharide for both HS and chondroitin sulfate (CS), fail to express either, and these knockout mice die before the 8-cell stage as the result of cytokinesis failure (Izumikawa et al. 2010). To identify the contributor of this lethality, specific glycanase treatments were performed. Treatment

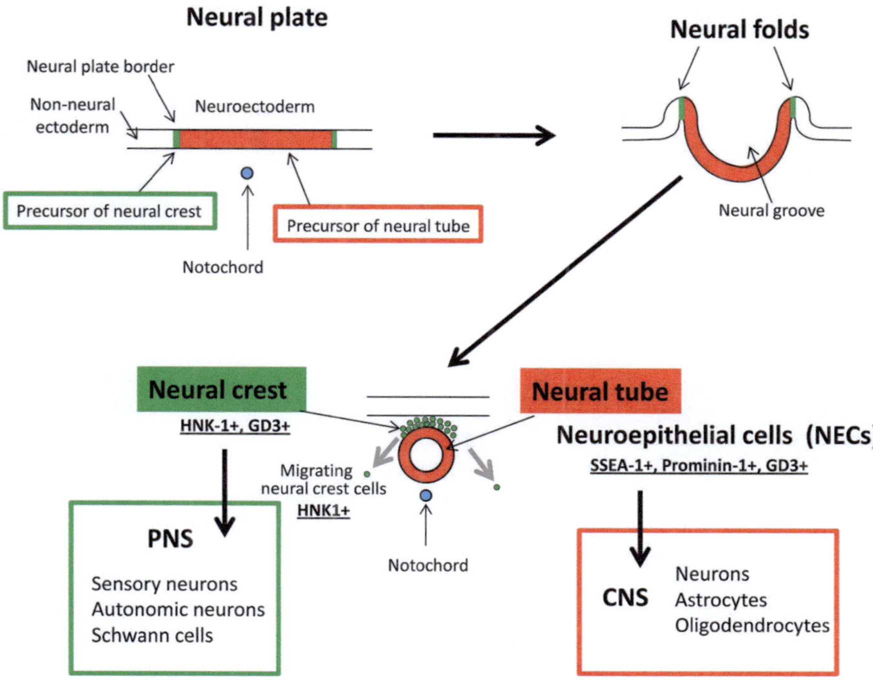

Fig. 9.6 Neural tube and neural crest formation. Cells in the neural plate undergo a series of divisions and form a neural groove that has neural folds on both sides. Cells in the neural fold are the future neural crest cells. The neural folds then approach each other in the median plane, become fused, and eventually form the neural tube. All the cells for the CNS are derived from neuroepithelial cells (NECs) of the neural tube. The PNS originates from neural crest cells. The precursors of neural crest cells reside in the dorsal neural tube, and these cells undergo epithelial to mesenchymal transition (EMT) and delaminate from the neural tube as neural crest cells. The known carbohydrate markers are *underlined*

of 2-cell embryos with chondroitinase ABC which degrades CS had marked effects on cell division were observed. At the same time, many heparitinase, which specifically degrade HS, treated embryos normally developed to blastocysts. Thus, CS is indispensable for embryonic cell division. These examples underscore the importance of glycoconjugates in embryonic cell division.

9.4 Neuroepithelial Cells and Radial Glial Cells

9.4.1 Neural Stem Cells in Development

Neuroepithelial cells (NECs) proliferate by repeating symmetric cell division at the wall of the neural tube, and as NECs accumulate the wall gradually becomes thicker. At first, an NEC elongates its fibers and becomes a radial glial cell (RGC) whose

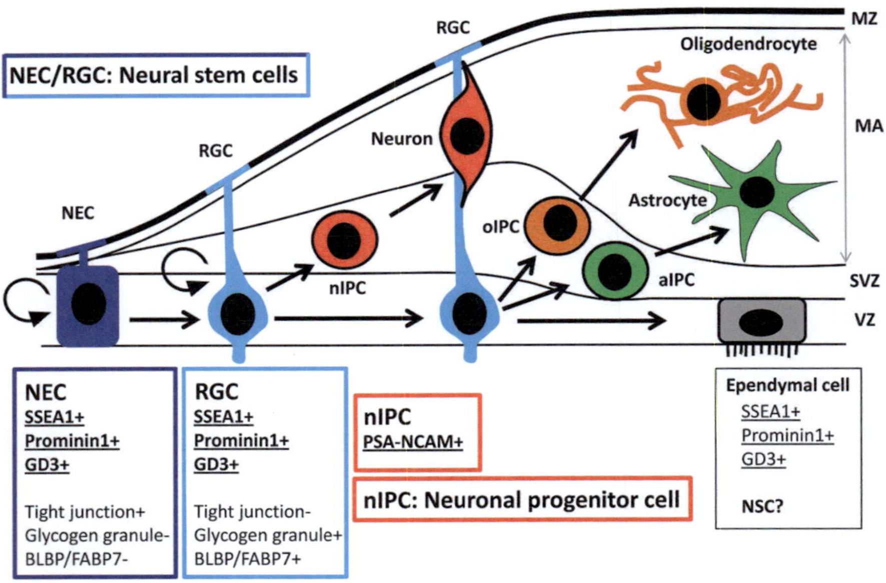

Fig. 9.7 Neuroepithelial cells (NECs) proliferate at the wall of the neural tube, and NECs elongate their fibers and become radial glial cells (RGCs). An RGC generates an RGC and an intermediate progenitor cell (IPC). *nIPC* neuronal intermediate progenitor (also called basal progenitor), *oIPC* oligodendrocyte IPC (also called oligodendrocyte precursor cell, OPC), *aIPC* astrocyte IPC, *MZ* marginal zone, *MA* mantle, *SVZ* subventricular zone, *VZ* ventricular zone. The known carbohydrate markers are *underlined*. Whether aIPCs are involved in this pattern is not well known

cell body lines the ventricular zone (VZ) and the apical surface meets the ventricles with the radial fibers reaching the pial surface (Fig. 9.7). Previously, RGCs were thought of as specialized glial cells whose function was to guide neuronal migration (Alvarez-Buylla et al. 2001; Fishell and Kriegstein 2003, 2005; Fujita 2003; Gotz and Huttner 2005; Miller and Gauthier 2007; Shimojo et al. 2011). Recently, RGCs were recognized as the precursors of neurons and glia. By asymmetric cell division, an RGC generates another RGC and an intermediate progenitor cell (IPC) or immature neuron (Malatesta et al. 2000; Miyata et al. 2001; Noctor et al. 2001). IPCs stay in the subventricular zone (SVZ) to proliferate and give rise to more neurons. Immature neurons migrate along with radial fibers into the cortical plate and then become mature neurons. At first, RGCs give rise to inner layer neurons and later to outer layer neurons. RGCs also give rise to oligodendrocytes and ependymal cells and can eventually differentiate into astrocytes. Both NECs and RGCs are considered NSCs (Franco and Muller 2013; Shimojo et al. 2011). The NSC niche is a specialized microenvironment that maintains stem cells in a multipotent and undifferentiated state. The NSC niche hosts a variety of stem/progenitor cells, such as NECs, RGCs and IPCs. Altogether, these versatile progenitors cooperate for neurogenesis and gliogenesis in the developing CNS (Fig. 9.7). In the following sections, some of the key glycoconjugate biomarkers are described.

Notch

Notch receptors are transmembrane proteins whose signaling has been shown to regulate a wide range of developmental processes (Hori et al. 2013; Koch et al. 2013). Notch signaling plays essential roles in neurogenesis, including inhibition of neurogenesis and oligodendrocyte differentiation, maintenance of the RGC pool, and promotion of astrocyte differentiation (de la Pompa et al. 1997; Gaiano and Fishell 2002). Notch signaling is activated by interaction with ligand molecules, such as *Delta-like1* (*Dll1*, *Delta* in *Drosophila*) or *Jagged* 1 (*Serrate* in *Drosophila*). Neuronal IPCs (nIPCs) or intermediate neuronal progenitor cells (INPCs) are known to be a source of Dll1 to activate Notch signaling in RGCs (Mizutani et al. 2007). nIPCs/INPCs provide intrinsic neuronal differentiation information to new neurons by themselves and by extrinsic inhibitory signals to maintain the stemness of RGCs. *Fringe* is a major regulator of Notch signaling, serving as a promoter of Delta–Notch signaling and as an inhibitor of *Serrate*–Notch signaling in *Drosophila wing* (Hou et al. 2012; Panin et al. 1997). In mammals, there are three *Fringe* genes (*Lfng, Mfng, Rfng*) expressed in different populations of cells in the developing cortex. *Lunatic fringe* (*Lfng*) is expressed in immature cells, presumed to be NECs and RGCs, in the VZ (Ishii et al. 2000; Kato et al. 2010). It is known that *O*-glycosylation of the Notch extracellular domain is essential for Notch activity by affecting protein folding, ligand interaction, and endocytosis of the Notch receptor (Okajima et al. 2008). The Notch receptor contains epidermal growth factor (EGF)-like repeats, which have *O*-fucose glycan modifications on the serine or threonine residues (Haines and Irvine 2003). These *O*-fucose glycans modulate protein–protein interactions and their resultant functional roles in regulating Notch signaling (Haines and Irvine 2003; Luther and Haltiwanger 2009; Stanley and Okajima 2010). The synthesis of the *O*-Fuc glycan is initiated by *O*-fucosyltransferase (OFUT) catalyzing the O-linked fucosylation of serine or threonine residues. Knockdown of *Drosophila* OFUT1 by RNA interference (RNAi) leads to defects in Notch signaling, indicating the importance of *O*-Fuc or the *O*-Fuc glycan in this process (Okajima and Irvine 2002). In cell culture, RNAi of OFUT1 inhibits both Delta–Notch and Serrate–Notch binding, whereas OFUT1 overexpression increases *Serrate*–Notch binding but inhibits Delta–Notch binding (Okajima et al. 2003). Deletion of OFUT1 in *Drosophila* prompts a severe Notch-like phenotype, exemplified by an overabundance of neurons due to failure of Notch-dependent lateral inhibition (Sasamura et al. 2003). Elimination of OFUT1 in mice causes the embryos to die in midgestation with defects in neurogenesis, somitogenesis, vasculogenesis, and cardiogenesis. The knockout mice present similar phenotypes as other mutants of Notch signaling molecules (Shi and Stanley 2003), suggesting that *O*-Fuc modification is conserved in various animal species. Interestingly, in addition to its role in glycosylation, OFUT 1 has been reported to have a distinct function as a molecular chaperone of Notch molecules (Okajima et al. 2005). *O*-Fuc residues are further modified by a series of glycosyltransferases, including β1-3*N*-acetylglucosaminyltransferase, β1-4galactosyltransferase, and α2-3sialyltransferase. *O*-Fuc glycan (SAα2-3Galβ1-4GlcNAcβ1-3Fuc-Ser/Thr) is synthesized by sequential addition of sugar residues, depending on the activities of these enzymes (Moloney et al. 2000). Intriguingly, *Fringe* protein, a promoter of *Delta* and an inhibitor of *Serrate*, has *N*-acetylglucosaminyl (GlcNAc) transferase activity and is required for modulation of Notch signaling (Bruckner et al. 2000; Moloney et al. 2000). Because the elongated *O*-Fuc glycans by *Fringe* leads to a higher affinity for Notch to *Delta* than to *Serrate*, the promoter activities of *Fringe* for *Delta* and inhibitor activities for *Serrate* are presumed to be modulated by the elongated O-Fuc glycans on Notch (Okajima et al. 2003). Recently it was reported that *Lfng*, which is distinctly expressed in the VZ, enhances the self-renewal of NSCs in the developing mouse brain (Kato et al. 2010). *Lfng* was also reported to be associated with neurogenesis in the chick spinal cord (Skaggs et al. 2011). These studies clearly indicate the importance of carbohydrate chains in the regulation of stem cell self-renewal and differentiation via Notch signaling.

9.4.2 Neuroepithelial Cells, Radial Glial Cells, and Intermediate Progenitor Cells

NECs

In the brain, neurons and glia originate from NSCs derived from the neuroectoderm. These cells have many epithelial cell characteristics and are known as NECs. Around E8, NECs undergo rapid proliferation by symmetric division to expand the progenitor pools (Smart 1973). From E9 to E10, the anterior portion of the neural tube, which later becomes the telencephalon, closes to form the lateral ventricle. Proliferative NECs are layered at the lateral ventricles as a pseudostratified neuroepithelium with epithelial apicobasal polarity. Tight junctions and adherent junctions are present at the most lateral end of the lateral plasma membrane. At the pial surface, NSCs make contact with the basal lamina (Aaku-Saraste et al. 1996; Graus-Porta et al. 2001; Lui et al. 2011; Smart 1973).

RGCs

NECs begin to transform into RGCs at E9.5. NECs lose some of their epithelial properties in favor of certain glial characteristics, but retain contacts with the ventricular and pial surfaces that give them their radial morphology. NEC-to-RGC transition is characterized by the loss of tight junctions, acquisition of glycogen storage granules, and the expression of astroglial genes, such as brain lipid-binding protein (BLBP) or fatty acid-binding protein 7 (FABP7), astrocyte-specific glutamate transporter (GLAST), and tenascin-C. RGCs still retain many NEC characteristics, such as adherent junctions, apical surface at ventricles, basal lamina contact, and expression of nestin, an NSC selective marker. During this period of development, the two cell types, NECs and RGCs, coexist. Although it was believed that mitotic cells in the VZ were the progenitors that generate neurons, astrocytes, and oligodendrocytes, more recent investigations have provided evidence that RGCs are the progenitors of most neurons, astrocytes, and oligodendrocytes in the CNS. The primary role of NECs is to expand the progenitor pool before transitioning to RGCs (Aaku-Saraste et al. 1996; Bruckner and Biesold 1981; Franco and Muller 2013; Hartfuss et al. 2001).

IPCs

Before their transformation to RGCs, only a small population of postmitotic neurons are generated directly from NECs. An RGC tends to divide asymmetrically and generates a RGC and a non-RGC daughter cell (Noctor et al. 2002). Only about 10 % of asymmetrically dividing RGCs are directly transformed into neurons (Attardo et al. 2008). Most RGCs divide into RGCs and IPCs. Unlike NEC and RGC, an IPC can undergo symmetric terminal division into two neurons. To generate more IPCs, certain IPCs can also undergo a limited number of additional symmetric divisions to paired IPCs (Noctor et al. 2004). The majority of RGCs can produce only neuronal or glial precursor cells (Malatesta et al. 2003). Occasionally, but rarely, RGCs host multipotent progenitor cells that generate both neurons and glia. The glial-specific progenitors typically generate either astrocytes or oligodendrocytes, but not both in vivo (McCarthy et al. 2001).

SSEA-1

SSEA-1 is expressed on NECs at early stages of development and the expression remains by E19 in the VZ and SVZ, where the NSC populations reside (Capela and Temple 2006; Hennen et al. 2011; Mai et al. 1998). This suggests a functional role for SSEA-1 in sustaining stem and progenitor cell growth. SSEA-1 can bind and regulate fibroblast growth factor 2 (FGF-2), which is known as a mitogen that maintains the stemness of NSCs (Dvorak et al. 1998; Jirmanova et al. 1999). In addition, the SSEA-1 epitope is also associated with chondroitin sulfate proteoglycan (CSPG) (Kabos et al. 2004), β1 integrin, glycolipids (Yanagisawa et al. 2005), lysosome-associated membrane protein 1 (LAMP-1) (Yagi et al. 2010a), extracellular matrix protein tenascin-C (Hanjan et al. 1982), phosphacan (Hanjan et al. 1982; Tole et al. 1995), and Wnt-1 (Capela and Temple 2006). Strong SSEA-1 expression can be observed during embryonic development on NSCs in neurogenic regions, such as the hippocampal primordium and the embryonic cerebral cortex; its expression remains clearly visible until E19 (Hennen et al. 2011; Mai et al. 1998). SSEA-1+ cells typically have bipolar morphology, radial orientation, and glial processes, and they resemble a subtype of RGCs (E12-E14) (Mai et al. 1998; Mo et al. 2007). In vitro experiments revealed that blockage of SSEA-1 by anti-SSEA-1 antibody inhibits cell migration from neurospheres, but does not affect cellular proliferation (von Holst et al. 2006; Yanagisawa et al. 2005). Recently, knockdown of FUT9 (a key enzyme for the biosynthesis of SSEA-1) in mouse NSCs was shown to downregulate Musashi-1 expression and NSC proliferation (Yagi et al. 2012). Musashi-1 plays a crucial role in maintaining the undifferentiated state of NSCs via activation of the Notch signaling pathway (Imai et al. 2001; Okano et al. 2005). SSEA-1 may regulate proliferation of NSCs via modulation of the expression of Musashi-1 (Yagi et al. 2012).

Prominin-1

Prominin-1, also known as CD133 or AC133 (the human homologue), is a pentaspan membrane glycoprotein originally identified as an antigen expressed on the apical surface of mouse NECs at E8.5 (Marzesco et al. 2005; Shmelkov et al. 2005; Weigmann et al. 1997). Prominin-1 is specifically associated with plasma membrane protrusions that have a microvilli-like structure on the apical surface of NECs (Weigmann et al. 1997). During development at E10.5–12.5, the apical plasma membrane protrusions containing prominin-1 are released into the lumen of the neural tube as a novel class of extracellular membrane particles (Marzesco et al. 2005). After E12.5, the release of prominin-1-containing extracellular particles is decreased (Marzesco et al. 2005; Yanagisawa et al. 2004a). At the same time, NEC proliferation decreases and NECs transit into RGCs. Prominin interacts with cholesterol and gangliosides in the plasma membrane to modulate the membrane microdomains (lipid rafts) at the membrane protrusions (Huttner and Zimmerberg 2001; Janich and Corbeil 2007; Roper et al. 2000). Analysis of mice deficient in prominin-1 revealed progressive degeneration of mature photoreceptors with complete loss of vision, but no other obvious abnormalities in brain development (Zacchigna et al. 2009). In prominin-1-deficient mice, upregulation of prominin-2, which is structurally related to prominin-1, was detected, and it seems that prominin-2 compensates for the loss of prominin-1.

Gangliosides

Expression of GD3 ganglioside (CD60a) in neural tubes early in development was detected using the GD3-specific monoclonal antibody (MAb) R24 (Rosner et al. 1992). Upon closer examination it was found to be expressed in NECs in neural tubes, in RGCs in the VZ of embryos, and in the SVZ of postnatal and adult rodents (Bannerman et al. 1996; Cammer and Zhang 1996a, b; Goldman et al. 1984; Nakatani et al. 2010). $GD3^+$ cells are also co-localized with SSEA-1 in the SVZ of mouse brains (Nakatani et al. 2010). In mouse neurosphere cultures, GD3 is the predominant ganglioside species (Nakatani et al. 2010; Yanagisawa et al. 2004b), accounting for more than 80 % of the total gangliosides. For this reason, it has been proposed that it can serve as a biomarker for mouse NSCs (Nakatani et al. 2010).

Heparin Sulfate Proteoglycans and Chondroitin Sulfate Proteoglycans

Proteoglycans, the major components of extracellular matrices (ECM), are a class of glycosylated proteins possessing covalently linked GAGs, sulfated carbohydrate chains made of repeating disaccharides. Proteoglycans are categorized into a number of subclasses, based on the components of disaccharides. For example, proteoglycans containing heparan sulfate GAGs are classified as heparin sulfate proteoglycans (HSPGs), whereas proteoglycans containing chondroitin sulfate GAGs are classified as CSPGs. Both HSPGs and CSPGs are known to be expressed in NSCs. (See Chap. 5 for more details about HSPGs and CSPGs.)

9.5 Neurogenesis

Neurons and astrocytes are generated in the CNS by a defined temporal sequence. At early developmental stages, a preplate consisting of the earliest-born neurons and possibly other cell types are formed between the VZ and meninges at the brain surface. The VZ is a densely packed cell layer formed by morphologically homogeneous RGCs, and the SVZ is a second proliferative layer. Newly generated neurons migrate radially out of the proliferative zones and form a new laminar structure. This preplate is subsequently split into the marginal zone and subplate by waves of migrating neurons. The neurons in the lower layers VI and V are born first, followed by those in layers IV, III, and II in the cortex. During development, the VZ becomes smaller, and after neurogenesis is completed, the VZ is replaced by an ependymal cell layer. Postnatally, most of the SVZ disappears except along the lateral wall of the lateral ventricles, where it is considered an NSC niche in the adult state (Franco and Muller 2013; Pinto and Gotz 2007; Qian et al. 2000).

9.5.1 Polysialic Acid–Neural Cell Adhesion Molecule

The polysialic acid (PSA) carbohydrate structure (Finne et al. 1983), carried exclusively by the neural cell adhesion molecule (NCAM), is expressed in neuronal precursor cells (nIPCs, INPs). PSA is a linear homopolymer containing up to 200 α2–8-linked sialic acid residues (SAα2-8SAα2-). Polysialyltransferases, ST8SiaII (also known as STX) and ST8SiaIV (also known as PST), are the responsible enzymes catalyzing the synthesis of PSA (Angata and Fukuda 2003; Kleene and Schachner 2004; Rutishauser and Landmesser 1996). PSA has interesting properties, including its highly negative charges, a high level of hydration, and an excellent ability to bind cations. Its remarkable structure enables PSA-NCAM to regulate myelination, axon guidance, synapse formation, and functional plasticity of the nervous system (Angata and Fukuda 2003; Aubert et al. 1995; Charles et al. 2000; Kleene and Schachner 2004; Seki and Rutishauser 1998). PSA-NCAM is prominently expressed during neural development; enzymatic deletion of PSA represses cell migration and induces premature neuronal differentiation as seen in the sprouting of axons, outgrowth of dendrites and axons, and dendritic branching (Durbec et al. 2001; Petridis et al. 2004; Yamamoto et al. 2000). Polysialyltransferase-deficient mice show developmental and behavioral defects, such as reduction of long-term potentiation and long-term depression, misguidance of mossy fibers, and ectopic synapse formation in the hippocampus (Angata et al. 2004; Eckhardt et al. 2000). In mouse NSC overexpressing PSA, cell migration is enhanced and oligodendrocyte genesis is suppressed (Franceschini et al. 2004). Thus, it is considered that the chemical structure of PSA-NCAM may modify cell fate.

9.5.2 9-O-Acetyl GD3

Ganglioside 9-O-acetyl GD3 (CD60b) was detected in neuroblasts during neural development using the JONES antibody (Blum and Barnstable 1987; Mendez-Otero et al. 1988). 9-O-acetyl GD3 is expressed in the SVZ and along the rostral migration stream (RMS) in both embryonic and adult brains (Mendez-Otero and Cavalcante 1996). Most of migrating neuroblasts expressing 9-O-acetyl GD3 also express PSA-NCAM (Miyakoshi et al. 2012). A more recent study casts some doubt on the importance of 9-O-acetyl GD3 in these studies. GD3 synthase knockout mice, in which GD3 and its downstream products, including 9-O-acetyl GD3, are missing, appear "grossly" normal in development (Yang et al. 2007). This raises the intriguing question whether the 9-O-acetyl sialic acid residue is conjugated with a protein and it functions in a similar manner as 9-O-acetyl GD3.

9.5.3 Gangliosides

During neuronal differentiation, the concentration of GD3, which is the predominant ganglioside in NSCs, is rapidly decreased. Concomitantly, the levels of GM1, GD1a, GD1b, and GQ1b continuously increase in young animals, reaching a plateau during adulthood (Hirschberg et al. 1996; Nakatani et al. 2010; Ngamukote et al. 2007). This pattern change follows closely with the upregulation of *N*-acetylgalactosaminyltransferase (GalNAcT) expression (Ngamukote et al. 2007). The dramatic changes in the expression profile of gangliosides during neuronal cell differentiation clearly reflect the biological needs at the particular stages during brain development (Fig. 9.3).

9.6 Gliogenesis

9.6.1 Oligodendrogenesis

Oligodendrocytes, the chief myelin-forming cells in the CNS, are derived from RGCs. The myelin structures provide efficient axon insulation and facilitate conduction of nerve impulses. At E12.5, the earliest oligodendrocyte progenitor cells (OPCs) are located in the developing cerebral cortex. The number of OPCs in the cortex increases between E16 and birth. However, most of the early generated oligodendrocytes disappear after birth. This suggests that most of the oligodendrocytes present in the adult cortex are generated at a later stage (Kriegstein and Alvarez-Buylla 2009; Rowitch and Kriegstein 2010). Many glial cell biomarkers are glycoconjugates and are described below.

A2B5

The first ganglioside antigen expressed in cells of glial lineage is the A2B5 antigen. A2B5 is a monoclonal antibody originally developed by Eisenbarth et al. using embryonic retina cells as the immunogen (Eisenbarth et al. 1979). The antigens recognized by the A2B5 monoclonal antibody have been established as the c-series gangliosides, including GQ1c, GT1c, and GT3 (Kasai and Yu 1983; Saito et al. 2001). These c-series gangliosides are abundant in fish brains and in mammalian embryonic, but not adult brains (Ando and Yu 1979; Freischutz et al. 1994, 1995; Rosner et al. 1988; Yu and Ando 1980). During development, the expression of c-series gangliosides is diminished in favor of the a- and b-series gangliosides, and the rate-limiting enzyme appears to be ST-III (Freischutz et al. 1994). Glial-restricted precursors (GRPs) have been recognized by the expression of A2B5 (Rao and Mayer-Proschel 1997). It is uncertain, however, whether GRPs exist in vivo.

NG2

Nerve/glial antigen 2 (NG2)/CSPG4 is one of the important CSPGs and was originally identified in rat (Stallcup 1981). The mouse homologue is also known as AG2. NG2 is a CSPG highly expressed in embryonic and adult brains (Jones et al. 2002). NG2+ cells are considered to be committed OPCs in developing brain. O-2A progenitor cells, glial precursor cells capable of differentiating into oligodendrocytes and Type 2 astrocytes, are positive for NG2 (Levine and Stallcup 1987; Raff et al. 1983b). O-2A progenitors exist in the ventricular germinal zones of the embryonic CNS and proliferate, migrate, and disseminate throughout the developing CNS (Richardson et al. 2011). Although the number of O-2A progenitors is decreased after birth, they are still found albeit in smaller numbers in the adult nervous system. O-2A progenitors are uniformly distributed throughout the CNS and are associated with axons where they generate myelinating oligodendrocytes (Dawson et al. 2003; Ffrench-Constant and Raff 1986a, b). Since O-2A cells exclusively generate oligodendrocytes during normal development, the term O-2A has been replaced as oligodendrocyte precursor (OLP) or OPCs. It has been reported that $NG2^+$ cells in postnatal mouse brain exhibit characteristics of NSCs, such as multipotency to differentiate into oligodendrocytes and astrocytes as well as neurons; this claim, however, has not been confirmed. More recently, Cre-lox fate mapping experiments revealed that embryonic $NG2^+$ cells generate mainly oligodendrocytes and some astrocytes, but not neurons, in the ventral zone (Zhu et al. 2011). None of the cells express either astrocyte or oligodendrocyte lineage markers, suggesting that at least two distinct types (either oligodendrocyte precursors or astrocyte precursors) of $NG2^+$ cells exist in the embryonic CNS. On the other hand, postnatal $NG2^+$ cells generate only oligodendrocytes in vivo (Zhu et al. 2011).

With respect to its functional roles, NG2 has been shown to have a high affinity for FGF-2 and platelet-derived growth factor-AA (PDGF-AA); both are important mitogens for OPCs (Goretzki et al. 1999). The high affinity between NG2 and growth factors is similar to that of HSPGs, which possess a strong affinity for FGF2. NG2 is required for the responsiveness of PDGFα-receptor to PDGF-induced cell proliferation or migration. Interestingly, NG2 knockout mice do not exhibit an obvious mutant phenotype during CNS development (Grako et al. 1999; Thallmair et al. 2006). However, the observation that mice deficient in the Olig2 basic helix–loop–helix (bHLH) transcription factor exhibit severe defects in $NG2^+$ cells in the developing CNS (Ligon et al. 2006) indicates that development of $NG2^+$ cells requires Olig transcription factors, especially Olig2.

O4 and O1

As oligodendrocyte development proceeds, unique GSLs appear on the oligodendrocyte plasma membrane and myelin. These GSLs include the O4 (sulfatide; HSO_3-3Galβ1-1′Cer) and O1 antigens (galactosylceramide; GalCer; Galβ1-1′Cer), which also have been used as specific markers to define immature and mature oligodendrocytes, respectively (Zhang 2001). The O1 and O4 antigens play important roles as modulators of oligodendrocyte development and function as well as major components of the myelin sheath to facilitate nerve conduction. (Please see Chap. 12.) A series of studies have clearly shown that knockout mice deficient in GalCer synthase or sulfatide synthase present severe neurological deficits, such as tremor, progressive ataxia, and reduction of nerve conduction velocity (Bosio et al. 1996; Coetzee et al. 1996; Honke et al. 2002). In these knockout mice, morphologically normal-appearing compact myelin is preserved, but paranodal loops are absent from the axon, and paranodal junctions are abnormal (Honke et al. 2002). The number of oligodendrocytes is increased in sulfatide knockout mice, indicating that the O4 antigen, sulfatide, is a critical molecule for the negative regulation of terminal differentiation of oligodendrocytes (Hirahara et al. 2004). GalCer expression factor-1, a rat homologue of hepatocyte growth factor-regulated tyrosine kinase substrate, has been cloned as an inducer of O1 antigen expression (Ogura et al. 1998). Overexpression of this molecule causes suppression of cell proliferation, causing dramatic change in morphology to become fibroblast-like in appearance (Ogura and Tai 2002). Although GalCer expression factor-1 may regulate the expression of O1 and O4 antigens during glial development, the function of GalCer expression factor-1 in NSC and glial precursor cells remains to be investigated.

9.6.2 Astrogliogenesis

The cell bodies of RGCs remain in the VZ throughout the period of neurogenesis and neuronal migration. At the end of cortical development, most RGCs lose their ventricular attachment and migrate toward the cortical plate by a process of somal translocation. Most RGCs transform into astrocytes. Some astrocytes may divide locally before terminal differentiation as a population of astrocyte IPCs is present in embryonic and postnatal stages (Hajos et al. 1981; Ichikawa et al. 1983). On the day of birth [postnatal day (P) 0], most astrocyte precursors are found in the inner half of the cortical width. On P4, the majority of astrocyte precursors are distributed in the outer half of the cortical width. The pattern of gliogenesis in the early postnatal rat thus shows an inside-out tendency, in analogy to neurogenesis (Kriegstein and Alvarez-Buylla 2009; Rowitch and Kriegstein 2010).

gp130

The cell surface glycoprotein gp130, also known as CD130, is a receptor component and signal transducer of interleukin (IL)-6 (Taga et al. 1989). This molecule mediates signaling activated by all of the eight members of the IL-6 family of cytokines: IL-6, IL-11, leukemia inhibitory factor (LIF), ciliary neurotrophic factor (CNTF), oncostatin M, cardiotrophin-1, cardiotrophin-like cytokines/novel neurotrophin-1/B-cell stimulating factor 3, and neuropoietin. The signaling pathways that are activated by the IL-6 family of cytokines via gp130 include the following: the Janus kinase (JAK)-signal transducer and activator of transcription 3 (STAT3) pathway, the Ras–MAPK pathway, and the phosphatidylinositol 3 kinase–Akt pathway (Fukuda and Taga 2005). The IL-6 family of cytokines induces astrocyte differentiation of NSCs via activation of gp130 and the JAK–STAT pathway (Bonni et al. 1997). Thus, gp130 is involved primarily in the induction of astrocytic differentiation. Cardiotrophin-1 is proposed to be a bona fide inducer of astrocytic differentiation via the gp130 pathway in the developing brain (Barondes et al. 1994). Deletion of gp130 results in reduction of the number of astrocytes in the developing mouse brain (Nakashima et al. 1999a). Astrocytic differentiation, however, is not regulated only by the IL-6 family of cytokines, gp130, and downstream JAK–STAT pathway signaling molecules. For instance, positive and negative cross talk between gp130 signaling and that of bone morphogenetic proteins (BMP) (Nakashima et al. 1999b), or Notch-hairy-enhancer of split (HES) signaling (Kamakura et al. 2004), as well as with neurogenin-2 (a bHLH transcription factor) (Sun et al. 2001), has been identified. Also, the epigenetic status of astrocytic genes in NSCs is critical for astrocyte differentiation via the gp130 pathway (Takizawa et al. 2001). In addition, gp130 is involved in maintenance of the proliferation of NSCs. CNTF maintains embryonic and adult NSCs in an undifferentiated state by blocking differentiation via gp130 signaling in cultured NSCs (Shimazaki et al. 2001). Conversely, CNTF lacking a secretory signal sequence is localized in the cytosol. Therefore, CNTF is not considered a secreted cytokine during brain development. Another member of the IL-6 family of cytokines, neuropoietins, was postulated to share the biological functions of CNTF (Derouet et al. 2004). Thus, NSC proliferation may be maintained by more than one IL-6 cytokine. Gp130 signaling has also been reported to support NSC survival via activation of the phosphatidylinositol 3 kinase–Akt pathway (Chang et al. 2004). Recently, unglycosylated gp130 present on the outer surface of the plasma membrane was found to be unable to form a heterodimer with the LIF receptor resulting in failure for signaling, because the unglycosylated gp130 could not be phosphorylated in response to LIF stimulation (Yanagisawa and Yu 2009). The above examples clearly show the N-glycans of gp130 are crucial for its activation, but not its cellular localization.

PtdGlc

A phosphoglycerolipid, phosphatidylglucoside (PtdGlc), is expressed in astrocytes and radial glia in rat embryonic brain (Nagatsuka et al. 2001). PtdGlc is localized in the lipid rafts, which are thought to operate as sorting platforms that bring together molecules for efficient cross talk that controls cellular signaling cascades, including those regulating cell proliferation and differentiation (Nagatsuka et al. 2003). A PtdGlc monoclonal antibody DIM21 has been developed that labels RGCs at E12.5-E14.5, astrocytes in late embryo to early postnatal stages, and RGC-like cells in the adult SVZ (Kinoshita et al. 2009a, b; Nagatsuka et al. 2006). In an in vitro study, the association of EGF receptors and PtdGlc-enriched lipid rafts was confirmed in NSCs, and PtdGlc-enriched lipid rafts were found to control NSC to astrocyte differentiation through EGF signaling (Kinoshita et al. 2009a). (See also Chap. 4.)

Gangliosides

GM3 and GD3 represent about 70 % and 10–20 %, respectively, of the total gangliosides in astrocytes bulk-isolated from neonatal rat brains (Asou et al. 1989; Murakami et al. 1999; Sbaschnig-Agler et al. 1988). Both Type 1 and Type 2 astrocytes express GM3. On the other hand, a recent study indicated that GD3 is expressed only in Type 2, but not in Type 1 astrocytes (Murakami et al. 1999). Type 2 astrocytes are known to express c-series gangliosides, which are recognized by mAb A2B5. The expression of GD3 and c-series gangliosides ($A2B5^+$) in Type 2 astrocytes indicates that these cells might have more similar properties as progenitor or immature cells than Type 1 astrocytes. mAb A2B5 has been recognized to identify GRPs (Rao and Mayer-Proschel 1997). On the other hand, Type 1 astrocytes are $GFAP^+/A2B5^-$, whereas Type 2 astrocytes express both GFAP and A2B5 (Raff et al. 1983a). For other cells of glial lineage, OPCs have been identified by $NG2^+/$ PDGFR-α^+ (Goretzki et al. 1999). In addition, immature oligodendrocytes express NG2 and can be identified by the phenotypic marker O4 (sulfatide), whereas mature oligodendrocytes express the O1 antigen (GalCer) (Fig. 9.4).

9.7 Adult NSCs and Niche

Although neurogenesis is mostly complete at the time of the development in most mammals, it continues to occur at a much slower pace and in a limited manner throughout the entire adult life. In the adult brain of mammals, neurogenesis persists primarily in two germinal zones, the SVZ of the lateral ventricles (Doetsch et al. 1997, 1999) and the subgranular zone (SGZ) in the dentate gyrus of the hippocampus (Seri et al. 2001; Suhonen et al. 1996).

9.7.1 SVZ

In the adult SVZ, four distinct cell types are present (Fig. 9.8). Type B cells are RGC-like cells and have been considered as NSCs. Type B NSCs are slow dividing (duration of cell cycle >15 days) and express GFAP. Type C cells are transient amplifying cells that are rapidly proliferating (duration of cell cycle about 13 h) and express the transcription factor Mash1. Type A cells are neuronal precursors that have already committed to differentiate into neurons, and these cells express PSA-NCAM on the cell surface (duration of cell cycle about 13 h) (Morshead 2004). Ependymal cells are lined on the wall of the ventricle and have multi-motile cilia, which are important for controlling the flow of cerebrospinal fluid (CSF). Multipotency of the ependymal cells has been reported (Johansson et al. 1999), although this is not settled (Chiasson et al. 1999; Doetsch et al. 1999; Laywell et al. 2000). Recently, ependymal cells were shown to be the most quiescent type of NSCs whose cell cycle is strictly regulated and reinitiated under specific circumstances. In certain restricted situations, a subpopulation of ependymal cells may develop into neurons, and these cells are considered as NSCs (Carlen et al. 2009; Coskun et al. 2008).

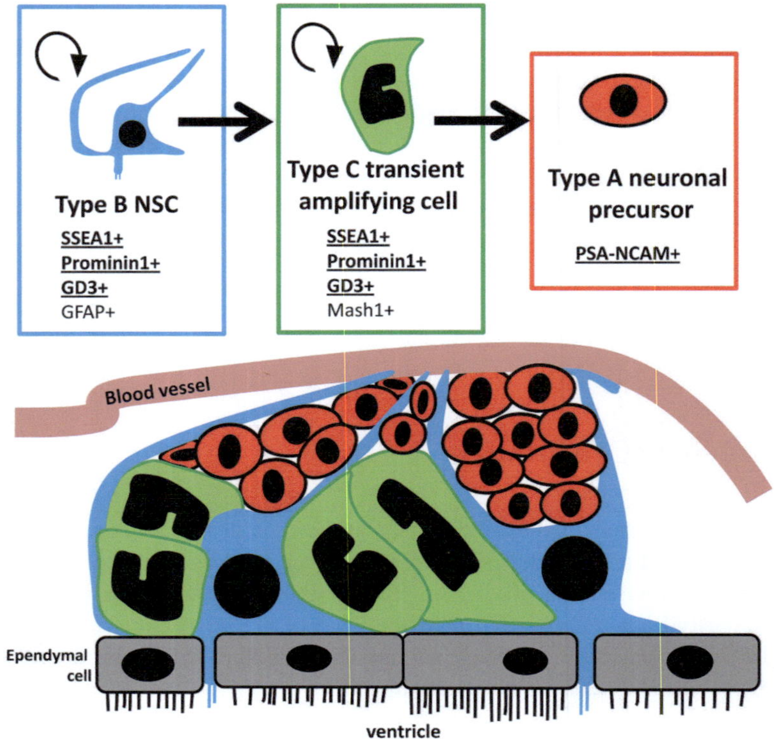

Fig. 9.8 Neural stem cell niche at the subventricular zone (SVZ) on the surface of the lateral ventricle in the adult mouse brain. Type B cells are radial glial cell (RGC)-like cells and have been considered as neural stem cells (NSCs). Type C cells are transient amplifying cells that are rapidly proliferating. Type A cells are neuronal precursors that have already committed to becoming neurons expressing PSA-NCAM on the cell surface. The known carbohydrate markers are *underlined*

9.7.2 SGZ

Five types of cells have been described in the SGZ (Filippov et al. 2003). Type 1 cells are considered quiescent neural progenitors that are RGC-like cells and largely equivalent to Type B NSCs in the SVZ. Type 2 cells express nestin, and this cell type has been classified into two cell populations: Type 2a cells are amplifying neural progenitors that are similar to Type C transient amplifying cells in the SVZ; Type 2b and 3 cells are neuroblasts that express PSA-NCAM (Encinas et al. 2006; Steiner et al. 2006). The other type of cells is mature granule neurons. Recently, it has been reported that $Mash1^+$ cells do not amplify and are therefore not Type 2a amplifying neural progenitors that can directly differentiate into early neuroblasts without mitosis (Lugert et al. 2012).

9.7.3 Glycoconjugates in Adult NSCs

Ganglioside GD3, SSEA-1, and prominin-1 are expressed in Type B NSCs in the SVZ and in Type 1 quiescent neural progenitors in the SGZ of the adult brain (Beckervordersandforth et al. 2010; Cammer and Zhang 1996a; Capela and Temple 2002; Nakatani et al. 2010; Walker et al. 2013). The intensity of prominin-1 expression in the SGZ is heterogeneous. Cells that do not express prominin-1 are not NSCs, but cells with intermediary or low levels of prominin-1 expression possess NSC properties. Analysis of cells deficient in prominin-1 indicates that there is no difference in the number of astrocytes, oligodendrocytes, neural precursor cells, or adult-born early postmitotic neurons, nor is there any difference in the ability for neurosphere formation.

9.7.4 Lectins

Lectins are carbohydrate-binding proteins that do not act enzymatically on their corresponding ligands. They are found in all kinds of organisms, including plants, microorganisms, animals, and humans. Each lectin specifically recognizes a monosaccharide or oligosaccharide structure and binds to glycoconjugates present in insoluble or soluble form (Sharon 2008; Sharon and Lis 1972). As neural and glial cells express various glycoconjugates, specific lectins can be used effectively for histochemical identification or sorting of specific cell types from heterogeneous NSC populations (Yanagisawa and Yu 2007). For example, the low expression of peanut agglutinin (PNA) ligand and the heat stable antigen (HSA, CD24a) in adult NSCs permit them to be effectively separated by negative selection (Rietze et al. 2001). PNA binds to the Galβ1-3GalNAc structure that is part of the ganglio-series ganglioside structure. For this reason, PNA is useful in recognizing those gangliosides. Another lectin, *Phaseolus vulgaris* erythroagglutinating lectin (PHA-E4), which binds to biantennary complex type N-glycans, has been used to isolate embryonic and adult NSCs by positive selection (Hamanoue et al. 2009), while *Ricinus communis* agglutinin (RCA), which binds to Galβ1-4GlcNAc-, has been used to detect Type A neuronal precursors (Kitada et al. 2011). Other lectins, such as *Agaricus bisporus* agglutinin (ABA) that shows specificity for Galβ1-3GalNAcα1, and PHA-E4 and wheat germ agglutinin (WGA) that show specificity for GlcNAc1-4GlcNAc recognize Type B NSCs and Type C transient amplifying cells in the SVZ as well as Type 1 quiescent neural progenitors and Type 2a amplifying neural progenitors in the SGZ (Kitada et al. 2011). These lectins are useful for the identification and purification of specific populations of NSCs.

9.8 Neural Crest Cells

The precursors of neural crest cells reside in the dorsal neural tube, and these cells undergo epithelial to mesenchymal transition (EMT) and detach from the neural tube and migrate during development to diverse locations (Fig. 9.6) (Anderson 1997; Sauka-Spengler and Bronner-Fraser 2008). Neural crest cells contain a population of neural crest stem cells (Bronner-Fraser and Fraser 1988; Morrison et al. 1999; Stemple and Anderson 1992). Neural crest stem cells are capable of self-renewal and have the multipotency to differentiate into Schwann cells upon induction with glial growth factor (Shah et al. 1994), autonomic neurons by induction with BMP (Shah et al. 1996), and smooth muscles by induction with transforming growth factor-β (Shah et al. 1996).

9.8.1 HNK-1

The human natural killer-1 (HNK-1) antigen (CD57) is a carbohydrate antigen whose structure has been established as HSO3-3GlcAβ1-3Galβ1-4GlcNAc- (Ariga et al. 1987; Chou et al. 1986). HNK-1 is distributed on the surface of neural crest cells and is required for their proper migration during development in avian, rat, and human (Bronner-Fraser 1987; Holley and Yu 1987; Nagase et al. 2003; Tucker et al. 1988). However, mouse neural crest cells are negative for HNK-1 expression by immunohistochemistry in fixed cryo-sections (Tucker et al. 1988). In a careful study using synthetic model compounds, the minimal carbohydrate unit for the HNK-1 epitope was shown to reside in the terminal disaccharide structure HSO3-3GlcAβ1-(Tokuda et al. 1998). A commonly used HNK-1 immunoreagent is an IgM monoclonal antibody, e.g., mouse mAb Leu 7, whose large molecular size (about 970 kDa) could sterically hinder its ability to cross-react with epitopes in fixed whole-mount tissues (Abo and Balch 1981). It is well known that fixed cells and living cells have far different staining patterns in studies using immunohistochemistry and flow cytometry. Loss of antigenicity with fixation could be caused by the conditions of immunohistochemical detection. Neural crest stem cells are isolated not only from the neural fold and neural tube but also from fetal peripheral nerve (Morrison et al. 1999) and fetal and postnatal gut (Bixby et al. 2002). Most recently Walters and colleagues found murine living neural crest stem cells do express HNK-1 (Walters et al. 2010). Less than half of murine HNK-1$^+$ cells express SRY (sex determining region Y)-box 10 (Sox10), known to be expressed in neural crest stem cells. Thus, the expression of HNK-1 alone is not sufficient to isolate a population of pure neural crest stem cells. The HNK-1 epitope is associated with a number of cell adhesion molecules (Jungalwala 1994; Kruse et al. 1984). Of particular interest is the fact that carrier molecules of the HNK-1 epitope can be a glycoprotein or a glycolipid. Among the glycoprotein antigens are L1, P0, MAG, and NCAM (Kruse et al. 1984), while the glycolipid antigens include just the two sulfated glucuronosyl glycolipids (SGGLs), sulfated glucuronosyl paragloboside (SGPG), and sulfated glucuronosyl lactosaminyl paragloboside (SGLPG) (Ariga et al. 1987; Chou et al. 1986). Interestingly, certain proteoglycans, e.g., CSPGs (Domowicz et al. 1995; Margolis et al. 1987; Pettway et al. 1996), cross-react with the HNK-1 antibody. Because of its wide distribution on various glycoconjugates, the HNK-1 epitope is expected to have important roles in neural development. So far, studies of brains from mice deficient in glucuronyltransferase P (GlcAT-P) or HNK-1 sulfotransferase, the two key enzymes of HNK-1 antigen synthesis, have not revealed any overt defect in brain development (Yamamoto et al. 2002). However, adult mice deficient in GlcAT-P or HNK-1 sulfotransferase exhibit reduced long-term potentiation and defective spatial memory formation, suggesting a functional role of the HNK-1 antigen in synaptic plasticity of the hippocampus, but not in brain development. Recently, HNK-1 expression in mouse embryonic NPCs was confirmed, and the HNK-1 epitope was present almost exclusively on tenascin-C (Yagi et al. 2010b; Yanagisawa et al. 2005).

9.8.2 PSA-NCAM

As a marker of neuronal precursor cells, PSA-NCAM is expressed not only in the CNS but also in the PNS. Sensory and autonomic neurons of rodents express PSA-NCAM (Boisseau et al. 1991; Stemple and Anderson 1992). Also PSA-NCAM expression is seen in the development of the enteric nervous system. PSA-NCAM$^+$ precursor cells from vagal, sacral, and rostral truncal regions of the neural crest migrate to the gut, stop at appropriate locations, proliferate and differentiate into the many different phenotypes of enteric neurons, form two ganglionated plexuses, and establish correct interconnections (Epstein et al. 1991; Heuckeroth et al. 1998; Le Douarin and Teillet 1973). The expression of high PSA-NCAM is restricted to early neuronal lineage cells derived from neural crest cells (Boisseau et al. 1991). In the rat gut, polysialyltransferases PST and STX are highly expressed at E14 to E18 and then downregulated postnatally (Faure et al. 2007). Approximately 30 % of neuron-committed cells in the myenteric layer express PSA-NCAM at E12. The number of PSA-NCAM$^+$ cells in the mesenteric plexus increases to 50 % at E14 and E16 and 80 % at E18 to E20 and then declines gradually during postnatal life. About 50 % of the cells committed to neuronal differentiation in the submucosal layer are PSA-NCAM$^+$ at E18 to E20. At P14 to P24, less than 10 % of the cells express PSA-NCAM in the submucosal plexus. In the development of the enteric nervous system, BMP enhances migration, neurite fasciculation, and clustering of neuronal cells via promotion of polysialylation of NCAM in the enteric nervous system formed from neural crest cells (Faure et al. 2007; Fu et al. 2006).

9.8.3 Other Glycoconjugates

Other glycoconjugate markers reported to be present in neural crest stem cells and neural crest-derived cells include GD3 in mouse neural crest cells (Stainier et al. 1991); SSEA-1 in cells committed to differentiating into sensory neurons in quail (Sieber-Blum 1989); B30 gangliosides, which are unidentified gangliosides recognized by the B30 antibody (one migrates between GM2 and GM1 and the other migrates between GD3 and GD1a on thin-layer chromatography) in mouse sensory neurons (Stainier et al. 1991); and O4 antigen (sulfatide) in Schwann cells and their precursors (Dong et al. 1999; Stemple and Anderson 1992).

9.9 Future Studies

Since their discovery, progress in the biology of NSCs has been made owing to their importance in the development of the nervous system. NSCs are characterized by their capacity for self-renewal and their ability to differentiate into neurons and glia (multipotency). Remarkably, they can be isolated not only from embryonic brains (Stemple and Anderson 1992) but from adult CNS tissue as well (Reynolds and Weiss 1992). NSCs cultured using the floating culture method with EGF and FGF-2 in a defined serum-free medium form neurospheres, which consist not only of NSCs, but of rather heterogeneous undifferentiated cell populations. A more homogenous population of cells can be prepared using monolayer, serum-free culture. NSCs from neurospheres or monolayer cultures can be induced to differentiate into multiple neural lineages upon growth factor withdrawal (Reynolds and Weiss 1992). The availability of relatively pure NSCs in culture has greatly enhanced current knowledge about the molecular mechanisms underlying cell fate determination and ultimately brain development. Moreover, cell reprogramming studies have indicated that lineage-restricted neuronal and glial precursors can display acquired properties that are not evident in vivo (Gabay et al. 2003; Kondo and Raff 2000; Palmer et al. 1999). For example, treatment of cells with fetal serum or BMPs can reprogram NG2 positive cells to generate NSCs containing reprogrammed multipotential stem cells that can differentiate into neurons, astrocytes, and oligodendrocytes (Kondo and Raff 2000). During reprogramming from OPCs to NSCs, chromatin remodeling and histone modifications occurred (Kondo and Raff 2004). The ability to manipulate NSC cell fate determination in vitro has greatly facilitated understanding of the properties and regulatory mechanisms of NSCs in the developing nervous system and adult brain that would have been difficult to decipher in vivo.

Glycoconjugates, including glycolipids and glycoproteins, are predominantly expressed on the cell surface. Because of their structural diversity, they have been used effectively as cell surface biomarkers for identification and isolation of specific cell types. During neural development and neuronal/glial cell differentiation, these glycoconjugates frequently undergo dramatic qualitative and quantitative changes that correlate with cellular changes. There is an urgent need to answer the question of whether these changes are merely consequences of differentiation needed to satisfy biological needs, such as cell–cell recognition, migration, and adhesion. More importantly, recent evidence has shed light on their roles in modulating signaling pathways during cellular differentiation and reprogramming. For example, we found that cell surface SSEA-1 modulates NSC proliferation mediated by the Notch signaling pathway and migration (Yagi et al. 2012; Yanagisawa et al. 2005). In addition, it was shown that GM3 can modulate EGF receptor function by inhibiting its tyrosine kinase activity. Most recently, we showed that GD3 associated with EGF receptor to modulate NSC proliferation (unpublished data). Additionally, GSLs have been shown to play an important role in the epithelial mesenchymal transition (EMT); changes in cell surface glycolipid expression by inhibition of glucosylceramide synthesis convert epithelial cells to a fibroblastic morphology (Guan et al. 2009). Conversely, overexpression of prominin in fibroblasts induces an epithelial cell-like phenotype with an abundance of microvilli-like protrusions (mesenchymal epithelial-like cell transition; MET) (Singh et al. 2013). These studies indicate that cell surface glycoconjugates may control cell fate in order to effect transdifferentiation of one cell type into another. Clearly this represents a fruitful area of future research.

Another critical area for future investigation is the basis of induced pluripotent stem cell (iPSC) generation. Cell surface glycoconjugates again occupy an important area for study. For example, in human fibroblasts, less than 1 % of the cells express SSEA-3 and SSEA-3$^+$ dermal fibroblast-enhanced iPSC generation, while no iPSCs could be generated from the SSEA-3$^-$ fraction (Reijo Pera et al. 2009; Wakao et al. 2011). SSEA-3$^+$ fibroblast and bone marrow stromal cells host a multipotent stem cell population that can generate the three germ layers without Yamanaka factors, such as Oct3/4, Sox2, c-Myc, and Klf4 (Kuroda et al. 2010). It clearly shows that SSEA-3 plays a crucial role during reprogramming of fibroblasts to stem cells in maintaining stemness. Although SSEA-1, SSEA-3, SSEA-4, GD3, and prominin-1 are all expressed in stem cells, mice deficient in one of these molecules show only subtle phenotypic abnormalities compared with the wild-type animals. Clearly, the biological function of one glycoconjugate can be substituted by another, albeit with less efficiency. The "biological redundancy" phenomenon governing cellular events needs to be further defined. Future studies in this regard should contribute greatly to regenerative and reparative biology.

Conflict of Interest The authors declare no conflicts of interest.

References

IUPAC-IUB Commission on Biochemical Nomenclature. The nomenclature of lipids. Recommendations (1976) Lipids. 1977;12(6):455–68.

Aaku-Saraste E, Hellwig A, Huttner WB. Loss of occludin and functional tight junctions, but not ZO-1, during neural tube closure-remodeling of the neuroepithelium prior to neurogenesis. Dev Biol. 1996;180(2):664–79.

Abo T, Balch CM. A differentiation antigen of human NK and K cells identified by a monoclonal antibody (HNK-1). J Immunol. 1981;127(3):1024–9.
Adewumi O, Aflatoonian B, Ahrlund-Richter L, Amit M, Andrews PW, Beighton G, et al. Characterization of human embryonic stem cell lines by the International Stem Cell Initiative. Nat Biotechnol. 2007;25(7):803–16.
Alvarez-Buylla A, Garcia-Verdugo JM, Tramontin AD. A unified hypothesis on the lineage of neural stem cells. Nat Rev Neurosci. 2001;2(4):287–93.
Anderson DJ. Cellular and molecular biology of neural crest cell lineage determination. Trends Genet. 1997;13(7):276–80.
Ando S, Yu RK. Isolation and characterization of two isomers of brain tetrasialogangliosides. J Biol Chem. 1979;254(23):12224–9.
Angata K, Fukuda M. Polysialyltransferases: major players in polysialic acid synthesis on the neural cell adhesion molecule. Biochimie. 2003;85(1–2):195–206.
Angata K, Long JM, Bukalo O, Lee W, Dityatev A, Wynshaw-Boris A, et al. Sialyltransferase ST8Sia-II assembles a subset of polysialic acid that directs hippocampal axonal targeting and promotes fear behavior. J Biol Chem. 2004;279(31):32603–13.
Ariga T, Kohriyama T, Freddo L, Latov N, Saito M, Kon K, et al. Characterization of sulfated glucuronic acid containing glycolipids reacting with IgM M-proteins in patients with neuropathy. J Biol Chem. 1987;262(2):848–53.
Asou H, Hirano S, Uyemura K. Ganglioside composition of astrocytes. Cell Struct Funct. 1989;14(5):561–8.
Attardo A, Calegari F, Haubensak W, Wilsch-Brauninger M, Huttner WB. Live imaging at the onset of cortical neurogenesis reveals differential appearance of the neuronal phenotype in apical versus basal progenitor progeny. PLoS One. 2008;3(6):e2388.
Aubert I, Ridet JL, Gage FH. Regeneration in the adult mammalian CNS: guided by development. Curr Opin Neurobiol. 1995;5(5):625–35.
Bannerman PG, Oliver TM, Xu Z, Shieh A, Pleasure DE. Effects of FGF-1 and FGF-2 on GD3 immunoreactive spinal neuroepithelial cells. J Neurosci Res. 1996;45(5):549–57.
Barondes SH, Castronovo V, Cooper DN, Cummings RD, Drickamer K, Feizi T, et al. Galectins: a family of animal beta-galactoside-binding lectins. Cell. 1994;76(4):597–8.
Beckervordersandforth R, Tripathi P, Ninkovic J, Bayam E, Lepier A, Stempfhuber B, et al. In vivo fate mapping and expression analysis reveals molecular hallmarks of prospectively isolated adult neural stem cells. Cell Stem Cell. 2010;7(6):744–58.
Bieberich E, MacKinnon S, Silva J, Yu RK. Regulation of apoptosis during neuronal differentiation by ceramide and b-series complex gangliosides. J Biol Chem. 2001;276(48):44396–404.
Bixby S, Kruger GM, Mosher JT, Joseph NM, Morrison SJ. Cell-intrinsic differences between stem cells from different regions of the peripheral nervous system regulate the generation of neural diversity. Neuron. 2002;35(4):643–56.
Blum AS, Barnstable CJ. O-acetylation of a cell-surface carbohydrate creates discrete molecular patterns during neural development. Proc Natl Acad Sci U S A. 1987;84(23):8716–20.
Boisseau S, Nedelec J, Poirier V, Rougon G, Simonneau M. Analysis of high PSA N-CAM expression during mammalian spinal cord and peripheral nervous system development. Development. 1991;112(1):69–82.
Bonni A, Sun Y, Nadal-Vicens M, Bhatt A, Frank DA, Rozovsky I, et al. Regulation of gliogenesis in the central nervous system by the JAK-STAT signaling pathway. Science. 1997;278(5337):477–83.
Bosio A, Binczek E, Stoffel W. Functional breakdown of the lipid bilayer of the myelin membrane in central and peripheral nervous system by disrupted galactocerebroside synthesis. Proc Natl Acad Sci U S A. 1996;93(23):13280–5.
Bronner-Fraser M. Perturbation of cranial neural crest migration by the HNK-1 antibody. Dev Biol. 1987;123(2):321–31.
Bronner-Fraser M, Fraser SE. Cell lineage analysis reveals multipotency of some avian neural crest cells. Nature. 1988;335(6186):161–4.
Bruckner G, Biesold D. Histochemistry of glycogen deposition in perinatal rat brain: importance of radial glial cells. J Neurocytol. 1981;10(5):749–57.

Bruckner K, Perez L, Clausen H, Cohen S. Glycosyltransferase activity of Fringe modulates Notch-Delta interactions. Nature. 2000;406(6794):411–5.

Cammer W, Zhang H. Carbonic anhydrase II in microglia in forebrains of neonatal rats. J Neuroimmunol. 1996a;67(2):131–6.

Cammer W, Zhang H. Ganglioside GD3 in radial glia and astrocytes in situ in brains of young and adult mice. J Neurosci Res. 1996b;46(1):18–23.

Capela A, Temple S. LeX/ssea-1 is expressed by adult mouse CNS stem cells, identifying them as nonependymal. Neuron. 2002;35(5):865–75.

Capela A, Temple S. LeX is expressed by principle progenitor cells in the embryonic nervous system, is secreted into their environment and binds Wnt-1. Dev Biol. 2006;291(2):300–13.

Carlen M, Meletis K, Goritz C, Darsalia V, Evergren E, Tanigaki K, et al. Forebrain ependymal cells are Notch-dependent and generate neuroblasts and astrocytes after stroke. Nat Neurosci. 2009;12(3):259–67.

Chang MY, Park CH, Son H, Lee YS, Lee SH. Developmental stage-dependent self-regulation of embryonic cortical precursor cell survival and differentiation by leukemia inhibitory factor. Cell Death Differ. 2004;11(9):985–96.

Charles P, Hernandez MP, Stankoff B, Aigrot MS, Colin C, Rougon G, et al. Negative regulation of central nervous system myelination by polysialylated-neural cell adhesion molecule. Proc Natl Acad Sci U S A. 2000;97(13):7585–90.

Chiasson BJ, Tropepe V, Morshead CM, van der Kooy D. Adult mammalian forebrain ependymal and subependymal cells demonstrate proliferative potential, but only subependymal cells have neural stem cell characteristics. J Neurosci. 1999;19(11):4462–71.

Chou DK, Ilyas AA, Evans JE, Costello C, Quarles RH, Jungalwala FB. Structure of sulfated glucuronyl glycolipids in the nervous system reacting with HNK-1 antibody and some IgM paraproteins in neuropathy. J Biol Chem. 1986;261(25):11717–25.

Coetzee T, Fujita N, Dupree J, Shi R, Blight A, Suzuki K, et al. Myelination in the absence of galactocerebroside and sulfatide: normal structure with abnormal function and regional instability. Cell. 1996;86(2):209–19.

Coskun V, Wu H, Blanchi B, Tsao S, Kim K, Zhao J, et al. CD133+ neural stem cells in the ependyma of mammalian postnatal forebrain. Proc Natl Acad Sci U S A. 2008;105(3):1026–31.

Dawson MR, Polito A, Levine JM, Reynolds R. NG2-expressing glial progenitor cells: an abundant and widespread population of cycling cells in the adult rat CNS. Mol Cell Neurosci. 2003;24(2):476–88.

de la Pompa JL, Wakeham A, Correia KM, Samper E, Brown S, Aguilera RJ, et al. Conservation of the Notch signalling pathway in mammalian neurogenesis. Development. 1997;124(6):1139–48.

Derouet D, Rousseau F, Alfonsi F, Froger J, Hermann J, Barbier F, et al. Neuropoietin, a new IL-6-related cytokine signaling through the ciliary neurotrophic factor receptor. Proc Natl Acad Sci U S A. 2004;101(14):4827–32.

Doetsch F, Caille I, Lim DA, Garcia-Verdugo JM, Alvarez-Buylla A. Subventricular zone astrocytes are neural stem cells in the adult mammalian brain. Cell. 1999;97(6):703–16.

Doetsch F, Garcia-Verdugo JM, Alvarez-Buylla A. Cellular composition and three-dimensional organization of the subventricular germinal zone in the adult mammalian brain. J Neurosci. 1997;17(13):5046–61.

Domowicz M, Li H, Hennig A, Henry J, Vertel BM, Schwartz NB. The biochemically and immunologically distinct CSPG of notochord is a product of the aggrecan gene. Dev Biol. 1995;171(2):655–64.

Dong Z, Sinanan A, Parkinson D, Parmantier E, Mirsky R, Jessen KR. Schwann cell development in embryonic mouse nerves. J Neurosci Res. 1999;56(4):334–48.

Durbec P, Cremer H. Revisiting the function of PSA-NCAM in the nervous system. Mol Neurobiol. 2001 Aug-Dec;24(1-3):53–64.

Dvorak P, Hampl A, Jirmanova L, Pacholikova J, Kusakabe M. Embryoglycan ectodomains regulate biological activity of FGF-2 to embryonic stem cells. J Cell Sci. 1998;111(Pt 19):2945–52.

Eckhardt M, Bukalo O, Chazal G, Wang L, Goridis C, Schachner M, et al. Mice deficient in the polysialyltransferase ST8SiaIV/PST-1 allow discrimination of the roles of neural cell adhesion

molecule protein and polysialic acid in neural development and synaptic plasticity. J Neurosci. 2000;20(14):5234–44.
Eisenbarth GS, Walsh FS, Nirenberg M. Monoclonal antibody to a plasma membrane antigen of neurons. Proc Natl Acad Sci U S A. 1979;76(10):4913–7.
Encinas JM, Vaahtokari A, Enikolopov G. Fluoxetine targets early progenitor cells in the adult brain. Proc Natl Acad Sci U S A. 2006;103(21):8233–8.
Epstein ML, Poulsen KT, Thiboldeaux R. Formation of ganglia in the gut of the chick embryo. J Comp Neurol. 1991;307(2):189–99.
Fang Y, Wu G, Xie X, Lu ZH, Ledeen RW. Endogenous GM1 ganglioside of the plasma membrane promotes neuritogenesis by two mechanisms. Neurochem Res. 2000;25(7):931–40.
Faure C, Chalazonitis A, Rheaume C, Bouchard G, Sampathkumar SG, Yarema KJ, et al. Gangliogenesis in the enteric nervous system: roles of the polysialylation of the neural cell adhesion molecule and its regulation by bone morphogenetic protein-4. Dev Dyn. 2007; 236(1):44–59.
Fenderson BA, Eddy EM, Hakomori S. Glycoconjugate expression during embryogenesis and its biological significance. Bioessays. 1990;12(4):173–9.
Fenderson BA, Zehavi U, Hakomori S. A multivalent lacto-N-fucopentaose III-lysyllysine conjugate decompacts preimplantation mouse embryos, while the free oligosaccharide is ineffective. J Exp Med. 1984;160(5):1591–6.
Ffrench-Constant C, Raff MC. The oligodendrocyte-type-2 astrocyte cell lineage is specialized for myelination. Nature. 1986a;323(6086):335–8.
Ffrench-Constant C, Raff MC. Proliferating bipotential glial progenitor cells in adult rat optic nerve. Nature. 1986b;319(6053):499–502.
Filippov V, Kronenberg G, Pivneva T, Reuter K, Steiner B, Wang LP, et al. Subpopulation of nestin-expressing progenitor cells in the adult murine hippocampus shows electrophysiological and morphological characteristics of astrocytes. Mol Cell Neurosci. 2003;23(3):373–82.
Finne J, Finne U, Deagostini-Bazin H, Goridis C. Occurrence of alpha 2-8 linked polysialosyl units in a neural cell adhesion molecule. Biochem Biophys Res Commun. 1983;112(2): 482–7.
Fishell G, Kriegstein A. Cortical development: new concepts. Neuron. 2005;46(3):361–2.
Fishell G, Kriegstein AR. Neurons from radial glia: the consequences of asymmetric inheritance. Curr Opin Neurobiol. 2003;13(1):34–41.
Franceschini I, Vitry S, Padilla F, Casanova P, Tham TN, Fukuda M, et al. Migrating and myelinating potential of neural precursors engineered to overexpress PSA-NCAM. Mol Cell Neurosci. 2004;27(2):151–62.
Franco SJ, Muller U. Shaping our minds: stem and progenitor cell diversity in the mammalian neocortex. Neuron. 2013;77(1):19–34.
Freischutz B, Saito M, Rahmann H, Yu RK. Activities of five different sialyltransferases in fish and rat brains. J Neurochem. 1994;62(5):1965–73.
Freischutz B, Saito M, Rahmann H, Yu RK. Characterization of sialyltransferase-IV activity and its involvement in the c-pathway of brain ganglioside metabolism. J Neurochem. 1995;64(1): 385–93.
Fu M, Vohra BP, Wind D, Heuckeroth RO. BMP signaling regulates murine enteric nervous system precursor migration, neurite fasciculation, and patterning via altered Ncam1 polysialic acid addition. Dev Biol. 2006;299(1):137–50.
Fujita S. The discovery of the matrix cell, the identification of the multipotent neural stem cell and the development of the central nervous system. Cell Struct Funct. 2003;28(4):205–28.
Fukuda S, Taga T. Cell fate determination regulated by a transcriptional signal network in the developing mouse brain. Anat Sci Int. 2005;80(1):12–8.
Furukawa K, Aixinjueluo W, Kasama T, Ohkawa Y, Yoshihara M, Ohmi Y, et al. Disruption of GM2/GD2 synthase gene resulted in overt expression of 9-O-acetyl GD3 irrespective of Tis21. J Neurochem. 2008;105(3):1057–66.
Gabay L, Lowell S, Rubin LL, Anderson DJ. Deregulation of dorsoventral patterning by FGF confers trilineage differentiation capacity on CNS stem cells in vitro. Neuron. 2003;40(3): 485–99.

Gaiano N, Fishell G. The role of notch in promoting glial and neural stem cell fates. Annu Rev Neurosci. 2002;25:471–90.

Goldman JE, Hirano M, Yu RK, Seyfried TN. GD3 ganglioside is a glycolipid characteristic of immature neuroectodermal cells. J Neuroimmunol. 1984;7(2–3):179–92.

Goretzki L, Burg MA, Grako KA, Stallcup WB. High-affinity binding of basic fibroblast growth factor and platelet-derived growth factor-AA to the core protein of the NG2 proteoglycan. J Biol Chem. 1999;274(24):16831–7.

Gotz M, Huttner WB. The cell biology of neurogenesis. Nat Rev Mol Cell Biol. 2005; 6(10):777–88.

Grako KA, Ochiya T, Barritt D, Nishiyama A, Stallcup WB. PDGF (alpha)-receptor is unresponsive to PDGF-AA in aortic smooth muscle cells from the NG2 knockout mouse. J Cell Sci. 1999;112(Pt 6):905–15.

Graus-Porta D, Blaess S, Senften M, Littlewood-Evans A, Damsky C, Huang Z, et al. Beta1-class integrins regulate the development of laminae and folia in the cerebral and cerebellar cortex. Neuron. 2001;31(3):367–79.

Guan F, Handa K, Hakomori SI. Specific glycosphingolipids mediate epithelial-to-mesenchymal transition of human and mouse epithelial cell lines. Proc Natl Acad Sci U S A. 2009; 106(18):7461–6.

Haines N, Irvine KD. Glycosylation regulates Notch signalling. Nat Rev Mol Cell Biol. 2003;4(10):786–97.

Hajos F, Woodhams PL, Basco E, Csillag A, Balazs R. Proliferation of astroglia in the embryonic mouse forebrain as revealed by simultaneous immunocytochemistry and autoradiography. Acta Morphol Acad Sci Hung. 1981;29(4):361–4.

Hamanoue M, Matsuzaki Y, Sato K, Okano HJ, Shibata S, Sato I, et al. Cell surface N-glycans mediated isolation of mouse neural stem cells. J Neurochem. 2009;110(5):1575–84.

Hanjan SN, Kearney JF, Cooper MD. A monoclonal antibody (MMA) that identifies a differentiation antigen on human myelomonocytic cells. Clin Immunol Immunopathol. 1982;23(2):172–88.

Hartfuss E, Galli R, Heins N, Gotz M. Characterization of CNS precursor subtypes and radial glia. Dev Biol. 2001;229(1):15–30.

Hennen E, Czopka T, Faissner A. Structurally distinct LewisX glycans distinguish subpopulations of neural stem/progenitor cells. J Biol Chem. 2011;286(18):16321–31.

Heuckeroth RO, Lampe PA, Johnson EM, Milbrandt J. Neurturin and GDNF promote proliferation and survival of enteric neuron and glial progenitors in vitro. Dev Biol. 1998;200(1):116–29.

Hirahara Y, Bansal R, Honke K, Ikenaka K, Wada Y. Sulfatide is a negative regulator of oligodendrocyte differentiation: development in sulfatide-null mice. Glia. 2004;45(3):269–77.

Hirschberg K, Zisling R, van Echten-Deckert G, Futerman AH. Ganglioside synthesis during the development of neuronal polarity. Major changes occur during axonogenesis and axon elongation, but not during dendrite growth or synaptogenesis. J Biol Chem. 1996;271(25): 14876–82.

Holley JA, Yu RK. Localization of glycoconjugates recognized by the HNK-1 antibody in mouse and chick embryos during early neural development. Dev Neurosci. 1987;9(2):105–19.

Honke K, Hirahara Y, Dupree J, Suzuki K, Popko B, Fukushima K, et al. Paranodal junction formation and spermatogenesis require sulfoglycolipids. Proc Natl Acad Sci U S A. 2002;99(7): 4227–32.

Hori K, Sen A, Artavanis-Tsakonas S. Notch signaling at a glance. J Cell Sci. 2013;126(Pt 10):2135–40.

Hou X, Tashima Y, Stanley P. Galactose differentially modulates lunatic and manic fringe effects on Delta1-induced NOTCH signaling. J Biol Chem. 2012;287(1):474–83.

Huttner WB, Zimmerberg J. Implications of lipid microdomains for membrane curvature, budding and fission. Curr Opin Cell Biol. 2001;13(4):478–84.

Ichikawa M, Shiga T, Hirata Y. Spatial and temporal pattern of postnatal proliferation of glial cells in the parietal cortex of the rat. Brain Res. 1983;285(2):181–7.

Imai T, Tokunaga A, Yoshida T, Hashimoto M, Mikoshiba K, Weinmaster G, et al. The neural RNA-binding protein Musashi1 translationally regulates mammalian numb gene expression by interacting with its mRNA. Mol Cell Biol. 2001;21(12):3888–900.

Ishii Y, Nakamura S, Osumi N. Demarcation of early mammalian cortical development by differential expression of fringe genes. Brain Res Dev Brain Res. 2000;119(2):307–20.

Izumikawa T, Kanagawa N, Watamoto Y, Okada M, Saeki M, Sakano M, et al. Impairment of embryonic cell division and glycosaminoglycan biosynthesis in glucuronyltransferase-I-deficient mice. J Biol Chem. 2010;285(16):12190–6.

Janich P, Corbeil D. GM1 and GM3 gangliosides highlight distinct lipid microdomains within the apical domain of epithelial cells. FEBS Lett. 2007;581(9):1783–7.

Jirmanova L, Pacholikova J, Krejci P, Hampl A, Dvorak P. O-linked carbohydrates are required for FGF-2-mediated proliferation of mouse embryonic cells. Int J Dev Biol. 1999;43(6):555–62.

Johansson CB, Momma S, Clarke DL, Risling M, Lendahl U, Frisen J. Identification of a neural stem cell in the adult mammalian central nervous system. Cell. 1999;96(1):25–34.

Jones LL, Yamaguchi Y, Stallcup WB, Tuszynski MH. NG2 is a major chondroitin sulfate proteoglycan produced after spinal cord injury and is expressed by macrophages and oligodendrocyte progenitors. J Neurosci. 2002;22(7):2792–803.

Jungalwala FB. Expression and biological functions of sulfoglucuronyl glycolipids (SGGLs) in the nervous system – a review. Neurochem Res. 1994;19(8):945–57.

Kabos P, Matundan H, Zandian M, Bertolotto C, Robinson ML, Davy BE, et al. Neural precursors express multiple chondroitin sulfate proteoglycans, including the lectican family. Biochem Biophys Res Commun. 2004;318(4):955–63.

Kamakura S, Oishi K, Yoshimatsu T, Nakafuku M, Masuyama N, Gotoh Y. Hes binding to STAT3 mediates crosstalk between Notch and JAK-STAT signalling. Nat Cell Biol. 2004;6(6):547–54.

Kasai N, Yu RK. The monoclonal antibody A2B5 is specific to ganglioside GQ1c. Brain Res. 1983;277(1):155–8.

Kato TM, Kawaguchi A, Kosodo Y, Niwa H, Matsuzaki F. Lunatic fringe potentiates Notch signaling in the developing brain. Mol Cell Neurosci. 2010;45(1):12–25.

Kawai H, Allende ML, Wada R, Kono M, Sango K, Deng C, et al. Mice expressing only monosialoganglioside GM3 exhibit lethal audiogenic seizures. J Biol Chem. 2001;276(10):6885–8.

Kinoshita MO, Furuya S, Ito S, Shinoda Y, Yamazaki Y, Greimel P, et al. Lipid rafts enriched in phosphatidylglucoside direct astroglial differentiation by regulating tyrosine kinase activity of epidermal growth factor receptors. Biochem J. 2009a;419(3):565–75.

Kinoshita MO, Shinoda Y, Sakai K, Hashikawa T, Watanabe M, Machida T, et al. Selective upregulation of 3-phosphoglycerate dehydrogenase (Phgdh) expression in adult subventricular zone neurogenic niche. Neurosci Lett. 2009b;453(1):21–6.

Kitada M, Kuroda Y, Dezawa M. Lectins as a tool for detecting neural stem/progenitor cells in the adult mouse brain. Anat Rec (Hoboken). 2011;294(2):305–21.

Kleene R, Schachner M. Glycans and neural cell interactions. Nat Rev Neurosci. 2004;5(3):195–208.

Koch U, Lehal R, Radtke F. Stem cells living with a Notch. Development. 2013;140(4):689–704.

Kondo T, Raff M. Oligodendrocyte precursor cells reprogrammed to become multipotential CNS stem cells. Science. 2000;289(5485):1754–7.

Kondo T, Raff M. Chromatin remodeling and histone modification in the conversion of oligodendrocyte precursors to neural stem cells. Genes Dev. 2004;18(23):2963–72.

Kriegstein A, Alvarez-Buylla A. The glial nature of embryonic and adult neural stem cells. Annu Rev Neurosci. 2009;32:149–84.

Kruse J, Mailhammer R, Wernecke H, Faissner A, Sommer I, Goridis C, et al. Neural cell adhesion molecules and myelin-associated glycoprotein share a common carbohydrate moiety recognized by monoclonal antibodies L2 and HNK-1. Nature. 1984;311(5982):153–5.

Kudo T, Fujii T, Ikegami S, Inokuchi K, Takayama Y, Ikehara Y, et al. Mice lacking alpha1,3-fucosyltransferase IX demonstrate disappearance of Lewis x structure in brain and increased anxiety-like behaviors. Glycobiology. 2007;17(1):1–9.

Kudo T, Ikehara Y, Togayachi A, Kaneko M, Hiraga T, Sasaki K, et al. Expression cloning and characterization of a novel murine alpha1, 3-fucosyltransferase, mFuc-TIX, that synthesizes the Lewis x (CD15) epitope in brain and kidney. J Biol Chem. 1998;273(41):26729–38.

Kuroda Y, Kitada M, Wakao S, Nishikawa K, Tanimura Y, Makinoshima H, et al. Unique multipotent cells in adult human mesenchymal cell populations. Proc Natl Acad Sci U S A. 2010;107(19):8639–43.

Laywell ED, Rakic P, Kukekov VG, Holland EC, Steindler DA. Identification of a multipotent astrocytic stem cell in the immature and adult mouse brain. Proc Natl Acad Sci U S A. 2000; 97(25):13883–8.

Le Douarin NM, Teillet MA. The migration of neural crest cells to the wall of the digestive tract in avian embryo. J Embryol Exp Morphol. 1973;30(1):31–48.

Levine JM, Stallcup WB. Plasticity of developing cerebellar cells in vitro studied with antibodies against the NG2 antigen. J Neurosci. 1987;7(9):2721–31.

Ligon KL, Kesari S, Kitada M, Sun T, Arnett HA, Alberta JA, et al. Development of NG2 neural progenitor cells requires Olig gene function. Proc Natl Acad Sci U S A. 2006;103(20): 7853–8.

Lugert S, Vogt M, Tchorz JS, Muller M, Giachino C, Taylor V. Homeostatic neurogenesis in the adult hippocampus does not involve amplification of Ascl1(high) intermediate progenitors. Nat Commun. 2012;3:670.

Lui JH, Hansen DV, Kriegstein AR. Development and evolution of the human neocortex. Cell. 2011;146(1):18–36.

Luther KB, Haltiwanger RS. Role of unusual O-glycans in intercellular signaling. Int J Biochem Cell Biol. 2009;41(5):1011–24.

Mai JK, Andressen C, Ashwell KW. Demarcation of prosencephalic regions by CD15-positive radial glia. Eur J Neurosci. 1998;10(2):746–51.

Malatesta P, Hack MA, Hartfuss E, Kettenmann H, Klinkert W, Kirchhoff F, et al. Neuronal or glial progeny: regional differences in radial glia fate. Neuron. 2003;37(5):751–64.

Malatesta P, Hartfuss E, Gotz M. Isolation of radial glial cells by fluorescent-activated cell sorting reveals a neuronal lineage. Development. 2000;127(24):5253–63.

Margolis RK, Ripellino JA, Goossen B, Steinbrich R, Margolis RU. Occurrence of the HNK-1 epitope (3-sulfoglucuronic acid) in PC12 pheochromocytoma cells, chromaffin granule membranes, and chondroitin sulfate proteoglycans. Biochem Biophys Res Commun. 1987;145(3): 1142–8.

Marzesco AM, Janich P, Wilsch-Brauninger M, Dubreuil V, Langenfeld K, Corbeil D, et al. Release of extracellular membrane particles carrying the stem cell marker prominin-1 (CD133) from neural progenitors and other epithelial cells. J Cell Sci. 2005;118(Pt 13):2849–58.

McCarthy M, Turnbull DH, Walsh CA, Fishell G. Telencephalic neural progenitors appear to be restricted to regional and glial fates before the onset of neurogenesis. J Neurosci. 2001; 21(17):6772–81.

Mendez-Otero R, Cavalcante LA. Expression of 9-O-acetylated gangliosides is correlated with tangential cell migration in the rat brain. Neurosci Lett. 1996;204(1–2):97–100.

Mendez-Otero R, Schlosshauer B, Barnstable CJ, Constantine-Paton M. A developmentally regulated antigen associated with neural cell and process migration. J Neurosci. 1988;8(2): 564–79.

Miller FD, Gauthier AS. Timing is everything: making neurons versus glia in the developing cortex. Neuron. 2007;54(3):357–69.

Miyakoshi LM, Todeschini AR, Mendez-Otero R, Hedin-Pereira C. Role of the 9-O-acetyl GD3 in subventricular zone neuroblast migration. Mol Cell Neurosci. 2012;49(2):240–9.

Miyata T, Kawaguchi A, Okano H, Ogawa M. Asymmetric inheritance of radial glial fibers by cortical neurons. Neuron. 2001;31(5):727–41.

Mizutani K, Yoon K, Dang L, Tokunaga A, Gaiano N. Differential Notch signalling distinguishes neural stem cells from intermediate progenitors. Nature. 2007;449(7160):351–5.

Mo Z, Moore AR, Filipovic R, Ogawa Y, Kazuhiro I, Antic SD, et al. Human cortical neurons originate from radial glia and neuron-restricted progenitors. J Neurosci. 2007;27(15):4132–45.

Moloney DJ, Shair LH, Lu FM, Xia J, Locke R, Matta KL, et al. Mammalian Notch1 is modified with two unusual forms of O-linked glycosylation found on epidermal growth factor-like modules. J Biol Chem. 2000;275(13):9604–11.

Morris-Wiman J, Brinkley LL. The role of the mesenchyme in mouse neural fold elevation. I. Patterns of mesenchymal cell distribution and proliferation in embryos developing in vitro. Am J Anat. 1990a;188(2):121–32.

Morris-Wiman J, Brinkley LL. The role of the mesenchyme in mouse neural fold elevation. II. Patterns of hyaluronate synthesis and distribution in embryos developing in vitro. Am J Anat. 1990b;188(2):133–47.

Morrison SJ, White PM, Zock C, Anderson DJ. Prospective identification, isolation by flow cytometry, and in vivo self-renewal of multipotent mammalian neural crest stem cells. Cell. 1999;96(5):737–49.

Morshead CM. Adult neural stem cells: attempting to solve the identity crisis. Dev Neurosci. 2004;26(2–4):93–100.

Murakami K, Asou H, Adachi T, Takagi T, Kunimoto M, Saito H, et al. Neutral glycolipid and ganglioside composition of type-1 and type-2 astrocytes from rat cerebral hemisphere. J Neurosci Res. 1999;55(3):382–93.

Muramatsu T, Muramatsu H. Carbohydrate antigens expressed on stem cells and early embryonic cells. Glycoconj J. 2004;21(1–2):41–5.

Nagase T, Sanai Y, Nakamura S, Asato H, Harii K, Osumi N. Roles of HNK-1 carbohydrate epitope and its synthetic glucuronyltransferase genes on migration of rat neural crest cells. J Anat. 2003;203(1):77–88.

Nagatsuka Y, Hara-Yokoyama M, Kasama T, Takekoshi M, Maeda F, Ihara S, et al. Carbohydrate-dependent signaling from the phosphatidylglucoside-based microdomain induces granulocytic differentiation of HL60 cells. Proc Natl Acad Sci U S A. 2003;100(13):7454–9.

Nagatsuka Y, Horibata Y, Yamazaki Y, Kinoshita M, Shinoda Y, Hashikawa T, et al. Phosphatidylglucoside exists as a single molecular species with saturated fatty acyl chains in developing astroglial membranes. Biochemistry. 2006;45(29):8742–50.

Nagatsuka Y, Kasama T, Ohashi Y, Uzawa J, Ono Y, Shimizu K, et al. A new phosphoglycerolipid, "phosphatidylglucose," found in human cord red cells by multi-reactive monoclonal anti-i cold agglutinin, mAb GL-1/GL-2. FEBS Lett. 2001;497(2–3):141–7.

Nakashima K, Wiese S, Yanagisawa M, Arakawa H, Kimura N, Hisatsune T, et al. Developmental requirement of gp130 signaling in neuronal survival and astrocyte differentiation. J Neurosci. 1999a;19(13):5429–34.

Nakashima K, Yanagisawa M, Arakawa H, Kimura N, Hisatsune T, Kawabata M, et al. Synergistic signaling in fetal brain by STAT3-Smad1 complex bridged by p300. Science. 1999b;284(5413): 479–82.

Nakatani Y, Yanagisawa M, Suzuki Y, Yu RK. Characterization of GD3 ganglioside as a novel biomarker of mouse neural stem cells. Glycobiology. 2010;20(1):78–86.

Ngamukote S, Yanagisawa M, Ariga T, Ando S, Yu RK. Developmental changes of glycosphingolipids and expression of glycogenes in mouse brains. J Neurochem. 2007;103(6):2327–41.

Noctor SC, Flint AC, Weissman TA, Dammerman RS, Kriegstein AR. Neurons derived from radial glial cells establish radial units in neocortex. Nature. 2001;409(6821):714–20.

Noctor SC, Flint AC, Weissman TA, Wong WS, Clinton BK, Kriegstein AR. Dividing precursor cells of the embryonic cortical ventricular zone have morphological and molecular characteristics of radial glia. J Neurosci. 2002;22(8):3161–73.

Noctor SC, Martinez-Cerdeno V, Ivic L, Kriegstein AR. Cortical neurons arise in symmetric and asymmetric division zones and migrate through specific phases. Nat Neurosci. 2004;7(2): 136–44.

Ogura K, Kohno K, Tai T. Molecular cloning of a rat brain cDNA, with homology to a tyrosine kinase substrate, that induces galactosylceramide expression in COS-7 cells. J Neurochem. 1998;71(5):1827–36.

Ogura K, Tai T. Molecular cloning and characterization of galactosylceramide expression factor-1 (GEF-1). Neurochem Res. 2002;27(7–8):779–84.

Okajima T, Irvine KD. Regulation of notch signaling by o-linked fucose. Cell. 2002;111(6): 893–904.

Okajima T, Matsuura A, Matsuda T. Biological functions of glycosyltransferase genes involved in O-fucose glycan synthesis. J Biochem. 2008;144(1):1–6.

Okajima T, Xu A, Irvine KD. Modulation of notch-ligand binding by protein O-fucosyltransferase 1 and fringe. J Biol Chem. 2003;278(43):42340–5.

Okajima T, Xu A, Lei L, Irvine KD. Chaperone activity of protein O-fucosyltransferase 1 promotes notch receptor folding. Science. 2005;307(5715):1599–603.

Okano H, Kawahara H, Toriya M, Nakao K, Shibata S, Imai T. Function of RNA-binding protein Musashi-1 in stem cells. Exp Cell Res. 2005;306(2):349–56.

Okuda T, Tokuda N, Numata S, Ito M, Ohta M, Kawamura K, et al. Targeted disruption of Gb3/CD77 synthase gene resulted in the complete deletion of globo-series glycosphingolipids and loss of sensitivity to verotoxins. J Biol Chem. 2006;281(15):10230–5.

Palmer TD, Markakis EA, Willhoite AR, Safar F, Gage FH. Fibroblast growth factor-2 activates a latent neurogenic program in neural stem cells from diverse regions of the adult CNS. J Neurosci. 1999;19(19):8487–97.

Panin VM, Papayannopoulos V, Wilson R, Irvine KD. Fringe modulates Notch-ligand interactions. Nature. 1997;387(6636):908–12.

Petridis AK, El-Maarouf A, Rutishauser U. Polysialic acid regulates cell contact-dependent neuronal differentiation of progenitor cells from the subventricular zone. Dev Dyn. 2004 Aug;230(4):675–84.

Pettway Z, Domowicz M, Schwartz NB, Bronner-Fraser M. Age-dependent inhibition of neural crest migration by the notochord correlates with alterations in the S103L chondroitin sulfate proteoglycan. Exp Cell Res. 1996;225(1):195–206.

Pinto L, Gotz M. Radial glial cell heterogeneity – the source of diverse progeny in the CNS. Prog Neurobiol. 2007;83(1):2–23.

Purves D, Lichtman JW. Principles of neural development. Sunderland, MA: Sinauer Associates; 1985.

Qian X, Shen Q, Goderie SK, He W, Capela A, Davis AA, et al. Timing of CNS cell generation: a programmed sequence of neuron and glial cell production from isolated murine cortical stem cells. Neuron. 2000;28(1):69–80.

Raff MC, Abney ER, Cohen J, Lindsay R, Noble M. Two types of astrocytes in cultures of developing rat white matter: differences in morphology, surface gangliosides, and growth characteristics. J Neurosci. 1983a;3(6):1289–300.

Raff MC, Miller RH, Noble M. A glial progenitor cell that develops in vitro into an astrocyte or an oligodendrocyte depending on culture medium. Nature. 1983b;303(5916):390–6.

Rao MS, Mayer-Proschel M. Glial-restricted precursors are derived from multipotent neuroepithelial stem cells. Dev Biol. 1997;188(1):48–63.

Reijo Pera RA, DeJonge C, Bossert N, Yao M, Hwa Yang JY, Asadi NB, et al. Gene expression profiles of human inner cell mass cells and embryonic stem cells. Differentiation. 2009;78(1): 18–23.

Reynolds BA, Weiss S. Generation of neurons and astrocytes from isolated cells of the adult mammalian central nervous system. Science. 1992;255(5052):1707–10.

Richardson WD, Young KM, Tripathi RB, McKenzie I. NG2-glia as multipotent neural stem cells: fact or fantasy? Neuron. 2011;70(4):661–73.

Rietze RL, Valcanis H, Brooker GF, Thomas T, Voss AK, Bartlett PF. Purification of a pluripotent neural stem cell from the adult mouse brain. Nature. 2001;412(6848):736–9.

Roper K, Corbeil D, Huttner WB. Retention of prominin in microvilli reveals distinct cholesterol-based lipid micro-domains in the apical plasma membrane. Nat Cell Biol. 2000;2(9):582–92.

Rosner H, al-Aqtum M, Rahmann H. Gangliosides and neuronal differentiation. Neurochem Int. 1992;20(3):339–51.

Rosner H, Greis C, Henke-Fahle S. Developmental expression in embryonic rat and chicken brain of a polysialoganglioside-antigen reacting with the monoclonal antibody Q 211. Brain Res. 1988;470(2):161–71.

Rowitch DH, Kriegstein AR. Developmental genetics of vertebrate glial-cell specification. Nature. 2010;468(7321):214–22.

Rutishauser U, Landmesser L. Polysialic acid in the vertebrate nervous system: a promoter of plasticity in cell–cell interactions. Trends Neurosci. 1996;19(10):422–7.

Saito M, Kitamura H, Sugiyama K. The specificity of monoclonal antibody A2B5 to c-series gangliosides. J Neurochem. 2001;78(1):64–74.
Sasamura T, Sasaki N, Miyashita F, Nakao S, Ishikawa HO, Ito M, et al. Neurotic, a novel maternal neurogenic gene, encodes an O-fucosyltransferase that is essential for Notch-Delta interactions. Development. 2003;130(20):4785–95.
Sauka-Spengler T, Bronner-Fraser M. A gene regulatory network orchestrates neural crest formation. Nat Rev Mol Cell Biol. 2008;9(7):557–68.
Sbaschnig-Agler M, Dreyfus H, Norton WT, Sensenbrenner M, Farooq M, Byrne MC, et al. Gangliosides of cultured astroglia. Brain Res. 1988;461(1):98–106.
Schoenwolf GC, Fisher M. Analysis of the effects of Streptomyces hyaluronidase on formation of the neural tube. J Embryol Exp Morphol. 1983;73:1–15.
Seki T, Rutishauser U. Removal of polysialic acid-neural cell adhesion molecule induces aberrant mossy fiber innervation and ectopic synaptogenesis in the hippocampus. J Neurosci. 1998; 18(10):3757–66.
Seri B, Garcia-Verdugo JM, McEwen BS, Alvarez-Buylla A. Astrocytes give rise to new neurons in the adult mammalian hippocampus. J Neurosci. 2001;21(18):7153–60.
Shah NM, Groves AK, Anderson DJ. Alternative neural crest cell fates are instructively promoted by TGFbeta superfamily members. Cell. 1996;85(3):331–43.
Shah NM, Marchionni MA, Isaacs I, Stroobant P, Anderson DJ. Glial growth factor restricts mammalian neural crest stem cells to a glial fate. Cell. 1994;77(3):349–60.
Sharon N. Lectins: past, present and future. Biochem Soc Trans. 2008;36(Pt 6):1457–60.
Sharon N, Lis H. Lectins: cell-agglutinating and sugar-specific proteins. Science. 1972; 177(4053):949–59.
Sheikh KA, Sun J, Liu Y, Kawai H, Crawford TO, Proia RL, et al. Mice lacking complex gangliosides develop Wallerian degeneration and myelination defects. Proc Natl Acad Sci U S A. 1999;96(13):7532–7.
Shi S, Stanley P. Protein O-fucosyltransferase 1 is an essential component of Notch signaling pathways. Proc Natl Acad Sci U S A. 2003;100(9):5234–9.
Shimazaki T, Shingo T, Weiss S. The ciliary neurotrophic factor/leukemia inhibitory factor/gp130 receptor complex operates in the maintenance of mammalian forebrain neural stem cells. J Neurosci. 2001;21(19):7642–53.
Shimojo H, Ohtsuka T, Kageyama R. Dynamic expression of notch signaling genes in neural stem/ progenitor cells. Front Neurosci. 2011;5:78.
Shmelkov SV, St Clair R, Lyden D, Rafii S. AC133/CD133/Prominin-1. Int J Biochem Cell Biol. 2005;37(4):715–9.
Sieber-Blum M. Commitment of neural crest cells to the sensory neuron lineage. Science. 1989;243(4898):1608–11.
Simpson MA, Cross H, Proukakis C, Priestman DA, Neville DC, Reinkensmeier G, et al. Infantile-onset symptomatic epilepsy syndrome caused by a homozygous loss-of-function mutation of GM3 synthase. Nat Genet. 2004;36(11):1225–9.
Singh RD, Schroeder AS, Scheffer L, Holicky EL, Wheatley CL, Marks DL, et al. Prominin-2 expression increases protrusions, decreases caveolae and inhibits Cdc42 dependent fluid phase endocytosis. Biochem Biophys Res Commun. 2013;434(3):466–72.
Skaggs K, Martin DM, Novitch BG. Regulation of spinal interneuron development by the Olig-related protein Bhlhb5 and Notch signaling. Development. 2011;138(15):3199–211.
Smart IH. Proliferative characteristics of the ependymal layer during the early development of the mouse neocortex: a pilot study based on recording the number, location and plane of cleavage of mitotic figures. J Anat. 1973;116(Pt 1):67–91.
Solter D, Knowles BB. Monoclonal antibody defining a stage-specific mouse embryonic antigen (SSEA-1). Proc Natl Acad Sci U S A. 1978;75(11):5565–9.
Stainier DY, Bilder DH, Gilbert W. The B30 ganglioside is a cell surface marker for neural crest-derived neurons in the developing mouse. Dev Biol. 1991;144(1):177–88.
Stallcup WB. The NG2 antigen, a putative lineage marker: immunofluorescent localization in primary cultures of rat brain. Dev Biol. 1981;83(1):154–65.

Stanley P, Okajima T. Roles of glycosylation in Notch signaling. Curr Top Dev Biol. 2010;92:131–64.

Steiner B, Klempin F, Wang L, Kott M, Kettenmann H, Kempermann G. Type-2 cells as link between glial and neuronal lineage in adult hippocampal neurogenesis. Glia. 2006;54(8):805–14.

Stemple DL, Anderson DJ. Isolation of a stem cell for neurons and glia from the mammalian neural crest. Cell. 1992;71(6):973–85.

Sugiura Y, Furukawa K, Tajima O, Mii S, Honda T. Sensory nerve-dominant nerve degeneration and remodeling in the mutant mice lacking complex gangliosides. Neuroscience. 2005;135(4):1167–78.

Suhonen JO, Peterson DA, Ray J, Gage FH. Differentiation of adult hippocampus-derived progenitors into olfactory neurons in vivo. Nature. 1996;383(6601):624–7.

Sun Y, Nadal-Vicens M, Misono S, Lin MZ, Zubiaga A, Hua X, et al. Neurogenin promotes neurogenesis and inhibits glial differentiation by independent mechanisms. Cell. 2001;104(3):365–76.

Susuki K, Baba H, Tohyama K, Kanai K, Kuwabara S, Hirata K, et al. Gangliosides contribute to stability of paranodal junctions and ion channel clusters in myelinated nerve fibers. Glia 2007;55(7):746–57.

Svennerholm L. Chromatographic separation of human brain gangliosides. J Neurochem. 1963;10:613–23.

Taga T, Hibi M, Hirata Y, Yamasaki K, Yasukawa K, Matsuda T, et al. Interleukin-6 triggers the association of its receptor with a possible signal transducer, gp130. Cell. 1989;58(3):573–81.

Takamiya K, Yamamoto A, Furukawa K, Yamashiro S, Shin M, Okada M, et al. Mice with disrupted GM2/GD2 synthase gene lack complex gangliosides but exhibit only subtle defects in their nervous system. Proc Natl Acad Sci U S A. 1996;93(20):10662–7.

Takizawa T, Nakashima K, Namihira M, Ochiai W, Uemura A, Yanagisawa M, et al. DNA methylation is a critical cell-intrinsic determinant of astrocyte differentiation in the fetâl brain. Dev Cell. 2001;1(6):749–58.

Thallmair M, Ray J, Stallcup WB, Gage FH. Functional and morphological effects of NG2 proteoglycan deletion on hippocampal neurogenesis. Exp Neurol. 2006;202(1):167–78.

Tokuda A, Ariga T, Isogai Y, Komba S, Kiso M, Hasegawa A, et al. On the specificity of anti-sulfoglucuronosyl glycolipid antibodies'. J Carbohyd Chem. 1998;17(4–5):535–46.

Tole S, Kaprielian Z, Ou SK, Patterson PH. FORSE-1: a positionally regulated epitope in the developing rat central nervous system. J Neurosci. 1995;15(2):957–69.

Tucker GC, Delarue M, Zada S, Boucaut JC, Thiery JP. Expression of the HNK-1/NC-1 epitope in early vertebrate neurogenesis. Cell Tissue Res. 1988;251(2):457–65.

von Holst A, Sirko S, Faissner A. The unique 473HD-Chondroitinsulfate epitope is expressed by radial glia and involved in neural precursor cell proliferation. J Neurosci. 2006;26(15):4082–94.

Wakao S, Kitada M, Kuroda Y, Shigemoto T, Matsuse D, Akashi H, et al. Multilineage-differentiating stress-enduring (Muse) cells are a primary source of induced pluripotent stem cells in human fibroblasts. Proc Natl Acad Sci U S A. 2011;108(24):9875–80.

Walker TL, Wierick A, Sykes AM, Waldau B, Corbeil D, Carmeliet P, et al. Prominin-1 allows prospective isolation of neural stem cells from the adult murine hippocampus. J Neurosci. 2013;33(7):3010–24.

Walters LC, Cantrell VA, Weller KP, Mosher JT, Southard-Smith EM. Genetic background impacts developmental potential of enteric neural crest-derived progenitors in the Sox10Dom model of Hirschsprung disease. Hum Mol Genet. 2010;19(22):4353–72.

Weigmann A, Corbeil D, Hellwig A, Huttner WB. Prominin, a novel microvilli-specific polytopic membrane protein of the apical surface of epithelial cells, is targeted to plasmalemmal protrusions of non-epithelial cells. Proc Natl Acad Sci U S A. 1997;94(23):12425–30.

Wu G, Fang Y, Lu ZH, Ledeen RW. Induction of axon-like and dendrite-like processes in neuroblastoma cells. J Neurocytol. 1998;27(1):1–14.

Wu G, Lu ZH, Kulkarni N, Amin R, Ledeen RW. Mice lacking major brain gangliosides develop parkinsonism. Neurochem Res. 2011;36(9):1706–14.

Wu G, Lu ZH, Xie X, Li L, Ledeen RW. Mutant NG108-15 cells (NG-CR72) deficient in GM1 synthase respond aberrantly to axonogenic stimuli and are vulnerable to calcium-induced apoptosis: they are rescued with LIGA-20. J Neurochem. 2001;76(3):690–702.

Yagi H, Saito T, Yanagisawa M, Yu RK, Kato K. Lewis X-carrying N-glycans regulate the proliferation of mouse embryonic neural stem cells via the Notch signaling pathway. J Biol Chem. 2012;287(29):24356–64.

Yagi H, Yanagisawa M, Kato K, Yu RK. Lysosome-associated membrane protein 1 is a major SSEA-1-carrier protein in mouse neural stem cells. Glycobiology. 2010a;20(8):976–81.

Yagi H, Yanagisawa M, Suzuki Y, Nakatani Y, Ariga T, Kato K, et al. HNK-1 epitope-carrying tenascin-C spliced variant regulates the proliferation of mouse embryonic neural stem cells. J Biol Chem. 2010b;285(48):37293–301.

Yamamoto N, Inui K, Matsuyama Y, Harada A, Hanamura K, Murakami F, et al. Inhibitory mechanism by polysialic acid for lamina-specific branch formation of thalamocortical axons. J Neurosci. 2000 Dec 15;20(24):9145–51.

Yamamoto S, Oka S, Inoue M, Shimuta M, Manabe T, Takahashi H, et al. Mice deficient in nervous system-specific carbohydrate epitope HNK-1 exhibit impaired synaptic plasticity and spatial learning. J Biol Chem. 2002;277(30):27227–31.

Yamashita T, Wada R, Sasaki T, Deng C, Bierfreund U, Sandhoff K, et al. A vital role for glycosphingolipid synthesis during development and differentiation. Proc Natl Acad Sci U S A. 1999;96(16):9142–7.

Yanagisawa M, Liour SS, Yu RK. Involvement of gangliosides in proliferation of immortalized neural progenitor cells. J Neurochem. 2004a;91(4):804–12.

Yanagisawa M, Nakamura K, Taga T. Roles of lipid rafts in integrin-dependent adhesion and gp130 signalling pathway in mouse embryonic neural precursor cells. Genes Cells. 2004b; 9(9):801–9.

Yanagisawa M, Taga T, Nakamura K, Ariga T, Yu RK. Characterization of glycoconjugate antigens in mouse embryonic neural precursor cells. J Neurochem. 2005;95(5):1311–20.

Yanagisawa M, Yu RK. The expression and functions of glycoconjugates in neural stem cells. Glycobiology. 2007;17(7):57R–74.

Yanagisawa M, Yu RK. N-glycans modulate the activation of gp130 in mouse embryonic neural precursor cells. Biochem Biophys Res Commun. 2009;386(1):101–4.

Yang CR, Liour SS, Dasgupta S, Yu RK. Inhibition of neuronal migration by JONES antibody is independent of 9-O-acetyl GD3 in GD3-synthase knockout mice. J Neurosci Res. 2007; 85(7):1381–90.

Yu RK, Ando S. Structures of some new complex gangliosides of fish brain. Adv Exp Med Biol. 1980;125:33–45.

Yu RK, Bieberich E, Xia T, Zeng G. Regulation of ganglioside biosynthesis in the nervous system. J Lipid Res. 2004;45(5):783–93.

Yu RK, Macala LJ, Taki T, Weinfield HM, Yu FS. Developmental changes in ganglioside composition and synthesis in embryonic rat brain. J Neurochem. 1988;50(6):1825–9.

Yu RK, Nakatani Y, Yanagisawa M. The role of glycosphingolipid metabolism in the developing brain. J Lipid Res. 2009;50(Suppl):S440–5.

Yu RK, Tsai YT, Ariga T. Functional roles of gangliosides in neurodevelopment: an overview of recent advances. Neurochem Res. 2012;37(6):1230–44.

Zacchigna S, Oh H, Wilsch-Brauninger M, Missol-Kolka E, Jaszai J, Jansen S, et al. Loss of the cholesterol-binding protein prominin-1/CD133 causes disk dysmorphogenesis and photoreceptor degeneration. J Neurosci. 2009;29(7):2297–308.

Zhang SC. Defining glial cells during CNS development. Nat Rev Neurosci. 2001;2(11):840–3.

Zhu X, Hill RA, Dietrich D, Komitova M, Suzuki R, Nishiyama A. Age-dependent fate and lineage restriction of single NG2 cells. Development. 2011;138(4):745–53.

Chapter 10
Gangliosides and Cell Surface Ganglioside Glycohydrolases in the Nervous System

Massimo Aureli, Maura Samarani, Valentina Murdica, Laura Mauri, Nicoletta Loberto, Rosaria Bassi, Alessandro Prinetti, and Sandro Sonnino

Abstract Gangliosides are a large group of complex lipids found predominantly on the outer layer of the plasma membranes of cells, and they are particularly concentrated in nerve endings. Their half-life in the nervous system is short, and their membrane composition and content are strictly connected to their metabolism. Their neobiosynthesis starts in the endoplasmic reticulum and is completed in the Golgi; catabolism occurs primarily in the lysosomes. However, the final content of gangliosides in the plasma membrane is affected by other cellular processes.

In this chapter structural changes in the oligosaccharide chains of gangliosides induced by the activity of glycohydrolases and in some cases by glycosyltransferases that are associated with plasma membranes are discussed. Some of the plasma membrane enzymes arise from fusion processes between intracellular fractions and the plasma membrane; however, other plasma membrane enzymes display a structure different from that of the intracellular enzymes. Several of these plasma membrane enzymes have been characterized and some of them seem to have a specific role in the nervous system.

Keywords Ganglioside • Glycosphingolipid • Glycohydrolases • Sphingolipid metabolism • Central nervous system • Neuronal differentiation • Neurodegeneration

---

M. Aureli • M. Samarani • V. Murdica • L. Mauri • N. Loberto • R. Bassi
A. Prinetti • S. Sonnino (✉)
Department of Medical Biotechnology and Translational Medicine, University of Milano, via Fratelli Cervi 93, 20090 Segrate (MI), Italy
e-mail: massimo.aureli@unimi.it; maura.samarani@unimi.it; valentina.murdica@unimi.it; laura.mauri@unimi.it; nicoletta.loberto@unimi.it; rosaria.bassi@unimi.it; alessandro.prinetti@unimi.it; sandro.sonnino@unimi.it

R.K. Yu and C.-L. Schengrund (eds.), *Glycobiology of the Nervous System*, Advances in Neurobiology 9, DOI 10.1007/978-1-4939-1154-7_10,

10.1 Gangliosides

Gangliosides are glycosphingolipids (GSLs) that contain one or more sialic acid residues. They are components of the external layer of all animal cell plasma membranes and are particularly abundant in brain where they are tenfold higher than in extra-nervous tissues, representing one-twelfth of the outer layer of glycerophospholipids. Neuronal gangliosides are lipids with strong amphiphilic character due to their acidic and, in general, large saccharide head group and double-tailed hydrophobic moiety. The lipid moiety of gangliosides, as well as that of all sphingolipids, is ceramide (Fig. 10.1a), which consists of a long-chain amino alcohol connected to a fatty acid by an amide linkage. As a peculiar characteristic of the nervous system, the amino alcohol can be either 2*S*,3*R*,4*E*,2-amino-1,3-dihydroxy-octadecene, known as sphingosine, or 2*S*,3*R*,4*E*,2-amino-1,3-dihydroxy-eicosene, known as eicosasphingosine. The ratio between the two long-chain alcohols is variable, with eicosasphingosine barely detectable in the brains of fetuses but progressively increasing with age to become the major species in the elderly. A few percent of the saturated species, sphinganine and eicosasphinganine, have also been identified (Valsecchi et al. 1996; Valsecchi et al. 1993).

In nonnervous tissue gangliosides exhibit heterogeneity in their acyl chains, with a very long fatty acid moiety often being a major component of the ceramide structure. In gangliosides of the nervous system, stearic acid is the most common fatty acid accounting for 90–95 % of the total fatty acid content. This characteristic of neuronal gangliosides may be necessary for neuronal membrane plasticity.

The oligosaccharide chain of neuronal gangliosides varies widely due to the neutral sugar content and number of sialic acid residues. This, together with the heterogeneity of the ceramide moiety, makes gangliosides a very large family of compounds. Table 10.1 shows the main ganglioside structures found in the human nervous system, together with trivial and commonly used abbreviations.

Sialic acid is the name that identifies all the derivatives of 5-amino-3,5-dideoxy-D-*glycero*-D-*galacto*-non-2-ulopyranosonic acid, or neuraminic acid. Among the sialic acids, the 5-*N*-acetyl (Fig. 10.1b), the 5-*N*-acetyl-9-*O*-acetyl (Fig. 10.1c), and the 5-*N*-glycolyl derivatives (Fig. 10.1d) are the most common. Healthy humans have only the first two sialic acids, in the ratio of 9:1 (Kamerling and Vliegenthart 1975). Also ganglioside lactones, containing polysialyl chains where the sialic acids are linked together with ketosidic and ester linkages (Fig. 10.1e), have been found in human brains (Riboni et al. 1986).

10.2 Gangliosides and Membrane Organization

Gangliosides are enriched in confined areas of the plasma membrane known as "lipid rafts" (Sonnino et al. 2006). Lipid rafts are enriched in sphingolipids and cholesterol with respect to glycerophospholipids and contain about 1–4 % of the

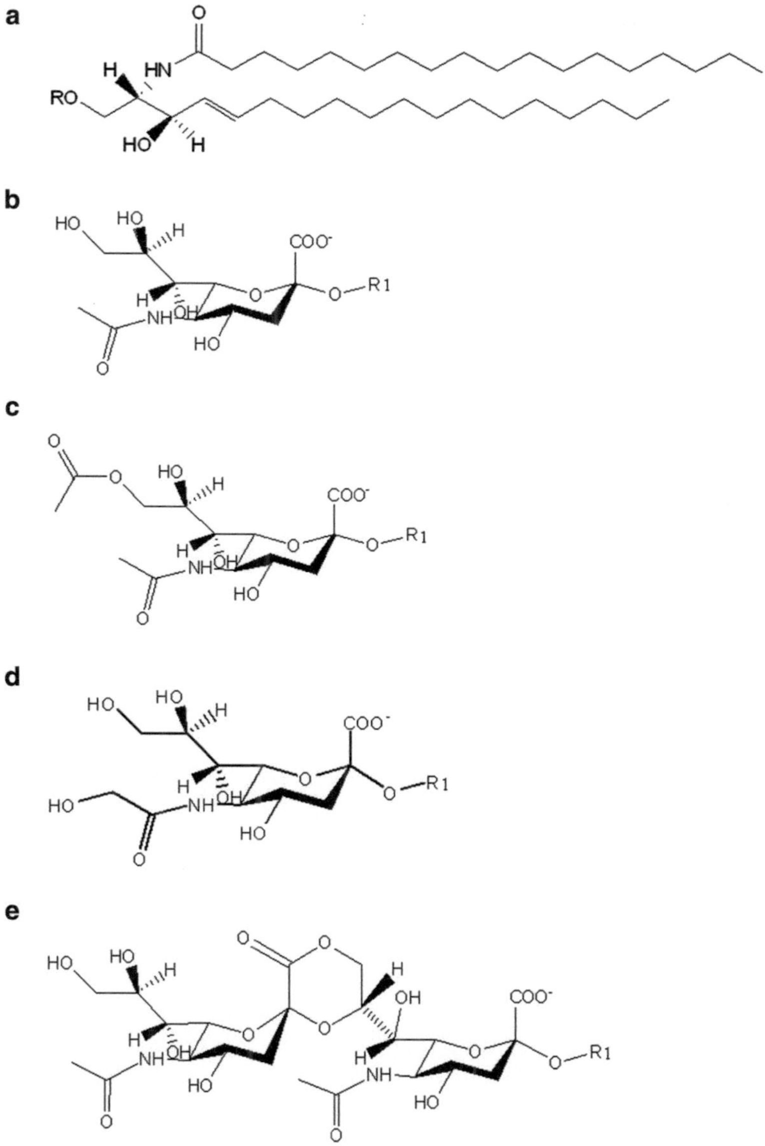

Fig. 10.1 Chemical structures of ceramide and the sialic acids. (**a**) Ceramide, a long-chain amino alcohol (sphingosine base) connected to a fatty acid by an amide linkage. R: phosphocholine, glucose, or oligosaccharides. (**b**) Sialic acid is the name that identifies all the derivatives of 5-acetamido-3,5-dideoxy-D-*glycero*-D-*galacto*-non-2-ulopyranosonic acid, or neuraminic acid. (**c**) 5-acetamido-9-*O*-acetyl-3,5-dideoxy-D-*glycero*-D-*galacto*-non-2-ulopyranosonic acid. (**d**) 3,5 dideoxy-5-glycolamido-D-*glycero*-D-*galacto*-non-2-ulopyranosonic acid. (**e**) Sialic acids linked together with ketosidic and ester linkage. R1: saccharides or oligosaccharide chain

Table 10.1 The main oligosaccharide series in the human nervous system

Structure of the oligosaccharide chain	Series
β-Gal-	Gal
β-Gal-(1-4)-β-Glc-	Lac
β-GalNAc-(1-4)-β-Gal-(1-4)-β-Glc-	Gg_3
β-Gal-(1-3)-β-GalNAc-(1-4)-β-Gal-(1-4)-β-Glc-	Gg_4
β-GalNAc-(1-4)-β-Gal-(1-3)-β-GalNAc-(1-4)-β-Gal-(1-4)-β-Glc-	Gg_5
β-Gal-(1-4)-β-GlcNAc-(1-3)-β-Gal-(1-4)-β-Glc-	nLc_4
The main gangliosides from the human nervous system	
Svennerholm nomenclature	IUPAC-IUB nomenclature
GM4	Neu5AcGalCer
GM3	II^3Neu5AcLacCer
GD3	$II^3(Neu5Ac)_2$LacCer
GM2	II^3Neu5AcGg_3Cer
GD2	$II^3(Neu5Ac)_2Gg_3$Cer
GM1	II^3Neu5AcGg_4Cer
3′-LM1	IV^3nLc_4Cer
GD1a	IV^3Neu5AcII^3Neu5AcGg_4Cer
GalNAc-GD1a	IV^3Neu5AcII^3Neu5AcGg_5Cer
GD1b	$II^3(Neu5Ac)_2Gg_4$Cer
GD1b-lactone	II^3[Neu5Ac-(2-8,1-9)-Neu5Ac]Gg_4Cer
GT1b	IV^3Neu5Ac$II^3(Neu5Ac)_2Gg_4$Cer
O-Acetyl-GT1b	IV^3Neu5AcII^3[Neu5,9Ac_2-(2-8)-Neu5Ac]Gg_4Cer
GQ1b	$IV^3(Neu5Ac)_2II^3(Neu5Ac)_2Gg_4$Cer
O-Acetyl-GQ1b	$IV^3(Neu5Ac)_2II^3(Neu5Ac)_2Gg_4$Cer

total protein content of the cell (Sonnino et al. 2006). Many lipid raft proteins are involved in cell signaling, and this led to the concept that ganglioside–protein interactions are instrumental in signal transduction and cell function. Protein properties might be affected by specific interactions but could also be modified by the physicochemical properties of the membrane which are determined by the lipid pattern, lipid amphiphilic and geometric properties, and lipid organization. Gangliosides, with their complex and bulky oligosaccharide structures, need a larger interfacial area than that required by glycerolipids. Phase separation with clustering of GSLs in a phospholipid bilayer is a spontaneous process driven by minimization of the interfacial free energy. The segregation of amphiphilic molecules with a bulky hydrophilic head group implies acquisition of a positive membrane curvature. The interfacial area increases with size of the oligosaccharide chain, along with a more positive membrane curvature and more pronounced segregation. The geometric properties of individual gangliosides inserted into the membrane depend primarily on the structural features of their oligosaccharides (Sonnino et al. 1994) and to a lesser extent on those of ceramide. Any variation in either the head group and/or the ceramide portion can have marked effects on the membrane surface, particularly on lipid rafts, where gangliosides are highly enriched. The dramatic effect on membrane curvature/organization seen when sialidase was used to convert the disialoganglioside GD1a to monosialoganglioside GM1, in an artificial membrane (Del Favero et al. 2011), supports the hypothesis that the effect of sequential hydrolysis of gangliosides to ceramide by coordination of plasma membrane-associated GSL hydrolases, as it occurs in senescent neurons and during neurodegeneration, should be much greater (Aureli et al. 2011c). Due to its two hydrocarbon chains, ceramide is very hydrophobic and almost insoluble in an aqueous solution. In a membrane it can be considered amphiphilic due to the primary hydroxyl group and the amide planar linkage located at the water–lipid interface. It is claimed that when a large amount of ceramide is produced from complex sphingolipids, it rapidly segregates forming macrodomains instrumental for cell signaling (Gulbins and Grassme 2002). Ceramide has a very high packing parameter suitable for negative curvature. The removal of the head group from sphingomyelin leads to ceramide-enriched endocytic vesicles, in artificial membranes (Holopainen et al. 2000). In a natural membrane this process would require rearrangement of the membrane, with exclusion of some components and sorting of others. In this context, the original lipid–protein interactions, or the forces exerted by the lipid environment on the protein conformation, would change with concomitant modifications of protein biological properties.

10.3 Metabolic Pathways of Gangliosides

The dynamic plasma membrane ganglioside content and pattern are determined by neobiosynthesis, catabolism, and complex trafficking into and out of the cell. Change in any of these pathways can lead to alterations in plasma membrane ganglioside content that can affect neuronal differentiation and neurodegeneration. A general scheme for GSL metabolism is shown in Fig. 10.2.

De novo biosynthesis of GSLs requires ceramide which is synthesized in the endoplasmic reticulum. In contrast to other cells, neurons have two different serine acyl-CoA acyltransferases: one specific for palmitoyl-CoA and the other for stearoyl-CoA. They are expressed in different ratios and in a spatiotemporal manner during neuronal development (Chigorno et al. 1997a) and are necessary for the first step in the synthesis of sphinganine and eicosasphinganine, respectively. The biosynthetic process leading to ceramide is shown in Fig. 10.3. Ceramide can be transported to the Golgi apparatus, the site of ganglioside synthesis, by either vesicular or non-vesicular (protein mediated) trafficking (Olayioye and Hausser 2012); neobiosynthesized gangliosides are transferred to the plasma membrane by a vesicle transport system, becoming external leaflet components. Catabolism of sphingolipids occurs in lysosomes, from which less complex products obtained in the degradation pathway can escape (Kolter and Sandhoff 2005). In neurons, only a minor portion of endocytic vesicles become lysosomes, while the rest rapidly reassociate with the plasma membrane. Among the more hydrophilic gangliosides, only a minor portion is released from the plasma membranes into the extracellular environment

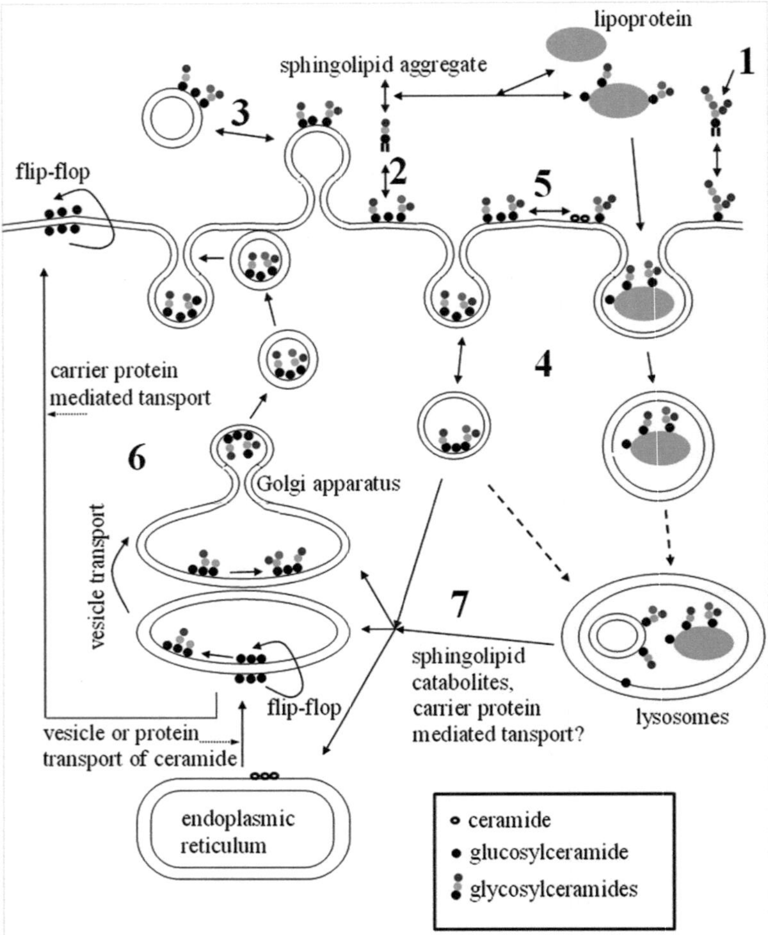

Fig. 10.2 Scheme depicting glycosphingolipid metabolism. Different metabolic pathways involved in changing plasma membrane glycosphingolipids composition. 1—plasma membrane uptake of extracellular glycolipids shed by different cells; 2—shedding of glycolipid monomers, some directly reenter the membrane, while others interact with the extracellular proteins or lipoproteins and are subsequently taken up by the cells and catabolized into lysosomes; 3—release of glycolipid-containing vesicles from the plasma membrane; 4—membrane endocytosis followed by sorting to lysosomes and lysosomal catabolism; 5—biosynthetic modifications by plasma membrane-associated glycosyltransferases and glycosidases; 6—neobiosynthesis of glycosphingolipids and their transport to the cell surface; 7—recycle in the biosynthetic pathway of partially catabolized molecules

(Chigorno et al. 2006). Sphingolipids present in the extracellular milieu are, at least in part, taken up by cells, becoming components of membranes where they may modify their composition, or they may be directly sorted to the lysosomes (Saqr et al. 1993). Finally, recent evidence indicates that both sphingolipid catabolic

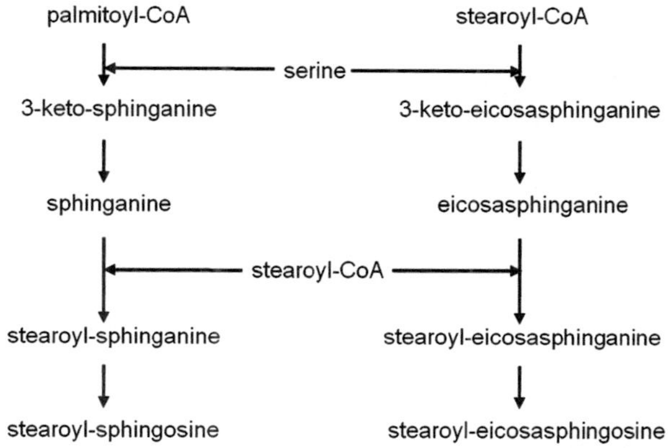

Fig. 10.3 Biosynthetic pathway for brain ceramide. Ceramide is synthesized in the endoplasmic reticulum. Neurons have two different serine acyl-CoA acyltransferases: one specific for palmitoyl-CoA and the other for stearoyl-CoA. They are expressed in different ratios and in a spatiotemporal manner during neuronal development and are necessary for the first step in the synthesis of sphinganine and eicosasphinganine, respectively

(hydrolases) and biosynthetic (transferases) enzymes are associated with the plasma membrane where they may act on membrane components (Aureli et al. 2011b).

While steps in the biosynthesis of GSLs are now well defined, little is known about how it is regulated. It is generally assumed that it is primarily regulated at the transcriptional level by control of expression of glycosyltransferases and/or transporter proteins. Indeed, changes in cellular GSL pattern such as that seen during neuronal development and upon oncogenic transformation and acquisition of drug resistance by tumor cells are paralleled by changes in expression of the corresponding glycosyltransferases. However, it is also possible that differences in intracellular metabolic flow and/or half-life of each GSL influence the GSL pattern (Veldman et al. 2002). In fact, regulation of intracellular trafficking may be as important as control of enzyme expression in determining the final GSL composition of the plasma membrane. Reliance of cells on neobiosynthesis and subsequent transport of the GSLs to the plasma membrane could require too much time to modify the plasma membrane GSL composition in response to extracellular signals. In addition, even though the half-life of GSLs may be short in neurons (Prinetti et al. 2001), it is long or very long in the majority of cells. In fibroblasts, where the half-life of gangliosides is long (Chigorno et al. 1997b), their major loss is by shedding into the extracellular environment (Chigorno et al. 2005), not catabolism. It has been determined that up to 7–8 % of the total cell sphingolipids are shed every day by cells in culture (Chigorno et al. 2006), making neobiosynthesis of GSLs a necessity.

The fact that some enzymes involved in GSL metabolism are found in the plasma membrane modifies the generalized concept that glycohydrolases are only in the lysosomes and glycosyltransferases are only in the Golgi. The association of enzymes needed for GSL metabolism with the plasma membrane, accumulating information about their activity on natural substrates in living cells, along with identification of the presence of pairs of enzymes, like sialidase and sialyltransferase, able to catalyze opposing reactions, suggest that changes in plasma membrane GSLs could occur rapidly in response to different conditions. These changes would be dependent only on kinetic properties of the enzymes, which can change very rapidly in response to ligand-triggered interactions.

10.4 Plasma Membrane-Associated Enzymes and Ganglioside Pattern

Enzymes involved in GSL metabolism that are associated with the plasma membrane include sialidase and sialyltransferase, β-hexosaminidase, β-galactosidase, and β-glucosidase (Fig. 10.4). In gangliosides sialic acid is usually linked by an α-linkage to the C3 position of galactose or to the C8 position of another sialic acid residue. Four different sialidases capable of catalyzing release of ganglioside sialic acid moieties have been identified. Neu1 is the lysosomal enzyme, known to be needed for catabolism of sialo-compounds; Neu2 is cytosolic and Neu4 is mitochondrial, but their role in vivo is not clear. Neu3 is associated with the plasma membrane and seems to be involved in functional processes. Small amounts of Neu1 and Neu2 have been also found in association with the plasma membrane.

Publications in the 1970s and 1980s described the association of sialidase (Schengrund and Rosenberg 1970; Tettamanti et al. 1972; Tettamanti et al. 1973; Tettamanti et al. 1975) and sialyltransferase (Preti et al. 1980) with synaptosomal membranes. These observations led to the hypothesis that a sialylation–desialylation cycle existed for gangliosides at the plasma membrane level and might be involved in the function of gangliosides in neurons. The sialylation–desialylation cycle

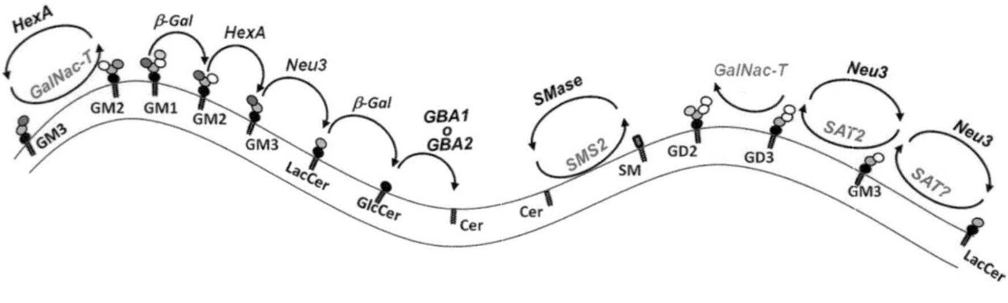

Fig. 10.4 Schematic representation of the glycosyltransferases and glycohydrolases associated with the cell surface. The presence of different glycosyltransferases and glycohydrolases at the cell plasma membrane allows the in situ modification of the cell surface glycolipid composition. The availability of a series of couples of enzymes catalyzing the same reaction in opposite directions extends the concept of cycle, originally reported for the sphingomyelin, to several other sphingolipids. HexA, hexosaminidase A; GalNac-T, UDP-N-UDP-N-acetylgalactosaminyltransferase; β-Gal, β-galactosidase; SMase, sphingomyelinase; SMS2

predates by 20 years the elucidation of the "sphingomyelin cycle" (Venable et al. 1995), in which the two enzymes sphingomyelinase and sphingomyelin synthase modulate cell proliferation and apoptosis through the level of ceramide. The existence of a plasma membrane-associated sialidase distinct from the lysosomal enzyme was first suggested by enzymatic and immunological studies (Schengrund et al. 1976; Miyagi et al. 1990a; Miyagi et al. 1990b; Schneider-Jakob and Cantz 1991; Kopitz et al. 1994), as well by metabolic studies (Riboni et al. 1991; Kopitz et al. 1997a); a membrane-bound sialidase was purified from human brain grey matter (Kopitz et al. 1997b) and from bovine brain (Hata et al. 1998; Oehler et al. 2002). In 1999, the existence of a specific membrane-associated sialidase, coded as Neu3, that was distinct from other known sialidases, was unambiguously proven by the cloning of its cDNA sequence for human (Wada et al. 1999) bovine (Miyagi et al. 1999) and mouse (Hasegawa et al. 2000).

Neu3 and gangliosides co-localize in Triton X-100-insoluble lipid rafts (Kalka et al. 2001). The nonrandom distribution of Neu3 at the cellular surface introduces the possibility that the biological effects of this enzyme might be due to the local reorganization of GSL-based signaling units. Remarkably, Neu3 modulates cell surface GSL composition by *trans* interactions, hydrolyzing substrates on the surface of neighboring cells (Papini et al. 2004).

In colon and renal cancer, this sialidase appears to be responsible for acting on GM3, thereby maintaining high cellular levels of lactosylceramide that can exert a Bcl-2-dependent antiapoptotic effect, contributing to survival of the cancer cells and subsequent tumor progression (Kakugawa et al. 2002; Ueno et al. 2006).

Neu3, together with plasma membrane-associated β-galactosidase and β-glucosidase, can act on gangliosides to produce bioactive ceramide at the cell surface of cultured human fibroblasts (Valaperta et al. 2006). The structure of plasma membrane-associated ß-galactosidase is still unknown; it could be the lysosomal enzyme, transferred to the plasma membrane by the fusion of lysosomes with the plasma membrane. Two distinct β-glucosidases are associated with the plasma membranes of fibroblasts. One is the lysosomal enzyme GBA1; the other, GBA2, displays a different structure. Expression of either Neu3, β-galactosidase, or β-glucosidase can affect the expression/activity of the others. Overexpression of Neu3 resulted in increased expression/activity of β-galactosidase and β-glucosidase. This resulted in an increase in ceramide and a shift from cell proliferation to cell death by apoptosis. Nevertheless the increase of cell surface ceramide was paralleled by a very scant reduction in ganglioside GM3, a substrate for Neu3, and the main ganglioside in fibroblasts. In fact, overexpression of Neu3 led to increased expression of GM3 synthase. This may be how cells avoid reduction of GM3 by the concomitant increase in Neu3, β-galactosidase, and β-glucosidase activities that combined catalyze conversion of GM3 to ceramide. The increased expression of sialyltransferase 1 (SAT1) increases its action on lactosylceramide depleting its availability for the biosynthesis of globotrihexosylceramide. Figure 10.4 shows relationships between several enzymes of GSL metabolism. As for Neu3, plasma membrane β-galactosidase and β-glucosidase display *trans* activity in living cells by acting on substrates on neighboring membranes in the absence of activator proteins or detergents (Papini et al. 2004; Aureli et al. 2009).

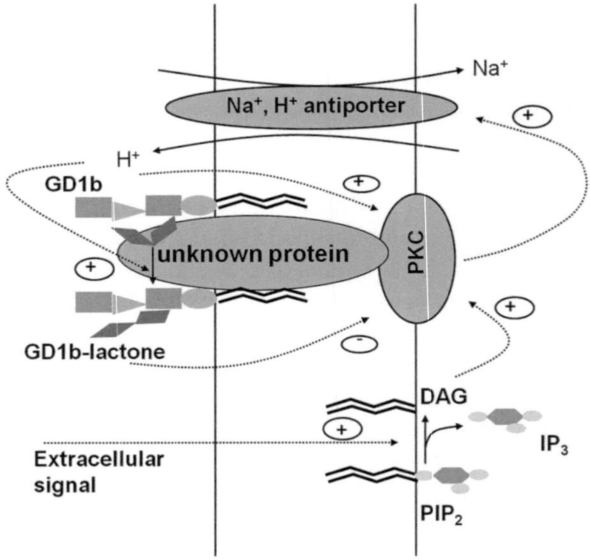

Fig. 10.5 Monolactone formation due to the acidic plasma membrane microenvironment. The catalytic protons are generated by the sodium/proton pump located at the cell surface and activated through the action of protein kinase C (PKC). The local acidic microenvironment is directly responsible for lactone formation

The presence of active β-hexosaminidase A in the external leaflet of the plasma membrane has also been found in cultured fibroblasts (Mencarelli et al. 2005). Immunological and biochemical characterization of the membrane-associated β-hexosaminidase indicated that this enzyme has the same structure as the lysosomal enzyme. This led to the hypothesis that a regulated fusion of lysosomes with the plasma membrane might be a general mechanism of repair for the plasma membrane (Reddy et al. 2001) and could also provide a way for lysosomal GSL-metabolizing enzymes to reach the cell surface where, together with specific and different membrane-associated enzymes, they could play an active role in remodeling the GSL composition of the plasma membrane.

Information is also available about in situ sialylation of cell surface gangliosides. The original report on the presence of a synaptosomal membrane sialyltransferase in calf brain (Preti et al. 1980) was confirmed by metabolic studies in chicken embryos (Matsui et al. 1986) and rat brains (Durrie et al. 1988; Durrie and Rosenberg 1989). More recently, it was shown that exposure to dexamethasone markedly increased GM3 synthesis, due to enhanced gene expression and increased enzyme activity of SAT1. Metabolic studies indicated that this event is localized at the plasma membrane (Iwamori and Iwamori 2005), thus confirming that glycolipid sialylation might occur outside the Golgi compartment, contributing to the local modulation of the cell surface GSL pattern.

A further possibility for modification of the oligosaccharide structure of gangliosides is by the lactonization of molecules containing a disialosyl residue, such as GD1b. Ganglioside lactones are present as minor components in vertebrate brains (Sonnino et al. 1983; Riboni et al. 1986). GD1b monolactone formation, in the presence of catalytic protons, has been studied in vitro (Bassi et al. 1989) (Fig. 10.5), and it has been shown that lactonization influences the conformation, aggregative

(Acquotti et al. 1987) and biological properties of GD1b (Bassi et al. 1991). GD1b is able to directly interact with several cellular proteins (Prinetti et al. 2000b) and to modulate several plasma membrane-associated protein kinase activities (Bassi et al. 1991), while GD1b-lactone does not or does so in a very reduced way (Bassi et al. 1991; Sonnino et al. 1990). This suggests that lactonization/delactonization might represent a localized event able to trigger specific ganglioside-mediated cellular events. In vivo GD1b lactonization occurs in neurons through a process that suggests the presence of a specific enzyme associated with the plasma membrane (Bassi et al. 1994).

10.5 Plasma Membrane Glycosphingolipid Hydrolases in the Nervous System

10.5.1 Sialidase Neu3

The first plasma membrane-associated enzyme identified that was involved in ganglioside catabolism was the sialidase Neu3 (NEU3, EC 3.2.1.18). It can be considered an ubiquitous enzyme as it is expressed albeit at different levels in the plasma membranes of most normal and pathological human tissues such as the human brain (Kopitz et al. 1994), normal colon, as well as colon rectal carcinoma tissues, hepatic tumors, and kidney carcinomas (Monti et al. 2002; Miyagi et al. 2008b; Miyagi et al. 2008a; Ueno et al. 2006; Kakugawa et al. 2002). In addition, its expression and activity were also found in normal and pathological cell lines such as erythroid and erythroleukemic cells (Venerando et al. 2002; Tringali et al. 2007b; Tringali et al. 2007a), fibroblasts (Chigorno et al. 1986), neurons, neuroblastoma cells (Schengrund and Repman 1982), breast ductal cancer T47D cells, colon carcinoma CaCo2 cells, colorectal adenocarcinoma HT29 cells, different types of ovarian cancer cells and cervix adenocarcinoma HeLa cells (Kakugawa et al. 2002).

Neu3 catalyzes the hydrolysis of both α2–8 and α2–3 external ketosidic linkages but is ineffective on inner α2–3 sialyl residues. An increase in Neu3 activity modifies cell surface ganglioside composition by catalyzing the conversion of polysialylated species of Gg4 and Gg3 to monosialoderivatives and GM3 to lactosylceramide. These changes have significant effects on neuronal differentiation and apoptosis in both normal and pathological cells (Kakugawa et al. 2002; Valaperta et al. 2007; Valaperta et al. 2006). In mouse and human neuroblastoma cells, the pharmacologically induced neuronal differentiation is accompanied by an increase in Neu3 expression and activity (Proshin et al. 2002). Neurite outgrowth can also be induced by transfection of cells with Neu3. In addition, an increase in Neu3 activity enhanced extension and/or branching of neurites promoted by exposure of neuroblastoma cells to 5-bromodeoxyuridine (Hasegawa et al. 2000). Conversely, in SK-N-MC neuroblastoma cells, inhibition of Neu3 activity resulted in the loss of neuronal differentiation markers (Kopitz et al. 1994; von Reitzenstein et al. 2001).

In cultured rat granule cells, Neu3 increased during differentiation and remained constant during aging (Aureli et al. 2011c). In cultured hippocampal neurons, Neu3 activity regulated the plasma membrane content of GM1 and was essential for axonal growth and regeneration after axotomy (Rodriguez et al. 2001). In these neurons, Neu3 activity is asymmetrically concentrated at the end of a single neurite and determines its axonal fate by a local increase in TrkA activity (Da Silva et al. 2005). However, reduction of Neu3 expression did not prevent induction of neuroblastoma cell differentiation (Valaperta et al. 2006). The nonrandom distribution of Neu3 at the cell surface (Kalka et al. 2001) introduces the possibility that the biological effects of this enzyme might be due to local reorganization of GSL-based signaling units, not just on cells with which Neu3 is associated, but on neighboring cells as well, due to *trans* interactions (Papini et al. 2004).

Taking into account the high concentration of gangliosides in the nervous system, and the very high and progressive increase of Neu3 and total sialyltransferase during neuronal differentiation, it is possible to hypothesize that a sialylation–desialylation cycle on the plasma membrane has a specific role during neuronal cell specialization, especially during cell stages in which it is necessary to build specialized membranes such as axonal protrusion and elongation, dendritic arborization, and synaptogenesis. These enzymes could also be essential for axon repair and/or synaptic function.

10.5.2 β-Glucocerebrosidases

At least three different β-glucocerebrosidases have been described: a β-glucocerebrosidase (GBA, GBA1, EC 3.2.1.45) sensitive to inhibition by conduritol B epoxide (CBE), associated primarily with lysosomes (Neufeld et al. 1996); a cytosolic β-glucosidase (GBA3, EC 3.2.1.21) (Daniels et al. 1981); and a non-lysosomal β-glucosylceramidase (GBA2, EC 3.2.1.45) (van Weely et al. 1993).

GBA2 is able to catalyze the hydrolysis of glucosylceramide both at the cell surface and in the endoplasmic reticulum (van Weely et al. 1993; Korschen et al. 2012). In fact, the cellular localization of GBA2 is still controversial, since it has been described as being associated with endosomal vesicles, the plasma membrane, and the endoplasmic reticulum. Analysis of data describing its localization indicates that it depends on the cell type and cell stage. Up to now, most of the information about GBA2 localization came from studies performed on cells overexpressing GBA2 as a fusion protein with green fluorescence protein (GFP); therefore, further investigation is needed to define its cellular topology. Database searching of GBA2 cDNA sequences revealed apparent orthologs of this enzyme in species ranging from Drosophila to Arabidopsis to vertebrates, indicating that the protein is highly conserved and suggesting its functional importance. Study of GBA2 expression and activity showed that in humans, this enzyme is abundant in the brain, heart, skeletal muscle, and kidney (Matern et al. 2001), whereas in mouse it is most abundant in the testis and brain (Yildiz et al. 2006).

A recent study using GBA2 knockout (KO) mice showed an abnormal accumulation of glucosylceramide in multiple tissues, including the brain, liver, and testis. The KO mice had normal bile acid metabolism and, apparently, no impairment in the CNS; however, the accumulated glucosylceramide led to decreased fertility due to formation of misshapen spermatozoa (Yildiz et al. 2006). More recent studies indicated that mutations in GBA2 caused autosomal recessive cerebellar ataxia with spasticity in humans (Martin et al. 2013; Hammer et al. 2013). Moreover, antisense morpholino oligonucleotides targeting the GBA2 orthologous gene in zebrafish led to abnormal motor behavior and axonal shortening/branching of motoneurons. This condition could be rescued by transfection with human wild-type mRNA for GBA2 but not with the mRNA containing the missense mutation found in GBA2 in patients affected by autosomal recessive cerebellar ataxia with spasticity. These data suggest a specific role for GBA2 in the control of the cellular glucosylceramide–ceramide balance that could be responsible for the onset of motoneuron defects (Martin et al. 2013). As mentioned before, homozygous GBA2 knockout mice showed no apparent neurological signs, liver dysfunction, or reduced viability when observed at 4 months of age even when an accumulation of glycolipid species was observed by mass spectrometry in the brain, liver, and testis (Yildiz et al. 2006). Longer times may be required to observe a neurological phenotype in these mice, as was found in other mouse models of hereditary spastic paraplegia. The latter may reflect the fact that the neurological signs are very subtle during the first months of life (Ferreirinha et al. 2004; Soderblom et al. 2010). Impairment of their nervous system could also be obscured by the different structure of mouse corticospinal tracts compared to humans or zebrafish or by a compensation for the loss in GBA2 activity by other GBA enzymes during early stages of development (Martin et al. 2013). It should be noted that the GBA2 knockout mice lacked only exons 5 to 10 and retained 50 % of normal glucosidase activity. On this basis it has been hypothesized that accumulation of glucosylceramide in the ER and/or plasma membrane did not reach the threshold needed to cause neurological symptoms.

It has also been shown that GBA2 activity increases more than threefold during neuronal differentiation (Aureli et al. 2011a). β-Glucosidase activity also increases during differentiation of murine neuronal stem cells (Aureli et al. 2011a). However, in primary neuronal cultures, cell surface β-glucosidase activity is due primarily to the GBA2 enzyme whose activity increases more than threefold during neuronal differentiation. While the CBE-sensitive β-glucosidase enzyme contributes to the total plasma membrane (PM) β-glucosidase activity, the CBE-sensitive β-glucosidase/GBA2 ratio is 0.4 in the first stage of neuronal differentiation and drops to 0.25 in fully differentiated neurons (Aureli et al. 2011c). Conversely, in murine neuronal stem cells, plasma membrane β-glucosidase activity was largely due to the CBE-sensitive β-glucosidase enzyme, with the CBE-sensitive β-glucosidase/GBA2 ratio 0.75 in precursors and 2.3 in differentiated cells. The different behaviors of these enzymes in the two different cellular models could reflect the fact that, as a result of serum-induced murine neuronal stem cell differentiation, the neuronal component in the differentiated cells accounted for only about 10 % of the cells, with about 70–80 % by glial cells, whose contribution in terms of the two different β-glucosidase activities could be very different than that of neurons. In fact, in nonneuronal cell lines such as human fibroblasts, we observed a ratio between CBE-sensitive β-glucosidase and GBA2 that was around 7 (Aureli et al. 2009). The lack of data on the activity of PM-associated glycohydrolases in cultures of astrocytes and oligodendrocytes indicates that future studies need to be done to assess this issue. Interestingly in studies of fibroblasts derived from patients affected by different variants of Gaucher disease (GD), the reduction in total GBA1 activity was paralleled by an increase in GBA2 activity and expression. This was particularly evident in fibroblasts from patients affected by type 2 GD (GD2), the most severe form of the neuronopathic type of GD (Aureli et al. 2012). Despite all of these observations, the link between GBA2 and neuronal differentiation as well as neurodegeneration is still unclear and merits further study.

Several years ago the presence of a cell surface CBE-sensitive β-glucosidase activity was described (Aureli et al. 2009). Studies of human fibroblasts derived from patients affected by GD showed that this plasma membrane-associated activity was significantly reduced with respect to fibroblasts from healthy patients permitting one to ascribe this activity to a deficiency in the GBA1 enzyme (Aureli et al. 2012).

Increased GBA1 and GBA2 activities on the plasma membrane of human fibroblasts and concomitant increase in ceramide are responsible for cell cycle arrest and apoptosis (Valaperta et al. 2006). This activity is due to both enzymes (Aureli et al. 2009).

A recent multicentre study demonstrated that mutations of GBA1 represent, to date, the most common genetic risk factor for Parkinson's disease (PD) (Sidransky et al. 2009). Importantly, a recent paper demonstrated that in neurons and brains from this type of PD patients, the lysosomal accumulation of glucosylceramide, the substrate for GBA1, directly influenced the abnormal lysosomal storage of α-synuclein oligomers resulting in a further inhibition of GBA1 activity (Mazzulli et al. 2011). These findings suggest for the first time that the bidirectional effect of decreased GBA1 activity and α-synuclein accumulation forms a positive feedback loop that may lead to self-propagating disease (Mazzulli et al. 2011). Up to now this process has just been described for the lysosomal function of GBA1, but due to its involvement in the production of apoptotic ceramide at the PM level, GBA1 associated with the cell PM could contribute to neuronal impairment in neurodegenerative diseases.

10.5.3 β-Galactosidases

Two different β-galactosidases involved in GSL metabolism have been described: β-galactocerebrosidase (β-Gal-ase, GALC EC 3.2.1.46), which catalyzes the hydrolysis of galactose from galactosylceramide, lactosylceramide, and galactosylsphingosine, and the β-galactosidase (GAL, EC 3.2.1.23), which catalyzes hydrolysis of the terminal galactose of GM1 (Li and Li 1999). As is well known, loss of function of β-galactosidase EC 3.2.1.46 is responsible for the disease globoid leukodystrophy (GLD, Krabbe disease), while deficiency of β-galactosidase EC 3.2.1.23 is the cause of GM1 gangliosidosis (Xu et al. 2010). Both these sphingolipidoses are characterized by an impairment of the CNS although the molecular bases are unclear.

In addition to the lysosomal enzymes, PM-associated β-galactosidase activity has been found in several cell lines (Aureli et al. 2011b). The identity of the protein, or proteins, responsible for the β-galactosidase activity present at the cell surface is still unknown. However, in living human fibroblasts, the presence of a β-galactosidase which displays a *trans* activity (on substrates belonging to the cell surface of neighboring cells) has been verified. It is active in the absence of detergents or activator proteins, suggesting that on the cell surface there is at least one enzyme with a β-galactocerebrosidase-like activity (Aureli et al. 2009). Using the same cells, it was shown that its expression is up-regulated by Neu3 overexpression and correlated with the onset of ceramide-mediated apoptosis (Valaperta et al. 2006). β-Galactosidase activity was measured during neuronal cell differentiation and aging, in both the total cell lysate and the plasma membrane fraction from rat cerebellar granule cells. Both activities were up-regulated during cell differentiation. As expected, β-galactosidase activity associated with the plasma membranes was much less than that found in the total cell homogenate. Total cell activity remained constant during differentiation and then increased fourfold during aging. In contrast, cell surface activity increased tenfold during differentiation and then doubled during neuronal senescence (Aureli et al. 2011c). Similar behavior was described for PM-associated β-galactosidase activity during neuronal differentiation of NSCs (Aureli et al. 2011a). β-Galactosidase activity has been proposed as a marker for senescence (Coates 2002; Dimri et al. 1995; Severino et al. 2000; Geng et al. 2010). The behavior of the PM-associated enzyme in rat cerebellar granule cells suggests that β-galactosidase activity could be used as hallmark of both neuronal differentiation and aging as well as of apoptosis in fibroblasts. Little is known regarding the functional role of PM-associated β-galactosidases. It has been hypothesized that they may act as cell surface receptors mediating various cell–cell and cell–matrix interactions responsible for cell migration, differentiation, and axonal protrusion (Evans et al. 1993; Huang et al. 1995). No data are available on their enzymatic properties.

10.5.4 β-Hexosaminidases

β-Hexosaminidase is a dimeric enzyme that exists in three different isoforms. There are two different subunits for β-hexosaminidase, α (528 residues) and β (556 residues), encoded by two different but evolutionarily related genes, HEX A and HEX B (Triggs-Raine et al. 2001). Each subunit has its own active site: the β-subunit hydrolyzes uncharged substrates, whereas the α-subunit catalyzes cleavage of GalNAc from negatively charged ones (Bearpark and Stirling 1978; Kytzia and

Sandhoff 1985). However, dimerization is necessary for the enzymes in order to become fully functional. Thus, the α- and β-subunits can form three different β-hexosaminidase isoenzymes: Hex A (αβ), Hex B (ββ), and Hex S (αα). Only the αβ heterodimer Hex A is able to remove β-linked nonreducing terminal GalNAc from ganglioside GM2 and GalNAc-GD1a, in the presence of the GM2 activator protein, a specific cofactor of Hex A (Kolter and Sandhoff 2006).

Genetic defects in either of the genes encoding the α- and β-subunits of Hex A or the GM2 activator protein can result in GM2 accumulation in neural tissue leading to one of the three forms of GM2-gangliosidosis: Tay–Sachs disease, due to defects in the α-subunit (TSD, OMIN 2728800); Sandhoff disease, characterized by defects in the β-subunit (SD, OMIN 268800); and the AB variant of Sandhoff disease (OMIN 272750), in which both the subunits are affected (Bateman et al. 2011). The massive neuronal accumulation of GM2 is accompanied by progressive neurological deterioration affecting motor, cerebral, and spinocerebellar functions.

The presence of active β-hexosaminidase A in the external leaflet of the plasma membrane was found in studies of cultured fibroblasts (Mencarelli et al. 2005). Immunological and biochemical characterization of the membrane-associated β-hexosaminidase indicated that the enzyme has the same structure as that in lysosomes. This suggests that a regulated fusion process between lysosomes and the plasma membrane might be responsible for transport of lysosomal enzymes to the cell surface where the enzymes could function in the remodeling of the glycolipid content and pattern on the external leaflet of the plasma membrane. During differentiation of murine neuronal stem cells, plasma membrane β-hexosaminidase increased its activity, reaching a maximum in fully differentiated cells (Aureli et al. 2011a). On the other hand, analysis of the PM-associated glycohydrolase activity in fibroblasts derived from patients affected by different variants of Gaucher disease indicated that the PM-associated β-hexosaminidase increased only in cells derived from patients affected by the most severe neuronopathic form of GD (GD2) (Aureli et al. 2012).

Analysis of epithelial cells indicated the presence on the cell surface of an UDP-GalNAc: LacCer/GM3/GD3 *N*-acetylgalactosaminyl transferase able to act on exogenous GM3 (Crespo et al. 2010). Currently, no data are available regarding its activity on endogenously synthesized GSLs. However, its ability to act on exogenous substrates could be important in the plasma membrane remodeling process. In fact, the glycolipid composition of the plasma membrane could also be remodeled by the uptake of glycolipids from the extracellular environment (e.g., from other cells, lipoproteins, or molecules shed by other cells). These compounds, depending on the "cellular request" and on their aggregation, could be endocytosed or become components of the cell surface directly or after modification by the action of PM-associated enzymes. The coexistence of β-hexosaminidase and β-hexosaminyl transferase activity on the cell surface supports the hypothesis that there is a glycolipid cycle occurring on the plasma membrane that can have important biophysical effects on the membrane itself and affect events that regulate the "cell social life."

10.6 Conclusions

During neuronal development dramatic changes occur in GSL content along with a parallel reorganization of PM lipid domains enriched in GSLs (Yu 1994; Prinetti et al. 2001; Yu et al. 2004). While the driving forces guiding these modifications are not completely understood, some information regarding the possibility of fine-tuning the cell PM GSL composition has been obtained regarding a synergy in activity by different plasma membrane glycohydrolases (Valaperta et al. 2006; Aureli et al. 2009).

In rat cerebellar granule cells, increases in total cell ceramide content (eightfold from the 2nd to the 17th day in culture) and in that belonging to the sphingolipid-enriched domains (tenfold from the 2nd to the 17th day in culture) were observed. A parallel reduction in the endogenous content of both sphingomyelin and gangliosides was observed in sphingolipid-enriched domains of senescent cells relative to fully differentiated neurons (Prinetti et al. 2000a). The increase of ceramide could be explained by the well-known ceramide–sphingomyelin cycle that is known to correlate with apoptotic phenomena (Venable et al. 1995). The increased activity of the PM-associated glycohydrolases during cell aging supports speculation that the augmented ceramide in the PM could be derived from cell surface catabolism of the glycosphingolipids, as found in human fibroblasts (Valaperta et al. 2006). As reported (Rodriguez et al. 2001; Da Silva et al. 2005), the sialidase Neu3 is able to influence extension and symmetry of axons in neuronal cells, possibly by inducing a local change in PM sphingolipid composition at the axonal cones. All these data support the idea that modulation of the activities of other PM-associated glycohydrolases during neuronal differentiation could affect differentiation itself and could also help define the curvature properties of specific areas of the PM (such as synapses or the negative curvature of the membrane near the axon protrusion) by a rapid in situ modification of GSL components. An example of the latter is provided by the geometry of synapses. Typically it is characterized by a succession of PM regions with negative or positive curvatures that correlate respectively to an enrichment of the area with simple sphingolipids or the enrichment of more complex GSLs (Sonnino et al. 1994; Brocca and Sonnino 1997). On the other hand, an aberrant increase in cell surface glycohydrolases can increase production of apoptotic ceramide and lead to the onset of neuronal impairment. For these reasons, the balance between glycosylation and de-glycosylation events at the cell surface could be a very important mechanism for maintaining the appropriate neuronal physiology.

Conflict of Interest All the authors declare that they have no conflict of interest.

References

Acquotti D, Fronza G, Riboni L, Sonnino S, Tettamanti G. Ganglioside lactones:1H-NMR determination of the inner ester position of GD1b-ganglioside lactone naturally occurring in human brain or produced by chemical synthesis. Glycoconj J. 1987;V4:119–27.

Aureli M, Bassi R, Loberto N, Regis S, Prinetti A, Chigorno V, et al. Cell surface associated glycohydrolases in normal and Gaucher disease fibroblasts. J Inherit Metab Dis. 2012;35:1081–91.

Aureli M, Gritti A, Bassi R, Loberto N, Ricca A, Chigorno V, et al. Plasma membrane-associated glycohydrolases along differentiation of murine neural stem cells. Neurochem Res. 2011a;37:1344–54.

Aureli M, Loberto N, Chigorno V, Prinetti A, Sonnino S. Remodeling of sphingolipids by plasma membrane associated enzymes. Neurochem Res. 2011b;36:1636–44.

Aureli M, Loberto N, Lanteri P, Chigorno V, Prinetti A, Sonnino S. Cell surface sphingolipid glycohydrolases in neuronal differentiation and aging in culture. J Neurochem. 2011c;116:891–9.

Aureli M, Masilamani AP, Illuzzi G, Loberto N, Scandroglio F, Prinetti A, et al. Activity of plasma membrane beta-galactosidase and beta-glucosidase. FEBS Lett. 2009;583:2469–73.

Bassi R, Chigorno V, Fiorilli A, Sonnino S, Tettamanti G. Exogenous gangliosides GD1b and GD1b-lactone, stably associated to rat brain P2 subcellular fraction, modulate differently the process of protein phosphorylation. J Neurochem. 1991;57:1207–11.

Bassi R, Riboni L, Sonnino S, Tettamanti G. Lactonization of GD1b ganglioside under acidic conditions. Carbohydr Res. 1989;193:141–6.

Bassi R, Riboni L, Tettamanti G. Cultured cerebellar granule cells, but not astrocytes, produce an ester of ganglioside GD1b, presumably GD1b monolactone, from exogenous GD1b. Biochem J. 1994;302(Pt 3):937–42.

Bateman KS, Cherney MM, Mahuran DJ, Tropak M, James MN. Crystal structure of beta-hexosaminidase B in complex with pyrimethamine, a potential pharmacological chaperone. J Med Chem. 2011;54:1421–9.

Bearpark TM, Stirling JL. A difference in the specificities of human liver N-acetyl-beta-hexosaminidases A and B detected by their activities towards glycosaminoglycan oligosaccharides. Biochem J. 1978;173:997–1000.

Brocca P, Sonnino S. Dynamic and spatial organization of surface gangliosides. Trends Glycosci Glycotech. 1997;9:433–45.

Chigorno V, Cardace G, Pitto M, Sonnino S, Ghidoni R, Tettamanti G. A radiometric assay for ganglioside sialidase applied to the determination of the enzyme subcellular location in cultured human fibroblasts. Anal Biochem. 1986;153:283–94.

Chigorno V, Giannotta C, Ottico E, Sciannamblo M, Mikulak J, Prinetti A, et al. Sphingolipid uptake by cultured cells: complex aggregates of cell sphingolipids with serum proteins and lipoproteins are rapidly catabolized. J Biol Chem. 2005;280:2668–75.

Chigorno V, Negroni E, Nicolini M, Sonnino S. Activity of 3-ketosphinganine synthase during differentiation and aging of neuronal cells in culture. J Lipid Res. 1997a;38:1163–9.

Chigorno V, Riva C, Valsecchi M, Nicolini M, Brocca P, Sonnino S. Metabolic processing of gangliosides by human fibroblasts in culture–formation and recycling of separate pools of sphingosine. Eur J Biochem. 1997b;250:661–9.

Chigorno V, Sciannamblo M, Mikulak J, Prinetti A, Sonnino S. Efflux of sphingolipids metabolically labeled with [1–3H]sphingosine, L-[3–3H]serine and [9,10–3H]palmitic acid from normal cells in culture. Glycoconj J. 2006;23:159–65.

Coates PJ. Markers of senescence? J Pathol. 2002;196:371–3.

Crespo PM, Demichelis VT, Daniotti JL. Neobiosynthesis of glycosphingolipids by plasma membrane-associated glycosyltransferases. J Biol Chem. 2010;285:29179–90.

Da Silva JS, Hasegawa T, Miyagi T, Dotti CG, Abad-Rodriguez J. Asymmetric membrane ganglioside sialidase activity specifies axonal fate. Nat Neurosci. 2005;8:606–15.

Daniels LB, Coyle PJ, Chiao YB, Glew RH, Labow RS. Purification and characterization of a cytosolic broad specificity beta-glucosidase from human liver. J Biol Chem. 1981;256:13004–13.

Del Favero E, Brocca P, Motta S, Rondelli V, Sonnino S, Cantu L. Nanoscale structural response of ganglioside-containing aggregates to the interaction with sialidase. J Neurochem. 2011;116:833–9.

Dimri GP, Lee X, Basile G, Acosta M, Scott G, Roskelley C, et al. A biomarker that identifies senescent human cells in culture and in aging skin in vivo. Proc Natl Acad Sci U S A. 1995;92:9363–7.

Durrie R, Rosenberg A. Anabolic sialosylation of gangliosides in situ in rat brain cortical slices. J Lipid Res. 1989;30:1259–66.
Durrie R, Saito M, Rosenberg A. Endogenous glycosphingolipid acceptor specificity of sialosyltransferase systems in intact Golgi membranes, synaptosomes, and synaptic plasma membranes from rat brain. Biochemistry. 1988;27:3759–64.
Evans MK, Robbins JH, Ganges MB, Tarone RE, Nairn RS, Bohr VA. Gene-specific DNA repair in xeroderma pigmentosum complementation groups A, C, D, and F. Relation to cellular survival and clinical features. J Biol Chem. 1993;268:4839–47.
Ferreirinha F, Quattrini A, Pirozzi M, Valsecchi V, Dina G, Broccoli V, et al. Axonal degeneration in paraplegin-deficient mice is associated with abnormal mitochondria and impairment of axonal transport. J Clin Invest. 2004;113:231–42.
Geng YQ, Guan JT, Xu XH, Fu YC. Senescence-associated beta-galactosidase activity expression in aging hippocampal neurons. Biochem Biophys Res Commun. 2010;396:866–9.
Gulbins E, Grassme H. Ceramide and cell death receptor clustering. Biochim Biophys Acta 2002;1585:139–45.
Hammer MB, Eleuch-Fayache G, Schottlaender LV, Nehdi H, Gibbs JR, Arepalli SK, et al. Mutations in GBA2 cause autosomal-recessive cerebellar ataxia with spasticity. Am J Hum Genet. 2013;92:245–51.
Hasegawa T, Yamaguchi K, Wada T, Takeda A, Itoyama Y, Miyagi T. Molecular cloning of mouse ganglioside sialidase and its increased expression in neuro2a cell differentiation. J Biol Chem. 2000;275:14778.
Hata K, Wada T, Hasegawa A, Kiso M, Miyagi T. Purification and characterization of a membrane-associated ganglioside sialidase from bovine brain. J Biochem (Tokyo). 1998;123:899–905.
Holopainen JM, Angelova MI, Kinnunen PK. Vectorial budding of vesicles by asymmetrical enzymatic formation of ceramide in giant liposomes. Biophys J. 2000;78:830–8.
Huang Q, Shur BD, Begovac PC. Overexpressing cell surface beta 1.4-galactosyltransferase in PC12 cells increases neurite outgrowth on laminin. J Cell Sci. 1995;108(Pt 2):839–47.
Iwamori M, Iwamori Y. Changes in the glycolipid composition and characteristic activation of GM3 synthase in the thymus of mouse after administration of dexamethasone. Glycoconj J. 2005;22:119–26.
Kakugawa Y, Wada T, Yamaguchi K, Yamanami H, Ouchi K, Sato I, et al. Up-regulation of plasma membrane-associated ganglioside sialidase (Neu3) in human colon cancer and its involvement in apoptosis suppression. Proc Natl Acad Sci U S A. 2002;99:10718–23.
Kalka D, von Reitzenstein C, Kopitz J, Cantz M. The plasma membrane ganglioside sialidase cofractionates with markers of lipid rafts. Biochem Biophys Res Commun. 2001;283:989–93.
Kamerling JP, Vliegenthart JF. Identification of O-cetylated N-acylneuraminic acids by mass spectrometry. Carbohydr Res. 1975;41:7–17.
Kolter T, Sandhoff K. Principles of lysosomal membrane digestion: stimulation of sphingolipid degradation by sphingolipid activator proteins and anionic lysosomal lipids. Annu Rev Cell Dev Biol. 2005;21:81–103.
Kolter T, Sandhoff K. Sphingolipid metabolism diseases. Biochim Biophys Acta. 2006;1758: 2057–79.
Kopitz J, Muhl C, Ehemann V, Lehmann C, Cantz M. Effects of cell surface ganglioside sialidase inhibition on growth control and differentiation of human neuroblastoma cells. Eur J Cell Biol. 1997a;73:1–9.
Kopitz J, Sinz K, Brossmer R, Cantz M. Partial characterization and enrichment of a membrane-bound sialidase specific for gangliosides from human brain tissue. Eur J Biochem. 1997b; 248:527–34.
Kopitz J, von Reitzenstein C, Muhl C, Cantz M. Role of plasma membrane ganglioside sialidase of human neuroblastoma cells in growth control and differentiation. Biochem Biophys Res Commun. 1994;199:1188–93.
Korschen HG, Yildiz Y, Raju DN, Schonauer S, Bonigk W, Jansen V, et al. The non-lysosomal beta-glucosidase GBA2 is a non-integral membrane-associated protein at the endoplasmic reticulum (ER) and Golgi. J Biol Chem. 2012;288:3381–93.

Kytzia HJ, Sandhoff K. Evidence for two different active sites on human beta-hexosaminidase A. Interaction of GM2 activator protein with beta-hexosaminidase A. J Biol Chem. 1985;260: 7568–72.

Li YT, Li SC. Enzymatic hydrolysis of glycosphingolipids. Anal Biochem. 1999;273:1–11.

Martin E, Schule R, Smets K, Rastetter A, Boukhris A, Loureiro JL, et al. Loss of function of glucocerebrosidase GBA2 is responsible for motor neuron defects in hereditary spastic paraplegia. Am J Hum Genet. 2013;92:238–44.

Matern H, Boermans H, Lottspeich F, Matern S. Molecular cloning and expression of human bile acid beta-glucosidase. J Biol Chem. 2001;276:37929–33.

Matsui Y, Lombard D, Massarelli R, Mandel P, Dreyfus H. Surface glycosyltransferase activities during development of neuronal cell cultures. J Neurochem. 1986;46:144–50.

Mazzulli JR, Xu YH, Sun Y, Knight AL, McLean PJ, Caldwell GA, et al. Gaucher disease glucocerebrosidase and alpha-synuclein form a bidirectional pathogenic loop in synucleinopathies. Cell. 2011;146:37–52.

Mencarelli S, Cavalieri C, Magini A, Tancini B, Basso L, Lemansky P, et al. Identification of plasma membrane associated mature beta-hexosaminidase A, active towards GM2 ganglioside, in human fibroblasts. FEBS Lett. 2005;579:5501–6.

Miyagi T, Sagawa J, Konno K, Handa S, Tsuiki S. Biochemical and immunological studies on two distinct ganglioside-hydrolyzing sialidases from the particulate fraction of rat brain. J Biochem (Tokyo). 1990a;107:787–93.

Miyagi T, Sagawa J, Konno K, Tsuiki S. Immunological discrimination of intralysosomal, cytosolic, and two membrane sialidases present in rat tissues. J Biochem (Tokyo). 1990b;107: 794–8.

Miyagi T, Wada T, Iwamatsu A, Hata K, Yoshikawa Y, Tokuyama S, et al. Molecular cloning and characterization of a plasma membrane-associated sialidase specific for gangliosides. J Biol Chem. 1999;274:5004–11.

Miyagi T, Wada T, Yamaguchi K. Roles of plasma membrane-associated sialidase NEU3 in human cancers. Biochim Biophys Acta. 2008a;1780:532–7.

Miyagi T, Wada T, Yamaguchi K, Shiozaki K, Sato I, Kakugawa Y, et al. Human sialidase as a cancer marker. Proteomics. 2008b;8:3303–11.

Monti E, Preti A, Venerando B, Borsani G. Recent development in mammalian sialidase molecular biology. Neurochem Res. 2002;27:649–63.

Neufeld EB, Cooney AM, Pitha J, Dawidowicz EA, Dwyer NK, Pentchev PG, et al. Intracellular trafficking of cholesterol monitored with a cyclodextrin. J Biol Chem. 1996;271:21604–13.

Oehler C, Kopitz J, Cantz M. Substrate specificity and inhibitor studies of a membrane-bound ganglioside sialidase isolated from human brain tissue. Biol Chem. 2002;383:1735–42.

Olayioye MA, Hausser A. Integration of non-vesicular and vesicular transport processes at the Golgi complex by the PKD-CERT network. Biochim Biophys Acta. 2012;1821:1096–103.

Papini N, Anastasia L, Tringali C, Croci G, Bresciani R, Yamaguchi K, et al. The plasma membrane-associated sialidase MmNEU3 modifies the ganglioside pattern of adjacent cells supporting its involvement in cell-to-cell interactions. J Biol Chem. 2004;279:16989–95.

Preti A, Fiorilli A, Lombardo A, Caimi L, Tettamanti G. Occurrence of sialyltransferase activity in the synaptosomal membranes prepared from calf brain cortex. J Neurochem. 1980;35: 281–96.

Prinetti A, Chigorno V, Prioni S, Loberto N, Marano N, Tettamanti G, et al. Changes in the lipid turnover, composition, and organization, as sphingolipid-enriched membrane domains, in rat cerebellar granule cells developing in vitro. J Biol Chem. 2001;276:21136–45.

Prinetti A, Chigorno V, Tettamanti G, Sonnino S. Sphingolipid-enriched membrane domains from rat cerebellar granule cells differentiated in culture. A compositional study. J Biol Chem. 2000a;275:11658–65.

Prinetti A, Marano N, Prioni S, Chigorno V, Mauri L, Casellato R, et al. Association of Src-family protein tyrosine kinases with sphingolipids in rat cerebellar granule cells differentiated in culture. Glycoconj J. 2000b;17:223–32.

Proshin S, Yamaguchi K, Wada T, Miyagi T. Modulation of neuritogenesis by ganglioside-specific sialidase (Neu 3) in human neuroblastoma NB-1 cells. Neurochem Res. 2002;27:841–6.
Reddy A, Caler EV, Andrews NW. Plasma membrane repair is mediated by Ca(2+)-regulated exocytosis of lysosomes. Cell. 2001;106:157–69.
Riboni L, Prinetti A, Bassi R, Tettamanti G. Cerebellar granule cells in culture exhibit a ganglioside-sialidase presumably linked to the plasma membrane. FEBS Lett. 1991;287:42–6.
Riboni L, Sonnino S, Acquotti D, Malesci A, Ghidoni R, Egge H, et al. Natural occurrence of ganglioside lactones. Isolation and characterization of GD1b inner ester from adult human brain. J Biol Chem. 1986;261:8514–9.
Rodriguez JA, Piddini E, Hasegawa T, Miyagi T, Dotti CG. Plasma membrane ganglioside sialidase regulates axonal growth and regeneration in hippocampal neurons in culture. J Neurosci. 2001;21:8387–95.
Saqr HE, Pearl DK, Yates AJ. A review and predictive models of ganglioside uptake by biological membranes. J Neurochem. 1993;61:395–411.
Schengrund CL, Repman MA. Density-dependent changes in gangliosides and sialidase activity of murine neuroblastoma cells. J Neurochem. 1982;39:940–7.
Schengrund CL, Rosenberg A. Intracellular location and properties of bovine brain sialidase. J Biol Chem. 1970;245:6196–200.
Schengrund CL, Rosenberg A, Repman MA. Ecto-ganglioside-sialidase activity of herpes simplex virus-transformed hamster embryo fibroblasts. J Cell Biol. 1976;70:555–61.
Schneider-Jakob HR, Cantz M. Lysosomal and plasma membrane ganglioside GM3 sialidases of cultured human fibroblasts. Differentiation by detergents and inhibitors. Biol Chem Hoppe Seyler. 1991;372:443–50.
Severino J, Allen RG, Balin S, Balin A, Cristofalo VJ. Is beta-galactosidase staining a marker of senescence in vitro and in vivo? Exp Cell Res. 2000;257:162–71.
Sidransky E, Nalls MA, Aasly JO, Aharon-Peretz J, Annesi G, Barbosa ER, et al. Multicenter analysis of glucocerebrosidase mutations in Parkinson's disease. N Engl J Med. 2009; 361:1651–61.
Soderblom C, Stadler J, Jupille H, Blackstone C, Shupliakov O, Hanna MC. Targeted disruption of the Mast syndrome gene SPG21 in mice impairs hind limb function and alters axon branching in cultured cortical neurons. Neurogenetics. 2010;11:369–78.
Sonnino S, Cantu L, Corti M, Acquotti D, Venerando B. Aggregative properties of gangliosides in solution. Chem Phys Lipids. 1994;71:21–45.
Sonnino S, Chigorno V, Valsecchi M, Bassi R, Acquotti D, Cantu L, et al. Relationship between the regulation of membrane enzyme activities by gangliosides and a possible ganglioside segregation in membrane microdomains. Indian J Biochem Biophys. 1990;27:353–8.
Sonnino S, Ghidoni R, Chigorno V, Masserini M, Tettamanti G. Recognition by two-dimensional thin-layer chromatography and densitometric quantification of alkali-labile gangliosides from the brain of different animals. Anal Biochem. 1983;128:104–14.
Sonnino S, Prinetti A, Mauri L, Chigorno V, Tettamanti G. Dynamic and structural properties of sphingolipids as driving forces for the formation of membrane domains. Chem Rev. 2006;106: 2111–25.
Tettamanti G, Morgan IG, Gombos G, Vincendon G, Mandel P. Sub-synaptosomal localization of brain particulate neuraminidose. Brain Res. 1972;47:515–8.
Tettamanti G, Preti A, Lombardo A, Bonali F, Zambotti V. Parallelism of subcellular location of major particulate neuraminidase and gangliosides in rabbit brain cortex. Biochim Biophys Acta. 1973;306:466–77.
Tettamanti G, Preti A, Lombardo A, Suman T, Zambotti V. Membrane-bound neuraminidase in the brain of different animals: behaviour of the enzyme on endogenous sialo derivatives and rationale for its assay. J Neurochem. 1975;25:451–6.
Triggs-Raine B, Mahuran DJ, Gravel RA. Naturally occurring mutations in GM2 gangliosidosis: a compendium. Adv Genet. 2001;44:199–224.
Tringali C, Anastasia L, Papini N, Bianchi A, Ronzoni L, Cappellini MD, et al. Modification of sialidase levels and sialoglycoconjugate pattern during erythroid and erytroleukemic cell differentiation. Glycoconj J. 2007a;24:67–79.

Tringali C, Lupo B, Anastasia L, Papini N, Monti E, Bresciani R, et al. Expression of sialidase Neu2 in leukemic K562 cells induces apoptosis by impairing Bcr-Abl/Src kinases signaling. J Biol Chem. 2007b;282:14364–72.

Ueno S, Saito S, Wada T, Yamaguchi K, Satoh M, Arai Y, et al. Plasma membrane-associated sialidase is up-regulated in renal cell carcinoma and promotes interleukin-6-induced apoptosis suppression and cell motility. J Biol Chem. 2006;281:7756–64.

Valaperta R, Chigorno V, Basso L, Prinetti A, Bresciani R, Preti A, et al. Plasma membrane production of ceramide from ganglioside GM3 in human fibroblasts. FASEB J. 2006;20:1227–9.

Valaperta R, Valsecchi M, Rocchetta F, Aureli M, Prioni S, Prinetti A, et al. Induction of axonal differentiation by silencing plasma membrane-associated sialidase Neu3 in neuroblastoma cells. J Neurochem. 2007;100:708–19.

Valsecchi M, Palestini P, Chigorno V, Sonnino S. Age-related changes of the ganglioside long-chain base composition in rat cerebellum. Neurochem Int. 1996;28:183–7.

Valsecchi M, Palestini P, Chigorno V, Sonnino S, Tettamanti G. Changes in the ganglioside long-chain base composition of rat cerebellar granule cells during differentiation and aging in culture. J Neurochem. 1993;60(1):193–6.

van Weely S, Brandsma M, Strijland A, Tager JM, Aerts JM. Demonstration of the existence of a second, non-lysosomal glucocerebrosidase that is not deficient in Gaucher disease. Biochim Biophys Acta. 1993;1181:55–62.

Veldman RJ, Klappe K, Hinrichs J, Hummel I, van der Schaaf G, Sietsma H, et al. Altered sphingolipid metabolism in multidrug-resistant ovarian cancer cells is due to uncoupling of glycolipid biosynthesis in the Golgi apparatus. FASEB J. 2002;16:1111–3.

Venable ME, Lee JY, Smyth MJ, Bielawska A, Obeid LM. Role of ceramide in cellular senescence. J Biol Chem. 1995;270:30701–8.

Venerando B, Fiorilli A, Croci G, Tringali C, Goi G, Mazzanti L, et al. Acidic and neutral sialidase in the erythrocyte membrane of type 2 diabetic patients. Blood. 2002;99:1064–70.

von Reitzenstein C, Kopitz J, Schuhmann V, Cantz M. Differential functional relevance of a plasma membrane ganglioside sialidase in cholinergic and adrenergic neuroblastoma cell lines. Eur J Biochem. 2001;268:326–33.

Wada T, Yoshikawa Y, Tokuyama S, Kuwabara M, Akita H, Miyagi T. Cloning, expression, and chromosomal mapping of a human ganglioside sialidase. Biochem Biophys Res Commun. 1999;261:21–7.

Xu YH, Barnes S, Sun Y, Grabowski GA. Multi-system disorders of glycosphingolipid and ganglioside metabolism. J Lipid Res. 2010;51:1643–75.

Yildiz Y, Matern H, Thompson B, Allegood JC, Warren RL, Ramirez DM, et al. Mutation of beta-glucosidase 2 causes glycolipid storage disease and impaired male fertility. J Clin Invest. 2006;116:2985–94.

Yu RK. Development regulation of ganglioside metabolism. Prog Brain Res. 1994;101:31–44.

Yu RK, Bieberich E, Xia T, Zeng G. Regulation of ganglioside biosynthesis in the nervous system. J Lipid Res. 2004;45:783–93.

Chapter 11
Role of Myelin-Associated Glycoprotein (Siglec-4a) in the Nervous System

Pablo H.H. Lopez

Abstract Myelin-associated glycoprotein (MAG) is a 100 kDa glycoprotein located at the innermost layer of myelin sheets that remains in intimate contact with the axonal membrane. It is selectively expressed by myelinating cells including Schwann cells in the peripheral nervous system (PNS) and oligodendrocytes in the central nervous system (CNS). Due to its selective location and its effects on neurons, it was originally thought to be involved in axon–glia communication. The generation of MAG-deficient mice greatly expanded our knowledge about the physiological role of MAG, which helped to establish its critical roles in the normal formation and maintenance of myelinated axons. Interest in MAG was revived when it was described as the first myelin-derived inhibitor of axon regeneration. Since then numerous publications have provided detailed information about its axonal receptors and their signaling pathways. Recently, the nurture role of MAG on neurons was confirmed. On the other hand, MAG mediates signals coming from the axons that strongly impact on oligodendrocytes and Schwann cells, highlighting the bidirectional nature of axon–glia communication. Overall, MAG is a critical component of axon–glia interactions with multiple functions in the biology of both neurons and glial cells.

Keywords MAG • Myelin-associated glycoprotein • Siglec-4a (sialic acid-binding immunoglobulin-type lectin 4a) • Myelin • Axon–glia interaction • Oligodendrocyte • Schwann cell • Axon regeneration • Neuroprotection • Gangliosides • NgR (Nogo-66 receptor)

P.H.H. Lopez (✉)
Laboratorio de Neurobiología, Instituto de Investigación Médica Mercedes y Martín Ferreyra, INIMEC-CONICET-Universidad Nacional de Córdoba, Córdoba, Argentina

Facultad de Psicología, Universidad Nacional de Córdoba, Córdoba, Argentina
e-mail: phhlopez@immf.uncor.edu

R.K. Yu and C.-L. Schengrund (eds.), *Glycobiology of the Nervous System*, Advances in Neurobiology 9, DOI 10.1007/978-1-4939-1154-7_11,

11.1 Introduction

MAG, also known as Siglec-4a, is a type I integral glycoprotein member of the immunoglobulin superfamily (IgSF) gene. It was first identified in 1973 by Richard Quarles and colleagues who isolated it as the major glycoprotein from myelin extracts (1 % of the total protein in the CNS, 0.1 % of the total protein in the PNS) (Everly et al. 1973). MAG is a cell surface member of the Siglec family, a subgroup of the IgSF that share structural and functional characteristics in addition to the common feature of recognition of sialic acid-bearing glycoconjugates (Crocker et al. 1998). It is encoded by a single gene that is conserved among vertebrates (Arquint et al. 1987). Together with the structurally related Schwann cell myelin protein (SMP, also known as Siglec-4b), they are the only two members of the Siglec family whose expression outside the hematopoietic system is restricted to the nervous system (Dulac et al. 1992). A third member of the family, Siglec-11, is expressed, but not restricted to, by the brain microglia (Wang et al. 2012; Hayakawa et al. 2005). Due to the fact that Siglec-4b and Siglec-11 expression is exclusive to avians and humans, respectively, this chapter will focus on the role of MAG in the nervous system, for which the extensive findings published during the past 40 years have helped to unmask many biological functions.

11.2 MAG Structure and Expression

The structure of MAG consists of an extracellular segment (which mediates binding to axons) containing five disulfide-bonded immunoglobulin-like domains, a single intramembrane segment, and a cytoplasmic domain (Lai et al. 1987; Arquint et al. 1987; Salzer et al. 1987). In the nervous system, MAG exists predominantly as two isoforms that arise from an alternative splicing that produces two proteins with identical extracellular and transmembrane domains but that differ in the length of their cytoplasmic tails and are identified as S-MAG (short) and L-MAG (long) (Lai et al. 1987; Salzer et al. 1987; Frail and Braun 1984). The difference in their cytoplasmic domains is traduced in unique signaling capacities that will be discussed later in this chapter. In addition, a soluble form of MAG containing only the extracellular domain as a result of proteolytic cleavage has been described (Tang et al. 1997b; Milward et al. 2008; Sato et al. 1982).

MAG is selectively expressed by myelinating cells including Schwann cells and oligodendrocytes in the nervous system and its expression correlates with the initiation of myelination during postnatal development (Quarles 1983). During myelination MAG appears to be expressed in cell bodies and processes as well as at the loose myelin layers of myelinating axons soon after completion of the first myelin loop. In mature CNS and PNS myelinated axons, MAG is expressed at the innermost layer of myelin sheets that remain in intimate contact with the axonal membrane (Trapp et al. 1989). In the PNS, MAG can also be found in paranodal loops and Schmidt–Lanterman incisures (strand of cytoplasm within the myelin sheath), as well as at the external surface of myelinating Schwann cells (Owens et al. 1989; Trapp et al. 1982; Trapp et al. 1989). More recent studies done using mice expressing a green fluorescent protein-tagged variant of MAG indicated that paranodal loops from most axons in the CNS also express MAG to some extent (Erb et al. 2006). MAG is also found in perisynaptic non-myelinating Schwann cells covering motor nerve terminals at neuromuscular junctions (Georgiou and Charlton 1999). Due to the specific periaxonal location and its effects on neurons, MAG was originally associated to axon–glia interactions (Johnson et al. 1989; Trapp et al. 1982; Attia et al. 1989).

MAG expression is regulated temporally and spatially in the nervous system. L-MAG predominates during CNS development and is the major protein during myelination, whereas S-MAG accumulates in later stages (Inuzuka et al. 1991; Ishiguro et al. 1991; Tropak et al. 1988). In contrast, S-MAG is the main isoform at all stages in the PNS while L-MAG remains as a minor constituent (Tropak et al. 1988; Inuzuka et al. 1991). MAG contains about 30 % by weight carbohydrates, consisting of heterogeneous N-linked oligosaccharides at eight extracellular sites (Everly et al. 1973; Quarles 1983; Tropak and Roder 1997, Sgroi et al. 1996). Most of them are of the complex type and negatively charged because of sialic acid and/or sulfate content, and many are bisected by N-acetylglucosamine (Yim et al. 1992; Noronha et al. 1989; Matthieu et al. 1975). It was reported that N-linked glycosylation of MAG and sialylation of MAG's expressing cells can modulate the structural conformation of the extracellular domain, probably by the folding back of the two outmost Ig-like domains (which includes the domain carrying the sialic acid-binding site) over the rest of the molecule (Tropak and Roder 1997; Attia et al. 1993). Another posttranslational modification of MAG includes phosphorylation of the cytoplasmic domains and palmitylation via an intramembranous thioester linkage (Pedraza et al. 1990; Edwards et al. 1988; Edwards et al. 1989; Afar et al. 1990; Agrawal et al. 1990).

11.3 Role of MAG in Axon–Myelin Interaction/Stability

Most of the knowledge about the functional role of MAG derives from studies performed on two different lines of MAG-deficient mice established for studying the role of MAG in myelination (Montag et al. 1994; Li et al. 1994). In the absence of MAG, mice were able to generate mature myelin in the nervous system with only subtle alterations, some of which are depicted in Fig. 11.1. These alterations included increased periaxonal spacing, reduced or loss of the cytoplasmic pericollar (cytoplasm in the innermost layer of myelin) in both the PNS and CNS, delayed myelination, redundant compact myelin (multiply myelinated axons) and disrupted compact myelin lamella (disorganized myelin), as well as an increased number of unmyelinated axons in the CNS (Fig. 11.1) (Li et al. 1994; Montag et al. 1994). On the other hand, young adult mice displayed altered maintenance of node of Ranvier

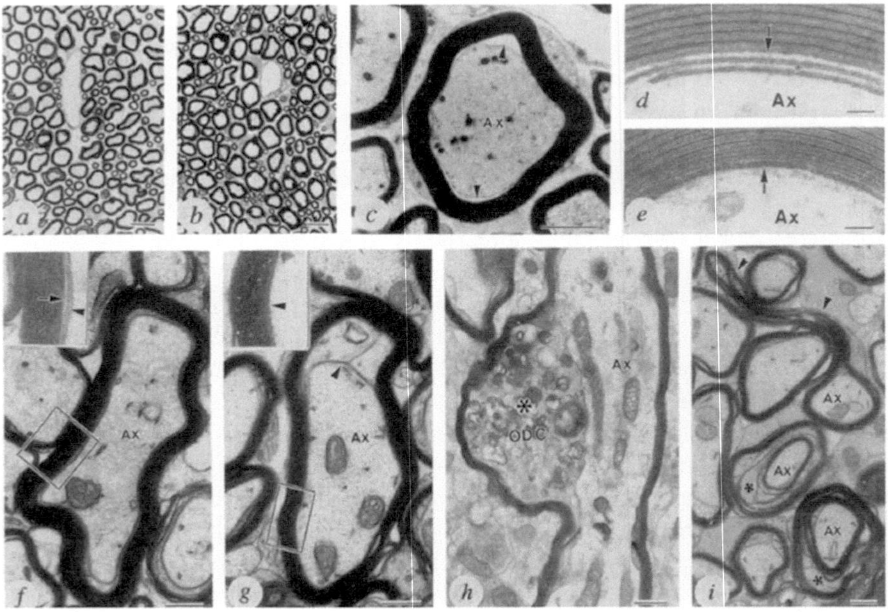

Fig. 11.1 MAG-deficient mice display subtle alterations in myelin morphology in the PNS (*upper panel*) and CNS (*lower panel*). (**a**, **b**) Light micrographs of 1 μm sections of spinal roots show normal gross myelination in 3-month-old Wt (**a**) and MAG-deficient (**b**) mice. (**c**) Electron micrograph (EM) of a myelinated axon (Ax) from spinal roots of a MAG-deficient mouse, displaying increased periaxonal space (*arrowheads*). (**d**, **e**) EMs of spinal roots from Wt (**d**) and MAG-deficient (**e**) mice. Loss of the normal 12–14 nm periaxonal space (*asterisk*) and the Schwann cell cytoplasmic collar (*arrow*) is observed in MAG-deficient mice. (**f**, **g**) EMs from optic nerve from Wt (**f**) and MAG-deficient (**g**) mice show that the normal periaxonal space (see *insert*, *arrowhead*) and cytoplasmic collar (*insert*, *arrow*) are reduced or missing in oligodendrocytes from a MAG-deficient mice (except in the mesaxon region, where the cytoplasmic collar is present, *arrowhead*). (**h**) Longitudinal section of optic nerves from MAG-deficient mice showing an oligodendrocyte (ODC) with focal disorganization of the periaxonal cytoplasmic collar (*asterisk*). (i) Transverse section of optic nerve from a MAG-deficient mouse shows disrupted compact myelin lamella (*asterisks*) and multiply compact myelin (*arrowheads*). Scale bars: (**a**, **b**) 10 μm; c, (**f**, **g**, **h**) 0.5 μm; (**d**, **e**) 0.05 μm. Reprinted with permission from Li, Tropak, Gerlai, ClapoV, Abramow-Newerly, Trapp, Peterson, and Roder, 1994, Nature 369, 747–750, 1994, Macmillan Publishers Ltd

organization (loss of myelin microvilli) associated with altered distribution of nodal and paranodal markers, although these alterations seemed to have a minor impact on the electrophysiological properties of nerves (Marcus et al. 2002; Weiss et al. 2001; Yin et al. 1998). Also original studies of MAG-deficient mice identified a late-onset axonal degeneration in the PNS, and further studies extended this observation to the CNS (Fruttiger et al. 1995; Pan et al. 2005; Nguyen et al. 2009). Interestingly, a recent time-course study revealed a constant rate of actively degenerating axons in both the PNS and CNS from adult MAG-deficient mice, which was hypothesized to be due to a loss of axonal stabilizing and protective effects of MAG (Nguyen et al. 2009). Mice engineered to express a truncated form of L-MAG only display alterations in axon–myelin integrity in the CNS similar to MAG-deficient mice, but do not display alterations in axon–myelin interactions or axon degeneration, emphasizing the role of S-MAG in maintaining the integrity of the PNS (Fujita et al. 1998).

Studies in aging MAG-deficient mice have shown sparse but clear structural alterations in oligodendrocytes characterized by dystrophic oligodendrocyte processes and cytoplasmic changes including the accumulation of vesicular material and granules (Weiss et al. 2000; Lassmann et al. 1997). Another important observation in MAG-deficient mice is the altered morphology of axons in the PNS, characterized by a reduced caliber and densely packed neurofilaments of myelinated axons (Yin et al. 1998). This was associated with reduced neurofilament phosphorylation as a consequence of decreased activities of extracellular signal-regulated kinases 1 and 2 (ERK1/2) and cyclin-dependent kinase-5 (cdk5) (Dashiell et al. 2002). The finding seems to be specifically associated with MAG since (a) neurons cultured on cell layers expressing MAG show increased expression and phosphorylation of neuronal cytoskeletal proteins and their associated kinases (Dashiell et al. 2002); (b) mice deficient in major myelin proteins displayed normal spacing of neurofilaments (Sanchez et al. 1996); and (c) human autoimmune demyelinating neuropathies characterized by the presence of high titers of anti-MAG antibodies in their sera are associated with reduced axon caliber (Lunn et al. 2002).

One important aspect to consider when interpreting the findings observed in MAG-deficient mice is that physical disruption of axon–myelin interactions that characterize these mice could affect other critical interactions beyond the ones established by MAG. Secondly, changes in the expression of other myelin proteins have been described in MAG-deficient mice, some of which could account for some compensatory mechanisms described in these mice (Li et al. 1994; Uschkureit et al. 2000; Pernet et al. 2008). Finally, the physiological functions of MAG can be underestimated due to molecular redundancy in the nervous system. Overall, these data emphasize the critical role of MAG in the maintenance of axon–glia interactions.

11.4 MAG as an Inhibitor of Axon Regeneration

As mentioned earlier, the identification of MAG as the first myelin-derived inhibitor of axon regeneration has generated a burst of research in the field. In the hunt for the causes of the contact inhibition CNS myelin had on neuritogenesis, MAG, isolated from myelin, was identified as a major inhibitor of neurite outgrowth using the motor neuron-like cell line NG108-15 (McKerracher et al. 1994). MAG was simultaneously described to inhibit neurite outgrowth from developing cerebellar granule cells and dorsal root ganglion (DRG) neurons (Mukhopadhyay et al. 1994). At the same time, MAG was identified as a sialic acid-dependent adhesion molecule of the IgG superfamily together with sialoadhesin and CD22 (Kelm et al. 1994). This study revealed a high specificity of MAG toward the terminal oligosaccharide NeuAcα2–3Galβ1–3GalNAc, which led to the identification of gangliosides GD1a and GT1b (glycosphingolipids bearing this terminal oligosaccharide structure) as the first functional axonal receptors (Yang et al. 1996; Vyas et al. 2002)

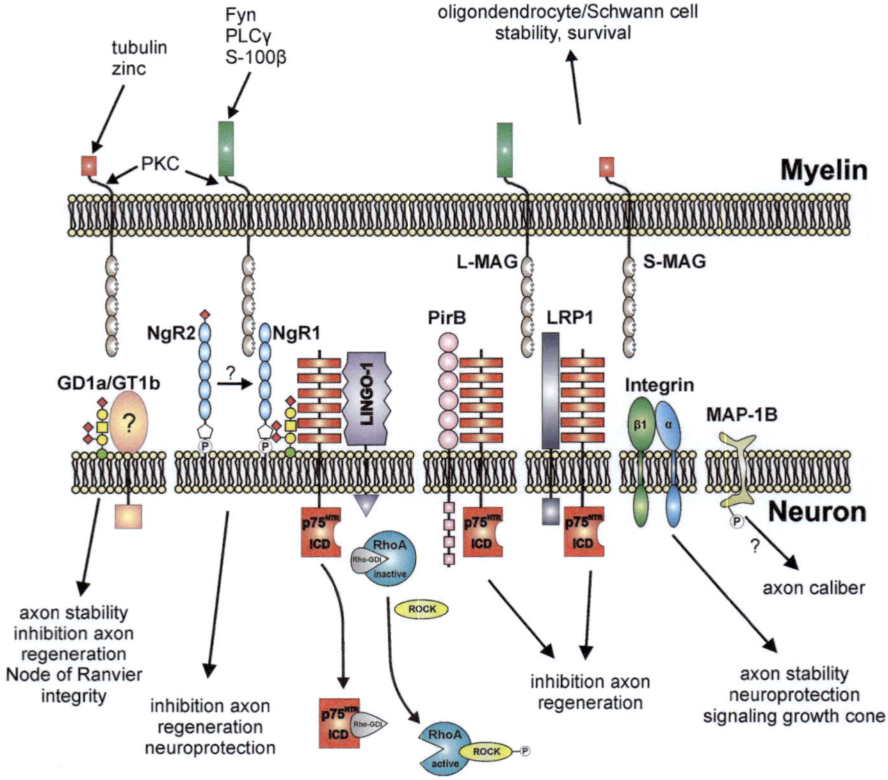

Fig. 11.2 Model for MAG-mediated signaling in axon–glia interactions. The figure summarizes knowledge about the complex signaling mechanisms associated with MAG. At present five functional neuronal receptors for MAG have been described including complex gangliosides GD1a and GT1b, Nogo receptors 1 and 2 (NgR1 and NgR2), paired immunoglobulin-like receptor B (PirB), β1-integrin, and LDL receptor-related protein 1 (LRP1). All of the receptors are associated with MAG-mediated inhibitory or repulsive signals of neurite outgrowth at the growth cone level. NgR1 and gangliosides are part of a multimeric signaling complex at the axonal membrane that includes the neurotrophin receptor $p75^{NTR}$ and Lingo-1. A signal for MAG-mediated inhibition of neurite outgrowth is transduced through $p75^{NTR}$, which binds to a Rho GDP dissociation inhibitor (Rho-GDI) and induces the subsequent activation of the small GTPase RhoA and its associated kinase ROCK. Gangliosides can also signal via $p75^{NTR}$-independent pathways by using an unidentified transducer molecule. NgR2 signaling mechanisms have not been resolved. PirB and LRP1 signal via $p75^{NTR}$-dependent pathways. Gangliosides, NgR receptors, and β1-integrin are also linked to stabilizing/neuroprotective effects of MAG on neurons. Gangliosides are additionally associated with interactions determining the node of Ranvier integrity. MAP-1B completes the list of MAG receptors. Although its functional role has not been elucidated, it was postulated that it could have a role in modulating the neuronal cytoskeleton, thereby controlling axonal caliber. MAG isoforms (S-MAG and L-MAG) are located at the oligodendrocyte/Schwann cell membrane, where they can signal via their cytoplasmic domain. While both isoforms can interact with PKC, S-MAG is able to bind tubulin and zinc through its specific cytoplasmic sequence. On the other hand, L-MAG containing a longer cytoplasmic sequence that can bind to Fyn, S-100β, and PLCγ. Activation of MAG at the oligodendrocyte membrane is associated with stability and survival of oligodendrocytes

(Fig. 11.2). Subsequently, it was demonstrated that the founding member of the NgR family named NgR1, a glycophosphatidylinositol-anchored protein acting as receptor for the myelin-derived inhibitors NogoA and OMgp, could also act as functional receptor for MAG (Liu et al. 2002; Domeniconi et al. 2002; Lauren et al. 2007) (Fig. 11.2). In contrast a second member of the NgR family, NgR2, interacts selectively with MAG in a sialic acid-dependent manner (Venkatesh et al. 2005) (Fig. 11.2). MAG interacts with gangliosides via the Ig-like domain 1 (critical residue is alanine-118), while Ig-like domains 4 and 5 are involved in recognition by NgRs (Lauren et al. 2007; Tang et al. 1997a; Cao et al. 2007; Robak et al. 2009). Gangliosides and NgRs are not likely to transduce signals on their own, but rather depend on specific carrier proteins (Schnaar and Lopez 2009). Research results indicate that the NgR1 transduces its inhibitory action on neurite outgrowth as part of a tripartite receptor complex via the neurotrophin receptor $p75^{NTR}$ or alternatively the orphan neurotrophin receptor TROY/TAJ, two members of the tumor necrosis factor receptor superfamily (Shao et al. 2005; Park et al. 2005; Wang et al. 2002; Wong et al. 2002). The complex is completed by the presence of Lingo-1, a transmembrane leucine-rich repeat protein that links $P75^{NTR}$ to NgR1 (Mi et al. 2004) (Fig. 11.2). An alternative receptor for Lingo-1, named AMIGO3, was recently identified (Ahmed et al. 2013). Upon receptor binding MAG can stimulate the regulated proteolytic cleavage of $p75^{NTR}$, which releases its cytoplasmic domain allowing its interaction with Rho-GDI (a Rho GDP dissociation inhibitor) and the subsequent activation of the small GTPase RhoA (Domeniconi et al. 2005; Niederost et al. 2002 (Fig. 11.2). Growth cone collapse and prevention of neurite extension are later achieved by a signaling cascade including Rho-associated protein kinase (ROCK) and its downstream effectors collapsin response mediator proteins 2 and 4 (CRMP2/4) and LIM domain kinase-1 (LIMK-1), which regulate microtubules and actin cytoskeleton, respectively (Mimura et al. 2006; Nagai et al. 2012; Hsieh et al. 2006). Gangliosides are able to signal in a $p75^{NTR}$-dependent manner via direct interaction with NgR1 and Lingo-1 (Williams et al. 2008; Saha et al. 2011) (Fig. 11.2). Gangliosides may also signal via $p75^{NTR}$-independent pathways by using other unidentified transducer molecules, although this is still controversial (Mehta et al. 2007; Tang et al. 1997a) (Fig. 11.2). Paired immunoglobulin-like receptor B (PirB), LDL receptor-related protein 1 (LRP1), and β1-integrin are three other functional receptors for MAG (Stiles et al. 2013; Goh et al. 2008; Atwal et al. 2008) (Fig. 11.2). PirB and LRP1 suppress axon regeneration in a $p75^{NTR}$-dependent manner (Fujita et al. 2011a; Fujita et al. 2011b; Stiles et al. 2013). β1-Integrin receptor has been associated with repulsive signaling events in neuronal growth cones acting through focal adhesion kinase (FAK) activation and independently of the NgR/Lingo-1/$p75^{NTR}$ pathway (Goh et al. 2008). The repertoire of axonal MAG receptors described so far includes microtubule-associated protein 1B, although a functional role for this receptor has not yet been described (Franzen et al. 2001) (Fig. 11.2). In addition MAG has been associated with the activation of downstream signaling pathways such as phosphatidylinositol-3,4,5-trisphosphate 3-phosphatase (PTEN), intracellular calcium signaling, and protein kinase C (PKC) that could cross talk with the above signaling cascades (Song et al. 1998; Wong et al. 2002; Wong et al. 2003). Finally, it is important to emphasize that MAG receptors can signal independently but in a synergistic manner to inhibit neurite outgrowth in a cell-type-specific combination (Mehta et al. 2007; Venkatesh et al. 2007) (Fig. 11.3).

Despite the vast in vitro evidence about the inhibitory role of MAG on axon regeneration, the in vivo role remains controversial. Initial studies using the optic nerve crush injury model failed to demonstrate improved regeneration in MAG-deficient mice (Bartsch et al. 1995). On the contrary, antibody-based strategies targeting MAG in the same optic nerve crush injury model resulted in modest regeneration (Wong et al. 2003). Also MAG-deficient mice crossbred with C57BL/WldS mutant mice, characterized by a delayed clearance of myelin debris, showed slightly improved regeneration in the PNS (Schafer et al. 1996). However, assessment of spinal cord regeneration in MAG-deficient mice was not conclusive, in part due to the presence of other myelin-derived and non-myelin-derived inhibitors of axon regeneration sharing functional receptors with MAG (Cafferty et al. 2010; Lee et al. 2010; Dickendesher et al. 2012). In conclusion, the presence of multiple inhibitors of axon regeneration sharing axonal receptors/pathways and compensatory expression of other myelin inhibitory proteins mask the in vivo role of MAG on axon regeneration.

Finally, there is evidence that MAG could promote neurite outgrowth of immature/young neurons and that the switch from stimulation to inhibition of neurite growth is associated with a reduction of cAMP levels in neurons (Mukhopadhyay et al. 1994; Cai et al. 2001; Hasegawa et al. 2004; Johnson et al. 1989). Moreover, outgrowth inhibition by MAG can be converted to neurite extension by inhibiting PKC activity, while the opposite effect can be achieved by inhibiting inositol 1,4,5-trisphosphate (IP3) (Hasegawa et al. 2004). Thus, a balance between cAMP/PKC and IP3 seems critical for bidirectional regulation of axon regeneration by MAG. These results could represent a redundancy rather than a functional role of MAG given that axons have already reached their targets by the time MAG expression starts in the nervous system.

11.5 Nurturing/Protective Properties of MAG on Neurons

As mentioned earlier, the generation of MAG-deficient mice unmasked a protective/stabilizing role of MAG on axons it ensheathes as evidenced by the finding of progressive late-onset axon degeneration and reduced axon caliber (Montag et al. 1994;

Fig. 11.3 (continued) or 10 μM Y-27632 (a specific inhibitor of Rho-associated protein kinase (ROCK)). After 24 h, the cultures were fixed and stained with anti-tubulin mAb. Representative fluorescence micrographs are presented as reverse grayscale images to enhance clarity (*bar*, 50 μm). Neurite outgrowth (mean ± S.E.) was quantified using image analysis and normalized with respect to the control. *Symbols* indicate statistical comparison with myelin-inhibited neurite outgrowth: $*p \leq 0.01$; $**p \leq 0.001$. This research was originally published by Mehta et al. (2007) with permission from © the American Society for Biochemistry and Molecular Biology

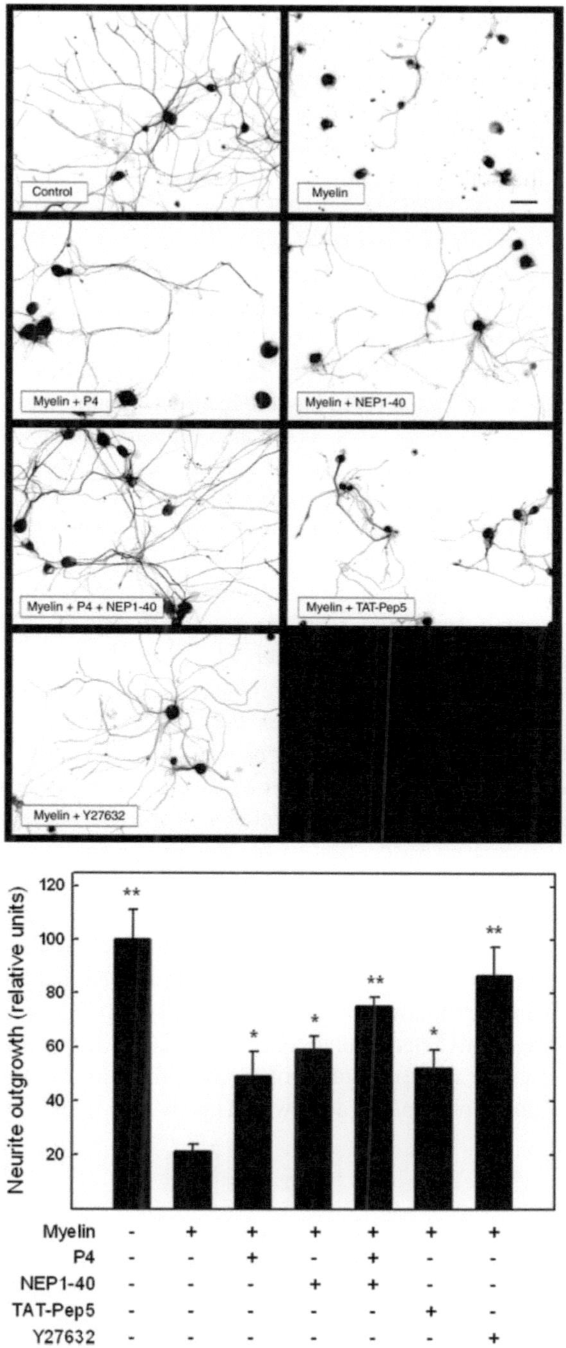

Fig. 11.3 MAG inhibition of neurite outgrowth from DRG neurons (DRGNs) is mainly via NgR and glycosphingolipids. DRGNs were cultured on control surfaces or the same surfaces adsorbed with detergent-extracted myelin proteins (*Myelin*). One hour after plating, cultures were treated with either 1 μM P4 (an inhibitor of ganglioside biosynthesis), 1 μM NEP1–40 (NgR blocking peptide), 200 nM of TAT-Pep5 (a cell-permeable peptide that blocks the intracellular association of $p75^{NTR}$ with the Rho GDP dissociation inhibitor, blocking its ability to activate RhoA),

Nguyen et al. 2009; Pan et al. 2005). Mice lacking complex gangliosides develop late-onset axonal degeneration, and further studies comparing neuropathology and behavioral deficits in single- and double-null mice for complex gangliosides and MAG supported the notion that gangliosides mediate axon stability in the nervous system (Pan et al. 2005; Sheikh et al. 1999) (Fig. 11.2). However, it was not until recently that the protective effects of MAG on axons were studied in in vitro settings. MAG was reported to stabilize axons from DRG neurons against vincristine, a potent microtubule-destabilizing agent used for cancer treatment that is also associated with human neuropathies due to its neurotoxicity (Nguyen et al. 2009). Further studies demonstrated that this effect requires gangliosides and involves ROCK-mediated activation of CRMP4, to induce microtubule stability (Mehta et al. 2010; Nagai et al. 2012) (Fig. 11.2). MAG also prevented axonal degeneration in response to neurotoxins such as acrylamide, granzyme B, and supernatants from activated cytotoxic T cells (Nguyen et al. 2009). In addition a protective role against axonal injury induced by acrylamide and the T cell-mediated inflammatory toxicity was confirmed in vivo using MAG-deficient mice (Nguyen et al. 2009; Jones et al. 2013). Interestingly, the neuroprotective role of MAG is not restricted to the axons that it engages; instead, MAG also seems to provide signals that contribute to neuron survival (Fig. 11.4). Thus, MAG protects hippocampal neurons in vitro from kainic acid-induced excitotoxicity predominantly via NgR and β1-integrin receptors and involves ROCK activation (Lopez et al. 2011) (Fig. 11.2). The fact that MAG inhibited neurite outgrowth in these neurons mainly via its interactions with gangliosides demonstrated that MAG can exert dual roles (neuroprotection and inhibition of neurite outgrowth) via different receptors in the same cell. These studies also showed increased susceptibility of MAG-deficient mice to kainic acid-induced seizure activity and increased susceptibility to excitotoxicity induced by intrastriatal injection of N-methyl-D-aspartate (Lopez et al. 2011). In the later model, pretreatment with a chimeric soluble form of MAG reduced brain damage. It is important to mention here that acute insults can induce the downregulation of MAG expression and disruption of axon–glial integrity, which ultimately highlights the dynamic nature of this interaction (Reimer et al. 2011; Xie et al. 2010). Overall, these studies emphasize the contribution of MAG to the nurturing/protective role of myelin on axons under physiological as well as pathological conditions of the nervous system.

11.6 MAG as a Functional Receptor in Oligodendrocytes

It has been proposed that the two isoforms of MAG present in the rodent nervous system, which differ in the length and sequence of their cytoplasmic domains, have different functions in the oligodendroglial cell cytoplasm. S-MAG and L-MAG share a common intracellular domain of 37 amino acids apposed to the plasma membrane (Lai et al. 1987). In addition, S-MAG contains a specific sequence of 10 amino acids that binds tubulin and zinc (Kursula et al. 2001; Kursula et al. 1999a)

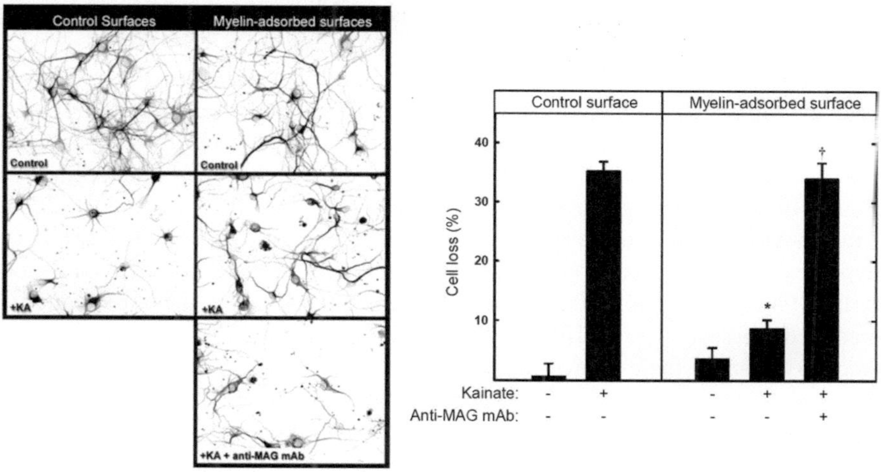

Fig. 11.4 MAG protects hippocampal neurons (HNs) from kainic acid-induced excitotoxicity in vitro. (*Left*) HNs were cultured onto control surfaces or the same surfaces adsorbed with detergent-extracted proteins from rat brain myelin. One hour after plating, some cultures were treated with 10 μg/mL of anti-MAG mAb. After 48 h, HNs were treated with 130 μM kainic acid (KA) to induce excitotoxicity. After an additional 24 h, cultures were incubated for 30 min with medium containing propidium iodide and then were fixed and stained with anti-tubulin mAb. Representative fluorescent micrographs are presented as reverse grayscale images to enhance clarity. (*Right*) Cell survival (mean ± SEM) was normalized with respect to control surface. Live cell counts from four microscopic fields from each of nine microwells from three independent experiments were averaged. Growth on myelin-adsorbed surfaces provided significant neuroprotection from KA (*$p<0.001$) compared to cells grown on control surfaces. Addition of anti-MAG antibody reversed this protection, resulting in cell loss that was not significantly different from KA-treated cells grown on control surfaces (*$p>0.7$). Reprinted with permission from Lopez, Ahmad, Mehta, Toner, Rowland, Zhang, Dorè, Schnaar. 2011, J Neurochem 116, 900–908, 2011, Johns Wiley and Sons Publishers Ltd. © 2011 International Society for Neurochemistry

(Fig. 11.2). On the other hand, L-MAG contains a specific cytoplasmic sequence of 54 amino acids that binds several signaling molecules including Fyn tyrosine kinase, phospholipase C-γ, and S100β (Kursula et al. 1999b; Jaramillo et al. 1994; Umemori et al. 1994) (Fig. 11.2). These findings suggest that MAG serves as a docking protein that allows the interaction between different signaling molecules. Several studies have addressed whether MAG can be phosphorylated in vivo and in vitro. The data obtained can be summarized as follows: (a) L-MAG and S-MAG are the major in vivo forms found in the CNS and PNS, respectively (Agrawal et al. 1990; Edwards et al. 1988; Edwards et al. 1989). (b) While both isoforms can be phosphorylated in vitro mainly on serine residues, L-MAG phosphorylation predominates on oligodendrocyte cultures and S-MAG is the major phosphorylated form in Schwann cell cultures (Agrawal et al. 1990; Yim et al. 1995). (c) Both isoforms contain potential phosphorylation sites on tyrosine residues, and tyrosine kinases v-fps and v-src can bind and phosphorylate MAG in vitro (Edwards et al. 1988; Jaramillo et al. 1994). (d) PKC can bind and phosphorylate in vitro both isoforms of MAG, while protein kinase A (PKA)-dependent phosphorylation is specific on L-MAG and inhibited by S100β (Kirchhoff et al. 1993; Kursula et al. 2000).

In addition, antibody-mediated cross-linking of MAG on cultured oligodendrocytes (to mimic axonal binding) was shown to trigger its redistribution to rafts and produce the activation of signaling cascades that includes dephosphorylation of serine and threonine residues in specific proteins and hyperphosphorylation of Fyn, a central integrator and mediator of axon–glia communication (Marta et al. 2004; Kramer-Albers and White 2011). Fyn signaling downstream of MAG has been hypothesized to play a critical role in the initiation of myelination in vivo (Umemori et al. 1994; Biffiger et al. 2000). An interesting functional consequence of antibody-mediated cross-linking of MAG on oligodendrocytes in vitro is an increased resistance to glutamate toxicity by mechanisms that remain to be elucidated (Irving et al. 2005). However, further studies are needed to fully understand the impact of signaling cascades downstream of MAG on oligodendrocytes.

11.7 Future Perspectives

Despite many years of prolific research in the field, our understanding of the biological roles of MAG remains incomplete. Part of this problem is related to the variety of molecular interactions associated with MAG and the multiple downstream signaling pathways associated with each receptor. In addition, the lack of a resolved crystal structure conspires against a comprehensive knowledge of its structural properties and functions. A future challenge will be to completely elucidate its biological roles under physiological and pathological conditions from a point of view that includes MAG as a critical player capable of supporting bidirectional communication between myelin and axons. These studies should include the complete dissection of the molecular mechanisms underlying its stabilizing effect on neurons, its modulatory role on the axon cytoskeleton, and its inhibitory effect on axon regeneration. It will also be crucial to study the neuronal survival properties of MAG in different experimental paradigms related to human diseases which, in addition to clarifying its protective role on axons, could eventually open a therapeutic opportunity to mitigate demyelinating as well as dysmyelinating diseases. Finally, efforts must be made to translate the discoveries in the field into a unifying view of MAG signaling.

Conflict of Interest The author declares that he has no conflict of interest.

References

Afar DE, Salzer JL, Roder J, Braun PE, Bell JC. Differential phosphorylation of myelin-associated glycoprotein isoforms in cell culture. J Neurochem. 1990;55(4):1418–26.

Agrawal HC, Noronha AB, Agrawal D, Quarles RH. The myelin-associated glycoprotein is phosphorylated in the peripheral nervous system. Biochem Biophys Res Commun. 1990;169(3):953–8.

Ahmed Z, Douglas MR, John G, Berry M, Logan A. AMIGO3 is an NgR1/p75 co-receptor signalling axon growth inhibition in the acute phase of adult central nervous system injury. PLoS One. 2013;8(4):e61878.

Arquint M, Roder J, Chia LS, Down J, Wilkinson D, Bayley H, et al. Molecular cloning and primary structure of myelin-associated glycoprotein. Proc Natl Acad Sci U S A. 1987;84(2): 600–4.

Attia J, Hicks L, Oikawa K, Kay CM, Dunn RJ. Structural properties of the myelin-associated glycoprotein ectodomain. J Neurochem. 1993;61(2):718–26.

Attia J, Tropak M, Johnson PW, Newerly-Abranow W, Pawson T, Roder JC et al. Modulated adhesion: a proposal for the role of myelin-associated glycoprotein in myelin wrapping. Clin Chem. 1989;35(5):717–20. Review.

Atwal JK, Pinkston-Gosse J, Syken J, Stawicki S, Wu Y, Shatz C, et al. PirB is a functional receptor for myelin inhibitors of axonal regeneration. Science. 2008;322(5903):967–70.

Bartsch U, Bandtlow CE, Schnell L, Bartsch S, Spillmann AA, Rubin BP, et al. Lack of evidence that myelin-associated glycoprotein is a major inhibitor of axonal regeneration in the CNS. Neuron. 1995;15(6):1375–81.

Biffiger K, Bartsch S, Montag D, Aguzzi A, Schachner M, Bartsch U. Severe hypomyelination of the murine CNS in the absence of myelin-associated glycoprotein and fyn tyrosine kinase. J Neurosci. 2000;20(19):7430–7.

Cafferty WB, Duffy P, Huebner E, Strittmatter SM. MAG and OMgp synergize with Nogo-A to restrict axonal growth and neurological recovery after spinal cord trauma. J Neurosci. 2010;30(20):6825–37.

Cai D, Qiu J, Cao Z, McAtee M, Bregman BS, Filbin MT. Neuronal cyclic AMP controls the developmental loss in ability of axons to regenerate. J Neurosci. 2001;21(13):4731–9.

Cao Z, Qiu J, Domeniconi M, Hou J, Bryson JB, Mellado W, et al. The inhibition site on myelin-associated glycoprotein is within Ig-domain 5 and is distinct from the sialic acid binding site. J Neurosci. 2007;27(34):9146–54.

Crocker PR, Clark EA, Filbin M, Gordon S, Jones Y, Kehrl JH, et al. Siglecs: a family of sialic-acid binding lectins. Glycobiology. 1998;8(2):v.

Dashiell SM, Tanner SL, Pant HC, Quarles RH. Myelin-associated glycoprotein modulates expression and phosphorylation of neuronal cytoskeletal elements and their associated kinases. J Neurochem. 2002;81(6):1263–72.

Dickendesher TL, Baldwin KT, Mironova YA, Koriyama Y, Raiker SJ, Askew KL, et al. NgR1 and NgR3 are receptors for chondroitin sulfate proteoglycans. Nat Neurosci. 2012;15(5):703–12.

Domeniconi M, Cao Z, Spencer T, Sivasankaran R, Wang K, Nikulina E, et al. Myelin-associated glycoprotein interacts with the Nogo66 receptor to inhibit neurite outgrowth. Neuron. 2002; 35(2):283–90.

Domeniconi M, Zampieri N, Spencer T, Hilaire M, Mellado W, Chao MV, et al. MAG induces regulated intramembrane proteolysis of the p75 neurotrophin receptor to inhibit neurite outgrowth. Neuron. 2005;46(6):849–55.

Dulac C, Tropak MB, Cameron-Curry P, Rossier J, Marshak DR, Roder J, et al. Molecular characterization of the Schwann cell myelin protein, SMP: structural similarities within the immunoglobulin superfamily. Neuron. 1992;8(2):323–34.

Edwards AM, Arquint M, Braun PE, Roder JC, Dunn RJ, Pawson T, et al. Myelin-associated glycoprotein, a cell adhesion molecule of oligodendrocytes, is phosphorylated in brain. Mol Cell Biol. 1988;8(6):2655–8.

Edwards AM, Braun PE, Bell JC. Phosphorylation of myelin-associated glycoprotein in vivo and in vitro occurs only in the cytoplasmic domain of the large isoform. J Neurochem. 1989;52(1): 317–20.

Erb M, Flueck B, Kern F, Erne B, Steck AJ, Schaeren-Wiemers N. Unraveling the differential expression of the two isoforms of myelin-associated glycoprotein in a mouse expressing GFP-tagged S-MAG specifically regulated and targeted into the different myelin compartments. Mol Cell Neurosci. 2006;31(4):613–27.

Everly JL, Brady RO, Quarles RH. Evidence that the major protein in rat sciatic nerve myelin is a glycoprotein. J Neurochem. 1973;21(2):329–34.

Frail DE, Braun PE. Two developmentally regulated messenger RNAs differing in their coding region may exist for the myelin-associated glycoprotein. J Biol Chem. 1984;259(23): 14857–62.

Franzen R, Tanner SL, Dashiell SM, Rottkamp CA, Hammer JA, Quarles RH. Microtubule-associated protein 1B: a neuronal binding partner for myelin-associated glycoprotein. J Cell Biol. 2001;155(6):893–8.

Fruttiger M, Montag D, Schachner M, Martini R. Crucial role for the myelin-associated glycoprotein in the maintenance of axon-myelin integrity. Eur J Neurosci. 1995;7(3):511–5.

Fujita N, Kemper A, Dupree J, Nakayasu H, Bartsch U, Schachner M, et al. The cytoplasmic domain of the large myelin-associated glycoprotein isoform is needed for proper CNS but not peripheral nervous system myelination. J Neurosci. 1998;18(6):1970–8.

Fujita Y, Endo S, Takai T, Yamashita T. Myelin suppresses axon regeneration by PIR-B/SHP-mediated inhibition of Trk activity. EMBO J. 2011a;30(7):1389–401.

Fujita Y, Takashima R, Endo S, Takai T, Yamashita T. The p75 receptor mediates axon growth inhibition through an association with PIR-B. Cell Death Dis. 2011b;2:e198.

Georgiou J, Charlton MP. Non-myelin-forming perisynaptic schwann cells express protein zero and myelin-associated glycoprotein. Glia. 1999;27(2):101–9.

Goh EL, Young JK, Kuwako K, Tessier-Lavigne M, He Z, Griffin JW, et al. Beta1-integrin mediates myelin-associated glycoprotein signaling in neuronal growth cones. Mol Brain. 2008;1:10.

Hasegawa Y, Fujitani M, Hata K, Tohyama M, Yamagishi S, Yamashita T. Promotion of axon regeneration by myelin-associated glycoprotein and Nogo through divergent signals downstream of Gi/G. J Neurosci. 2004;24(30):6826–32.

Hayakawa T, Angata T, Lewis AL, Mikkelsen TS, Varki NM, Varki A. A human-specific gene in microglia. Science. 2005;309(5741):1693.

Hsieh SH, Ferraro GB, Fournier AE. Myelin-associated inhibitors regulate cofilin phosphorylation and neuronal inhibition through LIM kinase and Slingshot phosphatase. J Neurosci. 2006;26(3): 1006–15.

Inuzuka T, Fujita N, Sato S, Baba H, Nakano R, Ishiguro H, et al. Expression of the large myelin-associated glycoprotein isoform during the development in the mouse peripheral nervous system. Brain Res. 1991;562(1):173–5.

Irving EA, Vinson M, Rosin C, Roberts JC, Chapman DM, Facci L, et al. Identification of neuroprotective properties of anti-MAG antibody: a novel approach for the treatment of stroke? J Cereb Blood Flow Metab. 2005;25(1):98–107.

Ishiguro H, Sato S, Fujita N, Inuzuka T, Nakano R, Miyatake T. Immunohistochemical localization of myelin-associated glycoprotein isoforms during the development in the mouse brain. Brain Res. 1991;563(1–2):288–92.

Jaramillo ML, Afar DE, Almazan G, Bell JC. Identification of tyrosine 620 as the major phosphorylation site of myelin-associated glycoprotein and its implication in interacting with signaling molecules. J Biol Chem. 1994;269(44):27240–5.

Johnson PW, Bramow-Newerly W, Seilheimer B, Sadoul R, Tropak MB, Arquint M, et al. Recombinant myelin-associated glycoprotein confers neural adhesion and neurite outgrowth function. Neuron. 1989;3(3):377–85.

Jones MV, Nguyen TT, Ewaleifoh O, Lebson L, Whartenby KA, Griffin JW, et al. Accelerated axon loss in MOG35-55 experimental autoimmune encephalomyelitis (EAE) in myelin-associated glycoprotein-deficient (MAGKO) mice. J Neuroimmunol. 2013;262(1–2):53–61.

Kelm S, Pelz A, Schauer R, Filbin MT, Tang S, de Bellard ME, et al. Sialoadhesin, myelin-associated glycoprotein and CD22 define a new family of sialic acid-dependent adhesion molecules of the immunoglobulin superfamily. Curr Biol. 1994;4(11):965–72.

Kirchhoff F, Hofer HW, Schachner M. Myelin-associated glycoprotein is phosphorylated by protein kinase C. J Neurosci Res. 1993;36(4):368–81.

Kramer-Albers EM, White R. From axon-glial signalling to myelination: the integrating role of oligodendroglial Fyn kinase. Cell Mol Life Sci. 2011;68(12):2003–12.

Kursula P, Lehto VP, Heape AM. The small myelin-associated glycoprotein binds to tubulin and microtubules. Brain Res Mol Brain Res. 2001;87(1):22–30.

Kursula P, Lehto VP, Heape AM. S100beta inhibits the phosphorylation of the L-MAG cytoplasmic domain by PKA. Brain Res Mol Brain Res. 2000;76(2):407–10.

Kursula P, Merilainen G, Lehto VP, Heape AM. The small myelin-associated glycoprotein is a zinc-binding protein. J Neurochem. 1999a;73(5):2110–8.

Kursula P, Tikkanen G, Lehto VP, Nishikimi M, Heape AM. Calcium-dependent interaction between the large myelin-associated glycoprotein and S100beta. J Neurochem. 1999b;73(4):1724–32.

Lai C, Brow MA, Nave KA, Noronha AB, Quarles RH, Bloom FE, et al. Two forms of 1B236/myelin-associated glycoprotein, a cell adhesion molecule for postnatal neural development, are produced by alternative splicing. Proc Natl Acad Sci U S A. 1987;84(12):4337–41.

Lassmann H, Bartsch U, Montag D, Schachner M. Dying-back oligodendrogliopathy: a late sequel of myelin-associated glycoprotein deficiency. Glia. 1997;19(2):104–10.

Lauren J, Hu F, Chin J, Liao J, Airaksinen MS, Strittmatter SM. Characterization of myelin ligand complexes with neuronal Nogo-66 receptor family members. J Biol Chem. 2007;282(8):5715–25.

Lee JK, Geoffroy CG, Chan AF, Tolentino KE, Crawford MJ, Leal MA, et al. Assessing spinal axon regeneration and sprouting in Nogo-, MAG-, and OMgp-deficient mice. Neuron. 2010;66(5):663–70.

Li C, Tropak MB, Gerlai R, Clapoff S, Bramow-Newerly W, Trapp B, et al. Myelination in the absence of myelin-associated glycoprotein. Nature. 1994;369(6483):747–50.

Liu BP, Fournier A, GrandPre T, Strittmatter SM. Myelin-associated glycoprotein as a functional ligand for the Nogo-66 receptor. Science. 2002;297(5584):1190–3.

Lopez PH, Ahmad AS, Mehta NR, Toner M, Rowland EA, Zhang J, et al. Myelin-associated glycoprotein protects neurons from excitotoxicity. J Neurochem. 2011;116(5):900–8.

Lunn MP, Crawford TO, Hughes RA, Griffin JW, Sheikh KA. Anti-myelin-associated glycoprotein antibodies alter neurofilament spacing. Brain. 2002;125(Pt 4):904–11.

Marcus J, Dupree JL, Popko B. Myelin-associated glycoprotein and myelin galactolipids stabilize developing axo-glial interactions. J Cell Biol. 2002;156(3):567–77.

Marta CB, Taylor CM, Cheng S, Quarles RH, Bansal R, Pfeiffer SE. Myelin associated glycoprotein cross-linking triggers its partitioning into lipid rafts, specific signaling events and cytoskeletal rearrangements in oligodendrocytes. Neuron Glia Biol. 2004;1(1):35–46.

Matthieu JM, Everly JL, Brady RO, Quarles RH. [35-S]sulfate incorporation into myelin clycoproteins; II. Peripheral nervous tissue. Biochim Biophys Acta. 1975;392(1):167–74.

McKerracher L, David S, Jackson DL, Kottis V, Dunn RJ, Braun PE. Identification of myelin-associated glycoprotein as a major myelin-derived inhibitor of neurite growth. Neuron. 1994;13(4):805–11.

Mehta NR, Lopez PH, Vyas AA, Schnaar RL. Gangliosides and Nogo receptors independently mediate myelin-associated glycoprotein inhibition of neurite outgrowth in different nerve cells. J Biol Chem. 2007;282(38):27875–86.

Mehta NR, Nguyen T, Bullen Jr JW, Griffin JW, Schnaar RL. Myelin-associated glycoprotein (MAG) protects neurons from acute toxicity using a ganglioside-dependent mechanism. ACS Chem Neurosci. 2010;1(3):215–22.

Mi S, Lee X, Shao Z, Thill G, Ji B, Relton J, et al. LINGO-1 is a component of the Nogo-66 receptor/p75 signaling complex. Nat Neurosci. 2004;7(3):221–8.

Milward E, Kim KJ, Szklarczyk A, Nguyen T, Melli G, Nayak M, et al. Cleavage of myelin associated glycoprotein by matrix metalloproteinases. J Neuroimmunol. 2008;193(1–2):140–8.

Mimura F, Yamagishi S, Arimura N, Fujitani M, Kubo T, Kaibuchi K, et al. Myelin-associated glycoprotein inhibits microtubule assembly by a Rho-kinase-dependent mechanism. J Biol Chem. 2006;281(23):15970–9.

Montag D, Giese KP, Bartsch U, Martini R, Lang Y, Bluthmann H, et al. Mice deficient for the myelin-associated glycoprotein show subtle abnormalities in myelin. Neuron. 1994;13(1):229–46.

Mukhopadhyay G, Doherty P, Walsh FS, Crocker PR, Filbin MT. A novel role for myelin-associated glycoprotein as an inhibitor of axonal regeneration. Neuron. 1994;13(3):757–67.

Nagai J, Goshima Y, Ohshima T. CRMP4 mediates MAG-induced inhibition of axonal outgrowth and protection against Vincristine-induced axonal degeneration. Neurosci Lett. 2012;519(1): 56–61.

Nguyen T, Mehta NR, Conant K, Kim KJ, Jones M, Calabresi PA, et al. Axonal protective effects of the myelin-associated glycoprotein. J Neurosci. 2009;29(3):630–7.

Niederost B, Oertle T, Fritsche J, McKinney RA, Bandtlow CE. Nogo-A and myelin-associated glycoprotein mediate neurite growth inhibition by antagonistic regulation of RhoA and Rac1. J Neurosci. 2002;22(23):10368–76.

Noronha AB, Hammer JA, Lai C, Kiel M, Milner RJ, Sutcliffe JG et al. Myelin-associated glycoprotein (MAG) and rat brain-specific 1B236 protein: mapping of epitopes and demonstration of immunological identity. J Mol Neurosci. 1989;1(3):159–70.

Owens GC, Bunge RP. Evidence for an early role for myelin-associated glycoprotein in the process of myelination. Glia. 1989;2(2):119–28.

Pan B, Fromholt SE, Hess EJ, Crawford TO, Griffin JW, Sheikh KA, et al. Myelin-associated glycoprotein and complementary axonal ligands, gangliosides, mediate axon stability in the CNS and PNS: neuropathology and behavioral deficits in single- and double-null mice. Exp Neurol. 2005;195(1):208–17.

Park JB, Yiu G, Kaneko S, Wang J, Chang J, He XL, et al. A TNF receptor family member, TROY, is a coreceptor with Nogo receptor in mediating the inhibitory activity of myelin inhibitors. Neuron. 2005;45(3):345–51.

Pedraza L, Owens GC, Green LA, Salzer JL. The myelin-associated glycoproteins: membrane disposition, evidence of a novel disulfide linkage between immunoglobulin-like domains, and posttranslational palmitylation. J Cell Biol. 1990;111(6 Pt 1):2651–61.

Pernet V, Joly S, Christ F, Dimou L, Schwab ME. Nogo-A and myelin-associated glycoprotein differently regulate oligodendrocyte maturation and myelin formation. J Neurosci. 2008;28(29):7435–44.

Quarles RH. Myelin-associated glycoprotein in development and disease. Dev Neurosci. 1983; 6(6):285–303.

Reimer MM, McQueen J, Searcy L, Scullion G, Zonta B, Desmazieres A, et al. Rapid disruption of axon-glial integrity in response to mild cerebral hypoperfusion. J Neurosci. 2011;31(49): 18185–94.

Robak LA, Venkatesh K, Lee H, Raiker SJ, Duan Y, Lee-Osbourne J, et al. Molecular basis of the interactions of the Nogo-66 receptor and its homolog NgR2 with myelin-associated glycoprotein: development of NgROMNI-Fc, a novel antagonist of CNS myelin inhibition. J Neurosci. 2009;29(18):5768–83.

Saha N, Kolev MV, Semavina M, Himanen J, Nikolov DB. Ganglioside mediate the interaction between Nogo receptor 1 and LINGO-1. Biochem Biophys Res Commun. 2011;413(1):92–7.

Salzer JL, Holmes WP, Colman DR. The amino acid sequences of the myelin-associated glycoproteins: homology to the immunoglobulin gene superfamily. J Cell Biol. 1987;104(4):957–65.

Sanchez I, Hassinger L, Paskevich PA, Shine HD, Nixon RA. Oligodendroglia regulate the regional expansion of axon caliber and local accumulation of neurofilaments during development independently of myelin formation. J Neurosci. 1996;16(16):5095–105.

Sato S, Quarles RH, Brady RO. Susceptibility of the myelin-associated glycoprotein and basic protein to a neutral protease in highly purified myelin from human and rat brain. J Neurochem. 1982;39(1):97–105.

Schafer M, Fruttiger M, Montag D, Schachner M, Martini R. Disruption of the gene for the myelin-associated glycoprotein improves axonal regrowth along myelin in C57BL/Wlds mice. Neuron. 1996;16(6):1107–13.

Schnaar RL, Lopez PH. Myelin-associated glycoprotein and its axonal receptors. J Neurosci Res. 2009;87(15):3267–76.

Sgroi D, Nocks A, Stamenkovic I. A single N-linked glycosylation site is implicated in the regulation of ligand recognition by the I-type lectins CD22 and CD33. J Biol Chem. 1996; 271(31):18803–9.

Shao Z, Browning JL, Lee X, Scott ML, Shulga-Morskaya S, Allaire N, et al. TAJ/TROY, an orphan TNF receptor family member, binds Nogo-66 receptor 1 and regulates axonal regeneration. Neuron. 2005;45(3):353–9.

Sheikh KA, Sun J, Liu Y, Kawai H, Crawford TO, Proia RL, et al. Mice lacking complex gangliosides develop Wallerian degeneration and myelination defects. Proc Natl Acad Sci U S A. 1999;96(13):7532–7.

Song H, Ming G, He Z, Lehmann M, McKerracher L, Tessier-Lavigne M, et al. Conversion of neuronal growth cone responses from repulsion to attraction by cyclic nucleotides. Science. 1998;281(5382):1515–8.

Stiles TL, Dickendesher TL, Gaultier A, Fernandez-Castaneda A, Mantuano E, Giger RJ, et al. LDL receptor-related protein-1 is a sialic-acid-independent receptor for myelin-associated glycoprotein that functions in neurite outgrowth inhibition by MAG and CNS myelin. J Cell Sci. 2013;126(Pt 1):209–20.

Tang S, Shen YJ, DeBellard ME, Mukhopadhyay G, Salzer JL, Crocker PR, et al. Myelin-associated glycoprotein interacts with neurons via a sialic acid binding site at ARG118 and a distinct neurite inhibition site. J Cell Biol. 1997a;138(6):1355–66.

Tang S, Woodhall RW, Shen YJ, DeBellard ME, Saffell JL, Doherty P, et al. Soluble myelin-associated glycoprotein (MAG) found in vivo inhibits axonal regeneration. Mol Cell Neurosci. 1997b;9(5–6):333–46.

Trapp BD, Andrews SB, Cootauco C, Quarles R. The myelin-associated glycoprotein is enriched in multivesicular bodies and periaxonal membranes of actively myelinating oligodendrocytes. J Cell Biol. 1989;109(5):2417–26.

Trapp BD, Quarles RH. Presence of the myelin-associated glycoprotein correlates with alterations in the periodicity of peripheral myelin. J Cell Biol. 1982;92(3):877–82.

Tropak MB, Johnson PW, Dunn RJ, Roder JC. Differential splicing of MAG transcripts during CNS and PNS development. Brain Res. 1988;464(2):143–55.

Tropak MB, Roder JC. Regulation of myelin-associated glycoprotein binding by sialylated cis-ligands. J Neurochem. 1997;68(4):1753–63.

Umemori H, Sato S, Yagi T, Aizawa S, Yamamoto T. Initial events of myelination involve Fyn tyrosine kinase signalling. Nature. 1994;367(6463):572–6.

Uschkureit T, Sporkel O, Stracke J, Bussow H, Stoffel W. Early onset of axonal degeneration in double (plp-/-mag-/-) and hypomyelinosis in triple (plp-/-mbp-/-mag-/-) mutant mice. J Neurosci. 2000;20(14):5225–33.

Venkatesh K, Chivatakarn O, Lee H, Joshi PS, Kantor DB, Newman BA, et al. The Nogo-66 receptor homolog NgR2 is a sialic acid-dependent receptor selective for myelin-associated glycoprotein. J Neurosci. 2005;25(4):808–22.

Venkatesh K, Chivatakarn O, Sheu SS, Giger RJ. Molecular dissection of the myelin-associated glycoprotein receptor complex reveals cell type-specific mechanisms for neurite outgrowth inhibition. J Cell Biol. 2007;177(3):393–9.

Vyas AA, Patel HV, Fromholt SE, Heffer-Lauc M, Vyas KA, Dang J, et al. Gangliosides are functional nerve cell ligands for myelin-associated glycoprotein (MAG), an inhibitor of nerve regeneration. Proc Natl Acad Sci U S A. 2002;99(12):8412–7.

Wang KC, Kim JA, Sivasankaran R, Segal R, He Z. P75 interacts with the Nogo receptor as a co-receptor for Nogo, MAG and OMgp. Nature. 2002;420(6911):74–8.

Wang X, Mitra N, Cruz P, Deng L, Varki N, Angata T, et al. Evolution of siglec-11 and siglec-16 genes in hominins. Mol Biol Evol. 2012;29(8):2073–86.

Weiss MD, Hammer J, Quarles RH. Oligodendrocytes in aging mice lacking myelin-associated glycoprotein are dystrophic but not apoptotic. J Neurosci Res. 2000;62(6):772–80.

Weiss MD, Luciano CA, Quarles RH. Nerve conduction abnormalities in aging mice deficient for myelin-associated glycoprotein. Muscle Nerve. 2001;24(10):1380–7.

Williams G, Wood A, Williams EJ, Gao Y, Mercado ML, Katz A, et al. Ganglioside inhibition of neurite outgrowth requires Nogo receptor function: identification of interaction sites and development of novel antagonists. J Biol Chem. 2008;283(24):16641–52.

Wong EV, David S, Jacob MH, Jay DG. Inactivation of myelin-associated glycoprotein enhances optic nerve regeneration. J Neurosci. 2003;23(8):3112–7.

Wong ST, Henley JR, Kanning KC, Huang KH, Bothwell M, Poo MM. A p75(NTR) and Nogo receptor complex mediates repulsive signaling by myelin-associated glycoprotein. Nat Neurosci. 2002;5(12):1302–8.

Xie W, Uchida H, Nagai J, Ueda M, Chun J, Ueda H. Calpain-mediated down-regulation of myelin-associated glycoprotein in lysophosphatidic acid-induced neuropathic pain. J Neurochem. 2010;113(4):1002–11.

Yang LJ, Zeller CB, Shaper NL, Kiso M, Hasegawa A, Shapiro RE, et al. Gangliosides are neuronal ligands for myelin-associated glycoprotein. Proc Natl Acad Sci U S A. 1996;93(2):814–8.

Yim SH, Quarles RH. Biosynthesis and expression of the myelin-associated glycoprotein in cultured oligodendrocytes from adult bovine brain. J Neurosci Res. 1992; 33(3):370–8.

Yim SH, Toda K, Goda S, Quarles RH. Comparison of the phosphorylation of myelin-associated glycoprotein in cultured oligodendrocytes and Schwann cells. J Mol Neurosci. 1995;6(1): 63–74.

Yin X, Crawford TO, Griffin JW, Tu P, Lee VM, Li C, et al. Myelin-associated glycoprotein is a myelin signal that modulates the caliber of myelinated axons. J Neurosci. 1998;18(6): 1953–62.

Chapter 12
Role of Galactosylceramide and Sulfatide in Oligodendrocytes and CNS Myelin: Formation of a Glycosynapse

Joan M. Boggs

Abstract The two major glycosphingolipids of myelin, galactosylceramide (GalC) and sulfatide (SGC), interact with each other by *trans* carbohydrate–carbohydrate interactions in vitro. They face each other in the apposed extracellular surfaces of the multilayered myelin sheath produced by oligodendrocytes and could also contact each other between apposed oligodendrocyte processes. Multivalent galactose and sulfated galactose, in the form of GalC/SGC-containing liposomes or silica nanoparticles conjugated to galactose and galactose-3-sulfate, interact with GalC and SGC in the membrane sheets of oligodendrocytes in culture. This interaction causes transmembrane signaling, loss of the cytoskeleton and clustering of membrane domains, similar to the effects of cross-linking by anti-GalC and anti-SGC antibodies. These effects suggest that GalC and SGC could participate in glycosynapses, similar to neural synapses or the immunological synapse, between GSL-enriched membrane domains in apposed oligodendrocyte membranes or extracellular surfaces of mature myelin. Formation of such glycosynapses in vivo would be important for myelination and/or oligodendrocyte/myelin function.

Keywords Carbohydrate–carbohydrate interactions • Glycosphingolipids • Membrane domains • Membrane rafts • Cytoskeleton • Signaling • Actin • Microtubules • Myelin basic protein • Silica nanoparticles • Liposomes

---

J.M. Boggs (✉)
Molecular Structure and Function Program, Research Institute, Hospital for Sick Children, 686 Bay St., Toronto, ON, Canada M5G OA4

Department of Laboratory Medicine and Pathobiology, University of Toronto, Toronto, ON, Canada M5G 1L5
e-mail: jmboggs@sickkids.ca

R.K. Yu and C.-L. Schengrund (eds.), *Glycobiology of the Nervous System*, Advances in Neurobiology 9, DOI 10.1007/978-1-4939-1154-7_12,

Abbreviations

Ab	Antibody
CGT	UDP-galactose:ceramide galactosyltransferase
CM-DiI	Cell tracker lipophilic red fluorescent dye
CNP	2′,3′-Cyclic nucleotide 3′-phosphodiesterase
CST	Galactosylceramide 3′-sulfotransferase
DIGs	Detergent-insoluble glycosphingolipid-enriched membrane domains
FA2H	Fatty acid 2-hydroxylase
Gal-BSA	Galactose conjugated to bovine serum albumin
GalC	Galactosylceramide
GlcC	Glucosylceramide
Glyco-nanoparticles	Where glyco=Gal, S-Gal, Glc, Man, silica nanoparticles conjugated to galactose, galactose-3-sulfate, glucose, or mannose
GPI	Glycosylphosphatidylinositol
GSLs	Glycosphingolipids
HFA	Hydroxy fatty acid form of GSL
KO	Gene knockout
LacC	Lactosylceramide
MAG	Myelin-associated glycoprotein
MAPK	Mitogen activated protein kinase (p42 and p44, 42 and 44 kDa isoforms)
MBP	Myelin basic protein
MCT	Lactate transporter
MGDG	Monogalactosyldiglyceride
MOG	Myelin/oligodendrocyte glycoprotein
NFA	Nonhydroxy fatty acid form of GSL
NMDA	*N*-methyl-D-aspartate
OLG	Oligodendrocyte
PLP	Proteolipid protein
SGC	Sulfatide sulfated form of GalC, galactosylceramide I^3-sulfate
SGG	Seminolipid sulfogalactosyldiglyceride, 3-sulfated form of MGDG
SM	Sphingomyelin

12.1 Requirement for Myelin Glycosphingolipids for Maintenance of the Myelin Sheath

Oligodendrocytes (OLGs), the cells that form myelin in the central nervous system (CNS), contain large amounts of the simple glycosphingolipids (GSLs), galactosylceramide (GalC) and sulfatide (SGC) (the sulfated form of GalC, galactosylceramide I^3-sulfate). Myelin is especially enriched in these two GSLs. GalC is 23 wt% and SGC is 4 wt% of the total lipid of myelin (Norton 1977). Myelin also contains low concentrations of two glyco-glycerolipids with the same sugar head groups, monogalactosyldiglyceride (MGDG) and sulfogalactosyldiglyceride (SGG). Their synthesis peaks at the time of most rapid myelination in the rat, and in cultured OLGs at the time of formation of membrane sheets, suggesting a role in myelin formation (Ishizuka and Inomata 1979; Pieringer et al. 1977; Burgisser et al. 1988; Shimomura and Kishimoto 1984).

The galacto and sulfated galactolipids are not essential for formation of myelin, but are essential for its maintenance and stability, to maintain axonal health, and for survival of the animal. Their absence, due to knocking out the genes encoding the enzymes that are necessary for their synthesis, uridine diphosphate-galactose:ceramide galactosyltransferase (CGT) and galactosylceramide 3′-sulfotransferase (CST), causes a more severe phenotype (Coetzee et al. 1996; Bosio et al. 1996; Honke et al. 2002) than knocking out the genes for several myelin-specific proteins (proteins that are much more abundant in myelin than other tissues) including proteolipid protein (PLP) (Boison and Stoffel 1994; Rosenbluth et al. 2009), which is the most abundant protein of myelin and 15 % of the mass of myelin, myelin-associated glycoprotein (MAG) (Roder 1994; Li et al. 1994), 2′,3′-cyclic nucleotide 3′-phosphodiesterase (CNP) (Edgar et al. 2009), and others. The only structural protein known to be even more essential for myelin formation and similarly essential for survival is myelin basic protein (MBP), which is the second most abundant protein in myelin (Readhead et al. 1990).

In the CGT and CST mutants, compact myelin is formed and is almost as thick as normal myelin. However, interaction of the paranodal loops with the axon is abnormal (Marcus et al. 2002, 2006; Hirahara et al. 2004). Delocalization of axolemmal proteins occurs around the node and paranode in these mutants (Marcus et al. 2002). The CGT-null mice have a more severe clinical phenotype than the CST-null mice, with nerve conduction deficits, paralysis, extensive myelin vacuolation and splitting at the intraperiod line, and early death. With age in the CST knockout (KO), the nodal structure also deteriorates, the amount of cytoplasm in myelin increases, and myelin vacuolar degeneration occurs (Marcus et al. 2006). Terminal differentiation of OLGs from the CST KO is enhanced in vitro and in vivo, suggesting that SGC is a negative regulator of OLG differentiation (Hirahara et al. 2004; Bansal and Pfeiffer 1989; Bansal et al. 1999). The number of terminally differentiated OLGs is also increased in vivo and this increase persists into adulthood in CST-null mice, due to increased proliferation and decreased apoptosis (Shroff et al. 2009; Honke 2013). In addition to changes observed in the CNS of the CST KO mouse, the number of Schmidt-Lanterman incisures was remarkably increased in the peripheral nervous system (PNS) (Hoshi et al. 2007).

Myelin galactosphingolipids contain either hydroxy fatty acids (HFA) or non-hydroxy fatty acids (NFA). The fatty acid 2-hydroxylase (FA2H) gene responsible for synthesis of hydroxy fatty acids is highly enriched in the brain, especially in OLGs (Edvardson et al. 2008). The myelin produced in the CGT mutant contains the hydroxy-fatty acid form of glucosylceramide (HFA-GlcC) (normally absent in OLGs) instead of GalC and SGC, and also increased ceramide and sphingomyelin

(SM) levels (Coetzee et al. 1996; Bosio et al. 1996). However, elimination of HFA-GlcC by knockout of the gene for uridine diphosphate-glucose:ceramide glucosyltransferase, targeted to OLGs in the CGT mutant, had no significant effect on phenotype, indicating that the HFA-GlcC produced does not compensate for the loss of GalC and SGC (Saadat et al. 2010).

Indeed, elimination of the HFA forms of myelin GSLs, by knocking out the fatty acid 2-hydroxylase (FA2H) gene, had little effect in young animals, although NFA-GalC was also decreased in amount (Zoller et al. 2008; Potter et al. 2011). Compact myelin formed and appeared normal and there was no significant effect on nerve conduction velocity, although some behavioral changes occurred. However, aged mice (18 months) had scattered axonal and myelin sheath degeneration in the spinal cord and even more pronounced loss of myelin sheaths in sciatic nerve, but not in the brain, indicating impairment of long-term maintenance of myelin. Progressive hind limb paralysis occurred at 22 months. These changes also occurred in mice in which FA2H was knocked out only in OLGs and Schwann cells (Fa2hflox/flox Cnp1-Cre mice) (Potter et al. 2011). Axonal degeneration started before myelin degeneration suggesting that the OLG/myelin HFA-GSLs were required for glial support of axonal function rather than myelin formation. The ratio of HFA to NFA-SGC increases in the brain with age (Shimomura and Kishimoto 1983). Mice overexpressing CGT had unstable myelin with progressive demyelination and a decreased ratio of HFA- to NFA-GalC (Fewou et al. 2005), suggesting that the HFA–NFA ratio may be important for myelin maintenance. However, MGDG was also increased in these mice and might also contribute to the myelin instability.

Several mutations of FA2H have been found in humans that cause decreased synthesis of the HFA-species of myelin GSLs, resulting in leukodystrophy and spastic paraplegia, with severity depending on the site of the mutation (Edvardson et al. 2008; Dick et al. 2010). These mutations affected only the CNS (contrary to the mouse) suggesting an additional FA2H is present in the PNS in humans. Compact myelin was formed initially in the CNS, but with age, thinning of the corpus callosum and the pons, as well as volume depletion of the cerebellum occurred, suggestive of demyelination, as found in the mouse. This is the first identified deficiency in the lipid composition of myelin that results in disease in humans.

If both CGT and FA2H were knocked out in the mouse, the mice lacked Sulf, GalC, HFA-GlcC, and HFA-SM. The amount of NFA-GlcC was also reduced and NFA-SM was significantly increased. However, the mice still formed compact myelin initially and there was no obvious phenotypic difference from the CGT KO (Meixner et al. 2011).

Myelin GSLs have primarily long chain C24:0 and 24:1 fatty acids (both HFA and NFA species). Knockout of ceramide synthase 2, which uses C22–24 acyl chain CoAs to synthesize the long fatty acid chain forms of ceramide, caused reduced levels of GalC, especially long chain NFA and HFA forms, in myelin, resulting in myelin degeneration and detachment at 5 months. The HFA-18:0-GalC species was elevated instead (Ben-David et al. 2011). Thus, fatty acid chain length in addition to the HFA–NFA ratio and carbohydrate moiety of GSLs is important for myelin maintenance.

The importance of glycolipids in formation of compact myelin has also been revealed by phylogenetic studies. Earthworm myelin, with only traces of glycolipids, and shrimp myelin with only GlcC, have a loosely wrapped, noncompacted form of myelin (Okamura et al. 1985; Kishimoto 1986). The space at the extracellular apposition of myelin teleost and higher vertebrates (Inouye and Kirschner 1990), correlates inversely with their proportions of galactolipids (Selivonchick and Roots 1976; Burgisser et al. 1986). In addition, a phylogenetically lower order of deep sea fish, the Gadiformes, whose myelin has an unusual glycolipid composition, including GlcC and its fatty acid ester, and galactoglycerolipids rather than galactosphingolipids, have thin loosely compacted myelin compared to more advanced species of fish (Tamai et al. 1992).

These studies indicate that galactolipids, especially galactosphingolipids, are necessary for maintenance of compact myelin structure with age, although they are not necessary for the process of myelination. They, including their HFA and long chain fatty acid species, are also necessary within compact myelin to sustain the myelinated axon and prevent neurodegeneration, as found also for myelin proteins that are not necessary for formation of compact myelin, such as PLP and MAG, but are necessary to prevent neurodegeneration with age.

12.2 Functions of Myelin Glycosphingolipids

The special properties of GSLs, and sphingolipids in general, could allow them to perform a number of roles in OLGs and myelin (Jackman et al. 2009). GSLs have longer, more saturated fatty acids than phospholipids, and can also participate in an intermolecular hydrogen bonding network (Boggs 1987; Boggs et al. 1984), which causes them to form more ordered membrane domains or rafts that may be involved in protein trafficking and signaling (Gielen et al. 2006). The hydroxyl group on the HFA species can also participate in intermolecular hydrogen bonding interactions (Boggs et al. 1984), and contribute to ordering and membrane domain formation. Interestingly, in the FA2H KO, the ratio of saturated 24:0 to unsaturated 24:1 species of Sulf and GalC was increased (Zoller et al. 2008), which might partially compensate for the loss of HFA species and help maintain ordered membrane domains.

The absence of some paranodal proteins in the CGT/CST mutant may be partly due to altered trafficking of these proteins to the plasma membrane resulting in abnormal paranodal loop formation. Indeed, the content of a number of proteins involved in regulation of cytoskeletal dynamics, energy metabolism, vesicular trafficking, or adhesion was altered, with some being increased and others decreased, in the myelin from the CGT and CST mutants (Fewou et al. 2010). Some proteins that were decreased in myelin were increased in the OLG cell body, suggesting transport from the OLG cell bodies into myelin may be differentially dysregulated in the absence of these GSLs. Similar loss of a lactate transporter (MCT) may occur in spermatocytes in the CST mutant due to impaired trafficking, since CST is also responsible for sulfation of SGG (also called seminolipid) in spermatocytes (Honke 2013). However, the increase in SM in the CGT/CST KO's may partially compensate for the loss of GSLs in raft formation and protein trafficking. The CGT/FA2H double knockout had significantly increased NFA-SM and still formed low density CHAPS-insoluble membrane domains (Meixner et al. 2011).

These GSLs, especially sulfatide, could also interact with various protein ligands. SGC has been described as a multifunctional lipid (Honke 2013). It interacts with a variety of cell adhesion molecules including fibrinogen, von Willebrand factor, P-selectin, thrombospondin and laminin and could interact with extracellular matrix proteins in the CNS. Impaired interaction of SGC with axonal proteins has been suggested to be involved in altered paranodal loop formation in the CGT/CST KO's. SGC, but not GalC, is one of the constituents in myelin responsible for myelin inhibition of CNS axonal outgrowth further suggestive of its interaction with an axonal constituent (Winzeler et al. 2011).

GSLs also organize specific proteins into signaling complexes. GalC is concentrated with cholesterol, GM1, the raft marker flotillin, caveolin, GPI-linked proteins, and kinases (Simons et al. 2000; Arvanitis et al. 2005; DeBruin and Harauz 2007; Gielen et al. 2006; Fitzner et al. 2006; Dupree and Pomicter 2010) in low density detergent-insoluble GSL-enriched fractions (DIGs), believed to come from membrane domains or rafts in OLG and myelin membranes. These DIGs from myelin also contain SGC but are not enriched in it compared to myelin (Arvanitis et al. 2005). They also contain some myelin proteins including MBP, which is a peripheral membrane protein bound to the cytoplasmic side of the membrane, and phosphorylated MBP (Arvanitis et al. 2005; DeBruin and Harauz 2007). The GalC and SGC in these membrane domains can also behave as receptors and receive extracellular signals that are then transmitted to the cytosol.

There is much evidence that GSLs and raft proteins are receptors for extracellular signals as indicated by the ability of antibodies to these constituents to trigger signaling. Antibodies, especially IgM antibodies, or IgG antibodies after cross-linking by second antibodies, or ligands such as cholera toxin B (for GM1), cause cross-linking of raft constituents and clustering of rafts, which may bring different signaling molecules in different raft domains into contact with each other and concentrate them into one domain (Sonnino et al. 2009; Harder and Simons 1999; Iwabuchi et al. 1998; Prinetti et al. 1999). These antibodies mimic natural ligands that are multivalent, such as agrin, which cross-links the acetylcholine receptor and is essential for neural synapse formation (Khan et al. 2001). Galectins, which bind Gal-GlcNAc branches of N-glycans of glycoproteins, are another type of multivalent ligand that can cross-link receptors in cells and regulate receptor signaling (Lajoie et al. 2009). Other multivalent interactions with cell receptors are due to *trans* interaction with a membrane domain in another cell, which allows for interactions between a number of constituents in the apposed membrane domains. This creates a multi-molecular complex with a multivalent cross-linking effect, such as the T cell receptor complex interacting with MHC proteins on antigen presenting cells, forming the immunological synapse (Vogt et al. 2002). Interestingly, agrin is also expressed in lymphocytes, and reorganizes lipid rafts in T cells and sets the threshold for T cell signaling (Khan et al. 2001). The immunological synapse is an example of an intermembrane intercellular interaction that triggers signaling in one or both cells. Similar synaptic interactions may occur in multilayered compact myelin, in which the extracellular surfaces face each other and are separated only by a small distance, and between the paranodal loops and the axon.

12.3 Involvement of OLG/Myelin GSLS in Signaling

Anti-GalC and anti-SGC antibodies (Abs) have long been known to have diverse effects on cultured mouse OLGs and Schwann cells suggesting that they can activate GSL-enriched signaling domains in these cells (Ranscht et al. 1987). The extracellular signals imparted by the Abs have effects on the cytoskeleton (Dyer 1993; Benjamins and Dyer 1990; Dyer et al. 1994) and affect OLG differentiation (Bansal et al. 1999; Bansal and Pfeiffer 1994, 1989), indicating that these signals are transmitted across the membrane to the cytoplasmic side. Anti-GalC and anti-SGC Abs caused redistribution of GalC and SGC on the extracellular surface over domains containing MBP on the cytoplasmic side in OLGs, and depolymerization of a lacy network of microtubules in the membrane sheets (Dyer 1993; Benjamins and Dyer 1990; Dyer et al. 1994). Anti-GalC Ab caused an influx of extracellular Ca^{2+}, rapid cycling of the phospholipid polar head groups, and a decrease in phosphorylation of MBP. Significantly, it did not have these effects on cultured OLGs from the *shiverer* mutant mouse, which lacks MBP, indicating that MBP is required to mediate the effects of anti-GalC Ab on OLGs (Dyer et al. 1994). Anti-SGC Ab inhibited differentiation of progenitor OLGs (Bansal et al. 1999; Bansal and Pfeiffer 1989) and downregulated gene expression in mature OLGs (Bansal and Pfeiffer 1994) consistent with the accelerated developmental time course of cultured OLGs from the CGT and CST mutants. This suggests that SGC is involved in signaling mechanisms that regulate differentiation (Hirahara et al. 2004; Bansal et al. 1999; Shroff et al. 2009).

12.4 Natural Ligands That Interact with GalC and SGC and Transmit Signals Across Apposed Membranes

The effects of anti-GalC and anti-SGC Abs on cultured OLGs suggest that there must be natural ligands that interact with these GSLs in signaling domains to confer signals that are transmitted across the membrane. These ligands may include axonal, OLG, or extracellular matrix proteins, but can also be glycolipids in apposed membranes. *Trans* carbohydrate–carbohydrate interactions between sugars, mediated by divalent cations, occur between simple sugars in crystals (Cook and Bugg 1975), and between multivalent polysaccharides in the presence of water. They mediate processes such as cellulose fibril formation and their subsequent cross-linking by xyloglucan (Gorshkova et al. 2010), as well as proteoglycan-mediated self-association of sponge cells (Bucior 2004; Bucior and Burger 2004). Interactions between single sugar molecules are disrupted by water, but the interactions between polymeric multivalent glycoconjugates in solution or in membrane lipid domains are stronger and persist in water. Furthermore, the water at the membrane surface is more ordered than bulk water and less likely to hydrate the carbohydrate groups (Vogler 1998). The forces between multivalent glycoconjugates have been measured by atomic force microscopy, surface plasmon resonance and other techniques and can be strong enough to cause specific adhesion of sponge cells into a multicellular organism (reviewed in Boggs et al. 2004, 2008a).

Membrane domains present glycolipids as a multivalent array of carbohydrate head groups. Hakomori's pioneering work on GSL-enriched signaling domains demonstrated that they interact by homotypic or heterotypic *trans* carbohydrate–carbohydrate interactions with each other across apposed cell membranes, similar to the neurosynapse or the immunological synapse, and he has termed these interacting domains a "glycosynapse" (Hakomori 1991, 2002). The interactions between these GSLs in GSL-enriched microdomains in apposed membranes trigger signals, probably due to noncovalent cross-linking or patching of the GSLs by the multivalent array of sugars, similar to the effect of Abs cross-linked by anti-Ig Abs, multivalent ligands, or multi-protein complexes in membrane domains. These signals are transmitted across the membrane to cytosolic signal transduction proteins (Iwabuchi et al. 1998; Prinetti et al. 1999).

Like several other GSLs in various cells, the myelin GSLs, GalC and SGC, can also participate in *trans* carbohydrate–carbohydrate interactions between apposed membranes (Hakomori 1991; Stewart and Boggs 1993). The HFA species interact more strongly than the NFA species (Stewart and Boggs 1993; Koshy et al. 1999). Bilayers of GalC only, or containing up to 15 mol% of a negatively charged phospholipid, are separated by a narrow space less than the diameter of a single water molecule, indicating an attractive interaction between them (Kulkarni et al. 1999). The adhesive forces between GalC and SGC bilayers have not been measured, but must be larger than for GalC alone, because heterotypic interactions between liposomes containing GalC and liposomes containing SGC were greater than their homotypic interactions (Stewart and Boggs 1993). Similar behavior was observed by mass spectrometry for GalC and SGC micelles in methanol/water (Koshy and Boggs 1996). If these GSLs are present only on the extracellular surface of myelin, the extracellular lipid must be about 40 mol% glycolipid, with cholesterol estimated to be also about 40 mol% (Inouye and Kirschner 1988). The extracellular surfaces of myelin, which face each other at the intraperiod line in the multilayered myelin sheath (Fig. 12.1c), would thus be covered with the simple sugars, galactose and sulfated galactose.

GalC and SGC are likely to act as ligands for each other in OLGs and myelin since *trans* interactions between them cause similar signaling effects as the Abs, as we showed by adding multivalent forms of galactose/sulfated galactose to cultured rat OLGs (Boggs et al. 2004, 2008a). Multivalent forms used included phospholipid/cholesterol liposomes containing GalC/SGC (Boggs and Wang 2001, 2004), a polyvalent form of galactose conjugated to bovine serum albumin (Gal-BSA) (Boggs and Wang 2004; Boggs et al. 2008b), and silica nanoparticles bearing

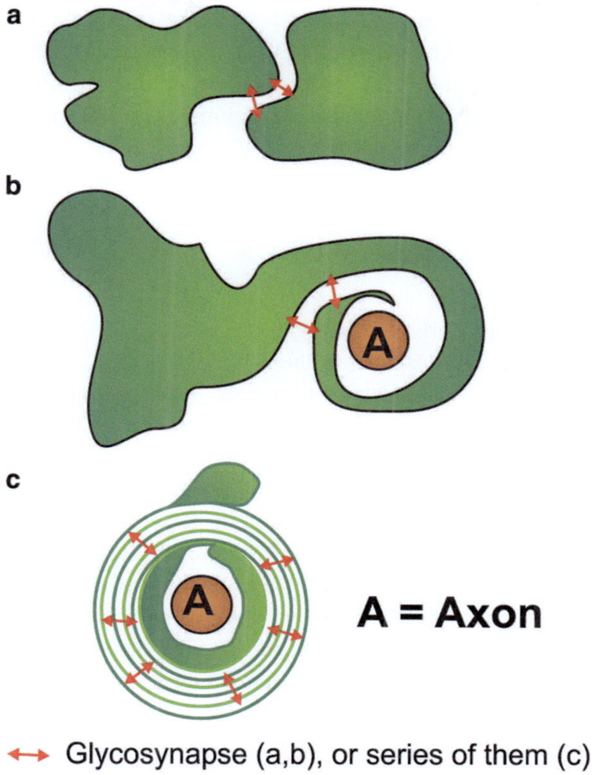

Fig. 12.1 Sites where glycosynapses between GalC/SGC-enriched domains in OL/myelin membranes might occur, resulting in signaling and depolymerization of the cytoskeleton. A *double headed orange arrow* represents glycosynapse formation between two apposed OLG membranes (panels **a**, **b**), or a series of glycosynapses between the multilayers of the myelin sheath (panel **c**). These glycosynapses are postulated to form transiently under certain conditions. (**a**) Glycosynapse between two different OLGs—could be a signal for process retraction or cessation of growth of a membrane sheet. (**b**) Glycosynapse between extracellular surfaces of membranes of a process of the same cell wrapping around an axon (tan, labeled A)—could be a signal for elimination of cytosol and formation of closely packed (compact) myelin layers. (**c**) Series of glycosynapses between the extracellular surfaces (at intraperiod line shown in *light green*) of compact myelin surrounding a nerve axon—could transmit extracellular or axonal signals throughout myelin layers. Major dense line where cytosolic surfaces are apposed is shown in *darker green*. The outer loop and inner loop containing cytoplasm are shown in *green*. Reprinted from Boggs et al. (2008a, 2010) with permission from Elsevier

galactose (Gal) and sulfated-galactose (SGal) (Boggs et al. 2010; Zhao et al. 2012). These glycoconjugates can be used to mimic interactions which are postulated to occur between GalC and SGC in apposed membranes, such as between the facing pairs of extracellular surfaces of myelin, or between OLGs or their processes in contact with each other (Fig. 12.1).

Although liposomes have the advantage that they can be made to resemble the extracellular surface of myelin or its GSL-enriched membrane domains, and therefore, the putative natural multivalent ligand, they have the disadvantage that they can affect cells not just by adhering to them, but possibly by exchange of lipids between the liposomes and the cell membranes or by internalization into the cell. The use of Gal-BSA or glyco-nanoparticles eliminates this possibility.

12.5 Binding and Effect of Multivalent Glyco-nanoparticles on OLGS

Dendrimers or silica nanoparticles conjugated with saccharides are multivalent glycoconjugates that have been used to mimic *trans* interactions between cell surface glycans (Seah et al. 2009). Azide-functionalized silica nanoparticles were conjugated with propargyl derivatives of sugars or fluorescent FITC-labeled sugars by copper-promoted azide–alkyne cycloaddition. We examined the binding and effect of these silica nanoparticles bearing galactose (Gal-nanoparticles), galactose-3-sulfate (SGal-nanoparticles), or a combination of galactose and galactose-3-sulfate (Gal/SGal-nanoparticles) on OLGs, as well as control glyco-nanoparticles bearing glucose (Glc-nanoparticles) or mannose (Man-nanoparticles) (Boggs et al. 2010; Zhao et al. 2012). Unglycosylated control particles were capped with propargyl alcohol to provide a hydroxyl-terminated nanoparticle (OH-nanoparticles). In each case, the nanoparticles carried about 78,000 carbohydrate groups per particle, and thus were highly multivalent. They bound specifically to microtiter plates coated with galactolipid, but not glucolipid, and bound to each other in a heterotypic fashion (Zhao et al. 2012). The binding did not require divalent cations. *Trans* interactions between many sugars require divalent cations (Hakomori 1991) as did the interactions between GalC and SGC in phosphatidylcholine/cholesterol liposomes (Stewart and Boggs 1993), but those between pure GalC bilayers and pure SGC bilayers did not (Boggs et al. 2000), similar to the glyco-nanoparticles where galactose and sulfated galactose are highly concentrated. The requirement for divalent cations for a lipid or lipid/protein composition resembling that of the extracellular surface of OLGs or myelin is not known. *Trans* interactions between the galactoglycerolipids of myelin, MGDG and SGG, have not been investigated but would probably occur since the interaction between Gal-nanoparticles and SGal-nanoparticles does not depend on a lipidic moiety. Furthermore, divalent cations have been shown to complex digalactosyldiglyceride membranes to each other (Webb et al. 1988).

The Gal- and SGal-nanoparticles also bound specifically to live or fixed OLGs (Fig. 12.2e, f) and did not bind to astrocytes present in the culture, which lack GalC

Fig. 12.2 (continued) fixation and staining with anti-MAG Ab (*red*); *yellow* indicates overlap. Similar results were obtained by addition of nanoparticles to fixed OLGs; (**e**) Gal/SGal-nanoparticles; (**f**) Gal-nanoparticles; (**g**) Cells were preincubated with anti-GalC O1 Ab prior to binding of fluorescent Gal/SGal-nanoparticles; binding was greatly diminished; (**h**) Cells were preincubated with anti-GalC O1 Ab prior to binding of Gal-nanoparticles; binding was not affected as expected for a heterotypic interaction of galactose with SGC. Scale bars = 20 μm. Panels **a–d**, **e**, **f**, and **h** are reprinted with permission from Zhao et al. (2012). Copyright (2012) American Chemical Society

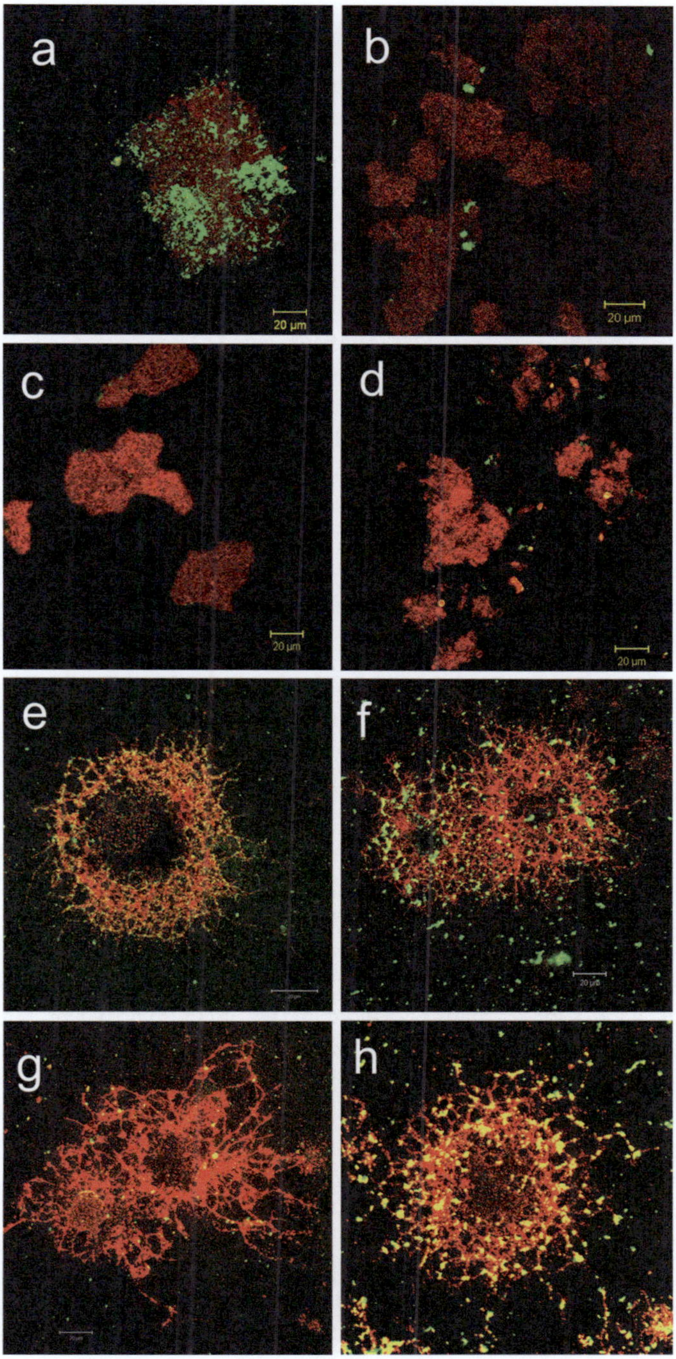

Fig. 12.2 (**a–d**) Fluorescence microscope images of fluorescein-labeled nanoparticles (*green*) bound to myelin fragments labeled with CD-DiI (*red*); *yellow* indicates overlap. (**a**) Gal/SGal-nanoparticles; (**b**) control OH-nanoparticles; (**c**) 4× excess nonfluorescent Gal/SGal nano particles followed by fluorescent Gal/SGal-nanoparticles; (**d**) myelin was preincubated with anti-GalC IgG Fab fragments followed by fluorescent Gal/SGal-nanoparticles. Bar = 20 μm. (**e–h**) Fluorescence microscope images of fluorescein-labeled nanoparticles (*green*) bound to live OLGs followed by

and SGC (Zhao et al. 2012). Pre-incubation of OLGs with the O1 anti-GalC monoclonal antibody greatly reduced binding of the Gal/SGal (Fig. 12.2g) or SGal-bearing particles to the OLGs, but had no effect on the binding of Gal-particles (Fig. 12.2h), as expected for a heterotypic interaction. The Gal/SGal nanoparticles also bound specifically to purified myelin fragments (Fig. 12.2a). Control nanoparticles bound much less to OLGs and myelin (shown for OH-nanoparticles and myelin in Fig. 12.2b). The binding was inhibited by an excess of nonfluorescent versions of Gal/SGal nanoparticles (Fig. 12.2c) and also by preincubation of myelin with monovalent anti-GalC IgG Fab fragments (Fig. 12.2d).

The Gal/SGal nanoparticles, GalC/SGC-containing liposomes and Gal-BSA all had sugar-specific and relatively similar effects on the distribution of GalC and MBP in cultured OLGs. They caused redistribution and clustering of GalC on the extracellular side, and MBP on the cytosolic side, into clusters of varying size such that the GalC domains usually overlaid the MBP domains (Boggs and Wang 2001, 2004; Boggs et al. 2010) (MBP clustering is shown for Gal-nanoparticles in panels d–f of Fig. 12.3, compared to untreated cells in panels a–c, and MBP or GalC clustering is shown for GalC/SGC-containing liposomes in Fig. 12.4c, g, compared to control liposome-treated cells in Fig. 12.4a, e). Pre-incubation of the OLGs with anti-GalC IgG Fab fragments prevented the effect of the Gal/SGal-nanoparticles on GalC and MBP distribution and OLG morphology (Boggs et al. 2010). Untreated mature cells have a complex cytoskeletal network with major veins and a lacy network of microtubules (Dyer 1993) (Fig. 12.4f). The Gal/SGal nanoparticles and GalC/SGC-containing liposomes also caused loss of the microtubular network (Fig. 12.4h) (Boggs and Wang 2001, 2004; Boggs et al. 2010).

The glyco-nanoparticles had a quantitatively greater effect on the OLGs than GalC/SGC-containing liposomes or Gal-BSA, with some effect seen after 6 h and a greater effect after overnight culture, whereas overnight culture was required in order to detect an effect of GalC/SGC-containing liposomes or Gal-BSA. At 8 days in culture, when the glyco-nanoparticles are added, the cell population typically consists of about 30 % mature cells with flat membrane sheets (resembling Fig. 12.3a–c) and the cells are almost completely GalC and MBP positive, even when less mature, with only thin processes instead of membrane sheets. However, after treatment with glyco-nanoparticles, more cells appeared less mature than in the control cell population, with many narrow processes, and only about 5 % appearing mature with membrane sheets. In many of these less mature GalC$^+$ cells, MBP staining was very low or absent, in contrast to untreated OLGs (Boggs et al. 2010). An example of this type of cell after treatment with Gal/SGal-nanoparticles, is shown in Fig. 12.3g–i. These effects, particularly the almost complete absence of MBP staining, were not commonly observed after liposome treatment (e.g., Fig. 12.4c). Since MBP appears relatively late in OLG differentiation (Bansal and Pfeiffer 1994), its loss in the treated cells suggests that dedifferentiation may have occurred. The Gal, SGal, or mixed Gal/SGal-nanoparticles had significantly more effect than Glc-, Man-, or unglycosylated-nanoparticles. Similarly, control liposomes with no GSL, or with GlcC or lactosylceramide (LacC) instead of GalC, had much less effect on GalC redistribution than those containing GalC and/or

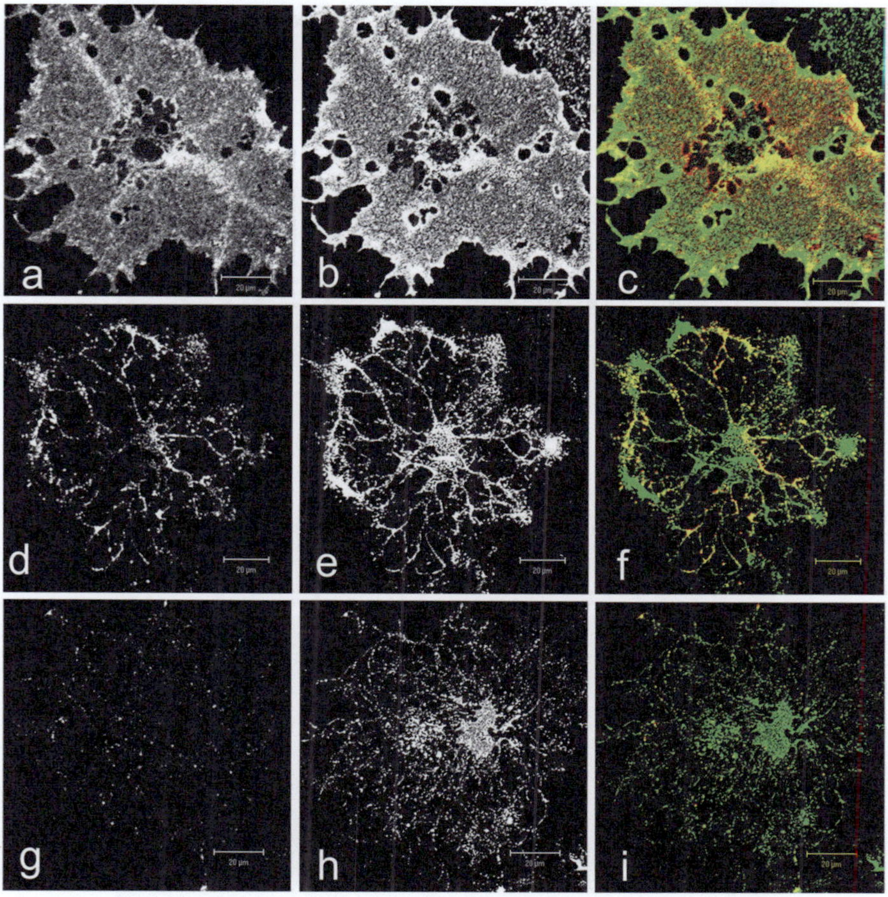

Fig. 12.3 Confocal microscope images of OLGs fixed and stained externally with monoclonal anti-GalC Ab (O1) (**b**, **e**, **h**); then stained internally with anti-MBP Ab (**a**, **d**, **g**). Merge is shown in (**c**, **f**, **i**) (MBP, *red*; GalC, *green*). Untreated OLGs (**a**–**c**); OLGs treated overnight with 2 μg/ml Gal-nanoparticles (**d**–**i**). Panels (**d**–**f**) show MBP and GalC redistribution/clustering in a more mature cell typical of that caused by GalC/SGC-containing liposomes. Panels (**g**–**i**) represent a cell which looks less mature and has lost most of its MBP; this occurs frequently after nanoparticle treatment but is much less frequent with liposome treatment. Bar = 20 μm. Reprinted from Boggs et al. (2010), with permission from Elsevier

SGC. BSA conjugated to glucose and mannose had significantly less effect than Gal-BSA (Boggs and Wang 2004). The nanoparticles bearing both Gal and SGal had a significantly greater effect on GalC redistribution than those with only Gal or SGC (Boggs et al. 2010). Similarly, liposomes containing both myelin GSLs had a greater effect than those with only SGC or GalC (Boggs et al. 2004).

Using liposomes, effects on clustering of other proteins and on the cytoskeleton were also determined. Similar clustering of GPI-linked proteins and of two transmembrane proteins, proteolipid protein (PLP) and myelin/OLG glycoprotein

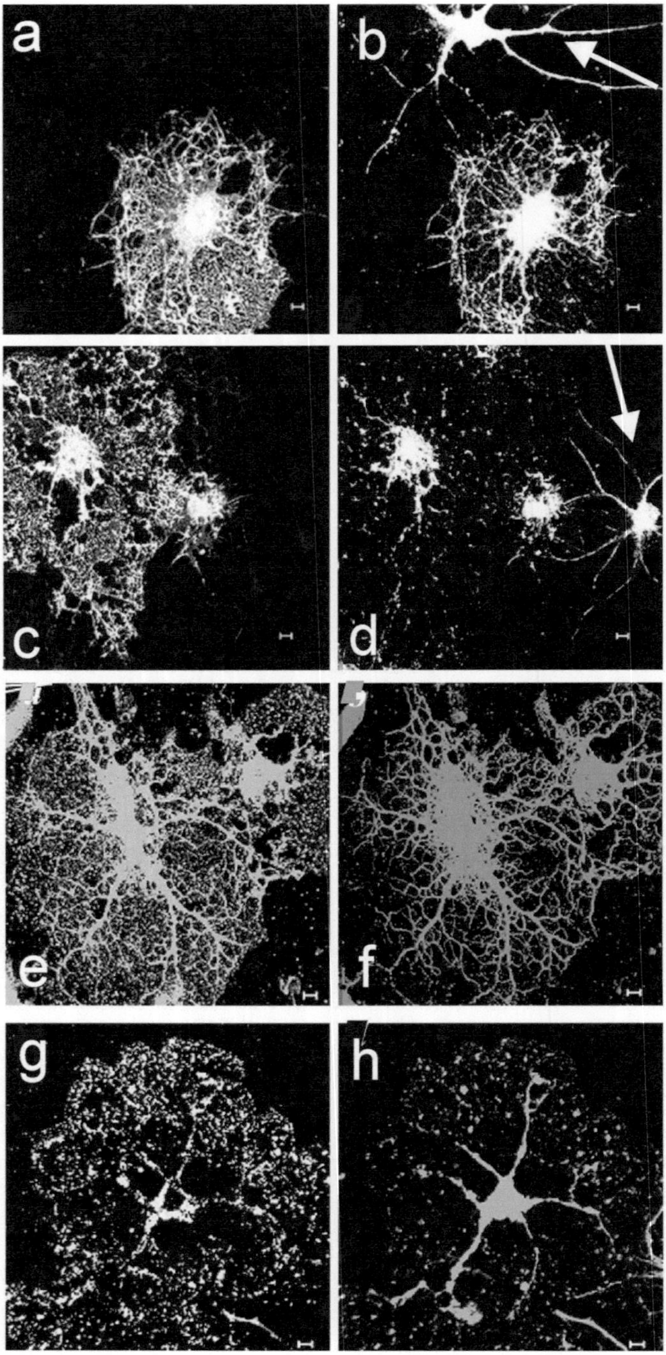

Fig. 12.4 Confocal microscope images of cultured OLGs treated overnight with control phospholipid/cholesterol liposomes (**a**, **b**, **e**, **f**) or GalC/SGC-containing phospholipid/cholesterol liposomes (**c**, **d**, **g**, **h**). (**a**, **c**) OLGs fixed, permeabilized, and stained with monoclonal anti-MBP and FITC-labeled second Ab and then (**b**, **d**) stained with Texas Red-phalloidin. The MBP-negative astrocytes present in the field in **b** above the OLG and in **d** to the far right of the OLG are stained only with phalloidin (*white arrows* in **b**, **d**). The GalC/SGC-containing liposomes caused

(MOG), occurred, suggesting that the domains which cluster are membrane rafts (Boggs and Wang 2004; Boggs et al. 2004, 2008a). Several proteins or phosphorylated proteins involved in signal transduction, MAPK, phosphorylated MBP, and some phospho-tyrosine-containing proteins also clustered with MBP and GalC (Boggs and Wang 2004), suggesting that these rafts are membrane signaling domains. Indeed, the GalC/SGC-containing liposomes also caused depolymerization of microtubules (Fig. 12.4h, compare to Fig. 12.4f), and actin filaments (Fig. 12.4d, compare to Fig. 12.4b) that form a lacy cytoskeletal network in the membrane sheets, indicating that the interaction of GalC/SGC-containing liposomes with the extracellular surface of the OLG caused transmission of a signal across the membrane (Boggs and Wang 2001, 2004). Note that the GalC/SGC-containing liposomes had no effect on the actin cytoskeleton of an astrocyte (MBP negative, white arrow) also present in the culture (Fig. 12.4d, compare to Fig. 12.4b). The Gal/SGal nanoparticles also caused loss of the microtubular network (Boggs et al. 2010), but their effect on actin has not been examined. Inhibition, using various reagents, of a number of kinases, such as Rho kinase, and phosphatases that are involved in regulation of the cytoskeleton prevented the liposome-mediated effects on the cytoskeleton (Boggs et al. 2008a, b).

These results show that a lipidic form of the sugar is not necessary for the effect; rather a multivalent form of the sugar, as found on the surface of GSL-containing liposomes, or bound to a polymer or nanoparticle, is sufficient. Finally, the effects are specific for Gal and SGal.

12.6 Receptors in OLGS Which Interact with Multivalent Gal/SGal by *Trans* Interactions

The receptor(s) in the OLG membrane which interact with the multivalent Gal/SGal presented by liposomes or polymers could be protein(s), but are likely to be GalC and SGC (and possibly also MGDG and SGG) for the following reasons: (1) GalC and SGC bind to each other by *trans* interactions across apposed surfaces and SGal-bearing nanoparticles bind to Gal-nanoparticles and to galactolipid in vitro (Hakomori 1991; Stewart and Boggs 1993; Zhao et al. 2012); (2) The effects of multivalent presentations of Gal/SGal on OLGs resemble the effects of anti-GalC/SGC Abs on

Fig. 12.4 (continued) depolymerization of the actin filaments only in the MBP-positive OLG and not in the astrocyte. (**e**, **g**) OLGs fixed, stained externally with polyclonal anti-GalC Ab and FITC-labeled second Ab and then (**f**, **h**) permeabilized and stained internally with monoclonal anti-α,β-tubulin Abs. Cell has large veins of microtubules and a lacy network of smaller microtubules in the membrane sheets (**f**). The GalC/SGC-containing liposomes caused loss of the lacy network of microtubules but the major veins remain (**h**). Scale bar = 20 μm. Panels **a**–**d** reprinted from Boggs and Wang (2001) with permission from John Wiley and Sons. Panels **e**–**h** reprinted from Boggs et al. (2004) with permission from Springer

OLGs reported by Dyer and Benjamins (Benjamins and Dyer 1990; Dyer 1993); (3) Fab fragments of anti-GalC IgG Ab, which had no effects on OLGs themselves, prevented the binding and effects of the Gal/SGal-nanoparticles on GalC and MBP distribution and OLG morphology (Boggs et al. 2010; Zhao et al. 2012); (4) The multivalent Gal/SGal did not bind to and had no effect on the cytoskeleton of astrocytes that were also present in the culture, and that lack these two GSLs, indicating a specific effect on GalC/SGC-containing OLGs (Boggs and Wang 2001; Boggs et al. 2004; Zhao et al. 2012); (5) Inhibition of GSL synthesis by treatment of OLGs with fumonisin B1 prevented the effect of liposomal GalC/SGC on MBP redistribution in the GalC/SGC-negative OLGs (Boggs et al. 2008a, b). These results support our suggestion that the natural ligand(s) for GalC and SGC in OLGs that are mimicked by anti-GalC/SGC Abs or multivalent Gal/SGal are, or include, a multivalent array of GalC and SGC (and possibly also MGDG and SGG) in apposed OL/myelin membranes. However, proteins with terminal galactose and galactose-3-sulfate moieties might also be able to serve as ligands for GalC and SGC; this has not been investigated.

We suggest that *trans* interactions between these GSLs clustered in apposed membrane domains can occur transiently under certain conditions between extracellular OLG membranes or the extracellular surfaces of compact myelin membranes. They cannot cause strong adhesion of these surfaces, but might contribute weak adhesive forces. Interestingly, in the PLP knockout mouse, compact myelin is formed with a smaller separation between the extracellular surfaces than in wild type mouse, although the myelin is more unstable and separates more easily under fixation conditions used for electron microscopy (Rosenbluth et al. 2009). Myelin particles have been shown to bind to cultured OLGs, and myelin particles bind to each other (Bakhti et al. 2013). The interaction between them is reduced, but still significant, if PLP is absent, indicating that although PLP contributes to adhesive interactions, other constituents, such as GalC and SGC, also contribute.

Loss of sialic acid on the OLG surface and downregulation of proteins with large extracellular domains, such as chondroitin sulfate proteoglycan 4 (NG2) and CD44, occur with maturation of OLGs and their production of myelin membranes. This prevents the electrostatic repulsion which occurs between other cell surfaces (Bakhti et al. 2013). Nevertheless, X-ray diffraction and electron microscopy indicate that the static separation of the extracellular surfaces of compact myelin is too great for the carbohydrate head groups of these GSLs to be in constant contact (Kirschner et al. 1989). However, they may come into transient contact under some conditions that cause protein clustering away from GSL-enriched domains, such as increased extracellular Ca^{2+} concentration (Hollingshead et al. 1981; Boggs et al. 2004, 2008a). Furthermore, membrane surfaces in multibilayers undulate, allowing the intermembrane separation to fluctuate (Niles et al. 1996) so that transient contact between GalC and SGC could occur. Transient, weak interactions between myelin GSLs, in addition to PLP, would allow myelin layers to slip by each other as the axon is ensheathed by growing OLG processes.

12.7 GalC/SGC Signaling Releases Cytoskeletal Restriction of Membrane Domains

Jasplakinolide, a reagent that stabilizes actin filaments, inhibited liposome-induced redistribution of all the membrane constituents that were clustered in its absence, including GalC, on the extracellular surface (Boggs and Wang 2004). It also prevented depolymerization of the microtubules. This result indicates that depolymerization of the actin filaments is required both for redistribution of the membrane constituents and for depolymerization of the microtubules. Thus, it is an early event in the transmembrane signaling mediated by multivalent Gal/SGal. The fact that jasplakinolide also prevented clustering of GalC on the extracellular surface suggests that it is not just individual GalC molecules that redistribute, but rather entire membrane domains/rafts that redistribute and coalesce (Fig. 12.5).

Since actin depolymerization is an early event following interaction of multivalent Gal/SGal with the OLG membrane, this interaction must first cause transmission of a signal across the membrane that affects actin. This initial signal may be Ca^{2+} entry, as found when anti-GalC Ab was added to OLGs (Dyer 1993; Benjamins and Dyer 1990; Paz Soldan et al. 2003), or a mechanical signal due to GalC/SGC cross-linking. Depolymerization of the cytoskeleton then allows for redistribution and coalescence of microdomains enriched in GalC, MBP, and the other membrane constituents examined. This sequence of events suggests that the cytoskeleton restricts lateral diffusion of these membrane constituents, either by binding to them or by binding to other transmembrane proteins. This is consistent with the membrane skeleton fence or picket fence model (Kusumi et al. 1999, 2012) as well as with studies indicating that lipids and transmembrane proteins undergo hop diffusion in compartmentalized membrane domains of 50–200 nm (Fujiwara et al. 2002; Marone et al. 2006). Upon depolymerization of the cytoskeleton, these domains are able to redistribute and coalesce into large clusters (Fig. 12.5). Multivalent Gal/SGal cross-linking or patching of the small membrane domains/rafts restrained within these compartments would facilitate this redistribution.

In some cells, such as T cells, receptor clustering causes actin polymerization or requires the actin cytoskeleton for clustering to occur (Harder and Simons 1999; Mitchell et al. 2009; Rodgers and Zavzavadjian 2001; Gomez-Mouton et al. 2001; Baumgartner et al. 2003). However, in other cells, receptor clustering increased on depolymerization of the cytoskeleton (Wang et al. 2001; Treanor et al. 2010; Hao and August 2005). Coclustering of separate GM1-containing rafts with GM3-containing rafts in fibroblasts increased significantly after actin depolymerization (Fujita et al. 2009). In B cells, activation by cross-linking of the B cell receptor caused rapid (15–30 s) global actin depolymerization, although actin repolymerization later recurred (Hao and August 2005), resulting in "corrals" around individual microclusters of B cell receptors that had formed (Treanor et al. 2011). Cross-linking of glycan moieties on retinal pigment epithelial cells with galectin-3 also caused depolymerization of the cytoskeleton, resulting in their failure to form pseudopodia and to attach and spread on a fibronectin-coated surface (Alge-Priglinger

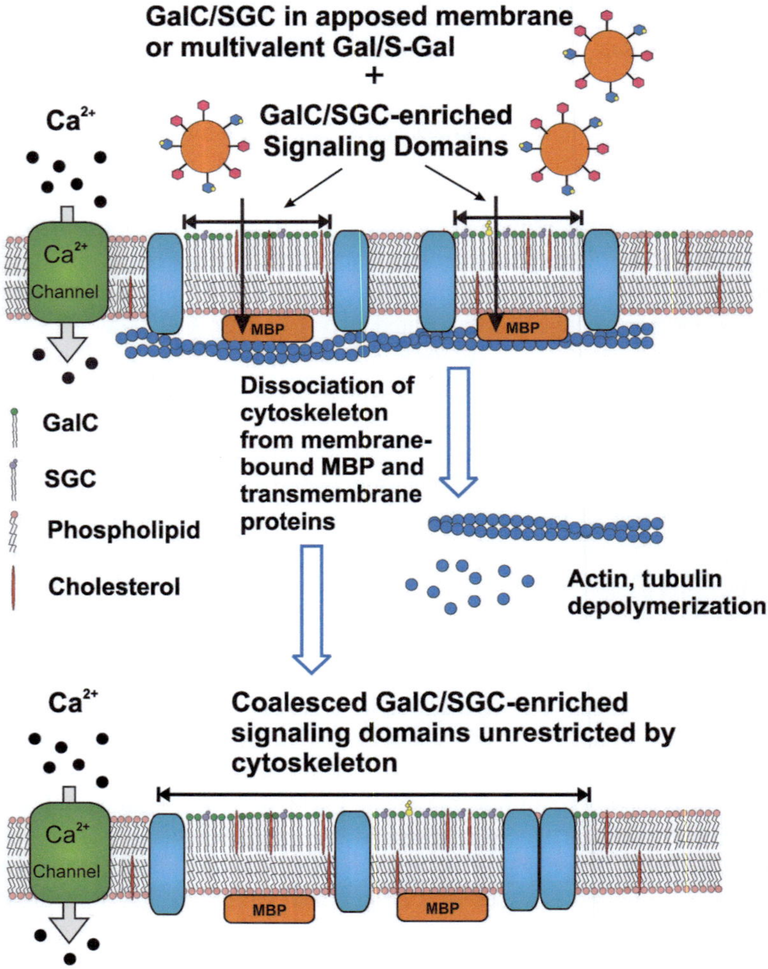

Fig. 12.5 Schematic of effects of multivalent Gal/SGal on OLG membrane sheets. GalC and SGC-enriched membrane domains (rafts) in OLG membrane sheets also contain MBP, a peripheral membrane protein on the cytoplasmic side, and are linked to the membrane skeleton (made up of linked *blue spheres*) via MBP (Boggs 2006; Harauz and Boggs 2013), transmembrane proteins and/or other membrane-actin binding proteins. Some transmembrane proteins (*blue oblongs*) bound to the membrane skeleton serve as picket fences, according to the hypothesis of Kusumi and colleagues (Kusumi et al. 1999, 2012; Fujiwara et al. 2002; Marone et al. 2006), which restrict lateral diffusion of both lipids and proteins in the membrane domains. *Upper panel*—binding of multivalent Gal/SGal (GalC/SGC-containing liposomes or glyco-nanoparticles conjugated to Gal and SGal, depicted by large *orange spheres* bearing Gal (*magenta hexagon*) and SGal (*blue hexagon*)), cross-links GalC and SGC molecules in the membrane domains, and triggers an initial signal, possibly Ca^{2+} entry, which causes dissociation of actin filaments and microtubules and their depolymerization. *Lower panel*—loss of the membrane skeleton permits lateral diffusion of the membrane domains so that they coalesce into larger clusters. The membrane domains may contain a number of other transmembrane proteins such as PLP and MOG, and signaling proteins such as MAPK (not shown), since they redistribute together with GalC, SGC, and MBP (Boggs and Wang 2004). MBP on the cytoplasmic side may be linked to GalC/SGC on the extracellular side via one

et al. 2011). An IgM that reacted with a raft constituent of neurons caused raft clustering and signaling, and coupled the rafts to microtubules, but caused F-actin networks to recede from the growth cone periphery (Xu et al. 2011). Thus, receptor cross-linking and clustering is coupled to dynamic reorganization of the cytoskeleton in many cells.

12.8 Role of MBP in Transmission of GalC/SGC-Mediated Signal

MBP on the cytosolic side may play an important role in transmission of the GalC/SGC-mediated signal to the cytoskeleton. Earlier studies by Dyer et al. (1994) showed that anti-GalC Ab did not cause effects in OLGs from the *shiverer* mutant mouse, which lacks MBP. Suppression of MBP synthesis in normal rat OLGs using MBP siRNA significantly inhibited the effect of GalC/SGC-containing liposomes on GalC redistribution in the MBP-negative OLGs (Boggs et al. 2008b) and on the cytoskeleton (Boggs et al. 2010). Coculture of OLGs with neurons induced dramatic lipid condensation or ordering, as detected by laurdan fluorescence, in OLG membranes and clustering of GalC-containing domains (Fitzner et al. 2006), which may be related to the GalC clustering induced by anti-GalC liposomes or multivalent Gal/SGal arrays. This effect of neurons was not observed in OLGs from *shiverer* mice indicating that MBP was required for this phenomenon also. CHAPS-insoluble membrane domains prepared from the OLGs contained more PLP and MBP after coculture with neurons, indicating protein redistribution in the membrane had also occurred.

MBP binds to and assembles actin filaments and microtubules and binds actin and microtubules to a lipid bilayer, and it may tether the cytoskeleton to the plasma membrane in OLGs (Boggs 2006; Harauz and Boggs 2013). Dynamic changes in co-localization of MBP with actin and tubulin, occurred in transfected N19-oligodendroglial cells during membrane ruffling stimulated by PMA, with enrichment of these proteins in the membrane ruffles, and in membrane domains resembling focal adhesion contacts induced by IGF-1 (Smith et al. 2012). MBP was coimmunoprecipitated with actin, tubulin, and signaling molecules from detergent extracts of primary OLGs (Boggs et al. 2014). Low density DIGs isolated from OLGs and myelin also contained actin and tubulin (Arvanitis et al. 2005; Taguchi et al. 2005; Marta et al. 2003) in addition to MBP. These studies indicate that OLG

Fig. 12.5 (continued) of these transmembrane proteins, since it is influenced by the GalC/SGC cross-linking and is necessary for transmission of the extracellular signal to the cytoskeleton (Boggs et al. 2008b; Dyer et al. 1994). The head groups of the lipids are depicted as GalC (*green*) SGC (*purple*), gangliosides (*yellow*), phospholipids (PL) (*pink*); cholesterol (*red rod*). Similar *trans* interactions between GalC/SGC-enriched domains in apposed OLG or myelin membranes are postulated to create a glycosynapse and have a similar signaling effect

membrane domains/rafts may be linked to the membrane skeleton via MBP, in addition to other proteins (Fig. 12.5). They also suggest that MBP may be linked to cytoskeletal proteins even in myelin. DIGs from myelin also contain the radial component, a junctional specialization within intermodal CNS myelin that passes through many layers of compact myelin (Karthigasan et al. 1994). It appears as a radial array by electron microscopy and has a similar appearance in isolated DIGs. It is made up of tight junctions that may control the ionic content of the extracellular space in myelin (Dyer 2002; Boggs et al. 2008a).

12.9 Role of Glycosynapses in OLGS or Myelin

GalC/SGC-enriched microdomains in the OL/myelin membrane may form glycosynapses at different sites and under certain conditions in OLG or myelin membranes. Glycosynapses could occur between apposed membranes of OLG processes in contact with each other, or between the extracellular surfaces of compact myelin (Fig. 12.1). If OLGs are in contact at high densities, or if OLG processes contact an already myelinated axon, this contact might cause process retraction requiring disruption of the cytoskeleton (Fig. 12.1a). Contact inhibition of oligodendrocyte progenitor cells occurs for cells grown at high density (Zhang and Miller 1996), but it is not known if this also occurs for more mature SGC and GalC-containing OLGs. However, in vivo and in vitro time lapse imaging shows that OLG processes often seem to withdraw after contact with a nearby process or myelinated axon, suggesting that they influence one another (Kirby et al. 2006; Ioannidou et al. 2012). This may be a mechanism to ensure uniform myelination. A signal applied to SGC via a glycosynapse between two OLG processes might be expected to inhibit differentiation and process extension (Hirahara et al. 2004) as observed with anti-SGC Ab (Bansal et al. 1999). The increased number and decreased apoptosis of OLGs in the CST-null mouse (Shroff et al. 2009) may be partly due to a failure of contact inhibition normally mediated by GalC–SGC interactions.

Dynamic regulation of the cytoskeleton is necessary for various stages of myelination (Bauer et al. 2009; Boggs et al. 2008a; Harauz and Boggs 2013). When mature myelinating OLGs initially ensheath axons, the first few layers of membrane around the axon contain cytosol (Remahl and Hildebrand 1990). Disruption of the cytoskeleton in these layers is necessary for the cytoplasmic surfaces to adhere and create compact myelin. GalC–SGC interactions could occur as the membrane sheets wrap around the nerve axon allowing GSLs in apposed surfaces to come into contact at least transiently and/or in localized domains and confer a signal for compaction (Fig. 12.1b). In CST-null mice, the processes which myelinate are thicker and retain more cytoplasm than in wild-type mice (Shroff et al. 2009), perhaps due to retention of the cytoskeleton in the absence of GalC–Sulf interactions. Formation of compact myelin requires close apposition between each pair of facing extracellular surfaces, and between each pair of facing cytoplasmic surfaces, which may be promoted by protein and GSL clustering. Fitzner et al. (2006) have shown that coculture of OLGs with neurons induced GalC ordering in OLG membranes, which was probably due to GalC clustering, and increased MBP distribution into DIGs. Thus, a neuronal signal inducing OLG processes to ensheath the axon could be followed by GalC–SGC contact between the apposed membranes, depolymerization of the cytoskeleton, and protein and GSL clustering, leading to elimination of cytosol and adhesion of the cytosolic surfaces.

Signaling resulting from *trans* GalC–SGC interactions may also occur in the mature myelin sheath (Fig. 12.1c). These signals may in turn be transmitted across the membrane to MBP and the cytoskeletal elements and junctions in the radial component in myelin, as found in cultured OLGs for anti-GalC antibody and GalC/sulfatide-containing liposomes, possibly by causing Ca^{2+} entry into the cytosolic domains. The accumulation of Ca^{2+} into the cytosolic space of compact myelin, mediated by *N*-methyl-D-aspartate (NMDA) receptors, has been detected within compact myelin (Micu et al. 2005). Subsequent effects on tight junctions or gap junctions between myelin layers may regulate their permeability (Morita et al. 1999; Dyer 2002) as occurs for intestinal epithelial cells (van Itallie and Anderson 2004). This process may allow for transmission of signals from the axon throughout compact myelin.

Communication between the myelin sheath and the axon may regulate both axonal and myelin function and is necessary to prevent neurodegeneration (Witt and Brady 2000; Edgar and Garbern 2004). Myelination affects the axon caliber, phosphorylation of axonal neurofilaments, the axonal cytoskeleton, and ion channel organization (Baba et al. 1999). Phosphorylation of MBP occurs in myelin in response to the nerve action potential (Murray and Steck 1984; Atkins and Sweatt 1999), and lipids and metabolic precursors are transferred from the axon to the myelin sheath (Chakraborty et al. 1999). The myelin sheath provides trophic support to the axon, which may compensate for its shielding of the axon from extracellular metabolic support (Nave and Trapp 2008; Simons and Lyons 2013). Lactate is one substance taken up by OLGs and myelin that may allow for such trophic support. MCT1, a lactate and pyruvate transporter, has been localized to OLGs and also shown to be in compact myelin by immunogold labeling, whereas a different lactate transporter, MCT2, is in axons. Lactate uptake causing a pH change was detected in both OLG cell bodies and processes aligned with axons (Rinholm, et al. 2011). Downregulation of MCT1 selectively in OLGs caused axonal damage and neuronal loss in animal and cell culture models, indicating a new mechanism by which OLGs support neurons and axons (Lee et al. 2012). It would be interesting to determine whether MCT1 is expressed normally in OLGs and myelin from the CGT/CST KO's since a lactate transporter in spermatocytes in the CST KO is suspected to be absent due to altered trafficking (Honke 2013).

The myelin sheath also produces ATP through generation of a proton gradient across the lamellae, and has been postulated to be a site of oxygen absorption and aerobic metabolism for the axons (Ravera et al. 2009). Neuronal activity also results in release of glutamate and ATP causing shifts in extracellular pH (Ro and Carson 2004; Butt et al. 2004). A Na^+/H^+ exchanger, Na^+/HCO_3^- cotransporter and carbonic anhydrase II are present in OLGs that can respond to shifts in pH of the extracellular space during neuronal activity (Ro and Carson 2004). pH microdomains in OLGs occur which differ in pH by over 0.1 pH unit. Similar pH fluctuations may be able to occur in the cytosolic spaces of myelin if the permeability of tight and gap junctions can be influenced by GSL-mediated signaling.

It appears to be necessary to have compact myelin containing all of its normal constituents surrounding an axon in order to provide trophic support to neurons, since neurodegeneration occurs in mutant mice in which one of several myelin proteins, such as PLP, 2′,3′-cyclic nucleotide 3′-phosphodiesterase (CNP), or myelin-associated glycoprotein (MAG), is eliminated (Edgar and Garbern; 2004; Edgar et al. 2009). Neurodegeneration also occurs in the CST, CGT, and FA2H KO's, even though a compact myelin sheath is formed initially in all cases. Participation of transient GalC and SGC interactions between the apposed extracellular surfaces of mature myelin might allow for transmission of signals throughout the myelin sheath regulating its metabolic activity, and thus facilitate myelin–axonal communication and trophic support of the axon. In support of this conclusion, deletion of SGC in the CST-null mouse decreased the caliber and shape of the axon with age (Marcus et al. 2006), suggesting that the lack of SGC decreased signaling from myelin to the axon.

12.10 Treatment of Demyelinating Disease by Stimulation of OLGS by GSL Cross-Linking

Interestingly, IgM antibody to GalC or SGC, produced by hybridoma cells implanted in the spinal cord of neonatal pups in vivo, or in an in vitro myelinating culture, resulted in formation of myelin with a large space between the extracellular surfaces and paranodal loops, without tight junctions between them (Rosenbluth et al. 1996; Rosenbluth and Moon 2003). Thus, the multivalent IgM Ab replaced normal interactions between the extracellular surfaces. The abnormal myelin was stable enough to persist at least 18 days; the effect at longer times was not examined. A similar IgG Ab, however, which would not be able to link the apposed extracellular surfaces, prevented myelin formation. These results indicate that some type of interaction between the extracellular surfaces is necessary for myelin to form and function, but that the spacing can be wider than normal as long as the surfaces are linked via a molecular interaction. IgM Abs produced in peripheral neuropathies and in multiple sclerosis, including anti-GalC and anti-SGC Abs (Ilyas et al. 2003), cause formation of myelin with similar wide-spaced lamellae (Jacobs and Scadding 1990). The Ab linking of apposed myelin lamellae could mediate signals that could have both positive and negative effects on myelin maintenance and function.

Anti-GalC and anti-SGC Abs produced in these diseases may also have signaling effects on OLGs in vivo, as found in the in vitro studies. Indeed, IgM antibodies to these and other myelin constituents have been found to increase Ca^{2+} entry into OLGs, inhibit apoptotic signaling and OLG differentiation, and ameliorate demyelinating disease in animal models (Warrington et al. 2007; Asakura et al. 1998;

Paz Soldan et al. 2003; Watzlawik et al. 2010). IgM monoclonal antibodies O1 and O4 bind to GalC and SGalC (in lipid bilayers of phosphatidylcholine with 6 % SGC and 16 % GalC) with unusually small apparent dissociation constants (K_D=0.9 nM) for natural antibodies, although they have similar properties to natural antibodies (Wittenberg et al. 2012). Such IgM antibodies for OLG/myelin constituents are now in clinical trials to treat multiple sclerosis (Wootla et al. 2013). Similarly, multivalent arrays of Gal and SGal, such as the Gal/SGal-nanoparticles used here, could have potential beneficial effects on myelination or myelin function, a possibility that warrants further investigation.

Acknowledgements Studies from the Boggs laboratory were supported by operating grants from the Multiple Sclerosis Society of Canada. My collaborator for the glyco-nanoparticle studies, Dr. Amit Basu, Brown University, Providence, Rhode Island, is gratefully acknowledged. Ms. Huimin Wang, Dr. Wen Gao, Dr. Yuanfang Liu, Mr. Jingsha Zhao, Dr. Yukie Hirahara, and Dr. Hyun-Joo Park are thanked for their excellent assistance with these studies.

Compliance with Ethics Requirements Joan Boggs declares that she has no conflict of interest. This article made use of cultured oligodendrocytes obtained from rats. All institutional and national guidelines for the care and use of laboratory animals were followed. This article does not contain any studies with human subjects.

References

Alge-Priglinger CS, Andre S, Schoeffl H, Kampik A, Strauss RW, Kernt M, et al. Negative regulation of RPE cell attachment by carbohydrate-dependent cell surface binding of galectin-3 and inhibition of the ERK-MAPK pathway. Biochimie. 2011;93:477–88.

Arvanitis DN, Min W, Gong Y, Meng YM, Boggs JM. Two types of detergent-insoluble, glycosphingolipid/cholesterol-rich membrane domains in isolated myelin. J Neurochem. 2005;94:1696–710.

Asakura K, Miller DJ, Pease LR, Rodriguez M. Targeting of IgM kappa antibodies to oligodendrocytes promotes CNS remyelination. J Neurosci. 1998;18:7700–8.

Atkins CM, Sweatt JD. Reactive oxygen species mediate activity-dependent neuron-glia signaling in output fibers of the hippocampus. J Neurosci. 1999;19:7241–8.

Baba H, Akita H, Ishibashi T, Inoue Y, Nakahira K, Ikenaka K. Completion of myelin compaction, but not the attachment of oligodendroglial processes triggers K channel clustering. J Neurosci Res. 1999;58:752–64.

Bakhti M, Snaidero N, Schneider D, Aggarwal S, Mobius W, Janshoff A, et al. Loss of electrostatic cell-surface repulsion mediates myelin membrane adhesion and compaction in the central nervous system. Proc Natl Acad Sci U S A. 2013;110:3143–8.

Bansal R, Pfeiffer SE. Reversible inhibition of oligodendrocyte progenitor differentiation by a monoclonal antibody against surface galactolipids. Proc Natl Acad Sci U S A. 1989;86:6181–5.

Bansal R, Pfeiffer SE. Regulation of gene expression in mature oligodendrocytes by the specialized myelin-like membrane environment: antibody perturbation in culture with the monoclonal antibody R-mAb. Glia. 1994;12:173–9.

Bansal R, Winkler S, Bheddah S. Negative regulation of oligodendrocyte differentiation by galactosphingolipids. J Neurosci. 1999;19:7913–24.

Bauer NG, Richter-Landsberg C, Ffrench-Constant C. Role of the oligodendroglial cytoskeleton in differentiation and myelination. Glia. 2009;57:1691–705.

Baumgartner W, Schutz GJ, Wiegand J, Golenhofen N, Drenckhahn D. Cadherin function probed by laser tweezer and single molecule fluorescence in vascular endothelial cells. J Cell Sci. 2003;116:1001–11.
Ben-David O, Pewzner-Jung Y, Brenner O, Laviad EL, Kogot-Levin A, Weissberg I, et al. Encephalopathy caused by ablation of very long acyl chain ceramide synthesis may be largely due to reduced galactosylceramide levels. J Biol Chem. 2011;286:30022–33.
Benjamins JA, Dyer CA. Glycolipids and transmembrane signaling in oligodendroglia. Ann N Y Acad Sci. 1990;605:90–100.
Boggs JM. Lipid intermolecular hydrogen bonding: influence on structural organization and membrane function. Biochim Biophys Acta. 1987;906:353–404.
Boggs JM. Myelin basic protein: a multifunctional protein. Cell Mol Life Sci. 2006;63:1945–61.
Boggs JM, Wang H. Effect of liposomes containing cerebroside and cerebroside sulfate on cytoskeleton of cultured oligodendrocytes. J Neurosci Res. 2001;66:242–53.
Boggs JM, Wang H. Co-clustering of galactosylceramide and membrane proteins in oligodendrocyte membranes on interaction with polyvalent carbohydrate and prevention by an intact cytoskeleton. J Neurosci Res. 2004;76:342–55.
Boggs JM, Koshy KM, Rangaraj G. Effect of fatty acid chain length, fatty acid hydroxylation, and various cations on the phase behavior of synthetic forms of cerebroside sulfate. Chem Phys Lipids. 1984;36:65–89.
Boggs JM, Menikh A, Rangaraj G. Trans interaction between galactosylceramide and cerebroside sulfate across apposed bilayers. Biophys J. 2000;78:874–85.
Boggs JM, Wang H, Gao W, Arvanitis D, Gong Y, Min W. A glycosynapse in myelin? Glycoconj J. 2004;21:97–110.
Boggs JM, Gao W, Hirahara Y. Myelin glycosphingolipids, galactosylceramide and sulfatide, participate in carbohydrate-carbohydrate interactions between apposed membranes and may form glycosynapses between oligodendrocyte or myelin membranes. Biochim Biophys Acta. 2008a;1780:445–55.
Boggs JM, Gao W, Hirahara Y. Signal transduction pathways involved on interaction of galactosylceramide/sulfatide-containing liposomes with cultured oligodendrocytes and requirement for myelin basic protein and glycosphingolipids. J Neurosci Res. 2008b;86:48–1458.
Boggs JM, Homchaudhuri L, Rangaraj G, Liu YF, Smith GST, Harauz G. Interaction of myelin basic protein with cytoskeletal and signaling proteins in cultured primary oligodendrocytes and N19 oligodendroglial cells. BioMed Central, in press, 2004.
Boggs JM, Gao W, Zhao J, Park H-J, Liu Y, Basu A. Participation of galactosylceramide and sulfatide in glycosynapses between oligodendrocyte or myelin membranes. FEBS Lett. 2010; 584:1771–8.
Boison D, Stoffel W. Disruption of the compacted myelin sheath of axons of the central nervous system in proteolipid protein-deficient mice. Proc Natl Acad Sci U S A. 1994;91:11709–13.
Bosio A, Binczek E, Stoffel W. Functional breakdown of the lipid bilayer of the myelin membrane in central and peripheral nervous system by disrupted galactocerebroside synthesis. Proc Natl Acad Sci U S A. 1996;93:13280–5.
Bucior I. Carbohydrate-carbohydrate interaction provides adhesion force and specificity for cellular recognition. J Cell Biol. 2004;165:529–37.
Bucior I, Burger MM. Carbohydrate-carbohydrate interaction as a major force initiating cell-cell recognition. Glycoconj J. 2004;21:111–23.
Burgisser P, Matthieu J-M, Waehneldt TV. Myelin lipids: a phylogenetic study. Neurochem Res. 1986;11:1261–72.
Burgisser P, Althaus H-H, Rohmann A, Neuhoff V. Lipid synthesis by oligodendrocytes from adult pig brain maintained in long-term culture. Neurochem Int. 1988;13:111–8.
Butt AM, Pugh M, Hubbard P, James G. Functions of optic nerve glia: axoglial signalling in physiology and pathology. Eye. 2004;18:1110–21.
Chakraborty G, Drivas A, Ledeen R. The phosphoinositide signaling cycle in myelin requires cooperative interaction with the axon. Neurochem Res. 1999;24:249–54.

Coetzee T, Fujita N, Dupree J, Shi R, Blight A, Suzuki K, et al. Myelination in the absence of galactocerebroside and sulfatide: normal structure with abnormal function and regional instability. Cell. 1996;86:209–19.

Cook WJ, Bugg CE. Calcium-carbohydrate bridges composed of uncharged sugars. Structure of a hydrated calcium bromide complex of α-fucose. Biochim Biophys Acta. 1975;389:428–35.

DeBruin LS, Harauz G. White matter rafting—membrane microdomains in myelin. Neurochem Res. 2007;32:213–28.

Dick KJ, Eckhardt M, Paisan-Ruiz C, Alshehhi AA, Proukakis C, Sibtain NA, et al. Mutation of FA2H underlies a complicated form of hereditary spastic paraplegia (SPG35). Hum Mutat. 2010;31:E1251–60.

Dupree JL, Pomicter AD. Myelin, DIGs, and membrane rafts in the central nervous system. Prostaglandins Other Lipid Mediat. 2010;91:118–29.

Dyer CA. Novel oligodendrocyte transmembrane signaling systems. Mol Neurobiol. 1993;7:1–22.

Dyer CA. The structure and function of myelin: from inert membrane to perfusion pump. Neurochem Res. 2002;27:1279–92.

Dyer CA, Philibotte TM, Wolf MK, Billings-Gagliardi S. MBP mediates extracellular signals that regulate microtubule stability in oligodendrocyte membrane sheets. J Neurosci Res. 1994;39:97–107.

Edgar JM, Garbern J. The myelinated axon is dependent on the myelinating cell for support and maintenance: molecules involved. J Neurosci Res. 2004;76:593–8.

Edgar JM, McLaughlin M, Werner HB, McCulloch MC, Barrie JA, Brown A, et al. Early ultrastructural defects of axons and axon-glia junctions in mice lacking expression of *Cnp1*. Glia. 2009;57:1815–24.

Edvardson S, Hama H, Shaag A, Gomori JM, Berger I, Soffer D, et al. Mutations in the fatty acid 2-hydroxylase gene are associated with leukodystrophy with spastic paraparesis and dystonia. Am J Hum Genet. 2008;83:643–8.

Fewou SN, Bussow H, Schaeren-Wiemers N, Vanier MT, Macklin WB, Gieselmann V, et al. Reversal of non-hydroxy;alpha-hydroxy galactosylceramide ratio and unstable myelin in transgenic mice overexpressing UDP-galactose:ceramide galactosyltransferase. J Neurochem. 2005;94:469–81.

Fewou SN, Fernandes A, Stockdale K, Francone VP, Dupree JL, Rosenbluth JR, et al. Myelin protein composition is altered in mice lacking either sulfated or both sulfated and non-sulfated galactolipids. J Neurochem. 2010;112:599–610.

Fitzner D, Schneider A, Kippert A, Mobius W, Willig KI, Hell SW, et al. Myelin basic protein-dependent plasma membrane reorganization in the formation of myelin. EMBO J. 2006;25:1–12.

Fujita A, Cheng J, Fujimoto T. Segregation of GM1 and GM3 clusters in the cell membrane depends on the intact actin cytoskeleton. Biochim Biophys Acta. 2009;1791:388–96.

Fujiwara T, Ritchie K, Murakoshi H, Jacobson K, Kusumi A. Phospholipids undergo hop diffusion in compartmentalized cell membrane. J Cell Biol. 2002;157:1071–81.

Gielen E, Baron W, Vandeven M, Steels P, Hoekstra D, Ameloot M. Rafts in oligodendrocytes: evidence and structure-function relationship. Glia. 2006;54:499–512.

Gomez-Mouton C, Abad JL, Mira E, Lacalle RA, Gallardo E, Jimenez-Baranda S, et al. Segregation of leading edge and uropod components into specific lipid rafts during T cell polarization. Proc Natl Acad Sci U S A. 2001;98:9642–7.

Gorshkova TA, Mikshina PV, Gurjanov OP, Chemikosova SB. Formation of plant cell wall supramolecular structure. Biochemistry (Mosc). 2010;75:159–72.

Hakomori S. Carbohydrate-carbohydrate interaction as an initial step in cell recognition. Pure Appl Chem. 1991;63:473–82.

Hakomori S. The glycosynapse. Proc Natl Acad Sci U S A. 2002;99:225–32.

Hao S, August A. Actin depolymerization transduces the strength of B-cell receptor stimulation. Mol Biol Cell. 2005;16:2275–84.

Harauz G, Boggs JM. Myelin management by the 18.5-kDa and 21.5-kDa myelin basic protein isoforms. J Neurochem. 2013;125:334–61.

Harder T, Simons K. Clusters of glycolipid and glycosylphosphatidylinositol-anchored proteins in lymphoid cells: accumulation of actin regulated by local tyrosine phosphorylation. Eur J Immunol. 1999;29:556–62.

Hirahara Y, Bansal R, Honke K, Ikenaka K, Wada Y. Sulfatide is a negative regulator of oligodendrocyte differentiation: development in sulfatide-null mice. Glia. 2004;45:269–77.

Hollingshead CJ, Caspar DLD, Melchior V, Kirschner DA. Compaction and particle segregation in myelin membrane arrays. J Cell Biol. 1981;89:631–44.

Honke K. Biosynthesis and biological function of sulfoglycolipids. Proc Jpn Acad Ser B Phys Biol Sci. 2013;89:129–38.

Honke K, Hirahara Y, Dupree J, Suzuki K, Popko B, Fukushima K, et al. Paranodal junction formation and spermatogenesis require sulfoglycolipids. Proc Natl Acad Sci U S A. 2002;99: 4227–32.

Hoshi T, Suzuki A, Hayashi S, Tohyama K, Hayashi A, Yamaguchi Y, et al. Nodal protrusions, increased Schmidt-Lanterman incisures, and paranodal disorganization are characteristic features of sulfatide-deficient peripheral nerves. Glia. 2007;55:584–94.

Ilyas AA, Chen ZW, Cook SD. Antibodies to sulfatide in cerebrospinal fluid of patients with multiple sclerosis. J Neuroimmunol. 2003;139:76–80.

Inouye H, Kirschner DA. Membrane interactions in nerve myelin: II. Determination of surface charge from biochemical data. Biophys J. 1988;53:247–60.

Inouye H, Kirschner DA. Phylogenetic aspects of myelin structure. In: Jeserich G, Althaus HH, Waehneldt TV, editors. Cellular and molecular biology of myelination. Berlin: Springer; 1990. p. 376–87.

Ioannidou K, Anderson KI, Strachan D, Edgar JM, Barnett SC. Time-lapse imaging of the dynamics of CNS glial-axonal interactions in vitro and ex vivo. PLoS One. 2012;7:e30775.

Ishizuka I, Inomata M. Sulphated glycoglycerolipids in rat brain: decrease and disappearance after developmental age. J Neurochem. 1979;33:387–8.

Iwabuchi K, Handa K, Hakomori S. Separation of "glycosphingolipid signaling domain" from caveolin-containing membrane fraction in mouse melanoma B16 cells and its role in cell adhesion coupled with signaling. J Biol Chem. 1998;273:33766–73.

Jackman N, Ishii A, Bansal R. Oligodendrocyte development and myelin biogenesis: parsing out the roles of glycosphingolipids. Physiology. 2009;24:290–7.

Jacobs JM, Scadding JW. Morphological changes in IgM paraproteinemic neuropathy. Acta Neuropathol (Berl). 1990;80:77–84.

Karthigasan J, Kosaras B, Nguyen J, Kirschner DA. Protein and lipid composition of the radial component-enriched CNS myelin. J Neurochem. 1994;62:1203–13.

Khan AA, Bose C, Yam LS, Soloski MJ, Rupp F. Physiological regulation of the immunological synapse by agrin. Science. 2001;292:1681–6.

Kirby BB, Takada N, Latimer AJ, Shin J, Carney TJ, Kelsh RN, et al. In vivo time-lapse imaging shows dynamic oligodendrocyte progenitor behavior during zebrafish development. Nat Neurosci. 2006;9:1506–11.

Kirschner DA, Inouye H, Ganser AL, Mann V. Myelin membrane structure and composition correlated: a phylogenetic study. J Neurochem. 1989;53:1599–609.

Kishimoto Y. Phylogenetic development of myelin glycosphingolipids. Chem Phys Lipids. 1986;42:117–28.

Koshy KM, Boggs JM. Investigation of the calcium-mediated interaction between the carbohydrate head groups of galactosylceramide and galactosylceramide I^3 sulfate by electrospray ionization mass spectrometry. J Biol Chem. 1996;271:3496–9.

Koshy KM, Wang J, Boggs JM. Divalent-cation mediated interaction between cerebroside sulfate and cerebrosides: an investigation of the effect of structural variations of lipids by electrospray ionization mass spectrometry. Biophys J. 1999;77:306–18.

Kulkarni K, Snyder DS, McIntosh TJ. Adhesion between cerebroside bilayers. Biochemistry. 1999;38:15264–71.

Kusumi A, Suzuki K, Koyasako K. Mobility and cytoskeletal interactions of cell adhesion receptors. Curr Opin Cell Biol. 1999;11:582–90.

Kusumi A, Fujiwara TK, Morone N, Yoshida KJ, Chadda R, Xie M, et al. Membrane mechanisms for signal transduction: the coupling of the meso-scale raft domains to membrane-skeleton-induced compartments and dynamic protein complexes. Stem Cell Dev Biol. 2012;23:126–44.
Lajoie P, Goetz JG, Dennis JW, Nabi IR. Lattices, rafts, and scaffolds: domain regulation of receptor signaling at the plasma membrane. J Cell Biol. 2009;185:381–5.
Lee Y, Morrison BM, Li Y, Lengacher S, Farah MH, Hoffman PN, et al. Oligodendroglia metabolically support axons and contribute to neurodegeneration. Nature. 2012;487:443–8.
Li C, Tropak MB, Gerlai R, Clapoff S, Abramow-Newerly W, Trapp B, et al. Disruption of the compacted myelin sheath of axons of the central nervous system in proteolipid protein-deficient mice. Proc Natl Acad Sci U S A. 1994;91:11709–13.
Marcus J, Dupree JL, Popko B. Myelin-associated glycoprotein and myelin galactolipids stabilize developing axo-glial interactions. J Cell Biol. 2002;156:567–77.
Marcus J, Honigbaum S, Shroff S, Honke K, Rosenbluth J, Dupree JL. Sulfatide is essential for the maintenance of CNS myelin and axon structure. Glia. 2006;53:372–81.
Marone N, Fujiwara T, Murase K, Kasai RS, Ike H, Yuasa S, et al. Three-dimensional reconstruction of the membrane skeleton at the plasma membrane interface by electron tomography. J Cell Biol. 2006;174:851–62.
Marta CB, Taylor CM, Coetzee T, Kim T, Winkler S, Bansal R, et al. Antibody cross-linking of myelin oligodendrocyte glycoprotein leads to its rapid repartitioning into detergent-insoluble fractions, and altered protein phosphorylation and cell morphology. J Neurosci. 2003;23: 5461–71.
Meixner M, Jungnickel J, Grothe C, Gieselmann V, Eckhardt M. Myelination in the absence of UDP-galactose:ceramide galactosyl-transferase and fatty acid 2-hydroxylase. BMC Neurosci. 2011;12:22.
Micu I, Jiang Q, Coderre E, Ridsdale A, Zhang L, Woulfe J, et al. NMDA receptors mediate calcium accumulation in myelin during chemical ischaemia. Nature. 2005;439:988–92.
Mitchell JS, Brown WS, Woodside DG, Vanderslice P, McIntrye BW. Clustering T cell GM1 lipid rafts increases cellular resistance to shear on fibronectin through changes in integrin affinity and cytoskeletal dynamics. Immunol Cell Biol. 2009;87:324–36.
Morita K, Sasaki H, Fujimoto K, Furuse M, Tsukita S. Claudin-11/OSP-based tight junctions of myelin sheaths in brain and Sertoli cells in testis. J Cell Biol. 1999;145:579–88.
Murray N, Steck AJ. Impulse conduction regulates myelin basic protein phosphorylation in rat optic nerve. J Neurochem. 1984;43:243–8.
Nave K-A, Trapp BD. Axon-glial signaling and the glial support of axon function. Annu Rev Neurosci. 2008;31:535–61.
Niles WD, Silvius JR, Cohen FS. Resonance energy transfer imaging of phospholipid vesicle interaction with a planar phospholipid membrane. J Gen Physiol. 1996;107:329–51.
Norton WT. Isolation and characterization of myelin. In: Morell P, editor. Myelin. New York: Plenum; 1977. p. 161–99.
Okamura N, Stoskopf M, Yamaguchi H, Kishimoto Y. Lipid composition of the nervous system of earthworms (Lumbricus terrestris). J Neurochem. 1985;45:1875–9.
Paz Soldan MM, Warrington AE, Bieber AJ, Ciric B, van Keulen V, Pease LR, et al. Remyelination-promoting antibodies activate distinct Ca^{2+} influx pathways in astrocytes and oligodendrocytes: relationship to the mechanism of myelin repair. Mol Cell Neurosci. 2003;22:14–24.
Pieringer J, Rao GS, Mandel P, Pieringer RA. The association of the sulphogalactosylglycerolipid of rat brain with myelination. Biochem J. 1977;166:421–8.
Potter KA, Kern MJ, Fullbright G, Bielawski J, Scherer SS, Yum SW, et al. Central nervous system dysfunction in a mouse model of FA2H deficiency. Glia. 2011;59:1009–21.
Prinetti A, Iwabuchi K, Hakomori S. Glycosphingolipid-enriched signaling domain in mouse neuroblastoma neuro2a cells. J Biol Chem. 1999;274:20916–24.
Ranscht B, Wood PM, Bates M, Bunge RP. Role of galactocerebroside in the formation of peripheral myelin. In: Althaus HH, Seifert W, editors. Glial-neuronal communication in development and regeneration. New York: Springer; 1987. p. 666–81.

Ravera S, Panfoli I, Calzia D, Aluigi MG, Bianchini P, Diaspro A, et al. Evidence for aerobic ATP synthesis in isolated myelin vesicles. Int J Biochem Cell Biol. 2009;41:1581–91.

Readhead C, Takasashi N, Shine HD, Saavedra R, Sidman R, Hood L. Role of myelin basic protein in the formation of central nervous system myelin. Ann N Y Acad Sci. 1990;605:280–5.

Remahl S, Hildebrand C. Relations between axons and oligodendroglial cells during initial myelination. II. The individual axon. J Neurocytol. 1990;19:883–98.

Rinholm JE, Hamilton NB, Kessaris N, Richardson WD, Bergersen LH, Attwell D. Regulation of oligodendrocyte development and myelination by glucose and lactate. J Neurosci. 2011;31: 538–48.

Ro H, Carson JH. pH microdomains in oligodendrocytes. J Biol Chem. 2004;279:37115–23.

Roder J. Myelination in the absence of myelin-associated glycoprotein. Nature. 1994;369: 747–50.

Rodgers W, Zavzavadjian J. Glycolipid-enriched membrane domains are assembled into membrane patches by associating with the actin cytoskeleton. Exp Cell Res. 2001;267:173–83.

Rosenbluth J, Moon D. Dysmyelination induced in vitro by IgM antisulfatide and antigalactocerebroside monoclonal antibodies. J Neurosci Res. 2003;71:104–9.

Rosenbluth J, Liang W-L, Liu Z, Guo D, Schiff R. Expanded CNS myelin sheaths formed *in situ* in the presence of an IgM antigalactocerebroside-producing hybridoma. J Neurosci. 1996; 16:2635–41.

Rosenbluth J, Schiff R, Lam P. Effects of osmolality on PLP-null myelin structure: implications re axon damage. Brain Res. 2009;1253:191–7.

Saadat L, Dupree JL, Kilkus J, Han X, Traka M, Proia RL, et al. Absence of oligodendroglial glucosylceramide synthesis does not result in CNS myelin abnormalities or alter the dysmyelinating phenotype of CGT-deficient mice. Glia. 2010;58:391–8.

Seah N, Santacroce PV, Basu A. Probing the lactose·GM3 carbohydrate-carbohydrate interaction with glycodendrimers. Org Lett. 2009;11:559–62.

Selivonchick DP, Roots BI. Variation in myelin lipid composition induced by change in environmental temperature of goldfish (Carassius aurattus L.). J Therm Biol. 1976;1:131–5.

Shimomura K, Kishimoto Y. An improved procedure for the quantitative determination and characterization of sulfatides in rat kidney and brain by high performance liquid chromatography. Biochim Biophys Acta. 1983;754:93–100.

Shimomura K, Kishimoto Y. Changes in monogalactosyl diacylglycerols, alkylgalactolipids, and cerebroside fatty acid esters in maturing brain measured by high-performance liquid chromatography. Biochim Biophys Acta. 1984;794:162–4.

Shroff SM, Pomicter AD, Chow WN, Fox MA, Colello RJ, Henderson SC, et al. Adult CST-null mice maintain an increased number of oligodendrocytes. J Neurosci Res. 2009;87:3403–14.

Simons M, Lyons DA. Axonal selection and myelin sheath generation in the central nervous system. Curr Opin Cell Biol. 2013;25:1–8.

Simons M, Kramer EM, Thiele C, Stoffel W, Trotter J. Assembly of myelin by association of proteolipid protein with cholesterol- and galactosylceramide-rich membrane domains. J Cell Biol. 2000;151:143–54.

Smith GST, Homchaudhuri L, Boggs JM, Harauz G. Classic 18.5- and 21.5-kDa myelin basic protein isoforms associate with cytoskeletal and SH3-domain proteins in the immortalized N19-oligodendroglial cell line stimulated by phorbol ester and IGF-1. Neurochem Res. 2012;37:77–1295.

Sonnino S, Prinetti A, Nakayama H, Yangida M, Ogawa H, Iwabuchi K. Role of very long fatty acid-containing glycosphingolipids in membrane organization and cell signaling: the model of lactosylceramide in neutrophils. Glycoconj J. 2009;26:615–21.

Stewart RJ, Boggs JM. The carbohydrate-carbohydrate interaction between galactosylceramide-containing liposomes and cerebroside sulfate-containing liposomes: dependence on the glycolipid ceramide composition. Biochemistry. 1993;32:10666–74.

Taguchi K, Yoshinaka K, Yoshino K, Yonezawa K, Maekawa S. Biochemical and morphologic evidence of the interaction of oligodendrocyte membrane rafts with actin filaments. J Neurosci Res. 2005;81:218–25.

Tamai Y, Kojima H, Saito S, Takayama-Abe K, Horichi H. Characteristic distribution of glycolipids in gadoid fish nerve tissues and its bearing on phylogeny. J Lipid Res. 1992;33:1351–9.

Treanor B, Depoil D, Gonzalez-Granja A, Barral P, Weber M, Dushek O, et al. The membrane skeleton controls diffusion dynamics and signaling through the B cell receptor. Immunity. 2010;32:187–99.

Treanor B, Depoil D, Bruckbauer A, Batista FD. Dynamic cortical actin remodeling by ERM proteins controls BCR microcluster organization and integrity. J Exp Med. 2011;208:1055–68.

van Itallie CM, Anderson JM. The molecular physiology of tight junction pores. Physiology. 2004;19:331–8.

Vogler EA. Structure and reactivity of water at biomaterial surfaces. Adv Colloid Interface Sci. 1998;74:69–117.

Vogt AB, Spindeldreher S, Kropshofer H. Clustering of MHC-peptide complexes prior to their engagement in the immunological synapse: lipid raft and tetraspan microdomains. Immunol Rev. 2002;189:135–51.

Wang J, Chen H, Brown EJ. L-plastin peptide activation of $\alpha v\beta 3$ mediated adhesion requires integrin conformational change and actin filament disassembly. J Biol Chem. 2001;276:14474–81.

Warrington AE, Bieber AJ, Ciric B, Pease LR, Van Keulen V, Rodriguez M. A recombinant human IgM promotes myelin repair after a single, very low dose. J Neurosci Res. 2007;85:967–76.

Watzlawik J, Holicky E, Edberg DD, Marks DL, Warrington AE, Wright BR, et al. Human remyelination promoting antibody inhibits apoptotic signaling and differentiation through Lyn kinase in primary rat oligodendrocytes. Glia. 2010;58:1782–93.

Webb MS, Tilcock CPS, Green BR. Salt-mediated interactions between vesicles of the thylakoid lipid digalactosyldiacylglycerol. Biochim Biophys Acta. 1988;938:323–33.

Winzeler AM, Mandemakers WJ, Sun MZ, Stafford M, Phillips CT, Barres BA. The lipid sulfatide is a novel myelin-associated inhibitor of CNS axon outgrowth. J Neurosci. 2011;31:6481–92.

Witt A, Brady ST. Unwrapping new layers of complexity in axon/glial relationships. Glia. 2000;29:112–7.

Wittenberg NJ, Im H, Xu X, Wootla B, Watzlawik J, Warrington AE, et al. High affinity binding of remyelinating natural autoantibodies to myelin-mimicking lipid bilayers revealed by nanohole surface Plasmon resonance. Anal Chem. 2012;84:6031–9.

Wootla B, Watzlawik JO, Denic A, Rodriguez M. The road to remyelination in demyelinating diseases: current status and prospects for clinical treatment. Expert Rev Clin Immunol. 2013;9:535–49.

Xu X, Warrington AE, Wright BR, Bieber AJ, van Keulen V, Pease LR, et al. A human IgM signals axon outgrowth: coupling lipid raft to microtubules. J Neurochem. 2011;119:100–12.

Zhang H, Miller RH. Density-dependent feedback inhibition of oligodendrocyte precursor expansion. J Neurosci. 1996;16:6886–95.

Zhao J, Liu Y, Park H-J, Boggs JM, Basu A. Carbohydrate-coated fluorescent silica nanoparticles as probes for the galactose/3-sulfogalactose carbohydrate-carbohydrate interaction using model systems and cellular binding studies. Bioconjug Chem. 2012;23:1166–73.

Zoller I, Meixner M, Hartmann D, Bussow H, Meyer R, Gieselmann V, et al. Absence of 2-hydroxylated sphingolipids is compatible with normal neural development but causes late-onset axon and myelin sheath degeneration. J Neurosci. 2008;28:9741–54.

Chapter 13
Glycosignaling: A General Review

Glyn Dawson

Abstract The concept of glycosignaling, in which neural cell-surface glycoconjugates form microdomains (Lipid Rafts) to facilitate the recruitment of signaling molecule components to form a transient signaling unit, is helping us understand the reason for glycoheterogeneity in the brain and is leadings to important translational efforts in medicine. In this review we first describe the origins of the concept of glycomicrodomains, how lipid heterogeneity might have relevance for the brain development, pathology and how the glycocalyx acts as a barrier in glia. After a discussion of how such microdomains are isolated and studied using modern technology such as nanoparticle labeling and molecular microscopy, we will present examples of how glycosignaling can function in such brain-specific situations as axonal growth and protein phosphorylation-mediated signaling.

Keywords Glycosphingolipids • Gangliosides • Proteoglycans • Glycoproteins • Microdomains • Lipid rafts • Neural stem cells • Quantum dots • Molecular microscopy • Protein phosphorylation

13.1 Introduction: Definition of "Glycosignaling"

The major membrane glycosphingolipids of brain contain typically up to four neutral sugars and up to five sialic acid residues within the ganglio series (GalNAcβ1–4Gal) and only trace amounts of the different core globo-series (Galα1–4-Gal), the neolactose series (Galβ1–4GlcNAc), or the lacto series (Galβ1–3GlcNAcβ1–4Gal) glycolipids (Hakomori 2008). Quantitatively dominant are the sulfated and

G. Dawson (✉)
Department of Pediatrics, MC4068, University of Chicago,
5841, S. Maryland Ave, Chicago, IL 60637, USA
e-mail: dawg@uchicago.edu

R.K. Yu and C.-L. Schengrund (eds.), *Glycobiology of the Nervous System*,
Advances in Neurobiology 9, DOI 10.1007/978-1-4939-1154-7_13,

non-sulfated galactosylceramides associated with the myelin sheath. Of the other glycoconjugates in brain, proteoglycans contain extensively sulfated uronic acid-GlcNAc/GalNAc repeating polymers and glycoproteins have extensively branched chains of oligosaccharides, which uniquely contain mannose and sialic acids. These are discussed extensively in other chapters. All these molecules appear to be involved in many types of cell recognition and signaling within the brain and there is increasing evidence for the importance of "glycosignaling" in all aspects of brain function.

The term "glycosignaling" was introduced by Hakomori et al. (1998) on the basis of their studies, mostly on non-CNS tissue. They described an assembly of glycosphingolipids and signal transducer molecules, which formed what they termed a "*glycosignaling domain*" (*GSD*). They ascribed to this GSD a functional role in cell interaction/adhesion, which involved second messenger signaling. An original difference from previously described sphingolipid-rich microdomains (Lipid rafts) was the absence of cholesterol and glycoprotein clusters from the GSDs. The GSL clustering could be imaged on EM using gold sol coated anti-polysialoganglioside antibody and a clustering of GSLs above a critical threshold was necessary for both antibody binding and cell adhesion. The glycosignaling concept has since been expanded to include glycoproteins and proteoglycans and the glyco-modification of lipids and proteins and many of these processes are associated with more conventional "Lipid Rafts" (Hakomori et al. 1998; Lingwood et al. 2008). Since isolated Lipid Rafts (LRs) also contain the elements of many G-protein signaling systems, there is also considerable interest in how the glycosignaling through lipid rafts (LRs) could influence the many signaling pathways critical for the functioning of the brain.

The original ideas of how glycosignaling systems might function in brain were laid out in a review by Saul Roseman (1970) which was based on many years of studying glycosyltransferases in developing chick brains. In this model, cell-surface glycosyltransferases interacted with cell surface glycolipid substrates to facilitate intercellular adhesion and recognition. This concept was refined by Hakomori/Prinetti (Hakomori 2008) whereby stimulation of GSDs by their ligands (complementary GSLs, cognins, antibodies, etc.) induces conformational changes in transducer molecules which activate positive or negative signals and then regulate such things as cell adhesion. They envisioned transducer molecules to include Src, and Ras kinase family proteins and Rho GTPases, all of which are active in brain. Evidence came from a mouse melanoma B16 cell which displays GM3-dependent adhesion to plates coated with Gg3 or anti-GM3 antibody. This resulted in enhanced c-Src phosphorylation, but there was no response when GM3 was absent or replaced with other gangliosides such as GM1 or GD1a, or asialo-glycolipids such as LacCer. From such data they concluded (Hakomori 2008) that many transducer molecules controlled cell motility and proliferation in response to various glyco-mediated stimuli. Thus, glycolipid-enriched microdomains are believed to be functionally involved in initiating or inhibiting signal transduction through interaction of their carbohydrate moiety with their ligands, hence the term "glycosignaling domain."

Mounting evidence suggests that glycosignaling operates in the heterogeneous mammalian CNS as it arises from undifferentiated neural stem cells (NSCs), with their high proliferative potential and capacity for self-renewal and multipotency (Yu and Yanagisawa 2007). Neuronal and Glial cells certainly express a sufficiently complex glycocalyx (made up from membrane-associated glycolipids, and membrane and extracellular glycoproteins and proteoglycans) to be critical participants in recognition, patterning, barrier formation and in dynamic signaling systems (Parker and Kohler 2009). A combination of a negatively charged polymer and a cluster of glycosphingolipids and glycoproteins plus heparan sulfates or a posttranslational modification to a protein could have a critical role in using glycosignaling to drive CNS development, function, and regeneration in either a positive or negative direction.

13.2 Isolation of Glycosignaling Complexes from Brain

Glycosphingolipids (GSLs) are enriched in membranes immersed in an insoluble matrix and glycosignaling complexes can be isolated by virtue of their insolubility in detergents such as 1 % Triton X-100 at 4 °C. The terms "detergent-insoluble material" (DIM) and "detergent-insoluble substrate attachment matrix" (DISAM) were originally applied to designate a special membrane compartment enriched in GSLs (Hakomori et al. 1998; Lingwood et al. 2008) but we typically now refer to all such preparations as a "Lipid Raft" (LR). For example (Dawson et al. 2012), one can extract 50 mg fresh weight of mouse cerebral hemisphere with 2 ml of 1 % Triton X-100/MES buffer–NaCI (pH 6.5) using 50 strokes of a Dounce homogenizer. Following centrifugation at 700×*g* to remove insoluble material such as myelin membranes, the extract is mixed with 2 ml of 1 % Triton X-100 in 80 % sucrose/MES buffer and placed in an ultracentrifugation tube. Five milliliter of 30 % sucrose/MES is layered on top of this fraction, followed by 3 ml of 5 % sucrose/MES and the samples ultracentrifuged at 39,000×*g* for 17 h. Typically, an opalescent band is seen in Fraction 4 and this band contains LR protein markers such as Flotillin-1 (Dawson et al. 2012) as well as almost all the sphingolipids. Sphingolipid modifying enzymes and other glycoproteins such a G-protein-coupled receptors may occur in this Fraction 4 or may translocate to fraction 4 after specific stimulation of the cell, for example with neurotransmitters. In brain, virtually all the gangliosides are present in the LR fraction, including those found only in fetal or pathological tissue such as GM3, GM2, GD3, and GD2. GM1 ganglioside is the cholera toxin receptor in brain but is also present in non-neural cells in a similar detergent-insoluble fraction along with virtually all other glycolipids. Lipid rafts can be visualized by transfecting cells (for example a human oligodendroglioma cell line HOG with green fluorescent protein-tagged GPI-anchored protein sequence which always localizes to Lipid Rafts (Fig. 13.1a, b)). The detergent-insolubility (and presence in insoluble LRs) of GM3 or GM1 in cells is usually interpreted to mean that these gangliosides are complexed with a cell adhesion system or

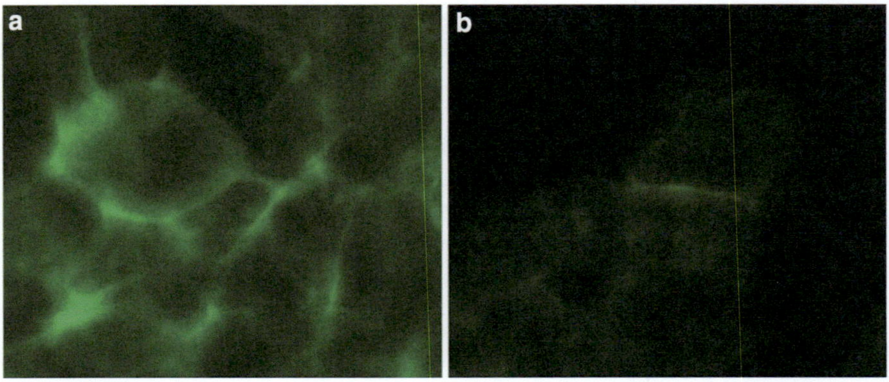

Fig. 13.1 Visualization of glycomicrodomains. GPI anchored proteins are exclusively localized to sphingolipid-enriched microdomains (LRs) and are involved in glycosignaling, for example the OMPR protein involved in forming a complex with gangliosides in regulating axonal growth (Saha et al. 2011) (as shown in Fig. 13.3). Fluorescent images were obtained after expression of LR-associated green fluorescent protein (GFP)-labeled GPI-anchored protein cDNA in neurotumor-derived HOG cells. These cells show increased concentration of GFP expression in plasma membranes (**a**), especially at adjacent cell membranes (**b**). These concentrations of GFP could represent aggregates of LRs corresponding to glycomicrodomains involved in cell adhesion and signaling

cytoskeletal system, since gangliosides are all soluble in aqueous detergent solutions such as Triton X-100.

Further evidence for LR-based glycosignaling in the brain comes from a study (Sekino-Suzuki et al. 2012) which showed that both R24 anti-GD3 antibody, and an antibody (GGR12), that specifically recognizes GD1b ganglioside, were able to induce Lyn tyrosine kinase activation in rat primary cerebellar granule cells. Immunoprecipitation studies on similar Triton X-100 LRs from these cells showed a complex containing Lyn, the transmembrane phosphoprotein Cbp and the physiological regulator of Src family kinases, Csk. The interaction of Cbp and Csk required the phosphorylation of Cbp at Tyr-314 by activated Lyn and they proposed (Sekino-Suzuki et al. 2012) that phosphorylated Cpb negatively regulated Lyn through the recruitment of Csk into LRs. Since both GD3 and GD1b contain the same Neu5Ac-(2–8)-Neu5Ac-(2–3)-Gal sequence and trigger the same response in neurons they suggest that this is the critical glycosignaling element in forming or stabilizing the LR complex. As discussed by Sonnino et al. (2013) it remains to be seen if these studies in cultured neurons have physiological significance in the brain.

13.3 Glycosignaling, Biosynthesis, and Brain Development

Notch signaling is critical for brain development and Notch is only active when glycosylated by an O-fucosylation step and the addition of GlcNAc to specific serine residues (Parker and Kohler 2009). Such glyco-posttranslational modifications of proteins are common and are likely initiators of glycosignaling in brain development and function since their presence is essential for recognition and this recognition triggers biological events. Regulation of such glycosylation (and therefore regulation of glycosignaling) can also occur at the level of nucleotide sugar availability, glycosyltransferase activation, or secretion and is therefore both complex and dynamic.

For example, gangliosides are attached through ceramides to LR membranes but free ceramides inhibit protein kinase B (Akt) through activation of a phosphatase. When ceramide is low (as in the *Fro/Fro* mouse and in many tumors) (Qin et al. 2012), Akt is activated (phosphorylated) together with the mTOR, p70D6k pathway to upregulate genes such as HAS2, which cause synthesis and secretion of the extracellular matrix proteoglycan, hyaluronic acid (Qin et al. 2012). This results in accelerated metastasis of a tumor if localized or impaired bone and brain development if occurring in all tissues, as in the *Fro/Fro* mouse (Qin et al. 2012). Additional control of such glycosignaling can be exerted by means of the many secreted and cell surface glycosidases, sulfatases, phosphatases, etc. (Aureli et al. 2012) acting on gangliosides and other cell surface glycoconjugates. Such modifications can also be shown to affect many critical developmental signaling systems such as the Wnt, EGF, and FGF pathways (Parker and Kohler 2009). Thus, glycosignaling must be thought of as an ongoing dynamic process, which involves many glycoconjugates and can impinge on many brain functions.

13.4 The Glycocalyx as a Barrier to Glycosignaling

The glycocalyx is a negatively charged barrier in brain because of the high content of lipid- and protein-linked sialic acids, sulfated lipids, and sulfated sugar and uronic acid-containing polymers (Parker and Kohler 2009; Sonnino et al. 2012) surrounding the cells. The glycocalyx is extensively produced by glial cells in the form of chondroitin and heparan sulfates and ceramide-galactosylsulfates, but neurons also produce lesser amounts of specific types of chondroitin sulfates. Both neurons and glia produce heparan sulfates, which are extensively involved in growth regulation via glycosignaling complexes with the EGF receptor. In contrast, gangliosides are synthesized by neurons. This negative surface charge difference between neurons and glia is most dramatically demonstrated by exposing brain slices containing differentiating, integrated brain cell types to intrinsically highly fluorescent nanoparticles. These consist of a CdSe/ZnS core/shell luminescent semiconductor or “quantum dot” (QD) approximately 6–10 nm in size (the same size as a typical lysosomal hydrolase) which are coated with a stabilizing/solubilizing polymer through dihydrolipoic acid (DHLA) residues (Fig. 13.2a). This polymer can be neutral (such as polyethyleneglycol) or negatively charged (CL4) (Walters et al. 2012; Boeneman et al. 2013). The Zn on the surface of the QDs is able to bind to histidine residues (H_6) incorporated into a peptide or lipopeptide cargo such as WGDap(Palmitoyl) VKIKKP$_9$GGH$_6$ (palm-1) or any His$_6$-tagged protein. The original GDap(Palmitoyl)

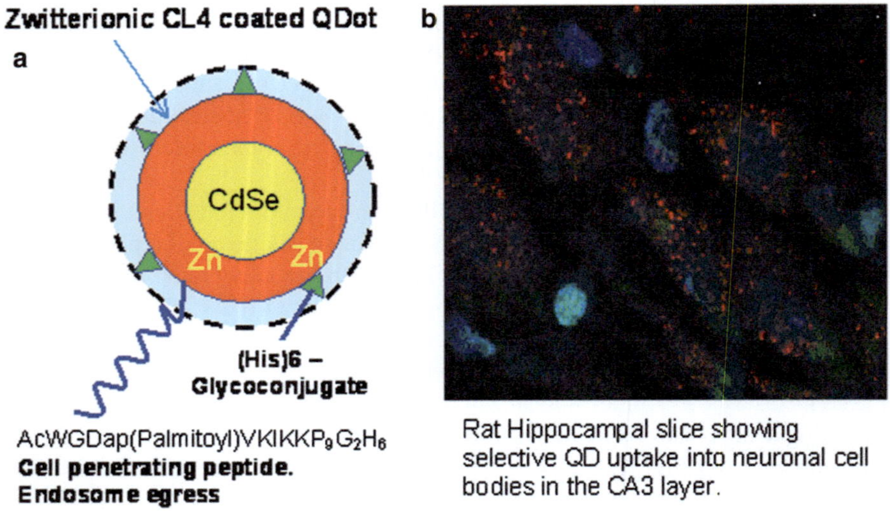

Fig. 13.2 Structure of Quantum dots showing site of attachment of His6-tagged peptides and proteins. (**a**) Quantum dot-Palm1 structure. This is a representation of a Cd/SeZnS Quantum dot coated with a dihydrolipoic acid (DHLA)-negatively charged (Zwitterionic) molecule (compact ligand 4 (CL4)). Membrane penetrating peptides such as $AcWDapVKIKKP_9G_2H_6$ or His_6-protein or His_6-glycoconjugate bind tightly to the Zn coat. These intensely fluorescent particles are 10–15 nm in diameter and can deliver various types of glycosignaling-modifying compounds to facilitate correction of brain mis-development or injury. (**b**) Confocal microscopic image showing QD delivery to hippocampal neurons. Rat hippocampal brain slices take up the highly fluorescent (625 nm) coated Quantum dots plus peptides initially into endosomes and then distribute it throughout the neuron. *Blue-staining* is DAPI (nor the intensely staining astrocyte nuclei with no QDs), *green staining* is Nissl body, neuron-specific, and *red staining* is Quantum dot–lipopeptide complexes

VKIKK sequence was based on the region of Ras-4B which is dynamically palmitoylated in vivo (Walters et al. 2012; Boeneman et al. 2013). The coating on the QD provides colloidal stability and Palm-1 uniquely allows the QDs to be taken up by cultured cells and readily exit the endosome into the soma (Walters et al. 2012; Boeneman et al. 2013). Electron microscopic images confirmed the endosomal egress only occurred in the presence of the Palm1 peptide, showing a much more dispersed cytosolic distribution of the CL4 QDs conjugated to Palm1 compared with CL4 QDs alone. Replacing a neutral polyampholyte coating (PEG) with a negatively heterocharged compact ligand, ("CL4"), was able to specifically target the palmitoylated peptide (palm-1) to neurons in the developing rat hippocampal brain slice cultures (Fig. 13.2b). There was no noticeable uptake of such Quantum dots by astrocytes, oligodendrocytes, and very little by microglia (identified by immunocytochemistry (Walters et al. 2012)). However, pretreatment of the cultures with chondroitinase ABC to reduce the surface negative charge, promoted uptake into oligodendrocytes. Thus, although neurons receive nutrients and signals by retrograde axonal transport they respond to external negatively charged glycosignals (sulfation, sialic acids, etc.), whereas glial cells are more resistant.

13.5 Glycosignaling Heterogeneity in Specific Brain Regions

Because glycoconjugates exist as both linear and branched structures and hydroxyl group modifications such as sulfation and acylation are common, the number of possible targets for glyco-binding proteins exceeds 7,000 (Lingwood et al. 2010). The importance of glycolipids to the CNS is amply demonstrated by the ability of bacteria and viruses to gain entry to cells by first binding to a specific carbohydrate sequence. Advances in technology have increased our understanding of how such glyco-recognition can be finely tuned. For example a ganglioside such as GD1a can be O-acetylated and contain either or both *N*-acetyl- and *N*-glycolyl-neuraminic acids and this affects where the GD1a is localized within the brain (Colsch et al. 2011) and which glycosignaling system it modifies.

A recent study using matrix-assisted laser desorption/ionization (MALDI) mass spectrometry (Colsch et al. 2011) showed that the qualitative localization of ganglioside species in the rat hippocampus depended both on the degree of glycosylation and type of sialylation (e.g., O-acetylation), and also the ceramide structure as defined by the presence of either sphingoid bases C18:1 or C20:1. Previous immunostaining with antibodies, lectins, or toxins to detect ganglioside molecules recognized the headgroup, but imaging mass spectrometry is able to recognize all components of a molecule, both oligosaccharide as well as ceramide core. The information obtained by molecular fragmentation produces "molecular microscopy" (Colsch et al. 2011) and enabled insights into modifications critical for glycosignaling. Thus, in the rat hippocampus, the molecular layer of the dentate gyrus (ML), is made up of three distinct layers, the inner molecular layer (IML), which contains gangliosides with C18-sphingosine exclusively, the middle molecular layer (MML), and the outer molecular layer (OML) where C20:1-sphingosine is the only sphingosine base in the gangliosides.

Intriguingly, the gangliosides which occur in embryonic development and reoccur in brain pathological conditions (GM3, GM2, GD3, and GD2) all contain predominantly C18-sphingosine and are predominantly present in cell layers made up of the pyramidal cell layer (Py) and the granular layer of the dentate gyrus (GL). In contrast, when the major mature brain gangliosides (GM1, GD1, and GT1) and GQ1s were mapped they were found to be predominantly in the substantia nigra, cerebral peduncle, hippocampus, and midbrain. However, C20:1 sphingosine GM1 was minimally present in the corpus callosum and midbrain and was mostly found in the hippocampus and the substantia nigra, Minor gangliosides GQ1 and the O-acetylated forms of GD1, GT1, and GQ1 gangliosides, contained both C18- and C20-sphingosines and their distribution was based on the degree of sialylation and acetylation of the oligosaccharide chains in the neuronal cell bodies. These are potentially exciting findings for glycosignaling (Colsch et al. 2011) since they suggest that the ceramide part of the glycosphingolipid may influence glycosignaling in the brain.

13.6 Evidence That Fatty Acids and Sphingosine Base Heterogeneity Can Affect Glycosignaling: Toxins

Although most glycolipids are typically firmly attached to membrane LR complexes (an exception being the "shedding" observed in cancer cells) the reactivity of the GSL sugars with binding proteins and antibodies does appear to be somewhat dependent upon the fatty acid composition. Most GSLs show extensive fatty acid heterogeneity (C16–24) with highly antigenic lipids such as galactosylceramides and sulfatides containing a wide range of fatty acids, about 50 % of which are alpha-hydroxylated and/or monounsaturated. This affects their reactivity, possibly because it affects their mobility in the plasma membrane lipid Raft (Lingwood 1999). A marked exception to this fatty acid heterogeneity rule are brain gangliosides which are almost exclusively C18:0 but do contain a mixture of C18 and C20 sphingoid bases. Gangliosides are very minor components of non-neural tissue and when they do occur, typically in cancerous cells, they show fatty acid heterogeneity typical of non-CNS glycolipids. So there is something unique to the fatty acid behavior of brain gangliosides, which affects the structure of microdomains (LRs) and most probably relates to some glycosignaling function. For example, fatty acid heterogeneity has been clearly shown to regulate the binding of toxins such as veratoxin to neutral glycolipids (GbOse3cer) (Lingwood 1999), so this phenomenon needs to be better investigated in brain.

13.7 Glycosignaling, Rho-GTPase, and Axonal Growth

The GalNAc transferase knockout mouse with absence of the major gangliosides appeared to have very little phenotype until Schnaar discovered that recovery from nerve injury was significantly improved and went on to show that sialidase treatment, which converted complex gangliosides to GM1, improved recovery from spinal cord injury (Mountney et al. 2010). Spinal cord injury typically results in lifelong loss of nerve function and morbidity (Barritt et al. 2006) and recent reports suggest that glycosignaling errors could contribute to this loss of function (Mountney et al. 2013). Studies have implicated the glycocalyx in inhibition of nerve injury recovery and both digestion of chondroitin sulfates and sialo-oligosaccharides has been of some therapeutic value in rat models of SC injury. Recently, Schnaar and associates (Mountney et al. 2013) showed that delivery of bacterial chondroitinase ABC or bacterial sialidase (but not the concomitant delivery of both) to the site of experimental spinal cord injuries resulted in improved spinal axon outgrowth, locomotor recovery, and cardiovascular reflex recovery.

The explanation for this is that CNS axons have the capacity to regenerate, but are inhibited from doing so by endogenous glycosylated-axon regeneration inhibitors, such as myelin-associated glycoprotein (MAG), Nogo, and oligodendrocyte-myelin glycoprotein in the myelin sheath and chondroitin sulfate proteoglycans

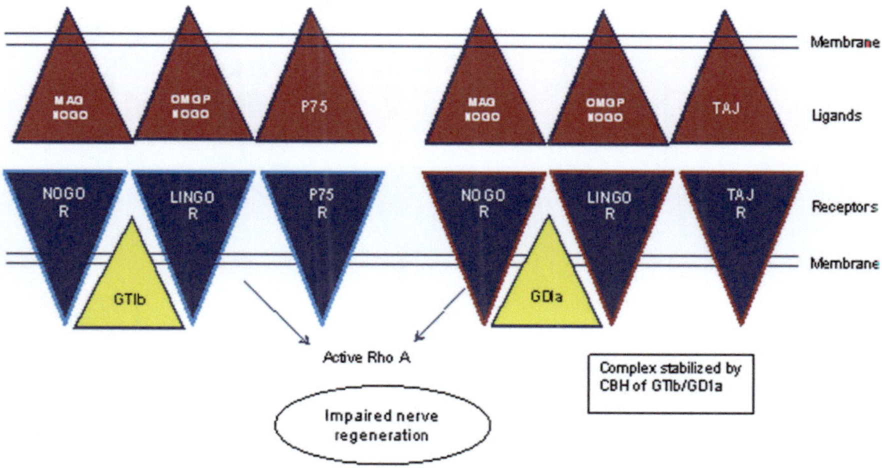

Fig. 13.3 Glycosignaling model for inhibition of nerve repair following injury. Gangliosides GT1b and GD1a (with terminal sialo 2–3 Gal) facilitate clustering of receptors for MAG, OMGP, p75 NGF, and most likely other glycoproteins to form a signaling complex which activates Rho A. In a 2-step procedure, GT1b/GD1a gangliosides associate with NgR1/LINGO-1 to form a functional receptor/co-receptor complex and this is followed by simultaneous recruitment of the transducer p75 and the myelin inhibitor Nogo-54. Once the initial GT1b/GD1a-mediated tripartite receptor/co-receptor complex is formed between NgR1/LINGO-1 and p75, the overall affinity of NgR1 for the Nogo-54 ligand is most likely enhanced and glycosignaling ensues [21]

(CSPGs) at the site of the glial scar. The glial glycoconjugates are believed to bind to complementary receptors on axons (possibly glyco sequences), thus preventing axonal regeneration. A simplified model is shown in Fig. 13.3. In this model, MAG, NOGO bind to NOGO receptors, OMGP/NOGO bind to LINGO receptors and p75/TAJ, bind to p75/TAJ receptors which are critically held together in a glycocluster by either of two gangliosides GTib/GD1a (Saha et al. 2011). Elimination of any of these components reverses some of the inhibition of nerve regeneration directed by activated RhoA. Thus, glycosignaling through this complex on the surface of the cell results in the negative signaling by RhoA inside the cell (Fig. 13.3).

13.8 Glycosignaling and Protein Phosphorylation

Yu and Yanagisawa (2007) presented a model in 2007 that described the involvement of a wide variety of glycoconjugates in signaling pathways critical for the proliferation and self-renewal of neural stem cells and which involved the regulation of protein phosphorylation. Our current view is summarized in Fig. 13.4. Such brain-derived cells the LRs are enriched in GD3 ganglioside, which activates a growth factor glycosignaling pathway, which in turn activates the Ras-MAPK pathway but not the Janus kinase (JAK-STAT) pathway. The latter is associated with

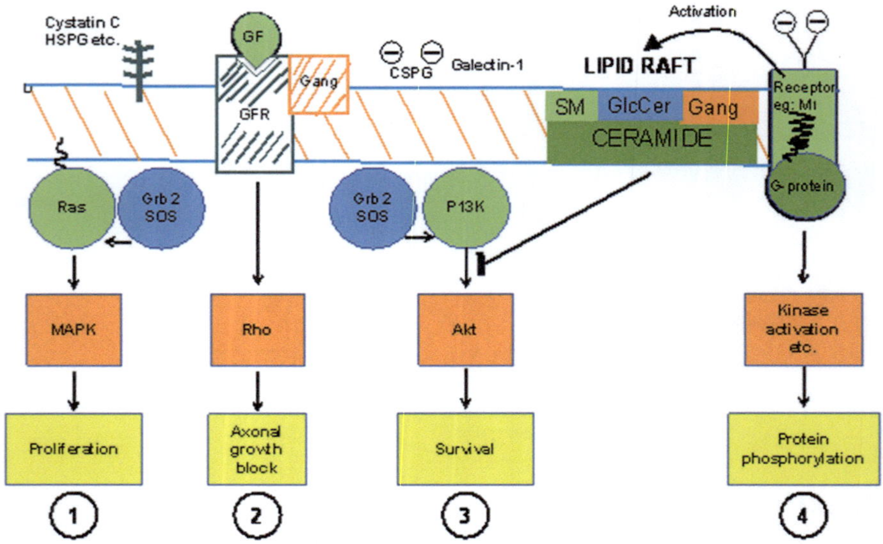

Fig. 13.4 Possible Glycosignaling pathways in Brain. (*1*) Heparan sulfate (HSPG) and chondroitin sulfate (CSPG) proteoglycans, proteins such as Cystalin C and Galectin-1, together with specific gangliosides promote a growth factor (GF) binding complex in a glycodomain which activates Ras and MAPK and promotes brain growth and development. (*2*) Activation of Rho GTPase which maintains synaptic fidelity and integrity by inhibition of axonal growth in the mature animal (see Fig. 13.3). (*3*) Activation of the phosphatidylinositol (P13K)/Akt pathway critically important for growth and development. This pathway can be negatively regulated by Ceramide (generated within a Lipid Raft whose integrity is maintained by glyco-glyco and hydrophobic interactions). Tumor cells accentuate this pathway and reduce ceramide either by glycosylation to form glucosylceramide (GlcCer) and more complex gangliosides Gang such as GD1a, increasing sphingomyelin (SM) by inhibition of sphingomyelinase or reducing de novo synthesis of ceramide. (*4*) G-Protein receptors (e.g., muscarinic M1) are activated following recruitment into Lipid Rafts where they activate palmitoylated/farnesylated modified G-proteins and promote signaling protein phosphorylation. This results in signaling under the control of complex glycosignaling pathways as discussed in the text

astrocyte differentiation and cell survival (Yu and Yanagisawa 2007). Other studies have emphasized a link between glycosignaling and protein phosphorylation, for example Shigatoxin binds to a LR glycolipid and this activates a tyrosine kinase (Katagiri et al. 1999). A similar phenomenon has been described in a neural-derived cell line, the GQ1b-dependent neuritogenesis of human neuroblastoma cell line, GOTO. This was associated with GQ1b-dependent ecto-type protein phosphorylation (Tsuji et al. 1992) of several cell surface proteins. The protein kinase inhibitor, K-252b (a non-membrane-permeable derivative of K-252a) inhibited both the GQ1b-dependent neuritogenesis as well as the GQ1b-stimulated phosphorylation, suggesting a direct coupling between the two cell events. Such events could be important in the CNS.

Another more complex glycosignaling system involves the non-sphingo-glycolipid. (PhGlc) and tyrosine phosphorylation. In this study (Kaneko et al. 2011) the novel lipid, phosphatidylglucoside (PhGlc), present in brain LRs, induced differentiation of cells as measured by the appearance of CD38 (a multifunctional ectoenzyme which acts as an NAD(+) glycohydrolase, an ADP-ribosyl cyclase, and a cyclic ADP-ribose hydrolase) and c-Myc downregulation. Reduction of endogenous cholesterol and dissociation of LRs with methyl-beta-cyclodextrin suppressed the associated tyrosine phosphorylation and signaling, suggesting that these had previously colocalized in the LR. Ligation of known components of LRs such as sphingomyelin and ganglioside GM1, with corresponding antibodies failed to induce differentiation or tyrosine phosphorylation, suggesting that PhGlc can directly glycosignal by inducing protein phosphorylation. More importantly, by using a specific anti-PhGlc antibody (Kaneko et al. 2011), they were able to show robust PhGlc staining in the two primary neurogenic regions of the adult rodent brain, the subventricular zone (SVZ) lining the lateral ventricle and the subgranular zone of the dentate gyrus. The staining pattern of PhGlc appeared to overlap that of glial fibrillary acidic protein, an adult neural stem cell marker and PhGlc expression overlapped with other proposed adult neural stem cell markers (Kaneko et al. 2011).

Thus, a glycosignaling hypothesis would propose that ceramide-enriched lipid platforms (LRs), and gangliosides contribute to receptor (for example CD38) activation to produce cADPR in response to receptor stimulation, (for example by muscarinic type 1 (M(1) agonists)). Thus, the M(1) receptor agonist, oxotremorine, should increased LR clustering on the membrane to form a complex of CD38 with LR components such as GM1, acid sphingomyelinase (ASMase), and ceramide, and the complex GSD should be abolished by LR disruptors, such as methyl-beta-cyclodextrin or filipin (Jia et al. 2008) There is experimental evidence for this. Further, fluorescence resonance energy transfer (FRET) showed the close proximity of ganglioside and CD38, and thus the formation of ceramide-enriched lipid macrodomains (LRs) appears to be crucial for agonist-induced activation of CD38 to produce cADPR. We can conclude that following translocation into lipid glycodomains transmembrane signaling through the G-protein-coupled M(1) receptor produces the second messenger cADPR (Jia et al. 2008).

However, the situation is more complex since the association of ganglioside and CD38 signaling also involves a role for sulfated glycoconjugates. This association was confirmed when the extracellular catalytic domain of CD38 was expressed as a fusion protein with maltose-binding protein, and then co-crystallized with one of its likely physiological brain ganglioside inhibitors (GT1b) in a 1:1 ratio per each asymmetric unit (Zhao et al. 2012). Three kinds of novel sulfated gangliosides structurally related to the Chol-1 (alpha-series) ganglioside GQ1b-alpha were tested and found to be potent inhibitors of the NADase activity of CD38 (Zhao et al. 2012). The disulfate of iso-GM1b was surprisingly found to be the most potent structure for both NADase inhibition and MAG-binding activity. Such sulfated gangliosides could bind to the internal sialic acid residues linked alpha2–3- to Gal as well as to siglec-dependent recognition sites with a terminal sialic acid residue. Thus, sialylated oligosaccharides binding to minor sulfated gangliosides could be a good model for carbohydrate–carbohydrate interactions at the cell membrane in the nervous system and much exciting work lies ahead in order to unravel the complex system we call "glycosignaling."

So the role of the sphingolipid/cholesterol microdomain glycolipid or other glycoconjugate is to create an environment (the lipid raft (LR)) which recruits proteins and facilitates protein interactions leading to kinase activation and biological effects in the brain. The evidence for such a mechanism in neural stem cells involving growth factors, integrins, and the Ras-MAPK pathway has been well-summarized (Yu and Yanagisawa 2007)) and the limitations in evidence duly noted. A summary scheme is presented in Fig. 13.4. Much of the evidence supporting these ideas comes from the immune system since many pathogens target glycosphingolipids in LRs but the results have relevance for the brain. Thus lactosylceramide (LacCer, CDw17) binds to *Candida albicans* and forms membrane microdomains together with the Src family tyrosine kinase Lyn. These LacCer-enriched membrane microdomains can mediate superoxide generation, migration, and phagocytosis, indicating that LacCer functions as a Pattern Recognition Receptor in innate immunity (Nakayama et al. 2013). Other studies in non-neural cells suggest that lactosylceramide may directly activate phospholipase A2 by a translocation mechanism (Nakamura et al. 2013). Thus we may conclude that glycosignaling in the brain is most likely all about recruiting signaling partners into microdomains with associated glycoconjugate and fatty acid/sphingosine heterogeneity offering the additional binding specificities necessary for the brain to function as an amazingly complex organ.

13.9 Lysosomal Storage Modifies Glycosignaling

The failure to degrade complex carbohydrates results in lysosomal storage diseases and often the destruction of the nervous system prenatally or before adolescence. This is the most dramatic demonstration of the power of quantitatively abnormal or unique glyco-structures to disrupt brain function. Although the pathology of these diseases suggests that physical accumulation of glycoconjugates can be a primary cause of neural cell death and resultant seizures and demyelination, it has been shown that critical glycosignaling systems involving ion channels etc. may be compromised by the storage of glycoconjugates and that this contributes to the pathology of these diseases (Futerman and van Meer 2004).

Since in lysosomal storage diseases there could be perturbations in late endocytic functions leading to abnormal lipid raft composition and trafficking (Futerman and van Meer 2004) we carried out a detailed lipidomics study of a pathological lysosomal storage disease in the brain of a mouse with a typical lysosomal storage disease (San Filippo Type3a (Dawson et al. 2012)). The major gangliosides GM1, GD1a, GD1b, and GT1a were C18/20 sphingosine with C18 fatty acid (as expected) and all were exclusively present in LRs. However, although the heparan sulfate-derived storage material was not in LRs (as expected), the abnormal gangliosides (GD2, GD3, and GM2 and GM3) were all in LRs. Thus, the glyco-part could be modifying the formation of complexes, which then disrupted normal glycosignaling in the brain, leading to seizures, blindness, and loss of all cortical functions.

13.10 Future Directions for Glycosignaling in the Brain

Any system involving modification of multiple membrane glycoproteins, proteoglycans, gangliosides, sulfated oligosaccharides, and glycosylated phosphoglycerides will take a while to understand, but no one doubts that the human brain has evolved into a very complex yet capable organ, so we look forward with eager anticipation to the secrets that will be revealed by the continued improvement and application of new technology to glycosignaling.

Conflict of Interest The author declares no conflict of interest.

References

Aureli M, Gritti A, Bassi R, Loberto N, Ricca A, Chigorno V, Prinetti A, Sonnino S. Plasma membrane-associated glycohydrolases along differentiation of murine neural stem cells. Neurochem Res. 2012;37:1344–54.

Barritt AW, Davies M, Marchand F, Hartley R, Grist J, Yip P, McMahon SB, Bradbury EJ. Chondroitinase ABC promotes sprouting of intact and injured spinal systems after spinal cord injury. J Neurosci. 2006;26:10856–67.

Boeneman K, Delehanty JB, Blanco-Canosa JB, Susumu K, Stewart MH, Oh E, Huston AL, Dawson G, Ingate S, Walters R, Domowicz M, Deschamps JR, Algar WR, DiMaggio S, Manono J, Spillmann CM, Thompson D, Jennings TL, Dawson PE, Mendintz I. Selecting improved peptidyl motifs for cytosolic delivery of disparate protein and nanoparticle. ACS Nano. 2013;7:3778–96.

Colsch B, Jackson SN, Dutta S, Woods AS. Molecular microscopy of brain gangliosides: illustrating their distribution in hippocampal cell layers. ACS Chem Neurosci. 2011;2:213–22.

Dawson G, Fuller M, Helmsey KM, Hopwood JJ. Abnormal gangliosides are localized in lipid rafts in Sanfilippo (MPS3a) mouse brain. Neurochem Res. 2012;37:1372–80.

Futerman AH, van Meer G. The cell biology of lysosomal storage disorders. Nat Rev Mol Cell Biol. 2004;5:554–65.

Hakomori SI. Structure and function of glycosphingolipids and sphingolipids: recollections and future trends. Biochim Biophys Acta. 2008;1780:325–46.

Hakomori S, Handa K, Iwabuchi K, Yamamura S, Prinetti A. New insights in glycosphingolipid function: "glycosignaling domain" a cell surface assembly of glycosphingolipids with signal transducer molecules, involved in cell adhesion coupled with signaling. Glycobiology. 1998;8:11–9.

Jia SJ, Jin S, Zhang F, Yi F, Dewey WL, Li PL. Formation and function of ceramide-enriched membrane platforms with CD38 during M1-receptor stimulation in bovine coronary arterial myocytes. Am J Physiol. 2008;295:1743–52.

Kaneko J, Kinoshita MO, Machida T, Shinoda Y, Nagatsuka Y, Hirabayashi Y. Phosphatidylglucoside: a novel marker for adult neural stem cells. J Neurochem. 2011;116:840–4.

Katagiri Y, Mori T, Nakamjima H, Katagiri C, Taguchi T, Takeda T, Kiyokawa N, Fujimoto J. Activation of Src family kinase induced by Shiga toxin binding to globotriasosyl ceramide (Gb3/CD77) in low density, detergent-insoluble microdomains. J Biol Chem. 1999;274:35278–82.

Lingwood CA. A holistic approach to glycolipid function: is the lipid moiety important? Trends Glycosci Glycotechnol. 1999;11:1–9.

Lingwood D, Reis J, Schwille P, Simons K. Plasma membranes are poised for activation of raft phase coalescence at physiological temperature. Proc Natl Acad Sci U S A. 2008;105: 10005–10.

Lingwood CA, Manis A, Mahfoud R, Khan F, Binnijngton B, Mylvaganam M. New aspects of the regulation of glycosphingolipid receptor function. Chem Phys Lipids. 2010;163:27–35.

Mountney A, Zahner MR, Lorenzini I, Oudega M, Schramm LP, Schnaar RL. Sialidase enhances recovery from spinal cord contusion injury. Proc Natl Acad Sci U S A. 2010;107:11561–6.

Mountney A, Zahner MR, Sturgill ER, Riley CJ, Aston JW, Oudega M, Schramm LP, Hurtado A, Schnaar R. Sialidase, chondroitinase ABC and combination therapy after spinal cord contusion injury. J Neurotrauma. 2013;30(3):181–90.

Nakamura H, Moriyama Y, Makiyama T, Emori S, Yamashita H, Yamazaki R, Murayama T. Lactosylceramide interacts with and activates cytosolic phospholipase A2α. J Biol Chem. 2013;288(32):23264–72.

Nakayama H, Ogawa H, Takamori K, Iwabuchi K. GSL-enriched membrane microdomains in innate immune responses. Arch Immunol Ther Exp (Warsz). 2013;61:217–28.

Parker RB, Kohler J. Regulation of intracellular signaling by extracellular glycan remodeling. ACS Chem Biol. 2009;5:35–46.

Qin J, Berdyshev E, Poirer C, Schwartz NB, Dawson G. Neutral sphingomyelinase 2 deficiency increases hyaluronan synthesis by up-regulation of Hyaluronan synthase 2 through decreased ceramide production and activation of Akt. J Biol Chem. 2012;287:13620–32.

Roseman S. The synthesis of complex carbohydrates by multiglycosyltransferase systems and their potential function in intercellular adhesion. In: Chemistry and metabolism of sphingolipids. Amsterdam: North Holland Publishing Co.; 1970. p. 270–97.

Saha N, Kolev MV, Semavina M, Himanen J, Nikolov DB. Ganglioside mediate the interaction between Nogo receptor 1 and LINGO-1. Biochem Biophys Res Commun. 2011;413:92–7.

Sekino-Suzuki N, Yuyama K, Miki T, Kaneda M, Suzuki H, Yamamoto N, Yamamoto T, Oneyama C, Okada M, Kasahara K. Involvement of gangliosides in the process of Cbp/PAG phosphorylation by Lyn in developing cerebellar growth cones. J Neurochem. 2012;124:514–22.

Sonnino S, Prioni S, Chigorno V, Prinetti A. Interactions between caveolin-1 and sphingolipids, and their functional relevance. Biochemical roles of eukaryotic cell surface macromolecules. Adv Exp Med Biol. 2012;749:97–115.

Sonnino S, Mauri L, Ciampa MG, Prinetti A. Gangliosides as regulators of cell signaling: ganglioside-protein interactions or ganglioside-driven membrane organization? J Neurochem. 2013;124:432–5.

Tsuji S, Yamashita T, Matsuda Y, Nagai Y. A novel glycosignaling system: GQ1b-dependent neuritogenesis of human neuroblastoma cell line, GOTO, is closely associated with GQ1b-dependent ecto-type protein phosphorylation. Neurochem Int. 1992;21:549–54.

Walters R, Kraig RP, Medintz I, Delehanty JB, Stewart MH, Susumu K, Huston AL, Dawson PE, Dawson G. Nanoparticle targeting to neurons in a rat hippocampal slice culture model. ASN Neuro. 2012;4:383–92.

Yu RK, Yanagisawa M. Glycosignaling in neural stem cells. J Neurochem. 2007;103:39–46.

Zhao J, Liu Y, Park H, Boggs JM, Basu A. Carbohydrate-coated fluorescent silica nanoparticles as probes for the galactose/3-Sulfogalactose carbohydrate–carbohydrate interaction using model systems and cellular binding studies. Bioconjug Chem. 2012;23:1166–73.

Chapter 14
Glycosphingolipids in the Regulation of the Nervous System

Koichi Furukawa, Yusuke Ohmi, Yuki Ohkawa, Orie Tajima, and Keiko Furukawa

Abstract The highest expression of gangliosides, sialic acid-containing glycosphingolipids (GSLs), is found in the nervous tissue of vertebrates. Changes in the profiles of gangliosides during the development of nervous tissues indicate that they are involved in the regulation of neurogenesis and synaptogenesis. Their distinct distribution patterns support the suggestion that they are involved in both the differentiation and function of neural cells. In addition to results of studies of GSLs done using biochemical, histopathological, and cell biological approaches, recent progress in the genetic engineering of glycosyltransferase genes has resulted in novel findings and concepts about their roles in the nervous system. Roles of GSLs in the regulation of signaling that determine cell fates in membrane microdomains such as lipid rafts have been extensively studied. In particular, gene targeting of glycosyltransferases in mice has enabled investigation of the in vivo functions of GSLs. The majority of abnormal phenotypes exhibited by knockout (KO) mice may reflect an abnormal structure and a resultant altered function of lipid rafts caused by alterations in their GSL composition. Generally speaking, abnormal phenotypes found in most KO mice were milder than expected, suggesting that the remaining GSLs compensate for the functions of those lost. There are also functions that

K. Furukawa (✉) • Y. Ohmi • Y. Ohkawa
Department of Biochemistry II, Nagoya University Graduate School of Medicine, 65 Tsurumai, Showa-ku, Nagoya 466-0065, Japan
e-mail: koichi@med.nagoya-u.ac.jp

O. Tajima
Department of Biomedical Sciences, College of Life and Health Sciences, Chubu University, 1200 Matsumoto-cho, Kasugai 487-8501, Japan

K. Furukawa
Department of Biochemistry II, Nagoya University Graduate School of Medicine, 65 Tsurumai, Showa-ku, Nagoya 466-0065, Japan

Department of Biomedical Sciences, College of Life and Health Sciences, Chubu University, 1200 Matsumoto-cho, Kasugai 487-8501, Japan

R.K. Yu and C.-L. Schengrund (eds.), *Glycobiology of the Nervous System*, Advances in Neurobiology 9, DOI 10.1007/978-1-4939-1154-7_14,

cannot be replaced by the remaining GSLs. Thus, there may be two modes of function of GSLs: one is nonspecific and can be carried out by multiple GSLs, the second mode is that in which the function of the missing GSL(s) cannot be compensated by others. Identification of natural ligands for individual GSLs is crucial in order to clarify the functions of each structure.

Keywords Glycosphingolipids • Microdomains • Gangliosides • Knockout • NGF • Lipid rafts

Abbreviations

GSLs	Glycosphingolipids
KO	Knockout
GlcCer	Glucosylceramide
LacCer	Lactosylceramide
GalCer	Galactosylceramide
DKO	Double KO
NGF	Nerve growth factor
CNS	Central nervous system
GPI	Glycosylphosphatidylinositol.

14.1 Introduction

Glycosphingolipids (GSLs) are unique amphipathic molecules consisting of a hydrophilic carbohydrate moiety and a hydrophobic lipid portion (Wiegandt 1985) majority of which is synthesized from glucosylceramide (GlcCer). The carbohydrate moiety can usually be classified into one of 4 major series. They are the ganglio-, globo-, lacto/neolacto-, and asialo-series. Lactosylceramide (LacCer), a common precursor of most GSLs is synthesized by addition of galactose (in a β1,4-linkage) to GlcCer. In addition, gala-series GSLs generated from galactosylceramide (GalCer) are also present. The GalCer-derived GSLs is limited in number as are their sites of expression. In addition to polymorphism in the carbohydrate moiety of GSLs, variability is seen in both their fatty acid and long-chain base composition. Thus, regulatory systems needed for proper expression of GSLs should be present in tissues in which they are found as well as at specific stages of cell/tissue differentiation.

Histopathological and biochemical studies of cells maintained in vitro and of various organs and tissues resulted in the identification of a number of biological functions for individual GSLs (Schengrund 1990; Yu et al. 1988). In particular,

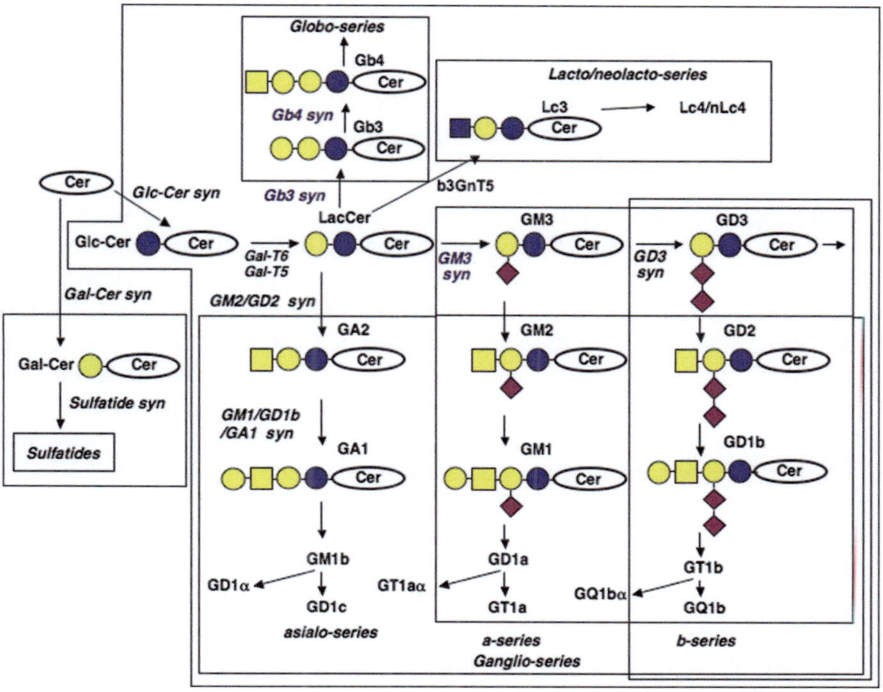

Fig. 14.1 Synthetic pathway and enzymes needed for GSL synthesis. The synthetic pathway and main enzymes responsible for GSL synthesis are shown. Deleted structures in individual knockout mice are indicated by *squares*

interrogation of neural cell lines sensitive to well-defined differentiation factors resulted in novel findings regarding mechanisms by which GSLs may affect cell function (Greene and Tischler 1976; Levi et al. 1988; Mutoh et al. 1995). Since the isolation of cDNA encoding the glycosyltransferase responsible for synthesis of GM2 and GD2 (Nagata et al. 1992), a number of other glycosyltransferase cDNAs have been isolated (Lloyd and Furukawa 1998). The synthetic pathways and enzymes for GSL synthesis are shown in Fig. 14.1. Isolation of the specific cDNAs facilitated studies of knockout (KO) animals, in which specific glycosyltransferase genes were genetically disrupted (Furukawa et al. 2001). Functions of missing GSLs were identified based on the phenotypes seen in the various glycosyltransferase gene KO mice (Furukawa et al. 2001). However, KO of specific glycosyltransferase genes in mice often resulted in milder abnormal phenotypes than expected (Furukawa et al. 2001, 2004).

In this chapter we summarize the roles of GSLs in the nervous system as identified using cultured cells and mice in which specific GSL synthase genes were knocked out, and discuss anticipated future research.

14.2 Glycosphingolipids in Cultured Cells

14.2.1 *Gangliosides Modulate Signals Transduced by Neurotrophic Factors/Receptors*

Gangliosides are highly expressed in nervous tissue, and have been considered to be involved in the regulation of their development, differentiation, and function (Schengrund 1990). A number of studies have reported that gangliosides play roles as neurotrophic factors in cultured neural cell lines and anti-apoptotic factors (Ferrari et al. 1983). Studies of a rat pheochromocytoma cell line PC12 demonstrated that gangliosides regulate proliferation (Fukumoto et al. 2000) and differentiation (Nishio et al. 2004) of cells by modulating cell signaling. Nerve growth factor (NGF) binds TrkA and triggers phosphorylation of the receptor and subsequent activation of Ras/Raf/MEK/Erk pathway leading to neurite extention (Vaudry et al. 2002). To our surprise, over-expression of ST8SIA1 (GD3 synthase) cDNA resulted in the constitutive activation of TrkA and Erk1/2, and consequently unresponsiveness to NGF. These cells showed continuous cell growth even after NGF treatment. GD3 synthase cDNA transfectant cells may represent the features of neural cells during early brain development, a time period in which GD3 is a dominant ganglioside (Yu et al. 1988). On the other hand, over-expression of GM1 synthase induced unresponsiveness to nerve growth factor (NGF). In this case, no phosphorylation of TrkA and subsequent signaling molecules could be found even after NGF treatment. In these transfectant cells, a dramatic alteration in the intracellular localization (inside to outside of lipid rafts) of the NGF receptor TrkA and other signaling molecules such as $p75^{NTR}$ and H-Ras was observed. Results of earlier studies had indicated that GM1 could act as a neurotrophic factor in the rescue of serum-deprived PC12 cells (Ferrari et al. 1983) as well as in the enhancement of neurite extension (Mutoh et al. 1995). Although the reason for the difference in the effects of GM1 (exogenous vs endogenous) on PC12 cells are not known, relative GM1 levels might be critical for the direction of cell signaling. The aberrant signaling seen in transfected cells is shown in Fig. 14.2. These observations are examples of the roles of lipid raft-associated gangliosides in the regulation of signal transduction.

14.2.2 *Essential Roles of GSLs for Development of Multicellular Organisms*

As shown by a UGCG (GlcCer synthase)-deficient cell line (Ichikawa et al. 1996), GSLs are not essential for the growth and survival of mammalian cells in culture. On the other hand, in GlcCer synthase-KO mice GSLs were needed for development at embyonic day E6.5–7.5. The GlcCer synthase-deficient mutant mice showed embryonal lethality by E7.5 accompanied by intensive apoptotic changes particularly in the ectoderm. This observation indicated that GSLs are essential for

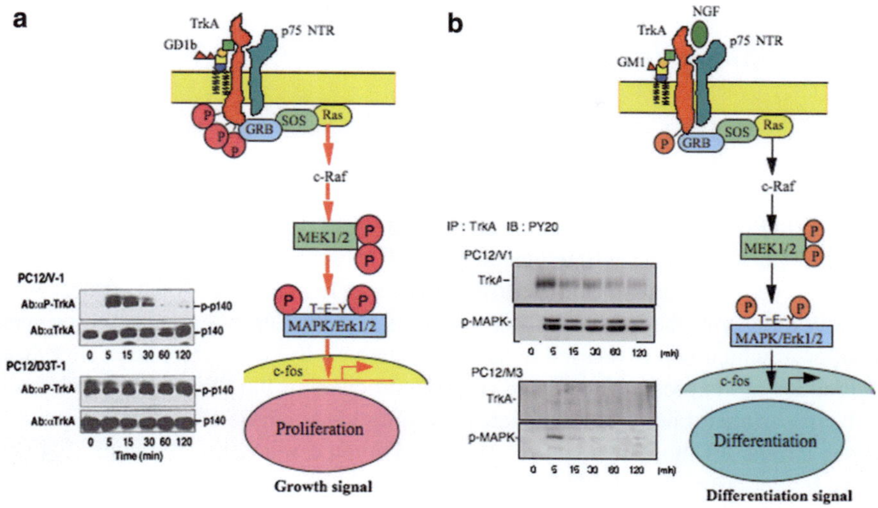

Fig. 14.2 Aberrant signaling seen in cells transfected with specific glycosyltransferase cDNAs. (**a**) Transfection of GD3 synthase cDNA into a PC12 cell line resulted in constitutive activation of TrkA and ERK1/2, which was accompanied by enhanced cell proliferation and unresponsiveness to NGF. (**b**) Transfection of PC12 cells with B3galt4 (GM1/GD1b synthase) cDNA resulted in the cells being unresponsive to NGF stimulation both in neurite extension and in TrkA/Erk1/2 activation

development of multicellular organisms (Yamashita et al. 1999). Conditional knockout (KO) of GlcCer synthase in mice, in which the gene was destroyed after birth, resulted in severe dysfunction of cerebellum and peripheral nerves that was associated with structural defects (Jennemann et al. 2005). The fact that the mice died within 3 weeks after birth again indicated that GSLs are indispensable for the maintenance and survival of organisms.

14.3 Impact of KO of Glycosyltransferase Genes Located at Stem Steps of GSL Synthesis

14.3.1 KO of ST3GAL5 (GM3 Synthase) in Mice

GM3 synthase is essential for the synthesis of GM3 from LacCer, and therefore, for the synthesis of all ganglio-series gangliosides. However, KO of the gene in mice resulted in no apparent abnormal phenotype except increased sensitivity to insulin (Yamashita et al. 2003). This phenotype is in dramatic contrast to that seen in patients with GM3 synthase deficiency (Simpson et al. 2004). Patients with the mutated GM3 synthase gene exhibit "infantile-onset symptomatic epilepsy" with growth and mental retardation. The basis for this marked difference between human and mouse is not known.

14.3.2 KO of UGT8 (GalCer Synthase) and GAL3ST1 (Sulfatide Synthase) in Mice

GalCer-derived GSLs were shown in studies of mutant mice lacking the GalCer synthase gene (Coetzee et al. 1996) to generate myelin that contained glucocerebroside. In addition to lacking GalCer these mice also lacked sulfatide (another myelin-enriched GSL) and seminolipid. The fact that these mice had severe neurological defects (Coetzee et al. 1996) indicated that GalCer and/or sulfatide has an important role in myelin. The mutants also showed male sterility (Fujimoto et al. 2000).

Sulfatide synthase is essential for synthesis of both sulfatides and seminolipids, neither of which was expressed in sulfatide synthase KO mice (Honke et al. 2002). Sulfatides are expressed mainly in oligodendrocytes in the CNS and in Schwann cells in peripheral nerves. As expected the sulfatide synthase KO mice had phenotypes similar to those of GalCer synthase-disrupted mice, but the abnormalities were milder (Honke et al. 2002).

14.3.3 LacCer Synthase KO Mice

B4galt6 (Lactosylceramide synthase, LacCer synthase) KO mice showed no definite abnormal phenotypes (Tokuda et al. 2013). In contrast, KO of B4galt5 resulted in severe defects in development, indicating that β1,4Gal-T5 might be the main LacCer synthase (Kumagai et al. 2009; Nishie et al. 2010). The phenotype resulting from disruption of the LacCer synthase was similar to that induced by KO of GlcCer synthase.

14.4 Compensation for Lost Functions by Remaining GSLS

Generally, disruption of ganglioside synthases resulted in milder phenotypes than expected. In particular, the fact that KO mice lacking the B4galnt1 (GM2/GD2 synthase) gene had an almost normal architecture of the central nervous system (CNS) at birth despite the lack of all complex gangliosides was quite surprising (Takamiya et al. 1996). No clear differences between KO and wild-type mice were found in brain morphology, myelination, and behavior. The only change observed was a reduction in nerve conductivity. Male infertility due to aspermatogenesis appeared to be the most serious phenotype (Takamiya et al. 1998). However, KO mice underwent neuronal degeneration that increased gradually with aging (Sugiura et al. 2005). The degenerative disorders were detected primarily in peripheral nerves and the dorsal horn of the spinal cord (Sugiura et al. 2005). These KO mice showed sensory nerve-dominant neurodegeneration, while another group using the same kind of KO mice reported Wallerian degeneration and abnormal neurological function in a motor-neuron-dominant manner (Sheikh et al. 1999; Chiavegatto et al. 2000). Morphological changes in synaptic vesicles and dendrites in the central terminals, and in glia processes indicated that compensatory modification of neural tissue took place after nerve degeneration, i.e., remodeling or regeneration. Furthermore, regeneration of resected hypoglossal nerves was strongly disturbed (Kittaka et al. 2008). These results indicated that complex gangliosides are not essential in morphogenesis, but important in the maintenance and repair of nerve tissue. Aberrant Ca^{2+}-regulating properties in the cerebellar neurons found in the same type of KO mice (Wu et al. 2001) may correspond with the neurological disorders described above. All these results indicate that increased GM3 and GD3 play important roles in compensating for the loss of more complex gangliosides as seen in GM2/GD2 synthase KO mice.

Despite the fact that genetic disruption of the GD3 synthase gene resulted in loss of all b-series and c-series gangliosides (Kawai et al. 1998; Okada et al. 2002), almost no apparent abnormalities were seen in either morphology or behavior. Although GD3 was reported to mediate the apoptotic signals mediated by Fas–Fas ligands (De Maria et al. 1997), sensitivity of thymocytes from these KO mice to apoptosis induced by anti-Fas antibody appeared unaffected. While no morphological abnormalities were detected in neural tissue of GD3 synthase KO mice, significantly reduced regenerative activity was found in hypoglossal nerve resection experiments (Okada et al. 2002). This indicated that b-series gangliosides have a crucial role in nerve regeneration, and supported observations indicating that b-series gangliosides were the most effective at stimulating rat hypoglossal nerve regeneration (Itoh et al. 2001).

14.5 Double KO Exhibited More Severe Phenotypes

As summarized in Fig. 14.3, profiles of GSL species in the individual KO mice support the hypothesis that remaining ones might compensate for the roles of lost GSLs. This could be due to two important possibilities. The first is that the total cell concentration of GSLs is strictly regulated so that their total concentration is similar despite variability in their composition. The second is that functions of some GSLs can be replaced by others. In order to evaluate the roles of each GSL we generated complex KO mice in which remaining GSLs were reduced as much as possible.

In double KO (DKO) mice lacking both GD3 synthase and GM2/GD2 synthase genes, animals only synthesized GM3 (Inoue et al. 2002), no obvious changes were detected at birth. However, they gradually died of unknown causes about 12 weeks after birth. They also exhibited refractory skin lesions on the face and neck. Reduced sensitivity to mechanical pain seemed to trigger the skin lesions. Neurodegenerative changes at a fairly young age may explain the dysfunction in the sensory system (Inoue et al. 2002). Audiogenic seizures and consequent sudden death by noise were observed in another group of the double DKO mice (Kawai et al. 2001). Interestingly, our DKO mice did not show the same response to noise. This difference

KO gene	Glc-Cer syn	GM3 syn	GD3 syn	GM2/GD2 syn	DKO
Lost structures	all GSLs	ganglio-series (a-, b-, c-)	b-series (and c-series)	all comlex gangliosides (inc. asialo-series)	all gangliosides except GM3
Remaining structures		asialo-series	a-series asialo-series	GM3, GD3 (and GT3)	GM3
	Gal-Cer sulfatides	Gal-Cer sulfatides n-GSLs	Gal-Cer sulfatides n-GSLs	Gal-Cer sulfatides n-GSLs	Gal-Cer sulfatides n-GSLs
	↓	↓	↓	↓	
	Emb. lethal	No apparent abnormalities	Mild phenotypes	Moderate phenotypes progressive	Severe phenotypes progressive

Fig. 14.3 Profiles of GSL species synthesized by various KO mice. GSLs found in the KO mice lines are shown. These structures may compensate for the role(s) of GSLs missing in the individual KO mice. DKO, double knockout of GD3 synthase and GM2/GD2 synthase genes. n-*GSLs* neutral glycosphingolipids, *Emb. Lethal* embyonal lethal

may be due to the different genetic backgrounds of the ES lines used. Phenotypic analyses of these mutant mice indicated that GM3 alone enabled them to undergo almost normal neurogenesis, birth, and development up to a certain point. However, the correct composition of gangliosides seemed to be essential for maintenance of intact morphology and function. Subsequent analyses to clarify the cause of neurodegeneration seen in the DKO mice revealed that in order to maintain integrity in the architecture and function of nervous tissue cells lipid rafts containing the appropriate composition of GSLs were needed (Ohmi et al. 2009, 2011).

Mice in which both GM3 synthase and GM2/GD2 synthase genes were knocked out were also generated (Yamashita et al. 2005) and used to examine the significance of asialo-series gangliosides. These animals eventually showed severe nerve degeneration, leading to an early death. The severity of this phenotype indicates that GM3 itself is essential for survival as well as maintenance of the CNS. Although GM3 has been reported to suppress EGF/EGFR-mediated signals by forming complexes with tetraspanin (Yoon et al. 2006), the specific effect of its loss in the DKO mice remains to be further clarified.

14.6 Response to Neurodegeneration by Modification of Gene Expression in the DKO mice

Changes in gene expression profiles in DKO mice were examined using DNA microarrays. The results indicated that an up- (or down-) regulation in expression of genes encoding proteins involved in inflammation- and immunological reaction-related events, and of those encoding proteins that react to inflammation and/or degeneration took place in KO brains. This grouping scheme is shown in Fig. 14.4. Up-regulation of expression of genes encoding proteins such as complement C4 and

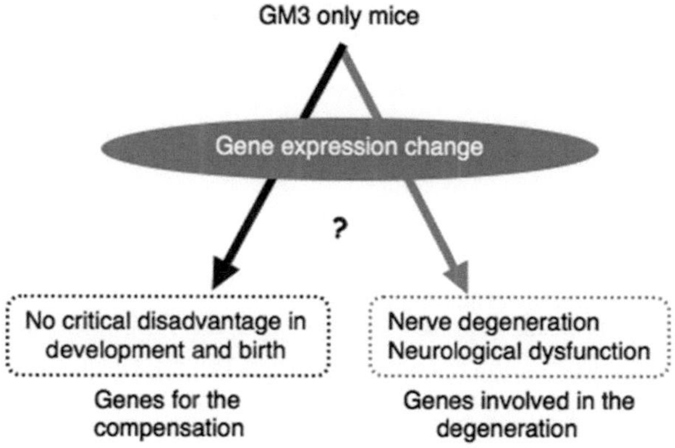

Fig. 14.4 Genes strongly upregulated or downregulated in nervous tissue of DKO mice. cDNA microarray analyses were performed to compare gene expression profiles between wild-type and DKO mouse brain tissue. One group seemed to include inflammation- and immunological reaction-related genes. The other group contained those regulated as a reaction to inflammation and/or degeneration in DKO brains. The former might be results of defects of gangliosides, and the latter might represent host reaction to inflammation and/or degeneration caused by ganglioside deficiency

C3a receptor 1, indicated that inflammation could be a major cause of the neurodegeneration seen. Support for this hypothesis was provided by results obtained in studies of triple KO mice lacking GM2/GD2 synthase, GD3 synthase, and C3 (Ohmi et al. 2009).

An increased expression of genes encoding anti-apoptotic proteins was also observed (Ohmi et al. 2011). Among upregulated genes in cerebellum of the DKO mice, was Wisp2/CCN5. This was of interest because expression of this gene in the CNS had not been previously reported. Its over-expression resulted in increased cell proliferation and neurite outgrowth upon serum withdrawal from cultured Neuro2a cells (Ohkawa et al. 2011). Integrin appears to be a receptor for secreted Wisp2/CCN5. The cDNA-transfected cells also exhibited resistance to H_2O_2-induced apoptosis. These results indicate that the Wisp2/CCN5 induced in neurons of DKO mice serves to protect them from neurodegeneration caused by ganglioside deficiency. Changes found in nervous tissue of DKO mice are summarized in Fig. 14.5.

14.7 Mechanisms by Which Gangliosides May Maintain the Integrity of the CNS

DKO of two major glycosyltransferase genes showed more severe neurodegeneration than those detected in single KO animals. Studies of potential mechanisms underlying neurodegeneration identified a role for gangliosides in regulation of the complement system (Inoue et al. 2002; Ohmi et al. 2009). Gangliosides in membrane microdomains were found to control the complement system and suppression of inflammation and neurodegeneration (Ohmi et al. 2012). Results of studies of

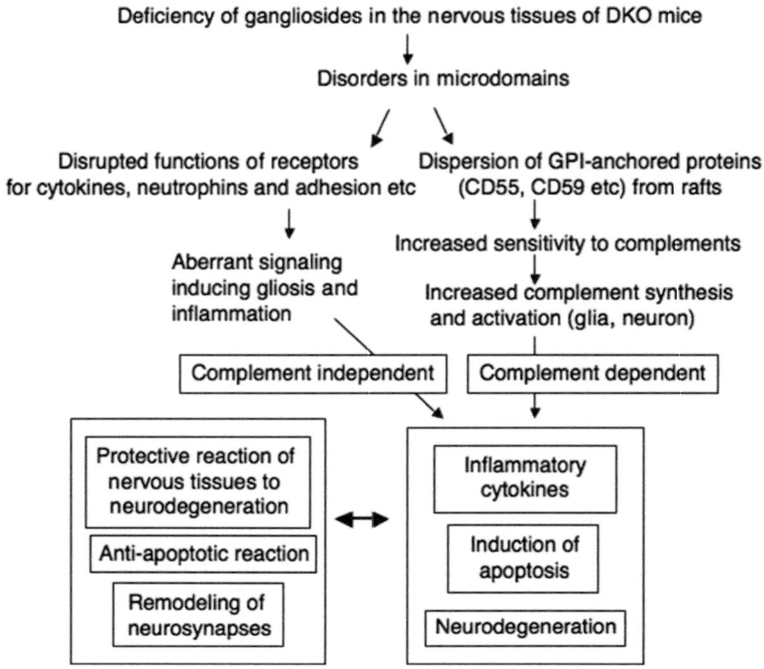

Fig. 14.5 Defects and reactions observed in nervous tissue of DKO mice

mice in which specific ganglioside synthases were knocked out indicated that the disruption of lipid rafts was accompanied by up-regulation of complement-related genes associated with proliferation of astrocytes and infiltration of microglia. Severity of the effect depended on the defects in ganglioside composition. Glycosylphosphatidylinositol (GPI)-anchored molecules such as DAF, CD59, and NCAM tended to disperse most severely from the raft fraction from DKO mice > GM2/GD2 synthase KO > GD3 synthase KO > WT. Even lipid raft markers such as flotillin-1 dispersed from the raft fractions in a similar order. These results indicated that the architecture of the lipid rafts was destroyed by deletion of gangliosides and that the degree of disruption depended on the severity of the change in GSL composition (Ohmi et al. 2011).

14.8 Potential Relationships Between the Role(s) of GSLS in the Regulation of the Architecture and Function of Lipid Rafts and Human Neurodegenerative Diseases

The inflammatory responses seen in nervous tissue in KO mice are similar to those seen in human neurodegenerative diseases such as Alzheimer and Parkinson's diseases (Rogers et al. 1992; Shen et al. 2001). Neurodegeneration in these diseases is frequently associated with autoimmune reactions. As shown in the triple KO mice described previously, immune suppression such as deprivation of complement components has been tried as a therapeutic approach (Sardi et al. 2011; Shen and Meri 2003). In addition, administration of antibodies or vaccination with disease-related proteins often alleviates pathological and clinical features associated with these neurodegenerative diseases (Delrieu et al. 2012; Shah and Federoff 2011).

The majority of studies on lipid rafts have been performed using cultured cell lines (Patra 2008; Simons and Gerl 2010), and not by experimental animals (Furukawa et al. 2007). On the other hand, many findings described in this chapter were substantially proven by analyzing molecular reactions in/near lipid rafts in the individual experimental systems with some done using brain tissue as the source of microdomains (Ohmi et al. 2009, 2011, 2012; Furukawa et al. 2011). Thus, it is reasonable to conclude that the majority of abnormal phenotypes observed in KO mice are due to disruption of the architecture and function of lipid rafts due to altered expression of GSLs.

14.9 Conclusions/Future Directions

We propose that there are two ways by which GSLs in lipid rafts function in the maintenance of structure and function of nervous tissue. The first is to help maintain the fundamental environment of the cell membrane, and the second is for each GSL to serve a particular function(s) that cannot be replaced by other glycoconjugates.

The availability of glycosyltransferase genes and mutant mice lacking them has provided a powerful tool for studying their role(s) in biological processes, such as development, cell growth, differentiation, and cell death. Interestingly, many of the novel findings obtained by analysis of the mutant mice are more complicated than expected, indicating that GSLs may have multiple functions. While some functions of the gangliosides can be compensated for by other molecules, others seem to be indispensable. In particular, it was quite surprising that GM3 synthase gene knockout mice (Yamashita et al. 2003) had no apparent morphological or behavioral abnormalities, suggesting that asialo-series gangliosides may have compensated for the loss of all GM3-derived gangliosides. Use of mice genetically engineered to lack specific glycosyltransferases should help clarify the role(s) of functional redundancy among GSLs. In cases where a specific role of a particular GSL cannot be replaced by other species it is possible that their function depends upon the interaction of the GSL with a specific ligand(s). Therefore, identification of these ligands is needed in order to understand how each GSL functions.

Finally, the question of where gangliosides/GSLs are located on the membranes of living cells remains to be interrogated. It may be possible to address this question as well as the question of whether they are located in close juxtaposition of the proteins with which they putatively interact using single molecule imaging with high spatiotemporal resolution (Suzuki et al. 2012, 2013).

Conflict of Interest The authors declare that they have no conflict of interest.

References

Chiavegatto S, Sun J, Nelson RJ, Schnaar RL. A functional role for complex gangliosides: motor deficits in GM2/GD2 synthase knockout mice. Exp Neurol. 2000;166(2):227–34.

Coetzee T, Fujita N, Dupree J, Shi R, Blight A, Suzuki K, et al. Myelination in the absence of galactocerebroside and sulfatide: normal structure with abnormal function and regional instability. Cell. 1996;86(2):209–19.

De Maria R, Lenti L, Malisan F, d'Agostino F, Tomassini B, Zeuner A, et al. Requirement for GD3 ganglioside in CD95- and ceramide-induced apoptosis. Science. 1997;277(5332):1652–5.

Delrieu J, Ousset PJ, Caillaud C, Vellas B. 'Clinical trials in Alzheimer's disease': immunotherapy approaches. J Neurochem. 2012;120 Suppl 1:186–93.

Ferrari G, Fabris M, Gorio A. Gangliosides enhance neurite outgrowth in PC12 cells. Brain Res. 1983;284(2–3):215–21.

Fujimoto H, Tadano-Aritomi K, Tokumasu A, Ito K, Hikita T, Suzuki K, et al. Requirement of seminolipid in spermatogenesis revealed by UDP-galactose: ceramide galactosyltransferase-deficient mice. J Biol Chem. 2000;275(30):22623–6.

Fukumoto S, Mutoh T, Hasegawa T, Miyazaki H, Okada M, Goto G, et al. GD3 synthase gene expression in PC12 cells results in the continuous activation of TrkA and ERK1/2 and enhanced proliferation. J Biol Chem. 2000;275(8):5832–8.

Furukawa K, Ohmi Y, Ohkawa Y, Tokuda N, Kondo Y, Tajima O. Regulatory mechanisms of nervous systems with glycosphingolipids. Neurochem Res. 2011;36(9):1578–86.

Furukawa K, Tajima O, Okuda T, Tokuda N, Furukawa K. Knockout mice and glycolipids. In: Kamerling JP, Boons GJ, Lee YC, Suzuki A, Taniguchi N, Voragen AGJ, editors. Comprehensive glycoscience from chemistry to systems biology. Oxford, UK: Elsevier; 2007. p. 149–57.

Furukawa K, Takamiya K, Okada M, Inoue M, Fukumoto S. Novel functions of complex carbohydrates elucidated by the mutant mice of glycosyltransferase genes. Biochim Biophys Acta. 2001;1525(1–2):1–12.

Furukawa K, Tokuda N, Okuda T, Tajima O. Glycosphingolipids in engineered mice: insights into function. Semin Cell Dev Biol. 2004;15(4):389–96.

Greene LA, Tischler AS. Establishment of a noradrenergic clonal line of rat adrenal pheochromocytoma cells which respond to nerve growth factor. Proc Natl Acad Sci U S A. 1976;73(7):2424–8.

Honke K, Hirahara Y, Dupree J, Suzuki K, Popko B, Fukushima K, et al. Paranodal junction formation and spermatogenesis require sulfoglycolipids. Proc Natl Acad Sci U S A. 2002;99(7):4227–32.

Ichikawa S, Sakiyama H, Suzuki G, Hidari KI, Hirabayashi Y. Expression cloning of a cDNA for human ceramide glucosyltransferase that catalyzes the first glycosylation step of glycosphingolipid synthesis. Proc Natl Acad Sci U S A. 1996;93(22):12654.

Inoue M, Fujii Y, Furukawa K, Okada M, Okumura K, Hayakawa T, et al. Refractory skin injury in complex knock-out mice expressing only the GM3 ganglioside. J Biol Chem. 2002;277(33):29881–8.

Itoh M, Fukumoto S, Iwamoto T, Mizuno A, Rokutanda A, Ishida HK, et al. Specificity of carbohydrate structures of gangliosides in the activity to regenerate the rat axotomized hypoglossal nerve. Glycobiology. 2001;11(2):125–30.

Jennemann R, Sandhoff R, Wang S, Kiss E, Gretz N, Zuliani C, et al. Cell-specific deletion of glucosylceramide synthase in brain leads to severe neural defects after birth. Proc Natl Acad Sci U S A. 2005;102(35):12459–64.

Kawai H, Allende ML, Wada R, Kono M, Sango K, Deng C, et al. Mice expressing only monosialoganglioside GM3 exhibit lethal audiogenic seizures. J Biol Chem. 2001;276(10):6885–8.

Kawai H, Sango K, Mullin KA, Proia RL. Embryonic stem cells with a disrupted GD3 synthase gene undergo neuronal differentiation in the absence of b-series gangliosides. J Biol Chem. 1998;273(31):19634–8.

Kittaka D, Itoh M, Ohmi Y, Kondo Y, Fukumoto S, Urano T, et al. Impaired hypoglossal nerve regeneration in mutant mice lacking complex gangliosides: down-regulation of neurotrophic factors and receptors as possible mechanisms. Glycobiology. 2008;18(7):509–16.
Kumagai T, Tanaka M, Yokoyama M, Sato T, Shinkai T, Furukawa K. Early lethality of beta-1, 4-galactosyltransferase V-mutant mice by growth retardation. Biochem Biophys Res Commun. 2009;379(2):456–9.
Levi A, Biocca S, Cattaneo A, Calissano P. The mode of action of nerve growth factor in PC12 cells. Mol Neurobiol. 1988;2(3):201–26.
Lloyd KO, Furukawa K. Biosynthesis and functions of gangliosides: recent advances. Glycoconj J. 1998;15(7):627–36.
Mutoh T, Tokuda A, Miyadai T, Hamaguchi M, Fujiki N. Ganglioside GM1 binds to the Trk protein and regulates receptor function. Proc Natl Acad Sci U S A. 1995;92(11):5087–91.
Nagata Y, Yamashiro S, Yodoi J, Lloyd KO, Shiku H, Furukawa K. Expression cloning of beta 1,4N-acetylgalactosaminyltransferase cDNAs that determine the expression of GM2 and GD2 gangliosides. J Biol Chem. 1992;267(17):12082–9.
Nishie T, Hikimochi Y, Zama K, Fukusumi Y, Ito M, Yokoyama H, et al. Beta4-galactosyltransferase-5 is a lactosylceramide synthase essential for mouse extra-embryonic development. Glycobiology. 2010;20(10):1311–22.
Nishio M, Fukumoto S, Furukawa K, Ichimura A, Miyazaki H, Kusunoki S, et al. Overexpressed GM1 suppresses nerve growth factor (NGF) signals by modulating the intracellular localization of NGF receptors and membrane fluidity in PC12 cells. J Biol Chem. 2004;279(32): 33368–78.
Ohkawa Y, Ohmi Y, Tajima O, Yamauchi Y, Furukawa K. Wisp2/CCN5 up-regulated in the central nervous system of GM3-only mice facilitates neurite formation in Neuro2a cells via integrin-Akt signaling. Biochem Biophys Res Commun. 2011;411(3):483–9.
Ohmi Y, Ohkawa Y, Yamauchi Y, Tajima O, Furukawa K. Essential roles of gangliosides in the formation and maintenance of membrane microdomains in brain tissues. Neurochem Res. 2012;37(6):1185–91.
Ohmi Y, Tajima O, Ohkawa Y, Mori A, Sugiura Y, Furukawa K. Gangliosides play pivotal roles in the regulation of complement systems and in the maintenance of integrity in nerve tissues. Proc Natl Acad Sci U S A. 2009;106(52):22405–10.
Ohmi Y, Tajima O, Ohkawa Y, Yamauchi Y, Sugiura Y, Furukawa K. Gangliosides are essential in the protection of inflammation and neurodegeneration via maintenance of lipid rafts: elucidation by a series of ganglioside-deficient mutant mice. J Neurochem. 2011;116(5):926–35.
Okada M, Itoh Mi M, Haraguchi M, Okajima T, Inoue M, Oishi H, et al. b-series Ganglioside deficiency exhibits no definite changes in the neurogenesis and the sensitivity to Fas-mediated apoptosis but impairs regeneration of the lesioned hypoglossal nerve. J Biol Chem. 2002;277(3): 1633–6.
Patra SK. Dissecting lipid raft facilitated cell signaling pathways in cancer. Biochim Biophys Acta. 2008;1785(2):182–206.
Rogers J, Cooper NR, Webster S, Schultz J, McGeer PL, Styren SD, et al. Complement activation by beta-amyloid in Alzheimer disease. Proc Natl Acad Sci U S A. 1992;89(21):10016–20.
Sardi F, Fassina L, Venturini L, Inguscio M, Guerriero F, Rolfo E, et al. Alzheimer's disease, autoimmunity and inflammation. The good, the bad and the ugly. Autoimmun Rev. 2011;11(2): 149–53.
Schengrund CL. The role(s) of gangliosides in neural differentiation and repair: a perspective. Brain Res Bull. 1990;24(1):131–41.
Shah S, Federoff HJ. Therapeutic potential of vaccines for Alzheimer's disease. Immunotherapy. 2011;3(2):287–98.
Sheikh KA, Sun J, Liu Y, Kawai H, Crawford TO, Proia RL, et al. Mice lacking complex gangliosides develop Wallerian degeneration and myelination defects. Proc Natl Acad Sci U S A. 1999;96(13):7532–7.

Shen Y, Lue L, Yang L, Roher A, Kuo Y, Strohmeyer R, et al. Complement activation by neurofibrillary tangles in Alzheimer's disease. Neurosci Lett. 2001;305(3):165–8.
Shen Y, Meri S. Yin and Yang: complement activation and regulation in Alzheimer's disease. Prog Neurobiol. 2003;70(6):463–72.
Simons K, Gerl MJ. Revitalizing membrane rafts: new tools and insights. Nat Rev Mol Cell Biol. 2010;11(10):688–99.
Simpson MA, Cross H, Proukakis C, Priestman DA, Neville DC, Reinkensmeier G, et al. Infantile-onset symptomatic epilepsy syndrome caused by a homozygous loss-of-function mutation of GM3 synthase. Nat Genet. 2004;36(11):1225–9.
Sugiura Y, Furukawa K, Tajima O, Mii S, Honda T. Sensory nerve-dominant nerve degeneration and remodeling in the mutant mice lacking complex gangliosides. Neuroscience. 2005;135(4):1167–78.
Suzuki KG, Kasai RS, Fujiwara TK, Kusumi A. Single-molecule imaging of receptor-receptor interactions. Methods Cell Biol. 2013;117:373–90.
Suzuki KG, Kasai RS, Hirosawa KM, Nemoto YL, Ishibashi M, Miwa Y, et al. Transient GPI-anchored protein homodimers are units for raft organization and function. Nat Chem Biol. 2012;8(9):774–83.
Takamiya K, Yamamoto A, Furukawa K, Yamashiro S, Shin M, Okada M, et al. Mice with disrupted GM2/GD2 synthase gene lack complex gangliosides but exhibit only subtle defects in their nervous system. Proc Natl Acad Sci U S A. 1996;93(20):10662–7.
Takamiya K, Yamamoto A, Furukawa K, Zhao J, Fukumoto S, Yamashiro S, et al. Complex gangliosides are essential in spermatogenesis of mice: possible roles in the transport of testosterone. Proc Natl Acad Sci U S A. 1998;95(21):12147–52.
Tokuda N, Numata S, Li X, Nomura T, Takizawa M, Kondo Y, et al. β4GalT6 is involved in the synthesis of lactosylceramide with less intensity than β4GalT5. Glycobiology. 2013;23(10):1175–83.
Vaudry D, Stork PJ, Lazarovici P, Eiden LE. Signaling pathways for PC12 cell differentiation: making the right connections. Science. 2002;296(5573):1648–9.
Wiegandt H. Gangliosides. In: Wiegandt H, editor. Glycolipids. Amsterdam: Elsevier; 1985. p. 199–260.
Wu G, Xie X, Lu ZH, Ledeen RW. Cerebellar neurons lacking complex gangliosides degenerate in the presence of depolarizing levels of potassium. Proc Natl Acad Sci U S A. 2001;98(1):307–12.
Yamashita T, Hashiramoto A, Haluzik M, Mizukami H, Beck S, Norton A, et al. Enhanced insulin sensitivity in mice lacking ganglioside GM3. Proc Natl Acad Sci U S A. 2003;100(6):3445–9.
Yamashita T, Wada R, Sasaki T, Deng C, Bierfreund U, Sandhoff K, et al. A vital role for glycosphingolipid synthesis during development and differentiation. Proc Natl Acad Sci U S A. 1999;96(16):9142–7.
Yamashita T, Wu YP, Sandhoff R, Werth N, Mizukami H, Ellis JM, et al. Interruption of ganglioside synthesis produces central nervous system degeneration and altered axon-glial interactions. Proc Natl Acad Sci U S A. 2005;102(8):2725–30.
Yoon SJ, Nakayama K, Hikita T, Handa K, Hakomori SI. Epidermal growth factor receptor tyrosine kinase is modulated by GM3 interaction with N-linked GlcNAc termini of the receptor. Proc Natl Acad Sci U S A. 2006;103(50):18987–91.
Yu RK, Macala LJ, Taki T, Weinfield HM, Yu FS. Developmental changes in ganglioside composition and synthesis in embryonic rat brain. J Neurochem. 1988;50(6):1825–9.

Chapter 15
Glycobiology of Ion Transport in the Nervous System

Martha C. Nowycky, Gusheng Wu, and Robert W. Ledeen

Abstract The nervous system is richly endowed with large transmembrane proteins that mediate ion transport, including gated ion channels as well as energy-consuming pumps and transporters. Transport proteins undergo N-linked glycosylation which can affect expression, location, stability, and function. The N-linked glycans of ion channels are large, contributing between 5 and 50 % of their molecular weight. Many contain a high density of negatively charged sialic acid residues which modulate voltage-dependent gating of ion channels. Changes in the size and chemical composition of glycans are responsible for developmental and cell-specific variability in the biophysical and functional properties of many ion channels. Glycolipids, principally gangliosides, exert considerable influence on some forms of ion transport, either through direct association with ion transport proteins or indirectly through association with proteins that activate transport through appropriate signaling. Examples of both pumps and ion channels have been

M.C. Nowycky (✉)
Department of Pharmacology and Physiology, RBHS, New Jersey Medical School, The State University of New Jersey, 185 South Orange Ave., Newark, NJ 07103, USA
e-mail: nowyckmc@njms.rutgers.edu

G. Wu
Department of Neurology and Neurosciences of Rutgers, RBHS, New Jersey Medical School, The State University of New Jersey, 185 South Orange Ave, Newark, NJ 07103, USA
e-mail: gwu@njms.rutgers.edu

R.W. Ledeen
Department of Neurology and Neurosciences of Rutgers, RBHS, New Jersey Medical School, The State University of New Jersey, 185 South Orange Ave., Newark, NJ 07103, USA

Department of Pharmacology and Physiology, RBHS, New Jersey Medical School, The State University of New Jersey, 185 South Orange Ave., Newark, NJ 07103, USA
e-mail: ledeenro@njms.rutgers.edu

R.K. Yu and C.-L. Schengrund (eds.), *Glycobiology of the Nervous System*, Advances in Neurobiology 9, DOI 10.1007/978-1-4939-1154-7_15,

revealed which depend on ganglioside regulation. While some of these processes are localized in the plasma membrane, ganglioside-regulated ion transport can also occur at various loci within the cell including the nucleus. This chapter will describe ion channel and ion pump structures with a focus on the functional effects of glycosylation on ion channel availability and function, and effects of alterations in glycosylation on nervous system function. It will also summarize highlights of the research on glycolipid/ganglioside-mediated regulation of ion transport.

Keywords Ion channels • Voltage-gated sodium channel • Transporters • GM1 and Ca^{2+} modulation • Sodium-calcium exchanger • TRP channels

Abbreviations

N'ase (also called sialidase)	Neuraminidase
TRP (TRPC3, TRPC5, TRPC6, TRPV1, TRPM8)	Transient receptor potential channel
AChR	Nicotinic acetylcholine receptor
CtxB	Cholera toxin B subunit
DRG	Dorsal root ganglion
ER	Endoplasmic reticulum
NCX	Sodium-calcium exchanger
PMCA	Plasma membrane Ca^{2+}-ATPase
SERCA	Sarco/endoplasmic reticulum Ca^{2+}-ATPase

15.1 Introduction

The transport of charged ions across impermeant lipid membranes is a critical physiological process in all cells. Pumps, transporters, and channels are found in the plasma membrane, as well as the membranes of organelles such as mitochondria, the endoplasmic reticulum (ER), and nucleus. Excitable cells, particularly neurons of the central and peripheral nervous system, are endowed with an unusually rich variety of voltage-gated and ligand-gated ion channels that are used for rapid electrical signaling and signal processing.

All transport proteins, and many of their auxiliary subunits, are glycosylated. As with most transmembrane proteins, changes in glycosylation affect both the expression and stability of proteins at the plasma membrane thereby changing the electrical properties of cells. In addition, the negative charges contributed by sialic acid residues in the glycan structures can affect the electrical potential near the pore of ion channels thereby influencing channel characteristics and behavior. In this chapter voltage-gated sodium channels will be used as exemplars to describe the functional effects of glycosylation on ion channel activity. We will focus on N-glycosylation of ion channels which is common to all transmembrane proteins. O-linked glycosylation has been described for a few channels (see e.g. Schwetz et al. 2011), however, it is much less well studied.

Glycolipids also contribute importantly to ion transport, gangliosides with their variety of oligosaccharide chains being the major species so engaged. As with glycoproteins, negatively charged sialic acid is crucial to such functional roles. These behave in many cases as modulatory agents through direct association with proteins that function as ion channels or pumps, while indirect modulation can also occur through association with proteins that influence ion transport downstream through appropriate signaling. Such processes are abundant in the plasma membrane but a growing number have been identified at various intracellular loci as well. The monosialo-ganglioside, GM1, has been the major focus of such activities, while the disialoganglioside, GD1a, has subsumed a supporting role as metabolic precursor to GM1. Our survey of such modulatory activities will include the especially prominent role that appears to have been assigned the GM1/GD1a duo in relation to Ca^{2+} transport.

15.2 Glycosylation of Ion Transport Proteins

15.2.1 Glycosylation of Voltage-Gated Na Channel

Voltage-gated Na channels are responsible for generating action potentials in neurons and other excitable cells. Structurally, they are representative of an ancient protein superfamily that includes voltage-gated channels selective for Na^+, Ca^{2+}, and K^+. The sodium channel was the first member of the family of voltage-gated ion channels to be sequenced and cloned [for historical review see Catterall 2012]. Early structural studies identified a large glycoprotein containing a major α subunit as well as a noncovalently associated β1 subunit and a disulfide-linked β2 subunit (Hartshorne and Catterall 1981, 1984; Barchi 1983). Currently, there are ten known isoforms of alpha subunits (proteins $Na_v1.1$ through $Na_v1.9$, genes SCN1A–SCN11A), and four isoforms of beta subunits (proteins β1–β4 and genes SCNB1–SCNB4). The β1 and β3 subunits are noncovalently associated with the main subunit, while the β2 and β4subunit are disulfide-linked (Isom et al. 1992).

Sodium channels, similar to essentially all membrane proteins, undergo "N-linked" glycosylation of an asparagine residue. The target sequence is Asn-X-Ser/Thr, where X is any amino acid other than proline. Both the main α subunit and auxiliary β subunits are heavily glycosylated. Glycosylation accounts for 15–40 % of the MW of the main pore-forming α subunit, depending on factors such as the channel isoform, cell type, and developmental or pathological state (see below). For example, the mature α subunit protein isolated from rat brain has a MW of 260 kDa (Schmidt and Catterall 1987). Inhibition of glycosylation with tunicamycin lowered

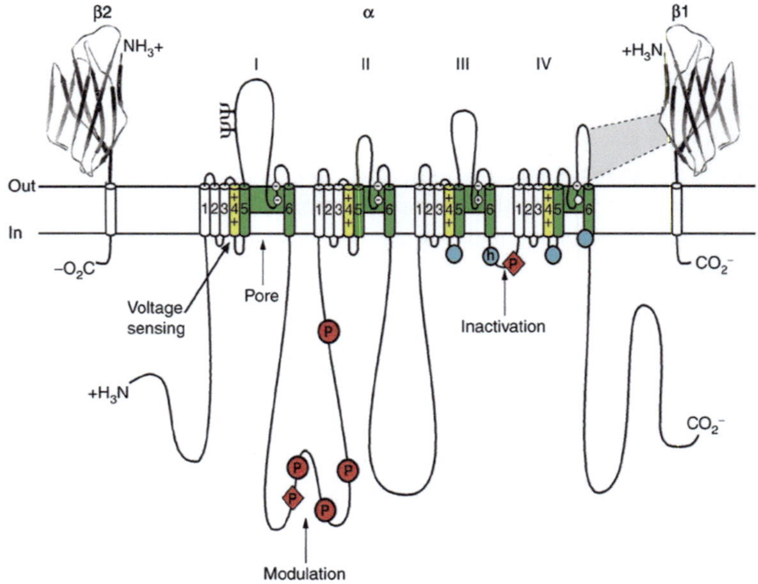

Fig. 15.1 Model of the primary structure of a voltage-gated sodium channel. The structure of the channel is illustrated with transmembrane α-helical regions represented as cylinders. S4 segments of domains I–IV are the primary voltage-sensing regions. S5–S6 with the re-entrant loop line the pore. Glycosylation sites on the α subunit are limited to the extracellular loop between S5 and S6 in domain I and indicated with Ψ. The lengths of lines are approximately proportional to the lengths of each extra- or intracellular poplypeptide segment. The extracellular domains of the β1 and β2 subunits are shown as immunoglobulin-like folds. Reprinted from Catterall (2012) with permission from John Wiley and Sons

the apparent MW to 203 kDa, a value close to that predicted for the ~2,000 amino acid sequence. The β subunits are also heavily glycosylated. Mutation of the putative glycosylation sites of the β1 subunit reduced the MW from 38 to 22 kDa, a value predicted for the amino acid sequence (Johnson et al. 2004).

The structure of the α subunit of voltage-gated Na channels is typical of the voltage-gated ion channel family: the subunit consists of 4 domains (I–IV) each with six alpha helical transmembrane segments (S1–S6) arranged to form a cylindrical structure in the membrane (Fig. 15.1). The four sets of S5–S6 segments form the inner wall of the pore, while the corresponding S1–S4 segments are arranged around them. The long extracellular loop between S5 and S6 is called the P loop (*P*ore) and is reentrant: it projects extracellularly, then dips half-way into the membrane before again emerging extracellularly. The S4 segment is the main voltage-sensor for activation or opening of voltage-gated channels, while the amino acids along the inward dip of the P loop are primary determinants of ion selectivity and conductance.

Despite the modular structure of the channel, all the putative glycosylation sites of the α subunit are restricted to the large extracellular S5–S6 loop of Domain I

(Bennett 2002). Deleting putative sites in the Domain I S5–S6 loop eliminated glycosylation (Bennett et al. 1997). When chimeric channels were formed by switching the Domain I S5–S6 loop of human isoform Nav1.4 (hSkM1), a heavily glycosylated isoform, with the loop from human Nav1.5 (hH1), an isoform with little glycosylation, both chimeric channels exhibited the glycosylation properties appropriate to the channel from which the loops were derived (Bennett 2002).

β subunits are single-pass transmembrane proteins of the Ig superfamily, with a short cytoplasmic C-terminus and large extracellular portion that has a V-type Ig-like fold (Fig. 15.1). The β1 subunit has 3–4 N-linked carbohydrate chains as determined by sequential treatment with neuraminidase and endoglycosidase (Messner and Catterall 1985). This corresponds well with the four putative extracellular glycosylation sites predicted by "sequence gazing" (Isom et al. 1992).

An unusual characteristic of ion channel glycosylation is the high sialic acid residue content of the glycan structures. In an analysis of carbohydrate content of the main subunit of the eel electroplax voltage-gated sodium channels, Miller et al. (1983) estimated that sialic acid residues made up 11.8 % of the alpha subunit MW and 39.7 % of total carbohydrate. Roberts and Barchi (1987) estimated that the main subunit molecule of the skeletal muscle sodium channel contained over 100 sialic acid residues. James and Agnew (1989) estimated a very high degree of polysialic acid for the same channel—~113 negative charges to the protein surface. The ratio of sialic acid residues to consensus glycosylation sites suggests that the terminal chains are well over ten sialosyl residues in length, potentially extending 10–30 nm into the extracellular environment (James and Agnew 1989). Polysialic acid in glycan structures of vertebrate proteins is rare and is most often associated with neural cell adhesion molecules (James and Agnew 1989; Rutishauser and Landmesser 1996). In electroplax membranes from *Torpedo*, the sodium channel is the only protein that is immunoreactive for sialic acid (James and Agnew 1989).

15.2.2 Functional Consequences of Ion Channel Glycosylation

The high degree of sialic acid content found in ion channels has consequences for their biophysical properties and function. Each sialic acid residue usually carries a negative charge at physiological pH. The cloud of negative charges created by a high density of sialic acid residues, particularly when localized just above the channel pore as it is for voltage-gated sodium channels, can have functional effects on voltage-dependent properties. Complete enzymatic removal of sialic acid from rat skeletal muscle sodium channels modified several parameters of voltage-dependent gating (Bennett et al. 1997). The voltage dependence of the time constants of channel activation and inactivation and the voltage for steady-state half-activation were ~10 mV more depolarized for neuraminidase-treated channels compared to controls. A similar shift was observed when channels were expressed in a sialylation-deficient cell line (lec2) or following deletion of likely glycosylation sites.

Desialylated channels were also less sensitive to the charge screening effects of external calcium. The authors concluded that sialic acid most likely contributed to the negative surface potential and altered the electric field sensed by channel gating elements (Bennett et al. 1997).

In a later study, the same group compared two human isoforms of sodium channels: Nav1.4 from skeletal muscle (hSkM1) and Nav1.5 (hH1) from heart (Bennett 2002). Native hSkM1 has numerous sialic acid residues while hH1 does not. Consistent with the predicted role of sialic acid, neuraminidase caused a depolarizing shift in channel gating of hSkM1 but not hH1. When the S5–S6 loop of domain I was exchanged between the two isoforms, the loop determined the degree of sialylation of the chimeric channels, indicating that the primary sequence of the S5–S6 linker of domain I to some extent regulates the composition of the glycan structure attached to the channel (Bennett 2002).

In addition to the effects of sialic acid residues attached to the main pore-forming subunit, Johnson et al. (2004) reported that glycosylation of the auxiliary β subunits also modifies the activation and inactivation range of voltage-gated Na channels. The fully sialylated β1 subunit caused a hyperpolarizing shift in the voltage range of gating for the cardiac and two neuronal α subunit isoforms. Mutation of the N-glycosylation sites abolished the effects of the β1 subunit. Interestingly, the β1 subunit had no effect on the gating properties of the heavily glycosylated skeletal muscle α subunit. The authors proposed a saturating electrostatic mechanism in which a spectrum of differentially sialylated α and β subunits could modulate channel gating and influence the excitability properties of cells (Johnson et al. 2004).

Variability in glycosylation of ion channels can have profound effects on the excitability characteristics of a cell. Stocker and Bennett (2006) proposed mechanisms by which cardiac voltage-gated sodium channel gating and subsequently cardiac rhythms are modulated by changes in channel-associated sialic acids. Changes in neuronal excitability and function have been reported for various regions of the nervous system. Isaev et al. (2007) reported that the activation and inactivation properties of voltage-gated Na channels from CA3 pyramidal cells were shifted towards depolarizing potentials by treatment of hippocampal slices by neuraminidase. This led to an increase in the action potential threshold due to enhanced steady-state inactivation. Neuraminidase treatment had powerful anticonvulsive action both in vitro and in acute and chronic in vivo models of epilepsy (Isaev et al. 2007; Isaeva et al. 2011). Large diameter dorsal root ganglion (DRG) cells (Aα and β) become hyperexcitable following chronic constriction injury. The hyperexcitability of injured DRG neurons was reduced substantially by neuraminidase treatment (Peng et al. 2004). Desialylation had no effect on normal intact neurons. Such effects may indicate plasticity in the glycosylation of ion channels.

Sialic acid residues modify the function of other voltage-gated channels in addition to Na channels. The channel gating of voltage-gated Kv1.1 (potassium) channels expressed in CHO cells was shifted in a depolarizing direction by tunicamycin and neuraminidase treatment (Thornhill et al. 1996). Kv3.1, a voltage-gated potassium channel expressed throughout the nervous system, was studied in its glycosylated (wild type), partially glycosylated (N220Q or N229Q) and unglycosylated (N220Q/N229Q) states (Hall et al. 2011). Wild-type Kv3.1 channel currents had faster activation and deactivation rates than the mutated forms. When channels were expressed in the B35 neuroblastoma cell line, cells with wild-type Kv3.1 channels migrated more rapidly in a wound-healing assay (Hall et al. 2011). Potassium channel activity can also be modulated by glycosylation of auxiliary subunits. Tunicamycin blocked the glycosylation of DPP10, an auxiliary subunit of the Kv4 channel and a key determinant of cardiac and neuronal excitability (Cotella et al. 2010). This abolished the effects of DPP10 on Kv4.3 inactivation and recovery from inactivation.

Glycosylation can also affect ion channels gated by ligands or other activators. Glycosylation affects the stability, assembly, and open times of the $GABA_A$ionotropic channel (Lo et al. 2010) Although more work needs to be done, the authors postulate that the N-glycans of at least one of the channel's glycosylation sites (N173) make stabilizing contacts with adjacent subunits or perhaps with chaperone proteins during the assembly process. A particularly intriguing observation is the role of glycosylation in regulation of TRPC6 and TRPC3, two closely related channels of the TRP family. TRPC6 has 2 glycosylation sites and is tightly regulated while TRPC3 has a single glycosylation site and is constitutively active. Mutation of the unique NX(S/T) motif in TRPC6 converted it to a constitutively active channel, while addition of the second glycosylation site to TRPC3 reduced TRPC3 basal activity (Dietrich et al. 2003). Glycosylation also regulates the ligand binding and gating properties of the TRP vanilloid 1 receptor (TRPV1; Wirkner et al. 2005). A mutant channel lacking the potential N-glycosylation site at position N604 had a depressed maximum of the dose–response curve for capsaicin, altered dependence of the capsaicin effect on extracellular pH, and decreased sensitivity to the antagonist capsazepine. Thus glycosylation affects both the gating, ligand binding, and pharmacology of TRP channels.

15.2.3 Plasma Membrane Expression

N-linked glycosylation plays an important role in the expression, insertion, and stability of many plasma membrane proteins. Expression and localization of ion channels is tightly regulated by multiple factors such as targeting to lipid rafts, anchoring to scaffolding proteins, and coassembly with essential or auxiliary subunits. Glycosylation, while important, is only one factor influencing expression. Consequently, the role of glycosylation varies for different ion channels, between cell types, and under specific physiological conditions.

As with most proteins expressed on the surface membrane, the number and density of voltage-gated Nachannels is decreased when glycosylation is inhibited by tunicamycin. In neuroblastoma cells, tunicamycin reduced the number of voltage-gated Na channels as measured by high-affnity saxitoxin binding to 20–28 % of control values over a 60 h period or by batrachotoxin-activated $^{22}Na^+$ influx (Waechter et al. 1983). In embryonic rat neocortical neurons, tunicamycin application decreased the voltage-clamped Na current to ~40 % of control values (Zona et al. 1990). The effect was much more rapid for neurons that were actively growing in culture (days 5–14 after dissociation) compared to more established neurons (days 20–40 after dissociation). Currents were reduced to 40 % within 24 h for more recently plated neurons compared with 68 h for neurons that were fully grown (Zona et al. 1990).

In squid giant fiber lobe neurons of the stellate ganglion, voltage-gated Na channels are present at high density in axons but are absent from its somata in vivo. This distribution is maintained in culture except for the appearance of low-level expression in cell bodies (Gilly et al. 1990). Tunicamycin disrupted the expression of Na channels in axonal membranes in vitro with no effect on low levels in the soma. Disruption of glycosylation did not affect voltage-gated Na channel turnover, axon viability, or K channel distribution, indicating that glycosylation had a specific effect on voltage-gated Na channel localization in this system (Gilly et al. 1990).

The effect of glycosylation on other channels is highly variable. Glycosylation has profound effects on the stability of the Shaker K channel, although it is not needed for expression (Khanna et al. 2001). In pulse chase experiments, the wild-type protein was stable with little degradation after 48 h, however, a mutant form with glycosylation sites removed (N259Q, N263Q) was rapidly degraded ($t_{1/2} \sim 18$ h). Glycosylation regulates efficient multimerization and transport of the TRPM8 channel (Erler et al. 2006). Similarly, mutation of two putative glycosylation sites within $\alpha 2\delta$, a subunit that regulates trafficking and function of voltage-gated calcium channels, decreased the number of functional charges in the plasma membrane (Sandoval et al. 2004). The pentameric nicotinic acetylcholine receptor (AChR) assembles when glycosylation is blocked, however, it is not inserted into the plasma membrane, remaining stuck in internal compartments (Sumikawa and Miledi 1989).

15.2.4 Variability of Glycosylation

The extent of glycosylation is highly variable not only for ion channel isoforms, but also for any given isoform expressed in different cells or at various developmental stages. Differences in glycosylation affect both channel expression levels and may manifest as profound alterations of biophysical properties. These, in turn, can affect neuronal excitability and function.

The degree of glycosylation can be due to differences in the primary sequence of an isoform. An example was provided earlier in the studies of the heavily glycosylated Nav1.4 channel compared to the lightly glycosylated Nav1.5 (Bennett 2002). Exchange of the domain I S5–S6 linker regions demonstrated that the extent of glycosylation was determined by the amino acids of the loop structure. Alternatively, the primary amino acid sequence distant from the glycosylation site may have an effect. Kv1.4 and Kv1.1 are two isoforms of a mammalian Shaker family channel involved in action potential repolarization. Block of N-glycosylation affected the protein trafficking, stability, and cell surface expression of Kv1.4, while Kv1.1 was unaffected. However, exchanging a trafficking pore region—a site distant from the glycosylation loop—from Kv1.4 into Kv1.1, caused changes in Kv1.1 comparable to those of Kv1.4 (Watanabe et al. 2004). The authors conclude that multiple regions of the protein must participate in proper folding, trafficking, and glycosylation of the channel.

Much of the variability in glycosylation is due to the expression and activity of the glycosylation machinery. N-linked glycosylation is a complex process that takes place within the ER and Golgi and involves numerous enzymes (see Chap. 3 for details). Briefly, the glycan structure is preformed in the endoplasmic reticulum attached to a lipid anchor. It is transferred to a target asparagine cotranslationally, while the peptide is synthesized. The N-linked glycan is further modified by removal of sugars (trimming) and re-addition of sugar residues (processing). Dozens of genes participate in this process. The enzymes are controlled through gene expression or through regulation or targeting of the proteins (Ohtsubo and Marth 2006).

Variability in the glycosylation of a single ion channel isoform can arise because of differences in expression of specific glycosylation enzymes between cells, at various stages of the developmental process, or in pathological conditions. A particularly striking example of the cell-type variability that can be observed is seen for cells derived from the various chambers of the heart. Montpetit et al. (2009) compared the expression of 239 genes coding for glycosyltransferases, glycosidases, and sugar nucleotide synthesis/transporter genes in four myocyte types: neonatal and adult atrium, and neonatal and adult ventricle. Of these, 110 glycogenes tested in mice were significantly differentially expressed among the four myocyte types (see Montpetit et al. 2009, their Fig. 15.1).

Stocker and Bennett (2006) found that the channels from neonatal atria, compared to adult atria or neonatal or adult ventricles, are much more heavily sialylated with approximately 15 more sialic acid residues attached to each alpha subunit. They showed that the difference is due to the expression of ST8sia2, a polysialyltransferase, that is expressed only in the neonatal atrium but not in adult atrium or in ventricles at any age (Montpetit et al. 2009). Comparison of action potential waveforms in neonatal atrial cells to those from ST8sia2 −/− mice showed distinct differences in time-to-peak and AP duration. These were consistent with the negative shift of gating of voltage-gated Na channels observed after ST8sia2 expression, while there was no effect on the level of expression of voltage-gated Na channels. The authors conclude that the expression of a single glycogene is sufficient to modulate cardiomyocyte excitability.

Similar changes in glycosylation and functional consequences also occur in neurons. The degree of sialylation can be developmentally regulated and accounts for variability in the biophysical properties of channels expressed at different stages of an animal's life. The tetrodotoxin-resistant voltage-gated Na channel isoform Nav1.9 exists in two glycosylated states in neonatal rat DRG neurons (but only in a less glycosylated isoform in adult DRG (Tyrrell et al. 2001)). Deglycosylation of Nav1.9 caused an 8 mV depolarizing shift in steady-state inactivation in the neonatal but not adult DRG (Tyrrell et al. 2001). Castillo et al. (1997) demonstrated a progressive shift in the gating characteristics of forebrain Na channels from P0, P15, and adult P30/P180 rats. The shifts in the midpoint potential of activation paralleled an increase in apparent size. Both were reversed by neuraminidase treatment, indicating the increased amount of sialylation during development. In addition to variability due to glycosylation of the main pore-forming subunit of voltage-gated Na channels, glycosylation of auxiliary subunits also contributes to the complexity of effects. For example, the β4 subunit is expressed in a 35 kDa form between P0 and P6, but shifts to a heavily glycosylated 38 kDa form on P7 (Zhou et al. 2012).

15.3 Glycolipids and Ion Transport

Glycolipids influence ion transport primarily through association with and modulation of transport-associated proteins, which are often glycoconjugates themselves. Two general approaches have been employed in such investigations: (a) study of endogenous glycolipids through manipulations such as genetic alteration or structurally specific perturbing agents, and (b) application of exogenous glycolipids to isolated transport systems or cultured cells containing the transport system. The implied assumption with the exogenous approach is that the applied glycolipid inserts into the membrane or associates with the isolated transporter in a manner corresponding to its natural topography. This is often the reality, in which case the observed glycolipid-induced effects are viewed as true manifestations of their physiological function. However, exogenous glycolipids are known to associate with cellular membranes in three distinct modes: (a) a loosely attached pool removable with serum, (b) a somewhat more tightly associated pool released by trypsin, and (c) a serum- and trypsin-stable component consisting of the membrane-inserted pool (Wu and Ledeen 1994). The latter fraction, normally a small portion of associated glycolipid, is the one most likely to mimic endogenous glycolipid in relation to natural function, although the other two pools can conceivably give rise to pharmacological effects of potential therapeutic interest. With those caveats in mind, we will recount examples where the exogenous approach has been employed, while emphasizing studies based on endogenous glycolipid function for which powerful research tools have become available.

In most cases the glycolipid under study has been one or another ganglioside, the negatively charged sialic acid being crucial to their role in mediating cation movement. GM1 ganglioside, the prototypic member of the ganglio-series (Fig. 15.2) is often considered in conjunction with GD1a, the other prominent member of the a-series (Ando and Yu 1979). A primary function of this disialoganglioside is that of metabolic precursor to GM1 by virtue of neuraminidase (N'ase, also called sialidase), most forms of which remove only the terminal sialic acid. These two glycolipids have received prominent attention in regard to Ca^{2+} transport mechanisms. GM1 is one of the few sialoglycoconjugates in nature resistant to most types of N'ase, a property that facilitates elevation of its concentration on membrane surfaces while retaining its negative charge. The fact that it binds with high affinity and relative selectivity to the B subunit of cholera toxin (CtxB) (Schengrund and Ringler 1989) has provided a useful tool for probing its functional roles as well as its location in and within specific cells.

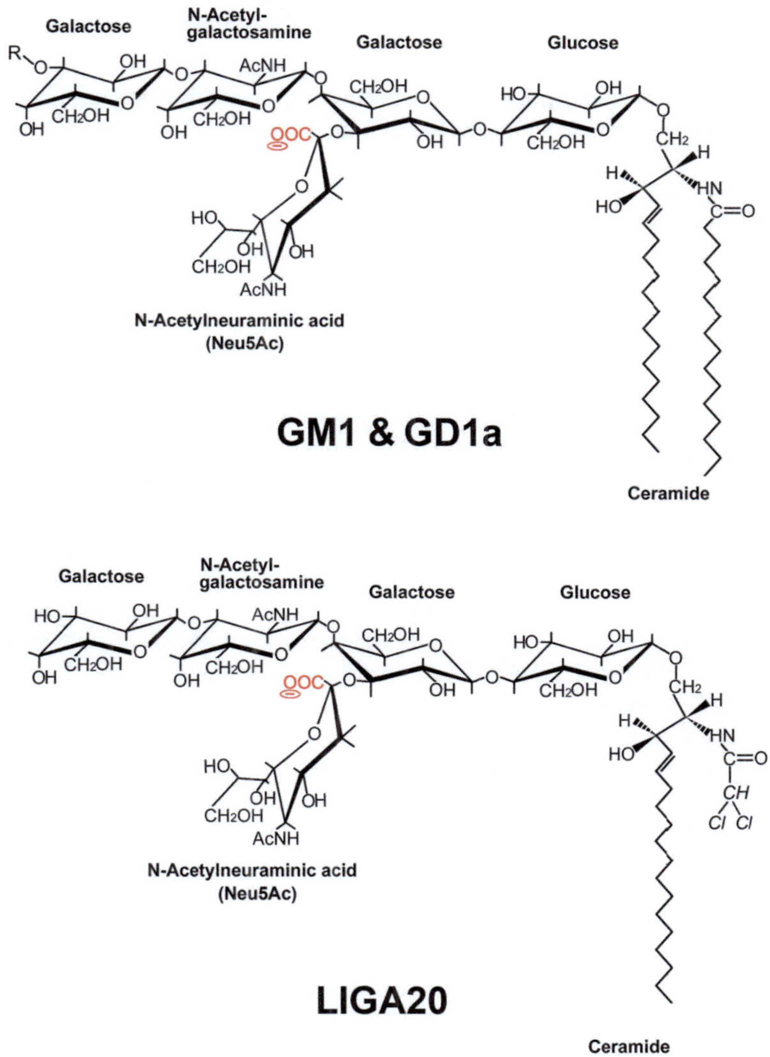

Fig. 15.2 Structure of GM1 ganglioside (R=H) and GD1a, a disialoganglioside in which R is an additional sialic acid (Neu5Ac). Also shown is LIGA20, a semi-synthetic, membrane permeable analog of GM1 in which the stearoyl unit (C18) of ceramide has been replaced with dichloroacetyl

A rare example of a neutral glycolipid influencing ion transport is that of glucosylceramide, which was shown to increase Ca^{2+} mobilization from intracellular stores in the ER, via activation of the ryanodine receptor (Lloyd-Evans et al. 2003; Korkotian et al. 1999). This property, proposed as an explanation for the pathophysiology of neuronopathic forms of Gaucher disease, was not shared with galactosylceramide and several other sphingolipids. Whether this is a normal function of glucosylceramide in nonpathological cells remains to be determined.

15.3.1 Ganglioside Modulation of Na^+ Transport

A role for GM1 in retaining neuronal conduction and excitability has been suggested in relation to its effect on Na^+ channels. In some neurological or neuroimmunological patients, clusters of voltage-gated Na^+ channels in nodes of Ranvier were shown to suffer damage by complement-mediated disruption through anti-GM1 antibodies, thought to represent disruption of axon-Schwann cell interactions at GM1 foci (Suzuki et al. 2007a). Mutant mice lacking the GM1 (ganglio) family of gangliosides due to disruption of the *B4galnt1* gene (GM2/GD2 synthase) were initially shown to have a slight reduction in the neural conduction velocity of the tibial nerve (Takamiya et al. 1996), whereas subsequent studies indicated altered paranodal junctions, broadened Na_v channel clusters, and aberrant Kv channel localization at the paranodes of peripheral motor nerves (Suzuki et al. 2007b). The fact that GD3 synthase gene knockout mice showed no loss of peripheral nerve conduction velocity (Handa et al. 2005) pointed to a-series gangliosides (GM1, GD1a) as the causative agents. These features further suggested that the GM1 contribution to Na_v channel function resided in maintenance of microdomain (raft) integrity. However, more intimate association of GM1 with Na_vchannels was suggested by the observation that current densities of both tetrodotoxin-sensitive and insentitive Na^+ channels were significantly decreased by CtxB (Qiao et al. 2008). That study provided evidence that endogenous GM1 plays a crucial role via modulation of Na_v channels in retaining the afferent conduction velocity of not only myelinated fibers of motor nerves but also of myelinated and unmyelinated fibers of visceral afferents. Sodium transport mediated by the antiporter, Na^+/K^+-ATPase was shown to be activated by nmolar concentrations of GM1, an effect that was diminished at higher concentrations (Leon et al. 1981). The fact that this regulatory property was shared with other ganglio-series gangliosides suggested the possible presence of N'ase in the employed crude membrane fraction that could have produced GM1.

15.3.2 Ganglioside Modulation of Ca^{2+} Transport at the Plasma Membrane

Efficient regulation of free intracellular Ca^{2+} is essential for maintaining viability and excitability of neurons, and GM1 ganglioside has been widely implicated in regulatory roles for this ion (Ledeen and Wu 2002). One approach to elucidating the relevant mechanisms has been to elevate its endogenous level on the cell surface with applied N'ase,which triggered Ca^{2+} influx in Neuro2a, B104, and B50 neuroblastoma cells but not N1A-103 or N18 cells (Wu and Ledeen 1991; Fang et al. 2000). Those cells experiencing an elevation of intracellular Ca^{2+} in this manner extended neurites which were subsequently shown to have axonal character (Wu et al. 1998a). Both N'ase-induced Ca^{2+} influx and neuritogenesis were blocked by CtxB, indicating that GM1 elevation, as opposed to other effects of N'ase, was key to the changes. Activation of a specific channel type by elevated GM1 was suggested based on blockade by low concentrations of amiloride, described as a specific inhibitor of low threshold voltage dependent T type channels (Tang et al. 1988). The physiological significance of these phenomena was suggested by the involvement of endogenous plasma membrane-localized ganglioside-reactive N'ase (Miyagi et al. 1999; Monti et al. 2000) that regulates axonal growth in Neuro2a cells (Hasegawa et al. 2000) and primary hippocampal neurons (Rodriguez et al. 2001).

As mentioned, some cell lines did not respond to N'ase with Ca^{2+} influx and axon outgrowth, but instead showed a different regulatory mechanism mediated by GM1. Contrary to the inhibitory effect of CtxB on Neuro2a cells, N18 cells responded to CtxB with Ca^{2+} influx (Masco et al. 1991; Carlson et al. 1994), the response being more robust if preceded by N'ase treatment (Fang et al. 2002). Thus, N'ase-mediated elevation of surface GM1 facilitated Ca^{2+} entry in both mechanisms but through activation of different channel types. Similar CtxB effects analogous to those in N18 cells were observed in primary neurons of both the PNS (Milani et al. 1992) and CNS (Wu et al. 1996), but not with microglia or oligodendrocytes (Nedelkoska and Benjamins 1998) or in Schwann cells (Skoff and Benjamins 1998). In a study with cerebellar granule neurons, CtxB-induce Ca^{2+} influx occurred during the first 7 days in culture after which CtxB inhibited Ca^{2+} influx. A somewhat similar developmental sequence was observed in NG108-15 cells, these responding to CtxB with Ca^{2+} influx only during the initial phase of axon outgrowth (Fang et al. 2002). In addition to this CtxB-mediated effect, NG108-15 cells also showed the above N'ase-induced Ca^{2+} influx, indicating coexistence of both GM1-regulated Ca^{2+} channels (Fang et al. 2002). Both mechanisms of Ca^{2+} influx resulted in axon-like neurite outgrowth, in contrast to dendrite-like processes that resulted from agents (e.g., retinoic acid, dibutyrylcAMP) that did not stimulate Ca^{2+} influx (Wu et al. 1998a). Paradoxically, elevation of cellular ganglioside through application of exogenous gangliosides resulted in Ca^{2+} influx that gave rise to dendrite-like processes (Wu et al. 1990; 1998a). Exogenous gangliosides were also shown to reduce intracellular Ca^{2+} elevated in Neuro2a cells by ionomycin (Wu and Ledeen 1994), suggesting ganglioside promotion of Ca^{2+} homeostasis as part of its neuroprotective mechanism (Nakamura et al. 1992).

Whereas Ca^{2+} influx induced by N'ase elevation of surface GM1 appeared to involve T type channels, the GM1-regulated mechanism activated by CtxB was eventually shown to involve the TRPC5 channel (Wu et al. 2007). TRPC5 is an isoform of the canonical subgroup of mammalian genes homologous to the transient receptor potential (TRP) family in *Drosophila* (Montell 2004). GM1 does not associate directly with this channel but rather with $\alpha5\beta1$ integrin heterodimers, these becoming cross-linked concurrently upon binding of CtxB to the ganglioside (Wu et al. 2007). Integrin cross-linking in this manner was shown to induce autophosphorylation of associated focal adhesion kinase, which in turn activated phospholipase $C\gamma$ and phosphoinositide-3 kinase. These effects were first revealed in NG108-15 cells following N'ase-induced elevation of cell surface GM1, which greatly enhanced the level of Ca^{2+} influx and neurite outgrowth induced by CtxB. This also promoted neurite outgrowth in murine cerebellar granule neurons without the need for N'ase pretreatment, this apparently being accomplished by upregulation of endogenous N'ase during neuronal differentiation. TRPC5 is prominently expressed in the soma of primary neurons and neuroblastoma cells only at an early stage of differentiation, consonant with selective activity of CtxB at that stage. The natural, endogenous cross-linking agent remained in doubt until studies with T cells in the immune system revealed that homodimeric galectin-1 exerts similar GM1 cross-linking as CtxB with similar TRPC5 Ca^{2+} channel activation (Wang et al. 2009a; Wu et al. 2011). Comparison of galectin-1 binding to GM1-deficient T cells vs wild type T cells suggested primary binding to GM1 (rather than glycoproteins) in that system (Wang et al. 2009a), consistent with the interaction of these two molecules in neuroblastoma cells and primary neurons (Kopitz et al. 1998; Gabius 2009). A schematic illustration has been presented (Ledeen et al. 2012) of homodimeric galectin-1 binding to the oligosaccharide structure of GM1 according to the Coulomb/van der Waals energy term obtained by computational interaction analysis (Siebert et al. 2003). Analogous to CtxB, anti-GM1 antibodies of the cross-linking IgM type were shown to induce similar Ca^{2+} changes (Quattrini et al. 2001) and neurite outgrowth (O'Hanlon et al. 2003) in neuroblastoma cells.

An additional example of GM1 modulation of Ca^{2+} influx, albeit indirectly, was seen in opioid activity of a certain type. Opioids are known to be capable of dual modulatory activities, as shown with action potential duration of sensory neurons (Shen and Crain 1989), neurotransmitter release in SK-N-SH cells (Keren et al. 1994), and calcium influx in NG108-15 cells (Jin et al. 1992). The excitatory mode, which promotes Ca^{2+} influx, was blocked by CtxB, implicating GM1 as facilitator of the excitatory response. This was verified in experiments showing conversion from inhibitory to excitatory mode by bath application of GM1 to CHO cells expressing the δ-opioid receptor (Wu et al. 1997a). The importance of the negatively charged carboxyl group of sialic acid was illustrated in loss of excitatory promotion by this modified GM1 (Wu et al. 1998b). Site-directed mutagenesis of the δ-opioid receptor involving replacement of the positively charged arginine residue at 192 with alanine also abolished the GM1-modulated excitatory response, suggesting this as the likely locus for charge-charge interaction of GM1 with this receptor (Wu et al. 1998b). This conformational interaction of GM1 with the δ-opioid receptor was seen as uncoupling of the receptor from G_i and facilitated coupling to G_s (Wu et al. 1997b). Neuronal Ca^{2+} influx stimulated by GM1 in this manner was postulated to occur through modulation of N-type Ca^{2+} channels (Keren et al. 1997).

Plasma membrane gangliosides have also been shown to influence Ca^{2+} efflux mechanisms, these often accounting for the neuroprotective activities observed with exogenously applied gangliosides (see above). Plasma membrane Ca^{2+}-ATPase (PMCA), the high affinity mechanism for extrusion of cytosolic Ca^{2+}, was studied in porcine brain synaptosomes and found to vary in response to different ganglioside structures in a manner reflecting the number of sialic acids: GD1b (two sialic acids) stimulated activity in contrast to GM1 (one sialic acid) which slightly reduced activity while asialo GM1 (no sialic acids) was strongly inhibitory (Zhao et al. 2004). Chain length was also considered important since GM2 and GM3 were both more inhibitory than GM1. The experimental procedure consisted of adding ganglioside to either synaptosomes or reconstituted proteo-liposomes containing purified synaptosomal PMCA followed by measurement of Ca^{2+} uptake, both procedures showing the same ganglioside modulatory effects. Purified PMCA was inactive due to delipidation during isolation, but was restored to full activity by reconstituting into liposomes containing phosphatidylcholine. Interestingly, a similar study with PMCA from pig erythrocytes gave very different results, all gangliosides being stimulatory up to sevenfold in the sequence: GD1b > GM1 > GM2 > GM3 = asialo-GM1 (Zhang et al. 2005). This difference was attributed to PMCA isoforms, PMCA1 and PMCA4 predominating in erythrocytes in contrast to PMCA2 and PMCA3 which are restricted to nerve cells. As these studies were carried out with applied gangliosides, it would be of interest to know whether the modulatory effect applied as well through in situ association with PMCA.

15.3.3 Ganglioside Modulation of Ca^{2+} Transport at Intracellular Loci

A number of studies on lysosomal storage disorders suggested a mechanistic link between ganglioside accumulation and disrupted Ca^{2+} homeostasis based on modulation of the sarco/endoplasmic reticulum Ca^{2+}-ATPase (SERCA) (Ginzburg et al. 2004). Thus a reduction in Ca^{2+}-uptake via the SERCA pump was observed in neurons and brain microsomes of the Hexb–/– mouse, a model of Sandhoff disease (Pelled et al. 2003). The latter study also demonstrated reduction in the rate of Ca^{2+} uptake in normal brain microsomes by exogenous GM2, to a lesser extent by GM1 and least by GM3. A more detailed follow up study with brain microsomes revealed the necessity of *N*-acetylneuraminic acid with a free carboxyl group for inhibitory activity (Ginzburg et al. 2008). That study further proposed that the GalNAc residue of GM2 and GM1 may be an additional structural requirement for SERCA inhibition. The oligosaccharides alone had no activity. Similar studies on skeletal muscle sarcoplasmic reticulum also reported an inhibitory effect of GM1 on SERCA activity but in contrast a positive modulatory effect by GM3 (Wang et al. 1999a). Using intrinsic and time-resolved fluorescence in addition to fluorescence quenching, the conformational changes observed indicated that GM1 could render the SERCA molecules less compact in the hydrophilic domain but more compact in the hydrophobic domain; GM3 on the other hand made the enzyme more compact in both the hydrophilic and hydrophobic domains. Further study of this system employing circular dichroism showed that both GM1 and GM3 reduced the α-helical content of the protein, with GM1 causing the stronger decrease; study of the proteo-liposomes containing this Ca^{2+}-ATPase using DPH as the probe showed that GM1 decreased membrane lipid fluidity while GM3 tended to increase it (Wang et al. 1999b). All the above SERCA studies were carried out with applied (exogenous) gangliosides, underscoring the desirability of ascertaining whether endogenous gangliosides have similar modulatory effects.

Nuclear GM1 was shown to have a prominent role in regulation of nuclear Ca^{2+} homeostasis through association with a sodium-calcium exchanger (NCX) in the inner nuclear membrane (Xie et al. 2002). This NCX, which mediates transfer of Ca^{2+} between the nucleoplasm and the luminal space of the nuclear envelope, is potently and specifically activated by GM1. Immunoblot analysis revealed an unusually tight association of GM1 with the nuclear NCX, so strong that it survived SDS-PAGE; this differed from the NCX of the plasma membrane for which a looser association with GM1 was suggested. A key feature proposed for the nuclear topology based on colocalization of GM1 and NCX in the inner nuclear membrane is interaction of the negative charge of N-acetylneuraminic acid of GM1 with the alternative splice region of the NCX loop containing positively charged amino acid(s) (Xie et al. 2004). For the plasma membrane, this NCX loop is seen as residing on the opposite side of the bilayer as GM1, thus accounting for the lower affinity association. Nuclear NCX is activated in developing neurons upon upregulation of GM1 synthesis, a process that also occurs in some but not all extraneural cell types (Xie et al. 2004). Use of a variety of cell types with and without nuclear NCX together with specific Ca^{2+} fluorescent indicators revealed transport of nucleoplasmic Ca^{2+} into the nuclear envelope followed by transfer to the ER lumen; in keeping with cytosolic Ca^{2+} flux through nuclear pores, the nuclear NCX/GM1 complex was shown to gate Ca^{2+} transfer from cytosol to ER, thus constituting an alternative mechanism to the SERCA pump for such transfer (Wu et al. 2009). As with plasma membrane NCX, the driving force for such transfer is the Na^{+} gradient created by Na^{+}/K^{+}-ATPase, and the latter transporter was shown to occur in the nuclear membrane suggesting concerted physiological coupling between these transporters (Galva et al. 2012). The relatively simple ganglioside pattern of the nuclear envelope includes GD1a which serves as metabolic precursor to GM1 owing to the presence of N'ase at the same locus (Wang et al. 2009b).

The nuclear NCX/GM1 complex was shown to serve a neuroprotective role in shielding the nucleus against prolonged elevation of nucleoplasmic Ca^{2+}, as seen in studies with mice lacking GM1 due to deletion of GM2/GD2 synthase [*B4galnt1(-/-)*]. Cultured cerebellar granule neurons from such mice were shown to have lost the ability possessed by wild-type cells to regulate Ca^{2+} homeostasis, resulting in apoptotic death when the cells were exposed to high K^{+} (Wu et al. 2001). This neuroprotective role was demonstrated in vivo with the above mice which showed enhanced susceptibility to kainate-induced seizures and neuronal apoptosis (Wu et al. 2005). Seizure activity and deterioration of pyramidal neurons in the CA3 of the hippocampus were significantly alleviated by intraperitoneal administration of LIGA20, a membrane permeable analog of GM1 (Fig. 15.2). This was coincident with LIGA20 entering the brain and brain cells, including neuronal nuclear membrane, with restoration of attenuated NCX activity. GM1 itself, with limited membrane permeability, showed little benefit when administered intraperitoneally.

The above reports on SERCA and nuclear NCX, examples of intracellular glycolipid regulation of ion transport, indicate the importance of considering the whole cell, not the plasma membrane alone. Additional examples of intracellular regulation are being revealed, such as the role of GM1 in modulating Ca^{2+} levels in the ER,

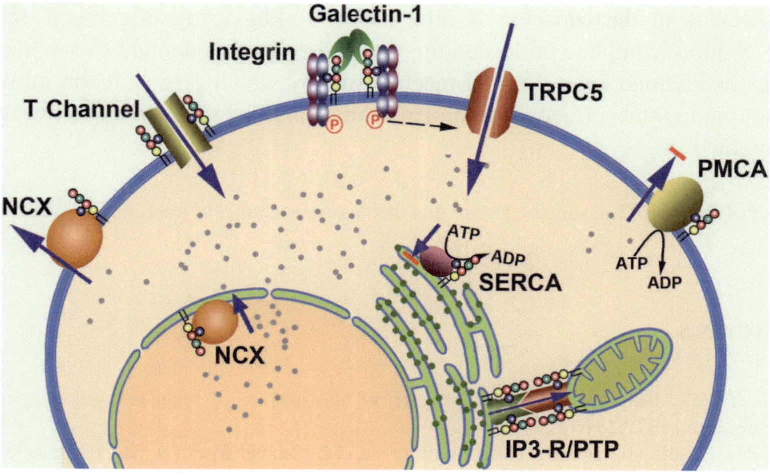

Fig. 15.3 Summary of GM1 modulatory roles for neuronal Ca^{2+}. For the plasma membrane, Ca^{2+} influx is promoted by GM1 via T type channels and by GM1 association with α5β1 integrin, cross-linking of which leads to TRPC5 channel activation. Calcium efflux is influenced positively in the plasma membrane by GM1 association with Na^{+}/Ca^{2+} exchanger (NCX) and negatively by GM1 association with plasma membrane Ca^{2+}-ATPase (PMCA). Intracellular mechanisms include GM1 association with NCX in the inner nuclear membrane that mediates transfer of Ca^{2+} from nucleoplasm to the nuclear envelope (and hence the ER); also inhibition of the SERCA pump. GM1 in the ER influences Ca^{2+} flux from that organelle to mitochondria

with potential for initiating a UPR-mediated apoptotic cascade (Tessitore et al. 2004). Further work in this area employing the GM1 gangliosidosis mouse model revealed GM1 accumulation in the raft fraction of mitochondria-associated ER membranes that influence Ca^{2+} flux between these organelles (Sano et al. 2009). The accumulated GM1 was shown to interact with the phosphorylated form of IP3 receptor-1, influencing the activity of that channel. It will be of interest to have more detail on the nature of this interaction and to know whether it occurs as well in non-pathological neurons. The numerous ways in which GM1 regulates neuronal Ca^{2+} homeostasis, as revealed to date, are summarized in Fig. 15.3.

15.4 Ion Transport and the Sugar Code

Considering the crucial role of ion transport in numerous aspects of nervous system functioning, it is not unexpected to find glycosylation in its multiple forms fulfilling a wide variety of essential modulatory roles. The relatively simple glycosylation patterns that characterize invertebrate nervous systems contrasts with the myriad arrays of protein and lipid glycoconjugates that occur and appear essential to the nervous systems of higher forms. The complexity of glycan configurations that mediate neural phenomena increases in tandem with complexity of evolutionary life forms. Carbohydrates, it has been pointed out, "…are second to no other class of biomolecules in the capacity of information coding using oligomers" (Gabius 2009). A more complete understanding of such coding as applied to ion transport awaits elucidation of the detailed mechanisms by which glycosylation influences channel/pump transport, conformational folding, and the stereochemistry of molecular interactions.

Conflict of Interest The authors declare that they have no conflict of interest.

References

Ando S, Yu RK. Isolation and characterization of two isomers of brain tetrasialogangliosides. J Biol Chem. 1979;254:12224–9.

Barchi RL. Protein components of the purified sodium channel from rat skeletal muscle sarcolemma. J Neurochem. 1983;40:1377–85.

Bennett ES. Isoform-specific effects of sialic acid on voltage-dependent Na^+ channel gating: functional sialic acids are localized to the S5-S6 loop of domain I. J Physiol. 2002;538:675–90.

Bennett E, Urcan MS, Tinkle SS, Koszowski AG, Levinson SR. Contribution of sialic acid to the voltage dependence of sodium channel gating. A possible electrostatic mechanism. J Gen Physiol. 1997;109:327–43.

Carlson RO, Masco D, Brooker G, Spiegel S. Endogenous ganglioside GM1 modulates L-type calcium channel activity in N18 neuroblastoma cells. J Neurosci. 1994;14:2272–81.

Castillo C, Diaz ME, Balbi D, Thornhill WB, Recio-Pinto E. Changes in sodium channel function during postnatal brain development reflect increases in the level of channel sialidation. Brain Res Dev Brain Res. 1997;104:119–30.

Catterall WA. Voltage-gated sodium channels at 60: structure, function and pathophysiology. J Physiol. 2012;590:2577–89.

Cotella D, Radicke S, Bortoluzzi A, Ravens U, Wettwer E, Santoro C, et al. Impaired glycosylation blocks DPP10 cell surface expression and alters the electrophysiology of I_{to} channel complex. Pflugers Arch. 2010;460:87–97.

Dietrich A, Mederos y Schnitzler M, Emmel J, Kalwa H, Hofmann T, Gudermann T. N-linked protein glycosylation is a major determinant for basal TRPC3 and TRPC6 channel activity. J Biol Chem. 2003;278:47842–52.

Erler I, Al-Ansary DM, Wissenbach U, Wagner TF, Flockerzi V, Niemeyer BA. Trafficking and assembly of the cold-sensitive TRPM8 channel. J Biol Chem. 2006;281(50):38396–404.

Fang Y, Wu G, Xie X, Lu Z-H, Ledeen RW. Endogenous GM1 ganglioside of the plasma membrane promotes neuritogenesis by two mechanisms. Neurochem Res. 2000;25:931–40.

Fang Y, Xie X, Ledeen RW, Wu G. Characterization of cholera toxin B subunit-induced Ca^{2+} influx in neuroblastoma cells: evidence for a voltage independent GM1-associated Ca^{2+} channel. J Neurosci Res. 2002;57:1–10.

Gabius H-J, editor. The sugar code. Fundamentals of glycosciences. Weinheim, Germany: Wiley-VHC; 2009.

Galva C, Artigas P, Gatto C. Nuclear Na^+K^+-ATPase plays an active role in nucleoplasmic Ca^{2+} homeostasis. J Cell Sci. 2012;125:6137–47.

Gilly WF, Lucero MT, Horrigan FT. Control of the spatial distribution of sodium channels in giant fiber lobe neurons of the squid. Neuron. 1990;5:663–74.

Ginzburg L, Kacher Y, Futerman AH. The pathogenesis of glycosphingolipid storage disorders. Semin Cell Dev Biol. 2004;15:417–31.

Ginzburg L, Li S-C, Li Y-T, Futerman AH. An exposed carboxyl group on sialic acid is essential for gangliosides to inhibit calcium uptake via the sarco/endoplasmic reticulum Ca^{2+}-ATPase: relevance to gangliosidoses. J Neurochem. 2008;104:140–6.

Hall MK, Cartwright TA, Fleming CM, Schwalbe RA. Importance of glycosylation on function of a potassium channel in neuroblastoma cells. PLoS One. 2011;6:e19317.

Handa Y, Ozaki N, Honda T, Furukawa K, Tomita Y, Inoue M, et al. GD3 synthase gene knockout mice exhibit thermal hyperalgesia and mechanical allodynia but decreased response to formalin-induced prolonged noxious stimulation. Pain. 2005;117:271–9.

Hartshorne RP, Catterall WA. Purification of the saxitoxin receptor of the sodium channel from rat brain. Proc Natl Acad Sci U S A. 1981;78:4620–4.

Hartshorne RP, Catterall WA. The sodium channel from rat brain. Purification and subunit composition. J Biol Chem. 1984;259:1667–75.

Hasegawa T, Yamaguchi K, Wada T, Takeda A, Itoyama Y, Miyagi T. Molecular cloning of mouse ganglioside sialidase and its increased expression in Neuro2a differentiation. J Biol Chem. 2000;275:8007–15.

Isaev D, Isaeva E, Shatskih T, Zhao Q, Smits NC, Shworak NW, et al. Role of extracellular sialic acid in regulation of neuronal and network excitability in the rat hippocampus. J Neurosci. 2007;27:11587–94.

Isaeva E, Lushnikova I, Savrasova A, Skibo G, Holmes GL, Isaev D. Effect of neuraminidase treatment on persistent epileptiform activity in the rat hippocampus. Pharmacol Rep. 2011;63: 840–4.

Isom LL, De Jongh KS, Patton DE, Reber BF, Offord J, Charbonneau H, et al. Primary structure and functional expression of the beta 1 subunit of the rat brain sodium channel. Science. 1992; 256:839–42.

James WM, Agnew WS. Alpha-(2–8)-polysialic acid immunoreactivity in voltage-sensitive sodium channel of eel electric organ. Proc R Soc Lond B Biol Sci. 1989;237:233–45.

Jin W, Lee NM, Loh HH, Thayer SA. Dual excitatory and inhibitory effects of opioids on intracellular calcium in neuroblastoma X glioma NG108-15 cells. Mol Pharmacol. 1992;42:1083–9.

Johnson D, Montpetit ML, Stocker PJ, Bennett ES. The sialic acid component of the beta1 subunit modulates voltage-gated sodium channel function. J Biol Chem. 2004;279:44303–10.

Keren O, Garty M, Sarne Y. Dual regulation by opioids of [^{3}H]norepinephrine release in the human neuroblastoma cell line SK-N-SH. Brain Res. 1994;646:319–23.

Keren O, Gafni M, Sarne Y. Opioids potentiate transmitter release from SK-N-SH human neuroblastoma cells by modulating N-type calcium channels. Brain Res. 1997;764:277–82.

Khanna R, Myers MP, Laine M, Papazian DM. Glycosylation increases potassium channel stability and surface expression in mammalian cells. J Biol Chem. 2001;276:34028–34.

Kopitz J, Von Reitzenstein C, Burchert M, Cantz M, Gabius H-J. Galectin-1 is a major receptor for ganglioside GM1, a product of the growth controlling activity of a cell surface ganglioside sialidase, on human neuroblastoma cells in culture. J Biol Chem. 1998;273:11205–11.

Korkotian E, Schwarz A, Pelled D, Schwarzmann G, Segal M, Futerman AH. Elevation of intracellular glucosylceramide levels results in an increase in endoplasmic reticulum density and in functional calcium stores in cultured neurons. J Biol Chem. 1999;274:21673–8.

Ledeen RW, Wu G. Ganglioside function in calcium homeostasis and signaling. Neurochem Res. 2002;27:637–47.

Ledeen RW, Wu G, André S, Bleich D, Huet G, Kaltner H, et al. Beyond glycoproteins as galectin-counterreceptors: tumor-effector T cell growth control via ganglioside GM1. Ann N Y Acad Sci. 2012;1253:206–21.

Leon A, Facci L, Toffano G, Sonnino S, Tettamanti G. Activation of (Na+, K+)-ATPase by nanomolar concentrations of GM1 ganglioside. J Neurochem. 1981;37:350–7.

Lloyd-Evans E, Pelled D, Riebeling C, Bodennec J, de-Morgan A, Waller H, Schiffmann R, Futerman AH. Glucosylceramide and glucosylsphingosine modulate calcium mobilization from brain microsomes via different mechanisms. J Biol Chem. 2003;278:23594–9.

Lo WY, Lagrange AH, Hernandez CC, Harrison R, Dell A, Haslam SM, et al. Glycosylation of β2 subunits regulates $GABA_A$ receptor biogenesis and channel gating. J Biol Chem. 2010; 285:31348–61.

Masco D, Van de Walle M, Spiegel S. Interaction of ganglioside GM1 with the B subunit of cholera toxin modulates growth and differentiation of neuroblastoma N18 cells. J Neurosci. 1991; 11:2443–52.

Messner DJ, Catterall WA. The sodium channel from rat brain. Separation and characterization of subunits. J Biol Chem. 1985;260:10597–604.

Milani D, Minozzi MC, Petrelli L, Guidolin D, Skaper SD, Spoerri PE. Interaction of ganglioside GM1 with the B subunit of cholera toxin modulates intracellular free calcium in sensory neurons. J Neurosci Res. 1992;33:466–75.

Miller JA, Agnew WS, Levinson SR. Principal glycopeptide of the tetrodotoxin/saxitoxin binding protein from electrophorus electricus: isolation and partial chemical and physical characterization. Biochemistry. 1983;22:462–70.

Miyagi T, Wada T, Iwamatsu A, Hata K, Yoshikawa Y, Tokuyama S, et al. Molecular cloning and characterization of plasma membrane-associated sialidase specific for gangliosides. J Biol Chem. 1999;274:5004–11.

Montell C. Exciting trips for TRPs. Nat Cell Biol. 2004;6:690–2.

Monti E, Bassi MT, Papini N, Riboni M, Manzoni M, Venerando B, et al. Identification and expression of NEU3, a novel human sialidase associated to the plasma membrane. Biochem J. 2000;81:284–96.

Montpetit ML, Stocker PJ, Schwetz TA, Harper JM, Norring SA, Schaffer L, et al. Regulated and aberrant glycosylation modulate cardiac electrical signaling. Proc Natl Acad Sci U S A. 2009;106:16517–22.

Nakamura K, Wu G, Ledeen RW. Protection of Neuro-2a cells against calcium ionophore cytotoxicity by gangliosides. J Neurosci Res. 1992;31:245–53.

Nedelkoska L, Benjamins JA. Binding of cholera toxin B subunit: a surface marker for murine microglia but not oligodendrocytes or astrocytes. J Neurosci Res. 1998;53:605–12.

O'Hanlon GM, Hirst TR, Willison HJ. Ganglioside GM1 binding toxins and human neuropathy-associated IgM antibodies differentially promote neuritogenesis in a PC12 assay. Neurosci Res. 2003;47:383–90.

Ohtsubo K, Marth JD. Glycosylation in cellular mechanisms of health and disease. Cell. 2006; 126:855–67.

Pelled D, Lloyd-Evans E, Riebeling C, Jeyakumar M, Platt FM, Futerman AH. Inhibition of calcium uptake via the sarco/endoplasmic reticulum Ca^{2+}-ATPase in a mouse model of Sandhoff disease and prevention by treatment with N-butyldeoxynojirimycin. J Biol Chem. 2003;278:29496–501.

Peng XQ, Zhang XL, Fang Y, Xie WR, Xie YK. Sialic acid contributes to hyperexcitability of dorsal root ganglion neurons in rats with peripheral nerve injury. Brain Res. 2004;1026(2): 185–93.

Qiao GF, Cheng ZF, Huo R, Sui XH, Lu YJ, Li BY. GM1 ganglioside contributes to retain the neuronal conduction and neuronal excitability in visceral and baroreceptor afferents. J Neurochem. 2008;106:1637–45.

Quattrini A, Lorenzetti I, Sciorati C, Corbo M, Previtali SC, Feltri ML, et al. Human IgM anti-GM1 autoantibodies modulate intracellular calcium homeostasis in neuroblastoma cells. J Neuroimmunol. 2001;114:213–9.

Roberts RH, Barchi RL. The voltage-sensitive sodium channel from rabbit skeletal muscle. Chemical characterization of subunits. J Biol Chem. 1987;262:2298–303.

Rodriguez JA, Piddini E, Hasegawa T, Miyagi T, Dotti CG. Plasma membrane ganglioside sialidase regulates axonal growth and regeneration in hippocampal neurons in culture. J Neurosci. 2001;21:8387–95.

Rutishauser U, Landmesser L. Polysialic acid in the vertebrate nervous system: a promoter of plasticity in cell-cell interactions. Trends Neurosci. 1996;19:422–7.

Sandoval A, Oviedo N, Andrade A, Felix R. Glycosylation of asparagines 136 and 184 is necessary for the alpha2delta subunit-mediated regulation of voltage-gated Ca^{2+} channels. FEBS Lett. 2004;576:21–6.

Sano R, Annunziata I, Patterson A, Moshiach S, Gomero E, Opferman J, et al. GM1 ganglioside accumulation at the mitochondria-associated ER membranes links ER stress to Ca^{2+}-dependent mitochondrial apoptosis. Mol Cell. 2009;36:500–11.

Schengrund C-L, Ringler NJ. Binding of *Vibrio cholera* toxin and the heat labile enterotoxin of *Escherichia coli* to GM1 and derivatives of GM1, and nonlipid, oligosaccharide polyvalent ligands. J Biol Chem. 1989;264:13233–7.

Schmidt JW, Catterall WA. Palmitylation, sulfation, and glycosylation of the alpha subunit of the sodium channel. Role of post-translational modifications in channel assembly. J Biol Chem. 1987;262:13713–23.

Schwetz TA, Norring SA, Ednie AR, Bennett ES. Sialic acids attached to O-glycans modulate voltage-gated potassium channel gating. J Biol Chem. 2011;286:4123–32.

Shen KF, Crain SM. Dual modulation of the action potential duration of mouse dorsal root ganglion neurons in culture. Brain Res. 1989;491:227–42.

Siebert HC, André S, Lu SY, Frank M, Kaltner H, van Kuik JA, et al. Unique conformer selection of human growth-regulatory lectin galectin-1 for ganglioside GM1 versus bacterial toxins. Biochemistry. 2003;42:14762–73.

Skoff AM, Benjamins JA. Antibodies to glycolipids and cholera toxin B subunit do not initiate Ca^{2+} signaling in rat Schwann cells. J Peripher Nerv Syst. 1998;3:19–27.

Stocker PJ, Bennett ES. Differential sialylation modulates voltage-gated Na^+ channel gating throughout the developing myocardium. J Gen Physiol. 2006;127:253–65.

Sumikawa K, Miledi R. Assembly and N-glycosylation of all ACh receptor subunits are required for their efficient insertion into plasma membranes. Brain Res Mol Brain Res. 1989;5: 183–92.

Susuki K, Rasband MN, Tohyama K, Koibuchi K, Okamoto S, Funakoshi K, et al. Anti-GM1 antibodies cause complement-mediated disruption of sodium channel clusters in peripheral motor nerve fibers. J Neurosci. 2007a;27:3956–67.

Susuki K, Baba H, Tohyama K, Kanai K, Kuwabara S, Hirata K, et al. Gangliosides contribute to stability of paranodal junctions and ion channel clusters in myelinated nerve fibers. Glia. 2007b;55:746–57.

Takamiya K, Yamamoto A, Furukawa K, Yamashiro S, Shin M, Okada M, et al. Mice with disrupted GM2/GD2 synthase gene lack complex gangliosides but exhibit only subtle defects in their nervous system. Proc Natl Acad Sci U S A. 1996;93:10662–7.

Tang C-M, Presser F, Morad M. Amiloride selectively blocks the low threshold (T) calcium channel. Science. 1988;240:213–5.

Tessitore A, del P Martin M, Sano R, Ma Y, Mann L, Ingrassia A, et al. GM1-ganglioside-mediated activation of the unfolded protein response causes neuronal death in a neurodegenerative gangliosidosis. Mol Cell. 2004;15:753–66.

Thornhill WB, Wu MB, Jiang X, Wu X, Morgan PT, Margiotta JF. Expression of Kv1.1 delayed rectifier potassium channels in Lec mutant Chinese hamster ovary cell lines reveals a role for sialidation in channel function. J Biol Chem. 1996;271:19093–8.

Tyrrell L, Renganathan M, Dib-Hajj SD, Waxman SG. Glycosylation alters steady-state inactivation of sodium channel Na_v 1.9/NaN in dorsal root ganglion neurons and is developmentally regulated. J Neurosci. 2001;21(24):9629–37.

Waechter CJ, Schmidt JW, Catterall WA. Glycosylation is required for maintenance of functional sodium channels in neuroblastoma cells. J Biol Chem. 1983;258:5117–23.

Wang Y, Tsui Z, Yang F. Antagonistic effect of ganglioside GM1 and GM3 on the activity and conformation of sarcoplasmic reticulum Ca^{2+}-ATPase. FEBS Lett. 1999a;457:144–8.

Wang Y, Tsui Z, Yang F. Mechanistic study of modulation of SR Ca^{2+}-ATPase activity by gangliosides GM1 and GM3 through some biophysical measurements. Glycoconj J. 1999b;16:781–6.

Wang J, Lu Z-H, Gabius H-J, Rolhowsky-Kochan C, Ledeen RW, Wu G. Cross-linking of GM1 ganglioside by galectin-1 mediates regulatory T cell activity involving TRPC5 channel activation: possible role in suppressing experimental autoimmune encephalomyelitis. J Immunol. 2009a;182:4036–45.

Wang J, Wu G, Miyagi T, Lu Z-H, Ledeen RW. Sialidase occurs in both membranes of the nuclear envelope and hydrolyzes endogenous GD1a. J Neurochem. 2009b;111:547–54.

Watanabe I, Zhu J, Recio-Pinto E, Thornhill WB. Glycosylation affects the protein stability and cell surface expression of Kv1.4 but not Kv1.1 potassium channels. A pore region determinant dictates the effect of glycosylation on trafficking. J Biol Chem. 2004;279:8879–85.

Wirkner K, Hognestad H, Jahnel R, Hucho F, Illes P. Characterization of rat transient receptor potential vanilloid 1 receptors lacking the N-glycosylation site N604. Neuroreport. 2005;16: 997–1001.

Wu G, Ledeen RW. Stimulation of neurite outgrowth in neuroblastoma cells by neuraminidase: putative role of GM1 ganglioside in differentiation. J Neurochem. 1991;56:95–104.

Wu G, Ledeen RW. Gangliosides as modulators of neuronal calcium. Prog Brain Res. 1994;101:101–12.

Wu G, Vaswani KK, Lu Z-H, Ledeen RW. Gangliosides stimulate calcium flux in Neuro-2A cells and require exogenous calcium for neuritogenesis. J Neurochem. 1990;55:484–91.

Wu G, Lu Z-H, Nakamura K, Spray DC, Ledeen RW. Trophic effect of cholera toxin B subunit in cultured cerebellar granule neurons: modulation of intracellular calcium by GM1 ganglioside. J Neurosci Res. 1996;44:243–54.

Wu G, Lu Z-H, Ledeen RW. Interaction of the δ-opioid receptor with GM1 ganglioside: conversion from inhibitory to excitatory mode. Brain Res Mol Brain Res. 1997a;44:341–6.

Wu G, Lu Z-H, Alfinito P, Ledeen RW. Opioid receptor and calcium channel regulation of adenylyl cyclase, modulated by GM1, in NG108-15 cells: competitive interactions. Neurochem Res. 1997b;22:1281–9.

Wu G, Fang Y, Lu Z-H, Ledeen RW. Induction of axon-like and dendrite-like processes in neuroblastoma cells. J Neurocytol. 1998a;27:1–14.

Wu G, Lu ZH, Wei TJ, Howells RD, Christoffers K, Ledeen RW. The role of GM1 ganglioside in regulating excitatory opioid effects. Ann N Y Acad Sci. 1998b;845:126–38.

Wu G, Xie X, Lu Z-H, Ledeen RW. Cerebellar neurons lacking complex gangliosides degenerate in the presence of depolarizing levels of potassium. Proc Natl Acad Sci U S A. 2001;98: 307–12.

Wu G, Lu ZH, Wang J, Wang Y, Xie X, Meyenhofer MF, et al. Enhanced susceptibility to kainate-induced seizures, neuronal apoptosis, and death in mice lacking gangliotetraosegangliosides: protection with LIGA20, a membrane-permeant analog of GM1. J Neurosci. 2005;25: 11014–22.

Wu G, Lu Z-H, Obukhov ASG, Nowycky MC, Ledeen RW. Induction of calcium influx through TRPC5 channels by cross-linking of GM1 ganglioside associated with α5β1 integrin initiates neurite outgrowth. J Neurosci. 2007;27:7447–58.

Wu G, Xie X, Lu Z-H, Ledeen RW. Sodium-calcium exchanger complexed with GM1 ganglioside in nuclear membrane transfers calcium from nucleoplasm to endoplasmic reticulum. Proc Natl Acad Sci U S A. 2009;106:10829–34.

Wu G, Lu Z-H, Gabius H-J, Ledeen RW, Bleich D. Ganglioside GM1 deficiency in effector T cells from NOD mice induces resistance to regulatory T-cell suppression. Diabetes. 2011;60: 2341–9.

Xie X, Wu G, Lu Z-H, Ledeen RW. Potentiation of a sodium-calcium exchanger in the nuclear envelope by nuclear GM1 ganglioside. J Neurochem. 2002;81:1185–95.

Xie X, Wu G, Lu Z-H, Rohowsky-Kochan C, Ledeen RW. Presence of sodium-calcium exchanger/GM1 complex in the nuclear envelope of non-neural cells: nature of the exchanger-GM1 interaction. Neurochem Res. 2004;29:2135–46.

Zhang J, Zhao Y, Duan J, Yang F, Zhang X. Gangliosides activate the phosphatase activity of the erythrocyte plasma membrane Ca^{2+}-ATPase. Arch Biochem Biophys. 2005;444:1–6.

Zhao Y, Fan X, Yang F, Zhang Z. Gangliosides modulate the activity of the plasma membrane Ca^{2+}-ATPase from porcine brain synaptosomes. Arch Biochem Biophys. 2004;427:204–12.

Zhou TT, Zhang ZW, Liu J, Zhang JP, Jiao BH. Glycosylation of the sodium channel β4 subunit is developmentally regulated and involves in neuritic degeneration. Int J Biol Sci. 2012;8: 630–9.

Zona C, Eusebi F, Miledi R. Glycosylation is required for maintenance of functional voltage-activated channels in growing neocortical neurons of the rat. Proc R Soc Lond B Biol Sci. 1990;239:119–27.

Chapter 16
O-GlcNAcylation of Neuronal Proteins: Roles in Neuronal Functions and in Neurodegeneration

Olof Lagerlöf and Gerald W. Hart

Abstract O-GlcNAc is the attachment of β-N-acetylglucosamine to the hydroxyl group of serine and threonine in nuclear and cytoplasmic proteins. It is generally not further elongated but exists as a monosaccharide that can be rapidly added or removed. Thousands of proteins involved in gene transcription, protein translation, and degradation as well as the regulation of signal transduction contain O-GlcNAc. Brain is one of the tissues where O-GlcNAc is most highly expressed and deletion of neuronal O-GlcNAc leads to death early in development. O-GlcNAc is also important for normal adult brain function, where dynamic processes like learning and memory at least in part depend on the modification of specific proteins by O-GlcNAc. Conversely, too much or too little O-GlcNAc on other proteins participates in neurodegenerative processes underlying diseases such as Alzheimer's and Parkinson's. In this chapter, we describe the expression and regulation of O-GlcNAc in the nervous system.

Keywords O-linked *N*-acetylglucosamine • O-GlcNAc • Learning and memory • Neurodegeneration • Alzheimer's disease • Signaling • Nutrient sensing • Post-translational modifications

16.1 Introduction

O-GlcNAc is the attachment of β-N-acetylglucosamine to the hydroxyl group of serine and threonine in nuclear and cytoplasmic proteins (Torres and Hart 1984). It is generally not further elongated but exists as a monosaccharide that can be rapidly

O. Lagerlöf • G.W. Hart (✉)
Department of Biological Chemistry, Johns Hopkins University School of Medicine, 725 N. Wolfe Street, Baltimore, MD 21205-2185, USA
e-mail: gwhart@jhmi.edu

R.K. Yu and C.-L. Schengrund (eds.), *Glycobiology of the Nervous System*, Advances in Neurobiology 9, DOI 10.1007/978-1-4939-1154-7_16,

added or removed (Hart et al. 2011). Thousands of proteins involved in gene transcription, protein translation, and degradation as well as the regulation of signal transduction contain O-GlcNAc (Trinidad et al. 2012; Alfaro et al. 2012). Brain is one of the tissues where O-GlcNAc is most highly expressed and deletion of neuronal O-GlcNAc leads to death early in development (Kreppel et al. 1997; O'Donnell et al. 2004). O-GlcNAc is also important for normal adult brain function, where dynamic processes like learning and memory at least in part depend on the modification of specific proteins by O-GlcNAc (Tallent et al. 2009; Rexach et al. 2012). Conversely, too much or too little O-GlcNAc on other proteins participates in neurodegenerative processes underlying diseases such as Alzheimer's and Parkinson's (Arnold et al. 1996; Liu et al. 2004; Yuzwa et al. 2012; Wang et al. 2010a, b; Marotta et al. 2012). In this chapter, we describe the expression and regulation of O-GlcNAc in the nervous system.

16.2 O-GlcNAc Is a Ubiquitous Monosaccharide That Cycles onto and off Serine and Threonine

16.2.1 O-GlcNAc Is Not Elongated to Yield Complex Oligosaccharides

O-GlcNAc is the covalent modification of nuclear and cytoplasmic proteins by β-N-acetylglucosamine (Torres and Hart 1984). O-GlcNAc is formed as a derivative of glucose through the hexosamine biosynthesis pathway (HBP). In the HBP, the oxygen on the second carbon of fructose-6-phosphate is exchanged for nitrogen forming GlcN-6-P, prior to acetylation of the nitrogen to yield GlcNAc-6-P. This is then coupled to the high-energy molecule uridine diphosphate (UDP), UDP-GlcNAc (Fig. 16.1). Upon modification of proteins by O-GlcNAc, the GlcNAc is cleaved from the UDP and attached in β-position to the hydroxyl group of serine or threonine (O-β-GlcNAc, O-GlcNAc). The reaction is catalyzed by the O-GlcNAc transferase, OGT. The removal of GlcNAc is catalyzed by the O-GlcNAc hydrolase, O-GlcNAcase (OGA).

Unlike "classical" O- and N-linked protein glycosylation the GlcNAc is generally not elongated but exists as a monosaccharide. In fact, when O-GlcNAc is artificially capped by galactose, its biological function is lost (Fang and Miller 2001). Although, O-GlcNAc is smaller than complex oligosaccharides, it is still much larger than many other protein modifications, such as protein methylation or protein phosphorylation (Hart et al. 2011).

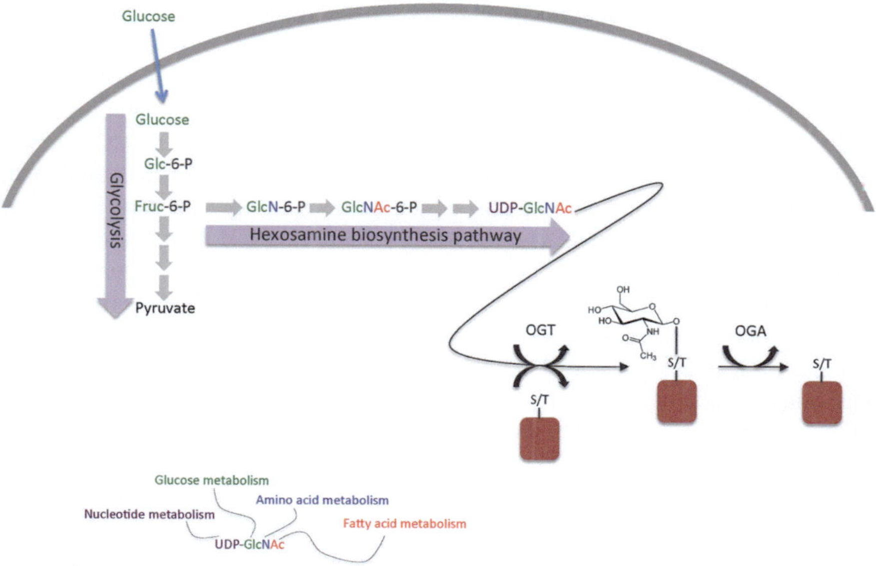

Fig. 16.1 O-GlcNAc cycling is regulated by two enzymes, OGT and OGA, and total O-GlcNAc levels depend on several metabolic pathways. The donor substrate for O-GlcNAc, UDP-GlcNAc, is produced by the hexosamine biosynthesis pathway (HBP). Many metabolites feed into the HBP and thereby modulate UDP-GlcNAc production. In the nucleus and cytosol of cells, the O-GlcNAc transferase (OGT) cleaves the UDP-GlcNAc and adds the GlcNAc to serine or threonine on proteins. The GlcNAc can then be removed from the protein by another nucleocytoplasmic protein, the O-GlcNAc hydrolase, OGA

16.2.2 O-GlcNAc Is Mostly Expressed on the Inside of Cells in Multicellular Organisms

O-GlcNAc is a highly conserved posttranslational modification. It has been found in evolutionary distinct clades like plantae, fungi, and animalia (Kreppel et al. 1997; Webster et al. 2009). In multicellular organisms, all types of cells investigated so far contain O-GlcNAc (Hart et al. 2011). O-GlcNAc has also been identified in some unicellular organisms, e.g. giardia—the oldest eukaryote, and inside several types of virus (Banerjee et al. 2009; Benko et al. 1988; Caillet-Boudin et al. 1989). Nonetheless, most studies on unicellular organisms fail to report the presence of β-O-GlcNAc. Protozoans modify proteins by O-linked GlcNAc but primarily on extracellular proteins and in α-linkage rather than in β-linkage. Yeasts appear to lack O-GlcNAc entirely. Bacteria are also largely devoid of cytoplasmic O-GlcNAc even though some bacterial proteins have been shown to carry O-linked GlcNAc (Schirm et al. 2004; Fredriksen et al. 2012). Interestingly, the bacterium *Clostridium novyi* exploits O-GlcNAc by encoding an O-GlcNAc transferase that modifies small G-proteins in the infected cell (Selzer et al. 1996; Hart et al. 2007, 2011).

O-GlcNAc is expressed almost exclusively on the inside of cells (Torres and Hart 1984). Until the discovery of O-GlcNAc protein glycosylation was known to occur only on proteins exposed to the extracellular matrix, or in cellular organelles topographically similar to the outside of the cell such as the endoplasmatic reticulum (ER) and the Golgi apparatus. In contrast, nearly all proteins that contain O-GlcNAc are expressed in the cytosolic or nuclear fraction of the cell. Proteins anchored to the cell membrane are modified with O-GlcNAc but usually only on parts stretching into the cytosol. This comes as no surprise as the O-GlcNAc transferase, OGT, is mainly nucleocytoplasmic rather than present in the Golgi or ER as other glycosyltransferases (there are at least two O-GlcNAc transferases with their active sites in the lumen of the ER, called, eOGTs, but these enzymes are distinct from the enzyme regulating nucleocytoplasmic O-GlcNAc. O-GlcNAc has been detected on extracellular domains of a handfull of proteins, e.g. Notch) (see Sect. 16.2.2; Alfaro et al. 2012). Also the hexosaminidase removing O-GlcNAc, OGA, is cytosolic and active at neutral pH (see Sect. 16.2.3). By comparison, cellular glycosidases breaking down glycoconjugates retrieved from the cell surface are primarily found in the lysosome and prefer an acidic milieu.

Importantly, the concentration of O-GlcNAc is not uniform across the cell. Some parts, like the nuclear membrane, are heavily modified whereas other parts, like the mitochondria, contain O-GlcNAc but to a lesser degree. All major organelles and other cytosolic substructures, e.g. the proteasome and ribosome, express O-GlcNAc (Holt and Hart 1986; Zhang et al. 2003; Zeidan et al. 2010). The precise level varies over time and is finely tuned to meet the conditions of the cell (see Sects. 16.2.3, 16.3 and 16.4).

16.2.3 O-GlcNAc Can Be Dynamically Attached and Removed

Whether or not a protein is modified by O-GlcNAc varies substantially over time. On many proteins, including the heat-shock protein αB-crystallin, the O-GlcNAc half-life is much shorter than the half-life of the peptide backbone (Roquemore et al. 1996). In fact, studies using selective inhibitors of the enzyme that removes O-GlcNAc from proteins, OGA, show that cycling rates are often on the order of minutes, making O-GlcNAc more akin to protein phosphorylation than "classical" protein glycosylation. "Classical" N- and O-linked glycosylation of proteins, glycosylation of proteins exposed to the extracellular matrix or within the secretory pathway, is, largely, stable once the mature glycan has been attached. There are examples of proteins, e.g. the nucleoporins that form pores through the nuclear membrane, where the O-GlcNAc does not appear to turnover faster than the protein itself (Holt et al. 1987; Miller et al. 1999).

It has been proposed that O-GlcNAc cycling works like a light switch with only two modes of operation—either "on" or "off." One argument in favor of this idea is the fact that there are only two enzymes that add and remove O-GlcNAc from proteins, OGT and OGA, respectively (see Sect. 16.2). Indeed, in some situations cells react by either elevating or suppressing global O-GlcNAc levels. For example, abundant nutrient supply leads to a general increase in O-GlcNAc and scant supply to a general decrease (see Sects. 16.2.2 and 16.4.3). Likewise, cellular stress is associated with raised O-GlcNAc throughout the cell (Zachara et al. 2004). All the same, early studies showed that activation of lymphocytes causes O-GlcNAc levels in the cytosol to go up while they go down in the nucleus (Kearse and Hart 1991). A recently developed FRET (fluorescence resonance energy transfer) reporter that measures OGT activity in real-time demonstrated further that during serum stimulation of transformed cell lines, OGT was activated manifold in some parts of the cytosol whereas in nearby areas OGT activity remained at baseline (Carillo et al. 2011). Work on the regulation of signal transduction by O-GlcNAc describes the same picture; upon stimulation, in a single pathway there can be proteins that become better substrates for OGT but also proteins that become worse substrates for OGT (Whelan et al. 2010). Thus, despite there being only two enzymes that add and remove O-GlcNAc, changes in O-GlcNAc can occur "locally" within the cell. For example, by forming dynamic multipartner complexes OGT and OGA can be directed towards select targets among a broader range of available substrates (see Sects 16.2.2–16.3; Whelan et al. 2008; Cheung and Hart 2008; Housley et al. 2009).

The spatiotemporal regulation of O-GlcNAc cycling is complex and occurs on many levels. While nutrients and stress can cause global changes in O-GlcNAc, binding partners to OGT and OGA tune O-GlcNAc occupancy locally. Below we will discuss in detail how O-GlcNAc cycling is controlled in the nervous system and how the dynamic modification of proteins by O-GlcNAc helps the brain to develop and respond to challenges in the environment.

16.3 O-GlcNAc Is Added to Proteins by OGT and Removed by OGA

16.3.1 Only Two Enzymes Regulate the Cycling of O-GlcNAc

O-GlcNAc exists as a monosaccharide on nuclear and cytoplasmic proteins and can cycle rapidly and repeatedly over the lifetime of the polypeptide chain (see Sect. 16.1). In a single cell, including neurons, thousands of proteins carry O-GlcNAc (Trinidad et al. 2012; see Sect. 16.3.2). Change in O-GlcNAc can happen globally throughout the cell but also locally on individual proteins or sites within a protein (see Sect. 16.2.3). In mammals, there are only two enzymes that add and remove O-GlcNAc. The O-GlcNAc transferase (OGT) adds O-GlcNAc to proteins (Haltiwanger et al. 1990; Kreppel et al. 1997). The O-GlcNAc glycosidase, O-GlcNAcase (OGA), removes O-GlcNAc from proteins (Dong and Hart 1994; Gao et al. 2001). As we will see in Sects. 16.4 and 16.5 loss or deregulation of O-GlcNAc cycling leads to severe developmental brain defects, impaired brain function in the adult and risk for many neurodegenerative disorders, e.g. Alzheimer's disease. In this section we will discuss how OGT and OGA are both promiscuous in order to accept a broad range of targets while at the same time specific to ensure that O-GlcNAc cycles on the correct site at the correct time and place.

16.3.2 O-GlcNAc Transferase; A Highly Conserved Glycosyltransferase Present in the Nucleus and Cytosol

In mammals, O-GlcNAc transferase (OGT) is encoded by a single gene. The gene is highly conserved and lies close to the centromeric region of the X chromosome (Xq13) (Shafi et al. 2000; Nolte and Muller 2002; Kreppel et al. 1997). It spans about 45 kilobase pairs (kb) and its locus is linked to Parkinsonian dystonia, a neurodegenerative movement disorder (Nolte and Muller 2002; Muller et al. 1998). In most organs there are five major OGT transcripts ranging from 4.2 kb to 9.5 kb. The transcripts undergo alternative splicing and two 4 kb transcripts may arise from an internal promoter. In brain the larger 9.5 and 6.4 kb transcripts, which include exons located 5′ of the internal promoter, dominate (Hanover et al. 2003; Nolte and Muller 2002). Most studies so far argue that the total expression of *OGT* is stable during most conditions. However, little is known about the regulation of the OGT gene and at least in neuroblastoma cells OGT mRNA increases after depriving the cells of glucose (Cheung and Hart 2008).

The protein encoded by *OGT* contains two major domains. The N-terminal half is comprised of several tetratricopeptide (TPR) repeats and the C-terminal half binds UDP-GlcNAc and harbors the glycosyltransferase activity. The TPR repeats form a flexible superhelix that can accommodate many protein–protein interactions (Lyer et al. 2003; Kreppel et al. 1997; Kreppel and Hart 1999; Jinek et al. 2004). The C-terminus exhibits a more compact structure and resembles members of the GT-B superfamily of glycosyltransferases (Gtfs) but adopts some unique folds as well (Lazarus et al. 2011). While most OGT-interacting proteins are believed to bind the TPR repeats the C-terminus includes regions that are necessary and sufficient for some interactions, e.g. to the mitogen-activated kinase (MAPK) p38 in neurons (Cheung and Hart 2008). The C-terminus also mediates translocation to the cell membrane upon insulin stimulation, probably via a cluster of lysines that pairs electrostatically with negatively charged phosphatylinositol (3,4,5)-triphosphate (PIP3) (Yang et al. 2008).

Catalysis occurs by an ordered sequential mechanism where OGT binds first UDP-GlcNAc and then the substrate. A hydroxyl group on the incoming substrate performs a direct nucleophilic attack on the anomeric carbon on GlcNAc, thereby inverting the glycosidic bond from UDP-α-GlcNAc to (serine/threonine-) O-β-GlcNAc. In the catalytic groove, there is a general base (histidine 498) and a general acid (probably lysine 842) that catalyze the nucleophilic attack by activating the hydroxyl group (Lazarus et al. 2011 Banerjee et al. 2013). In vitro experiments have shown that if the TPR domain is removed from OGT, the C-terminus alone can modify peptides with O-GlcNAc. In contrast, for protein substrates to be modified, the TPR domain is required (Kreppel and Hart 1999; Lyer and Hart 2003). It has been hypothesized based on computer simulations that the TPRs induce a conformational change in the substrate that enables the O-GlcNAc site to dock at the catalytic groove. Most likely, OGT goes through a conformational change as well; the TPR helix pivots around the intervening region between the TPRs and the C-terminus exposing the entrance to the groove (Trinidad et al. 2012; Lazarus et al. 2011.

Three major isoforms of OGT have been described (Hanover et al. 2003). All share the same catalytic domain but differ in their N-terminus. Nucleocytoplasmic OGT (ncOGT) is the full-length variant and includes 11–12 TPRs, depending on the species. Mitochondrial OGT (mOGT) starts with a mitochondrial targeting sequence, followed by a membrane-spanning region and continues with the last nine TPRs found in ncOGT. The third, and shortest, isoform is soluble OGT (sOGT). It contains only three TPRs. According to most studies, only ncOGT is present in total brain lysate (Kreppel et al. 1997; Marz et al. 2006). However, sOGT may become upregulated in older animals and little is known about whether mOGT or sOGT is present in specific regions of the brain or in specific subcellular compartments (Liu et al. 2012). Apart from mOGT, which is anchored to the inner membrane of the mitochondria, almost all OGT activity is found in the nucleocytoplasm or as a nonintegral membrane protein associated with the cytosolic face of cell membranes (Hanover et al. 2003; Haltiwanger et al. 1990)

In vivo, OGT oligomerizes into a dimer or trimer (Kreppel et al. 1997; Marz et al. 2006). OGT was first purified from rat liver and described as a heterotrimer consisting of two ncOGT and one 78 kDa unit. The 78 kDa unit is enriched in certain tissues and may correspond to sOGT, but may also be a proteolytic fragment of ncOGT (Haltiwanger et al. 1992; Kreppel et al. 1997). The major form of OGT expressed in brain is more likely a homodimer of ncOGT (Marz et al. 2006). Dimerization occurs over an evolutionary conserved hydrophobic region in the TPR domain (TPR 6). Dimerization is stable even in very high salt concentrations and probably not subject to posttranslational regulation (Kreppel and Hart 1999; Jinek et al. 2004).

OGT is exquisitely regulated adding O-GlcNAc only to particular sites at any given time and place. In vitro, OGT exhibits sequence specificity. If purified OGT is mixed with UDP-GlcNAc and a peptide that contains several possible O-GlcNAc sites, often only one or a few sites become significantly modified. Likewise, substrate peptides derived from different proteins are often modified with different efficiency (Kreppel and Hart 1999). Also UDP-GlcNAc levels influence peptide substrate specificity (Kreppel and Hart 1999; Shen et al. 2012). Nevertheless, there is no absolute substrate consensus sequence for OGT. Indeed, the catalytic site of OGT interacts primarily with the peptide backbone of the substrate and not particular side chains (Lazarus et al. 2011). In vivo, O-GlcNAc sites concentrate on disordered regions of proteins and close to proline and valine, the so called PVS motifs (Alfaro et al. 2012; Trinidad et al. 2012; Hart et al. 2011). Presumably, the reason is that such regions can be made to fit OGT's catalytic groove more easily.

Primary sequence plays a role in determining the major O-GlcNAc sites on a given protein. However, in cells, whether a particular protein will be modified

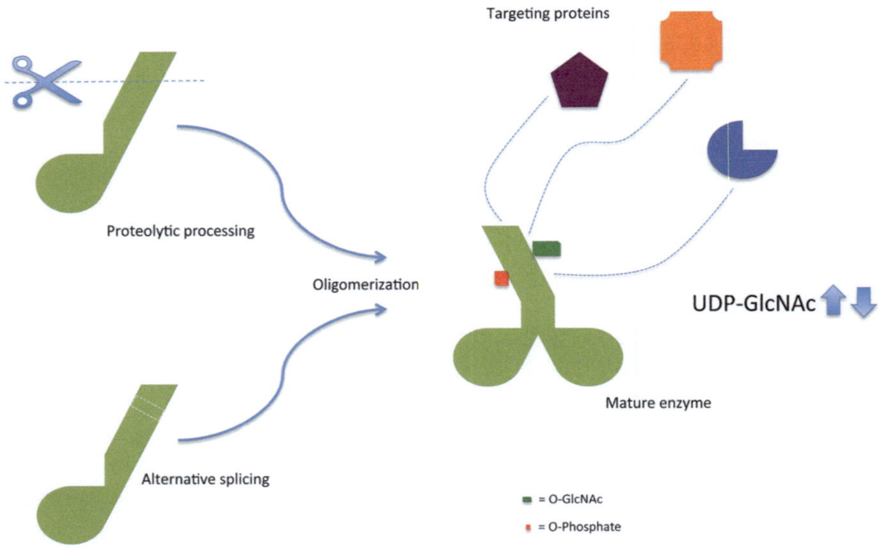

Fig. 16.2 The regulation of the O-GlcNAc transferase, OGT, is complex. In cells, OGT forms an oligomer that interacts with many other proteins. The composition of the oligomer depends on what OGT isoforms are transcribed and probably also on proteolytic processing. The interacting proteins direct OGT to its substrates at specific points in time and space. OGT activity can also be regulated by posttranslational modifications such as O-GlcNAc and O-phosphate and UDP-GlcNAc abundance

depends also on OGT-binding proteins. OGT operates as a holoenzyme, where its interacting proteins direct OGT to its substrates (Whelan et al. 2008; Cheung and Hart 2008; Cheung et al. 2008; Housley et al. 2009; Marz et al. 2006; Yang et al. 2002). For example, in neurons, neurofilaments become O-GlcNAcylated only after activated p38 binds OGT (Cheung et al. 2008). OGT-interacting proteins can enhance OGT activity towards peptide substrates as well (Marz et al. 2006). Moreover, OGT is multiple phosphorylated and it modifies itself with O-GlcNAc (Kreppel et al. 1997; Song et al. 2008; Whelan et al. 2008). Phosphorylation of OGT can both activate OGT, e.g. CaMKIV during neuronal depolarization, and alter OGT's substrate specificity (Whelan et al. 2008; Song et al. 2008; Bullen and Hart Forthcoming paper). OGT is strongly inhibited by free UDP but, unlike most other glycosyltransferases that utilize a UDP-bound sugar as donor, OGT does not require divalent cations (Haltiwanger et al. 1990). For a summary of the regulation of OGT, see Fig. 16.2.

OGT has emerged as a key cellular nutrient sensor. The donor for O-GlcNAc, UDP-GlcNAc, is produced by the hexosamine biosynthesis pathway, the HBP (see Sect. 16.2.1). UDP-GlcNAc is an abundant high-energy small molecule and ranges in concentrations within cells from 0.1 to 1 mM. Almost all metabolic pathways feed into the HBP and contribute to the production of UDP-GlcNAc, including fatty acids, nitrogen and 2–5 % of all cellular glucose. A rich energy supply elevates

UDP-GlcNAc, and, subsequently, protein O-GlcNAcylation whereas a scarce supply is associated with reduced levels of both UDP-GlcNAc and O-GlcNAc (Hart et al. 2011; Hart et al. 2007). Also in the brain fasting has been shown to decrease O-GlcNAc generally, a change reversed upon re-feeding (Liu et al. 2004; Li et al. 2006). Interestingly, depending on the substrate, the K_m for UDP-GlcNAc can vary more than 20-fold but most become better substrates in the presence of higher concentrations of UDP-GlcNAc (Shen et al. 2012; Kreppel and Hart 1999). Therefore, although metabolic flux is associated with global changes in O-GlcNAc, some proteins are more sensitive to nutrient supply than others. In Sects. 16.4 and 16.5 we will discuss how OGT as a nutrient sensor is involved in the pathology behind neurodegenerative disease.

16.3.3 O-GlcNAcase; A Cytosolic O-β-GlcNAc Hydrolase with Neutral pH Optima

In accord with the O-GlcNAc transferase, the enzyme that removes O-GlcNAc from proteins, the O-GlcNAcase (OGA), is expressed from a single gene. The gene was first cloned as an antigen associated with meningioma and it is localized to the long arm of chromosome 10 (10q24), a locus tightly linked to late-onset Alzheimer's disease (Heckel et al. 1998; Bertram et al. 2000). In addition, alternatively spliced OGA transcripts have been associated with sporadic cases of Alzheimer's disease (Twine et al. 2011).

OGA hydrolyses GlcNAc from proteins using substrate-assisted catalysis. Two aspartates activate the acetamido group on GlcNAc's second carbon to perform a nucleophilic attach on its anomeric carbon. This results in the displacement of GlcNAc from the protein (Macauley et al. 2005; Dennis et al. 2006). Contrary to many other hexosaminidases, OGA has neutral pH optima and is not inhibited by GalNAc. Nor does it remove GalNAc from proteins but is specific for β-linked O-GlcNAc (Dong and Hart 1994). OGA is also genetically and immunogenically distinct from other glycosidases (Gao et al. 2001).

OGA has been identified in all tissues so far investigated. It is expressed in the nucleus and cytoplasm of cells and the highest expression is found in brain (Gao et al. 2001). OGA is highly conserved but contains a middle, intrinsically unfolded region that exhibits more variability (Gao et al. 2001; Heckel et al. 1998; Butkinaree et al. 2008). Like OGT, OGA consists of two major domains. The glycosidase domain is C-terminal. The N-terminus contains a histone acetyl transferase (HAT) domain. While the HAT domain was reported to be functionally active (Toleman et al. 2004; Toleman et al. 2006), other groups have not been able to repeat this finding (Butkinaree et al. 2008). Caspase 3 cleaves OGA between the glycosidase domain and the HAT domain during apoptosis without any loss of O-GlcNAcase activity (Butkinaree et al. 2008).

Very little is known about the regulation of OGA. It is believed that OGA forms multipartner complexes that direct OGA to its substrates, much like how OGT is regulated (Hart et al. 2011; see above). Recently, highly specific pharmacological inhibitors of OGA have been developed (Yuzwa et al. 2008; Yuzwa et al. 2012). In Sects. 16.4 and 16.5 we will learn more about the way in which OGA is important for the brain's ability to learn and form memories and serves as a drug target for neurodegenerative disorders.

16.4 O-GlcNAc Is Highly Expressed in the Nervous System

16.4.1 O-GlcNAc is Found Throughout the Brain

Brain is one of the organs where O-GlcNAc is the most abundant. OGT and OGA are expressed in comparatively high levels in the brain (Kreppel et al. 1997; Lubas et al. 1997; Gao et al. 2001). Both enzymes are present across brain regions, including cortex, the cerebellum, and subcortical nuclei such as the amygdala (Liu et al. 2012; Rexach et al. 2012). Within neurons, O-GlcNAc has been identified in all subcellular structures investigated, albeit to differing degrees. Biochemical fractionation was used to show that O-GlcNAc transferase activity was present in synapses (Cole and Hart 2001). Immunoelectron microscopy indicated that OGT and O-GlcNAc were present in both the post- and presynapse. In the presynapse, OGT is concentrated on synaptic vesicles storing neurotransmitter (Akimoto et al. 2003). Most research on O-GlcNAc in the brain has focused on O-GlcNAc's role in neurons of the central nervous system. Less is known about its expression in neurons of the peripheral nervous system or in glia.

16.4.2 Thousands of Neuronal Proteins Are Modified by O-GlcNAc

One difficulty in studying the role of O-GlcNAc for brain function is its vast abundance. Technical breakthroughs have allowed identification of O-GlcNAc sites *en masse* by coupling selective enrichment of O-GlcNAc-modified proteins with electron capture dissociation (ECD) or electron transfer dissociation (ETD) mass spectrometry (types of mass spectrometry that can fragment peptides without losing the GlcNAc) (Wang et al. 2010a, b; Trinidad et al. 2012). According to some estimates, about 40 % of all neuronal proteins are modified by O-GlcNAc (Trinidad et al. 2012). Some proteins, like bassoon, which is important for neurotransmitter release, have more than a dozen O-GlcNAc sites. On other proteins, e.g. CaMKII, only a single site has been detected (Trinidad et al. 2012). As discussed in Sect. 16.2.2 O-GlcNAc sites concentrate on structurally disordered regions of proteins and there

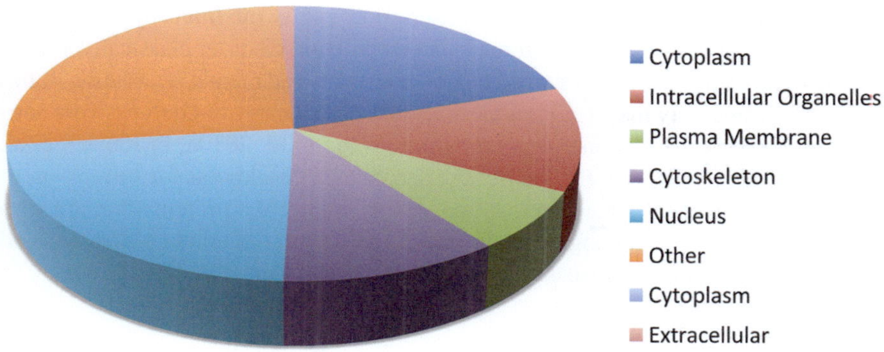

Fig. 16.3 O-GlcNAc modified proteins in neurons belong to all functional classes. In neurons, more than a 1,000 proteins carry O-GlcNAc. The O-GlcNAc-modified proteins include chromatin, transcription factors, cytoskeletal proteins, and enzymes like kinases. Many of these are present in the synapse. The figure shows the distribution of O-GlcNAc in brain across cell compartments, modified from Alfaro et al. Proc Natl Acad Sci USA. (2012) 109(19):7280–5

is some preference for a sequence context rich in proline and valine. For any particular site, O-GlcNAc is usually present in substochiometric levels (Hart et al. 2007). The absolute occupancy is dynamic and depends on the activity of the neuron, possibly due to CaMK-dependent stimulation of OGT (Khidekel et al. 2007; Rexach et al. 2010; Song et al. 2008).

Proteins modified by O-GlcNAc belong to all functional classes (Fig. 16.3). Many O-GlcNAc proteins are shared across cell types and underlie constitutive cellular functions such as transcription, translation, and protein degradation. Nevertheless, the role played by O-GlcNAc on these factors is often neuron-specific. For example, in neurons, the common transcription factor cyclic AMP-response element binding protein (CREB) is modified by O-GlcNAc on serine 40 upon cell depolarization. Once modified, CREB is prevented from binding to its coactivator CRTC (CREB-regulated transcription coactivator). This inhibits CREB and leads to an inactivation of several transcription pathways involved in synaptic plasticity (see Sect. 16.4.2; Rexach et al. 2012). Other O-GlcNAc proteins are particular to neurons, including certain scaffolding proteins, signaling proteins, and cytoskeletal proteins. Many of these are proteins found in the synapse. One interesting case is CaMKII-α. CaMKII-α is modified by O-GlcNAc in a small region known to contain one phosphorylation site that can activate the enzyme and another phosphorylation site that deactivates it (Trinidad et al. 2012). The regulation of CaMKII-α activity is crucial for synaptic events that underlie learning and memory (Lisman et al. 2002). Therefore, elucidation of how O-GlcNAc may fine tune CaMKII-α activity is important not only to our understanding of synapse biology but also higher-order brain function. Moreover, in the synapse, there is a significant overrepresentation of kinases, the enzymes that add O-phosphate to proteins, among the proteins that are modified by O-GlcNAc (Trinidad et al. 2012).

Several proteins that contribute to the development of neurocognitive disease carry O-GlcNAc. The transcription factor methyl CpG binding protein 2 (MeCP2) coordinates activity-dependent gene transcription and is modified by O-GlcNAc. Loss of MeCP2 causes Rett syndrome, a developmental disorder classically associated with repetitive and stereotypical hand movements and mental retardation. It has been shown that only the MeCP2 molecules that do not carry O-GlcNAc were activated in response to neuronal depolarization (Wang et al. 2010a, b; Rexach et al. 2012). Seemingly on MeCP2, O-GlcNAc serves as a checkpoint for turning on or off activity-dependent gene transcription. Other O-GlcNAc proteins that are intimately involved in neurocognitive disease include tau (Alzheimer's disease) and α-synuclein (Parkinson's disease) (Arnold et al. 1996; Wang et al. 2010a, b).

16.4.3 O-GlcNAc Regulates Diverse Cellular Processes

Proteins modified by O-GlcNAc regulate diverse cellular processes such as gene transcription, protein translation and degradation and signal transduction. The function of the O-GlcNAc modification depends on the specific protein that is modified as well as the site on that protein that is modified (Hart et al. 2007; Hart et al. 2011). Here we will briefly discuss different mechanisms by which O-GlcNAc underlies normal cell function.

Gene Transcription

RNA polymerase II. O-GlcNAc may directly affect basal transcription by inactivating RNA polymerase II. RNA polymerase II in the so-called pre-initiation complex is heavily modified by O-GlcNAc in its carboxyl terminal domain (CTD). However, the RNA polymerase II in the so-called elongation complex is devoid of O-GlcNAc. Instead, upon transcription initiation the CTD of RNA polymerase II becomes phosphorylated. It has been suggested that the loss of O-GlcNAc is required before it can be phosphorylated and thereby activated (Kelly et al. 1993; Comer and Hart 1999; Ranuncolo et al. 2012).

Histones. Histones are modified by multiple posttranslational modifications, including O-GlcNAc (Sakabe et al. 2010). The combination of these modifications is hypothesized to form a "code" that either facilitates or prevents the access of transcription factors to DNA. The O-GlcNAc on serine 112 enhances the (mono-) ubiquitination of histone 2B thereby activating gene transcription (Fujiki et al. 2011).

Transcription factors. O-GlcNAc takes advantage of several different mechanisms to regulate the activity of transcription factors toward gene transcription (Ozcan et al. 2010). O-GlcNAc activates NeuroD by promoting its shuttling from the cytosol to the nucleus and protects p53 by saving it from ubiquitin-mediated degradation (Andrali et al. 2007). O-GlcNAc on serine 228 on Oct4 does not affect its general activity but influences its promoter specificity (Jang et al. 2012).

Protein Translation

O-GlcNAc plays several distinct roles during the translation of messenger RNA to polypeptide chains. OGT and OGA bind very tightly to the ribosome and overexpression of OGT was shown to facilitate ribosome assembly. Several ribosomal proteins are modified by O-GlcNAc, including the mammalian target of rapamycin (mTOR) pathway protein S6 (Zeidan et al. 2010). Also associated translational factors are modified by O-GlcNAc. The eukaryotic initiation factor 2 (eIF2) initiates translation by forming a complex with p67. It has been suggested that the O-GlcNAc on p67 is required for p67's ability to protect eIF2 from phosphorylation and thereby inactivation (Datta et al. 1989. Ray et al. 1992).

Protein and Vesicle Trafficking

The subcellular distribution of multiple proteins has been shown to depend on O-GlcNAc (Sayat et al. 2008; Geng et al. 2012). For instance, E-cadherin is a cell junction protein that, among other things, is important for inhibitory synapse formation (Fiederling et al. 2011). If the O-GlcNAc on E-cadherin's cytosolic tail is not removed, E-cadherin becomes trapped in the endoplasmatic reticulum (ER) and cannot be transported further to the Golgi apparatus (Geng et al. 2012). O-GlcNAc may also play a role in the transport of microvesicles within cells, e.g. vesicles that mediate the release of neurotransmitter. There is no direct evidence showing that O-GlcNAc regulates microvesicle trafficking per se. However, proteins such as adaptor protein complex 2 (AP-2) that mediates clathrin-dependent endocytosis and *N*-ethylmaleimide-sensitive fusion protein (NSF), which is an ATPase involved in vesicle fusion, are modified by O-GlcNAc (Clark et al. 2008).

Protein Degradation

O-GlcNAc is intimately involved in the control of protein degradation. O-GlcNAc sites often fall in regions with high PEST-scores (the Pro-Glu-Ser-Thr sequence is associated with rapid degradation of proteins) and an increase in global O-GlcNAc leads to an increased ubiquitination of proteins whereas a decrease in global O-GlcNAc decreases protein ubiquitination (Hart et al. 2007; Guinez et al. 2008). In addition, O-GlcNAc can directly inhibit the proteasome by modifying the 26S and 19S proteasomes (Zang et al. 2003).

Signaling and the Crosstalk Between O-GlcNAc and O-Phosphate

An emerging theme in the regulation of signal transduction pathways is the crosstalk between O-GlcNAc and O-phosphate. Many O-phosphorylated proteins are also O-GlcNAc proteins (Trinidad et al. 2012). O-GlcNAc and O-phosphate can directly and reciprocally inhibit each other by sharing the same site, as in the case of threonine 58 in the trans-activation domain of the transcription factor c-myc (Chou et al. 1995a; Chou et al. 1995b). On other proteins the crosstalk is indirect and can be both negative and positive. Removal of O-GlcNAc from threonine 57/ serine 58 on CaMKIV, for example, prevents the phosphorylation of threonine 200, the main activation site on CaMKIV. At the same time, loss of O-GlcNAc on serine 189 facilitates threonine 200 phosphorylation (Dias et al. 2009). In fact, on a global level, the occupancy of most dynamic phosphorylation sites is affected by acute inhibition of O-GlcNAc cycling (Wang et al. 2008). As a group, kinases, the enzymes that add O-phosphate to proteins, are more often modified by O-GlcNAc than other kinds of proteins (Dias et al. 2012; Trinidad et al. 2012). Furthermore, as we discussed in Sect. 16.2.2, phosphorylation of OGT can activate OGT and influence its substrate specificity.

16.5 O-GlcNAc Is Essential for Brain Function

In Sects. 16.1–16.4 we have learned that O-GlcNAc dynamically modifies a vast array of neuronal proteins involved in many cellular processes, such as gene transcription and signal transduction. In the following section we will discuss how the regulation of neuronal function by O-GlcNAc underlies normal brain function and contributes to neurocognitive disease (summarized in Fig. 16.4).

16.5.1 Early and Late Brain Development Depends Upon O-GlcNAc Cycling

In vertebrates, O-GlcNAc is essential for normal development. In many different cell types, it has been shown that perturbations of O-GlcNAc cycling interrupt the cell cycle by preventing cytokinesis (Slawson et al. 2005; Slawson et al. 2008; Wang et al. 2010a, b). In the embryo, this leads to failure of stem cell proliferation and subsequent death at the single cell stage (Shafi et al. 2000). Studies taking advantage of partial depression of OGT expression, where cell proliferation is preserved, indicated that renewal of stem cell pluripotency and stem cell differentiation depends upon O-GlcNAc. In fact, two members of the core pluripotency network in embryonic stem cells, Oct4 and Sox2, are modified by O-GlcNAc. Loss of O-GlcNAc on threonine 228 of Oct4 alters its promoter specificity disrupting transcription of dozens of genes, including some that are involved in neuronal differentiation (e.g. Nanog) (Jang et al. 2012; Wang et al. 2012).

Later in development, O-GlcNAc plays a role in brain morphogenesis and patterning. Loss of OGT decreases brain size, while overexpression of OGT blurs the organization and distinction of hind-, mid-, and forebrain regions in zebra fish (Webster et al. 2009). Likewise, when OGT was deleted from neurons specifically

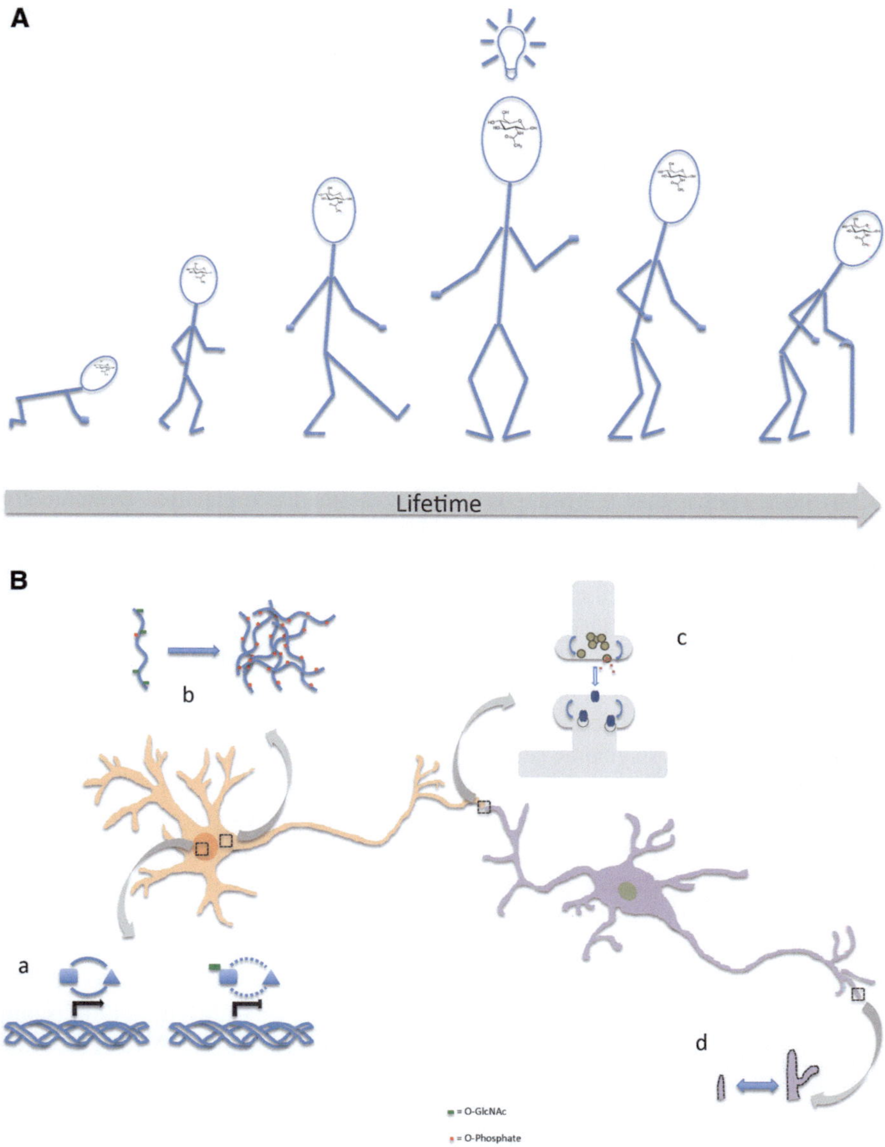

Fig. 16.4 O-GlcNAc underlies normal brain development and function and contributes to the development of neurocognitive disease by regulating diverse cellular processes. (**A**) O-GlcNAc is expressed in the brain from early development until late in life. O-GlcNAc is essential for brain development and modulates many normal functions in the adult, e.g. learning and memory. Disturbed O-GlcNAc cycling may contribute to the pathology behind many neurodegenerative diseases. How O-GlcNAc impacts neuron function is exemplified in (**B**). (*a*) When the transcription factor CREB is modified by O-GlcNAc, it cannot bind the coactivator CRTC. This leads to the repression of many genes that are involved in synaptic plasticity. (*b*) Tau is normally modified by O-GlcNAc. In Alzheimer's disease, tau loses its O-GlcNAc and precipitates into cytotoxic aggregates. (*c*) It is believed that dynamic changes in synaptic function underlie learning and memory. At presynaptic terminals, O-GlcNAc may regulate synaptic plasticity by affecting the recycling of neurotransmitter-containing vesicles, possibly through affecting the function of the vesicle-binding protein synapsin I. In the postsynapse, the O-GlcNAc effect on learning and memory may be related to modulation of neurotransmitter receptor trafficking. (*d*) One mechanism by which O-GlcNAc impacts normal development is its regulation of axonal growth and branching

by crossing floxed OGT mice with mice expressing Cre recombinase under the synapsin I promoter, viability until term was reduced. The pups that survived failed to develop any locomotor activity and died within 10 days (O'Donnell et al. 2004). In postmitotic neurons, the O-GlcNAc glycosylation of CREB underlies both axonal and dendritic growth and may, at least partly, explain the impaired development of brain function (Rexach et al. 2012). O-GlcNAc is also known to affect axonal branching (Francisco et al. 2009).

16.5.2 O-GlcNAc Underlies Learning and Memory

Learning and memory are fundamental properties of the brain. Over the past decades it has been shown that the brain is a highly malleable organ that dynamically responds to challenges in the environment. It is believed that memories are encoded by neurons through changes in their synaptic communications with other neurons, so-called synaptic plasticity. A prevailing model for how such changes occur is long-term potentiation (LTP) and depression (LTD), occurring mainly at the postsynapse (see, e.g., Hanley 2008; Lynch 2004). Multiple disorders involving mental retardation such as Fragile X and Rett syndrome have been linked to proteins underlying synaptic plasticity (Verpelli and Sala 2012; Grant 2012). OGT is present in the synapse and many postsynaptic proteins involved in LTP and LTD carry O-GlcNAc (see Sects. 16.3.1–2 and 16.4.2–3). Using electrophysiology, it was further established that O-GlcNAc cycling is necessary for normal expression of LTP; pharmacological inhibition of OGT diminished LTP whereas OGA inhibition elevated LTP (Tallent et al. 2009). The effect may relate to the synaptic relocation of the α-amino-3-hydroxy-5-methyl-4-isoxazoleproprionic (AMPA) receptor, the glutamate neurotransmitter receptor responsible for most fast synaptic transmission in the central nervous system (CNS). O-GlcNAc modifies CaMKII in close proximity to a phosphorylation site that regulates its activity (see Sect. 16.3.2). CaMKII, in turn, is one of the major pathways that can induce the synaptic insertion of the AMPA receptor (Tallent et al. 2009; Shepherd and Huganir 2007). Interestingly, an OGT-dependent increase in AMPA receptor conductivity was absent in AMPA receptors where the main CaMKII phosphorylation site was mutated to alanine (Kanno et al. 2010).

O-GlcNAc cycling is important for dynamic synaptic function at the presynapse. Acute OGA inhibition in vivo decreases paired-pulse facilitation (PPF). PPF is a measurement of vesicle release of neurotransmitter. O-GlcNAc may regulate PPF via synapsin I. The presynaptic protein synapsin I is extensively modified by O-GlcNAc and inhibition of OGA increases the phosphorylation of synapsin I (Tallent et al. 2009).

The studies investigating the effect of O-GlcNAc on learning and memory have so far relied heavily upon pharmacological manipulation of OGT and OGA. Unfortunately, some of the drugs used are known to cause off-target effects in cells. This may explain some apparent contradictions in published results (e.g. Tallent et al. 2009 argues that O-GlcNAc does not affect basal synaptic transmission whereas Kanno et al. 2010 does see a large effect on basal AMPA receptor function).

For future research, it will be important to develop mouse models of O-GlcNAc function and focus on mechanisms for the way in which O-GlcNAc may underlie learning and memory. One good example is a recent study that compared wild-type CREB with an O-GlcNAc loss-of-function mutant of CREB. This study overexpressed these proteins in vivo and showed that the O-GlcNAc mutant enhanced CREB-dependent transcription of several synaptic plasticity associated genes as well as short-term memory in mice (Rexach et al. 2012).

16.5.3 Impaired O-GlcNAc Cycling Contributes to Neurodegenerative Disease

Neurodegenerative disease is a collection of disorders that are characterized by general cognitive decline and loss of neurons and neuronal synapses. They become more common with age but early-onset variants exist (Koffie et al. 2011; Schulz-Schaeffer 2010). Impaired O-GlcNAc cycling is implicated in the development of several types of neurodegenerative disease, including Alzheimer's, Parkinsonism, and Huntington's (Arnold et al. 1996; Dias and Hart 2007; Wang et al. 2012; Hart et al. 2011). The loci for the genes encoding OGT and OGA are linked to Parkinsonian dystonia and Alzheimer's disease, respectively. In addition, OGA transcripts are alternatively spliced in the brain in patients with sporadic Alzheimer's (see Sects. 16.2.2–16.3).

Two pathological hallmarks of Alzheimer's disease are neurofibrillary tangles and amyloid plaques (Koffie et al. 2011; Dias and Hart 2007). The neurofibrillary tangles consist of paired helical filaments of tau, a protein that is important for microtubuli stability. Tau is extensively modified by O-GlcNAc, including in the region that binds to microtubuli (Arnold et al. 1996; Wang et al. 2010a, b). Results from both in vitro and in vivo experiments indicated that increasing O-GlcNAc on tau protected it from filament precipitation (Yuzwa et al. 2012). O-GlcNAc may save tau from precipitating either directly or indirectly by protecting it from hyperphosphorylation (Yuzwa et al. 2012; Yuzwa et al. 2008). Alzheimer's patients present with hyperphosphorylated and hypo-O-GlcNAc glycosylated tau. No O-GlcNAc can be found on the tau tangles isolated from these patients (Liu et al. 2004). Importantly, in mouse models of neurodegenerative disease resulting from expressing mutant forms of tau prone to precipitate, elevating global O-GlcNAc, including O-GlcNAc on tau, by selectively inhibiting OGA pharmacologically, slowed symptom progression (Yuzwa et al. 2012). Interestingly, the main contributor amyloid plaques, amyloid precursor protein (APP), is also modified by O-GlcNAc (Griffith et al. 1995).

Alzheimer's disease is intimately associated with dysregulated glucose metabolism and insulin resistance (Biessels et al. 2006; de la Monte and Wands 2008). Due to the regulation of O-GlcNAc by nutrient supply and metabolic hormones, OGT has emerged as a key nutrient sensor in cells (see Sect. 16.2.2). Much of the toxicity associated with excessive intake of glucose is mediated by the hexosamine biosynthesis pathway, the same pathway that produces the O-GlcNAc donor substrate UDP-GlcNAc. Many forms of insulin resistance are also related to OGT

(Hart et al. 2007). Overexpression of OGT can independently lead to insulin resistance (Yang et al. 2008). From this perspective, impaired O-GlcNAc signaling represents a model, and pharmacological target, for how metabolic dysfunction may result in neurodegenerative disease.

O-GlcNAc may also be involved in neurodegenerative diseases other than Alzheimer's. For example, in transgenic models where a polyglutamine expansion of huntingtin (the protein that causes Huntington's disease) was overexpressed, deletion of OGT attenuated neuron loss while deletion of OGA exacerbated neuron loss (Wang et al. 2012). In addition, α-synuclein, the protein found in lesions overrepresented in Parkinson's disease, carries O-GlcNAc in the region known to induce self-aggregation (Wang et al. 2010a, b). It should also be noted that O-GlcNAc in other cell types becomes elevated from many different types of stress and that deletion of OGT makes cells more vulnerable to stress (Zachara et al. 2004; Kazemi et al. 2010).

16.6 Summary and Outlook

Every time we think, act, and feel information is passed between neurons in the brain. Neurons are extraordinary cells with complex morphology and a lifespan that matches the lifespan of the organism. The posttranslational modification O-GlcNAc, the attachment of β-N-acetylglucosamine to serine and threonine, is expressed in all compartments of the neuron. By the help of only two enzymes, OGT and OGA, O-GlcNAc regulates thousands of proteins underlying diverse cellular processes such as gene transcription and signal transduction. The big challenge for the future will be to derive a mechanistic picture for how specific O-GlcNAc sites affect protein function and thereby synaptic transmission. Recently, several new techniques have been developed that improve and simplify the detection of O-GlcNAc. Nevertheless, more tools such as site-specific antibodies for O-GlcNAc are needed to make the O-GlcNAc field more accessible to the broader neuroscience research community. Understanding the way O-GlcNAc works in the nervous system will not only afford fundamental principles of brain function but also provide novel targets for the treatment of neurocognitive disease.

Conflicts of Interest The authors declare no conflicts of interest.

References

Akimoto Y, Comer FI, Cole RN, Kwakami AKH, Hirano H, Hart GW. Localization of the O-GlcNAc transferase and O-GlcNAc-modified proteins in rat cerebellar cortex. Brain Res. 2003;966(2):194–205.

Alfaro JF, Gong CX, Monroe ME, Aldrich JT, Clauss TR, Purvine SO, et al. Tandem mass spectrometry identifies many mouse brain O-GlcNAcylated proteins including EGF domain-specific O-GlcNAc transferase targets. Proc Natl Acad Sci U S A. 2012;109(19):7280–5.

Andrali SS, Qian Q, Ozcan S. Glucose mediates the translocation of NeuroD1 by O-linked glycosylation. J Biol Chem. 2007;282(21):15589–96.
Arnold SC, Johnson GVW, Cole RN, Dong DLY, Lee M, Hart GW. The microtubule-associated protein tau is extensively modified with O-linked N-acetylglucosamine. J Biol Chem. 1996;271(46):28741–4.
Banerjee S, Robbins PW, Samuelson J. Molecular characterization of nucleocytosolic O-GlcNAc transferases of Giardia lamblia and Cryptosporidium parvum. Glycobiology. 2009;19(4): 331–6.
Banerjee PS, Hart GW, Cho JW. Chemical approaches to study O-GlcNAcylation. Chem Soc Rev. 2013;42(10):4345–57.
Benko DM, Haltiwanger RS, Hart GW, Gibson W. Virion basic phosphoprotein from human cytomegalovirus contains O-linked N-acetylglucosamine. Proc Natl Acad Sci U S A. 1988; 85(8):2573–7.
Bertram L, Blacker D, Mullin K, Keeney D, Jones J, Basu S, et al. Evidence for genetic linkage of Alzheimer's disease to chromosome 10q. Science. 2000;290(5500):2302–3.
Biessels GJ, Staekenborg S, Brunner E, Brayne C, Scheltens P. Risk of dementia in diabetes mellitus: a systematic review. Lancet Neurol. 2006;5(1):64–74.
Bullen, Hart. AMPK regulates OGT substrate specificity. Forthcoming paper.
Butkinaree C, Cheung WD, Park S, Park K, Barbr M, Hart GW. Characterization of beta-N-acetylglucosaminidase cleavage by caspase-3 during apoptosis. J Biol Chem. 2008;283(35): 23557–66.
Caillet-Boudin ML, Strecker G, Michalski JC. O-linked GlcNAc in serotype-2 adenovirus fibre. Eur J Biochem. 1989;184(1):205–11.
Carillo LD, Froemming JA, Mahal LK. Targeted in vivo O-GlcNAc sensors reveal discrete compartment-specific dynamics during signal transduction. J Biol Chem. 2011;286(8): 6650–8.
Cheung WD, Hart GW. AMP-activated protein kinase and p38 MAPK activate O-GlcNAcylation of neuronal proteins during glucose deprivation. J Biol Chem. 2008;283(19):13009–20.
Cheung WD, Sakabe K, Housley MP, Dias WB, Hart GW. O-linked beta-N-acetylglucosaminyltransferase substrate specificity is regulated by myosin phosphatase targeting and other interacting proteins. J Biol Chem. 2008;283(49):33935–41.
Chou TY, Dang CV, Hart GW. Glycosylation of the c-Myc transactivation domain. Proc Natl Acad Sci U S A. 1995a;92(10):4417–21.
Chou TY, Hart GW, Dang CV. c-Myc is glycosylated at threonine 58, a known phosphorylation site and a mutational hot spot in lymphomas. J Biol Chem. 1995b;270(32):18961–5.
Clark PM, Dweck JF, Mason DE, Hart CR, Buck SB, Peters EC, et al. Direct in-gel fluorescence detection and cellular imaging of O-GlcNAc-modified proteins. J Am Chem Soc. 2008;130(35):11576–7.
Cole RN, Hart GW. Cytosolic O-glycosylation is abundant in nerve terminals. J Neurochem. 2001;79(5):1080–9.
Comer FI, Hart GW. O-GlcNAc and the control of gene expression. Biochim Biophys Acta. 1999;1473(1):161–71.
Datta B, Ray MK, Chakrabarti D, Wylie DE, Gupta NK. Glycosylation of eukaryotic peptide chain initiation factor 2 (eIF-2)-associated 67-kDa polypeptide (p67) and its possible role in the inhibition of eIF-2 kinase-catalyzed phosphorylation of the eIF-2 alpha-subunit. J Biol Chem. 1989;264(34):20620–4.
de la Monte SM, Wands JR. Alzheimer's disease is type 3 diabetes-evidence reviewed. J Diabetes Sci Technol. 2008;2(6):1101–13.
Dennis RJ, Taylor EJ, Macauley MS, Stubbs KA, Turkenburg JP, Hart SJ, et al. Structure and mechanism of a bacterial beta-glucosaminidase having O-GlcNAcase activity. Nat Struct Mol Biol. 2006;13(4):365–71.
Dias WB, Hart GW. O-GlcNAc modification in diabetes and Alzheimer's disease. Mol Biosyst. 2007;3(11):766–72.

Dias WB, Cheung WD, Wang Z, Hart GW. Regulation of calcium/calmodulin-dependent kinase IV by O-GlcNAc modification. J Biol Chem. 2009;284(32):21327–37.
Dias WB, Cheung WD, Hart GW. O-GlcNAcylation of kinases. Biochem Biophys Res Commun. 2012;422(2):224–8.
Dong DL, Hart GW. Purification and characterization of an O-GlcNAc selective N-acetyl-beta-D-glucosaminidase from rat spleen cytosol. J Biol Chem. 1994;269(30):19321–30.
Fang B, Miller MW. Use of galactosyltransferase to assess the biological function of O-linked N-acetyl-d-glucosamine: a potential role for O-GlcNAc during cell division. Exp Cell Res. 2001;263(2):243–53.
Fiederling A, Ewert R, Andreyeva A, Jungling K, Gottmann K. E-cadherin is required at GABAergic synapses in cultured cortical neurons. Neurosci Lett. 2011;501(3):167–72.
Francisco H, Kollins K, Varghis N, Vocadlo D, Vosseller K, Gallo G. O-GLcNAc post-translational modifications regulate the entry of neurons into an axon branching program. Dev Neurobiol. 2009;69(2–3):162–73.
Fredriksen L, Mathiesen G, Moen A, Bron PA, Kleerebezem M, Eijsink VG, et al. The major autolysin Acm2 from Lactobacillus plantarum undergoes cytoplasmic O-glycosylation. J Bacteriol. 2012;194(2):325–33.
Fujiki R, Hashiba W, Sekine H, Yokoyama A, Chikanishi T, Ito S, et al. GlcNAcylation of histone H2B facilitates its monoubiquitination. Nature. 2011;480(7378):557–60.
Gao Y, Wells L, Comer FI, Parker GJ, Hart GW. Dynamic O-glycosylation of nuclear and cytosolic proteins: cloning and characterization of a neutral, cytosolic beta-N-acetylglucosaminidase from human brain. J Biol Chem. 2001;276(13):9838–45.
Geng F, Zhi W, Anderson RA, Leber B, Andrews DW. Multiple post-translational modifications regulate E-cadherin transport during apoptosis. J Cell Sci. 2012;125(Pt 11):2615–25.
Grant SG. Synaptopathies: diseases of the synaptome. Curr Opin Neurobiol. 2012;22(3):522–9.
Griffith LS, Mathes M, Schmitz B. Beta-amyloid precursor protein is modified with O-linked N-acetylglucosamine. J Neurosci Res. 1995;41(2):270–8.
Guinez C, Mir AM, Dehennaut V, Cacan R, Harduin-Lepers A, Michalski JC, et al. Protein ubiquitination is modulated by O-GlcNAc glycosylation. FASEB J. 2008;22(8):2901–11.
Haltiwanger RS, Holt GD, Hart GW. Enzymatic addition of O-GlcNAc to nuclear and cytoplasmic proteins. Identification of a uridine diphospho-N-acetylglucosamine:peptide beta-N-acetylglucosaminyltransferase. J Biol Chem. 1990;265(5):2563–8.
Haltiwanger RS, Blomberg MA, Hart GW. Glycosylation of nuclear and cytoplasmic proteins. Purification and characterization of a uridine diphospho-N-acetylglucosamine:polypeptide beta-N-acetylglucosaminyltransferase. J Biol Chem. 1992;267(13):9005–13.
Hanley JG. AMPA receptor trafficking pathways and links to dendritic spine morphogenesis. Cell Adh Migr. 2008;2(4):276–82.
Hanover JA, Song Y, Lubas WB, Shin SH, Ragano-Caracciola M, Kochran J, et al. Mitochondrial and nucleocytoplasmic isoforms of O-linked GlcNAc transferase encoded by a single mammalian gene. Arch Biochem Biophys. 2003;409(2):287–97.
Hart GW, Housley MP, Slawson C. Cycling of O-linked beta-N-acetylglucosamine on nucleocytoplasmic proteins. Nature. 2007;26(7139):1017–22.
Hart GW, Slawson C, Ramirez-Correa G, Lagerlof O. Cross talk between O-GlcNAcylation and phosphorylation: roles in signaling, transcription, and chronic disease. Annu Rev Biochem. 2011;80:825–58.
Heckel D, Comtesse N, Brass N, Blin N, Zang KD, Meese E. Novel immunogenic antigen homologous to hyaluronidase in meningioma. Hum Mol Genet. 1998;7(12):1859–72.
Holt GD, Hart GW. The subcellular distribution of terminal N-acetylglucosamine moieties. Localization of a novel protein-saccharide linkage, O-linked GlcNAc. J Biol Chem. 1986;26(17):8049–57.
Holt GD, Snow CM, Senior A, Haltiwanger RS, Gerace L, Hart GW. Nuclear pore complex glycoproteins contain cytoplasmically disposed O-linked N-acetylglucosamine. J Cell Biol. 1987; 104(5):1157–64.

Housley MP, Udeshi ND, Rodgers JT, Shabanowitz J, Puigserver P, Hunt DF, et al. A PGC-1alpha-O-GlcNAc transferase complex regulates FoxO transcription factor activity in response to glucose. J Biol Chem. 2009;284(8):5148–57.
Jang H, Kim TW, Yoon S, Choi SY, Kang TW, Kim SY, et al. O-GlcNAc regulates pluripotency and reprogramming by directly acting on core components of the pluripotency network. Cell Stem Cell. 2012;11(1):62–74.
Jinek M, Rehwinkel J, Lazarus BD, Izaurralde E, Hanover JA, Conti E. The superhelical TPR-repeat domain of O-linked GlcNAc transferase exhibits structural similarities to importin alpha. Nat Struct Mol Biol. 2004;11(10):1001–7.
Kanno T, Yaguchi T, Nagata T, Mukasa T, Nishizaki T. Regulation of AMPA receptor trafficking by O-glycosylation. Neurochem Res. 2010;35(5):782–8.
Kazemi Z, Chang H, Haserodt S, McKen C, Zachara NE. O-linked beta-N-acetylglucosamine (O-GlcNAc) regulates stress-induced heat shock protein expression in a GSK-3beta-dependent manner. J Biol Chem. 2010;285(50):39096–107.
Kearse KP, Hart GW. Lymphocyte activation induces rapid changes in nuclear and cytoplasmic glycoproteins. Proc Natl Acad Sci U S A. 1991;88(5):1701–5.
Kelly WG, Dahmus ME, Hart GW. RNA polymerase II is a glycoprotein. Modification of the COOH-terminal domain by O-GlcNAc. J Biol Chem. 1993;268(14):10416–24.
Khidekel N, Ficarro SB, Clark PM, Bryan MC, Swaney LD, Rexach JE, et al. Probing the dynamics of O-GlcNAc glycosylation in the brain using quantitative proteomics. Nat Chem Biol. 2007;3(6):339–48.
Koffie RM, Hyman BT, Spires-Jones TL. Alzheimer's disease: synapses gone cold. Mol Neurodegener. 2011;6(1):63.
Kreppel LK, Hart GW. Regulation of a cytosolic and nuclear O-GlcNAc transferase. Role of the tetratricopeptide repeats. J Biol Chem. 1999;274(45):32015–22.
Kreppel LK, Blomberg MA, Hart GW. Dynamic glycosylation of nuclear and cytosolic proteins. Cloning and characterization of a unique O-GlcNAc transferase with multiple tetratricopeptide repeats. J Biol Chem. 1997;272(14):9308–15.
Lazarus MB, Nam Y, Jian J, Sliz P, Walker S. Structure of human O-GlcNAc transferase and its complex with a peptide substrate. Nature. 2011;469(7331):564–7.
Li X, Lu F, Wang JZ, Gong CX. Concurrent alterations of O-GlcNAcylation and phosphorylation of tau in mouse brains during fasting. Eur J Neurosci. 2006;23(8):2078–86.
Lisman J, Schulman H, Cline H. The molecular basis of CaMKII function in synaptic and behavioural memory. Nat Rev Neurosci. 2002;3(3):175–90.
Liu F, Iqbal K, Grundke-Iqbal I, Hart GW, Gong CX. O-GlcNAcylation regulates phosphorylation of tau: a mechanism involved in Alzheimer's disease. Proc Natl Acad Sci U S A. 2004; 101(29):10804–9.
Liu Y, Li X, Yu Y, Shi J, Liang Z, Run X, et al. Developmental regulation of protein O-GlcNAcylation, O-GlcNAc transferase and O-GlcNAcase in mammalin brain. PLoS One. 2012;7(8):e43724.
Lubas WA, Frank DW, Krause M, Hanover JA. O-Linked GlcNAc transferase is a conserved nucleocytoplasmic protein containing tetratricopeptide repeats. J Biol Chem. 1997;272(14): 9316–24.
Lyer SPN, Hart GW. Roles of the tetratricopeptide repeat domain in O-GlcNAc transferase targeting and protein substrate specificity. J Biol Chem. 2003;278(27):24608–16.
Lyer SPN, Akimoto Y, Hart GW. Identification and cloning of a novel family of coiled-coil domain proteins that interact with O-GlcNAc transferase. J Biol Chem. 2003;278(7):5399–409.
Lynch MA. Long-term potentiation and memory. Physiol Rev. 2004;84(1):87–136.
Macauley MS, Whitworth GE, Debowski AW, Chin D, Vocadlo DJ. O-GlcNAcase uses substrate-assisted catalysis: kinetic analysis and development of highly selective mechanism-inspired inhibitors. J Biol Chem. 2005;280(27):25313–22.
Marotta NP, Cherwien CA, Abeywardana T, Pratt MR. O-GlcNAc modification prevents peptide-dependent acceleration of α-synuclein aggregation. Chembiochem. 2012;13(18): 2665–70.

Marz P, Stetefeld J, Bendfeldt K, Nitsch C. Reinstein, Shoeman RL et al. Ataxin-10 interacts with O-linked beta-N-acetylglucosamine transferase in the brain. J Biol Chem. 2006;281(29): 20263–70.

Miller MW, Caracciolo MR, Berlin WK, Hanover JA. Phosphorylation and glycosylation of nucleoporins. Arch Biochem Biophys. 1999;367(1):51–60.

Muller U, Steinberger D, Nemeth AH. Clinical and molecular genetics of primary dystonias. Neurogenetics. 1998;1(3):165–77.

Nolte D, Muller U. Human O-GlcNAc transferase (OGT): genomic structure, analysis of splice variants, fine mapping in Xq13.1. Mamm Genome. 2002;13(1):62–4.

O'Donnell N, Zachara NE, Hart GW, Marth JD. Ogt-dependent X-chromosome-linked protein glycosylation is a requisite modification in somatic cell function and embryo viability. Mol Cell Biol. 2004;24(2):1680–90.

Ozcan S, Andrali SS, Cantrell JE. Modulation of transcription factor function by O-GlcNAc modification. Biochim Biophys Acta. 2010;1799(5–6):353–64.

Ranuncolo SM, Ghosh S, Hanover JA, Hart GW, Lewis BA. Evidence of the involvement of O-GlcNAc-modified human RNA polymerase II CTD in transcription in vitro and in vivo. J Biol Chem. 2012;287(28):23549–61.

Ray MK, Datta B, Chakraborty A, Chattopadhyay A, Meza-Keuthen S, Gupta NK. The eukaryotic initiation factor 2-associated 67-kDa polypeptide (p67) plays a critical role in regulation of protein synthesis initiation in animal cells. Proc Natl Acad Sci U S A. 1992;89(2):539–43.

Rexach JE, Rogers CJ, Yu SH, Tao J, Sun YE, Hsieh-Wilson LC. Quantification of O-glycosylation stoichiometry and dynamics using resolvable mass tags. Nat Chem Biol. 2010;6(9):645–51.

Rexach JE, Clark PM, Mason DE, Neve RL, Peters EC, Hsieh-Wilson LC. Dynamic O-GlcNAc modification regulates CREB-mediated gene expression and memory formation. Nat Chem Biol. 2012;89(3):253–61.

Roquemore EP, Chevrier MR, Cotter RJ, Hart GW. Dynamic O-GlcNAcylation of the small heat shock protein alpha B-crystallin. Biochemistry. 1996;35(11):3578–86.

Sakabe K, Wang Z, Hart GW. Beta-N-acetylglucosamine (O-GlcNAc) is part of the histone code. Proc Natl Acad Sci U S A. 2010;107(46):19915–20.

Sayat R, Leber B, Grubac V, Wiltshire L, Persad S. O-GlcNAc-glycosylation of beta-catenin regulates its nuclear localization and transcriptional activity. Exp Cell Res. 2008;314(15):2774–87.

Schirm M, Kalmokoff M, Aubry AP, Thibault P, Sandoz M, Logan SM. Flagellin from Listeria monocytogenes is glycosylated with beta-O-linked N-acetylglucosamine. J Bacteriol. 2004; 186(20):6721–7.

Schulz-Schaeffer WJ. The synaptic pathology of alpha-synuclein aggregation in dementia with Lewy bodies, Parkinson's disease and Parkinson's disease dementia. Acta Neuropathol. 2010;120(2):131–43.

Selzer J, Hofmann F, Rex G, Wilm M, Mann M, Just I, Aktories K. Clostridium novyi alpha-toxin-catalyzed incorporation of GlcNAc into Rho subfamily proteins. J Biol Chem. 1996;271(41): 25173–7.

Shafi R, Lyer SPN, Ellies LG, O'Donnell N, Marek KW, Chui D, et al. The O-GlcNAc transferase gene resides on the X chromosome and is essential for embryonic stem cell viability and mouse ontogeny. Proc Natl Acad Sci U S A. 2000;97(11):5735–9.

Shen DL, Gloster TM, Yuzwa SA, Vocadlo DJ. Insights into O-linked N-acetylglucosamine ([0-9] O-GlcNAc) processing and dynamics through kinetic analysis of O-GlcNAc transferase and O-GlcNAcase activity on protein substrates. J Biol Chem. 2012;287(19):15395–408.

Shepherd JD, Huganir RL. The cell biology of synaptic plasticity: AMPA receptor trafficking. Annu Rev Cell Dev Biol. 2007;23:613–43.

Slawson C, Zachara NE, Vosseller K, Cheung WD, Lane MD, Hart GW. Perturbations in O-linked beta-N-acetylglucosamine protein modification cause severe defects in mitotic progression and cytokinesis. J Biol Chem. 2005;280(38):32944–56.

Slawson C, Lakshmanan T, Knapp S, Hart GW. A mitotic GlcNAcylation/phosphorylation signaling complex alters the posttranslational state of the cytoskeletal protein vimentin. Mol Biol Cell. 2008;19(10):4130–40.

Song M, Kim HS, Park JM, Kim SH, Kim IH, Ryu SH, et al. o-GlcNAc transferase is activated by CaMKIV-dependent phosphorylation under potassium chloride-induced depolarization in NG-108-15 cells. Cell Signal. 2008;20(1):94–104.

Tallent MK, Varghis N, Skorobogatko Y, Hernandez-Cuebas L, Whelan K, Vocadlo DJ, et al. In vivo modulation of O-GlcNAc levels regulates hippocampal synaptic plasticity through interplay with phosphorylation. J Biol Chem. 2009;284(1):174–81.

Toleman C, Paterson AJ, Whisenhunt TR, Kudlow JE. Characterization of the histone acetyltransferase (HAT) domain of a bifunctional protein with activable O-GlcNAcase and HAT activities. J Biol Chem. 2004;279(51):53665–73.

Toleman CA, Paterson AJ, Kudlow JE. The histone acetyltransferase NCOAT contains a zinc finger-like motif involved in substrate recognition. J Biol Chem. 2006;281(7):3918–25.

Torres CR, Hart GW. Topography and polypeptide distribution of terminal N-acetylglucosamine residues on the surfaces of intact lymphocytes. Evidence for O-linked GlcNAc. J Biol Chem. 1984;259(5):3308–17.

Trinidad JC, Barkan DT, Gulledge BF, Thalhammer A, Sali A, Shoepfer R, et al. Global identification and characterization of both O-GlcNAcylation and phosphorylation at the murine synapse. Mol Cell Proteomics. 2012;11(8):215–29.

Twine NA, Janitz K, Wilkins MR, Janitz M. Whole transcriptome sequencing reveals gene expression and splicing differences in brain regions affected by Alzheimer's disease. PLoS One. 2011;6(1):e16266.

Verpelli C, Sala C. Molecular and synaptic defects in intellectual disability syndromes. Curr Opin Neurobiol. 2012;22(3):530–6.

Wang Z, Gucek M, Hart GW. Cross-talk between GlcNAcylation and phosphorylation: site-specific phosphorylation dynamics in response to globally elevated O-GlcNAc. Proc Natl Acad Sci U S A. 2008;105(37):13793–8.

Wang Z, Udeshi ND, O'Malley M, Shabanowitz J, Hunt DF, Hart GW. Enrichment and site mapping of O-linked N-acetylglucosamine by a combination of chemical/enzymatic tagging, photochemical cleavage, and electron transfer dissociation mass spectrometry. Mol Cell Proteomics. 2010a;9(1):153–60.

Wang Z, Namrata UD, Slawson C, Compton PD, Sakabe K, Cheung WD, et al. Extensive crosstalk between O-GlcNAcylation and phosphorylation regulates cytokinesis. Sci Signal. 2010b; 3(104):ra2.

Wang P, Lazarus BD, Forsythe ME, Love DC, Krause MW, Hanover JA. O-GlcNAc cycling mutants modulate proteotoxicity in Caenorhabditis elegans models of human neurodegenerative diseases. Proc Natl Acad Sci U S A. 2012;109(43):17669–74.

Webster DM, Teo CF, Sun Y, Wloga D, Gay S, Klonowski KD, et al. O-GlcNAc modifications regulate cell survival and epiboly during zebrafish development. BMC Dev Biol. 2009;9:28.

Whelan SA, Lane MD, Hart GW. Regulation of the O-linked beta-N-acetylglucosamine transferase by insulin signaling. J Biol Chem. 2008;283(31):21411–7.

Whelan SA, Dias WB, Thiruneelakantapillai L, Lane MD, Hart GW. Regulation of insulin receptor substrate 1 (IRS-1)/AKT kinase-mediated insulin signaling by O-Linked beta-N-acetylglucosamine in 3T3-L1 adipocytes. J Biol Chem. 2010;285(8):5204–11.

Yang X, Zhang F, Kudlow JE. Recruitment of O-GlcNAc transferase to promoters by corepressor mSin3A: coupling protein O-GlcNAcylation to transcriptional repression. Cell. 2002; 110(1):69–80.

Yang X, Ongusaha PP, Miles PD, Havstad JC, Zhang F, So WV, et al. Phosphoinositide signalling links O-GlcNAc transferase to insulin resistance. Nature. 2008;451(7181):964–9.

Yuzwa SA, Macauley MS, Heinonen JE, Shan X, Dennis RJ, He Y, et al. A potent mechanism-inspired O-GlcNAcase inhibitor that blocks phosphorylation of tau in vivo. Nat Chem Biol. 2008;4(8):483–90.

Yuzwa SA, Shan X, Macauley MS, Clark T, Skorobogatko Y, Vosseller K, et al. Increasing O-GlcNAc slows neurodegeneration and stabilizes tau against aggregation. Nat Chem Biol. 2012;8(4):393–9.

Zachara NE, O'Donnell N, Cheung WD, Mercer JJ, Marth JD, Hart GW. Dynamic O-GlcNAc modification of nucleocytoplasmic proteins in response to stress. A survival response of mammalian cells. J Biol Chem. 2004;279(29):30133–42.

Zeidan Q, Wang Z, De Maio A, Hart GW. O-GlcNAc cycling enzymes associate with the translational machinery and modify core ribosomal proteins. Mol Biol Cell. 2010;21(12):1922–36.

Zhang F, Su K, Yang X, Bowe DB, Paterson AJ, Kudlow JE. O-GlcNAc modification is an endogenous inhibitor of the proteasome. Cell. 2003;115(6):715–25.

Chapter 17
N-Glycosylation in Regulation of the Nervous System

Hilary Scott and Vladislav M. Panin

Abstract Protein N-glycosylation can influence the nervous system in a variety of ways by affecting functions of glycoproteins involved in nervous system development and physiology. The importance of N-glycans for different aspects of neural development has been well documented. For example, some N-linked carbohydrate structures were found to play key roles in neural cell adhesion and axonal targeting during development. At the same time, the involvement of glycosylation in the regulation of neural physiology remains less understood. Recent studies have implicated N-glycosylation in the regulation of neural transmission, revealing novel roles of glycans in synaptic processes and the control of neural excitability. N-Glycans were found to markedly affect the function of several types of synaptic proteins involved in key steps of synaptic transmission, including neurotransmitter release, reception, and uptake. Glycosylation also regulates a number of channel proteins, such as TRP channels that control responses to environmental stimuli and voltage-gated ion channels, the principal determinants of neuronal excitability. Sialylated carbohydrate structures play a particularly prominent part in the modulation of voltage-gated ion channels. Sialic acids appear to affect channel functions via several mechanisms, including charge interactions, as well as other interactions that probably engage steric effects and interactions with other molecules. Experiments also indicated that some structural features of glycans can be particularly important for their function. Since glycan structures can vary significantly between different cell types and depend on the metabolic state of the cell, it is important to analyze glycan functions using in vivo approaches. While the complexity of the nervous system and intricacies of glycosylation pathways can create serious obstacles for in vivo experiments in vertebrates, recent studies have

H. Scott • V.M. Panin (✉)
Department of Biochemistry and Biophysics, Texas A&M University,
2128 TAMU, College Station, TX 77843, USA
e-mail: panin@tamu.edu

R.K. Yu and C.-L. Schengrund (eds.), *Glycobiology of the Nervous System*, Advances in Neurobiology 9, DOI 10.1007/978-1-4939-1154-7_17,

indicated that more simple and experimentally tractable model organisms like *Drosophila* should provide important advantages for elucidating evolutionarily conserved functions of N-glycosylation in the nervous system.

Keywords Glycosylation • Sialylation • N-Glycan • Neural transmission • Neural excitability • Ion channel • *Drosophila*

Abbreviations

β4GalNAcTA	β1,4-*N*-acetylgalactosaminyltransferase A
AMPA	α-Amino-3-hydroxy-5-methyl-4-isoxazolepropionic acid
ASIC	Acid-sensing ion channel
CDGs	Congenital disorders of glycosylation
ConA	Concanavalin A
CSAS	CMP-sialic acid synthetase
DSiaT	*Drosophila* sialyltransferase
GABA	γ-Aminobutyric acid
GalNAc	*N*-Acetylgalactosamine
GnTI	*N*-Acetylglucosaminyltransferase I
iGluR	Ionotropic glutamate receptor
LacNAc	*N*-Acetyllactosamine
nAChR	Nicotinic acetylcholine receptor
NCAM	Neural cell adhesion molecule
NMDA	*N*-Methyl-D-aspartate
NMJ	Neuromuscular junction
Para	Paralytic
PSA	Polysialic acid
Sia	Sialic acid(s)
SV2	Synaptic vesicle protein 2
TRP	Transient receptor potential

17.1 Introduction

N-Linked glycan modifications of proteins exist in all three domains of life, Eukarya, Bacteria, and Archaea (Abu-Qarn et al. 2008). N-Glycosylation is especially abundant in eukaryotic cells, where it represents one of the most frequent and ubiquitous posttranslational protein modifications (Stanley et al. 2009; Moremen et al. 2012). In human cells, the majority of N-glycosylation sequon-containing proteins likely acquire N-glycans in the secretory pathway (Apweiler et al. 1999). Although N-glycosylation is not a prerequisite for the viability of mammalian cells in cultured conditions (Gottlieb et al. 1975; Stanley et al. 1975), the biosynthesis of N-glycans appears to be essential for cell communication as defects in N-glycosylation result in embryonic lethality (Ioffe and Stanley 1994; Metzler et al. 1994). The best known functions of N-glycosylation concentrate on promoting protein folding and mediating quality control within the secretory pathway inside the cell (Helenius and Aebi 2001). While biological functions of N-glycans outside of the cell are significantly less understood, they are involved in many essential processes, including cell communication and adhesion. It is more challenging to study these functions because they are less amenable to cell culture approaches and require in vivo analyses that are commonly complicated by pleiotropic effects and complex regulation of glycosylation pathways. The repertoire of N-glycan structures present on a protein can be very heterogeneous at the tissue and cellular level. Their biosynthesis is intimately linked to cell metabolism, reflecting a dynamic read-out of a physiological state of the cell (Dennis et al. 2009).

Many extracellular functions of N-glycans depend on interactions with specific lectins, proteins that bind particular carbohydrate structures (Varki et al. 2009). Glycoprotein-lectin interactions are known to affect a multitude of cell adhesion and signaling processes. These interactions are also involved in building a functional molecular landscape of cell surfaces (Sharon 2007; Dennis et al. 2009). Moreover, N-glycosylation can promote glycoprotein functions via stabilizing steric interactions that protect from proteolysis ((Wittwer and Howard 1990), reviewed in (Wormald and Dwek 1999)). All these functional outcomes of N-glycosylation are pertinent to the development and physiology of different organs and tissues, including the nervous system. In this review, we will focus on several novel paradigms of neural functions of N-glycans. Our goal is not to provide an extensive review of experimental data in this field. Instead, we will concentrate on the discussion of a number of recent studies that unraveled some interesting functional mechanisms underlying these paradigms.

17.2 N-Glycosylation in Neural Development

The critical involvement of N-glycosylation in development of the nervous system is evident from the studies of human congenital disorders of glycosylation (CDGs). They revealed that genetic defects in the N-glycosylation pathway are almost always associated with severe neurological abnormalities (reviewed in (Freeze et al. 2012)). Gene inactivation experiments in mice have shed light on the in vivo functions of several key glycosyltransferase genes and their glycan products in the nervous system (Lowe and Marth 2003). For example, brain-specific inactivation of GlcNAcT-I, a glycosyltransferase that mediates the biosynthesis of hybrid and complex N-linked carbohydrates, was found to result in severe neurological defects, including abnormal locomotion, tremors, and paralysis (Ye and Marth 2004). However, pleiotropic effects of glycosylation on the development and physiology commonly obstruct conclusive analyses and interpretation of phenotypes produced by knockouts that affect core structures. On the other hand, mutations affecting more specialized and some terminal structures of glycans have proven to be more amenable to study. Phenotypes of such mutations demonstrated the involvement of certain N-glycan structures in specific regulatory events. Thus, genetic inactivation of ST8Sia II and ST8Sia IV polysialyltransferases that modify N-glycans of the neural cell adhesion molecule (NCAM) with polysialic acid (PSA) unveiled the prominent role of PSA in the nervous system (Weinhold et al. 2005; Angata et al. 2007; Hildebrandt et al. 2009). PSA is a long polymer of α2,8-linked sialic acid residues that can be attached to the termini of glycans on some glycoproteins, including N-glycans of NCAM. The PSA structure was shown to regulate brain development, neurite outgrowth and targeting, and to affect synaptic plasticity, learning, and memory (for reviews on the structure and functions of NCAM-PSA see (Muhlenhoff et al. 1998, 2009; Rutishauser 2008; Colley 2010)). Remarkably, the most severe phenotypes associated with PSA deficiency, including early postnatal lethality, defects in major axonal tracts and progressive hydrocephalus, result from the gain-of-function effect of NCAM that lacks proper PSA modification, and the genetic inactivation of NCAM rescues all gross morphological defects in the brain of PSA-deficient mice (Weinhold et al. 2005). Another notable example of a N-glycan structure that plays a specialized role in neural development is represented by poly-*N*-acetyllactosamine oligosaccharides (PLN). The synthesis of PLN in the developing olfactory system depends on the activity of β1,3-*N*-acetylglucosaminyltransferase (β3GnT2) that initiates and participates in the elongation of PLN on the terminal β1-linked galactose residues of N-glycans (Zhou et al. 1999). Targeted genetic inactivation of the β3GnT2 glycosyltransferase results in numerous abnormalities in the olfactory system in mice, including defects of axonal guidance and failure of glomeruli formation. The primary cause of this phenotype appears to be the hypoglycosylation of adenylyl cyclase 3. This enzyme generates cAMP, a key signaling molecule that functions in olfactory axon targeting, and the loss of PLN dramatically downregulates the activity of adenylyl cyclase and the production of cAMP (Henion et al. 2011).

These examples likely correspond to just the tip of the iceberg of numerous yet unknown important roles of N-glycosylation in the development of the nervous system. The nervous system is regulated by a broad spectrum of N-glycosylated proteins, including cell surface and extracellular matrix (ECM) glycoproteins participating in cell adhesion and signaling (Kleene and Schachner 2004; Dityatev et al. 2010). Intriguingly, the functions of a number of these glycoproteins are affected by N-glycosylation outside of the nervous system. Thus, *β1,6*-branching GlcNAc modifications were found to modulate cell adhesion and cell motility by affecting the functions of laminin 332 and α3β1 integrins (Zhao et al. 2006; Kariya et al. 2008). Another example of a carbohydrate structure that can markedly affect molecular interactions is *α2,6*-sialylation. It was found to regulate the functions of α4β1 integrins and receptor protein tyrosine phosphatase CD45 by modifying their conformation or interactions with functionally important partners in immune system cells (Amano et al. 2003; Woodard-Grice et al. 2008). Laminins, integrins, and receptor protein tyrosine phosphatases also function in the nervous system, affecting neuronal migration, axonal growth and myelination, neuromuscular junction development, neuronal survival, etc. (Wang et al. 2009; Barros et al. 2010; Tan et al. 2011), and glycosylation of these proteins may be implicated in these neural functions. Several studies demonstrated that *α2,6*-linked sialic acids play an important role as negative regulators of galectin binding, which revealed a paradigm that is expected to be pertinent in many cellular and molecular contexts (reviewed in (Zhuo and Bellis 2011)). These and other examples suggest that similar glycan-dependent regulatory mechanisms may operate in the nervous system, and they need to be explored.

Several novel mechanisms implicating N-glycosylation in the modulation of neural transmission have been recently elucidated, and are discussed in more detail below.

17.3 N-Glycans in Neural Physiology

17.3.1 Glycans in Synaptic Transmission

Recent research revealed a connection between mutations in the gene encoding glutamine-fructose-6-phosphate transaminase 1 (GFPT1) and a group of congenital myasthenic syndromes (CMS, e.g. OMIM 608931) characterized by hereditary defects in synaptic transmission at neuromuscular junctions (Engel 2012). GFPT1 mediates the first, rate-limiting step in the synthesis of hexosamine needed for glycan biosynthesis (Senderek et al. 2011). The importance of protein glycosylation for different aspects of synaptic transmission was unveiled by a number of studies (see reviews (Kleene and Schachner 2004; Dityatev et al. 2010; Dani and Broadie 2012)). However, the molecular and cellular bases for the effects seen on synaptic transmission are complex and not well understood. Combined, these observations indicate that glycosylation in general is required for normal synaptic functions.

N-Glycosylation controls the function of many key players in synaptic processes and its effect on synaptic physiology is multifaceted. For instance, the function of synaptic vesicle protein 2 (SV2), ubiquitously present at vertebrate synapses, was found to depend on its N-glycosylation. Targeted gene inactivation in mice demonstrated the importance of SV2 for neural transmission as the deletion of two out of three existing SV2 isoforms resulted in postnatal lethality due to severe seizures. Notably, no developmental defects were found in the brains of these mutants, which indicated that SV2 proteins function mainly in synaptic physiology (Janz et al. 1999). It was suggested that they mediate a novel maturation step of primed synaptic vesicles, which potentiates responsiveness of synaptic vesicles to Ca^{2+} regulation. All SV2 isoforms have multiple N-linked glycan chains attached to their intravesicular loop. The most ubiquitous SV2 isoform, SV2a, has three glycosylation sites, and the removal of all of them inhibits the synaptic targeting of SV2a along with its function (Chang and Sudhof 2009). These results suggest that N-glycans are required for proper folding and trafficking of SV2 within the neuron.

More recent analyses of SV2 mutants lacking individual glycosylation sites indicated that single N-glycans are partially dispensable and their function is redundant for the proper sorting of SV2a to synaptic vesicles (Kwon and Chapman 2012). Similar approaches were used to examine the role of N-glycosylation in the regulation of two other major glycoproteins of synaptic vesicles, synaptotagmin 1 and synaptophysin. It was found that the role of glycosylation in glycoprotein sorting to synaptic vesicles can range from dispensable (synaptotagmin 1) to essential (synaptophysin) (Kwon and Chapman 2012). These results illustrated that glycans can play highly individualized regulatory roles that are tailored for a particular glycoprotein and its specific function in the nervous system.

Another type of prominent players in synaptic transmission is represented by neurotransmitter receptors that function as ligand-gated channel proteins and mediate communication among neurons within the nervous system, or between neurons and muscles at neuromuscular junctions. Neurotransmitter receptors are commonly glycosylated, having several N-glycans attached to their extracellular domains. Substantial evidence indicating the functional importance of these carbohydrate modifications has started to emerge. Thus, a number of studies have demonstrated that N-linked carbohydrate chains are involved in the function of nicotinic acetylcholine receptors (nAChRs). nAChR proteins correspond to founding members of the pentameric ligand-gated super family of ion channels, that also includes serotonin, γ-aminobutyric acid (GABA), and glycine receptors (Chen 2010). nAChRs regulate postsynaptic responses at neuromuscular junctions and a variety of synaptic connections in the brain. These receptors are implicated in diverse neural functions, including the processing of sensory information and learning and memory (Miwa et al. 2011). Results of studies on the involvement of N-glycosylation in the function of nAChRs indicate that glycosylation affects functional properties of the receptors. It was proposed that N-glycans can promote the local folding of some functional protein domains, without influencing interactions between receptor subunits or their cell surface expression (Gehle et al. 1997; Chen et al. 1998). Using *Torpedo* nAChRs as a model system, experiments revealed that N-glycosylation is implicated in receptor modulation, as receptors with mutated glycosylation sites have abnormal conductance and desensitization (the rate of current decay) (Nishizaki 2003). Interestingly, the pharmacological application of concanavalin A (ConA) to in vitro assays of wild type receptors mimicked the effect of mutations affecting N-glycosylation. ConA is a lectin that binds N-linked glycans, and thus its effect on nAChRs was interpreted as evidence that the glycans may function as a modulating "lid" at the channel pore, and the lack of sugar chains or the inhibition of its movement by lectin binding caused the decreased rate of desensitization (Nishizaki 2003). More recent experiments indicated that carbohydrate modifications of nAChRs can influence their surface expression and cholinergic agonist-dependent gating. However, glycosylation does not change their binding affinity for the agonists or the stability of folded receptors (Dacosta et al. 2005; Dellisanti et al. 2007).

The role of N-glycosylation of ionotropic glutamate receptors (iGluRs) was also analyzed. iGluRs mediate fast transmission at the majority of excitatory synapses within the mammalian nervous system, and they play essential roles in synaptic plasticity and extrasynaptic modulation of neurons (reviewed in (Traynelis et al. 2010)). These receptors form tetrameric complexes that function as ligand-gated ion channels. iGluRs encompass large subfamilies of AMPA, kainate, and NMDA receptors (Traynelis et al. 2010). The majority of these receptors appear to be N-glycosylated, with consensus glycosylation sites in their amino-terminal domains involved in receptor assembly and modulation, as well as in their ligand-binding domains (Partin et al. 1993; Everts et al. 1997; Everts et al. 1999; Mah et al. 2005). While the presence of glycans at these sites has not been well characterized, N-glycans were found to affect desensitization of AMPA and kainite receptors (Hollmann et al. 1994; Everts et al. 1997). At the same time, N-glycosylation is not generally required for iGluR function since the synthesis, transport, and subunit assembly of functional receptors on the plasma membrane are not significantly affected by the lack of glycosylation (Sumikawa et al. 1988; Everts et al. 1997; Gill et al. 2009). In agreement with these data, crystallization studies of kainate receptors showed that N-linked sugar chains are not directly involved in ligand binding and subunit association of iGluRs (Armstrong et al. 1998; Nanao et al. 2005). In contrast to the AMPA and kainite receptors, functional expression of NMDA-type receptors was found to be dramatically downregulated by inhibition of N-glycosylation. This effect was shown to be associated with a specific reduction in expression of the NR1 subunit, suggesting that glycans are required for its folding and/or trafficking (Everts et al. 1997).

Some specialized carbohydrate structures that can be present on N-linked glycans were found to be involved in the functional modulation of glycoproteins participating in neural transmission. These structures include HNK-1 and sialic acid (discussed below). The HNK-1 glycoepitope (initially discovered on *h*uman *n*atural *k*iller cells (Abo and Balch 1981)) was shown to be involved in regulation of the AMPA-type receptor subunit GluR2. The HNK-1 epitope can be also present on some glycolipids. Using these glycolipids, two research groups independently demonstrated that the HNK-1 epitope represents a sulfated glucuronic acid linked to *N*-acetyllactosamine on the nonreducing termini of oligosaccharides (HSO3–3GlcAβ1–3Galβ1–4GlcNAc) (Chou et al. 1986; Ariga et al. 1987). The expression of HNK-1 is highly enriched in the nervous system. Genetic inactivation of enzymes responsible for the biosynthesis of this epitope (glucuronyltransferase GlcAT-P, sulfotransferase HNK-1 ST, and β4-galactosyltransferase-2) lead to neurological phenotypes in mice, including reduced long-term potentiation in hippocampal CA1 synapses, electrophysiological abnormalities of hippocampal interneurons, defects in neural plasticity, and learning and memory, which suggest that HNK-1 is important for synaptic functions (Senn et al. 2002; Yamamoto et al. 2002; Gurevicius et al. 2007; Yoshihara et al. 2009). It was found that the HNK-1 epitope downregulates endocytosis of the AMPA glutamate receptor subunit GluR2 and stabilizes its expression on neuronal plasma membranes. Moreover, the presence of HNK-1 promotes the interaction between GluR2 and *N*-cadherin, which probably regulates the stability of GluR2 on the cell surface at synaptic connections (Morita et al. 2009). The HNK-1 epitope is present on a number of glycoproteins involved in intracellular adhesion, cell migration, and synaptic plasticity (reviewed in (Kleene and

Schachner 2004; Yanagisawa and Yu 2007)). More recently this epitope was found to be expressed on a tenascin-C spliced variant and involved in the regulation of mouse neural stem cells (Yagi et al. 2010).

Synaptic transmission can be significantly influenced by the function of neurotransmitter transporters, synaptic proteins essential for control of the concentration of neurotransmitters in the synaptic cleft. The SLC6 (solute carrier) family of membrane proteins includes a subfamily of transporters that mediate the translocation of neurotransmitters across the plasma membrane by coupling it to the cotransport of Na^+ and Cl^- (reviewed in (Kristensen et al. 2011)). The members of this subfamily include the transporters for serotonin (5-hydroxytryptamine, or 5-HT), dopamine, norepinephrine, GABA, and glycine. All these transport proteins appear to be N-glycosylated at the large extracellular loop 2 region, suggesting that this modification is functionally important. While removal of this glycosylation by mutagenesis or glycosidase treatment reduces the number of transporters at the cell surface, it usually does not have a strong effect on ligand binding and transporter activity. This reduction in transporter amount was attributed to a decrease in protein stability or a disruption in trafficking of nonglycosylated transporters to the plasma membrane (Tate and Blakely 1994; Olivares et al. 1995; Melikian et al. 1996; Nguyen and Amara 1996; Martinez-Maza et al. 2001; Li et al. 2004; Kristensen et al. 2011). N-Glycosylation of GAT1, the predominant GABA transporter in the brain, was found to promote both, the stability of the protein and its trafficking to the cell surface. Moreover, N-glycans were found to be important for GABA-uptake activity of the transporter, with sialic acids appearing to play an essential part in this regulation as the absence of sialylation slowed down the kinetics of the GABA transport cycle and reduced the apparent affinity of GAT1 for extracellular Na^+ (Cai et al. 2005; Hu et al. 2011). Interestingly, a nonsynonymous single nucleotide polymorphism (SNP) in the human *SLC6A4* gene encoding the serotonin transporter (hSERT) creates an ectopic glycosylation site (K201N) that was found to enhance glycosylation of hSERT with a concomitant increase in the level of transporter expression and activity. Although it is not yet known whether this SNP is associated with a neurological phenotype, by analogy to a well-studied polymorphism that also changes the expression level of hSERT, it was suggested that the K201N allele may affect personality traits and psychiatric disease susceptibility (Rasmussen et al. 2009).

17.3.2 N-Glycosylation Regulates Ion Channels in Vertebrate Neurons

N-Glycosylation can be an important modulator of ion channels in the nervous system. In general, glycans can regulate channels via at least three different mechanisms: (1) by promoting their folding and trafficking to the cell surface, (2) by affecting their stability and distribution on the cell surface (e.g., via regulating protein endocytosis and/or recycling at the plasma membrane), and (3) by changing

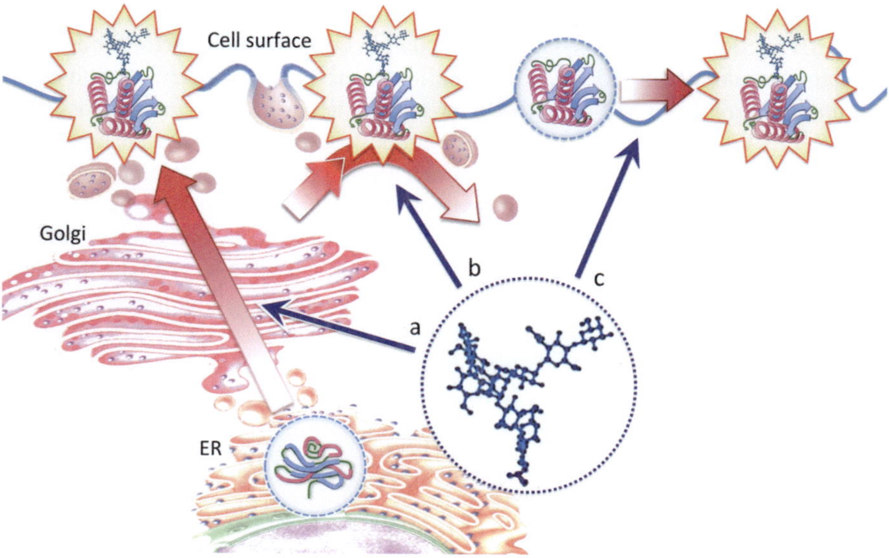

Fig. 17.1 Main effects of protein N-glycosylation. N-glycans can potentiate glycoprotein functions by facilitating protein folding and trafficking to the cell surface (**a**), promoting protein stability on the cell surface via regulation of protein uptake and recycling to the plasma membrane (**b**), and by enhancing protein activity via changing protein biophysical properties (**c**). N-Glycosylation is sketched as a single generic N-glycan (not to scale). The number of glycans can be different for distinct proteins, while glycan structures can vary and have different effects on protein functions

their molecular properties and thus potentiating channel functions (e.g., affecting biophysical characteristics and/or functional interactions with other molecules) (Fig. 17.1). Outcomes of the first two mechanisms impinge on the control of the number of channels on the cell surface. The effect of N-glycosylation on channel cell surface expression was demonstrated for several types of neuronal channels, including acid-sensing channels (e.g., ASIC1a and 1b (Kadurin et al. 2008; Jing et al. 2012)) and voltage-gated ion channels (e.g., potassium channels Kv1.3, Kv1.4, and HERG (Gong et al. 2002; Watanabe et al. 2004; Zhu et al. 2012), and calcium channels Cav3.2 (Weiss et al. 2013)). The effect on biophysical properties is frequently mediated by glycans attached to channel pore loops that can influence channel gating. Channel pore N-glycan modifications can effectively modulate the function of the TRPM8 channel, a member of a large family of transient receptor potential (TRP) ion channels playing essential roles in sensory physiology. TRPM8 glycosylation was found to cause a marked shift in the voltage dependence of channel activation (Pertusa et al. 2012). TRPM8 is expressed in sensory neurons that respond to cold (Mckemy et al. 2002; Peier et al. 2002). The N-linked glycans affect the temperature threshold of TRPM8 activation, and therefore they can function as critical molecular determinants that establish cold sensitivity in primary sensory neurons (Pertusa et al. 2012). Notably, the membrane localization of channels in this case appears to be unaffected by glycosylation and therefore the effect of glycans is concentrated on the regulation of channel biophysical properties (Pertusa et al. 2012). Similarly, glycosylation was found to be an essential factor for agonist-mediated regulation of TRPV1 (TRP Vanilloid Type 1), a nonspecific cation channel that functions as a key sensor of pain-sensing nerve fibers. A nonglycosylated mutant TRPV1 (N604T) was shown to be properly expressed on the plasma membrane; however, it did not undergo sustained regulation by capsaicin and had substantially altered desensitization properties (Veldhuis et al. 2012). While glycans affect the biophysical properties of several TRP channels (e.g., TRPC3 and TRPC6 (Dietrich et al. 2003; Wirkner et al. 2005)), glycosylation can also promote the function of TRP channels by regulating their expression and subcellular localization and thus influencing the number of available functional channels (TRPV4 and TRPV5 (Chang et al. 2005; Xu et al. 2006)). These different mechanisms mediated by N-glycosylation do not appear to be mutually exclusive. They could operate at the same time, while one of them could become more prominent, depending on particular molecular and cellular contexts.

17.3.3 N-Glycans and Interactions with Lectins

Exogenous lectins that interact with N-linked glycan structures were found to have a strong modulatory effect on some neurotransmitter receptors in pharmacological assays. As mentioned above, ConA can bind to N-glycans of nAChRs and influence desensitization of wild-type receptors in a way that mimics the effect of mutations that eliminate N-glycosylation sites (Nishizaki 2003). The modulatory effect of ConA was also demonstrated for iGluR subfamilies of AMPA, kainate, and NMDA receptors (Traynelis et al. 2010). ConA exerts a pronounced effect on kainate receptors by inhibiting their desensitization (Partin et al. 1993; Everts et al. 1997, 1999). Experiments indicate that ConA can interact with N-glycans attached to the amino-terminal domain of iGluRs and affect receptor conformational changes. This action appears to depend on the conformational state of the channel, since agonist-induced desensitization prior to ConA application eliminates the effect (Everts et al. 1997, 1999; Fay and Bowie 2006). Some other lectins, such as wheat germ agglutinin, soybean agglutinin, and succinyl-ConA, were also shown to potentiate kainate receptors (Thio et al. 1993; Yue et al. 1995). The in vitro effects of these lectins suggested that glycans may play a specialized role in the modulation of receptors in vivo (Everts et al. 1997; Nanao et al. 2005), while some endogenous, yet unknown lectins can potentially bind to these glycans and regulate the function of neurotransmitter receptors. A related mechanism of lectin-dependent regulation has recently been described for the Ca^{2+} TRPV5 channel in renal epithelial cells. Retention of TRPV5 on the cell surface is an essential regulatory process in the control of channel function. This regulation is mediated by Klotho, a humoral factor with glycosidase activity that appears to directly modify channel glycans, which in turn potentiates interactions with galectin and facilitates cell surface retention of the channel (Chang et al. 2005; Cha et al. 2008; Leunissen et al. 2013). However, the regulation of TRPV5 is not fully understood, and it is likely mediated by the converging effects of several mechanisms, also including sialylation that appears to work in parallel to promote lipid raft-mediated internalization of the channel (Leunissen et al. 2013). It will be important to investigate whether lectin-mediated regulation can also operate in the nervous system to regulate channels involved in neural transmission.

17.3.4 N-Glycans in Regulation of Voltage-Gated Ion Channels and Membrane Excitability

A large group of voltage-gated ion channels represents principal regulators of cell excitability. Glycosylation can affect cell excitability of neurons by modulating the function of various members of this channel superfamily, including channels that regulate membrane permeability for Na^+, K^+, and Ca^{2+} ions (e.g., (Recio-Pinto et al. 1990; Zhang et al. 1999; Bennett 2002; Gong et al. 2002; Watanabe et al. 2003; Johnson et al. 2004; Watanabe et al. 2007; Schwetz et al. 2010; Weiss et al. 2013)). In mammals, N-glycosylation of voltage-gated Na^+ and K^+ channels was found to be regulated developmentally and in a cell-specific manner in the heart and the nervous system, suggesting that glycans participate in setting the distinct levels of excitability required in different cells and at various developmental stages (Castillo et al. 1997; Tyrrell et al. 2001; Schwalbe et al. 2008; Montpetit et al. 2009).

In addition to the direct effects of channel glycans, N-glycosylation can influence ion channels indirectly, in a molecule nonautonomous manner, by regulating other glycoproteins that control channel functions. For example, glycosylation of auxiliary subunits that interact with channels can promote cell surface localization and modify channel biophysical properties (Johnson et al. 2004; Cotella et al. 2010). The nonautonomous effect of N-glycans can be potentially pertinent for regulation of many channels; however, this possibility remains largely unexplored.

Numerous studies of channel glycosylation have concentrated on sialylated glycans (reviewed in (Ednie and Bennett 2012)). Sialylated carbohydrate chains are negatively charged and can participate in electrostatic interactions with ions and other charged groups located on the cell surface, thus potentially affecting channel functions. Vertebrate voltage-gated Na^+ channels are heavily decorated with sialylated structures. It was estimated that up to 30 % of Na^+ channel molecular mass is represented by carbohydrate chains, with sialic acids (Sia) comprising nearly 50 % of channel glycans (Miller et al. 1983; Elmer et al. 1985; Messner and Catterall 1985; James and Agnew 1987; Roberts and Barchi 1987). Electrophysiological assays indicated that sialylated glycans can markedly affect the gating properties of Na^+ channels (Recio-Pinto et al. 1990; Bennett et al. 1997; Zhang et al. 1999; Cronin et al. 2005). This effect varies significantly for different channels, and it can also be isoform- and subunit-specific (Bennett 2002; Johnson et al. 2004; Johnson and Bennett 2006).

The role of sialylation in the modulation of vertebrate voltage-gated Na^+ channels has been generally explained by the electrostatic effect of the large negative charge provided by the numerous Sia residues present in the vicinity of the channel pore. Remarkably, more than 100 Sia residues can be attached to a channel protein, with the majority of them being incorporated as PSA structures (Miller et al. 1983; James and Agnew 1987; Zuber et al. 1992). The specific role of PSA in the regulation of voltage-gated Na^+ channels was uncovered by analyses of mouse mutant cardiomyocytes that had genetically inactivated ST8Sia II polysialyltransferase, an enzyme involved in PSA biosynthesis. The ST8Sia II deficiency was found to cause defects in cell excitability and channel gating, including abnormal action potentials with a significantly broader waveform and a delayed peak, considerable depolarizing shifts of gating curves, and compromised fast inactivation of channels (Montpetit et al. 2009).

While the effect of PSA was confirmed by several studies, a line of evidence suggested that sialylation can also affect voltage-gated channels via mechanisms that cannot be attributed to PSA or the significant charge of numerous Sia residues attached to channel glycans. These data indicated that Sia can play a more specific role in the modulation of channel functions. Thus, electrophysiological analyses of the cardiac sodium channel in cell culture revealed that its function can be affected by some "functional" Sia residues rather than by the total charge of channel sialylation (Stocker and Bennett 2006). Furthermore, experiments with rat hippocampal organotypic slice cultures suggested that PSA does not always influence the function of voltage-gated Na^+ channels, since treatment with Endo-N sialidase, a glycosidase that specifically removes PSA, was found to have no apparent impact on intracellularly recorded action potentials and evoked synaptic transmission (Muller et al. 1996). Additionally, the effects of PSA and non-PSA Sia residues on the function of α-$Na_V1.4$ channels were found to be distinct when they were analyzed using mutant Chinese hamster ovary (CHO) cell lines with defects in sialylation or polysialylation pathways. The loss of Sia and PSA in these mutant cells results in opposite shifts of voltage-dependent activation and steady-state inactivation of α-$Na_V1.4$, while only the loss of Sia has a significant effect on recovery from fast inactivation (Ahrens et al. 2011). Finally, unnatural Sia residues with *N*-acetyl groups changed to *N*-pentanoyl or *N*-propanoyl structures, when introduced metabolically, were found to have a notable effect on conductance properties of the Kv3.1 voltage-gated K+ channel. Collectively, these data suggest that sialylation can modulate channels through specific steric effects, in addition to its role in electrostatic interactions (Hall et al. 2011).

It is worth noting that most studies on the role of glycosylation in the regulation of ion channels and synaptic glycoproteins have been performed in vitro or in cell culture using transgenic approaches in various types of heterologous cells (for example, using frog oocytes (Everts et al. 1997; Gehle et al. 1997; Nishizaki 2003), different mammalian cell cultures (Bennett 2002; Dellisanti et al. 2007; Watanabe et al. 2007; Hu et al. 2011; Gurba et al. 2012), or in vitro reconstituted lipid membranes (Recio-Pinto et al. 1990; Castillo et al. 2003; Cronin et al. 2005)). It is important to keep in mind that the structure of glycosylation and its functional implications can vary significantly between different cell types, and between cultured cells and neural cells in vivo. Furthermore, the glycosylation of multisubunit protein complexes could also depend on a particular combination of subunits expressed by the cell (e.g., the glycosylation of $GABA_A$ $\beta3$ subunits can be affected by the coexpression of

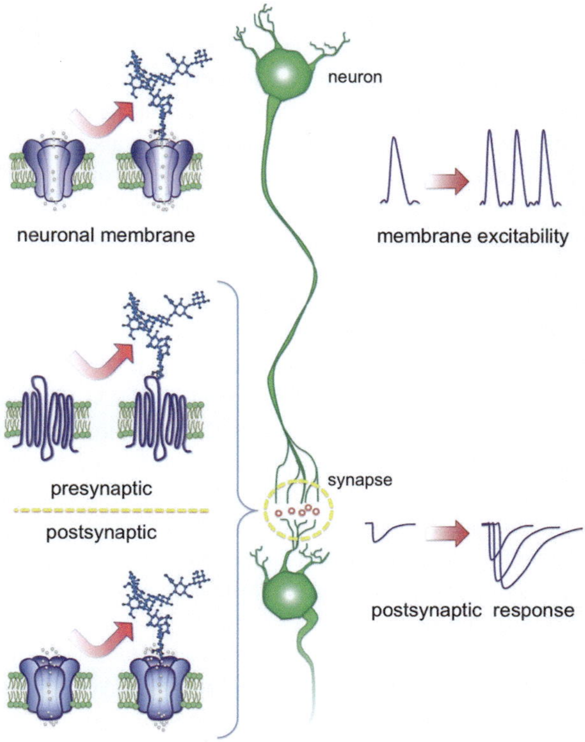

Fig. 17.2 N-Glycosylation can affect neural transmission by modulating voltage-gated ion channels that generate action potentials and determine neuronal excitability, and by influencing synaptic transmission via impact on the function of synaptic proteins, such as synaptic vesicle proteins and neurotransmitter receptors. N-Glycosylation is sketched as a generic N-glycan. Glycans can also include some specific modifications, such as polysialylation and the HNK-1 epitopes (not shown). Modified from (Scott and Panin 2014)

other receptor subunits (Gurba et al. 2012)). Therefore, it is important to exercise caution when interpreting data from in vitro and cell culture experiments in terms of mechanisms that operate in vivo. Nevertheless, taken together, experimental data clearly indicate that glycosylation can substantially influence the function of glycoproteins playing key roles in neural transmission (Fig. 17.2). This influence can be dissimilar for distinct types of factors regulating the nervous system. Moreover, even within the same family of related proteins (e.g., iGluRs) glycosylation can underlie distinct modulatory mechanisms that can also depend on the structure and location of carbohydrate chains. These effects of N-glycosylation potentially create an extra layer of regulatory processes that control neural physiology.

17.3.5 In Vivo Functions of Sialylated N-glycans

Studies that investigate the function of sialic acids in vivo remain relatively scarce. The biological importance of sialylation of voltage-gated Na^+ channels was most unambiguously demonstrated in the context of cardiac functions. Analyses of cardiomyocytes with defective channel sialylation (using mouse genetic models with diminished sialylation or glycosidase-treated rat cardiomyocytes) suggested that abnormal channel sialylation can result in cardiac excitability phenotypes and heart failure (Ufret-Vincenty et al. 2001; Stocker and Bennett 2006; Montpetit et al. 2009). Murine models were also used to examine the role of channel sialylation in the nervous system. These experiments analyzed neural excitability after treatment with glycosidases to remove sialylated glycans, as well as upon inhibition of endogenous neuraminidases that trim sialic acids from carbohydrate chains in vivo. It was found that glycoprotein sialylation can significantly affect the excitability of neural networks and influence seizure threshold in kindling epilepsy models. These studies suggested that sialic acids can effectively modulate voltage-gated Na^+ channels in brain neurons (Tyrrell et al. 2001; Isaev et al. 2007, 2011; Isaeva et al. 2010).

Recent analyses of the mouse model of Angelman syndrome revealed the possibility that an abnormal sialylation of cell surface proteins plays a key role in the etiology of the syndrome (Condon et al. 2013). This neurological genetic disorder is caused by the maternal loss of the E3 ubiquitin ligase Ube3a and is associated with motor dysfunction, mental retardation, speech impairment, seizures, and a high prevalence of autism (Williams et al. 2006). Loss of Ube3a causes defects in synaptic development and function, including a deficit in experience-dependent synaptic plasticity and decreased plasma membrane localization of AMPA receptors at excitatory synapses (Jiang et al. 1998; Dindot et al. 2008; Yashiro et al. 2009; Greer et al. 2010). Intriguingly, the Ube3a defect also causes a dramatic reduction of glycoprotein sialylation due to the structural and homeostatic disruption of the Golgi apparatus, which indicated that the deficiency of glycoprotein sialylation likely underlies the pathobiological mechanism of Angelman syndrome (Condon et al. 2013).

Although a number of in vivo experiments indicate that glycoprotein sialylation can significantly influence the excitability of neural networks, it remains unknown whether this effect is primarily due to the sialylation of voltage-gated channels or some other glycoproteins. It is challenging to address this question in vertebrates because of the complexity of the nervous system, intricacies of glycosylation pathways, potential functional redundancy of glycosylation genes, as well as the ubiquity of sialylation that affects a panoply of glycoconjugates in the majority of vertebrate cells. With its power of genetic approaches, a spectrum of well-established neurobiological approaches, and simplified glycosylation pathways, *Drosophila* has recently emerged as a promising model for elucidating conserved genetic and molecular mechanisms of neural glycosylation.

17.4 N-Glycosylation Regulates the Nervous System of *Drosophila*

17.4.1 Drosophila *Mutations Affecting N-Glycosylation*

Recent glycoproteomic approaches characterized in detail the totality of *Drosophila* N-glycosylated proteins, identifying more than 450 glycoproteins expressed in the head (Koles et al. 2007; Vandenborre et al. 2010; Baycin-Hizal et al. 2011).

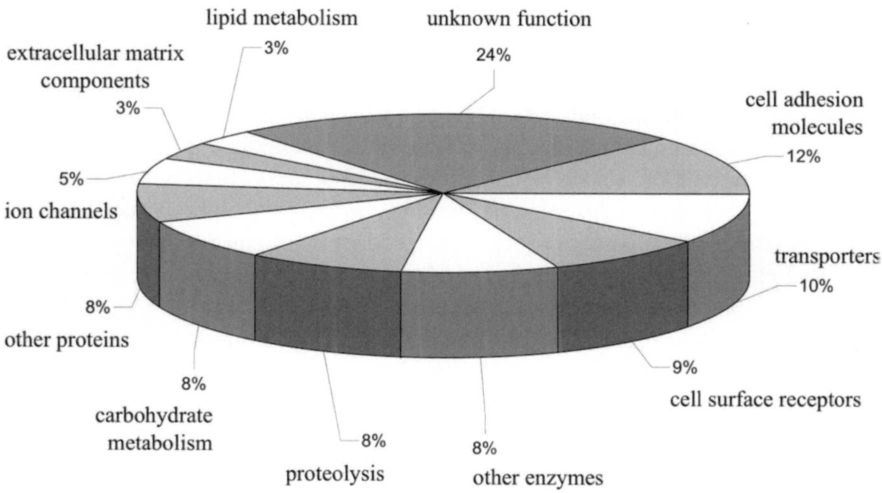

Fig. 17.3 Distribution of the different protein classes among N-glycosylated proteins identified in *Drosophila* head by glycoproteomics approaches. Figure adapted with permission from Koles et al. (2007)

These proteins comprise ion channels, transporters, cell surface receptors, cell adhesion molecules, molecules involved in proteolysis and carbohydrate metabolism, and some other protein families, including a large proportion of proteins with unknown functions (Fig. 17.3) (Koles et al. 2007; Baycin-Hizal et al. 2011). The repertoire of N-glycan structures in *Drosophila* is different from that in mammalian organisms. Detailed mass spectrometry analyses of the *Drosophila* N-glycome revealed that paucimannose and high mannose structures dominate the spectrum of N-glycosylation (Aoki et al. 2007; Koles et al. 2007). In contrast to mammalian N-glycans that are represented by abundant complex structures (Antonopoulos et al. 2011), complex and hybrid-type oligosaccharides that correspond to more processed mature structures represent only 12 % of the total *Drosophila* N-glycan profile (Aoki et al. 2007). Nevertheless, these minor glycan species play prominent roles in the nervous system, suggesting that their functions are evolutionarily conserved (Schachter 2010). The importance of these glycans for the nervous system was revealed in a number of studies that analyzed mutants with defects in the N-glycosylation pathway. Thus, genetic inactivation of the *MGAT1* gene that encodes GlcNAcT I, a key enzyme in the production of processed N-glycan structures, was found to result in severe neurological phenotypes, including locomotor abnormalities, significantly decreased life span, and the “fused lobes” phenotype, a developmental defect affecting a specialized brain structure involved in memory formation, the mushroom bodies (Sarkar et al. 2006, 2010). *MGAT1* mutants have prominent synaptic defects, including overgrowth of neuromuscular junctions and abnormal synaptic vesicle cycling. *MGAT1* mutant synapses have disrupted extracellular synaptomatrix and the accumulation of Mind the gap, a lectin-like extracellular matrix protein of the synaptic cleft. They also show a decreased expression of several key markers of functional synaptic morphology, such as Bruchpilot, a presynaptic active zone protein, and GLURIIB, a postsynaptic iGluR subunit B (Parkinson et al. 2013). Mutations of *fused lobes* cause cell-lethal phenotype in mosaic clones of olfactory projection neurons and result in mushroom body defects similar to those found in *MGAT1* mutants (the mushroom body *lobes* become *fused*) (Boquet et al. 2000; Sekine et al. 2013). *Fused lobes* encodes Golgi β-N-acetylglucosaminidase that inhibits the biosynthesis of hybrid and complex N-glycans and concomitantly promotes the production of paucimannose structures (Leonard et al. 2006). Downregulation of *sugar-free frosting,* a gene encoding a *Drosophila* homolog of SAD kinase that regulates secretory flux through the Golgi, inhibits synthesis of the HRP glycoepitope (α3-linked core fucose) and increases the amount of hybrid and complex N-glycan structures. *Sugar-free frosting* mutations lead to neuromuscular junction defects in larvae and locomotor abnormalities in adult flies (Baas et al. 2011). Meigo, a putative nucleotide sugar transporter, appears to specifically regulate the targeting of neurite projections in the olfactory system by affecting N-glycosylation of ephrin (Sekine et al. 2013). Deficiency of β1,4-N-acetylgalactosaminyltransferase A (β4GalNAcTA), a glycosyltransferase potentially involved in the biosynthesis of complex and hybrid N-glycans, results in prominent neurological phenotypes, including defects of locomotion, reduction in the number of synaptic boutons at neuromuscular junctions and decreased frequency of spontaneous release of neurotransmitters (Haines and Irvine 2005; Haines and Stewart 2007; Nakamura et al. 2012). Taken together, these examples highlight the notion that protein N-glycosylation plays important and specific functions in the *Drosophila* nervous system, and that these functions require the structural diversity of N-glycans. These data also indicate an intriguing possibility that many genes involved in the N-glycosylation pathway could be associated with evolutionarily conserved mechanisms that regulate the nervous system in a wide range of animals, from arthropods to mammals.

17.4.2 Sialylated N-Glycans Control Neural Excitability in **Drosophila**

Sialylated glycans represent less than 0.1 % of the total glycan profile of the *Drosophila* N-glycome. As a result, they can only be unambiguously detected and analyzed by the most sensitive glycomic approaches, such as multidimensional mass spectrometry (Aoki et al. 2007; Koles et al. 2007). Despite the fact that sialylation is so scarce, it has a prominent function in the nervous system of *Drosophila*, which was revealed by analysis of mutant phenotypes of the *Drosophila sialyltransferase* (*DSiaT*) and *CMP-sialic acid synthetase* (*CSAS*) genes that play key roles in the sialylation pathway (Koles et al. 2009). Unlike mammalian organisms that have 20 different sialyltransferases, *Drosophila* possesses only one sialyltransferase, DSiaT, which significantly simplifies the in vivo analysis of sialylation functions (Koles et al. 2009). DSiaT shows a close evolutionary relationship to the

ST6Gal family of mammalian sialyltransferases; it modifies glycoproteins by attaching α*2,6*-linked sialic acids to LacNAc termini of N-glycans (Koles et al. 2004; Repnikova et al. 2010). The expression of DSiaT is dynamic and largely restricted to subsets of fully differentiated CNS neurons during development and in adult flies, which indicates that the pattern of sialylation is tightly controlled in a cell-specific and developmentally regulated manner (Koles et al. 2009; Repnikova et al. 2010; Islam et al. 2013). The expression of CSAS, an enzyme generating the CMP-sialic acid sugar donor for sialylation, is similarly restricted, which can partially explain the low overall amount of sialylated glycans present in *Drosophila* (Koles et al. 2007; Repnikova et al. 2010; Islam et al. 2013), even when DSiaT was ectopically expressed throughout the CNS (North et al. 2006).

Genetic inactivation of the sialylation pathway in vivo revealed that sialylated N-glycans play a prominent and specific role in the regulation of the nervous system. Targeted deletion of *DSiaT* results in a significantly shortened life span, locomotion abnormalities, and temperature-sensitive paralysis phenotype. *DSiaT* mutant larvae have structural and physiological defects in their neuromuscular junction synaptic connections. Electrophysiological assays of *DSiaT* mutants indicated that DSiaT activity is required for normal neuronal excitability and specifically affects the function of Para, the main voltage-gated Na^+ channel in *Drosophila* (Repnikova et al. 2010). Similar phenotypes result from *CSAS* mutations that are predicted to also block the sialylation pathway (Islam et al. 2013). Interestingly, the paralysis phenotype of *CSAS* mutants can be significantly ameliorated by an extra gene copy of *para*, which suggests that sialylation potentially controls the number of functional voltage-gated channels on the cell surface (Islam et al. 2013). Moreover, the genetic interactions between *DSiaT* and *β4GalNAcTA* indicated that sialic acids may function as masking residues hindering the recognition of LacNAc termini of glycans by some endogenous lectins (Nakamura et al. 2012). While further experiments are required to test these intriguing hypotheses, taken together, these results reveal an important novel, nervous system-specific function for α*2,6*-sialylated N-glycans in the regulation of neural transmission. It is tempting to speculate that this regulatory role corresponds to one of the most ancient evolutionarily conserved functions of sialylation in metazoan organisms. This intriguing hypothesis requires further investigation.

17.5 Conclusions

N-Glycosylation can affect glycoproteins by a number of mechanisms, e.g., by facilitating protein folding and stability, supporting trafficking, participating in interactions with other molecules, including lectins, as well as by mediating electrostatic and steric effects on protein dynamics and conformation. In the nervous system, many key players of neural transmission bear N-linked carbohydrate modifications. The roles of these modifications usually depend on molecular and cellular contexts and can vary from nonessential effects to obligatory requirements for protein functions (Table 17.1). This broad range of possible effects is expected

Table 17.1 Examples of effects of N-glycosylation on glycoproteins involved in neural physiology. Modified from (Scott and Panin 2014)

Glycoprotein	Function in the nervous system	Role of N-glycosylation	References
SV2 (synaptic vesicle protein 2)	Major synaptic vesicle protein, controls maturation step of primed synaptic vesicles	Required for proper folding and trafficking to synapses	(Chang and Sudhof 2009; Kwon and Chapman 2012)
Synaptophysin	Major synaptic vesicle protein, regulates the kinetics of synaptic vesicle endocytosis	Required for synaptic localization	(Kwon and Chapman 2012)
Nicotinic acetylcholine receptors (nAChRs)	Ligand-gated cation channels, regulate postsynaptic responses to neurotransmitter acetylcholine, regulates diverse brain functions	Regulates desensitization and channel gating	(Chen et al. 1998; Gehle et al. 1997; Nishizaki 2003)
Ionotropic glutamate receptors (iGluRs)	Ligand-gated ion channels, regulate fast transmission at the majority of excitatory synapses	Affects maximal currents and desensitization of AMPA and kainite receptors Required for folding or trafficking of NMDA receptors	(Everts et al. 1999; Everts et al. 1997; Partin et al. 1993; Thio et al. 1993; Yue et al. 1995)
		HNK-1 structure downregulates endocytosis of AMPA GluR2 subunit and promotes receptors' stability on neuronal membranes	(Morita et al. 2009; Senn et al. 2002; Yamamoto et al. 2002; Yoshihara et al. 2009)
Neurotransmitter transporters	Major determinants of synaptic signaling, mediate uptake of neurotransmitters and regulate synaptic concentration of neurotransmitters	Promotes protein stability and trafficking, increases the number of transporters at the cell surface Sialylated glycans can affect the kinetics of GABA transporter activity and affinity for Na^+	(Hu et al. 2011; Kristensen et al. 2011; Li et al. 2004; Martinez-Maza et al. 2001; Melikian et al. 1996; Nguyen and Amara 1996; Olivares et al. 1995; Tate and Blakely 1994)
Acid-sensing channels (ASICs)	Acidosis-activated cation channels. Play roles in pain, neurological and psychiatric diseases, potential mechanosensory function in sensory neurons	Effect on cell surface expression of ASIC1a and ASIC1b	(Jing et al. 2012; Kadurin et al. 2008)
TRP channels (transient receptor potential ion channels)	Playing essential roles in sensory physiology	Affects the temperature threshold of TRPM8 activation in response to cold. Affects biophysical properties of TRPC3, TRPC6 and TRPV1. Affects expression and subcellular localization of TRPV4 and TRPV5.	(Chang et al. 2005; Dietrich et al. 2003; Pertusa et al. 2012; Wirkner et al. 2005; Xu et al. 2006)

Voltage-gated ion channels	Control cell excitability, generate action potentials, affect neural transmission		
Potassium channels		Affects cell surface expression and stability of Kv11.1, Kv1.3, Kv12.2, and Kv1.4. Affects gating of Kv1.1, Kv1.5, Kv12.2, I_{sK} Affects trafficking and gating of Kv1.2 Affects simulated action potentials for Kv1.1 and Kv1.2 *Sialylation:* Affects gating of Kv1.1, Kv1.5, Kv3.1 Gating of *Drosophila* Shaker channel expressed heterologously in mammalian cells is affected by N-glycans and sialylation	(Freeman et al. 2000; Gong et al. 2002; Hall et al. 2011; Johnson and Bennett 2008; Noma et al. 2009; Schwetz et al. 2010; Sutachan et al. 2005; Thornhill et al. 1996; Watanabe et al. 2003; Watanabe et al. 2004; Watanabe et al. 2007; Zhu et al. 2009; Zhu et al. 2012)
Calcium channels		Controls activity and cell surface expression of Cav3.2, affects glucose-dependent potentiation	(Weiss et al. 2013)
Sodium channels		Affects gating of Nav1.4 and Nav1.5, Alters steady-state inactivation of Nav1.9 *Sialylation:* Affects gating of Nav1.4, and Nav1.5, electroplax channel. Sialylation of Nav beta(2) subunit affects gating of Nav1.5. Affects functional properties of *Drosophila* Nav Para (unknown if the effect is direct)	(Bennett et al. 1997; Cronin et al. 2005; Johnson et al. 2004; Recio-Pinto et al. 1990; Repnikova et al. 2010; Stocker and Bennett 2006; Zhang et al. 1999)

to create a full gamut of states of neural transmission that can be controlled by glycosylation pathways (Fig. 17.2). Collectively, these data suggest that N-glycans can function in vivo as potent regulators of synaptic transmission and excitability of neural circuits, while also providing an important link between neural transmission and metabolism. These data also pose a number of outstanding questions about molecular, cellular, and genetic mechanisms that can underlie the glycan-mediated neural regulation in vivo, as well as a potential involvement of neural N-glycosylation in the pathobiology of neurological disorders. Obtaining answers to these challenging but fundamentally important questions is expected to require a combination of in vitro and in vivo approaches, and should be facilitated by studies using genetically tractable model organisms with simplified glycosylation pathways and a decreased complexity of the nervous system.

Conflict of Interest The authors declare that they have no conflict of interest.

Acknowledgements We are grateful to Dr. Mark Zoran for stimulating discussions, Dr. Linda Baum and Dr. Mark Lehrman for their inspiration to review the topics discussed in the paper; Dr. Daria Panina for comments on the manuscript. We thank all members of the Panin laboratory for helpful discussions. This work was supported in part by NIH grant NS075534 to V.M.P.

Ethical and Biosafety Standards Policy: Research experiments in the Panin laboratory have been approved by the Institutional Biosafety Committee of Texas A&M University (Permit IBC2013-053).

References

Abo T, Balch CM. A differentiation antigen of human NK and K cells identified by a monoclonal antibody (HNK-1). J Immunol. 1981;127(3):1024–9.

Abu-Qarn M, Eichler J, Sharon N. Not just for Eukarya anymore: protein glycosylation in Bacteria and Archaea. Curr Opin Struct Biol. 2008;18(5):544–50.

Ahrens J, Foadi N, Eberhardt A, Haeseler G, Dengler R, Leffler A, Muhlenhoff M, Gerardy-Schahn R, Leuwer M. Defective polysialylation and sialylation induce opposite effects on gating of the skeletal Na+ channel NaV1.4 in Chinese hamster ovary cells. Pharmacology. 2011;87(5–6): 311–7.

Amano M, Galvan M, He J, Baum LG. The ST6Gal I sialyltransferase selectively modifies N-glycans on CD45 to negatively regulate galectin-1-induced CD45 clustering, phosphatase modulation, and T cell death. J Biol Chem. 2003;278(9):7469–75.

Angata K, Huckaby V, Ranscht B, Terskikh A, Marth JD, Fukuda M. Polysialic acid-directed migration and differentiation of neural precursors are essential for mouse brain development. Mol Cell Biol. 2007;27(19):6659–68.

Antonopoulos A, North SJ, Haslam SM, Dell A. Glycosylation of mouse and human immune cells: insights emerging from N-glycomics analyses. Biochem Soc Trans. 2011;39(5):1334–40.

Aoki K, Perlman M, Lim JM, Cantu R, Wells L, Tiemeyer M. Dynamic developmental elaboration of N-linked glycan complexity in the Drosophila melanogaster embryo. J Biol Chem. 2007;282(12):9127–42.

Apweiler R, Hermjakob H, Sharon N. On the frequency of protein glycosylation, as deduced from analysis of the SWISS-PROT database. Biochim Biophys Acta. 1999;1473(1):4–8.

Ariga T, Kohriyama T, Freddo L, Latov N, Saito M, Kon K, Ando S, Suzuki M, Hemling ME, Rinehart Jr KL, et al. Characterization of sulfated glucuronic acid containing glycolipids reacting with IgM M-proteins in patients with neuropathy. J Biol Chem. 1987;262(2):848–53.

Armstrong N, Sun Y, Chen GQ, Gouaux E. Structure of a glutamate-receptor ligand-binding core in complex with kainate. Nature. 1998;395(6705):913–7.

Baas S, Sharrow M, Kotu V, Middleton M, Nguyen K, Flanagan-Steet H, Aoki K, Tiemeyer M. Sugar-free frosting, a homolog of SAD kinase, drives neural-specific glycan expression in the Drosophila embryo. Development. 2011;138(3):553–63.

Barros CS, Franco SJ, Muller U. Extracellular matrix: functions in the nervous system. Cold Spring Harb Perspect Biol. 2010;3(1):a005108.

Baycin-Hizal D, Tian Y, Akan I, Jacobson E, Clark D, Chu J, Palter K, Zhang H, Betenbaugh MJ. GlycoFly: a database of Drosophila N-linked glycoproteins identified using SPEG–MS techniques. J Proteome Res. 2011;10(6):2777–84.

Bennett ES. Isoform-specific effects of sialic acid on voltage-dependent Na+ channel gating: functional sialic acids are localized to the S5-S6 loop of domain I. J Physiol. 2002;538(3):675–90.

Bennett E, Urcan MS, Tinkle SS, Koszowski AG, Levinson SR. Contribution of sialic acid to the voltage dependence of sodium channel gating: a possible electrostatic mechanism. J Gen Physiol. 1997;109(3):327–43.

Boquet I, Hitier R, Dumas M, Chaminade M, Preat T. Central brain postembryonic development in Drosophila: implication of genes expressed at the interhemispheric junction. J Neurobiol. 2000;42(1):33–48.

Cai G, Salonikidis PS, Fei J, Schwarz W, Schulein R, Reutter W, Fan H. The role of N-glycosylation in the stability, trafficking and GABA-uptake of GABA-transporter 1. Terminal N-glycans facilitate efficient GABA-uptake activity of the GABA transporter. FEBS J. 2005;272(7): 1625–38.

Castillo C, Diaz ME, Balbi D, Thornhill WB, Recio-Pinto E. Changes in sodium channel function during postnatal brain development reflect increases in the level of channel sialidation. Brain Res Dev Brain Res. 1997;104(1–2):119–30.

Castillo C, Thornhill WB, Zhu J, Recio-Pinto E. The permeation and activation properties of brain sodium channels change during development. Brain Res Dev Brain Res. 2003;144(1):99–106.

Cha SK, Ortega B, Kurosu H, Rosenblatt KP, Kuro OM, Huang CL. Removal of sialic acid involving Klotho causes cell-surface retention of TRPV5 channel via binding to galectin-1. Proc Natl Acad Sci U S A. 2008;105(28):9805–10.

Chang WP, Sudhof TC. SV2 renders primed synaptic vesicles competent for Ca2+ -induced exocytosis. J Neurosci. 2009;29(4):883–97.

Chang Q, Hoefs S, Van Der Kemp AW, Topala CN, Bindels RJ, Hoenderop JG. The beta-glucuronidase klotho hydrolyzes and activates the TRPV5 channel. Science. 2005; 310(5747):490–3.

Chen L. In pursuit of the high-resolution structure of nicotinic acetylcholine receptors. J Physiol. 2010;588(Pt 4):557–64.

Chen D, Dang H, Patrick JW. Contributions of N-linked glycosylation to the expression of a functional alpha7-nicotinic receptor in Xenopus oocytes. J Neurochem. 1998;70(1):349–57.

Chou DK, Ilyas AA, Evans JE, Costello C, Quarles RH, Jungalwala FB. Structure of sulfated glucuronyl glycolipids in the nervous system reacting with HNK-1 antibody and some IgM paraproteins in neuropathy. J Biol Chem. 1986;261(25):11717–25.

Colley KJ. Structural basis for the polysialylation of the neural cell adhesion molecule. Adv Exp Med Biol. 2010;663:111–26.

Condon KH, Ho J, Robinson CG, Hanus C, Ehlers MD. The Angelman syndrome protein Ube3a/E6AP is required for Golgi acidification and surface protein sialylation. J Neurosci. 2013;33(9):3799–814.

Cotella D, Radicke S, Bortoluzzi A, Ravens U, Wettwer E, Santoro C, Sblattero D. Impaired glycosylation blocks DPP10 cell surface expression and alters the electrophysiology of Ito channel complex. Pflugers Arch. 2010;460(1):87–97.

Cronin NB, O'reilly A, Duclohier H, Wallace BA. Effects of deglycosylation of sodium channels on their structure and function. Biochemistry. 2005;44(2):441–9.

Dacosta CJ, Kaiser DE, Baenziger JE. Role of glycosylation and membrane environment in nicotinic acetylcholine receptor stability. Biophys J. 2005;88(3):1755–64.

Dani N, Broadie K. Glycosylated synaptomatrix regulation of trans-synaptic signaling. Dev Neurobiol. 2012;72(1):2–21.

Dellisanti CD, Yao Y, Stroud JC, Wang ZZ, Chen L. Crystal structure of the extracellular domain of nAChR alpha1 bound to alpha-bungarotoxin at 1.94 A resolution. Nat Neurosci. 2007;10(8):953–62.

Dennis JW, Nabi IR, Demetriou M. Metabolism, cell surface organization, and disease. Cell. 2009;139(7):1229–41.

Dietrich A, Mederos Y, Schnitzler M, Emmel J, Kalwa H, Hofmann T, Gudermann T. N-linked protein glycosylation is a major determinant for basal TRPC3 and TRPC6 channel activity. J Biol Chem. 2003;278(48):47842–52.

Dindot SV, Antalffy BA, Bhattacharjee MB, Beaud et al. The Angelman syndrome ubiquitin ligase localizes to the synapse and nucleus, and maternal deficiency results in abnormal dendritic spine morphology. Hum Mol Genet. 2008;17(1):111–8.

Dityatev A, Schachner M, Sonderegger P. The dual role of the extracellular matrix in synaptic plasticity and homeostasis. Nat Rev Neurosci. 2010;11(11):735–46.

Ednie AR, Bennett ES. Modulation of voltage-gated ion channels by sialylation. Compr Physiol. 2012;2(2):1269–301.

Elmer LW, O'brien BJ, Nutter TJ, Angelides KJ. Physicochemical characterization of the alpha-peptide of the sodium channel from rat brain. Biochemistry. 1985;24(27):8128–37.

Engel AG. Current status of the congenital myasthenic syndromes. Neuromuscul Disord. 2012;22(2):99–111.

Everts I, Villmann C, Hollmann M. N-Glycosylation is not a prerequisite for glutamate receptor function but Is essential for lectin modulation. Mol Pharmacol. 1997;52(5):861–73.

Everts I, Petroski R, Kizelsztein P, Teichberg VI, Heinemann SF, Hollmann M. Lectin-induced inhibition of desensitization of the kainate receptor GluR6 depends on the activation state and can be mediated by a single native or ectopic N-linked carbohydrate side chain. J Neurosci. 1999;19(3):916–27.

Fay AM, Bowie D. Concanavalin-A reports agonist-induced conformational changes in the intact GluR6 kainate receptor. J Physiol. 2006;572(Pt 1):201–13.

Freeze HH, Eklund EA, Ng BG, Patterson MC. Neurology of inherited glycosylation disorders. Lancet Neurol. 2012;11(5):453–66.

Gehle VM, Walcott EC, Nishizaki T, Sumikawa K. N-glycosylation at the conserved sites ensures the expression of properly folded functional ACh receptors. Brain Res Mol Brain Res. 1997;45(2):219–29.

Gill MB, Vivithanaporn P, Swanson GT. Glutamate binding and conformational flexibility of ligand-binding domains are critical early determinants of efficient kainate receptor biogenesis. J Biol Chem. 2009;284(21):14503–12.

Gong Q, Anderson CL, January CT, Zhou Z. Role of glycosylation in cell surface expression and stability of HERG potassium channels. Am J Physiol Heart Circ Physiol. 2002;283(1): H77–84.

GottliebC,BaenzigerJ,KornfeldS.Deficienturidinediphosphate-N-acetylglucosamine:glycoprotein N-acetylglucosaminyltransferase activity in a clone of Chinese hamster ovary cells with altered surface glycoproteins. J Biol Chem. 1975;250(9):3303–9.

Greer PL, Hanayama R, Bloodgood BL, Mardinly AR, Lipton DM, Flavell SW, Kim TK, Griffith EC, Waldon Z, Maehr R, Ploegh HL, Chowdhury S, Worley PF, Steen J, Greenberg ME. The Angelman Syndrome protein Ube3A regulates synapse development by ubiquitinating arc. Cell. 2010;140(5):704–16.

Gurba KN, Hernandez CC, Hu N, Macdonald RL. GABRB3 mutation, G32R, associated with childhood absence epilepsy alters alpha1beta3gamma2L gamma-aminobutyric acid type A (GABAA) receptor expression and channel gating. J Biol Chem. 2012;287(15):12083–97.

Gurevicius K, Gureviciene I, Sivukhina E, Irintchev A, Schachner M, Tanila H. Increased hippocampal and cortical beta oscillations in mice deficient for the HNK-1 sulfotransferase. Mol Cell Neurosci. 2007;34(2):189–98.

Haines N, Irvine KD. Functional analysis of Drosophila beta1,4-N-acetlygalactosaminyltransferases. Glycobiology. 2005;15(4):335–46.

Haines N, Stewart BA. Functional roles for beta1,4-N-acetlygalactosaminyltransferase-A in Drosophila larval neurons and muscles. Genetics. 2007;175(2):671–9.

Hall MK, Reutter W, Lindhorst T, Schwalbe RA. Biochemical engineering of the N-acyl side chain of sialic acids alters the kinetics of a glycosylated potassium channel Kv3.1. FEBS Lett. 2011;585(20):3322–7.

Helenius A, Aebi M. Intracellular functions of N-linked glycans. Science. 2001;291(5512): 2364–9.

Henion TR, Faden AA, Knott TK, Schwarting GA. beta3GnT2 maintains adenylyl cyclase-3 signaling and axon guidance molecule expression in the olfactory epithelium. J Neurosci. 2011;31(17):6576–86.

Hildebrandt H, Muhlenhoff M, Oltmann-Norden I, Rockle I, Burkhardt H, Weinhold B, Gerardy-Schahn R. Imbalance of neural cell adhesion molecule and polysialyltransferase alleles causes defective brain connectivity. Brain. 2009;132(Pt 10):2831–8.

Hollmann M, Maron C, Heinemann S. N-glycosylation site tagging suggests a three transmembrane domain topology for the glutamate receptor GluR1. Neuron. 1994;13(6):1331–43.

Hu J, Fei J, Reutter W, Fan H. Involvement of sialic acid in the regulation of gamma–aminobutyric acid uptake activity of gamma-aminobutyric acid transporter 1. Glycobiology. 2011;21(3): 329–39.

Ioffe E, Stanley P. Mice lacking N-acetylglucosaminyltransferase I activity die at mid-gestation, revealing an essential role for complex or hybrid N-linked carbohydrates. Proc Natl Acad Sci U S A. 1994;91(2):728–32.

Isaev D, Isaeva E, Shatskih T, Zhao Q, Smits NC, Shworak NW, Khazipov R, Holmes GL. Role of extracellular sialic acid in regulation of neuronal and network excitability in the rat hippocampus. J Neurosci. 2007;27(43):11587–94.

Isaev D, Zhao Q, Kleen JK, Lenck-Santini PP, Adstamongkonkul D, Isaeva E, Holmes GL. Neuroaminidase reduces interictal spikes in a rat temporal lobe epilepsy model. Epilepsia. 2011;52(3):e12–5.

Isaeva E, Lushnikova I, Savrasova A, Skibo G, Holmes GL, Isaev D. Blockade of endogenous neuraminidase leads to an increase of neuronal excitability and activity-dependent synaptogenesis in the rat hippocampus. Eur J Neurosci. 2010;32(11):1889–96.

Islam R, Nakamura M, Scott H, Repnikova E, Carnahan M, Pandey D, Caster C, Khan S, Zimmermann T, Zoran MJ, Panin VM. The role of Drosophila cytidine monophosphate-sialic acid synthetase in the nervous system. J Neurosci. 2013;33(30):12306–15.

James WM, Agnew WS. Multiple oligosaccharide chains in the voltage-sensitive Na channel from electrophorus electricus: evidence for alpha-2,8-linked polysialic acid. Biochem Biophys Res Commun. 1987;148(2):817–26.

Janz R, Goda Y, Geppert M, Missler M, Sudhof TC. SV2A and SV2B function as redundant Ca2+ regulators in neurotransmitter release. Neuron. 1999;24(4):1003–16.

Jiang YH, Armstrong D, Albrecht U, Atkins CM, Noebels JL, Eichele G, Sweatt JD, Beaud et al. Mutation of the Angelman ubiquitin ligase in mice causes increased cytoplasmic p53 and deficits of contextual learning and long-term potentiation. Neuron. 1998;21(4): 799–811.

Jing L, Chu XP, Jiang YQ, Collier DM, Wang B, Jiang Q, Snyder PM, Zha XM. N-glycosylation of acid-sensing ion channel 1a regulates its trafficking and acidosis-induced spine remodeling. J Neurosci. 2012;32(12):4080–91.

Johnson D, Bennett ES. Isoform-specific effects of the beta(2) subunit on voltage-gated sodium channel gating. J Biol Chem. 2006;281(36):25875–81.

Johnson D, Montpetit ML, Stocker PJ, Bennett ES. The sialic acid component of the beta(1) subunit modulates voltage-gated sodium channel function. J Biol Chem. 2004;279(43):44303–10.

Kadurin I, Golubovic A, Leisle L, Schindelin H, Grunder S. Differential effects of N-glycans on surface expression suggest structural differences between the acid-sensing ion channel (ASIC) 1a and ASIC1b. Biochem J. 2008;412(3):469–75.

Kariya Y, Kato R, Itoh S, Fukuda T, Shibukawa Y, Sanzen N, Sekiguchi K, Wada Y, Kawasaki N, Gu J. N-Glycosylation of laminin-332 regulates its biological functions. A novel function of the bisecting GlcNAc. J Biol Chem. 2008;283(48):33036–45.

Kleene R, Schachner M. Glycans and neural cell interactions. Nat Rev Neurosci. 2004; 5(3):195–208.

Koles K, Irvine KD, Panin VM. Functional characterization of Drosophila sialyltransferase. J Biol Chem. 2004;279(6):4346–57.

Koles K, Lim JM, Aoki K, Porterfield M, Tiemeyer M, Wells L, Panin V. Identification of N-glycosylated proteins from the central nervous system of Drosophila melanogaster. Glycobiology. 2007;17(12):1388–403.

Koles K, Repnikova E, Pavlova G, Korochkin LI, Panin VM. Sialylation in protostomes: a perspective from Drosophila genetics and biochemistry. Glycoconj J. 2009;26(3):313–24.

Kristensen AS, Andersen J, Jorgensen TN, Sorensen L, Eriksen J, Loland CJ, Stromgaard K, Gether U. SLC6 neurotransmitter transporters: structure, function, and regulation. Pharmacol Rev. 2011;63(3):585–640.

Kwon SE, Chapman ER. Glycosylation is dispensable for sorting of synaptotagmin 1 but is critical for targeting of SV2 and synaptophysin to recycling synaptic vesicles. J Biol Chem. 2012;287(42):35658–68.

Leonard R, Rendic D, Rabouille C, Wilson IB, Preat T, Altmann F. The Drosophila fused lobes gene encodes an N-acetylglucosaminidase involved in N-glycan processing. J Biol Chem. 2006;281(8):4867–75.

Leunissen EH, Nair AV, Bull C, Lefeber DJ, Van Delft FL, Bindels RJ, Hoenderop JG. The epithelial calcium channel TRPV5 is regulated differentially by klotho and sialidase. J Biol Chem. 2013;288:29238–46.

Li LB, Chen N, Ramamoorthy S, Chi L, Cui XN, Wang LC, Reith ME. The role of N-glycosylation in function and surface trafficking of the human dopamine transporter. J Biol Chem. 2004;279(20):21012–20.

Lowe JB, Marth JD. A genetic approach to Mammalian glycan function. Annu Rev Biochem. 2003;72:643–91.

Mah SJ, Cornell E, Mitchell NA, Fleck MW. Glutamate receptor trafficking: endoplasmic reticulum quality control involves ligand binding and receptor function. J Neurosci. 2005; 25(9):2215–25.

Martinez-Maza R, Poyatos I, Lopez-Corcuera B, Núñez E, Gimenez C, Zafra F, Aragon C. The role of N-glycosylation in transport to the plasma membrane and sorting of the neuronal glycine transporter GLYT2. J Biol Chem. 2001;276(3):2168–73.

Mckemy DD, Neuhausser WM, Julius D. Identification of a cold receptor reveals a general role for TRP channels in thermosensation. Nature. 2002;416(6876):52–8.

Melikian HE, Ramamoorthy S, Tate CG, Blakely RD. Inability to N-glycosylate the human norepinephrine transporter reduces protein stability, surface trafficking, and transport activity but not ligand recognition. Mol Pharmacol. 1996;50(2):266–76.

Messner DJ, Catterall WA. The sodium channel from rat brain. Separation and characterization of subunits. J Biol Chem. 1985;260(19):10597–604.

Metzler M, Gertz A, Sarkar M, Schachter H, Schrader JW, Marth JD. Complex asparagine-linked oligosaccharides are required for morphogenic events during post-implantation development. EMBO J. 1994;13(9):2056–65.

Miller JA, Agnew WS, Levinson SR. Principal glycopeptide of the tetrodotoxin/saxitoxin binding protein from Electrophorus electricus: isolation and partial chemical and physical characterization. Biochemistry. 1983;22(2):462–70.

Miwa JM, Freedman R, Lester HA. Neural systems governed by nicotinic acetylcholine receptors: emerging hypotheses. Neuron. 2011;70(1):20–33.

Montpetit ML, Stocker PJ, Schwetz TA, Harper JM, Norring SA, Schaffer L, North SJ, Jang-Lee J, Gilmartin T, Head SR, Haslam SM, Dell A, Marth JD, Bennett ES. Regulated and aberrant glycosylation modulate cardiac electrical signaling. Proc Natl Acad Sci U S A. 2009;106(38): 16517–22.

Moremen KW, Tiemeyer M, Nairn AV. Vertebrate protein glycosylation: diversity, synthesis and function. Nat Rev Mol Cell Biol. 2012;13(7):448–62.

Morita I, Kakuda S, Takeuchi Y, Itoh S, Kawasaki N, Kizuka Y, Kawasaki T, Oka S. HNK-1 glyco-epitope regulates the stability of the glutamate receptor subunit GluR2 on the neuronal cell surface. J Biol Chem. 2009;284(44):30209–17.

Muhlenhoff M, Eckhardt M, Gerardy-Schahn R. Polysialic acid: three-dimensional structure, biosynthesis and function. Curr Opin Struct Biol. 1998;8(5):558–64.

Muhlenhoff M, Oltmann-Norden I, Weinhold B, Hildebrandt H, Gerardy-Schahn R. Brain development needs sugar: the role of polysialic acid in controlling NCAM functions. Biol Chem. 2009;390(7):567–74.

Muller D, Wang C, Skibo G, Toni N, Cremer H, Calaora V, Rougon G, Kiss JZ. PSA-NCAM is required for activity-induced synaptic plasticity. Neuron. 1996;17(3):413–22.

Nakamura M, Pandey D, Panin VM. Genetic interactions between Drosophila sialyltransferase and beta1,4-N-acetylgalactosaminyltransferase-a genes indicate their involvement in the same pathway. G3 (Bethesda). 2012;2(6):653–6.

Nanao MH, Green T, Stern-Bach Y, Heinemann SF, Choe S. Structure of the kainate receptor subunit GluR6 agonist-binding domain complexed with domoic acid. Proc Natl Acad Sci U S A. 2005;102(5):1708–13.

Nguyen TT, Amara SG. N-linked oligosaccharides are required for cell surface expression of the norepinephrine transporter but do not influence substrate or inhibitor recognition. J Neurochem. 1996;67(2):645–55.

Nishizaki T. N-glycosylation sites on the nicotinic ACh receptor subunits regulate receptor channel desensitization and conductance. Brain Res Mol Brain Res. 2003;114(2):172–6.

North SJ, Koles K, Hembd C, Morris HR, Dell A, Panin VM, Haslam SM. Glycomic studies of Drosophila melanogaster embryos. Glycoconj J. 2006;23(5–6):345–54.

Olivares L, Aragon C, Gimenez C, Zafra F. The role of N-glycosylation in the targeting and activity of the GLYT1 glycine transporter. J Biol Chem. 1995;270(16):9437–42.

Parkinson W, Dear ML, Rushton E, Broadie K. N-glycosylation requirements in neuromuscular synaptogenesis. Development. 2013;140(24):4970–81.

Partin KM, Patneau DK, Winters CA, Mayer ML, Buonanno A. Selective modulation of desensitization at AMPA versus kainate receptors by cyclothiazide and concanavalin A. Neuron. 1993;11(6):1069–82.

Peier AM, Moqrich A, Hergarden AC, Reeve AJ, Andersson DA, Story GM, Earley TJ, Dragoni I, Mcintyre P, Bevan S, Patapoutian A. A TRP channel that senses cold stimuli and menthol. Cell. 2002;108(5):705–15.

Pertusa M, Madrid R, Morenilla-Palao C, Belmonte C, Viana F. N-glycosylation of TRPM8 ion channels modulates temperature sensitivity of cold thermoreceptor neurons. J Biol Chem. 2012;287(22):18218–29.

Rasmussen TN, Plenge P, Bay T, Egebjerg J, Gether U. A single nucleotide polymorphism in the human serotonin transporter introduces a new site for N-linked glycosylation. Neuropharmacology. 2009;57(3):287–94.

Recio-Pinto E, Thornhill WB, Duch DS, Levinson SR, Urban BW. Neuraminidase treatment modifies the function of electroplax sodium channels in planar lipid bilayers. Neuron. 1990; 5(5):675–84.

Repnikova E, Koles K, Nakamura M, Pitts J, Li H, Ambavane A, Zoran MJ, Panin VM. Sialyltransferase regulates nervous system function in Drosophila. J Neurosci. 2010; 30(18):6466–76.

Roberts RH, Barchi RL. The voltage-sensitive sodium channel from rabbit skeletal muscle. Chemical characterization of subunits. J Biol Chem. 1987;262(5):2298–303.

Rutishauser U. Polysialic acid in the plasticity of the developing and adult vertebrate nervous system. Nat Rev Neurosci. 2008;9(1):26–35.

Sarkar M, Leventis PA, Silvescu CI, Reinhold VN, Schachter H, Boulianne GL. Null mutations in Drosophila N-acetylglucosaminyltransferase I produce defects in locomotion and a reduced life span. J Biol Chem. 2006;281(18):12776–85.

Sarkar M, Iliadi KG, Leventis PA, Schachter H, Boulianne GL. Neuronal expression of Mgat1 rescues the shortened life span of Drosophila Mgat1 1 null mutants and increases life span. Proc Natl Acad Sci U S A. 2010;107(21):9677–82.

Schachter H. Mgat1-dependent N-glycans are essential for the normal development of both vertebrate and invertebrate metazoans. Semin Cell Dev Biol. 2010;21(6):609–15.

Schwalbe RA, Corey MJ, Cartwright TA. Novel Kv3 glycoforms differentially expressed in adult mammalian brain contain sialylated N-glycans. Biochem Cell Biol. 2008;86(1):21–30.

Schwetz TA, Norring SA, Bennett ES. N-glycans modulate K(v)1.5 gating but have no effect on K(v)1.4 gating. Biochim Biophys Acta. 2010;1798(3):367–75.

Sekine SU, Haraguchi S, Chao K, Kato T, Luo L, Miura M, Chihara T. Meigo governs dendrite targeting specificity by modulating ephrin level and N-glycosylation. Nat Neurosci. 2013; 16(6):683–91.

Senderek J, Muller JS, Dusl M, Strom TM, Guergueltcheva V, Diepolder I, Laval SH, Maxwell S, Cossins J, Krause S, Muelas N, Vilchez JJ, Colomer J, Mallebrera CJ, Nascimento A, Nafissi S, Kariminejad A, Nilipour Y, Bozorgmehr B, Najmabadi H, Rodolico C, Sieb JP, Steinlein OK, Schlotter B, Schoser B, Kirschner J, Herrmann R, Voit T, Oldfors A, Lindbergh C, Urtizberea A, Von Der Hagen M, Hubner A, Palace J, Bushby K, Straub V, Beeson D, Abicht A, Lochmuller H. Hexosamine biosynthetic pathway mutations cause neuromuscular transmission defect. Am J Hum Genet. 2011;88(2):162–72.

Senn C, Kutsche M, Saghatelyan A, Bosl MR, Lohler J, Bartsch U, Morellini F, Schachner M. Mice deficient for the HNK-1 sulfotransferase show alterations in synaptic efficacy and spatial learning and memory. Mol Cell Neurosci. 2002;20(4):712–29.

Sharon N. Lectins: carbohydrate-specific reagents and biological recognition molecules. J Biol Chem. 2007;282(5):2753–64.

Stanley P, Narasimhan S, Siminovitch L, Schachter H. Chinese hamster ovary cells selected for resistance to the cytotoxicity of phytohemagglutinin are deficient in a UDP-N-acetylglucosamine–glycoprotein N-acetylglucosaminyltransferase activity. Proc Natl Acad Sci U S A. 1975;72(9):3323–7.

Stanley P, Schachter H, Taniguchi N. N-Glycans. In: Varki A, Cummings RD, et al., editors. Essentials of glycobiology. Harbor, NY: Cold Spring Harbor Laboratory Press; 2009.

Stocker PJ, Bennett ES. Differential sialylation modulates voltage-gated Na+ channel gating throughout the developing myocardium. J Gen Physiol. 2006;127(3):253–65.

Sumikawa K, Parker I, Miledi R. Effect of tunicamycin on the expression of functional brain neurotransmitter receptors and voltage-operated channels in Xenopus oocytes. Brain Res. 1988; 464(3):191–9.

Tan CL, Kwok JC, Patani R, Ffrench-Constant C, Chandran S, Fawcett JW. Integrin activation promotes axon growth on inhibitory chondroitin sulfate proteoglycans by enhancing integrin signaling. J Neurosci. 2011;31(17):6289–95.

Tate CG, Blakely RD. The effect of N-linked glycosylation on activity of the Na(+)- and Cl(-)-dependent serotonin transporter expressed using recombinant baculovirus in insect cells. J Biol Chem. 1994;269(42):26303–10.

Thio LL, Clifford DB, Zorumski CF. Concanavalin A enhances excitatory synaptic transmission in cultured rat hippocampal neurons. Synapse. 1993;13(1):94–7.

Traynelis SF, Wollmuth LP, Mcbain CJ, Menniti FS, Vance KM, Ogden KK, Hansen KB, Yuan H, Myers SJ, Dingledine R. Glutamate receptor ion channels: structure, regulation, and function. Pharmacol Rev. 2010;62(3):405–96.

Tyrrell L, Renganathan M, Dib-Hajj SD, Waxman SG. Glycosylation alters steady-state inactivation of sodium channel Nav1.9/NaN in dorsal root ganglion neurons and is developmentally regulated. J Neurosci. 2001;21(24):9629–37.

Ufret-Vincenty CA, Baro DJ, Lederer WJ, Rockman HA, Quinones LE, Santana LF. Role of sodium channel deglycosylation in the genesis of cardiac arrhythmias in heart failure. J Biol Chem. 2001;276(30):28197–203.

Vandenborre G, Van Damme EJ, Ghesquiere B, Menschaert G, Hamshou M, Rao RN, Gevaert K, Smagghe G. Glycosylation signatures in Drosophila: fishing with lectins. J Proteome Res. 2010;9(6):3235–42.

Varki A, Etzler ME, Cummings RD, Esko JD. Discovery and classification of glycan-binding proteins. In: Varki A, Cummings RD, et al., editors. Essentials of glycobiology. Harbor, NY: Cold Spring; 2009.

Veldhuis NA, Lew MJ, Abogadie FC, Poole DP, Jennings EA, Ivanusic JJ, Eilers H, Bunnett NW, Mcintyre P. N-glycosylation determines ionic permeability and desensitization of the TRPV1 capsaicin receptor. J Biol Chem. 2012;287(26):21765–72.

Wang PS, Wang J, Xiao ZC, Pallen CJ. Protein-tyrosine phosphatase alpha acts as an upstream regulator of Fyn signaling to promote oligodendrocyte differentiation and myelination. J Biol Chem. 2009;284(48):33692–702.

Watanabe I, Wang HG, Sutachan JJ, Zhu J, Recio-Pinto E, Thornhill WB. Glycosylation affects rat Kv1.1 potassium channel gating by a combined surface potential and cooperative subunit interaction mechanism. J Physiol. 2003;550(1):51–66.

Watanabe I, Zhu J, Recio-Pinto E, Thornhill WB. Glycosylation affects the protein stability and cell surface expression of Kv1.4 but Not Kv1.1 potassium channels. A pore region determinant dictates the effect of glycosylation on trafficking. J Biol Chem. 2004;279(10):8879–85.

Watanabe I, Zhu J, Sutachan JJ, Gottschalk A, Recio-Pinto E, Thornhill WB. The glycosylation state of Kv1.2 potassium channels affects trafficking, gating, and simulated action potentials. Brain Res. 2007;1144:1–18.

Weinhold B, Seidenfaden R, Rockle I, Muhlenhoff M, Schertzinger F, Conzelmann S, Marth JD, Gerardy-Schahn R, Hildebrandt H. Genetic ablation of polysialic acid causes severe neurodevelopmental defects rescued by deletion of the neural cell adhesion molecule. J Biol Chem. 2005;280(52):42971–7.

Weiss N, Black SA, Bladen C, Chen L, Zamponi GW. Surface expression and function of Cav3.2T-type calcium channels are controlled by asparagine-linked glycosylation. Pflugers Arch. 2013;465(8):1159–70.

Williams CA, Beaud et al, Clayton-Smith J, Knoll JH, Kyllerman M, Laan LA, Magenis RE, Moncla A, Schinzel AA, Summers JA, Wagstaff J. Angelman syndrome 2005: updated consensus for diagnostic criteria. Am J Med Genet A. 2006;140(5):413–8.

Wirkner K, Hognestad H, Jahnel R, Hucho F, Illes P. Characterization of rat transient receptor potential vanilloid 1 receptors lacking the N-glycosylation site N604. Neuroreport. 2005;16(9):997–1001.

Wittwer AJ, Howard SC. Glycosylation at Asn-184 inhibits the conversion of single-chain to two-chain tissue-type plasminogen activator by plasmin. Biochemistry. 1990;29(17):4175–80.

Woodard-Grice AV, Mcbrayer AC, Wakefield JK, Zhuo Y, Bellis SL. Proteolytic shedding of ST6Gal-I by BACE1 regulates the glycosylation and function of alpha4beta1 integrins. J Biol Chem. 2008;283(39):26364–73.

Wormald MR, Dwek RA. Glycoproteins: glycan presentation and protein-fold stability. Structure. 1999;7(7):R155–60.

Xu H, Fu Y, Tian W, Cohen DM. Glycosylation of the osmoresponsive transient receptor potential channel TRPV4 on Asn-651 influences membrane trafficking. Am J Physiol Renal Physiol. 2006;290(5):F1103–9.

Yagi H, Yanagisawa M, Suzuki Y, Nakatani Y, Ariga T, Kato K, Yu RK. HNK-1 epitope-carrying tenascin-C spliced variant regulates the proliferation of mouse embryonic neural stem cells. J Biol Chem. 2010;285(48):37293–301.

Yamamoto S, Oka S, Inoue M, Shimuta M, Manabe T, Takahashi H, Miyamoto M, Asano M, Sakagami J, Sudo K, Iwakura Y, Ono K, Kawasaki T. Mice deficient in nervous system-specific carbohydrate epitope HNK-1 exhibit impaired synaptic plasticity and spatial learning. J Biol Chem. 2002;277(30):27227–31.

Yanagisawa M, Yu RK. The expression and functions of glycoconjugates in neural stem cells. Glycobiology. 2007;17(7):57R–74.

Yashiro K, Riday TT, Condon KH, Roberts AC, Bernardo DR, Prakash R, Weinberg RJ, Ehlers MD, Philpot BD. Ube3a is required for experience-dependent maturation of the neocortex. Nat Neurosci. 2009;12(6):777–83.

Ye Z, Marth JD. N-glycan branching requirement in neuronal and postnatal viability. Glycobiology. 2004;14(6):547–58.

Yoshihara T, Sugihara K, Kizuka Y, Oka S, Asano M. Learning/memory impairment and reduced expression of the HNK-1 carbohydrate in beta4-galactosyltransferase-II-deficient mice. J Biol Chem. 2009;284(18):12550–61.

Yue KT, Macdonald JF, Pekhletski R, Hampson DR. Differential effects of lectins on recombinant glutamate receptors. Eur J Pharmacol. 1995;291(3):229–35.

Zhang Y, Hartmann HA, Satin J. Glycosylation influences voltage-dependent gating of cardiac and skeletal muscle sodium channels. J Membr Biol. 1999;171(3):195–207.

Zhao Y, Nakagawa T, Itoh S, Inamori K, Isaji T, Kariya Y, Kondo A, Miyoshi E, Miyazaki K, Kawasaki N, Taniguchi N, Gu J. N-acetylglucosaminyltransferase III antagonizes the effect of N-acetylglucosaminyltransferase V on alpha3beta1 integrin-mediated cell migration. J Biol Chem. 2006;281(43):32122–30.

Zhou D, Dinter A, Gutierrez Gallego R, Kamerling JP, Vliegenthart JF, Berger EG, Hennet T. A beta-1,3-N-acetylglucosaminyltransferase with poly-N-acetyllactosamine synthase activity is structurally related to beta-1,3-galactosyltransferases. Proc Natl Acad Sci U S A. 1999; 96(2):406–11.

Zhu J, Yan J, Thornhill WB. N-glycosylation promotes the cell surface expression of Kv1.3 potassium channels. FEBS J. 2012;279(15):2632–44.

Zhuo Y, Bellis SL. Emerging role of alpha2,6-sialic acid as a negative regulator of galectin binding and function. J Biol Chem. 2011;286(8):5935–41.

Zuber C, Lackie PM, Catterall WA, Roth J. Polysialic acid is associated with sodium channels and the neural cell adhesion molecule N-CAM in adult rat brain. J Biol Chem. 1992;267(14): 9965–71.

Chapter 18
Roles of Carbohydrates in the Interaction of Pathogens with Neural Cells

Cara-Lynne Schengrund

Abstract Numerous pathogens that can affect neural function utilize oligosaccharide–protein interactions as a first step in the infection process. The variability in carbohydrate structures as well as the presence of carbohydrate binding receptors on the surface of cells provides a plethora of potential binding sites for viruses, bacteria, and bacterial toxins. This chapter discusses scenarios for how carbohydrates may affect the ability of infectious agents to interact with neural cells, provides examples of problems that may result from development of antibodies to carbohydrate antigens found on pathogens that are similar to epitopes expressed on mammalian cells, and presents approaches either in use or under consideration for translational uses of this information.

Keywords Bacterial toxins • Molecular mimicry • Glycodendrimers • Multivalency • Latency • Lipid rafts • Peptide mimetics

Abbreviations

BBB	Blood–brain barrier
Bgp	Borrelia GAG-binding protein
BoNT	Botulinum neurotoxin
CaMKII	Ca^{2+}/calmodulin-dependent kinase II
CNS	Central nervous system
CTx	Cholera toxin
DC-SIGN	Dendritic cell-specific intercellular molecule-grabbing nonintegrin

C.-L. Schengrund (✉)
Department of Biochemistry and Molecular Biology, The Pennsylvania State University College of Medicine, Hershey, PA 17033, USA
e-mail: cxs8@psu.edu

R.K. Yu and C.-L. Schengrund (eds.), *Glycobiology of the Nervous System*, Advances in Neurobiology 9, DOI 10.1007/978-1-4939-1154-7_18,

GABA	Gamma-aminobutyric acid
GAG	Glycosaminoglycan
Gb3	Globotriaosylceramide
GBS	Guillain–Barre syndrome
GSL	Glycosphingolipid
H	Hemagglutinin
HIV	Human immunodeficiency virus
Iv	Intravenous
MßCD	Methyl-ß-cyclodextrin
SNAP-25	Synaptic vesicle-associated protein with a mass of 25 kilodaltons
SV2	Synaptic vesicle glycoprotein 2
TeNT	Tetanus neurotoxin
TNFα	Tumor necrosis factor-alpha
TTC	Carboxyl terminal half of the heavy chain of tetanus toxin
VAMP-1-2	Vesicle-associated membrane protein-1-2 or synaptobrevin-1 and -2

In previous chapters you learned about the various types of glycoconjugates expressed in the nervous system and the almost limitless diversity they can display when considered from a structural point of view. Also discussed were some of the myriad number of problems that may occur as a result of incorrect glycosylation. The variability in carbohydrate structures as well as the presence of carbohydrate-binding receptors on the surface of cells provides a plethora of potential binding sites for viruses, bacteria, and bacterial toxins. This chapter will present (1) scenarios for how carbohydrates may affect the ability of infectious agents to interact with neural cells, (2) examples of problems that may result from development of antibodies to carbohydrate antigens found on pathogens that are similar to epitopes found on mammalian cells, and (3) approaches either in use or under consideration for translational uses of this information.

18.1 Bacterial Toxins: Lessons Learned

Cholera toxin (CTx), the heat labile enterotoxin of *Escherichia coli*, and Shiga-like toxin, causative agents for the symptoms of cholera, travelers' diarrhea, and hemolytic-uremic syndrome, respectively, are well-studied examples of carbohydrate-binding agents. Except for Shiga-like toxin produced by *E. coli* O157:H7 that can also infect the CNS (Hamano et al. 1993), these toxins do not induce neuronal disease. CNS involvement induced by Shiga-like toxin 2 is thought to result from inflammatory responses in the brain initiated by Shiga-like toxin binding to globotriaosylceramide (Gb3, sugar compositions of sphingolipids discussed are shown in Table 18.1) on brain endothelial cells, not neurons and glia. Resultant neuronal death appears to

Table 18.1 Carbohydrate composition of oligosaccharides discussed

Name	Carbohydrate composition
Gb3	Gal(α1–4)Gal(ß1–4)Glcß1–
GM3	SA(α2–3)Gal(ß1–4)Glcß1–
GM1	Gal(ß1–3)GalNAc(ß1–4)[SA(α2–3)]Gal(ß1–4)Glcß1–
GD3	SA(α2–8)SA(α2–3)Gal(ß1–4)Glcß1–
GD2	GalNAc(ß1–4)[SA(α2– 8)(SA(α2–3)]Gal(ß1–4)Glcß1–
GD1b	Gal(ß1–3)GalNAc(ß1–4)[SA(α2–8)SA(α2–3)]Gal(ß1–4)Glcß1–
GT1b	SA(α2–3)Gal(ß1–3)GalNAc(ß1–4)[SA(α2–8)SA(α2–3)]Gal(ß1–4)Glcß1–
GQ1b	SA(α2–8)SA(α2–3)Gal(ß1–3)GalNAc(ß1–4)[SA(α2–8)SA(α2–3)] Gal(ß1–4)Glcß1–
LLG-3[a]	8-OMe-Neu5Ac(α2–11)Neu5Gc(α2–3)Gal(ß1–4)Glcß1–
Heparan sulfate[b]	[GlcNAc(ß1–4)GlcUA/IdoA]$_n$-linkage tetrasaccharide
Dermatan sulfate	[IdoA/GlcUA(α1–3)GalNac]$_n$-linkage tetrasaccharide
Hyaluronic acid	[GlcUA(ß1–3)GlcNAc]$_n$
PGL-1 *M. leprae*	3,6-di-O-methylGlc(α1–4)2,3-di-O-methylrhamnose(ß1–2)3-O-methylmannose

[a]Neu refers to neuraminic acid while SA refers to sialic acid the general name used when not specifically indicating substituents

[b]Heparin and heparan sulfate have the same disaccharide repeating unit but heparin has more *N*-acetyl groups, and fewer N and O-sulfates. Dermatan sulfate contains predominantly iduronic acid while hyaluronic acid is not sulfated. For details about GAGs see Chap. 5

be caused by inflammatory responses as seen in the up-regulation of mRNA for interleukin-1ß and tumor necrosis factor-alpha (TNFα, Takahashi et al. 2008). Regardless of the effects of these toxins on the CNS, identification of their glycosphingolipid receptors promoted our understanding of infection by other carbohydrate-binding pathogens. Each of the three toxins can be described as being of the AB_5 type. This indicates that the binding subunit is made up of five identical polypeptide chains while the A subunit, or a part of it, mediates activity of the toxin. In the case of cholera toxin, the five identical binding subunits were shown to each contain components of a single binding site that was formed by the intersection of one peptide with the next, resulting in a binding subunit able to adhere to five receptors (Merritt et al. 2002). While individual protein–carbohydrate interactions are often of low affinity, the adherence of multiple binding subunits present on the AB_5 toxins to multiple carbohydrate residues present on either a single or multiple receptors was shown to result in a much higher binding affinity, one that was greater than the sum of the individual interactions (Lee and Lee 2000). This observation provided an explanation for why “multivalent” carbohydrate inhibitors, termed glycodendrimers, were observed to be more effective inhibitors of the binding of AB_5 toxins than the free oligosaccharide (e.g. Thompson and Schengrund 1997; Kitov et al. 2000) and provided the basis for understanding how pathogens may use cell surface carbohydrates as high-affinity binding sites.

As indicated above protein–sugar interactions are often strengthened as a result of the protein adhering to multiple carbohydrates supporting the hypothesis that when the receptor is a glycosphingolipid (GSL) with just a single oligosaccharide

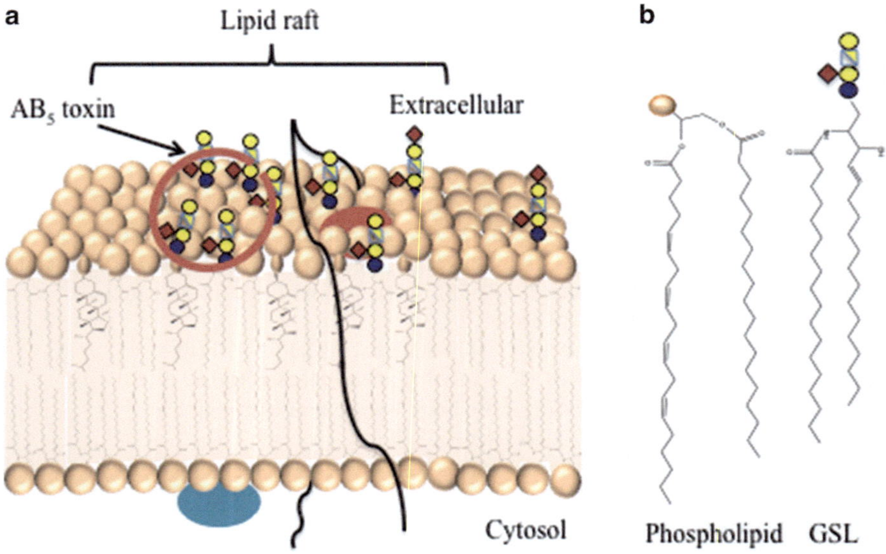

Fig. 18.1 (**a**) Schematic of the lipid raft portion on the outer surface of a cell's plasma membrane. The increased concentration of glycosphingolipids are indicated by linked, *colored circles* and diamonds, possible adherence of an AB_5 toxin to a number of carbohydrates is indicated by the round *red structure*, a transmembrane-raft-associated protein by the *black line* and surface-associated proteins by the elipses. Due to the increase in cholesterol and sphingolipids such as gangliosides present in lipid rafts, lateral mobility within the membrane is reduced relative to that in the nonraft, phospholipid-enriched portions of the membrane. This reflects the ring structure of cholesterol and the presence of sphingosine with its single *trans* double bond and the fact that the fatty acid linked to it to form the ceramide portion of sialylated glycosphingolipids such as gangliosides is usually saturated. (**b**) Comparison of the hydrocarbon chain composition frequently seen in phospholipids relative to that seen in glycosphingolipids such as gangliosides. Note the *cis* double bonds in the unsaturated fatty acid often found at the 2 position in a phospholipid. The *trans* double bond and saturated fatty acid found in ceramides (Cer) associated with gangliosides allow for tighter packing of hydrocarbon chains than seen when unsaturated fatty acids are present. Sugars on the GSL chain are glc (●), gal (○), galNAc (▧), and sialic acid (◆)

residue its concentration on the cell surface might affect binding by the pathogen. Hanashima et al (2008) found this applied to the binding of Shiga-like toxin to its receptor, Gb3. It adhered to wild-type Vero cells where Gb3 was present in lipid rafts but did not bind mutated Vero cells whose lipid rafts have a lower density of Gb3. For a schematic of a lipid raft, an area of the membrane enriched in cholesterol, glycosphingolipids, and signal transduction molecules (Brown and London 2000; Simons and Toomre 2000) see Fig. 18.1.

The *Clostridial* neurotoxins, tetanus (TeNT) and botulinum (BoNT), causative agents of tetani and botulism, respectively, recognize carbohydrate moieties on gangliosides as receptor components. Tetanus toxin induces the spastic paralysis associated with tetani, by acting on neurons in the CNS to block release of the inhibitory neurotransmitters, γ-aminobutyric acid (GABA) and glycine. BoNT induces the flaccid paralysis associated with botulism, by inhibiting acetylcholine release at the neuromuscular junction. While tetanus toxin exists as a single serotype, eight serotypes of BoNT (A-H plus numerous subtypes) have been identified (e.g. Dover et al. 2014). Serotypes A, B, and E are most often seen in human botulism while serotypes C and D are found in animals. Regardless of serotype each toxin consists of a heavy chain and a light chain linked together by a disulfide bond. The heavy chain of TeNT has two carbohydrate-binding sites that adhere to the carbohydrate portion of gangliosides GD1b and GT1b (Habermann and Dreyer 1986; Svennerholm's nomenclature 1980) while the heavy chain of BoNT has either one site that in the case of BoNT serotype A adheres to GT1b and GD1a (Yowler and Schengrund 2004) or two distinct binding sites as seen in BoNT/C, one preferring GD1a and GT1b, the other GD1b over GT1b (Karalewitz et al. 2012). These examples highlight the fact that different serotypes may express different ganglioside specificities. In addition to gangliosides, proteins are also a part of the receptor complex needed for BoNT and possibly for TeNT binding. Proteins identified are synaptotagmin I and II for BoNT serotypes B and G (e.g. Dong et al. 2007), and synaptic vesicle glycoprotein 2 (SV2) for BoNT serotypes A (Mahrhold et al. 2006; Dong et al. 2006), D (Peng et al. 2011), E (Dong et al. 2008), and F (Rummel et al. 2009) and a 15 kDa membrane bound glycoprotein for TeNT (Herreros et al. 2000). In the case of BoNT serotype E, it has been hypothesized that the glycosyl group on Asp-573 of the SV2 is needed for appropriate binding (Kwon and Chapman 2012).

Interestingly, Synaptotagmin 1, has been found in lipid rafts isolated from rat brain synaptosomes (Gil et al. 2005). The fact that all SV2 proteins interact directly with synaptotagmin 1 (Yao et al. 2010) suggests that they too are associated with lipid rafts. Since lipid rafts are enriched in glycosphingolipids (Brown and London 2000; Simons and Toomre 2000) their presence in the same lipid rafts as the protein coreceptors would provide an optimum environment for BoNT adherence. Despite the possible colocalization of both ganglioside and protein receptors in lipid rafts, analysis of the effect of disruption of lipid rafts using methyl-ß-cyclodextrin (MßCD) on susceptibility of SNAP-25 in N2a murine neuroblastoma cells, indicated that disruption of rafts prior to addition of BoNT/A resulted in enhanced cleavage of SNAP-25 by BoNT/A relative to that seen in control cells not exposed to MßCD (Petro et al. 2006). Once internalized and released into the cytosol the toxin light chains, Zn^{2+} endoproteases, block neurotransmission by catalyzing cleavage of one of the synaptic SNARE proteins, syntaxin-1 (BoNT/C), SNAP-25 (BoNT/A,C, and E), or VAMP-1-2 (synaptobrevin-1 and -2, BoNT/B,C,D, and F; TeTx). For a comprehensive review see Bercsenyi et al. (2013).

While many people are immunized against TeNT, the same is not true for BoNT even though it is the deadliest of all known biological substances (Singh 1999). While its lethality has allowed BoNT to be thought of as a possible bioterrorism agent, its effectiveness at inhibiting transmitter release at the neuromuscular junction has resulted in its use for treating nerve, muscle, and gland hyperactivity disorders caused by altered behavior of cholinergic neurons (Peng et al. 2012). Clinical use has shown that people can be injected a number of times with a specific serotype (generally BoNT/A) before developing an immune response sufficient to affect its efficacy as a drug. This is when the availability of different serotypes is a positive as in some instances it has been possible to use a different serotype and continue treatment. For an evaluation of the different BoNTs available for clinical use see Hallett et al. (2013). As the properties of the various domains of the toxin (binding, catalytic, and translocation) have become more completely understood, investigations into their use in the development of new therapeutics are underway (Chaddock 2013). The clinical significance of BoNT has influenced thought about immunizing people against accidental exposure, as that would eliminate its pharmacological uses. See Table 18.2 for examples of pathogens that may cause neural problems. Pathogens discussed below provide examples of the various roles glycoconjugates have in their interactions with host cells.

18.2 Bacterial Infections

Borrelia burgdorferi, the spirochete causative agent of Lyme disease is transmitted by bites from the Ixodes ticks that carry it (Burgdorfer et al. 1982). The spirochetes proliferate in the guts of the ticks, become disseminated, and are then injected into humans when the infected ticks feed on their blood. While *B. burgdorferi* can infect a variety of tissues, when it infects the nervous system it is most often seen clinically as meningitis, cranial neuritis, and radiculoneuritis (inflammation of nerve roots and their accompanying peripheral nerves; Halperin et al. 2007). The ability of the bacteria to infect the CNS means that treatment requires drugs that can cross the BBB. The spirochete has been shown to adhere to extracellular matrix glycosaminoglycans (sulfated and/or carboxylated carbohydrates generally linked to a protein core, see Chap. 1) via one of its cell surface glycosaminoglycan (GAG)-binding proteins. It has been hypothesized that pathogen binding to extracellular matrix GAGs serves to concentrate it on the cell surface thereby enhancing its ability to interact with additional receptors and infect cells (Kurre et al. 1999). The recombinant *B*orrelia *GAG-binding p*rotein identified as Bgp in studies of strain N40 D10/E9 (Parveen and Leong 2000) was shown to adhere to heparin and to inhibit binding of intact *B. burgdorferi*. Another GAG-binding protein found on *B. burgdorferi* was shown to adhere to dermatan sulfate and heparin (Fischer et al. 2006) as well as laminin (Verma et al. 2009). In addition to its ability to adhere to glycosaminoglycans, earlier studies done in three different labs showed that the bacteria bound to Gal-Cer and some studies also indicated that different strains adhered to additional GSLs (Garcia-Monco et al. 1992; Backenson et al. 1995; Kaneda et al. 1997). It was also found that when virulent low-passage strains of bacteria were serially subcultured, high passage strains did not bind as well to Gal-Cer and could not infect mice (Kaneda et al. 1997). While this observation indicates the need for interaction with GSLs, the question of whether its ability to interact with GSLs is mediated by one of the GAG-binding proteins or an as yet unidentified protein has not been addressed.

Table 18.2 Examples of diseases caused by pathogen-induced utilization of cell surface carbohydrates/lectins in the nervous system[a]

Agent	Cells infected	Disease	Ligand/anchor
Toxins			
1. Botulinum neurotoxin	Neurons (PNS)	Botulism	GT1b
2. Tetanus neurotoxin	Neurons (CNS)	Tetani	GT1b
3. Shiga toxin-2	Myelin sheaths	Neuronal apoptosis	Gb3-Cer
Bacteria			
4. *Borrelia burgdorferi*	Neurons/Glia	Meningitis and radicular pain	Dermatan sulfate/heparin/ Gal-cer
5. *Listeria monocytogenes*	Meninges/neurons	Meningitis/ meningoencephalitis	Heparin/heparin sulfate
6. *Mycobacterium Leprae*	Schwann cells	Leprosy	Laminin α2 binds trisaccharide on PGL-1
7. *Staphylococcus aureus*	Microglia	Meningitis/brain abscess	Heparin/heparin sulfate Staphylococcus aureus (methicillin-resistant)
Protozoa			
8. *Plasmodium falciparum*	Microvasculature	Cerebral malaria	Heparin
Epitope mimicry			
9. *Campylobacter jejuni*	Peripheral neurons	Peripheral neuropathy	Gangliosides
10. Group A streptococci	Neurons	Sydenham chorea	GlcNAc, Gangliosides
Viruses			
11. Adenovirus	Neurons	Keratoconjunctivitis	GD1a
12. Ebola	Astrocytes	Ebola	Mannose-binding lectins
13. HIV	Microglia/neurons	AIDS	Glc-Cer containing GSLs plus CD4, CCR5/ CXCR4
14. Influenza H5 N1	Neurons/microglia	Ataxia, tremor, bradykinesia	α2-3/α2-6 sialic acid
15. JC Polyomavirus	Oligodendroglia/ astrocytes	Leukoencephalopathy	α2-6/α2-3 sialic acid
16. Rabies	Neurons	Encephalitis/paralysis	Nicotinic cholinergic receptor
17 Varicella zoster	Neurons	Latent—shingles	Heparin/mann-6-P receptors
18. West Nile fever	Neurons	Encephalitis/flaccid paralysis	DC-SIGNR
Yeast			
19. *Cryptococcus neoformans*	Endothelial cells between vessels and neuropil	Meningoencephalitis	Hyaluronic acid

[a]References for the above are 1. Bercsenyi et al. (2013); 2. Herreros et al. (2000); 3. Fujii et al. (1998), Takahashi et al. (2008); 4. Kaneda et al. (1997) and Fischer et al. (2006); 5. Disson and Lecuit (2012); 6. Ng et al. (2000) and Rambukkana et al. (2002); 7. Naesens et al. (2009); 8. Boyle et al. (2010) and Ramos et al. (2013); 9. Rees et al. (1995) and Usuki et al. (2006); 10. Kirvan et al. (2003) and Cunningham (2012); 11. Nilsson et al. (2011); 12. Denizot et al. 2012); 13. Harouse et al. (1995) and Puri et al. (1998); 14. Simon et al. (2011); 15. Komagome et al. (2002) and Neu et al. (2011); 16. Rustici et al. (1989); 17. Zhu et al. (1995) and Jacquet et al. (1998); 18. Davis et al. (2006); and 19. Chang et al. (2004) and Huang et al. (2011)

18.3 Neural Problems Induced by Mimicry Between Pathogen and Host

Infection by *Campylobacter jejuni*, a major cause of bacterial diarrhea in the U.S., is perhaps the best known example of the negative effect expression of antibodies to bacterial cell wall oligosaccharides may have on neural function. After apparent recovery from infection by *C. jejuni* a number of patients develop Guillain–Barre syndrome (GBS, Rees et al. 1995), characterized by peripheral neuropathy. In response to the question of how an enteric bacterium that causes diarrhea can affect the nervous system, the answer is that upon infection the body produces antibodies to the lipo-oligosaccharide expressed on the bacterial wall of *C. jejuni*, and in some instances those antibodies recognize gangliosides associated with neural cells. Their adherence induces the neurological symptoms associated with GBS. Injection of the lipo-oligosaccharide into rabbits was shown to induce neuropathy similar to acute motor axonal neuropathy as well as anti-ganglioside antibodies. To hasten recovery from GBS, plasmapheresis is used to dilute the antibody concentration (Meena et al. 2011).

Another bacteria whose negative effects are enhanced by this type of molecular mimicry are group A streptococci (*Streptococcus pyogenes*). With respect to carbohydrates, studies of rheumatic heart disease indicate that infection by group A streptococci induces expression of antibodies to the group A carbohydrate, *N*-acetyl-ß-D-glucosamine (GlcNAc). The antibodies produced can also recognize an epitope found on lysoganglioside GM1 (GM1 lacking a fatty acyl residue, Kirvan et al. 2003). Binding of these antibodies to neuronal cell surface GM1 results in activation of Ca^{2+}/calmodulin –dependent kinase II (CaMKII) as well as increased release of dopamine. These changes correlate with the symptoms associated with Sydenham chorea, a disorder of the CNS characterized by involuntary movements and changes in mood. Initial results have shown that iv administration of immunoglobulins ameliorated movement problems but more research is needed in this area (Cunningham 2012).

18.4 Viral Infection

Some viruses adhere to carbohydrates on the surface of their target cells while others are bound by lectins expressed on the surface of their target cells. Human immunodeficiency virus (HIV) is perhaps the most widely recognized human pathogen that utilizes carbohydrates in its binding to human target cells. HIV-induced dementia affects a number of people with HIV (20–30 %, Albright et al. 1999) and the CNS effects range from cognitive impairment to central diabetes insipidus (Banks et al. 2001). Gp120 shed from the surface of HIV virus particles is able to cross the BBB as is intact HIV-1 which can be taken up by brain endothelial cells and move across the BBB (Banks et al. 2001). Primary cells infected are brain microglial cells. Gp120 can induce neuronal cell death by apoptosis (Corasaniti et al. 2001) and induce formation of reactive oxygen species leading to neurodegeneration (Maccarrone et al. 2002). Different strains of HIV-1 have been shown to require glucocerebroside-containing GSLs in addition to CD4, CCR5, and CXCR4 in order for effective gp120/gp41-mediated fusion of the virus with its target cells to occur (Harouse et al. 1995). Initial evidence for this was provided by the observation that anti-Gal-Cer antibodies were able to inhibit HIV-1 entry into glioma and human neuroblastoma cells (Harouse et al. 1991). Subsequent studies indicated that when synthesis of Glc-Cer containing GSLs was inhibited, gp120/gp41-mediated fusion of virus to target cells was inhibited. However, addition of Gb3 to the cells resulted in recovery of fusion (Puri et al. 1998) indicating that GSLs were required. This observation was used to develop effective multivalent oligosaccharide inhibitors of HIV-1 fusion with its target cells (e.g. Rosa Borges et al. 2010).

Another example of viral adherence to target cell glycoconjugates is provided by human polyoma JC virus. It is a double-stranded DNA virus that when present in immunocompromised individuals can cause progressive multifocal leukoencephalopathy due to its infection of oligodendroglia (Stettner et al. 2009). Of its three capsid proteins (VP1-3), VP_1 was shown to have a primary role in mediating viral attachment to cells (Goldmann et al 1999). VP_1 was found to adhere to neoglycoproteins containing terminal α2-3- or α2-6-linked sialic acid moieties with contribution from an internal α2-6-linked one as well. It was also shown to bind GM3, GD2, GD3, GD1b, GT1b, and GQ1b, gangliosides found on the plasma membranes of brain cells (Bullens et al. 2003). Importance of its adherence to GT1b was confirmed when pretreatment of JC virus with GT1b was found to inhibit its ability to infect IMR-32 cells by about 80 % (Komagome et al 2002). While many are nonneuronal, JC virus is just one of a number of viruses that recognize sialic acid as part of their binding site (for a general review see Neu et al. 2011).

Interestingly, while most people think of influenza virus as infecting the respiratory system, the strain responsible for the 1918 Spanish influenza pandemic induced neurological symptoms as does the very pathogenic avian influenza virus H5 N1, currently considered a serious pandemic threat. The virus has been shown to infect CNS neurons and microglia and to induce neurodegeneration (Jang et al. 2009). After binding to its receptor on cranial nerves it is transported axonally to the brain stem. While the hemagglutinin molecules on human influenza viruses tend to prefer terminal α2-6 linked sialic acid residues as part of their binding site, the H5 of H5 N1 can adhere to both α2-6 and α2-3 linked sialosyl moieties. The latter may enable it to bind not only to glycoproteins but to gangliosides such as GD1a present on the termini of cranial nerve endings (Simon et al. 2011). The presence of terminal sialic acid residues on both glycoproteins and glycolipids on the outer surface of cell plasma membranes makes them relatively accessible pathogen-binding sites. The presence of plasma membrane-associated sialidase able to catalyze cleavage of sialyl residues on the plasma membrane of cells permits the cells to modulate sialic acid expression thereby modulating availability of those binding sites (Pshezhetsky and Hinek 2011).

Cell surface lectins that can bind mannose residue(s) present on viral cell surface oligosaccharides function as attachment sites for a number of viruses such as West Nile (Davis et al. 2006), Dengue (Fuchs et al. 2010), and Ebola (Brudner et al. 2013). All three of these viruses may affect the CNS causing problems ranging from encephalitis and flaccid paralysis to widespread cell apoptosis. In the CNS, mannose-binding lectins, able to recognize carbohydrates on Ebola virions, are found on astrocytes while dendritic cell-specific intercellular molecule-grabbing nonintegrin (DC-SIGN) lectins that bind mannosyl-containing oligosaccharides on all three viruses are found on perivascular cells. In contrast to Ebola and West Nile viruses, Dengue is also bound by the mannose macrophage receptor (for a review discussing CNS effects of emerging viruses see Denizot et al. 2012).

18.5 Dormancy in the CNS

When a virus infects the CNS and is not cleared by the body it is possible that it will remain in a dormant or latent form (either not undergoing replication or replication is minimal). Varicella zoster, a human alphaherpesvirus best known as the causative agent of chickenpox, provides an example of such behavior. It interacts with cells by first binding to heparan sulfate proteoglycan. In order for it to enter target cells it must also be bound by a mannose 6-phosphate receptor (Zhu et al. 1995). While the symptoms associated with chickenpox are generally resolved within a few weeks, the virus is not entirely cleared from the body. Instead it remains in a latent state in the cell bodies of certain neurons. It may then reappear many years later, frequently in stressed or immunocompromised individuals, in the form of the painful rash characteristic of shingles and/or as encephalomyelitis. Interestingly in immunocompetent patients meningitis was the CNS presentation seen more often than encephalomyelitis (Pahud et al. 2011). While a shingles vaccine has been developed the problem of latency upon infection of the CNS is significant and becomes more so when people become less immunocompetent, a problem that appears to become more prevalent with increasing age.

18.6 Potential Treatment/Uses

Knowledge about the roles that carbohydrates can have in the interaction of pathogens with their target cells as well as their mechanism of action has led to development of approaches to prevent their interaction with cells, and to the use of some as therapeutic agents.

As information about binding sites for different pathogens has been acquired, it has been used to develop inhibitors of their binding. The finding that "multivalent" carbohydrates tend to be bound by carbohydrate-binding proteins more tightly than monovalent ones provided an explanation for the observation that a dendrimer core

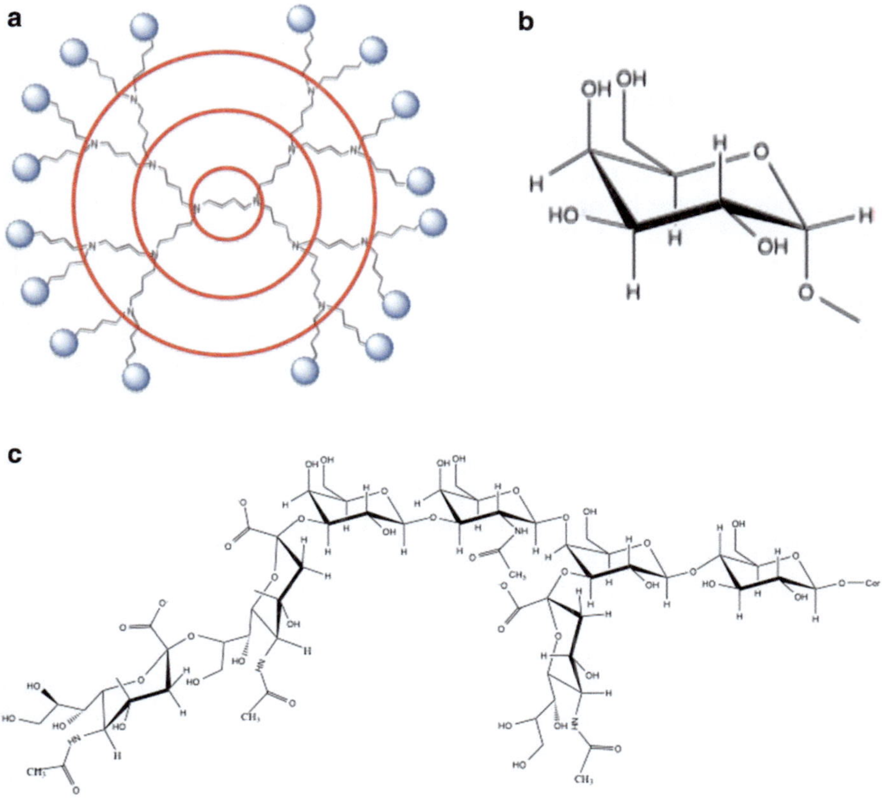

Fig. 18.2 Schematic showing a third generation PAMAM dendrimer derivatized with a residue at the ends of each of its 16 termini (**a**). Each circle indicates a generation. Carbohydrates linked to the ends of each dendrimer can vary from a simple galactose moiety (**b**) to something more complex such as the oligosaccharide portion of GQ1b (**c**)

derivatized with multiple oligosaccharide GM1 molecules was a more effective inhibitor of the binding of cholera toxin to GM1 than monovalent oligo-GM1 (Schengrund and Ringler 1989, see Fig. 18.2 for a schematic of one type of dendrimer). Crystallographic data indicating that Shiga-like toxins contained a pentameric binding subunit with three carbohydrate-binding sites per subunit arranged so that all 15 binding sites are on the same surface (Ling et al. 1998) was used in development of decavalent "starfish" dendrimers. These were comprised of a core molecule of glucose derivatized with five dimers of Gb3, one of which was linked to each glucose hydroxyl (Kitov et al. 2000). IC_{50}s obtained for Shiga-like toxins 1 and 2 binding to the dendrimers were close to 10^{-10} and 10^{-9}, respectively. Similar studies have indicated that multivalent presentations of the oligosaccharide portion of GM3 or Gb3 inhibited binding of HIV to target cells (Rosa Borges et al. 2010), and multibranched dendritic polymers derivatized with mannosyl residues inhibited binding of Ebola virus to DC-SIGN (Lasala et al. 2003).

While use of these multivalent ligands has confirmed the need for specific carbohydrates for adherence of the pathogen to its target cells, the use of these compounds to treat neural infections is problematic. For those using more complex carbohydrates as ligands, there is the need to obtain the carbohydrates in sufficient quantities and at costs that would make their use feasible. Another is targeting them to appropriate sites. To overcome some of the difficulties encountered using chemical methods to synthesize oligosaccharides, chemoenzymatic approaches are being used by a number of investigators (e.g. Champion et al. 2009; Muthana et al. 2009). In order to obtain enzymes capable of catalyzing a specific reaction investigators have used molecular approaches to engineer more active enzymes or enzymes with altered specificity. This approach was used to synthesize the neuritogenic starfish ganglioside LLG-3 (Rich and Withers 2012). To avoid the need to synthesize oligosaccharides, natural products have been evaluated for their effectiveness at inhibiting pathogen–host cell interactions. An example of this approach was the observation that cyanovirin-N, a high mannose oligosaccharide-binding protein found in blue-green algae, could adhere to sugars on the surface of HIV and Ebola thereby reducing their ability to bind and infect target cells and eliminating the need for "multivalent" oligosaccharide inhibitors (Barrientos et al. 2003).

In addition to the use of multivalent ligands to inhibit binding of pathogens bound by cell surface lectins, investigators are looking at the possibility of using recombinant lectins as possible drugs. An example of their potential effectiveness was provided by the observation that an increase in the concentration of serum recombinant human mannose binding lectin by sevenfold or more allowed mice to survive after injection with lethal concentrations of Ebola virus (Michelow et al. 2011).

Instead of dealing with the complexities of obtaining quantities of specific oligosaccharides, or preparing recombinant lectins, investigators have started looking for peptide mimetics. The following two examples indicate the range of possibilities for this approach. In one, a phage display library was used to identify peptides that could bind to a mAb (14G2a) that recognizes GD2 (Horwacik et al. 2011). GD2 is an antigen associated with most neuroblastomas as well as other cancers. For children with neuroblastoma, anti-GD2 antibodies are used for diagnosis and to follow treatment response. Effective peptide mimetics of GD2 should make it possible to use them instead of GD2 for preparation of anti-GD2 antibodies. Since GD2 is one of the gangliosides identified as a binding site for JC virus, it is possible that this peptide or others mimicking the oligosaccharide portions of other GSLs might by effective inhibitors of pathogen–cell interactions.

Studies of the antibody response to strain HS_{19} of *C. jejuni* that led to the molecular mimicry seen in Guillain–Barre syndrome indicated that the antibodies recognized the oligosaccharide portion of GD3 reflecting the fact that the terminal 3 sugar moieties of GD3 are also part of the bacterial lipo-oligosaccharide. The presence of GD3 on the Schwann cell surface and in the nodes of Ranvier of the sciatic nerve (Usuki et al. 2006) makes it an available target when the body responds by producing that particular anti-lipo-oligosaccharide antibody. Using phage display Usuki et al. (2010) identified a peptide that when injected into rats is able to restore peripheral nerve function in those previously induced to develop Guillain–Barre syndrome by injection of the GD3-like lipo-oligosaccharide. This observation supports the hypothesis that the peptide might provide an additional approach for treating people with Guillain–Barre syndrome. Identification of effective peptide mimetics provides an approach for obtaining quantities of specific mimetics at prices that may be more realistic than those currently associated with obtaining similar quantities of specific carbohydrates.

A different approach has been used to inhibit cell to cell spread of influenza virus. In order for newly synthesized virions to be released from the cell surface its sialidase (N1) catalyzes cleavage of sialic acid residues from both the newly synthesized virions and the cell surface thereby removing binding sites for the hemagglutinin (H). This releases newly formed virions from the cell surface, a necessary step if they are to move and infect more cells. While N1 associated with different strains of human influenza was found to preferentially catalyze cleavage of α2-3 sialic acid linkages, it could also catalyze cleavage of those linked α2-6 albeit somewhat more slowly (Mochalova et al. 2007). The requirement for sialidase activity led to analysis of its sialic acid-binding site, knowledge of which was then used in development of inhibitors of its action. Inhibitors of N1 currently in use, Relenza and Tamiflu, were developed by modifying sialic acid with either a positively charged guanidinium or ammonium substituent on C4. The problem with these drugs, especially Tamiflu, is that resistant strains are starting to appear. To counteract this, mechanism-based difluoro-inhibitors having the same substituents at C4 but with fluorines at C2 and 3 were developed and found effective at suppressing viral replication and prolonging survival of mice exposed to the virus intranasally (Kim et al. 2013).

Targeting drugs to the CNS requires their transport across the blood–brain barrier (BBB) in which tight junctions between cerebral endothelial cells restricts access. Research has shown that this problem can be circumvented by linking the drug to a vector that can bind to luminal surface receptors on cerebral endothelial cells and be transcytosed across the BBB. The fact that the carboxyl-terminal portion of the heavy chain of tetanus toxin (TTC) can be retrogradely transported and enter motor neurons has led to use of TTC as a carrier of proteins such as superoxide dismutase, and glial-derived neurotrophic factor and brain derived neurotrophic factor to them. This approach is seen as a possible means for delivering drugs needed for treatment of motor neuron diseases resulting in muscle paralysis (Calvo et al. 2013). In addition to the use of TTC, antibodies and peptides have also been used to target drug-carrying nanoparticles across the BBB. For example, the anti-transferrin antibody, OX26, has been found to be an effective vector for this purpose (e.g. Bao et al. 2012) as has the GM1 binding peptide, G23, although it is targeted to the lungs as well (Stojanov et al. 2012). While much of the work done with nanoparticles has been aimed at targeting cancer therapeutics, as microorganisms become more resistant to drugs currently available, and new neural pathogens evolve, it may become necessary to develop new drugs, such as those that inhibit the carbohydrate–protein interactions needed for infection, as well as approaches for targeting them to the CNS.

As our understanding of how viruses interact with cells has progressed, investigators have started to look at how viruses able to infect neural cells could be used to deliver genes to patients suffering from neurological diseases. The identification of different cell receptors for adenoviruses has allowed for development of specific adenovirus-based gene delivery systems (Cupelli and Stehle 2011). Adeno-associated viral vectors have proven useful as depending upon viral capsid type, they can (1) infect a number of different CNS cell types, (2) they induce relatively few side-affects, (3) gene transfer is efficient, and (4) transgene expression is long-lasting (for a recent review see Lentz et al. 2012). In addition to using viruses as gene delivery systems, a harmless strain of *E. coli* having a mutation that resulted in truncation of its lipopolysaccharide (LPS) was genetically manipulated to express specific glycosyl transferases able to catalyze addition of saccharides to the outer core region of the LPS to yield oligosaccharides recognized by the Shiga-like toxin receptor. When the modified bacteria were used to inhibit infection in mice by Shiga toxin *E. coli* expressing bacteria they were found to be 100 % effective (Paton et al. 2010). While the modified *E.coli* were used to treat an enteric infection, the two examples presented indicate that as more is learned about how various pathogens act, we may be able to use them to help treat/prevent disease. Interestingly this type of receptor mimicry is being considered as a possible therapeutic approach for treating agents considered as biothreats (Thomas 2010).

From the foregoing discussion, it can be seen that a number of pathogens initiate their interaction with target cells by either binding to a cell surface carbohydrate residue or by having a carbohydrate on their surface bound by a cell surface lectin. In many instances these interactions were strengthened by the fact that they are "multivalent" and that information has been used in the development of effective inhibitors of their binding. As more is learned about steps involved in the binding and release of pathogens, the use of ligands targeted to block specific steps in the infectious process should increase as should our ability to use this knowledge in the treatment of people with neural diseases.

Compliance with Ethics Requirements Cara-Lynne Schengrund declares that she has no conflict of interest and that she has used no human subjects in work cited that was done in her laboratory.

References

Albright AV, Shieh JT, Itoh T, Lee B, Pleasure D, O'Connor MJ. Microglia express CCR5, CXCR4, and CCR3, but of these, CCR5 is the principal coreceptor for human immunodeficiency virus type 1 dementia isolates. J Virol. 1999;73:205–13.

Backenson PB, Coleman JL, Benach JL. Borrelia burgdorferi shows specificity of binding to glycosphingolipids. Infect Immun. 1995;63:2811–7.

Banks WA, Freed EO, Wolf KM, Robinson SM, Franko M, Kumar VB. Transport of human immunodeficiency virus type 1 pseudoviruses across the blood-brain barrier: role of envelope proteins and adsorptive endocytosis. J Virol. 2001;75:4681–91.

Bao H, Jin X, Li L, Lv F, Liu T. OX26 modified hyperbranched polyglycerol-conjugated poly(lactic-co-glycolic acid) nanoparticles: synthesis, characterization and evaluation of its brain delivery ability. J Mater Sci Mater Med. 2012;23:1891–901.

Barrientos LG, O'Keefe BR, Bray M, Sanchez A, Gronenborn AM, Boyd MR. Cyanovirin-N binds to the viral surface glycoprotein, GP1,2 and inhibits infectivity of Ebola virus. Antiviral Res. 2003;58:47–56.

Bercsenyi K, Giribaldi F, Schiavo G. The elusive compass of Clostridial neurotoxins: deciding when and where to go? In: Rummel A, Binz T, editors. Botulinum Neurotoxins, Curr Top Microbiol Immunol. 2013;364:91–113.

Boyle MJ, Richards JS, Gilson PR, Chai W, Beeson JG. Interactions with heparin-like molecules during erythrocyte invasion by Plasmodium falciparum merozoites. Blood. 2010;115: 4559–68.

Brown DA, London E. Structure and function of sphongolipid- and cholesterol-rich membrane rafts. J Biol Chem. 2000;275:17221–4.

Brudner M, Karpel M, Lear C, Chen L, Yantosca LM, Scully C, et al. Lectin-dependent enhancement of Ebola virus infection via soluble and transmembrane C-type lectin receptors. PLoS One. 2013;8:e60838. doi:10.1371/journal.pone.0060838. Epub 2013 Apr 2.

Bullens RW, O'Hanlon GM, Wagner E, Molenaar PC, Furukawa K, Furukawa K, et al. Roles of complex gangliosides at the neuromuscular junction. Ann N Y Acad Sci. 2003;998:401–3.

Burgdorfer W, Barbour AG, Hayes SF, Benach JL, Grunwaldt E, Davis JP. Lyme disease-a tick-borne spirochetosis? Science. 1982;216:1317–9.

Calvo AC, Zaragoza P, Osta R. Gene therapy based on fragment C of tetanus toxin in ALS: a promising neuroprotective strategy for the bench to the bedside. Gene therapy—tools and potential applications. 2013; pp. 249–68. DOI: 10.5772/52896.

Chaddock J. Transforming the domain structure of botulinum neurotoxins into novel therapeutics. Curr Top Microbiol Immunol. 2013;364:287–306.

Champion E, André I, Moulis C, Boutet J, Descroix K, Morel S, et al. Design of alpha-transglucosidases of controlled specificity for programmed chemoenzymatic synthesis of antigenic oligosaccharides. J Am Chem Soc. 2009;131:7379–89.

Chang YC, Stins MF, McCaffery MJ, Miller GF, Pare DR, Dam T, et al. Cryptococcal yeast cells invade the central nervous system via transcellular penetration of the blood-brain barrier. Infect Immun. 2004;72:4985–95.

Corasaniti MT, Maccarrone M, Nistico R, Malorni W, Rotiroti D, Bagetta G. Exploitation of the HIV-1 coat glycoprotein, gp120, in neurodegenerative studies in vivo. J Neurochem. 2001;79:1–8.

Cunningham MW. Streptococcus and rheumatic fever. Curr Opin Rheumatol. 2012;24:408–16.

Cupelli K, Stehle T. Viral attachment strategies: the many faces of adenoviruses. Curr Opin Virol. 2011;1:84–91.

Davis CW, Nguyen HY, Hanna SL, Sánchez MD, Doms RW, Pierson TC. West Nile virus discriminates between DC-SIGN and DC-SIGNR for cellular attachment and infection. J Virol. 2006;80:1290–301.

Denizot M, Neal JW, Gasque P. Encephalitis due to emerging viruses: CNS innate immunity and potential therapeutic targets. J Infect. 2012;65:1–16.

Disson O, Lecuit M. Targeting of the central nervous system by Listeria monocytogenes. Virulence. 2012;3:213–21.

Dong M, Yeh F, Tepp WH, Dean C, Johnson EA, Janz R, et al. SV2 is the protein receptor for botulinum neurotoxin A. Science. 2006;312:592–6.

Dong M, Tepp WH, Liu H, Johnson EA, Chapman ER. Mechanism of botulinum B and G entry into hippocampal neurons. J Cell Biol. 2007;179:1511–22.

Dong M, Liu H, Tepp WH, Johnson EA, Janz R, Chapman ER. Glycosylated SV2A and SV2B mediate the entry of botulinum neurotoxin E into neurons. Mol Biol Cell. 2008;19:5226–37.

Dover N, Barash JR, Hill KK, Xie G, Arnon SS. Molecular characterization of a novel botulinum neurotoxin type H gene. J Infect Dis. 2014;209:192–202.

Fischer JR, LeBlanc KT, Leong JM. Fibronectin binding protein BBK32 of the Lyme disease spirochete promotes bacterial attachment to glycosaminoglycans. Infect Immun. 2006;74:435–41.

Fuchs A, Lin TY, Beasley DW, Stover CM, Schwaeble WJ, Pierson TC, et al. Direct complement restriction of flavivirus infection requires glycan recognition by mannose-binding lectin. Cell Host Microbe. 2010;8:186–95.

Fujii J, Kinoshita Y, Yamada Y, Yutsudo T, Kita T, Takeda T, et al. Neurotoxicity of intrathecal Shiga toxin 2 and protection by intrathecal injection of anti-Shiga toxin 2 antiserum in rabbits. Microb Pathog. 1998;25:139–46.

Garcia-Monco JC, Fernandez-Villar B, Rogers RC, Szczepanski A, Wheeler CM, Benach JL. Borrelia burgdorferi and other related spirochetes bind to galactocerebroside. Neurology. 1992;42:1341–8.

Gil C, Soler-Jover A, Blasi J, Aguilera J. Synaptic proteins and SNARE complexes are localized in lipid rafts from rat brain synaptosomes. Biochem Biophys Res Commun. 2005;329: 117–24.

Goldmann C, Petry H, Frye S, Ast O, Ebitsch S, Jentsch KD, et al. Molecular cloning and expression of major structural protein VP1 of the human polyomavirus JC virus: formation of virus-like particles useful for immunological and therapeutic studies. J Virol. 1999;73:4465–9.

Habermann E, Dreyer F. Clostridial neurotoxins: handling and action at the cellular and molecular level. Curr Top Microbiol Immunol. 1986;129:93–179.

Hallett M, Albanese A, Dressler D, Segal KR, Simpson DM, Truong D, et al. Evidence-based review and assessment of botulinum neurotoxin for the treatment of movement disorders. Toxicon. 2013;67:94–114.

Halperin JJ, Shapiro ED, Logigian E, Belman AL, Dotevall L, Wormser GP, et al. Practice parameter: treatment of nervous system Lyme disease (an evidence-based review): report of the Quality Standards Subcommittee of the American Academy of Neurology. Neurology. 2007; 69:91–102.

Hamano S, Nakanishi Y, Nara T, Seki T, Ohtani T, Oishi T, et al. Neurological manifestations of hemorrhagic colitis in the outbreak of Escherichia coli O157:H7 infection in Japan. Acta Paediatr. 1993;82:454–8.

Hanashima T, Miyake M, Yahiro K, Iwamaru Y, Ando A, Morinaga N, et al. Effect of Gb3 in lipid rafts in resistance to Shiga-like toxin of mutant Vero cells. Microb Pathog. 2008;45:124–33.

Harouse JM, Bhat S, Spitalnik SL, Laughlin M, Stefano K, Silberberg DH, et al. Inhibition of entry of HIV-1 in neural cell lines by antibodies against galactosyl ceramide. Science. 1991;253: 320–3.

Harouse JM, Collman RG, González-Scarano F. Human immunodeficiency virus type 1 infection of SK-N-MC cells: domains of gp120 involved in entry into a CD4-negative, galactosyl ceramide/3′ sulfo-galactosyl ceramide-positive cell line. J Virol. 1995;69:7383–90.

Herreros J, Lalli G, Montecucco C, Schiavo G. Tetanus toxin fragment C binds to a protein present in neuronal cell lines and motoneurons. J Neurochem. 2000;74:1941–50.

Horwacik I, Kurciński M, Bzowska M, Kowalczyk AK, Czaplicki D, Koliński A, et al. Analysis and optimization of interactions between peptides mimicking the GD2 ganglioside and the monoclonal antibody 14G2a. Int J Mol Med. 2011;28:47–57.

Huang SH, Long M, Wu CH, Kwon-Chung KJ, Chang YC, Chi F, Lee S, Jong A. Invasion of Cryptococcus neoformans into human brain microvascular endothelial cells is mediated through the lipid rafts-endocytic pathway via the dual specificity tyrosine phosphorylation-regulated kinase 3 (DYRK3). J Biol Chem. 2011;286:34761–9.

Jacquet A, Haumont M, Chellun D, Massaer M, Tufaro F, Bollen A, et al. The varicella zoster virus glycoprotein B (gB) plays a role in virus binding to cell surface heparan sulfate proteoglycans. Virus Res. 1998;53:197–207.

Jang H, Boltz D, Sturm-Ramirez K, Shepherd KR, Jiao Y, Webster R, et al. Highly pathogenic H5N1 influenza virus can enter the central nervous system and induce neuroinflammation and neurodegeneration. Proc Natl Acad Sci U S A. 2009;106:14063–8.

Kaneda K, Masuzawa T, Yasugami K, Suzuki T, Suzuki Y, Yanagihara Y. Glyco sphingolipid-binding protein of Borrelia burgdorferi sensu lato. Infect Immun. 1997;65:3180–5.

Karalewitz AP, Fu Z, Baldwin MR, Kim JJ, Barbieri JT. Botulinum neurotoxin serotype C associates with dual ganglioside receptors to facilitate cell entry. J Biol Chem. 2012;287:40806–16.

Kim JH, Resende R, Wennekes T, Chen HM, Bance N, Buchini S, et al. Mechanism-based covalent neuraminidase inhibitors with broad-spectrum influenza antiviral activity. Science. 2013;340:71–5.

Kirvan CA, Swedo SE, Heuser S, Cunningham MW. Mimicry and autoanti-body-mediated neuronal cell signaling in Sydenham chorea. Nat Med. 2003;9:914–20.
Kitov PI, Sadowska JM, Mulvey G, Armstrong GD, Ling H, Pannu NS, et al. Shiga-like toxins are neutralized by tailored multivalent carbohydrate ligands. Nature. 2000;403:669–72.
Komagome R, Sawa H, Suzuki T, Suzuki Y, Tanaka S, Atwood WJ, et al. Oligosaccharides as receptors for JC virus. J Virol. 2002;76:12992–3000.
Kurre P, Kiem HP, Morris J, Heyward S, Battini JL, Miller AD. Efficient transduction by an amphotropic retrovirus vector is dependent on high-level expression of the cell surface virus receptor. J Virol. 1999;73:495–500.
Kwon SE, Chapman ER. Glycosylation is dispensable for sorting of synaptotagmin 1 but is critical fro targeting of SV2 and synaptophysin to recycling synaptic vesicles. J Biol Chem. 2012; 287:35658–68.
Lasala F, Arce E, Otero JR, Rojo J, Delgado R. Mannosyl glycodendritic structure inhibits DC-SIGN-mediated Ebola virus infection in cis and in trans. Antimicrob Agents Chemother. 2003;47:3970–2.
Lee RT, Lee YC. Affinity enhancement by multivalent lectin-carbohydrate interaction. Glycoconj J. 2000;17:543–51.
Lentz TB, Gray SJ, Samulski RJ. Viral vectors for gene delivery to the central nervous system. Neurobiol Dis. 2012;48:179–88.
Ling H, Boodhoo A, Hazes B, Cummings MD, Armstrong GD, Brunton JL, et al. Structure of the shiga-like toxin I B-pentamer complexed with an analogue of its receptor Gb3. Biochemistry. 1998;37:1777–88.
Maccarrone M, Navarra M, Catani V, Corasaniti MT, Bagetta G, Finazzi-Agrò A. Cholesterol-dependent modulation of the toxicity of HIV-1 coat protein gp120 in human neuroblastoma cells. J Neurochem. 2002;82:1444–52.
Mahrhold S, Rummel A, Bigalke H, Davletov B, Binz T. The synaptic vesicle protein 2C mediates the uptake of botulinum neurotoxin A into phrenic nerves. FEBS Lett. 2006;580:2011–4.
Meena AK, Khadilkar SV, Murthy JMK. Treatment guidelines for Guillain-Barre Syndrome. Ann Indian Acad Neurol. 2011;14(1):S73–81.
Merritt EA, Zhang Z, Pickens JC, Ahn M, Hol WG, Fan E. Characterization and crystal structure of a high-affinity pentavalent receptor-binding inhibitor for cholera toxin and E coli heat-labile enterotoxin. J Am Chem Soc. 2002;124:8818–24.
Michelow IC, Lear C, Scully C, Prugar LI, Longley CB, Yantosca LM, et al. High-dose mannose-binding lectin therapy for Ebola virus infection. Infect Dis. 2011;203:175–9.
Mochalova L, Kurova V, Shtyrya Y, Korchagina E, Gambaryan A, Belyanchikov I. Oligosaccharide specificity of H1N1 virus neuraminidases. Arch Virol. 2007;152:2047–57.
Muthana S, Cao H, Chen X. Recent progress in chemical and chemoenzymatic synthesis of carbohydrates. Curr Opin Chem Biol. 2009;13:573–81.
Naesens R, Ronsyn M, Druwé P, Denis O, Ieven M, Jeurissen A. Central nervous system invasion by community-acquired meticillin-resistant Staphylococcus aureus. J Med Microbiol. 2009; 58:1247–51.
Neu U, Bauer J, Stehle T. Viruses and sialic acids: rules of engagement. Curr Opin Struct Biol. 2011;21:610–8.
Ng V, Zanazzi G, Timpl R, Talts JF, Salzer JL, Brennan PJ, Rambukkana A. Role of the cell wall phenolic glycolipid-1 in the peripheral nerve predilection of Mycobacterium leprae. Cell. 2000;103:511–24.
Nilsson EC, Storm RJ, Bauer J, Johansson SM, Lookene A, Angstrom J, et al. The GD1a glycan is a cellular receptor for adenoviruses causing epidemic keratoconjunctivitis. Nat Med. 2011; 17:105–9.
Pahud BA, Glaser CA, Dekker CL, Arvin AM, Schmid DS. Varicella zoster disease of the central nervous system: epidemiological, clinical, and laboratory features 10 years after the introduction of the varicella vaccine. J Infect Dis. 2011;203:316–23.
Parveen N, Leong JM. Identification of a candidate glycosaminoglycan-binding adhesin of the Lyme disease spirochete Borrelia burgdorferi. Mol Microbiol. 2000;35:1220–34.

Paton AW, Morona R, Paton JC. Bioengineered bugs expressing oligosaccharide receptor mimics: toxin-binding probiotics for treatment and prevention of enteric infections. Bioeng Bugs. 2010;1(3):172–7.

Peng L, Tepp WH, Johnson EA, Dong M. Botulinum neurotoxin D uses synaptic vesicle protein SV2 and gangliosides as receptors. PLoS Pathog. 2011;7(3):e1002008.

Peng CZ, Morris Jr JG, Rodriguez RL, Shukla AW, Tapia-Núñez J, Okun MS. Emerging opportunities for serotypes of botulinum neurotoxins. Toxins (Basel). 2012;4:1196–222.

Petro KA, Dyer MA, Yowler BC, Schengrund C-L. Disruption of lipid rafts enhances activity of botulinum neurotoxin serotype A. Toxicon. 2006;48:1035–45.

Pshezhetsky AV, Hinek A. Where catabolism meets signalling: neuraminidase 1 as a modulator of cell receptors. Glycoconj J. 2011;28:441–52.

Puri A, Hug P, Jernigan K, Barchi J, Kim HY, Hamilton J, et al. The neutral glycosphingolipid globotriaosylceramide promotes fusion mediated by a CD4-dependent CXCR4-utilizing HIV type 1 envelope glycoprotein. Proc Natl Acad Sci U S A. 1998;95:14435–40.

Rambukkana A, Zanazzi G, Tapinos N, Salzer JL. Contact-dependent demyelination by Mycobacterium leprae in the absence of immune cells. Science. 2002;296:927–31.

Ramos TN, Bullard DC, Darley MM, McDonald K, Crawford DF, Barnum SR. Experimental cerebral malaria develops independently of endothelial expression of intercellular adhesion molecule-1 (icam-1). J Biol Chem. 2013;288:10962–6.

Rees JH, Soudain SE, Gregson NA, Hughes RAC. *Campylobacter jejuni* infection and Guillain–Barré syndrome. N Engl J Med. 1995;333:1374–9.

Rich JR, Withers SG. A chemoenzymatic total synthesis of the neurogenic starfish ganglioside LLG-3 using an engineered and evolved synthase. Angew Chem Int Ed Engl. 2012;51: 8640–3.

Rosa Borges A, Wieczorek L, Johnson B, Benesi AJ, Brown BK, Kensinger RD, et al. Multivalent dendrimeric compounds containing carbohydrates expressed on immune cells inhibit infection by primary isolates of HIV-1. Virology. 2010;408:80–8.

Rummel A, Hafner K, Mahrhold S, Darashchonak N, Holt M, Jahn R, et al. Botulinum neurotoxins C, E and F bind gangliosides via a conserved binding site prior to stimulation-dependent uptake with botulinum neurotoxin F utilizing the three isoforms of SV2 as a second receptor. J Neurochem. 2009;110:1942–54.

Rustici M, Santucci A, Lozzi L, Petreni S, Spreafico A, Neri P. A monoclonal antibody to a synthetic fragment of rabies virus glycoprotein binds ligands of the nicotinic cholinergic receptor. J Mol Recognit. 1989;2:51–5.

Schengrund CL, Ringler NJ. Binding of Vibrio cholera toxin and the heat-labile enterotoxin of Escherichia coli to GM1, derivatives of GM1, and nonlipid oligosaccharide polyvalent ligands. J Biol Chem. 1989;264:13233–7.

Simon O, Lacour A, Delval A, Beaume A, Vermersch P. Recurrent multiple cranial nerve palsy and anti-GD1a antibodies. Muscle Nerve. 2011;43:447–8.

Simons K, Toomre D. Lipid rafts and signal transduction. Nat Rev Mol Cell Biol. 2000;1:31–9.

Singh BR. Introduction: biomedical and toxico-chemical aspects of botulinum neurotoxins. J Toxicol Toxin Rev. 1999;18:vii–x.

Stettner MR, Nance JA, Wright CA, Kinoshita Y, Kim WK, Morgello S. SMAD proteins of oligodendroglial cells regulate transcription of JC virus early and late genes coordinately with the Tat protein of human immunodeficiency virus type 1. J Gen Virol. 2009;90:2005–14.

Stojanov K, Georgieva JV, Brinkhuis RP, van Hest JC, Rutjes FP, Dierckx RA, et al. In vivo biodistribution of prion- and GM1-targeted polymersomes following intravenous administration in mice. Mol Pharm. 2012;9:1620–7.

Svennerholm L. Ganglioside designation. Adv Exp Med Biol. 1980;125:11.

Takahashi K, Funata N, Ikuta F, Sato S. Neuronal apoptosis and inflammatory responses in the central nervous system of a rabbit treated with Shiga toxin-2. J Neuroinflammation. 2008;5:11. doi:10.1186/1742-2094-5-11.

Thomas RJ. Use of receptor mimicry as a possible therapeutic treatment for biothreat agents. Bioeng Bugs. 2010;1:17–30.

Thompson JP, Schengrund CL. Oligosaccharide-derivatized dendrimers: defined multivalent inhibitors of the adherence of the cholera toxin B subunit and the heat labile enterotoxin of E. coli to GM1. Glycoconj J. 1997;14:837–45.

Usuki S, Thompson SA, Rivner MH, Taguchi K, Shibata K, Ariga T, et al. Molecular mimicry: sensitization of Lewis rats with Campylobacter jejuni lipopolysaccharides induces formation of antibody toward GD3 ganglioside. J Neurosci Res. 2006;83:274–84.

Usuki S, Taguchi K, Gu YH, Thompson SA, Yu RK. Development of a novel therapy for Lipo-oligosaccharide-induced experimental neuritis: use of peptide glycomimics. J Neurochem. 2010;113:351–62.

Verma A, Brissette CA, Bowman A, Stevenson B. Borrelia burgdorferi BmpA is a lamininbinding protein. Infect Immun. 2009;77:4940–6.

Yao J, Nowack A, Kensel-Hammes P, Gardner RG, Bajjalieh SM. Cotrafficking of SV2 and synaptotagmin at the synapse. J Neurosci. 2010;30:5569–78.

Yowler BC, Schengrund CL. Botulinum neurotoxin A changes conformation upon binding to ganglioside GT1b. Biochemistry. 2004;43:9725–31.

Zhu Z, Gershon MD, Ambron R, Gabel C, Gershon AA. Infection of cells by varicella zoster virus: inhibition of viral entry by mannose 6-phosphate and heparin. Proc Natl Acad Sci U S A. 1995;92:3546–50.

Chapter 19
Glycoconjugate Changes in Aging and Age-Related Diseases

Susumu Ando

Abstract The significance of glycosphingolipids and glycoproteins is discussed in their relation to normal aging and pathological aging, aging with diseases. Healthy myelin that looks stable is found to be gradually degraded and reconstructed throughout life for remodeling. An exciting finding is that myelin P0 protein is located in neurons and glycosylated in aging brains. In pathological aging, the roles of glycosphingolipids and glycoproteins as risk factors or protective agents for Alzheimer's and Parkinson's diseases are discussed. Intensive studies have been performed aiming to remove the risks from and to restore the functional deficits of the brain. Some of them are expected to be translated to therapeutic means.

Keywords Aging • Alzheimer's disease • Parkinson's disease • Ganglioside • Glycoprotein • Cholinergic • Remyelination • Sialylcholesterol • Warfarin

Abbreviations

APP	Amyloid precursor protein
Aβ	Amyloid beta
BBB	Blood–brain barrier
BDNF	Brain-derived neurotrophic factor
CBA	Glucocerebrosidase
CNS	Central nervous system
GnT-III	*N*-acetylglucosaminyltransferase III
LCB	Long-chain base

S. Ando (✉)
Tokyo Metropolitan Institute of Gerontology, Tokyo 173-0015, Tokyo, Japan

Soka Royal Care Center, Saitama 340-0001, Japan
e-mail: andos1008@nifty.com

R.K. Yu and C.-L. Schengrund (eds.), *Glycobiology of the Nervous System*, Advances in Neurobiology 9, DOI 10.1007/978-1-4939-1154-7_19,

MAP5	Microtubule-associated protein 5
NCAM	Neural cell adhesion molecule
NFTs	Neurofibrillary tangles
OPC	Oligodendrocyte precursor cell
PD	Parkinson's disease
P-gp	P-glycoprotein
PSA	Polysialic acid
SAP	Serum amyloid P

19.1 Preface

Aging is defined as the detrimental process of a body composed of somatic cells after cessation of reproductive activity. The body as a short-term vehicle for passing on the genes to the next generation is evolutionally disposable as based on the disposable soma theory of aging (Kirkwood 1977). The theory means that the body must take natural courses to death. Individuals, however, do not always fulfill their maximum life spans, but frequently suffer from various diseases injurious to their health. In this sense, aging is classified into two types, aging without disease, physiological aging and aging with disease, pathological aging. In this article, glycoconjugate changes in the brain that occur during physiological aging are described first, followed by discussion of their possible participation in the pathophysiology and pathogenesis of age-related neurological diseases. Kobata (2011) introduced the word "glycogerontology" for the field covering the roles of glycoconjugates in relation to gerontology and geriatrics. The physiology and pathology of glycosphingolipids (GSLs) and glycoproteins, will be reviewed here with special reference to their roles in glycogerontology.

19.2 Glycoconjugates in Aging Brains

19.2.1 Age-Related Changes of Glycosphingolipids in the Brain

GSLs are composed of neutral and acidic glycolipids, and the latter are divided into two groups, sulfatides and sialic acid-containing GSLs or gangliosides. The sialyloligosaccharide structures of gangliosides show great diversity as a result of the combination of neutral oligosaccharides of various chain lengths and different numbers of sialic acids attached to the core neutral sugars (Ando 1983) as shown in Fig. 19.1. Because of the complexity in their carbohydrate structure, it is hypothesized that gangliosides play crucial roles in cellular events (Cantù et al. 2011; Ohmi et al. 2012; Yu et al. 2012), and as a result they have attracted the most attention of

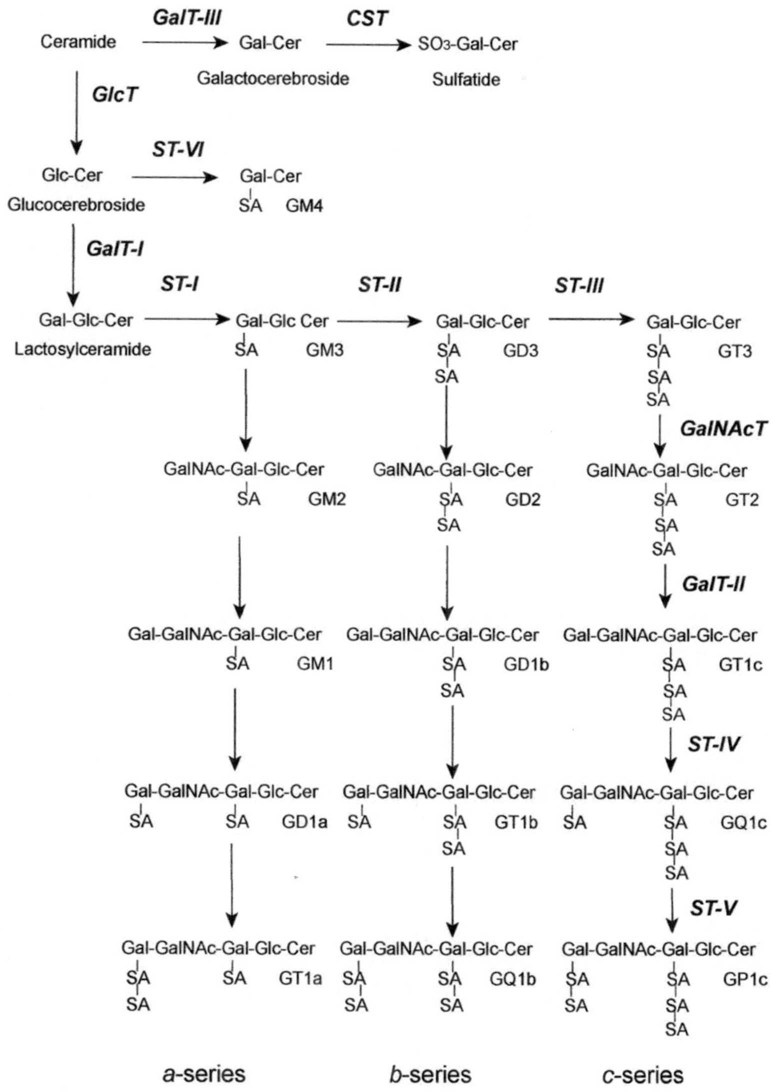

Fig. 19.1 Pathways for formation of sulfatide and gangliosides. Transferases are typed in *bold-italic*. *Cer* ceramide, *CST* galactosylceramide sulfotransferase, *Gal* galactose, *GalNAc* *N*-acetylgalactosamine, *Glc* glucose, *GalT-I* galactosyltransferase I, *GalT-II* galactosyltransferase II, *GalT-III* galactosyltransferase III, *GlcT* glucosyltransferase, *SA* sialic acid. ST-I, II, III, IV, V, and VI correspond to sialyltransferases I (GM3 synthase), II (GD3 synthase), III, IV, V, and VI, respectively

any of the GSLs from researchers in aging research. The supposition is that alterations in the concentration and composition of gangliosides in aging neural membranes should be correlated with age-related functional changes in the brain.

Age-related compositional changes of gangliosides in the human brain were first reported by (Suzuki 1965). In the frontal cortices, GD1a, which is predominant in new-born brains, decreases with age. GM1 also tends to decrease. On the other hand, GD1b and GT1b both increase and by age 30 reach adult levels which normally remain essentially constant until a person is near 90 years of age. These trends in ganglioside pattern changes were supported by the data reported by Segler-Stahl et al. (1983). Similarly, gangliosides GD1b, GT1b, and GQ1b in *b*-series gangliosides (Fig. 19.1) were shown to increase with age, while GM1 and GD1a decrease (Svennerholm et al. 1989). As GM1 is the major component of myelin (Ledeen and Yu 1982), *b*-series gangliosides in neural components other than myelin may continue to increase to reduce the relative contents of GM1 in the brain after the period of active myelination. Alternatively, the decrease of GM1 in advanced age may indicate some disintegration of myelin structure. Disruption of the myelin sheath has been reported to occur in aged brain (Peters et al. 2001). The myelin components were shown to never be stable but to be dynamically turned over in physiological aging (Ando et al. 2003). Probably as the consequence of the remodeling of the composition of myelin, cognitive declines with aging were hypothesized to correlate with the altered white matter tracts (Aine et al. 2011; Schulze et al. 2011).

19.2.2 Age-Related Changes of Glycosphingolipids in Synapses

The brain is composed of neurons, glias, and blood vessels. The ganglioside distribution patterns were shown biochemically to be distinct for neural cell types as well as their subcellular fractions (Ando 1983). Immunohistochemical identification of the regional distribution of major gangliosides in the rat brain (Kotani et al. 1993), indicated that a specific ganglioside or set of gangliosides might be responsible for particular neuronal functions. To elucidate their biological roles, it will be important to obtain information on the ganglioside contents and composition of particular cells and subcellular components. In respect to neurons, nerve endings or synapses are functional elements for neurotransmission, and gangliosides in synaptic membranes are thought to affect efficiency of acetylcholine release (Ando 2012; Ando et al. 2004; Tanaka et al. 1997) and long-term potentiation at synapses (Furuse et al. 1998; Wieraszko and Seifert 1986). Waki et al. (1994) developed excellent methods for quantitative isolation of gangliosides from membrane preparations and for their accurate quantitation using gas chromatography–mass spectrometry. Using this method the ganglioside content and composition of mouse brain synaptic plasma membranes was shown to remain constant from adult to senescence. This result seems to be reasonable. Functional units such as synaptic plasma membranes may retain the structure required for their function until senescence, even though the density of the units decreases with age resulting in the changes in ganglioside content and composition observed in tissues or whole brains as described in Sect. 19.2.1.

During the juvenile period of brain development, the content of GD1a was shown to increase and then decrease in the human frontal cortex (Suzuki 1965). GD1a found in cerebral microsomal fractions may be considered a marker for dendritic arborization (Yusuf and Dickerson 1978). Synaptic membrane preparations from mouse brains showed high concentrations of GD1a after birth and reduced levels of GD1a at 6 months of age (Waki et al. 1994). The transient increase in GD1a seemed to coincide with the temporary increase in synaptic density in infant brains (Huttenlocher 1979), and its decrease might correspond to synapse elimination during brain development. GD1a levels then remained constant throughout adult life. These observations indicate that GD1a expression may be related to arborization and synaptogenesis that occur in the initial formation of the neuronal network.

19.2.3 Age-Related Changes in Myelin GSLs

Many studies have examined the effects of age on various brain regions. Most of them, however, were focused on specific areas and the data seemed to be too fragmented to make valid comparisons. Upon comparing results obtained for the composition of 16 automatically segmented measures of the human brain Walhovd et al. (2005) found distinct age changes in different brain structures. Two representative age responses observed were that the volume of white matter showed a curvilinear relationship with age, while the volume of gray matter was reduced in a linear fashion. In the case of cerebral white matter, the volume increased early in life, and decreased during senescence following the steady-state adult period. The main components of white matter are axons and myelin sheaths. While no changes were found in the diameters of axons with age, the numbers of myelin lamellae increased in aged monkeys (Peters et al. 2001). These morphometric observations (Peters et al. 2001; Walhovd et al. 2005) indicate that some changes may continuously take place in myelin components across one's life span.

Age-related changes in the glycosphingolipids composition of white matter were determined for the human brain (Svennerholm et al. 1994). The three major GSLs, cerebrosides, sulfatides, and gangliosides were found to decrease (μmole per gram tissue weight) with advancing age starting at ~20 years. The same group published somewhat different age-related changes for these glycolipids in separate papers. Cerebroside content in white matter remained constant till 90 years and then sharply declined (Svennerholm et al. 1991). Another paper reported that GM1 in white matter increase in the early developmental age until 1 year and remained at a constant level in the adult (Vanier et al. 1971). These discrepancies might come from their research designs in which glycoconjugates were quantified based on fresh tissue weights. Cellular composition and even tissue water content may change with aging. To avoid such effects analyses should be performed on distinct components of the brain and GSL content related to dry weight.

Norton and Poduslo (1973) isolated myelin fractions from rat brains and found age-related changes in lipid composition along with myelination. The rate of accumulation of cerebroside in the brain paralleled that of myelin. The content of cerebroside in myelin increased during the active myelinogenesis period till 30 days of age, and then remained at constant until rats were 425 days. On the other hand, the content of sulfatide in myelin increased more slowly than that of cerebroside even during active myelinogenesis, and continued to increase gradually during adult life. Sulfatide is produced from cerebroside by the action of galactosylceramide sulfotransferase (Fig. 19.1), and the enzyme is thought to remain active in the remodeling of mature myelin. The biological function of sulfatide was revealed using a mouse model incapable of synthesizing it (Ishibashi et al. 2002). The sulfatide-deficient mice were normal at birth, developed neurological deficits after 6 weeks of life, and survived to >1 year of age. Immunohistochemical studies indicated that sulfatide was not necessary for initial cluster formation, but was essential for proper localization of axonal proteins such as Na^+ and K^+ channels as well as the maintenance of these proteins around nodes.

Yu and Iqbal (1979) found that gangliosides GM1 and GM4 were concentrated in the myelin fraction isolated from the human brain. Since myelin constitutes the bulk of the oligodendrocyte plasma membranes, GM1 and GM4 may be synthesized in the cell bodies and incorporated into myelin. This leads to the question of why the ganglioside composition of myelin is quite dissimilar to that of parent oligodendrocytes whose ganglioside pattern is complex and rather similar to that of neurons. Saito and Yu (1992) performed an in vitro experiment in which incubation of the myelin fraction obtained from young rats with sialidase was found to produce a ganglioside pattern similar to that observed in in vivo maturation of myelin. They suggested that myelin-associated sialidase may play a role in the processing of gangliosides in myelin membranes. This is an interesting hypothesis in light of another suggesting a role for the sialidase-GM1 interaction in stabilization of the multilamellar structure of myelin sheaths (Saito and Yu 1993). The activities of membrane-bound sialidase were determined for synaptic plasma membranes and nuclear membranes, and their age-related changes examined. The enzyme activities decreased by one-sixth in synaptic plasma membranes (Saito et al. 1995) and by one-third in nuclear membranes (Saito et al. 2002) in old animals.

In my laboratory, the lipid composition of myelin fractions isolated from mouse brains at different ages was examined (Ando 1985). Age-related changes of cerebroside and sulfatide were a little different from those reported for rats (Norton and Poduslo 1973). Our data showed that the contents of cerebroside and sulfatide as expressed per dry weight of myelin both remained at near constant levels from weaning to old age (20–820 days of age), and increased, almost suddenly, at the oldest old age of 1,055 days. With respect to gangliosides in mouse myelin, the content continued to increase from young to old age, and suddenly decreased at the last stage of life. The changes seen in myelin GSLs at the extreme old age may indicate occurrence of an abrupt metabolic disruption in myelin. Similar age-changes in myelin glycolipids at the final stage of life were also reported for human brains (Walhovd et al. 2005; Svennerholm et al. 1991).

To understand the age-related changes in myelin composition metabolic turnover rates for myelin components were measured (Ando et al. 2003). Myelin has a tightly compacted multi-membrane structure containing the least volume of cytosole among membranes, and was not expected to be actively metabolized, so short-term labeling experiments seemed inappropriate. In long-term monitoring studies, depending upon precursors used, the turnover rates of radioisotope-labeled myelin components were reported to vary 6–167 days as half-lives for phosphatidylcholine (Sun and Sun 1979). These big discrepancies may reflect the fact that different tracers recycle at different rates. Our in vivo deuterium-labeling method was shown to eliminate the contribution of the recycling or reuse of labels so that more reliable turnover rates of myelin components could be determined (Ando et al. 2003). Turnover rates of myelin components as half-lives were calculated from decay curves of initially labeled molecules. In fact, very long half-lives of up to 1 year could be measured. Individual components of myelin in the mouse brain were found to be metabolized at separate rates, and their turnover rates were affected differently by aging. Turnover rates of GSLs were calculated from the incorporation rates or disappearance rates of their labeled neutral sugar moieties. Cerebroside and GM1 appeared to be rapidly incorporated into myelin in infant brains as were cholesterol and phospholipids, and their turnover rates varied with aging, decreasing during young and adult periods and rebounding in the senescence stage. The accelerated metabolism of myelin in old age may explain the curious compositional changes observed at the last stage of a mouse's life (Ando 1985) as well as in human brains (Walhovd et al. 2005; Svennerholm et al. 1991). The metabolism of GM1 appeared to be composed of two compartments in myelin, one with a short half-life of 52 days and another with a long half-life of 131 days, which correspond to rapidly and slowly exchanging pools, respectively (Ando et al. 2003). The concept that metabolic turnover of membrane components occurs continuously in mature myelin and the finding that turnover rates change with age may provide a better understanding of the mechanism underlying myelin aging.

19.2.4 Alterations in the Lipid Portions of Glycosphingolipids with Aging

GSLs are inserted into the outer layer of neural membranes through their ceramide portions. Ceramide is composed of a fatty acid linked to sphingosine via an amide linkage. It is known that the ceramide portion plays a number of biological roles. It may control the interaction of the saccharide portion with external ligands (Kannagi et al. 1982), and regulate the aggregative properties and surface dynamics of the GSLs (Yohe et al. 1976). In a model system using synaptosomes and liposomes increasing concentrations of gangliosides in the membranes were shown to increase membrane fluidity or decrease membrane microviscosity (Ando et al. 1986). Gangliosides may affect membrane physicochemical characteristics such as fluidity through the homophilic interaction of the sialylsaccharides and the heterophilic interaction of the saccharides with other ligands, and further through the lipidic interaction of their ceramide portions with other lipids in membranes. Recently, Furukawa's group (Ohmi et al. 2012) showed, using mice expressing mutant ganglioside synthases, that glycolipid-enriched microdomain/rafts architecture was destroyed by ganglioside deficiency. Others generated mutant mice defective in their ability to synthesize ceramide containing very long-chain fatty acids (C22-C24) due to ablation of ceramide synthase 2 (Ben-David et al. 2011). The mice had reduced levels of both nonhydroxy-C22-C24- and 2-hydroxy-C22-C24-galactosylceramide, and developed brain lesions. Myelin degeneration and detachment occurred in the brain. The mice also exhibited abnormal motor behavior and histological abnormalities such as vacuolization and astrogliosis.

Age-related changes in the ceramide portion of GSLs were studied in order to learn more about the probable physiological roles of ceramides. Changes in fatty acid composition were studied using samples from human brains (Mansson et al. 1978; Svennerholm and Ställberg-Stenhagen 1968). The adult fatty acid composition of cerebrosides and sulfatides in regard to degree of unsaturation and total percentage of C22-C26 acids was reached at 2 years of age, but the percentage of odd-numbered fatty acids continued to increase up to about 10–15 years. Fatty acid changes occurred in cerebrosides first and in sulfatides later. The delay in sulfatides may reflect the fact that cerebrosides are precursors in the synthesis of sulfatides (Norton and Poduslo 1973). Stearic acid was reported to be the major component in the ceramide portions of gangliosides in human brains (Mansson et al. 1978). The proportion of stearic acid was found to decrease from about 94 to 86 % in both GM1 from white matter and GT1 from the cerebral cortex upon aging. These data were obtained from studies of three brains, one from a 77 year old (Svennerholm and Ställberg-Stenhagen 1968), one from a 71 year old and one from an 89 year old (Mansson et al. 1978).

Long-chain bases (LCBs) or sphingosines in the ceramide portion of GSLs in the brain are composed of two major species, d18:1 and d20:1. The compositional changes in LCB with aging were first observed in human brains (Mansson et al. 1978). The averaged values for the molar proportion of d20:1 were shown to increase from about 20 to 70 % with advancing age. Age-related changes in the LCBs of gangliosides were studied in detail using rat forebrains (Palestini et al. 1990). The d18:1 LCB, predominant at 3 days (91–96 %), diminished with age and at 2 years was 73, 65, 61, 59, and 45 % of the total for GD1a, GM1, GT1b, GD1b, and GQ1b, respectively. The content of d20:1 LCB, low at birth (4–9 %), increased with age in all gangliosides and at 2 years 27–55 % of the total. Molecular species of all gangliosides carrying d18:1 LCB were virtually devoid of C20 fatty acid. Analysis of the ceramide portions of gangliosides isolated from synaptosomes and myelin fractions of rat brains of different ages indicated that the fatty acid composition did not undergo appreciable changes (Palestini et al. 1993). Large age-related changes in LCB composition were observed in all gangliosides in both synaptosomal and myelin fractions. The steady increase in the proportion of d20:1 LCB observed in the two subcellular fractions appeared to coincide with the age-related changes reported for gangliosides isolated from rat whole brains.

Sugiura et al. (2008) provided a good answer for the question of whether the different molecular species would show different distribution patterns in the brain. Using imaging mass spectrometry, they found that gangliosides containing d18:1 or d20:1 LCB were differentially distributed in mouse brain. While the d18:1-species was widely distributed throughout the frontal brain, the d20:1-species selectively localized along the entorhinal-hippocampal projections, especially in the molecular layer of the dentate gyrus. The finding of developmental- and age-related accumulation of the d20:1 species in the hippocampal formation provided evidence that changes in the ganglioside molecular species may contribute to the process of brain aging.

19.2.5 Age-Related Changes in the Carbohydrate Structure of Glycoproteins

The glycan moieties of glycoproteins are known to play crucial roles not only in modulating the property of the stem glycoproteins, but also in regulating various molecular recognition processes (Kobata 1992). In the developing brain, nervous system glycans have been implicated as important mediators of adhesive interactions among neural cells (Schachner and Martini 1995). Age-related changes in glycoproteins in the central nervous system (CNS) have been infrequently documented in the literature, but evidence for them has been summarized in review articles (Kobata 2011; Sato and Endo 2010). The myelin glycoprotein P0 is one that has been studied.

Originally using a goat anti-P0 antibody and immunofluorescence, the myelin P0 glycoprotein was found to be located exclusively in the myelin of peripheral nerves, but not in CNS myelin (Ishaque et al. 1980). Subsequently, Endo's group (Sato et al. 1999) was able to show that P0 was present in the spinal cord of the rat. They further demonstrated by immunohistochemical and immunocytochemical analyses that CNS neuronal cells expressed P0 (Sato and Endo 2000). To explore whether any age-related changes occurred in the glycans of CNS P0, glycoproteins obtained from the brain or spinal cord of 9-week old and 29-month old rats were separated by electrophoresis and stained with *Lens culinaris* agglutinin (Sato et al. 1999). This lectin specifically binds to the P0 glycan. While the glycoprotein patterns of spinal cords showed marked differences between the two age groups, samples from brains did not. Nonglycosylated P0 molecules present in the young spinal cord were replaced with glycosylated ones during aging. As it was reported that the glycan moiety of P0 plays an important role in cell–cell adhesion (Yazaki et al. 1992), the appearance of glycosylated P0 may function in the remodeling of neural structures that occurs with aging. A study on age-related changes in the glycan structure of peripheral nerve myelin glycoprotein P0 indicated that P0 from adult rats contained high-mannose and/or hybrid-type oligosaccharides not seen in P0 from 5-day-old nerves (Brunden 1992).

Following the finding of expression of P0 in the mammalian CNS, its cellular localization in rat spinal cord was examined by immunohistochemical and immunocytochemical methods using a polyclonal anti-P0 antibody (Sato and Endo 2000). Nissl staining-positive motor neurons and sensory neurons showed strong reactivity with the antibody. The neuronal localization of P0 was confirmed by double immunofluorescence labeling using the anti-P0 antibody and an anti-neurofilament monoclonal antibody. Further analysis indicated that neurons expressing P0 in the spinal cord of 30 month-old rats had a different morphology than those of 14 week-old rats. The number of neurons stained in the old rats was about 80 % of that in the young rats, while the average size of neurons in the old rats was about 62 % of that in young rats. These observations indicate that both cell number and average size of neurons in the spinal cord decreased with age supporting the hypothesis that the activity and survival of the neurons might be regulated or changed by age-related changes in glycosylation of P0.

Endo's lab (Sato et al. 2006), using 2D-electrophresis using Concanavalin A staining, surveyed the contents of cytosolic glycoproteins in rat cerebral cortices and found several spots increased in the aged brains. The glycoprotein that was most prominently increased was identified as cathepsin D. It was detected in the cytosolic fractions of aged rats, but not in those of young adult rats. Cathepsin D in the microsomal fractions did not show age-related changes. The increase of cytosolic cathepsin D during aging was not due to disruption of the lysosomal membrane, because other lysosomal enzymes did not increase in the cytosolic fractions. The level of cathepsin D transcripts in aged rats was 1.6 times higher than in the young adults. Cathepsin D is known to digest neurofilaments and tau (Bednarski and Lynch 1996) which are components of the cytoskeleton of neuronal cells. The enhanced expression of cathepsin D may facilitate cytoskeletal degradation leading to morphological changes and functional loss of neurons in aged brains.

Sialic acid is present in both glycoproteins and gangliosides. It attaches to the nonreducing terminals of sugar chains by α2-3, α2-6, or α2-8 ketosidic linkage. A homopolymer of α2-8-linked sialic acid moieties (polysialic acid, PSA) is found on neural cell adhesion molecules (NCAM), and is known to modulate the adhesion property of NCAM and to regulate neurite outgrowth and cell migration (Brusés and Rutishauser 1998). PSA-NCAM (embryonic form of NCAM) was shown to be present in adult brain regions where neuronal regeneration occurred but its expression decreased during aging (Seki and Arai 1991). PSA-NCAM expression is lost after the cessation of neuronal cell migration and synapse formation, but can be retained by axons capable of synaptic remodeling.

The histochemical distribution of sialyl α2-3 galactose and sialyl α2-6 galactose was examined in the rat hippocampus using *Maackia amurensis* lectin to label the former, and *Sambucus sieboldiana* for the latter, and electron microscopy to visualize them (Sato et al. 2001). Both lectins stained the plasma membranes of pyramidal cells and synapses. The staining intensity by both the lectins of synapses was reduced in 30-month-old rats. In addition to staining glycoproteins, the *Maackia amurensis* lectin could also bind all of the series of gangliosides possessing sialyl α2-3 galactose linkages. The *Sambucus sieboldiana* lectin could bind the α-series of gangliosides possessing sialyl α2-6 galactose. Expression of sialyl α2-6 galactose decreased in the aged brain.

19.3 Glycoconjugates and Age-Related Diseases

19.3.1 Gangliosides and the Pathology of Alzheimer's Disease

An interesting question addressed by Mizutani's group was whether the histopathological changes that occur in Alzheimer's brains were the results of accelerated aging (Mizutani and Kasahara 1997; Yamada et al. 1998). In advancing physiological or normal aging, morphometric measurements of brain volumes in comparison with intracranial volumes showed a very slow progression of brain atrophy and an insignificant correlation between the rate of atrophy and age. In contrast, similar measurements with Alzheimer's brains revealed intensive atrophy occurred (Yamada et al. 1998). The hippocampal atrophy observed in Alzheimer's brains was distinct from that seen in control, nondemented brains (Mizutani and Kasahara 1997). In the hippocampal formation of control brains, no atrophy was observed in the hippocampus and parahippocampus. In fact, in all cases with Alzheimer's disease the stratum lacunosum-radiatum was decreased in thickness, and this was accompanied by loss of myelin and fibrillary gliosis. Interestingly, the numbers of senile plaques and neurofibrillary tangles (NFTs) and the degrees of neuronal loss in the pyramidal layer varied from case to case. The severer degeneration in the parahippocampus including the entorhinal cortex than in the hippocampus indicates that degeneration of the entorhinal cortex may induce degeneration of the hippocampus because the perforant pathway in the hippocampus originates from the entorhinal cortex. Entorhinal cortex dysfunction was detected in early Alzheimer's disease by positron emission tomography (Eustache et al. 2001). In regard to cognitive dysfunction, the concept of "synaptic pathology," in which synaptic loss is the major correlate of cognitive deficits, was proposed by Terry et al. (1991; Masliah and Terry 1993), and supported by others (Blennow et al. 1996; Heinonen et al. 1995; Scheff et al. 1993). Based on the observations made using Alzheimer's brains, a dementia animal model was successfully generated by inducing synapse-specific lesions in the entorhinal cortex (Ando et al. 2002). This supported the synaptic pathology hypothesis.

Alzheimer's disease is an age-related disorder, but is not the result of advanced normal aging. The progressive cognitive dysfunction seen in Alzheimer's disease is characterized by the accumulation of senile plaques and NFTs, and the degeneration of neurons in brain regions such as the hippocampal formation. A recent review of the literature on the correlation of Alzheimer's neuropathologic changes with cognitive status stresses that the severity of cognitive impairment correlates best with the burden of neocortical NFTs (Nelson et al. 2012). NFTs are related to neuronal loss. In contrast, from the standpoint of the pathogenesis of Alzheimer's disease, generation of amyloid β (Aβ)-peptides is thought to be at the beginning of a cascade that leads to the disease. Recent studies indicate that assembly of Aβ-peptides into Aβ-oligomers or protofibrils, can cause cognitive declines by disrupting synaptic function (Dahlgren et al. 2002; Matsumura et al. 2011; O'Nuallian et al. 2010). Wild type Aβ monomers assemble first into protofibrils and then amyloid fibrils to form senile plaques. Aβ-peptides deposit in diffuse plaques, the earliest stage of senile plaques. Yanagisawa et al. (1995) isolated the diffuse plaque fraction as a "light Aβ" from Alzheimer's disease and Down's syndrome brains and found that Aβ associated with diffuse plaques bound to ganglioside GM1 in a noncovalent fashion. Yanagisawa's group hypothesized that GM1-bound Aβ acted as a seed in the initiation of amyloid fibril formation (Kakio et al. 2001). Another in vitro experiment showed that Aβ selectively bound to membranes containing GM1 and that no Aβ binding was observed with GM1-free membranes (Choo-Smith et al. 1997). Studies of brains from monkeys indicated that GM1-Aβ was formed in early endosomes and transported to late endosomes for degradation (Kimura and Yanagisawa 2007). It is thought that impaired recycling in the endosome pathway results in accumulation of GM1-Aβ in endosomes and that its subsequent transport to the cell surface where it acts as a seed for amyloid fibril formation. Furthermore, it was shown that a toxic soluble Aβ assembly was formed in a GM1 dependent manner through incubation of soluble Aβ with neuronal membranes prepared from aged mouse brains (Yamamoto et al. 2007). These observations may help to define the mechanism underlying the plaque-independent neuronal death seen in Alzheimer's disease.

19.3.2 Expression of Unusual Gangliosides in the Alzheimer's Brain Indicative of Brain Plasticity

Expression of *c*-series gangliosides in Alzheimer's brains was reported by our laboratory (Takahashi et al. 1991). Since the *c*-series of gangliosides with trisialyl residues were structurally identified (Ando and Yu 1979), the metabolic pathway for their synthesis from GT3 (Fig. 19.1) was established (Yu et al. 2004). The *c*-series gangliosides were found to be expressed in embryonic brains and designated as fetal antigens (Hirabayashi et al. 1988). Anti-*c*-series ganglioside antibodies were found to label neuritic elements composing senile plaques as well as perivascular amyloid deposits (Takahashi et al. 1991). Positive staining with the antibodies was not observed in brains of nondemented individuals, except for those containing a small number of senile plaques. NFTs were immunolabeled with the monoclonal antibody A2B5, which had been shown to recognize the *c*-series ganglioside, GQ1c (Emory et al. 1987). Subsequently A2B5 was shown to react with gangliosides other than those in the *c*-series, and with sulfatides (Majocha et al. 1989) and even with glycoproteins having α2,8-trisialic acid units (Inoko et al. 2010). In this context, our specific anti-*c*-series ganglioside antibodies definitely demonstrated the presence of the fetal gangliosides in Alzheimer's brains. This finding was strengthened by the evidence that another fetal antigen, microtubule-associated protein 5 (MAP5), colocalized with *c*-series gangliosides in NFTs (Takahashi et al. 1991). The expression of fetal antigens such as *c*-series gangliosides and MAP5 may indicate that neuronal regeneration (cell proliferation and sprouting) occurs along with neuronal degeneration in the Alzheimer's brain. Evidence for the regeneration in the Alzheimer's brain was also provided by the observation of increased dendritic

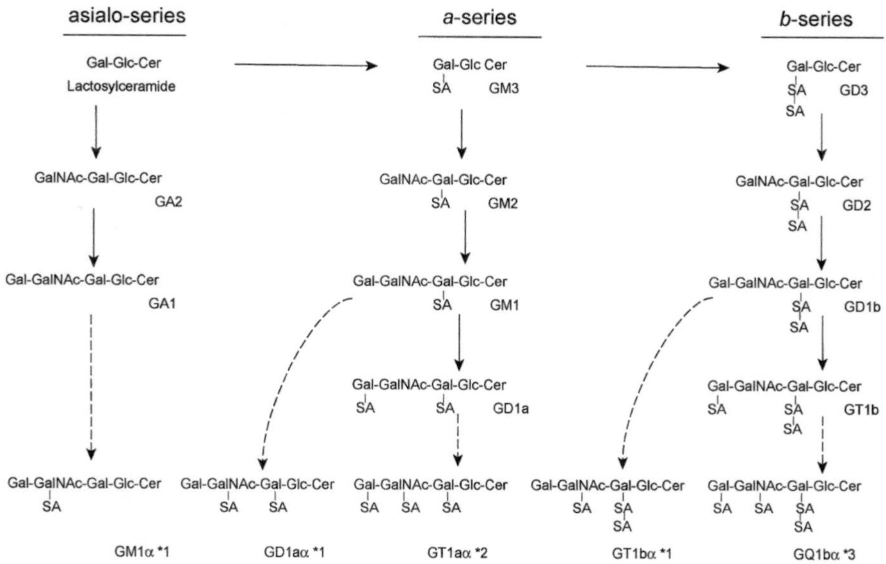

Fig. 19.2 Synthetic pathways for α-series gangliosides (Chol-1α antigens). Dashed lines indicate sialylation reactions catalyzed by sialyltransferase VII that catalyzes formation of sialyl 2-6 *N*-acetylgalactosamine linkages. Alpha-series of gangliosides marked by *1, *2, and *3 were characterized as Chol-1α by Irie et al. (1996), Ando et al. (1992) and Hirabayashi et al. (1992), respectively

sprouting and arborization (Probst et al. 1983; Scheibel and Tomiyasu 1978). These findings are consistent with the concept that neuronal regeneration does occur while degeneration proceeds (Lopez-Toledano and Shelanski 2004; Uchida 2010). As the brain has innate plasticity, it is possible that development of synapses achieved by various interventions may help protect the brain from degenerative insults and to recover from their damage (Ando 2012).

19.3.3 Functional Significance of α-Series Gangliosides in Alzheimer's Disease

Whittaker and his colleagues (Richardson et al. 1982) found that cholinergic-specific antigens in the Chol-1 family, comprised of Chol-1α, Chol-1β, and Chol-1γ, were gangliosides. We isolated and characterized two molecular species corresponding to Chol-1α, termed GT1aα and GQ1bα as illustrated in Fig. 19.2 (Ando et al. 1992; Hirabayashi et al. 1992). The "α" in the abbreviations is assigned to the unique branching structure of a sialic acid residue linked α2-6 to the *N*-acetylgalactosamine in the gangliotetraose backbone (Taki et al. 1986). Immunostaining indicated that Chol-1α was present in cholinergic nuclei such as the septal nucleus (Irie et al. 1994). To examine the involvement of gangliosides in the function of cholinergic synapses, a series of monoclonal antibodies against gangliosides were tested for their ability to suppress release of acetylcholine. Only the anti-Chol-1α antibody, GGR-41, affected acetylcholine release (Ando et al. 2004). To ascertain its mechanism of action, choline uptake and acetylcholine synthesis were measured using synaptosomes in the presence of GGR-41. Both choline uptake and acetylcholine synthesis were inhibited in a dose-dependent manner by GGR-41. When Chol-1α was added to a synaptosomal fraction, it accelerated high affinity choline uptake into synaptosomes and this resulted in enhancement of acetylcholine synthesis. These observations led to the question of whether Chol-1α gangliosides participated physiologically in the cognitive function of the brain. To answer this, GGR-41 was continuously infused into the rat brain septal area in order to disrupt the septohippocampal cholinergic pathway (Ando et al. 2004). This resulted in a reduction in the learning ability of the rats similar to that seen in rats given mecamylamine, a nicotinic cholinergic receptor antagonist. Memory retention was also severely impaired in rats infused with GGR-41. Chol-1α is thought to localize with nicotinic acetylcholine receptors because it was originally found in torpedo electric organs composed of pure nicotinic nerve terminals (Richardson et al. 1982). In sum, Chol-1α functions to accelerate acetylcholine turnover in cholinergic synapses and serve as a cholinergic marker.

Expression of Chol-1α antigens is known to be developmentally regulated as they appear at the time of cholinergic synapse formation in the rat brain (Derrington and Borroni 1990). Neurogenesis and neuronal regeneration are enhanced in cholinergic lesioned brains (Ho et al. 2009) and by environmental stimulation in aged brains (Nakamura et al. 1999). Expression of fetal *c*-series gangliosides in Alzheimer's brains may also reflect neuronal plastic responses (Section 19.3.2). To examine possible alterations in ganglioside metabolism in relation to Alzheimer's disease, animal models with disrupted ganglioside biosynthesis have been developed (Ariga et al. 2010, 2011, 2013; Bernardo et al. 2009; Oikawa et al. 2009). Ganglioside changes were analyzed in brains from double transgenic (Tg) mice that coexpressed amyloid precursor protein with the Swedish mutation and presenilin-1 with a deletion of exon 9 (Ariga et al. 2010). No significant changes were detected in the concentration and composition of major gangliosidesin brains from the double-Tg mice. In contrast, expression of cholinergic gangliosides such as GT1aα and GQ1bα (Chol-1α antigens) increased. Their increased expression may reflect cholinergic neuronal regeneration response to damages induced by Aβ. The above double-Tg mice were cross-bred with GD3S(St8sia1)$^{-/-}$ mice to generate mice deficient in GD3-synthase responsible for synthesis of *b*-series gangliosides (Fig. 19.1) (Bernardo et al. 2009). In the triple-Tg brains all the *b*-series gangliosides including GD3 were absent, while GM1 and GD1a were increased. Surprisingly, triple-Tg mice showed memory performance similar to that of wild-type control and GD3S$^{-/-}$ mice, while the double-Tg mice exhibited cognitive impairments. Consistent with normalized cognition, Aβ plaques were almost eliminated. These results indicate that *b*-series gangliosides may be one of the causes of Aβ accumulation. Remarkably in the triple-Tg brain, the concentration of GT1aα was elevated, while no expression of GQ1bα was observed. Thus, the elevated cholinergic ganglioside GT1aα may

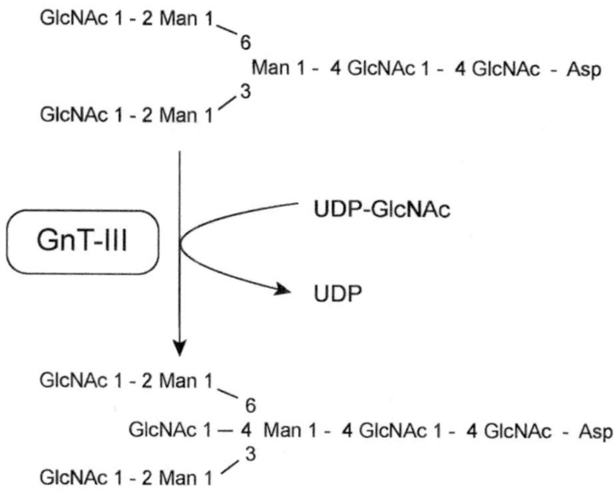

Fig. 19.3 Bisecting GlcNAc residues in N-glycans synthesized by *N*-acetylglucosaminyltransferase III (GnT-III)

contribute to memory retention (Ariga et al. 2013). The double-Tg mice expressing human amyloid precursor protein having the Swedish and London mutations were crossbred with GM2-synthase knockout mice (Oikawa et al. 2009). The mutant mice expressing a large amount of GM3 but not GM1 showed a remarkable increase of Aβ deposition in vascular tissues (amyloid angiopathy). This observation may indicate another mechanism exists for Aβ deposition than that covered by the hypothesis that GM1-bound Aβ is involved in amyloid fibril formation (Kakio et al. 2001) (Sect. 19.3.1). These results indicate that the significance of gangliosides relevant to the physiology and pathology of the brain cannot be deduced simply from studies of the loss-of-function or gain-of-function of each ganglioside. Further comprehensive studies are needed.

19.3.4 Glycoproteins: In Connection with Alzheimer's Disease

Core N-glycosylation and N-glycans were reported to modulate the synthesis of amyloid precursor protein (Pahlsson et al. 1992), suggesting that N-glycosylation status might affect the metabolic pathway of amyloid precursor protein (APP). Analysis of the N-glycan structures determined of mutant amyloid precursor proteins (Swedish type and London type) produced by transfected C17 cells showed that they contained higher contents of bisecting *N*-acetylglucosamine (GlcNAc) residues than normal APP (Akasaka-Manya et al. 2008). To examine the reason for overexpression of the bisecting structure, expression of *N*-acetylglucosaminyltransferase III (GnT-III)-mRNA (GnT-III is the enzyme responsible for synthesizing bisecting GlcNAc residues) (Fig. 19.3) was measured and found to be increased in Alzheimer's brains (Akasaka-Manya et al. 2010).

Incubation of Neuro2a cells with Aβ42 increased GnT-III gene expression. In a separate setting, Neuro2a cells transfected with a GnT-III expression vector downregulated Aβ production. Fiala et al. (2007) reported that blood monocytes exposed to Aβ peptides upregulated transcription of GnT-III and increased Aβ clearance by phagocytosis. Thus, upregulation of GnT-III appears to exert two effects: inhibition of Aβ production in neurons and enhancement of its clearance by monocytes. Both responses can be protective against the further progression of Alzheimer's disease.

The possible mechanisms by which bisecting GlcNAc residues could reduce Aβ production include the following (Akasaka-Manya et al. 2010): (1) Addition of bisecting GlcNAc may affect the conformation of the APP, thereby inducing changes in its susceptibility to α-, β-, and/or γ-secretases. (2) The increase of bisecting GlcNAc on the APP leads to changes in its N-glycan structure such as diminished degrees of elongation, branching, and sialylation. (3) Bisecting GlcNAc affects trafficking of the APP and, as a result, its susceptibility to secretases. This is supported by the report that localization and trafficking of APP is affected by its glycan modifications (McFarlane et al. 1999). (4) The bisecting GlcNAc directly affects secetase activities. Support for this possibility was provided the observations that (a) enzyme activities of α- and β-secretases were significantly increased and decreased, respectively, in GnT-III-transfected Neuro2a cells (Akasaka-Manya et al. 2010); (b) Western blot analysis indicated that changes in N-glycan structures were present in the secretases (TACE and BACE). These findings may account for changes in secretase activities and resultant inhibition of Aβ formation.

19.3.5 Glucocerebrosidase Gene Mutations as a Risk Factor for Parkinson's Disease

Some patients with Gaucher disease were found to develop parkinsonism (Mitsui et al. 2009; Rhouma et al. 2012; Sunwoo et al. 2011). Gaucher disease is an autosomal recessive, lysosomal storage disease caused by mutations in the β-glucocerebrosidase gene (Beutler and Grabowski 1995). Glucocerebrosidase (GBA) is a lysosomal enzyme that catalyzes the hydrolysis of glucocerebroside to ceramide and glucose. Homozygous mutations in the GBA that affect its activity result in accumulation of glucocerebroside in various tissues. Heterozygous loss of function mutations at the GBA locus are a potent risk factor for Parkinson's disease (PD) (Lwin et al. 2004) as evidenced by the finding of multiple cases of parkinsonism among Gaucher disease carriers (Goker-Alpan et al. 2004).

An international collaborative study was done to ascertain the frequency of GBA mutations in ethnically diverse patients with PD (Sidransky et al. 2009). Sixteen centers participated, including five from the Americas, six from Europe, two from Israel, and three from Asia. Two GBA mutations, L444P and N370S, were found in 15.3 % of Ashkenazi Jewish patients with PD, and in 3.2 % of non-Ashkenazi patients indicating a strong association between GBA mutations and PD. When a complete sequence analysis of the variants was carried out for a large cohort of

European, mostly French, patients with PD (Lesage et al. 2011), the results revealed that carrier frequency in the non-Ashkenazi Jewish populations was 7 %, much higher than that, the 3.2 %, reported by the above 16 centers' study (Sidransky et al. 2009). These results indicate that limited screening might miss more than half of the mutant alleles. They also showed that GBA mutations were significantly more frequent (odds ratio = 6.98, 95 % confidence interval 2.54–19.21; $p = 0.00002$) in PD than in controls. After the 16 centers' study, resequencing of GBA was performed for Japanese patients with PD. The frequency of pathogenic variants in the heterozygous state was shown to be 9.4 % in Parkinson's patients compared to 0.37 % in controls (odds ratio, 28.0). Mutations in the GBA gene are hypothesized to accelerate the pathogenesis of PD (Clark et al. 2007). Support for this was provided by the observation that GBA carriers had a 2.5 year earlier age of onset of PD compared to noncarriers. Although genetic research in the past delineated many mutations that cause PD, such as those in genes encoding E3 ubiquitin ligase parkin (PARK2), leucine-rich repeat kinase 2 (LRRK2), and alpha-synuclein (Bras et al. 2008), research on GBA mutations supports the idea that they are the most common genetic risk factor for PD (Lesage et al. 2011; Neumann et al. 2009).

To address the question of what deleterious effects are caused by GBA mutations, cerebral metabolic activity was assessed in carriers of the GBA mutation both with and without parkinsonism using positron emission tomography (Kono et al. 2010). All GBA mutation carriers had significantly decreased glucose metabolic rates in the supplemental motor area (cortex region anterior to the primary motor cortex), and the carriers with parkinsonism showed additional hypometabolism. The hypometabolism in the cortex region may be related to the clinical manifestation of parkinsonism. Measurement of GBA activity showed it decreased significantly in brains of Parkinson's patients carrying heterozygous GBA mutation (Gegg et al. 2012). The greatest deficiency was found in the substantia nigra (58 % decrease) known to be affected in PD. In PD the brain loses dopaminergic neurons which results in severe reduction in the dopamine content of the striatum (Hornykiewicz 1966). Protein levels of GBA were also reduced, indicating that the lowered expression of the enzyme as well as its decreased activity could contribute to its deficiency (Gegg et al. 2012). Immunofluorescence studies on brain tissues from patients with PD associated with GBA mutations showed that the enzyme was present in 75 % of Lewy bodies while GBA-positive Lewy bodies were found in only 4 % of the subjects without mutations (Goker-Alpan et al. 2010).

Possible mechanisms for the link between GBA gene mutations and PD have been speculated. Mutations in α-synuclein are known to result in aberrant aggregation of the protein, which is associated with neuronal death. The aggregated polymers are proposed to be necessary prerequisites for Lewy body formation seen in PD (Ishizawa et al. 2003). Functional loss of GBA in primary cultured neurons was found to compromise lysosomal protein degradation, cause accumulation of α-synuclein, and result in neurotoxicity through aggregation-dependent mechanisms (Mazulli et al. 2011). These results indicate that increased glucosylceramide, the GBA substrate, directly influences amyloid formation of α-synuclein. Increased glucosylceramide due to a deficiency of GBA is speculated to disrupt the membrane binding of α-synuclein, enhancing its aggregation in the cytoplasm (DePaolo et al. 2009). Another theory about the relationship of GBA to PD involves its interference with the clearance of mutated proteins. Since most mutations in GBA are missense mutations, the protein likely becomes aberrantly folded and as a result undergoes parkin-mediated poly-ubiquitination and subsequent proteasome-mediated degradation. The presence of mutant GBA might cause build-up of other parkin substrates, causing endoplasmic reticulum-stress and eventual apoptosis of the neurons (Westbroek et al. 2011). Finally, a physical linkage between α-synuclein and GBA was verified using immunoprecipitation and immunofluorescence (Yap et al. 2011).

19.3.6 Gangliosides: In Relation to the Pathogenesis of Parkinson's Disease

The symptoms of PD are manifested after dopaminergic innervation of the striatum is lost as a result of degeneration of dopaminergic neurons in the substantia nigra (Hornykiewicz 1966). Over two decades ago GM1 was hypothesized to act as a neurotrophic factor for dopaminergic neurons (Schneider et al. 1992). A comprehensive review on degenerative diseases and the therapeutic potentials of gangliosides are described in Chap. 20. A causative aspect of reduced gangliosides in PD is partly discussed here. Recently Using a genetic mouse model of Parkinson's disease in which major gangliosides were depleted Ledeen's group (Wu et al. 2011) proved that GM1 was involved in the pathogenesis of parkinsonism. The knockout mice were generated by disrupting the B4galnt1 gene for GM2/GD2 synthase thereby eliminating synthesis of GM2, GD2, and all gangliotetraose gangliosides. The B4galnt1$^{-/-}$ mice manifested clinical parkinsonism and the pathological loss of dopaminergic neurons in the substantia nigra. The symptoms of parkinsonism were largely attenuated by administration of a GM1 analogue, LIGA-20, developed by Costa's group (Manev et al. 1990). This finding supports the hypothesis that GM1 is involved in the pathogenesis of PD. Ledeen's group (Wu et al. 2012) also found that heterozygous mice with one defective allele for the B4galnt1 gene displayed virtually the same degree of parkinsonism as the knockout mice. Interestingly, the levels of GM1 and GD1a, *a*-series gangliosides (Fig. 19.1), decreased in the heterozygous mice by 43 % and 46 %, respectively, while those of *b*-series gangliosides did not significantly decrease. Loss of dopaminergic neurons was evident in the heterozygous mice, and dopamine levels in the striatum decreased progressively as the animals aged. Treatment with LIGA-20 increased the number of tyrosine hydroxylase-positive neurons in the substantia nigra and levels of dopamine in the striatum, indicating recovery of dopaminergic neurons. Combined these observations indicate that the heterozygous mouse model carrying the mutant B4galnt1 gene will be quite useful in future studies of PD.

In human PD, the number of dopaminergic neurons in the substantia nigra, identified by staining for tyrosine hydroxylase, decreased by 40 % compared to that in controls (Wu et al. 2012). Quantification of dopaminergic neurons expressing

GM1 in 11 Parkinson's patients revealed a noticeable GM1 deficiency: 19.7 % of dopaminergic neurons in the patients compared to 61.8 % in controls. This provides an interesting parallel to the deficit of GM1 observed in the heterozygous mouse model discussed above. GM1 staining with cholera toxin B-FITC (specific for GM1) was significantly decreased in both nondopaminergic neurons as well as dopaminergic neurons in Parkinson's brain sections, indicating the widespread abnormal expression of gangliosides in the diseased brain. Not addressed is whether the decreased GM1 is the cause for or result of PD. So far no mutations of genes involved in ganglioside metabolism have been identified as risk factors for PD.

Anti-GM1 ganglioside antibodies have been found in the sera of patients with neurological diseases such as lower motor neuron syndromes (Pestronk et al. 1990; Sunwoo et al. 2011), amyotrophic lateral sclerosis (Pestronk et al. 1988; Rhouma et al. 2012), and Alzheimer's (Chapman et al. 1988; Saito et al. 2002). In an early survey of a limited number of parkinsonian demented patients high titers of anti-GM1 antibodies were found (Saito et al. 2002). Subsequently Zappia et al. (Zappia et al. 2002) studied a large group of PD patients, most of whom had a tremor-dominant form of the disease, and found that more than one-quarter had sera with increased levels of IgM anti-GM1 antibodies. It is hypothesized that increased anti-GM1 antibodies could affect the function of ganglioside GM1 in dopaminergic neurons. Anti-ganglioside immune responses are speculated to contribute to axonal damage via a T cell-mediated mechanism in multiple sclerosis (Pender et al. 2003; McFarlane et al. 1999). When patients with lower motor neuron syndromes and high serum titers of IgM anti-GM1 antibodies were subjected to repeated plasma exchanges (Pestronk et al. 1994; Beutler and Grabowski 1995), removal of the antibodies brought about progressive improvement in muscle strength. These findings indicate that anti-ganglioside antibodies may act as a risk factor for neuronal dysfunction or disease pathogenesis.

19.4 Strategies for Anti-Aging and Prevention of Age-Related Diseases

Studies of glycoconjugates expression and aging have provided much information (Sects. 19.2 and 19.3), and some may provide the basis for development of future therapeutics. Possibilities include use of the knowledge for inducing remyelination, clearance of Aβ from the brain, and enhancement of innate neuronal plasticity.

19.4.1 Myelin Repair and Remyelination

The fact that the volume of white matter decreases in senescence (Walhovd et al. 2005) indicates that enhanced demyelination occurs in aging brains and more severely in demyelinating disease brains (Zhang et al. 2011). Cognitive declines appear to correlate with altered myelination of nerve tracts (Aine et al. 2011; Schulze et al. 2011). While disruption of mature myelin is going on, remyelination takes place (Peters and Sethares 2003). This is supported by observations indicating that the number of myelin lamellae increase (Peters et al. 2001) and intermodal lengths of myelin get shorter with advancing age (Bowley et al. 2010). To enhance myelin repair or remyelination following demyelination, promoters of remyelination have been examined as potential therapeutics. The age-associated inefficiency of remyelination (Shields et al. 1999) is known to be due to impairment of both recruitment and differentiation of oligodendrocyte progenitor cells (OPC) (Sim et al. 2002). Myelin synthesis is preceded by downregulation of OPC differentiation inhibitors such as PSA-NCAM (Shen et al. 2008). This downregulation is epigenetically controlled by recruitment of histone deacetylases to promoter regions, indicating that efficient remyelination requires deacetylation of nucleosomal histones. Epigenetic control of remyelination is potentially an important therapeutic target. Another promising intervention to enhance remyelination is hormonal stimulation of OPC (Calzà et al. 2010; Fernandez et al. 2004). Thyroid hormone was shown to regulate proliferation and differentiation of OPC and induce myelinating oligodendrocytes. Actually, systematic administration of thyroid hormone enhanced myelination in an experimental allergic encephalomyelitis model (Fernandez et al. 2004) as well as in a cuprizone-induced demyelination model (Franco et al. 2008). Using an in vivo deuterium labeling method (Ando et al. 2003) thyroxine was found to enhance incorporation of ganglioside GM1 into mature myelin of adult mice (Ando et al. 1984).

19.4.2 Acceleration of Aβ Clearance

Aging is the major risk factor for neurodegenerative diseases such as Alzheimer's disease and PD (Sect. 19.3) with accumulation of aberrant precipitates of amyloid-β peptide in brains from the former and α-synuclein in those from the latter. Clearance of those toxic compounds from the brain is driven by an active transport system at the blood–brain barrier (BBB) and blood–cerebrospinal fluid barrier. In the transport system at the barriers, the receptor low-density lipoprotein receptor-related protein 1 takes up the toxic compounds at the abluminal sites of capillary endothelial cells, and P-glycoprotein (P-gp) at the luminal sites of the cells excretes them into the blood stream. Decreased function of P-gp was observed in Alzheimer's patients (Vogelgesang et al. 2004) and in Parkinson's patients (Bartels et al. 2008). The expression of P-gp decreases in the BBB with age (Silverberg et al. 2010). As the reduced clearance of amyloid-β across the BBB is thought to enhance amyloid accumulation, a decrease in P-gp could be a risk factor for developing Alzheimer's disease. In this context, development of a model to upregulate expression of P-gp and enhance clearance of amyloid-β could provide a new method for treating Alzheimer's disease. Support for this idea was provided by both in vitro (Abuznait et al. 2011) and in vivo studies (Hartz et al. 2010).

Serum amyloid P (SAP) is also thought to be involved in the pathogenesis of Alzheimer's disease. SAP, a member of the pentraxin family, is a glycoprotein secreted by the liver into the blood stream. The main functions of SAP are to recognize carbohydrates, nuclear substances, and amyloid fibrils (Agrawal et al. 2009). SAP does not cross the BBB, however, neurons produce it in the brain, and its production is upregulated in Alzheimer's brains (Yasojima et al. 2000). SAP is assumed to contribute to the pathogenesis of Alzheimer's disease by binding to amyloid plaques and entering neurons to induce apoptotic cell death (Ulbanyi et al. 2003). Although the hallmarks of Alzheimer's disease include amyloid plaques and neurofibrillary tangles, a significant population of individuals who have plaques and tangles have no signs of cognitive impairment. Crawford et al. (2012) measured SAP levels in the hippocampus and cerebral cortex of post mortem samples obtained from Alzheimer's patients and nondemented individuals with Alzheimer's neuropathology, and found that the latter had no significant difference in SAP levels compared to normal controls while the former had increased SAP levels. The lack of dementia in individuals with Alzheimer's neuropathology and low levels of SAP, a marker of inflammation, may be due to reduced inflammatory responses. From the therapeutic point of view, reduction of the increased serum SAP might be beneficial for Alzheimer's patients. Pepys et al. (2002) invented a drug, R-1-[6- [R-2-carboxy-pyrrolidin-1-yl]-6-oxo-hexanoyl] pyrrolidine-2-carboxylic acid that is a competitive inhibitor of SAP binding to amyloid fibrils. Kolstoe et al. (2009) administered it to patients with probable Alzheimer's disease for 12 weeks, and found that the SAP concentrations in serum and cerebrospinal fluid were significantly reduced. Their trial was too short to provide measurable neurological or cognitive effects.

19.4.3 Gene-Transfer of Enzymes Responsible for Glycoconjugates

Loss of function mutations at the GBA locus is a known risk factor for PD (Lwin et al. 2004) (see Sect. 19.3.5). Sardi et al. (2011) developed mouse models of Gaucher disease including homozygote ($Gba1^{D409V/D409V}$) and heterozygotes ($Gba1^{D409V/+}$ and $Gba1^{+/-}$) that showed loss of GBA activity. GBA activity was reduced in homozygous mice (19 % of that of wild type) and heterozygous (59 % for $Gba1^{D409V/+}$and 54 % for $Gba1^{+/-}$). Adeno-associated virus-mediated expression of exogenous GBA in the hippocampus of homozygous mice brought about remarkable amelioration of both the neuropathology and memory deficits. Sardi et al. (2013) developed a transgenic mouse model overexpressing A53T α-synuclein and using it found that augmented striatal expression of exogenous GBA in the mice reduced the levels of both cytosolic soluble α-synuclein and membrane-associated striatal α-synuclein.

The observation that PSA-NCAM (see Sect. 19.2.5) was expressed concurrently with sprouting of injured or axotomized Purkinje cells in glial scars (Dusart et al. 1999) led to interrogation of whether induction of PSA expression in damaged tissues in adult CNS could enhance neuronal regeneration. To test the therapeutic potential of PSA, a viral vector encoding polysialyltransferase was employed to transfect scar astrocytes and sustain expression of high levels of PSA (El Maarouf et al. 2006). This resulted in substantial growth of axonal processes through the spinal injury site. Induced expression of PSA may be a promising strategy for promoting neural tissue repair. Similar genetic manipulations using viral vectors encoding GnT-III (see Sect. 19.3.4) could be applied to Alzheimer's brains to inhibit amyloid β production and to enhance the clearance of amyloid β as previously suggested (Akasaka-Manya et al. 2010). A possible alternative would be to use the curcumin derivative, bisdemethoxycurcumin found to induce expression of GnT-III (Fiala et al. 2007).

19.4.4 Ganglioside Analogues and Environmental Effectors of Neuronal Plasticity

Gangliosides, sialic acid-containing GSLs, are important mediators of neuronal function, age-related neuronal dysfunction, and Alzheimer's pathology. Studies have shown that gangliosides may mimic nerve growth factor neurotrophic activity thereby restoring cholinergic parameters (Cuello et al. 1989; Ferrari et al. 1995; Mutoh et al. 1995; Rabin and Mocchetti 1995). The beneficial effects of ganglioside GM1 were shown in the experimental models of glutamate neurotoxicity (Favaron et al. 1988; Vaccarino et al. 1987; Wu et al. 2005), brain and spinal cord lesions (Geisler et al. 1991; Oderfeld-Nowak et al. 1993), and cerebral ischemia (Seren et al. 1990). Although gangliosides have been hypothesized to serve as neuroprotective agents capable of reducing brain and spinal cord damages in disease conditions, they were proven to attenuate neuronal apoptosis induced by serum deprivation, ionomycin, or cyclosporine A in an in vitro system (Ryu et al. 1999). However, gangliosides did not attenuate neuronal necrosis induced by exposure of cultured cortical cells to excitotoxins, and even increased necrosis induced by oxidative stress. Thus, gangliosides seem to rescue neurons from becoming apoptotic.

Two decades ago, GM1 was reported to activate the TrkA (tropomyosin-related kinase A) receptor (Ferrari et al. 1995; Mutoh et al. 1995; Rabin and Mocchetti 1995). Later it was shown to activate TrkC by inducing release of neurotrophin-3 (Rabin et al. 2002). In contrast to GM1 which induces neurotrophin-3 but not BDNF, a synthetic analogue of GM1, LIGA20 (Manev et al. 1990), was found to increase secretion of BDNF by stimulating TrkB in NIH-3T3 fibroblasts. When the same amount of GM1 and LIGA20 were given orally to rats, the concentration of LIGA20 in the brain was 50-fold higher than that of GM1 (Polo et al. 1994). This may account for the superior potency of LIGA20 compared to natural GM1.

LIGA20 is similar in structure of lyso-GM1 (Manev et al. 1990). Although LIGA20 appears to be membrane permeable, it was hypothesized that smaller molecules would be more suitable for transportation across the BBB. To address this sialylcholesterol was synthesized (Sato et al. 1987). Effects of the two stereoisomers, α- and β-sialylcholesterols, on cholinergic synapses were examined (Tanaka and Ando 1996) to determine whether they could ameliorate age-related decrements in synaptic functions observed in studies of mouse brain synaptosomes (Tanaka et al. 1996). While both isomers enhanced acetylcholine release from synaptosomes, their underlying mechanisms differed. Alpha-sialylcholesterol increased depolarization-induced influx of calcium ions to enhance acetylcholine release, while not affecting choline uptake. On the other hand, β-sialylcholesterol activated high-affinity choline uptake, resulting in enhancement of acetylcholine synthesis followed by augmentation of acetylcholine release. The β-isomer had no effect on calcium ion influx. These results imply that the two isomers of sialylcholesterol modulate the synaptic functions in different ways, and that their effects on cholinergic synaptic transmission are synergistic. The results also indicate that the actions of the synthetic sialylcholesterols on synapses mimic those of naturally occurring gangliosides (Ando et al. 1998). The action of the α-isomer on calcium channels appears to be similar to that observed using ganglioside GQ1b. In contrast, the β-isomer activation of the choline transporter was similar to the effect of the Chol-1α ganglioside (see Sect. 19.3.3). The similarity of the effects of both gangliosides and sialylcholesterols on synaptic action is summarized in a review (Ando 2012).

Neuronal plasticity, a crucial requirement for adapting to new environments, developing neuronal integrity, and restoring damaged neural architecture, are central themes in basic and clinical neuroscience. Age-related deterioration in brain functions or damage caused by neurodegenerative diseases or various insults can be partly restored by innate neuronal plasticity. Stimulation of plasticity can be considered the ultimate strategy for maintaining brain function during aging and for its efficient recovery from neuronal damage. Synaptic plasticity was enhanced as revealed by increased synaptogenesis and cognitive improvement when aged rats were put in an enriched environment (Ando 2012; Saito et al. 1994). In a mouse model of PD, physical activity and environmental enrichment stimulated oligodendrocyte differentiation in the substantia nigra, again indicating their positive effect on cellular plasticity (Klaissle et al. 2012). Neuronal plasticity is thought to be regulated by at least two groups of molecules: neurotrophins and adhesion molecules. The effect of activity on plasticity may be explained by the observation that expression of BDNF and NCAM were upregulated in the spinal cords of rats given exercise (Macias et al. 2002). Drugs can also affect neuronal plasticity. Warfarin is routinely prescribed for patients with atrial fibrillation or cerebral infarction, but since it is also a vitamin K antagonist and may cause vitamin K-deficiency it must be used with care. The first indication that vitamin K was needed by the nervous system was the observation by Sundaram and Lev (1988) that young mice fed warfarin had decreased brain concentrations of sulfatide due to reduced activity of galactocerebroside sulfotransferase, needed for its synthesis (Fig. 19.1). Lifelong intake of a low-vitamin K diet was shown to impair cognitive performance by old rats. They also had low levels of gangliosides in the pons medulla and midbrain (Carrié et al. 2011). Sulfatide contents were not different between high- and low-dosed groups in old rats. In humans, fetal exposure to warfarin is known to induce warfarin embryopathy identified by symptoms such as blindness and mental retardation due to optic and cerebral atrophies (Hall et al. 1980). Patients with early stage Alzheimer's disease were found to have significantly lower intakes of vitamin K compared to age- and gender-matched controls (Presse et al. 2008). With regard to brain plasticity or responsiveness, both positive and negative environmental risk factors such as enriched environment and warfarin must be considered.

19.5 Concluding Remarks

In Sect. 19.2, age-related changes in glycoconjugates, mostly GSLs and glycoproteins, in the central nervous system are summarized.

In Section **3**, new findings on the pathogenesis of age-related neurodegenerative diseases such as Alzheimer's and Parkinson's are discussed. Ganglioside GM1 may function as a seed for amyloid fibril formation, and the expression of Chol-1α gangliosides in the Alzheimer's model mouse brain is thought to be a compensatory response to restore cholinergic function. Evidence suggests that mutations in the gene for GBA and partial deficits in *a*-series gangliosides may contribute to PD.

In section **4**, possible strategies for anti-aging and the prevention of age-related dementia are discussed, including those to (1) enhance remyelination, (2) enhance clearance of amyloid-β, and (3) express the GBA gene. To ameliorate the functional deficits in aging and diseased brains, ganglioside-analogues, LIGA20 and sialylcholesterols, have been tested. Alpha- and β-stereoisomers of the latter are found to activate calcium ion channels and high-affinity choline uptake at presynapses, respectively. Finally, environmental factors (eg physical activity and drugs) affecting neuronal plasticity in relation to glycoconjugates are discussed.

Conflict of Interest The author declares no conflicts of interest in preparing this article.

References

Abuznait AH, Cain C, Ingram D, Burk D, Kaddoumi A. Up-regulation of P-glycoprotein reduces intracellular accumulation of beta-amyloid: investigation of P-glycoprotein as a novel therapeutic target for Alzheimer's disease. J Pharm Pharmacol. 2011;63:1111–8.

Agrawal A, Singh PP, Bottazzi B, Garlanda C, Montovani A. Pattern recognition by pentraxins. Adv Exp Med Biol. 2009;653:98–116.

Aine CJ, Sanfratello L, Adair JÇ, Knoefel JE, Caprihan A, Stephen JM. Development and decline of memory functions in normal, pathological and healthy successful aging. Brain Topogr. 2011;24:323–39.

Akasaka-Manya K, Manya H, Sakurai Y, Wojczyk BS, Spitalnik SL, Endo T. Increased bisecting and core-fucosylated N-glycans on mutant human amyloid precursor proteins. Glycoconj J. 2008;25:775–86.

Akasaka-Manya K, Manya H, Sakurai Y, Wojczyk BS, Kozutsumi Y, Sato Y, et al. Protective effect of-glycan bisecting GlcNAc residues on β–amyloid production in Alzheimer's disease. Glycobiology. 2010;20:99–106.

Ando S. Review: gangliosides in the nervous system. Neurochem Int. 1983;5:507–37.
Ando S. Biochemistry of brain aging. Nihon Rinsho. 1985;43:1399–403 (in Japanese).
Ando S. Neuronal dysfunction with aging and its amelioration. Proc Jpn Acad Ser B Phys Biol Sci. 2012;88:266–82.
Ando S, Yu RK. Isolation and characterization of two isomers of brain tetrasialogangliosides. J Biol Chem. 1979;254:12224–9.
Ando S, Tanaka Y, Ono Y, Kon K. Incorporation rate of GM1 ganglioside into mouse brain myelin: effect of aging and modification by hormones and other compounds. Adv Exp Med Biol. 1984;174:241–8.
Ando S, Tanaka Y, Kon K. Membrane aging of the brain synaptosomes with special reference to gangliosides. In: Tettamanti G et al., editors. Gangliosides and neural plasticity. Padva: Liviana Press; 1986. p. 23–30.
Ando S, Hirabayashi Y, Kon K, Inagaki F, Tate S, Whittaker VP. A trisialoganglioside containing a sialyl-α2,6-N-acetylgalactosamine residue is a cholinergic-specific antigen, Chol-1α. J Biochem. 1992;111:287–90.
Ando S, Tanaka Y, Waki H, Kon K, Iwamoto M, Fukui F. Gangliosides and sialylcholesterol as modulators of synaptic functions. Ann N Y Acad Sci. 1998;845:232–9.
Ando S, Kobayashi S, Waki H, Kon K, Fukui F, Tadenuma T, et al. Animal model of dementia induced by entorhinal damage and partial restoration of cognitive deficits by BDNF and carnitine. J Neurosci Res. 2002;70:519–27.
Ando S, Tanaka Y, Toyoda Y, Kon K. Turnover of myelin lipids in aging brain. Neurochem Res. 2003;28:5–13.
Ando S, Tanaka Y, Kobayashi S, Fukui F, Iwamoto M, Waki H, et al. Synaptic function of cholinergic-specific Chol-1α ganglioside. Neurochem Res. 2004;29:857–67.
Ariga T, Yanagisawa M, Wakada C, Ando S, Buccafusco JJ, McDonald MP, et al. Ganglioside metabolism in a transgenic mouse model of Alzheimer's disease: expression of Chol-1α antigens in the brain. ASN Neuro. 2010;2:e00044.
Ariga T, Wakada C, Yu RK. The pathological roles of ganglioside metabolism in Alzheimer's disease: effect of gangliosides on neurogenesis. Int J Alzheimers Dis. 2011;2011:193618.
Ariga T, Itokazu Y, McDonald MP, Hirabayashi Y, Ando S, Yu RK. Brain gangliosides of a transgenic mouse model of Alzheimer's disease with deficiency in GD3-synthase: expression of elevated levels of a cholinergic-specific ganglioside, GT1aα. ASN Neuro. 2013;5:141–8.
Bartels AL, Willemsen ATM, Kortekaas R, de Jong BM, de Vries R, de Klerk O, et al. Decreased blood–brain barrier P-glycoprotein function in the progression of Parkinson's disease, PSP and MSA. J Neural Transm. 2008;115:1001–9.
Bednarski E, Lynch G. Cytosolic proteolysis of tau by cathepsin D in hippocampus following suppression of cathepsins B and L. J Neurochem. 1996;67:1846–55.
Ben-David O, Pewzner-Jung Y, Brenner O, Laviad EL, Kogot-Levin A, Weissberg I, et al. Encephalopathy caused by ablation of very long acyl chain ceramide synthesis may be largely due to reduced galactosylceramide levels. J Biol Chem. 2011;286:30022–33.
Bernardo A, Harrison FE, McCord M, Zhao J, Bruchey A, Davies SS, et al. Elimination of GD3 synthase improves memory and reduces amyloid-β plaque load in transgenic mice. Neurobiol Aging. 2009;30:1777–91.
Beutler E, Grabowski G. Gaucher disease. In: Scriver CR, Beaudet al, Sly WS, Valle D, editors. The metabolic and molecular basis of inherited disease. New York, NY: McGraw-Hill; 1995. p. 2641–70.
Blennow K, Bogdanovic N, Alafuzoff I, Ekman R, Davidsson P. Synaptic pathology in Alzheimer's disease: relation to severity of dementia, but not to senile plaques, neurofibrillary tangles, or the apo E4 allele. J Neural Transm. 1996;103:603–18.
Bowley MP, Cabral H, Rosene DL, Peters A. Age changes in myelinated nerve fibers of the cingulate bundle and corpus callosum in the rhesus monkey. J Comp Neurol. 2010;518:3046–64.
Bras J, Singleton A, Cookson MR, Hardy J. Potential role of ceramide metabolism in Lewy body disease. FEBS Lett. 2008;275:5767–73.

Brunden KR. Age-dependent changes in the oligosaccharide structure of the major myelin glycoprotein, P0. J Neurochem. 1992;58:1659–66.

Brusés JL, Rutishauser U. Reguration of neural cell adhesion molecule polysialylation: evidence for nontranscriptional control and sensitivity to an intracellular pool of calcium. J Cell Biol. 1998;140:1177–86.

Calzà L, Fernandez M, Giardino L. Cellular approaches to central nervous system remyelination stimulation: thyroid hormone to promote myelin repair via endogenous stem and precursor cells. J Mol Endocrinol. 2010;44:13–23.

Cantù L, Del Favero E, Sonnino S, Prinetti A. Gangliosides and the multiscale modulation of membrane structure. Chem Phys Lipids. 2011;164:796–810.

Carrié I, Bélanger E, Portoukalian J, Rochford J, Ferland G. Lifelong low-phylloquinone intake is associated with cognitive impairments in old rats. J Nutr. 2011;141:1495–501.

Chapman J, Sela BA, Wertman E, Michaelson DM. Antibodies to ganglioside GM1 in patients with Alzheimer's disease. Neurosci Lett. 1988;86:235–40.

Choo-Smith L-P, Garzon-Rodriguez W, Glabe CG, Surewicz WK. Acceleration of amyloid fibril formation by specific binding of Aβ-(1-40) peptide to ganglioside-containing membrane vesicles. J Biol Chem. 1997;272:22987–90.

Clark LN, Ross BM, Wang Y, Mejia-Santana H, Harris J, Louis ED, et al. Mutations on the glucocerebrosidase gene are associated with early-onset Parkinson disease. Neurology. 2007;69:1270–7.

Crawford JR, Bjorklund NL, Taglialatela G, Gomer RH. Brain serum amyloid P levels are reduced in individuals that lack dementia while having Alzheimer's disease neuropathology. Neurochem Res. 2012;37:795–801.

Cuello AC, Garofalo L, Kenigsberg RL, Maysinger D. Gangliosides potentiate in vivo and in vitro effects of nerve growth factor on central cholinergic neurons. Proc Natl Acad Sci U S A. 1989;86:2056–60.

Dahlgren KN, Manelli AM, Stine Jr WB, Baker LK. Oligomeric and fibrillar species of amyloid- β peptides differentially affect neuronal viability. J Biol Chem. 2002;277:32046–53.

DePaolo J, Goker-Alpan O, Samaddar T, Lopez G, Sidransky E. The association between mutations in the lysosomal protein glucocerebrosidase and parkinsonism. Mov Disord. 2009;24:1571–8.

Derrington EA, Borroni E. The developmental expression of the cholinergic-specific antigen Chol-1 in the central and peripheral nervous system.of the rat. Dwvelop. Brain Res. 1990;52:131–40.

Dusart I, Morel MP, Wehrlé R, Sotelo C. Late axonal sprouting of injured Purkinje cells and its temporal correlation with permissive changes in the glial scar. J Comp Neurol. 1999;408:399–418.

El Maarouf A, Petridis AK, Rutishauser U. Use of polisialic acid in repair of the central nervous system. Proc Natl Acad Sci U S A. 2006;103:16989–94.

Emory CR, Ala TA, Frey WH. Ganglioside monoclonal antibody (A2B5) labels Alzheimer's neurofibrillary tangles. Neurology. 1987;37:768–72.

Eustache F, Desgranges B, Giffard B, de la Sayette V, Barom J-C. Entorhinal cortex disruption causes memory deficit in early Alzheimer's disease as shown by PET. Neuroreport. 2001;12:683–5.

Favaron M, Manev H, Alho H, Bertolino M, Ferret B, Guidotti A, et al. Gangliosides prevent glutamate and kinate neurotoxicity in primary neuronal cultures of neonatal rat cerebellum and cortes. Proc Natl Acad Sci U S A. 1988;85:7351–5.

Fernandez M, Giuliani A, Pirondi S, D'Intino G, Giardino L, Aloe L, et al. Thyroid hormone administration enhances remyelination in chronic demyelinating inflammatory disease. Proc Natl Acad Sci U S A. 2004;101:16363–8.

Ferrari G, Anderson B, Stephens B, Kaplan D, Greene L. Prevention of apoptotic neuronal death by GM1 ganglioside. involvement of Trk neurotrophin receptors. J Biol Chem. 1995;270:3074–80.

Fiala M, Liu PT, Espinosa-Jeffrey A, Rosenthal MJ, Bernard G, Ringman JM, et al. Innate immunity and transcription of MGAT-III and toll-like receptors in Alzheimer's disease patients are improved by bisdemetoxycurcumin. Proc Natl Acad Sci U S A. 2007;104:12849–54.

Franco PG, Silverstroff L, Soto EF, Pasquini JM. Thyroid hormones promote differentiation of oligodendrocyte progenitor cells and improve remyelination after cuprizone-induced demyelination. Exp Neurol. 2008;212:458–67.

Furuse H, Waki H, Kaneko K, Fujii S, Miura M, Sasaki H, et al. Effect of the mono- and tetra-sialogangliosides, GM1 and GQ1b, on long-term potentiation in the CA1 hippocampal neurons of the guinea pig. Exp Brain Res. 1998;123:307–14.

Gegg ME, Burke D, Heales SJR, Cooper JM, Hardy J, Wood NW, et al. Glucocerebrosidase deficiency in substantia nigra of Parkinson disease brains. Ann Neurol. 2012;72:455–63.

Geisler FH, Dorsey FC, Coleman WP. Recovery of motor function after spinal-cord injury: a randomized placebo-controlled trial with GM-1 ganglioside. N Engl J Med. 1991;324:1829–38.

Goker-Alpan O, Schiffmann R, LaMarca ME, Nussbaum RL, Mclnerney-Leo A, Sidransky E. Parkinsonism among Gaucher disease carriers. J Med Genet. 2004;41:937–40.

Goker-Alpan O, Stubblefield BK, Giasson BI, Sidransky E. Glucocerebrosidase is present in α-synuclein inclusions in Lewy body disorders. Acta Neuropathol. 2010;120:641–9.

Hall JG, Pauli RM, Wilson KM. Maternal and fetal sequelae of anticoagulation during pregnancy. Am J Med. 1980;68:122–40.

Hartz AMS, Miller DS, Bauer B. Restoring blood–brain barrier P-glycoprotein reduces brain amyloid-β in a mouse model of Alzheimer's disease. Mol Pharmacol. 2010;77:715–23.

Heinonen O, Soininen H, Sorvari H, Kosunen O, Paljarvi L, Koivisto E, et al. Loss of synaptophysin-like immunoreactivity in the hippocampal formation is an early phenomenon in Alzheimer's disease. Neuroscience. 1995;64:375–84.

Hirabayashi Y, Hirota M, Matsumoto M, Tanaka H, Obata K, Ando S. Development changes of C-series polysialogangliosides in chick brains revealed by mouse monoclonal antibodies M6704 and M7103 with different epitope specificities. J Biochem. 1988;104:973–9.

Hirabayashi Y, Nakao T, Irie F, Whittaker VP, Kon K, Ando S. Structural characterization of a novel cholinergic neuron-specific ganglioside in bovine brain. J Biol Chem. 1992;267: 12973–8.

Ho NF, Han SP, Dawe GS. Effect of voluntary running on adult hippocampal neurogenesis in cholinergic lesioned mice. BMC Neurosci. 2009;10:57.

Hornykiewicz O. Metabolism of brain dopamine in human parkinsonism: Neurochemical and clinical aspects. In: Costa E, Yahr MD, editors. Biochemistry and pharmacology of the brain ganglia. New York, NY: Raven; 1966. p. 171–81.

Huttenlocher PR. Synaptic density in human frontal cortex—developmental changes and effects of aging. Brain Res. 1979;163:195–205.

Inoko E, Nishimura Y, Tanaka H, Takahashi T, Furukawa K, Kitajima K, et al. Developmental stage-dependent expression of an alpha 2,8-trisialic acid unit on glycoproteins in mouse brain. Glycobiology. 2010;20:916–28.

Irie F, Hashikura T, Tai T, Seyama Y, Hirabayashi Y. Distribution of cholinergic neuron-specific gangliosides (GT1aα and GQ1bα) in the rat central nervous system. Brain Res. 1994;665: 161–6.

Irie F, Kurono S, Li YT, Seyama Y, Hirabayashi Y. Isolation of three novel cholinergic neuron-specific gangliosides from bovine brain and their vitro syntheses. Glycoconj J. 1996;13: 177–86.

Ishaque A, Roomi MW, Szymanska I, Kowalski S, Eylar EH. The P0 glycoprotein of peripheral nerve myelin. Can J Biochem. 1980;58:913–21.

Ishibashi T, Dupree JL, Ikenaka K, Hirahara Y, Honke K, Peles E, et al. A myelin galactolipid, sulfatide, is essential for maintenance of ion channels on myelinated axon but not essential for initial cluster formation. J Neurosci. 2002;22:6507–14.

Ishizawa T, Mattila P, Davies P, Wang D, Dickson DW. Colocalization of tau and alpha-synuclein epitopes in Lewy bodies. J Neuropathol Exp Neurol. 2003;62:389–97.

Kakio A, Nishimoto S, Yanagisawa K, Kozutsumi Y, Matsuzaki K. Cholesterol-dependent formation of GM1 ganglioside-bound amyloid β-protein, and endogenous seed for Alzheimer amyloid. J Biol Chem. 2001;276:24985–90.

Kannagi R, Nudelmann E, Hakomori SI. Possible role of ceramide in defining structure and function of membrane glycolipids. Proc Natl Acad Sci U S A. 1982;79:3470–4.

Kimura N, Yanagisawa K. Endosomal accumulationof GM1 ganglioside-bound amyloid β-protein in neurons of aged monkey brains. Neuroreport. 2007;18:1669–73.

Kirkwood TB. Evolution of aging. Nature. 1977;270:301–4.

Klaissle P, Lesemann A, Huehnchen P, Hermann A, Storch A, Steiner B. Physical activity and environmental enrichment regulate the generation of neural precursors in the adult mouse substantia nigra in a dopamine-dependent manner. BMC Neurosci. 2012;13:132.

Kobata A. Structures and functions of the sugar chains of glycoproteins. Eur J Biochem. 1992;209:483–501.

Kobata A. Glycobiology in the field of gerontology (glycogerontology). Adv Exp Med Biol. 2011;705:411–29.

Kolstoe SE, Ridha BH, Bellotti V, Wang N, Robinson CV, Crutch SJ, et al. Molecular dissection of Alzheimer's disease neuropathology by depletion of serum amyloid P component. Proc Natl Acad Sci U S A. 2009;106:7619–23.

Kono S, Ouchi Y, Terada T, Ida H, Suzuki M, Miyajima H. Functional brain imaging in glucocerebrosidase mutation carriers with and without parkinsonism. Mov Disord. 2010;25:1823–9.

Kotani M, Kawashima I, Ozawa H, Terashima T, Tai T. Differential distribution of major gangliosides in rat central nervous system detected by specific monoclonal antibodies. Glycobiology. 1993;3:137–46.

Ledeen RW, Yu RK. Gangliosides: structure, isolation, and analysis. Meth Enzymol. 1982;83: 139–92.

Lesage S, Anheim M, Condroyen C, Pollak P, Durif F, Dupuits C, et al. Large-scale screening of the Gaucher's disease-related glucocerebrosidase gene in Europeans with Parkinson's disease. Hum Mol Genet. 2011;20:202–10.

Lopez-Toledano MA, Shelanski ML. Neurogenic effect of beta-amyloid peptide in the development of neural stem cells. J Neurosci. 2004;24:5439–44.

Lwin A, Orvisky E, Goker-Alpan O, LaMarca ME, Sidransky E. Glucocerebrosidase mutations in subjects with parkinsonism. Mol Genet Metab. 2004;81:70–3.

Macias M, Fehr S, Dwornik A, Sulejczak D, Wiater M, Czarkowska-Bauch J, et al. Exercise increases mRNA levels for adhesion molecules N-CAM and L1 correlating with BDNF response. Neuroreport. 2002;13:2527–30.

Majocha RE, Jungalwala FB, Rodenrys A, Marotta CA. Monoclonal antibody to embryonic CNS antigen A2B5 provides evidence for the involvement of membrane components at sites of Alzheimer degeneration and detects sulfatides as well as gangliosides. J Neurochem. 1989; 53:953–61.

Manev H, Favaron M, Vicini S, Guidotti A, Costa E. Glutamate-induced neuronal death in primary cultures of cerebeller granule cells: protection by synthetic derivatives of endogenous sphingolipids. J Pharmacol Exp Ther. 1990;252:419–27.

Mansson J-E, Vanier M-T, Svennerholm L. Changes in the fatty acid and sphingosine composition of the major gangliosides of human brain with age. J Neurochem. 1978;30:273–5.

Masliah E, Terry RD. Role of synaptic pathology in the mechanisms of dementia in Alzheimer's disease. Clin Neurosci. 1993;1:192–8.

Matsumura S, Shinoda K, Yamada M, Yokojima S, Inoue M, Ohnishi T, et al. Two distinct amyloid β-protein (A β) assembly pathways leading to oligomers and fibrils identified by combined fluorescence correlation spectroscopy, morphology, and toxicity analysis. J Biol Chem. 2011;286:11555–62.

Mazulli JR, Xu Y-H, Sun Y, Knight AL, McLean PJ, Caldwell GA, et al. Gaucher's disease glucocerebrosidase and α-synuclein form a bidirectional pathogenic loop in synucleinopathies. Cell. 2011;146:37–52.

McFarlane I, Georgopoulou N, Coughlan CM, Gillian AM, Breen KC. The role of the protein glycosylation state in the control of cellular transport of the amyloid β-precursor protein. Neuroscience. 1999;90:15–25.

Mitsui J, Mizuta I, Toyoda A, Ashida R, Takahashi Y, Goto J, et al. Mutations for Gaucher disease confer high susceptibility to Parkinson disease. Arch Neurol. 2009;66:571–6.

Mizutani T, Kasahara M. Hippocampal atrophy secondary to entorhinal cortical degeneration in Alzheimer-type dementia. Neurosci Lett. 1997;222:119–22.

Mutoh T, Tokuda A, Miyadai T, Hamaguchi M, Fujiki N. Ganglioside GM1 binds to the Trk protein and regulates receptor function. Proc Natl Acad Sci U S A. 1995;92:5087–91.

Nakamura H, Kobayashi S, Ohashi Y, Ando S. Age-changes of brain synapses and synaptic plasticity in response to an enriched environment. J Neurosci Res. 1999;56:307–15.

Nelson PT, Alafuzoff I, Bigio EH, Bouras C, Braak H, Caims NJ, et al. Correlation of Alzheimer disease neuropathologic changes with cognitive status: a review of the literature. J Neuropathol Exp Neurol. 2012;71:362–81.

Neumann J, Bras J, Deas E, O'Sullivan SS, Parkkinen L, Robin H, et al. Glucocerebrosidase mutations in clinical and pathologically proven Parkinson's disease. Brain. 2009;132:1783–94.

Norton WT, Poduslo SE. Myelination in rat brain: changes in myelin composition during brain maturation. J Neurochem. 1973;21:759–73.

O'Nuallian B, Freir DB, Nicoll AJ, Risse E, Ferguson N, Herron CE, et al. Aβ dimers rapidly form stable synaptotoxic protofibrils. J Neurosci. 2010;30:14411–9.

Oderfeld-Nowak B, Casamenti F, Pepeu G. Gangliosides in the repair of brain cholinergic neurons. Acta Biochim Pol. 1993;40:395–404.

Ohmi Y, Ohkawa Y, Yamauchi Y, Tajima O, Furukawa K, Furukawa K. Essential roles of gangliosides in the formation and maintenance of membrane microdomains in brain tissues. Neurochem Res. 2012;37:1185–91.

Oikawa N, Yamaguchi K, Ogino K, Taki T, Yuyama K, Yamamoto N, et al. Gangliosides determine the amyloid pathology of Alzheimer's disease. Neuroreport. 2009;20:1043–6.

Pahlsson P, Shakin-Eshleman SH, Spitalnik SL. *N*-Linked glycosylation of β-amyloid precursor protein. Biochem Biophys Res Commun. 1992;189:1667–73.

Palestini P, Masserini M, Sonnino S, Giuliani A, Tettamanti G. Changes in the ceramide composition of rat forebrain gangliosides with age. J Neurochem. 1990;54:230–5.

Palestini P, Masserini M, Fiorilli A, Calappi E, Tettamanti G. Age-related changes in the ceramide composition of the major gangliosides present in rat brain subcellular fractions enriched in plasma membranes of neuronal and myelin origin. J Neurochem. 1993;61:955–60.

Pender MP, Csurhes PA, Wolfe NP, Hooper KD, Good MF, McCombe PA, et al. Increased circulating T cell reactivity to GM3 and GQ1b gangliosides in primary progressive multiple sclerosis. J Clin Neurosci. 2003;10:63–6.

Pepys MB, Herbert J, Hutchinson WL, Tennent GA, Lachmann HJ, Gallimore JR, et al. Targeted pharmacological depletion of serum amyloid P component for treatment of human amyloidosis. Nature. 2002;417:254–9.

Pestronk A, Adams RN, Clawson L, Cornblath D, Kuncl RW, Griffin D, et al. Serum antibodies to GM1 ganglioside in amyotrophic lateral sclerosis. Neurology. 1988;38:1457–61.

Pestronk A, Chaudhry V, Feldman EL, Griffin JW, Cornblath DR, Denys EH, et al. Lower motor neuron syndromes defined by patterns of weakness, nerve conduction abnormalities, and high titers of antiglycolipid antibodies. Ann Neurol. 1990;27:316–26.

Pestronk A, Lopate G, Kornberg AJ, Elliott JL, Blume G, Yee WC, et al. Distal lower motor neuron syndrome with high-titer serum IgM anti-GM1 antibodies: improvement following immunotherapy with monthly plasma exchange and intravenous cyclophosphamide. Neurology. 1994;44:2027–31.

Peters A, Sethares C. Is there remyelination during aging of the primate central nervous system? J Comp Neurol. 2003;460:238–54.

Peters A, Sethares C, Killiany R. Effects of age on the thickness of myelin sheath in monkey primary visual cortex. J Comp Neurol. 2001;435:241–8.

Polo A, Kirschner G, Guidotti A, Costa E. Brain content of GSLs after oral administration of monosialoganglioside GM1 and LIGA20 to rats. Mol Chem Neuropathol. 1994;21:41–53.
Presse N, Shatenstein B, Kergoat MJ, Ferland G. Low vitamin K intakes in community-dwelling elders at an early stage of Alzheimer's disease. J Am Diet Assoc. 2008;108:2095–9.
Probst A, Basler V, Bron B, Ulrich J. Neuritic plaques in senile dementia of Alzheimer type: a Golgi analysis in the hippocampal region. Brain Res. 1983;268:249–54.
Rabin SJ, Mocchetti I. GM1 ganglioside activates the high-affinity nerve growth factor receptor TrkA. J Neurochem. 1995;65:347–54.
Rabin SJ, Bachis A, Mocchetti I. Gangliosides activate Trk receptors by inducing the release of neurotrophins. J Biol Chem. 2002;277:49466–72.
Rhouma FB, Kallel F, Kefi R, Cherif W, Nagara M, Azaiez H, et al. Adult Gaucher disease in southern Tunisia: report of three cases. Diagn Pathol. 2012;7:4.
Richardson PJ, Walker JH, Jones RT, Whittaker VP. Identification of a cholinergic-specific antigen Chol-1 as a ganglioside. J Neurochem. 1982;38:1605–14.
Ryu BR, Choi DW, Hartley DM, Costa E, Jou I, Gway BJ. Attenuation of cortical neuronal apoptosis by gangliosides. J Pharmacol Exp Ther. 1999;290:811–6.
Saito M, Yu RK. Role of myelin-associated neuraminidase in the ganglioside metabolism of rat brain myelin. J Neurochem. 1992;58:83–7.
Saito M, Yu RK. Possible role of myelin-neuraminidase in membrane adhesion. J Neurosci Res. 1993;36:127–32.
Saito S, Kobayashi S, Ohashi Y, Igarashi M, Komiya Y, Ando S. Decreased synaptic density in aged brains and its prevention by rearing under enriched environment as revealed by synaptophysin content. J Neurosci Res. 1994;39:57–62.
Saito M, Tanaka Y, Tang C-P, Yu RK, Ando S. Characterization of sialidase activity in mouse synaptic plasma membranes and its age-related changes. J Neurosci Res. 1995;40:401–6.
Saito M, Hagita H, Ito M, Ando S, Yu RK. Age-dependent reduction in sialidase activity of nuclear membranes from mouse brain. Exp Gerontol. 2002;37:937–41.
Sardi SP, Clarke J, Kinnecom C, Tamsett TJ, Li L, Stanek LM, et al. CNS expression of glucocerebrosidase corrects α-synuclein pathology and memory in a mouse model of Gaucher-related synucleinopathy. Proc Natl Acad Sci U S A. 2011;108:12101–6.
Sardi SP, Clarke J, Kinnecom C, Viel C, Chan M, Tamsett TJ, Treleaven CM, et al. Augmenting CNS glucocerebrosidase activity as a therapeutic strategy for parkinsonism and other Gaucher-related synucleinopathies. Proc Natl Acad Sci U S A. 2013;110:3537–42.
Sato Y, Endo T. Alterations with age of the neurons expressing P0 in the rat spinal cord. Neurosci Lett. 2000;281:41–4.
Sato Y, Endo T. Alteration of brain glycoproteins during aging. Geriatr Gerontol Int. 2010;10: S32–40.
Sato S, Fujita S, Furukata K, Ogura H, Yoshimura S, Itoh M, et al. Synthesis of 2-(5-cholesten-3β-yloxy)glycosides of N-acetyl-D-neuraminic acid derivatives. Chem Pharm Bull. 1987;35: 4043–8.
Sato Y, Kimura M, Yasuda C, Nakao Y, Tomita M, Kobata A, Endo T. Evidence for the presence of major peripheral myelin glycoprotein P0 in mammalian spinal cord and a change of its glycosylation state during aging. Glycobiology. 1999;9:655–60.
Sato Y, Akimoto Y, Kawakami H, Hirano H, Endo T. Location of sialylglycoconjugates containing the Siaα2-3Gal and Siaα2-6Gal groups in the rat hippocampus and the effect of aging on their expression. J Histochem Cytochem. 2001;49:1311–9.
Sato Y, Suzuki Y, Ito E, Shimazaki S, Ishida M, Yamammoto T, et al. Identification and characterization of an increased glycoprotein in aging: age-associated translocation of cathepsin D. Mech Ageing Dev. 2006;127:771–8.
Schachner M, Martini R. Glycans and the modulation of neural-recognition molecule function. Trends Neurosci. 1995;18:183–91.
Scheff SW, Sparks DL, Price DA. Quantitative assessment of synaptic density in the entorhinal cortex in Alzheimer's disease. Ann Neurol. 1993;34:356–61.

Scheibel AB, Tomiyasu U. Dendritic sprouting in Alzheimer's presenile dementia. Exp Neurol. 1978;60:1–8.

Schneider JS, Pope A, Simpson K, Taggart J, Smith MG, DiStetano L. Recovery from experimental parkinsonism in primates with GM1 ganglioside treatment. Science. 1992;256:843–6.

Schulze ET, Geary EK, Susmaras TM, Paliga JT, Maki PM, Little DM. Anatomical correlates of age-related working memory declines. J Aging Res. 2011;2011:606871.

Segler-Stahl K, Webster JC, Brunngraber EG. Changes in the concentration and composition of human brain gangliosides with aging. Gerontology. 1983;29:161–8.

Seki T, Arai Y. The persistent expression of a highly polysialylated NCAM in the dentate gyrus of the adult rat. Neurosci Res. 1991;12:503–13.

Seren MS, Rubini R, Lazzaro A, Zanoni R, Fiori MG, Leon A. Protective effects of a monosialoganglioside derivative following transitory forebrain ischemia in rats. Stroke. 1990;21:1607–12.

Shen S, Sandoval J, Swiss VA, Li J, Dupree J, Franklin RJM. Age-dependent epigenetic control of differentiation inhibitors is critical for remyelination efficiency. Nat Neurosci. 2008;11: 1024–34.

Shields SA, Gilson JM, Blakemore WF, Franklin RJ. Remyelination occurs as extensively but slowly in old rats compared to young rats following gliotoxin-induced CNS demyelination. Glia. 1999;28:77–83.

Sidransky E, Nalls MA, Aasly JO, Aharon-Perez J, Annesi G, Barbose ER, et al. Multi-center analysis of glucocerebrosidase mutations in Parkinson disease. N Engl J Med. 2009;361: 1651–61.

Silverberg GD, Messier AA, Miller MC, Machan JT, Majmudar SS, Stopa EG, et al. Amyloid efflux transporter expression at the blood–brain barrier declines in normal aging. J Neuropathol Exp Neurol. 2010;69:1034–43.

Sim FJ, Zhao C, Penderis J, Franklin RJ. The age-related decrease in CNS remyelination efficiency is attributable to an impairment of both oligodendrocyte progenitor recruitment and differentiation. J Neurosci. 2002;22:2451–9.

Sugiura Y, Shimma S, Konishi Y, Yamada MK, Setou M. Imaging mass spectrometry technology and application of ganglioside study: visualization of age-dependent accumulation of C20-ganglioside molecular species in the mouse hippocampus. PLoS One. 2008;3(9):e3232.

Sun GY, Sun KL. Metabolism of arachidonyl phosphoglycerides in mouse brain subcellular fractions. J Neurochem. 1979;32:1053–9.

Sundaram KS, Lev M. Warfarin administration reduces synthesis of sulfatides and other sphingolipids in mouse brain. J Lipid Res. 1988;29:1475–9.

Sunwoo M-K, Kim S-M, Lee S, Lee RH. Parkinsonism associated with glucocerebrosidase mutation. J Clin Neurol. 2011;7:99–101.

Suzuki K. The pattern of mammalian brain gangliosides—III. Regional and developmental differences. J Neurochem. 1965;12:969–79.

Svennerholm L, Ställberg-Stenhagen S. Changes in the fatty acid composition of cerebrosides and sulfatides of human nervous tissue with age. J Lipid Res. 1968;9:215–25.

Svennerholm L, Boström K, Fredman P, Månsson J-E, Rosengren B, Rynmaek B-M. Human brain gangliosides: developmental changes from early fetal stage to advanced age. Biochim Biophys Acta. 1989;1005:109–17.

Svennerholm L, Boström K, Helander CG, Jungbjer B. Membrane lipids in the aging human brain. J Neurochem. 1991;56:2051–9.

Svennerholm L, Bostöm K, Jungbjer B, Olsson L. Membrane lipids of adult human brain: lipid composition of frontal and temporal lobe in subjects of age 20 to 100 years. J Neurochem. 1994;63:1802–11.

Takahashi H, Hirokawa K, Ando S, Obata K. Immunohistological study on brains of Alzheimer's disease using antibodies to fetal antigens, C-series gangliosides and microtubule-associated protein 5. Acta Neuropathol. 1991;81:626–31.

Taki T, Hirabayashi Y, Ichikawa H, Ando S, Kon K, Tanaka Y, et al. A ganglioside of rat ascites hepatoma AH7974F cells: occurrence of a novel disialoganglioside (GD1aα) with a unique *N*-acetylneuraminyl-α2,6-*N*-acetylgalactosamine structure. J Biol Chem. 1986;261:3075–8.

Tanaka Y, Ando S. Modulation of cholinergic synaptic functions by sialylcholesterol. Glycoconj J. 1996;13:321–6.

Tanaka Y, Hasegawa A, Ando S. Impaired synaptic functions with aging as characterized by decreased calcium influx and acetylcholine release. J Neurosci Res. 1996;43:63–70.

Tanaka Y, Waki H, Kon K, Ando S. Gangliosides enhance KCl-induced Ca^{2+} influx and acetylcholine release in brain synaptosomes. Neuro Report. 1997;8:2203–7.

Terry RD, Masliah E, Salmon DP, Butters N, DeTeresa R, Hill R, et al. Physical basis of cognitive alterations in Alzheimer's disease: synaptic loss in the major correlate of cognitive impairment. Ann Neurol. 1991;30:572–80.

Uchida Y. Molecular mechanism of regeneration in Alzheimer's disease brain. Geriatr Gerontol Int. 2010;10:S158–68.

Ulbanyi Z, Laszlo L, Yomasi TB, Toth E, Mekes E, Sass M, et al. Serum amyloid P component induces neuronal apoptosis and beta-amyloid immunoreactivity. Brain Res. 2003;988:67–77.

Vaccarino F, Giodotti A, Costa E. Ganglioside inhibition of glutamate-mediated protein kinase C translocation in primary cultures of cerebellar neurons. Proc Natl Acad Sci U S A. 1987;84: 8707–11.

Vanier MT, Holm M, Öhman R, Svennerholm L. Developmental profiles of gangliosides in human and rat brain. J Neurochem. 1971;18:581–92.

Vogelgesang S, Warzok RW, Cascorbi I, Kunert-Keil C, Schroeder E, Kroemer HK, et al. The role of P-glycoprotein in cerebral amyloid angiopathy; implications for the early pathogenesis of Alzheimer's disease. Curr Alzheimer Res. 2004;1:121–5.

Waki H, Kon K, Tanaka Y, Ando S. Facile methods for isolation of gangliosides in a small scale: age-related changes of gangliosides in mouse brain synaptic plasma membranes. Anal Biochem. 1994;222:156–62.

Walhovd KB, Fjell AM, Reinvang I, Lundervold A, Dale AM, Eilertsen DE, et al. Effects of age on volumes of cortex, white matter and subcortical structures. Neurobiol Aging. 2005;26: 1261–70.

Westbroek W, Gustafson AM, Sidransky E. Exploring the link between glucocerebrosidase mutations and parkinsonism. Trends Mol Med. 2011;17:485–93.

Wieraszko A, Seifert W. Evidence for the functional role of monosialoganglioside GM1 in synaptic transmission in rat hippocampus. Brain Res. 1986;371:305–13.

Wu G, Lu Z-H, Wang J, Wang Y, Xie X, Meyenhofer MF, et al. Enhanced susceptibility to kainite-induced seizures, neuronal apoptosis, and death in mice lacking gangliotetraose gangliosides: protection with LIGA20, a membrane-permeant analogue GM1. J Neurochem. 2005;25: 11014–22.

Wu G, Lu Z-H, Kulkarni N, Amin R, Ledeen RW. Mice lacking major gangliosides develop parkinsonism. Neurochem Res. 2011;36:1706–14.

Wu G, Lu Z-H, Kulkarni N, Ledeen RW. Deficiency of ganglioside GM1 correlates with Parkinson's disease in mice and humans. J Neurosci Res. 2012;90:1997–2008.

Yamada S, Mizutani T, Asano T, Enomoto M, Sakata M, Esaki Y, et al. Age-related brain atrophy with a constant cortical thickness in the normal elderly. Neuropathology. 1998;18:276–83.

Yamamoto N, Matsubara E, Maeda S, Minagawa H, Takashima A, Maruyama W, et al. A ganglioside-induced toxic soluble A β assembly, its enhanced formation from A β bearing the Arctic mutation. J Biol Chem. 2007;282:2646–55.

Yanagisawa K, Odaka A, Suzuki N, Ihara Y. GM1 ganglioside-bound amyloid β-protein (A β): a possible form of preamyloid in Alzheimer's disease. Nat Med. 1995;1:1062–6.

Yap TL, Grushus JM, Velayati A, Westbroek W, Goldin E, Moaven N, et al. α-Synuclein interacts with glucocerebrosidase providing a molecular link between Parkinson and Gaucher diseases. J Biol Chem. 2011;286:28080–8.

Yasojima K, Schwab C, McGeer EG, McGeer PL. Human neurons generate C-reactive protein and amyloid P: upreguration in Alzheimer's disease. Brain Res. 2000;887:80–9.

Yazaki T, Miura M, Asou H, Kitamura K, Toya S, Uyemura K. Glycopeptide of P0 protein inhibits hemophilic cell adhesion: competition assay with transformants and peptides. FEBS Lett. 1992;307:361–6.

Yohe HC, Roark DE, Rosenberg A. C20 sphingosine as a relevant factor in determining aggregative properties of gangliosides. J Biol Chem. 1976;251:7083–7.

Yu RK, Iqbal K. Sialosylgalactosyl ceramide as a specific marker for human myelin and oligodendroglial perikarya : gangliosides of human myelin oligodendroglia and neurons. J Neurochem. 1979;32:293–300.

Yu RK, Bieberich E, Xia T, Zeng G. Regulation of ganglioside biosynthesis in the nervous system. J Lipid Res. 2004;45:783–93.

Yu RK, Tsai YT, Ariga T. Functional roles of gangliosides in neurodevelopment: an overview of recent advances. Neurochem Res. 2012;37:1230–44.

Yusuf HKM, Dickerson JWT. Disialoganglioside GD1a of rat brain subcellular particles during development. Biochem J. 1978;174:655–7.

Zappia M, Crescibene L, Bosco D, Arabia G, Nicoletti G, Bagalia A, et al. Anti-GM1 ganglioside antibodies in Parkinson's disease. Acta Neurol Scand. 2002;106:54–7.

Zhang J, Kramer EG, Asp L, Dutta DJ, Navrazhina K, Pham T, et al. Promoting myelin repair and return of function in multiple sclerosis. FEBS Lett. 2011;585:3813–20.

Chapter 20
Gangliosides and Glycolipids in Neurodegenerative Disorders

J.S. Schneider

Abstract Glycolipids and gangliosides play important roles in maintaining the functional integrity of the nervous system. However, surprisingly little is known about how glycolipids and gangliosides in particular participate in various neurodegenerative processes. For example, it has been known for a long time that administration of gangliosides and in particular, GM1 ganglioside, can ameliorate damage to the central and peripheral nervous systems and can mitigate effects of a variety of neurodegenerative processes. What is not known is the extent to which dysfunctional biosynthesis or metabolism of gangliosides may be involved in various neurodegenerative disorders and if alterations observed reflect an intrinsic disease-related process or represent the response of the brain to a degenerative process. This chapter briefly reviews recent advances in the study of glycolipids and gangliosides and their potential participation in a variety of neurodegenerative disorders including Parkinson's disease, Alzheimer's disease, Huntington's disease and the potential link between Gaucher disease and Parkinson's disease.

Keywords Gangliosides • Glycolipids • Parkinson's disease • Huntington's disease • Alzheimer's disease • Gaucher disease

Abbreviations

AD Alzheimer's disease
CNS Central nervous system
Ca Calcium

J.S. Schneider, Ph.D. (✉)
Department of Pathology, Anatomy and Cell Biology, Thomas Jefferson University, Philadelphia, PA 19107, USA
e-mail: jay.schneider@jefferson.edu

R.K. Yu and C.-L. Schengrund (eds.), *Glycobiology of the Nervous System*, Advances in Neurobiology 9, DOI 10.1007/978-1-4939-1154-7_20,

mg/kg	Milligram per kilogram
TH	Tyrosine hydroxylase
HVA	Homovanillic acid
MPTP	1-Methyl-4-phenyl 1,2,3,6-tetrahydropyridine
UPDRS	Unified Parkinson's Disease Rating Scale
APP	Amyloid precursor protein
Aβ	Amyloid-beta
BACE1	Beta-secretase 1
DARP-32	Dopamine- and cAMP-regulated neuronal phosphoprotein
GBA1	Beta-glucosidase 1

Glycolipids are broadly defined as lipids with a carbohydrate attached. There are a wide variety of glycolipids found in biological systems. Of particular relevance for the nervous system is a subset of glycolipids known as glycosphingolipids that contain sphingosine. A further subset of glycosphingolipids of particular relevance to the structure and function of the nervous system are cerebrosides and gangliosides. Glycosphingolipids are present on cell surface membranes and are particularly abundant in the brain (Hirabayashi 2012). Cerebrosides comprise a group of glycosphingolipids consisting of ceramide and sugar residues, the major forms being glucocerebrosides and galactocerebrosides. Gangliosides are sialic acid-containing glycosphingolipids that are integral components of cell surface membranes and are highly enriched in the central nervous system. The major ganglioside species in brain are a- and b-series gangliosides and consist of the major gangliosides GM1, GD1a, GD1b, and GT1b. GM1 ganglioside, a main component of membrane signaling domains (lipid rafts), is particularly important in the central nervous system (CNS) as it plays important roles in neuronal development and survival and modulates a wide variety of cellular functions through modulation of cell signaling mechanisms. GM1 has been shown to exert neurotrophic or neuroprotective effects under a variety of circumstances and influences numerous cellular activities mediated at the level of the plasma membrane as well as intracellularly, where it influences Ca^{2+} homeostasis, mitochondrial function, and lysosomal integrity, among other processes (Hakomori and Igarashi 1993; Allende and Proia 2002; Shield et al. 2006; Wei et al. 2009).

This brief review focuses on recent advances in the description and understanding of the role of glycolipids and in particular, gangliosides and glucocerebrosides, in neurodegenerative disorders. This review focuses on Parkinson's disease, Alzheimer's disease, and Huntington's Disease and the accumulating evidence for a link between Parkinson's disease and Gaucher disease.

20.1 Gangliosides and Parkinson's Disease

GM1 ganglioside has long been suggested as a potential agent for the treatment of Parkinson's disease, although the precise reasons for the efficacy of GM1 in animal models of Parkinson's disease and in Parkinson's disease patients is still unclear.

Toffano et al. (1983) first described the ability of GM1 ganglioside to enhance recovery of the nigrostriatal projection system. In this initial study, administration of GM1 ganglioside (30 mg/kg, beginning the second day after surgery) to rats that received unilateral hemitransections of ascending nigrostriatal dopaminergic fibers significantly increased striatal tyrosine hydroxylase (TH) activity, homovanillic acid (HVA) content, TH immunohistochemical staining ipsilateral to the hemitransection and reduced apomorphine-induced rotations. These data were the first demonstration of improvement of biochemical, morphological, and behavioral parameters associated with damage to the nigrostriatal pathway with GM1 treatment (Toffano et al. 1983). Over the next couple of decades, numerous studies were performed showing that administration of GM1 ganglioside at least partially reversed detrimental effects of various types of insults to the nigrostriatal dopamine system including mechanical transection (Agnati et al. 1983; Toffano et al. 1983, 1984), 1-methyl-4-phenyl-1,2,3,6-tetrahydropyridine (MPTP) toxin-induced lesions (Hadjiconstantinou et al. 1986; Schneider and Yuwiler 1989; Hadjiconstantinou and Neff 1988), and 6-hydroxydopamine-induced lesions (Tilson et al. 1988) to the nigrostriatal dopamine system, as well as ameliorate age-related dopaminergic changes in brain. Results in normal aged animals are particularly interesting in that GM1 administration restored a variety of dopaminergic markers in the striatum and substantia nigra to levels approximate to those seen in young animals (Goettl et al. 1999, 2003). Although GM1 administration had little effect on several measures of motor function assessed in aged animals (Goettl et al. 2001), improvements in cognitive functioning (i.e., spatial learning and memory) were described (Fong et al. 1997). Although relatively few papers have examined age-related changes in ganglioside content in brain, in rats, GM1 levels in whole brain were reported to be fairly consistent between 3 and 24 months of age, while GD1a and GT1b levels decreased (Aydin et al. 2000). These data need to be viewed cautiously as there are significant regional differences in ganglioside content in brain that might be obscured by examining whole brain expression. Regional expression of gangliosides examined in human brain showed variation by region and age (between 4 months and 80/90 years of age), with GM1 levels significantly decreased with advanced age in the frontal cortex and moderately decreased in hippocampus (Kracun et al. 1992b), with a possible shift in expression from the a- to b-series gangliosides with aging (see Chap. 19).

While most of the experimental work with GM1 and Parkinson's disease models was performed with rodents, studies using nonhuman primate models of Parkinson's disease also showed significant improvements with GM1 treatment. In MPTP-treated monkeys, GM1 treatment resulted in improved motor and cognitive functioning, compared to saline-treated lesion control animals, and also resulted in small but statistically significant increases in striatal dopamine levels and increased density of tyrosine hydroxylase (TH)-positive fibers (Schneider et al. 1992). Other studies showed that GM1 treatment resulted in significant increases in TH protein levels in residual dopaminergic neurons in the substantia nigra of MPTP-treated monkeys (Herrero et al. 1993).

These promising preclinical research findings led to clinical studies of GM1 treatment in Parkinson's disease patients. All clinical studies using GM1 were approved by the appropriate institutional ethics committees, performed in accordance with the ethical standards laid down in the 1964 Declaration of Helsinki, and written informed consent was obtained from all subjects prior to enrollment. Following a small phase I safety study (Schneider et al. 1995b), a randomized, double-blind, placebo-controlled trial was conducted with 45 Parkinson's disease patients ($N=22$ GM1; $N=23$ placebo) over a 16-week study period (Schneider et al. 1998). At the study initiation visit, subjects received intravenous infusion of 1,000 mg GM1 or placebo and then received two subcutaneous injections per day (GM1 100 mg/2 ml per injection or placebo) for the remainder of the 16 week study period. The primary efficacy measure was change from baseline in the Unified Parkinson's Disease Rating Scale (UPDRS) motor score, assessed during a practically defined off period (that is, in the morning after at least 12 h from the last dose of any anti-Parkinson medication). The GM1-treated group showed significant improvements in UPDRS motor scores compared to subjects receiving placebo (Schneider et al. 1998). The treatment effect size at week 16 was -6.79 ± 1.24 ($p<0.0002$). Secondary measures of activities of daily living and performance of timed motor tests also showed significant effects in favor of the GM1-treated subjects (Schneider et al. 1998). GM1 was well tolerated and no serious adverse effects were noted. This brief duration study suggested that GM1 could have a symptomatic effect in Parkinson's disease patients. Although the precise mechanisms underlying this response are not known, the effects observed in this study are consistent with preclinical data demonstrating an ability of GM1 to enhance TH expression and dopamine synthesis in residual neurons in animal models of Parkinson's disease (Herrero et al. 1993; Schneider et al. 1995a).

A group of subjects who completed the study described above consented to continue to receive GM1 and to be followed clinically for up to 5 years (Schneider et al. 2010). In this open extension study (26 subjects at start; 13 from the prior placebo group and 13 from the prior GM1-treated group), subjects were administered a total daily dose of 200 mg GM1 by subcutaneous injection (two injections of 100 mg GM1 per day). Subjects were evaluated during practically defined off periods at 6 month intervals over the course of the study. Changes in UPDRS motor scores and activities of daily living scores were assessed as were timed tests of motor function. In subjects who received GM1 during the previous randomized study, regression modeling showed a slow linear increase of the average UPDRS motor score over time, with an estimated annual rate of increase of approximately 1.2 points (95 % confidence interval, CI: −0.1 to 2.5, $p=0.06$) (Schneider et al. 2010). However, at the end of 5 years of continuous GM1 use, the estimated average UPDRS motor score (for the ten subjects who continued through the entire 5 year period) was still lower than that at baseline prior to randomization into the main randomized trial, suggesting that improvements achieved during the initial study were maintained during the open extension period. In subjects previously randomized to receive placebo but who received GM1 during the open-extension study, regression modeling also showed a change in the average UPDRS motor scores over time. Over the first 2 years of the extension study, motor scores decreased by about 3–4 points per year on average while in later years, the scores increased slightly and approached pre-randomization levels by the fifth year. However, for nine of the ten subjects followed for the full 5 year period, no appreciable progression of motor symptoms was observed. This relatively small open label study suggested that chronic use of GM1 in PD patients was safe (no serious adverse events reported over the course of the study) and may be beneficial in Parkinson's disease patients. However, without a control group, it is not possible to come to any firm conclusions regarding the long-term efficacy of GM1 as a potential disease modifying agent in Parkinson's disease.

In order to better assess efficacy of GM1 as a potential disease modifying therapy for Parkinson's disease, a randomized, controlled, delayed start trial of GM1 in Parkinson's disease was conducted (Schneider et al. 2013). Due to the complex nature of Parkinson's disease, there are several challenges in the clinical assessment of disease progression and the demonstration of disease modification, since many drugs with potential disease modifying effects may also exert symptomatic effects. Since we had previously shown that GM1 may potentially have both symptomatic and disease modifying effects on Parkinson's disease, a study design, the delayed start study, previously suggested to possibly differentiate between these aspects of drug response (Olanow et al. 2008, 2009; D'Agostino 2009) was used. In the delayed start trial design, subjects are initially randomized to either an early start group (i.e., randomized to receive GM1 and to continue to receive GM1 for the duration of the trial (out to week 120)) or a delayed start group (randomized to receive placebo but switched to GM1 after 24 weeks). In addition, a separate group of standard-of-care subjects ("Comparison group") was recruited according to the same criteria used for recruitment of the treatment groups and these subjects were assessed longitudinally to provide descriptive information regarding disease progression. The primary outcome measure in this study was change from baseline UPDRS motor scores assessed during a practically defined off period. Subjects were also evaluated during their best on period (i.e., typically at least 1 h. after taking their standard anti-Parkinson medication). At week 24, the early-start (GM1) group showed significant improvement in UPDRS motor scores compared to a significant worsening of scores in the delayed-start (placebo) group (Fig. 20.1). The early-start group also showed a sustained benefit compared to the delayed-start group at the end of the study (week 120) (Fig. 20.1). Following an initial improvement in UPDRS motor scores in the early-start group, this measure of symptom severity changed little over the course of the study and by the end of the observation period, these scores were still lower (i.e., improved) than they were at study baseline (Schneider et al. 2013). The standard-of-care comparison group showed a pattern of symptom progression different from both the early-start subjects who received GM1 throughout the study and the delayed-start subjects after they started receiving GM1 after week 24 (Schneider et al. 2013). Both early-start and delayed-start groups showed significant symptom worsening during washout (assessed at 1–2 years after the last use of GM1) (Fig. 20.1). The results of this study supported the previous report that GM1 use for a relatively short period of time was superior to placebo for improving motor symptoms and showed that extended GM1 use (up to 120 weeks) resulted in a lower than expected rate of symptom progression (also

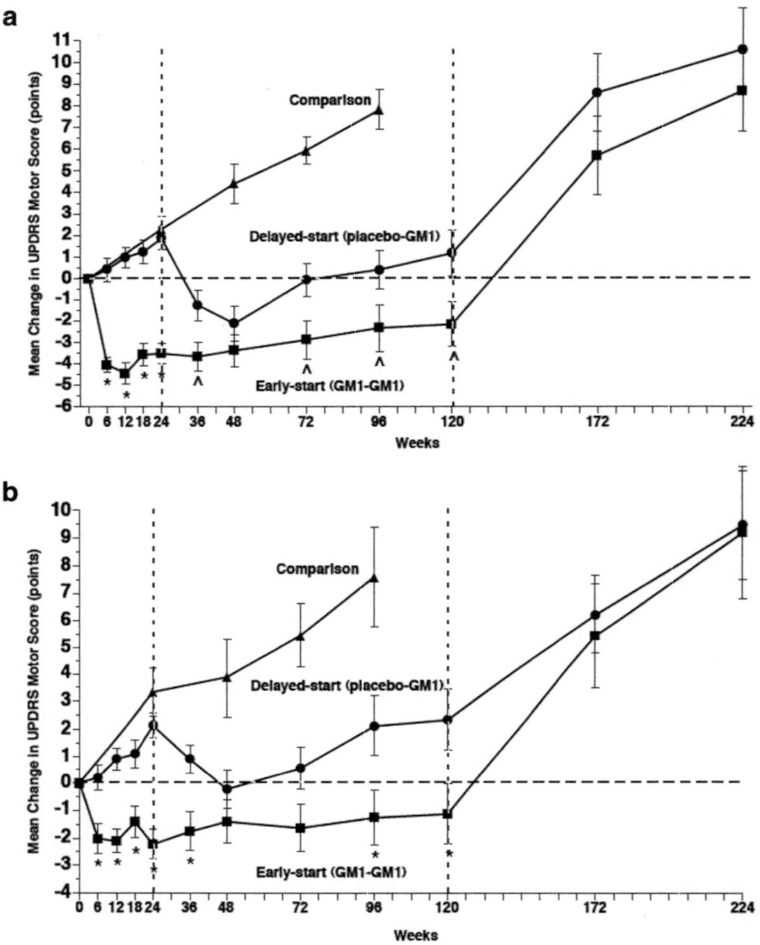

Fig. 20.1 Changes in unified Parkinson's disease rating scale (UPDRS) motor subsection scores. (**a**) The mean (±SE) change from baseline (observed scores) in Early-start and Delayed-start study subjects and in the standard-of-care Comparison group, assessed in the practically defined "off" condition. The *dashed vertical line* at week 24 indicates the end of study Phase I. The *dashed vertical line* at week 120 indicates the end of study Phase II. The *horizontal dashed line* indicates baseline level. An increase of score indicates symptom worsening; a decrease in score indicates symptom improvement. *$p < 0.0001$ Early-start vs. Delayed-start; ^$p < 0.05$ Early-start vs. Delayed-start. (**b**) The mean (±SE) change from baseline in Early-start and Delayed-start study subjects and in the standard-of-care Comparison group, assessed in the best "on" condition. The *dashed vertical line* at week 24 indicates the end of study Phase I. The *dashed vertical line* at week 120 indicates the end of study Phase II. The *horizontal dashed line* indicates baseline level. *$p < 0.01$ Early-start vs. Delayed-start. [Reprinted from Schneider JS, Gollomp SM, Sendek S, Colcher A, Cambi F, and Du W (2013) A randomized, controlled, delayed start trial of GM1 ganglioside in treated Parkinson's disease patients. *Journal of the Neurological Sciences* **324**:140–148

suggested in the previous open label study). Thus, the data from this relatively small study suggested that GM1 may have symptomatic and potentially disease modifying effects on Parkinson's disease.

Despite the positive effects of GM1 on preclinical models of Parkinson's disease and positive effects of GM1 in Parkinson's disease patients, the mechanisms underlying these effects remain somewhat obscure. We have suggested that at least in part, the apparent neuroprotective/neurorestorative effects of GM1 may involve modulation of lipid raft structure/function by altering the GM1 content of the rafts that could exert significant influence on a variety of signaling pathways (Schneider et al. 2013). Further, several Parkinson's disease-relevant proteins such as alpha synuclein, LRRK2, parkin, and PINK1 associate with lipid rafts and co-localize with GM1, potentially influencing neurodegeneration in Parkinson's disease (Martinez et al. 2007; Fallon et al. 2002; Hatano et al. 2007). One reason why GM1 may seemingly work so well in Parkinson's disease is that administration of GM1 to Parkinson's disease patients may represent GM1 replacement therapy, that is, restores a deficiency in GM1 levels (see Chap. 15). This would only be true if there is in fact a decrease in GM1 levels in the Parkinson's disease brain, and specifically, in the nigrostriatal dopamine system. Recent data from our lab (Kidd et al. 2012) and others (Wu et al. 2012) suggest that GM1 levels may be lower than normal in the substantia nigra of Parkinson's disease patients compared to age matched controls. This may be due to lower levels of GM1 in residual neurons. Additionally, we have made the novel observation that expression of key glycosyltransferase genes *B3Galt4* (the gene encoding the glycosyltransferase necessary for conversion of GM2 into GM1) and *St3gal2* (the gene encoding the sialyltransferase necessary for conversion of GM1 into GD1A) are decreased in the substantia nigra in Parkinson's disease brains compared with age-matched controls (Kidd et al. 2012). We have also recently observed increases in gene expression in Parkinson's disease substantia nigra for at least some of the enzymes responsible for the production of both a- and b-series gangliosides (i.e. *St3Gal5*, *St8sia3*, see Chap. 9 for metabolic pathways), compared to samples from neurologically intact, age-matched controls (Mettil et al. 2013). These findings together suggest that in addition to a loss of GM1 in the Parkinson's disease substantia nigra, there may also be an imbalance between a- and b-series gangliosides. More work is needed in order to understand the extent to which there is indeed a defect in GM1 expression in the Parkinson's disease brain and if so, the extent to which this defect is specific to the nigrostriatal dopamine system.

20.2 Gangliosides in Alzheimer's Disease

Several lines of evidence have linked Alzheimer's disease (AD) with aberrant lipid homeostasis and with abnormal expression of gangliosides. There have been a number of reports of decreased ganglioside concentrations in various regions of the AD brain dating back to the late 1960s, but some of these findings have been contradictory and lacked experimental consistency in the way in which gangliosides were measured and in the pathological confirmation of AD and AD subtypes. Gottfries et al. (1983) described a decreased expression of brain gangliosides in AD brain but the pattern of ganglioside expression was not evaluated. Kalanj et al. (1991) and Kracun et al. (1992a) reported that in AD brain, all ganglio-series gangliosides (i.e., GM1, GD1a, GD1b, and GT1b) were decreased in temporal and frontal cortices as well as in the nucleus basalis of Meynert, while gangliosides GM2, GD3, and GM3 were elevated in frontal and parietal cortex. Svennerholm and Gottfries (1994) described significant decreases in gangliosides in frontal and temporal cortices, caudate nucleus, and hippocampus in AD type I cases (early onset form) and more restricted loss of gangliosides in temporal cortex, hippocampus and frontal white matter in AD type II (late onset) cases. These authors suggested that based on a diminished yield of synaptosomes in AD type I brains that there was marked loss of synapses and neuronal processes in the AD type I brain. Studies examining CSF from AD patients showed increased GM1 in CSF in early onset AD compared to late onset AD, suggested by the authors to indicate more severe neurodegeneration in type I vs. type II AD patients (Blennow et al. 1991).

Recent evidence, however, suggests that more than simple measurement of ganglioside levels in AD brain may be needed to better appreciate the role that gangliosides might play in the pathophysiology of AD. More complex alterations in the lipid profile in brain may be associated with AD, with region-specific lipid anomalies potentially linked to AD pathogenesis (Chan et al. 2012). Aberrant lipid homeostasis has been suggested to play a role in AD as the neuronal lipid composition regulates activity of key proteins such as APP, BACE1 and presenilin that control Aβ levels (Chan et al. 2012). Gangliosides (GM1 in particular) have been suggested to modulate the pathogenic potential of Aβ by influencing its aggregation properties. A number of studies have suggested that the interaction of Aβ with GM1 results in GM1-bound Aβ that acts as a seed for the Aβ fibrillogenesis in the AD brain (Yanagisawa 2007). Recent data further suggest that Aβ is preferentially incorporated into GM1-rich membrane regions (i.e., lipid rafts) where the peptides undergo a conformational shift that disrupts membrane stability and promotes peptide–peptide interaction and oligomer formation (Haughey et al. 2010). Although the precise role of gangliosides and other lipids in AD remains to be determined, it is clear that alterations in sphingolipid metabolism and expression likely play an important role in the pathological processes contributing to AD. The role of GM1 ganglioside in particular in AD remains to be elucidated as some reports suggest that GM1 administration may be at least partially neuroprotective in model systems (Kreutz et al. 2011) and in humans (Svennerholm et al. 2002).

20.3 Gangliosides and Huntington's Disease

Desplats et al. (2007) reported abnormal expression levels of various genes encoding glycosyltransferases and sialyltransferases that are involved in the biosynthesis of various gangliosides in the striatum of Huntington's disease transgenic mice (R6/1 mice) and in postmortem caudate nucleus tissue from Huntington's disease patients. In particular, increased expression of *St8sia1* (encoding GD3 synthase) and decreased expression of *St8sia2*, *St8sia3* (GD3 synthase) and *B4galnt1* (GM2/GD2 synthase) was found in the striatum of R6/1 transgenic Huntington's disease mice. The mRNA expression levels of *St3gal5* (encoding GM3 synthase), *St3gal2* (encoding GM1b/GD1a/GT1b synthase), and *St6galnac5* (*SiaT7e*) were not affected in the R6/1 transgenic mice (Desplats et al. 2007). Ganglioside analysis showed a significant decrease in GM1 levels in striatum from R6/1 transgenic mice (Desplats et al. 2007). The gene expression of *St3gal5, St8sia3*, *B4galnt1*, and *St3gal2* were significantly decreased in the caudate nucleus from HD patients. However, in human Huntington's disease caudate, an overall decrease in ganglioside levels was reported, compared to control subjects with the exception of a specific increase in GD3 levels (Desplats et al. 2007). This could be significant as GD3 may be an apoptogenic ganglioside (Scorrano et al. 1999) and increased levels of GD3 in Huntington's disease caudate could contribute to apoptotic neurodegeneration. Although these data are from a very small sample of Huntington's patients (i.e., three cases), they do suggest altered ganglioside biosynthesis and expression associated with Huntington's disease.

Following up on these data, Maglione et al. (2010) described decreased expression of *B3galt4* (GM1/GD1b synthase) mRNA and reduced levels of GM1 ganglioside in cell and animal models of Huntington's disease as well as in fibroblasts isolated from Huntington's disease patients (Maglione et al. 2010). They further reported that the presence of mutant huntingtin protein affected ganglioside metabolism in a transgenic mouse Huntington's disease model, with reduced GM1, GD1a, and GT1b levels noted in striatum and cortex (Maglione et al. 2010). Administration of GM1 to a knock-in cell model of Huntington's disease restored GM1 levels and protected cells from apoptosis. GM1 administration also promoted AKT activation and resulted in phosphorylation of mutant huntingtin protein which may support neuroprotection and perhaps decrease toxicity of the mutant protein (Maglione et al. 2010). Importantly, this study demonstrated that relatively small alterations of GM1 content in the plasma membrane could cause significant changes in cell susceptibility to apoptosis.

As a follow up to the findings that levels of GM1 may be decreased in the HD brain (as well as in HD models), Di Pardo et al. (2012) demonstrated that administering GM1 by intraventricular infusion in a mouse HD model ((YAC)128 mice) restored normal motor behavior, increased striatal expression of DARP-32, a protein highly enriched in intrinsic striatal neurons, and increased phosphorylation of huntingtin at serine 13 and serine 16, suggested a potential role for GM1 in the treatment of HD.

20.4 Role of Glycolipids in the Link Between Parkinson's Disease and Gaucher Disease

A possible connection between Gaucher disease and Parkinson's disease has recently been suggested, although the molecular link between the two diseases is still somewhat obscure (Goldin 2010) (see also Chap. 19). The occurrence of Parkinson's disease in some type 1 (non-neuronopathic) Gaucher disease individuals and their first degree relatives suggested *GBA1* mutations might be a genetic risk factor for idiopathic Parkinson's (Bultron et al. 2010). It has recently been shown that compared to the general population, there is an approximate 20-fold increase in lifetime risk of developing Parkinson's disease in patients with type 1 Gaucher disease (Bultron et al. 2010). Gaucher disease is caused by mutations in the *GBA1* gene that encodes the enzyme glucocerebrosidase which degrades glycosphingolipids to ceramide and glucose (Beutler and Grabowski 2001). *GBA1* mutations result in lysosomal accumulation of glucocerebroside (glucosylceramide). In type II Gaucher disease, increased levels of ganglioside GD3 have been reported in brain and in cerebrospinal fluid (Gornati et al. 2002); some studies have reported increases in expression of gangliosides GM2, GM3, GM1, and GD3 in the CNS in Gaucher disease while others have indicated either no change or decreases in gangliosides (Gonzalez-Sastre et al. 1974; Gornati et al. 2002; Conradi et al. 1984; Kaye et al. 1986). However, most of the reports of altered ganglioside expression were in association with type II or infantile forms of Gaucher disease. The extent to which CNS ganglioside expression is altered in type I Gaucher is unclear. However, ganglioside GM3 is reported to be strikingly elevated in plasma of type 1 Gaucher disease patients (Ghauharali-van der Vlugt et al. 2008).

A further link between Gaucher disease and Parkinson's disease is the presence of α-synuclein inclusions in brains of patients with type I Gaucher disease and Parkinsonism. Such patients also have Lewy body pathology and loss of substantia nigra dopaminergic neurons (Wong et al. 2004). Brain samples from patients with Gaucher disease or from Gaucher disease carriers with Parkinsonism all showed *GBA1* mutations and Lewy body pathology, with α-synuclein and glucocerebrosidase detected in Lewy body inclusions (Shachar et al. 2011; Goker-Alpan et al. 2010). It has also recently been shown that increased intracellular glucosylceramide (GlcCer) levels (GlcCer accumulates in affected tissues in Gaucher disease due to glucocerebrosidase deficiency) enhance the formation of toxic α-synuclein assemblies which may lead to neurodegeneration (Mazzulli et al. 2011).

20.5 Concluding Remarks

Although there have been tremendous advances in the understanding of the pathophysiology of numerous neurodegenerative disorders, there is still much that is not known about the mechanisms that initiate and drive neurodegeneration. This is particularly true in Parkinson's disease and Alzheimer's disease. Recent data though suggest that alterations in ganglioside biosynthesis and/or metabolism may be involved in numerous neurodegenerative disorders. Further research is now needed in order to better understand the nature of the changes in brain gangliosides in these disorders and to understand the extent to which these changes are either the cause or effect of neurodegenerative processes.

Ethics Statement Some of the research reported in this chapter was supported by NIH grant NS038681. The author declares that he has no conflict of interest.

References

Agnati LF, Fuxe K, Calza L, Benfenati F, Cavicchioli L, Toffano G, Goldstein M. Gangliosides increase the survival of lesioned nigral dopamine neurons and favour the recovery of dopaminergic synaptic function in striatum of rats by collateral sprouting. Acta Physiol Scand. 1983;119:347–63.

Allende ML, Proia RL. Lubricating cell signaling pathways with gangliosides. Curr Opin Struct Biol. 2002;12:587–92.

Aydin M, Cengiz S, Agachan B, Yilmaz H, Isbir T. Age-related changes in GM1, GD1a, GT1b components of gangliosides in Wistar albino rats. Cell Biochem Funct. 2000;18:41–5.

Beutler E, Grabowski G. Gaucher disease. In Scriver CR, Beaudet AL, Sly WS, Valle D (eds). The Metabolic and Molecular Bases of Inherited Disease. New York: McGraw-Hill. 2001; pp. 3635–3668.

Blennow K, Davidsson P, Wallin A, Fredman P, Gottfries CG, Karlsson I, Mansson JE, Svennerholm L. Gangliosides in cerebrospinal fluid in 'probable Alzheimer's disease'. Arch Neurol. 1991;48:1032–5.

Bultron G, Kacena K, Pearson D, Boxer M, Yang R, Sathe S, Pastores G, Mistry PK. The risk of Parkinson's disease in type 1 Gaucher disease. J Inherit Metab Dis. 2010;33:167–73.

Chan RB, Oliveira TG, Cortes EP, Honig LS, Duff KE, Small SA, Wenk MR, Shui G, Di Paolo G. Comparative lipidomic analysis of mouse and human brain with Alzheimer disease. J Biol Chem. 2012;287:2678–88.

Conradi NG, Sourander P, Nilsson O, Svennerholm L, Erikson A. Neuropathology of the Norrbottnian type of Gaucher disease. Morphological and biochemical studies. Acta Neuropathol. 1984;65:99–109.

D'Agostino Sr RB. The delayed-start study design. N Engl J Med. 2009;361:1304–6.

Desplats PA, Denny CA, Kass KE, Gilmartin T, Head SR, Sutcliffe JG, Seyfried TN, Thomas EA. Glycolipid and ganglioside metabolism imbalances in Huntington's disease. Neurobiol Dis. 2007;27:265–77.

Di Pardo A, Maglione V, Alpaugh M, Horkey M, Atwal RS, Sassone J, Ciammola A, Steffan JS, Fouad K, Truant R, Sipione S. Ganglioside GM1 induces phosphorylation of mutant huntingtin and restores normal motor behavior in Huntington disease mice. Proc Natl Acad Sci U S A. 2012;109:3528–33.

Fallon L, Moreau F, Croft BG, Labib N, Gu WJ, Fon EA. Parkin and CASK/LIN-2 associate via a PDZ-mediated interaction and are co-localized in lipid rafts and postsynaptic densities in brain. J Biol Chem. 2002;277:486–91.

Fong TG, Neff NH, Hadjiconstantinou M. GM1 ganglioside improves spatial learning and memory of aged rats. Behav Brain Res. 1997;85:203–11.

Ghauharali-van der Vlugt K, Langeveld M, Poppema A, Kuiper S, Hollak CE, Aerts JM, Groener JE. Prominent increase in plasma ganglioside GM3 is associated with clinical manifestations of type I Gaucher disease. Clin Chim Acta. 2008;389:109–13.

Goettl VM, Wemlinger TA, Duchemin AM, Neff NH, Hadjiconstantinou M. GM1 ganglioside restores dopaminergic neurochemical and morphological markers in aged rats. Neuroscience. 1999;92:991–1000.

Goettl VM, Wemlinger TA, Colvin AE, Neff NH, Hadjiconstantinou M. Motoric behavior in aged rats treated with GM1. Brain Res. 2001;906:92–100.

Goettl VM, Zhang H, Burrows AC, Wemlinger TA, Neff NH, Hadjiconstantinou M. GM1 enhances dopaminergic markers in the brain of aged rats. Exp Neurol. 2003;183:665–72.

Goker-Alpan O, Stubblefield BK, Giasson BI, Sidransky E. Glucocerebrosidase is present in alpha-synuclein inclusions in Lewy body disorders. Acta Neuropathol. 2010;120:641–9.

Goldin E. Gaucher disease and parkinsonism, a molecular link theory. Mol Genet Metab. 2010;101:307–10.

Gonzalez-Sastre F, Pampols T, Sabater J. Infantile Gaucher's disease: a biochemical study. Neurology. 1974;24:162–7.

Gornati R, Berra B, Montorfano G, Martini C, Ciana G, Ferrari P, Romano M, Bembi B. Glycolipid analysis of different tissues and cerebrospinal fluid in type II Gaucher disease. J Inherit Metab Dis. 2002;25:47–55.

Gottfries CG, Adolfsson R, Aquilonius SM, Carlsson A, Eckernas SA, Nordberg A, Oreland L, Svennerholm L, Wiberg A, Winblad B. Biochemical changes in dementia disorders of Alzheimer type (AD/SDAT). Neurobiol Aging. 1983;4:261–71.

Hadjiconstantinou M, Neff NH. Treatment with GM1 ganglioside restores striatal dopamine in the 1-methyl-4-phenyl-1,2,3,6-tetrahydropyridine-treated mouse. J Neurochem. 1988;51:1190–6.

Hadjiconstantinou M, Rossetti ZL, Paxton RC, Neff NH. Administration of GM1 ganglioside restores the dopamine content in striatum after chronic treatment with MPTP. Neuropharmacology. 1986;25:1075–7.

Hakomori S, Igarashi Y. Gangliosides and glycosphingolipids as modulators of cell growth, adhesion, and transmembrane signaling. Adv Lipid Res. 1993;25:147–62.

Hatano T, Kubo S, Imai S, Maeda M, Ishikawa K, Mizuno Y, Hattori N. Leucine-rich repeat kinase 2 associates with lipid rafts. Hum Mol Genet. 2007;16:678–90.

Haughey NJ, Bandaru VV, Bae M, Mattson MP. Roles for dysfunctional sphingolipid metabolism in Alzheimer's disease neuropathogenesis. Biochim Biophys Acta. 2010;1801:878–86.

Herrero MT, Perez-Otano I, Oset C, Kastner A, Hirsch EC, Agid Y, Luquin MR, Obeso JA, Del Rio J. GM-1 ganglioside promotes the recovery of surviving midbrain dopaminergic neurons in MPTP-treated monkeys. Neuroscience. 1993;56:965–72.

Hirabayashi Y. A world of sphingolipids and glycolipids in the brain—novel functions of simple lipids modified with glucose. Proc Jpn Acad Ser B Phys Biol Sci. 2012;88:129–43.

Kalanj S, Kracun I, Rosner H, Cosovic C. Regional distribution of brain gangliosides in Alzheimer's disease. Neurol Croat. 1991;40:269–81.

Kaye EM, Ullman MD, Wilson ER, Barranger JA. Type 2 and type 3 Gaucher disease: a morphological and biochemical study. Ann Neurol. 1986;20:223–30.

Kidd SK, Mettil W, Anderson DW, Schneider JS. Ganglioside regulation in the human substantia nigra and its relation to Parkinson's disease. Program No. 754.16. 2012 Neuroscience Meeting Planner. New Orleans, LA: Society for Neuroscience, Online; 2012.

Kracun I, Kalanj S, Talan-Hranilovic J, Cosovic C. Cortical distribution of gangliosides in Alzheimer's disease. Neurochem Int. 1992a;20:433–8.

Kracun I, Rosner H, Drnovsek V, Vukelic Z, Cosovic C, Trbojevic-Cepe M, Kubat M. Gangliosides in the human brain development and aging. Neurochem Int. 1992b;20:421–31.

Kreutz F, Frozza RL, Breier AC, de Oliveira VA, Horn AP, Pettenuzzo LF, Netto CA, Salbego CG, Trindade VM. Amyloid-beta induced toxicity involves ganglioside expression and is sensitive to GM1 neuroprotective action. Neurochem Int. 2011;59:648–55.

Maglione V, Marchi P, Di Pardo A, Lingrell S, Horkey M, Tidmarsh E, Sipione S. Impaired ganglioside metabolism in Huntington's disease and neuroprotective role of GM1. J Neurosci. 2010;30:4072–80.

Martinez Z, Zhu M, Han S, Fink AL. GM1 specifically interacts with alpha-synuclein and inhibits fibrillation. Biochemistry. 2007;46:1868–77.

Mazzulli JR, Xu YH, Sun Y, Knight AL, McLean PJ, Caldwell GA, Sidransky E, Grabowski GA, Krainc D. Gaucher disease glucocerebrosidase and alpha-synuclein form a bidirectional pathogenic loop in synucleinopathies. Cell. 2011;146:37–52.

Mettil W, Kidd SK, Anderson DW, Schneider JS. Decreased GM1 ganglioside expression and altered ganglioside biosynthetic pathways in the brain of Parkinson's disease patients. Program No. 806.09. 2013 Neuroscience Meeting Planner. San Diego, CA: Society for Neuroscience, Online; 2013.

Olanow CW, Hauser RA, Jankovic J, Langston W, Lang A, Poewe W, Tolosa E, Stocchi F, Melamed E, Eyal E, Rascol O. A randomized, double-blind, placebo-controlled, delayed start study to assess rasagiline as a disease modifying therapy in Parkinson's disease (the ADAGIO study): rationale, design, and baseline characteristics. Mov Disord. 2008;23:2194–201.

Olanow CW, Rascol O, Hauser R, Feigin PD, Jankovic J, Lang A, Langston W, Melamed E, Poewe W, Stocchi F, Tolosa E. A double-blind, delayed-start trial of rasagiline in Parkinson's disease. N Engl J Med. 2009;361:1268–78.

Schneider JS, Yuwiler A. GM1 ganglioside treatment promotes recovery of striatal dopamine concentrations in the mouse model of MPTP-induced parkinsonism. Exp Neurol. 1989;105:177–83.
Schneider JS, Pope A, Simpson K, Taggart J, Smith MG, DiStefano L. Recovery from experimental parkinsonism in primates with GM1 ganglioside treatment. Science. 1992;256:843–6.
Schneider JS, Kean A, DiStefano L. GM1 ganglioside rescues substantia nigra pars compacta neurons and increases dopamine synthesis in residual nigrostriatal dopaminergic neurons in MPTP-treated mice. J Neurosci Res. 1995a;42:117–23.
Schneider JS, Roeltgen DP, Rothblat DS, Chapas-Crilly J, Seraydarian L, Rao J. GM1 ganglioside treatment of Parkinson's disease: an open pilot study of safety and efficacy. Neurology. 1995b;45:1149–54.
Schneider JS, Roeltgen DP, Mancall EL, Chapas-Crilly J, Rothblat DS, Tatarian GT. Parkinson's disease: improved function with GM1 ganglioside treatment in a randomized placebo-controlled study. Neurology. 1998;50:1630–6.
Schneider JS, Sendek S, Daskalakis C, Cambi F. GM1 ganglioside in Parkinson's disease: results of a five year open study. J Neurol Sci. 2010;292:45–51.
Schneider JS, Gollomp SM, Sendek S, Colcher A, Cambi F, Du W. A randomized, controlled, delayed start trial of GM1 ganglioside in treated Parkinson's disease patients. J Neurol Sci. 2013;324:140–8.
Scorrano L, Petronilli V, Di Lisa F, Bernardi P. Commitment to apoptosis by GD3 ganglioside depends on opening of the mitochondrial permeability transition pore. J Biol Chem. 1999; 274:22581–5.
Shachar T, Lo Bianco C, Recchia A, Wiessner C, Raas-Rothschild A, Futerman AH. Lysosomal storage disorders and Parkinson's disease: Gaucher disease and beyond. Mov Disord. 2011;26: 1593–604.
Shield AJ, Murray TP, Board PG. Functional characterisation of ganglioside-induced differentiation-associated protein 1 as a glutathione transferase. Biochem Biophys Res Commun. 2006;347:859–66.
Svennerholm L, Gottfries CG. Membrane lipids, selectively diminished in Alzheimer brains, suggest synapse loss as a primary event in early-onset form (type I) and demyelination in late-onset form (type II). J Neurochem. 1994;62:1039–47.
Svennerholm L, Brane G, Karlsson I, Lekman A, Ramstrom I, Wikkelso C. Alzheimer disease—effect of continuous intracerebroventricular treatment with GM1 ganglioside and a systematic activation programme. Dement Geriatr Cogn Disord. 2002;14:128–36.
Tilson HA, Harry GJ, Nanry K, Hudson PM, Hong JS. Ganglioside interactions with the dopaminergic system of rats. J Neurosci Res. 1988;19:88–93.
Toffano G, Savoini G, Moroni F, Lombardi G, Calza L, Agnati LF. GM1 ganglioside stimulates the regeneration of dopaminergic neurons in the central nervous system. Brain Res. 1983;261: 163–6.
Toffano G, Savoini G, Aporti F, Calzolari S, Consolazione A, Maura G, Marchi M, Raiteri M, Agnati LF. The functional recovery of damaged brain: the effect of GM1 monosialoganglioside. J Neurosci Res. 1984;12:397–408.
Wei J, Fujita M, Nakai M, Waragai M, Sekigawa A, Sugama S, Takenouchi T, Masliah E, Hashimoto M. Protective role of endogenous gangliosides for lysosomal pathology in a cellular model of synucleinopathies. Am J Pathol. 2009;174:1891–909.
Wong K, Sidransky E, Verma A, Mixon T, Sandberg GD, Wakefield LK, Morrison A, Lwin A, Colegial C, Allman JM, Schiffmann R. Neuropathology provides clues to the pathophysiology of Gaucher disease. Mol Genet Metab. 2004;82:192–207.
Wu G, Lu ZH, Kulkarni N, Ledeen RW. Deficiency of ganglioside GM1 correlates with Parkinson's disease in mice and humans. J Neurosci Res. 2012;90:1997–2008.
Yanagisawa K. Role of gangliosides in Alzheimer's disease. Biochim Biophys Acta. 2007;1768: 1943–51.

Chapter 21
Glycosidases: Inborn Errors of Glycosphingolipid Catabolism

Hisashi Ashida and Yu-Teh Li

Abstract Glycosphingolipids (GSLs) are information-rich glycoconjugates that occur in nature mainly as constituents of biomembranes. Each GSL contains a complex carbohydrate chain linked to a ceramide moiety that anchors the molecule to biomembranes. In higher animals, catabolism of GSLs takes place in lysosomes where sugar chains in GSLs are hydrolyzed by exo-glycosidases to cleave a sugar residue from the non-reducing end of a sugar chain. Inborn errors of GSL-catabolism, collectively called sphingolipidoses or GSL-storage diseases, are caused by the deficiency of exo-glycosidases responsible for the degradation of the specific sugar residues at the non-reducing termini in GSLs. This chapter briefly discusses glycone, anomeric, linkage, and aglycone specificities of exo-glycosidases and some of the historical landmarks on their associations with the chemical pathology of the five best known sphingolipidoses: GM1 gangliosidosis, GM2 gangliosidosis (Tay–Sachs disease), Fabry disease, Gaucher disease, and Krabbe disease.

Keywords Exo-glycosidases • Inborn errors of glycosphingolipid catabolism • Sphingolipid storage diseases • GM1 gangliosidosis • GM2 gangliosidosis • Tay–Sachs disease • Fabry disease • Gaucher disease • Krabbe disease

H. Ashida
Faculty of Biology-Oriented Science and Technology, Kinki University,
Kinokawa-shi, Wakayama 649-6493, Japan
e-mail: ashida@waka.kindai.ac.jp

Y.-T. Li (✉)
Department of Biochemistry and Molecular Biology, Tulane University School of Medicine,
New Orleans, LA 70112, USA
e-mail: yli1@tulane.edu

R.K. Yu and C.-L. Schengrund (eds.), *Glycobiology of the Nervous System*,
Advances in Neurobiology 9, DOI 10.1007/978-1-4939-1154-7_21,

Abbreviations

4MU	4-Methylumbelliferyl
CNS	Central nerve system
GALC	Galactocerebrosidase
GalSph	Galactosylsphingosine
GlcSph	Glucosylsphingosine
GM2-AP	GM2 activator protein
GSL	Glycosphingolipid
Hex A	β-Hexosaminidase A
Hex B	β-Hexosaminidase B
PNS	Peripheral nerve system
TGM2	Taurine-conjugated GM2
TSD	Tay–Sachs disease

21.1 Introduction

Sugar chains of the three major mammalian glycoconjugates, glycosphingolipids (GSLs), glycoproteins, and proteoglycans, are synthesized in the endoplasmic reticulum and the Golgi by sequential addition of sugar residues by a wide variety of glycosyltransferases, and subsequently transported to their destinations, such as cell surface or extracellular environments (Varki et al. 2009). It has been well established that homeostasis of cellular constituents is maintained by lysosomal degradation through autophagy (Klionsky 2007) and that lysosomes contain a myriad of acid hydrolases capable of digesting such biomolecules as nucleic acids, proteins, lipids, and glycoconjugates (de Duve 1983). GSLs are information-rich glycoconjugates that occur in nature, mainly as constituents of biomembranes. Each GSL contains a complex carbohydrate chain linked to a ceramide moiety that anchors the molecule to biomembranes. Except for sulfatide (galactosylceramide-3-sulfate), lysosomal degradation of GSLs in higher animals is by sequential hydrolysis of their sugar chains by exo-glycosidases. All glycosidases referred to in this chapter are exo-glycosidases; an exo-glycosidase cleaves a sugar residue from the non-reducing end of a sugar chain. Inborn errors of GSL-catabolism, collectively called sphingolipidoses or GSL-storage diseases, are caused by the deficiency of glycosidases responsible for the degradation of a specific sugar residue at the non-reducing termini of GSLs. With the exception of Krabbe disease, the deficiency of a glycosidase results in the lysosomal accumulation of the GSLs, as well as other glycoconjugates that are the substrates of the missing enzyme. The disorders are characterized by the excessive accumulation of undegraded GSLs and other partially degraded glycoconjugates in tissues as well as their excretion in the urine. The five best-known sphingolipidoses associated with the deficiency of lysosomal glycosidases are GM1 gangliosidosis, Tay–Sachs disease (TSD; GM2 gangliosidosis), Fabry disease, Gaucher disease, and Krabbe disease. With the exception of Fabry disease, an X-linked disorder, the other four are autosomal recessive disorders. Since clinical and pathological descriptions, gene mutations, and therapy for these disorders have been reviewed in detail in "The Online Metabolic and Molecular Bases of Inherited Diseases," this chapter focuses on some of the historical landmarks for glycosidases associated with the pathogenesis of GSL-storage disorders.

21.2 Specificities of Glycosidases

The primary structure of a sugar chain is determined by (1) the anomeric configurations of different sugar residues and their sequential arrangement and (2) the linkages between the two neighboring sugar residues in the sugar chain. One of the most important considerations in the enzymatic degradation of a sugar chain is the specificities of the glycosidases. The four basic types of glycosidase specificity are glycone specificity, anomeric specificity, linkage specificity, and aglycone specificity (Li and Li 1999).

21.2.1 Glycone Specificity

As shown in Fig. 21.1, the β-galactopyranoside at the non-reducing end of the sugar chain in GM1 is the glycone recognized by β-galactosidases. In general, glycone specificities of glycosidases are quite strict; however, they are not absolute. Some glycosidases are not able to distinguish structurally related glycones. For example, the recombinant glycosidase encoded by the glycoside hydrolase gene *bglMKg* was found to exhibit β-glucosidase, β-fucosidase and β-galactosidase activities (Wierzbicka-Wos et al. 2013). Another widely known example is the dual specificity of β-hexosaminidases. β-Hexosaminidases are able to cleave both β-linked *N*-acetylglucosaminide (GlcNAc) and β-linked *N*-acetylgalactosaminide (GalNAc) (Robinson and Stirling 1968; Woollen et al. 1961). Apparently the configurations of the hydroxyl groups at the C-4 of GlcNAc and GalNAc are irrelevant for the active site of β-hexosaminidases. Although enzymes that cleave exclusively the β-linked GlcNAc or β-linked-GalNAc are rare, a neutral β-*N*-acetylglucosaminidase isolated from rat spleen was found to be highly selective in cleaving *O*-β-GlcNAc-bearing glycopeptides (Dong and Hart 1994).

21.2.2 Anomeric Specificity

Without exception, the anomeric specificities of glycosidases are absolute. For example, α-linked galactopyranosides are refractory to β-galactosidases while β-linked galactopyranosides are refractory to α-galactosidases.

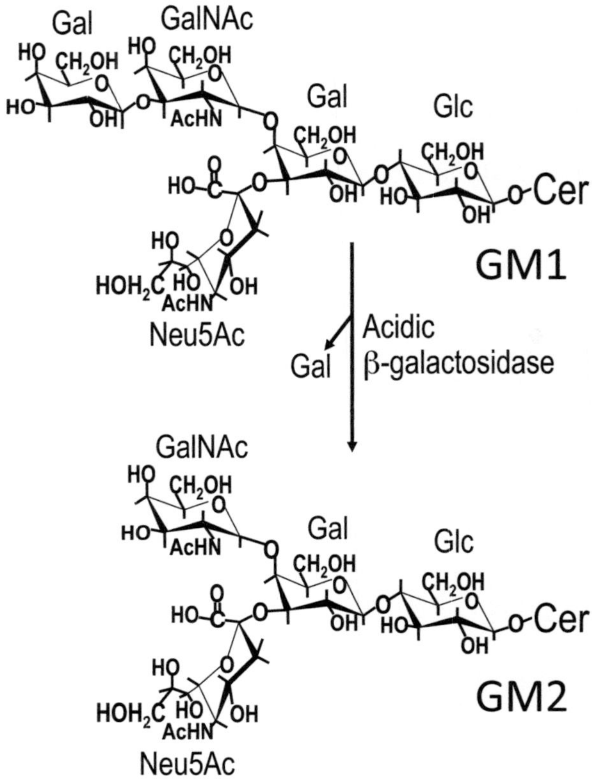

Fig. 21.1 Enzymatic conversion of GM1 to GM2. Conformational representation is used for each monosaccharide

21.2.3 Linkage Specificity

Most glycosidases do not have stringent linkage specificity. However, some glycosidases do exhibit varying degrees of preference toward a specific glycosidic linkage (Tyagarajan et al. 1996). For example, the sialidase from *Macrobdella* leech only cleaves the Neu5Acα2,3Gal linkage among various sialosyl linkages (Chou et al. 1994).

21.2.4 Aglycone Specificity

Aglycone specificities of glycosidases are the least obvious and their existence is often overlooked. Glycosidases do exhibit aglycone specificities. For example, GM1, LacCer and GalCer all contain β-linked Gal as the glycone, but their aglycones are very different. The acidic β-galactosidase capable of hydrolyzing the terminal Gal from GM1 not only recognizes the terminal Gal residue but also recognizes the aglycone part of GM1 moiety from the penultimate sugar GalNAc to the ceramide (see Fig. 21.1). However, this acidic β-galactosidase is not able to hydrolyze the Gal residue in GalCer (Suzuki et al. 2013). Similarly, the β-galactosidase (galactocerebrosidase) that hydrolyzes the Gal reside in GalCer is unable to hydrolyze the terminal Gal residue in GM1 (Wenger et al. 2013). Although acidic β-galactosidase and galactocerebrosidase are two distinct β-galactosidases, both of them are able to cleave LacCer. Thus, glycosidases are capable of distinguishing structurally related sugar chains in glycoconjugates through their aglycone specificities. The aglycone specificities of glycosidases are quite complex and their biological significances are still not well understood. It should be emphasized that although synthetic glycosides, such as 4-methylumbelliferyl (4MU)- and *p*-nitrophenyl-glycosides are convenient substrates for assaying glycosidase activities in vitro, these substrates only possess correct glycone moieties and anomeric configurations. Their aglycones are profoundly different from those found in glycoconjugates. Therefore, the activity of a glycosidase determined by using a synthetic substrate does not necessarily represent the activity toward natural substrates.

21.3 Detergents and Activator Proteins for the Hydrolysis of Sugar Chains in GSLs by Glycosidases

Due to their amphipathic nature, a GSL may exist in aqueous medium as a monolayer, micelles or bilayer vesicles. GSLs are poor substrates for water-soluble glycosidases. In general, in vitro hydrolysis of sugar residues in GSLs by glycosidases requires the presence of a detergent. Bile salts and Triton X-100 are most commonly used detergents to assist with the enzymatic hydrolysis. Under the same conditions, GSLs with shorter sugar chains are usually more resistant to glycosidases than those with longer ones. Depending on the glycosidase and the nature of the GSL substrate, certain detergents can be more effective than others in promoting hydrolysis. Moreover, the same detergent may not exert a similar stimulatory effect on the same functional glycosidase isolated from different sources. The effects of detergents on the in vitro hydrolysis of sugar chains in GSLs by glycosidases are still not well understood. Furthermore, in vitro hydrolyses of GSLs by glycosidases in the presence of detergents may not correlate with their in vivo activities. Since bile salts do not exist in tissues other than the liver and digestive organs; the in vivo degradation of amphipathic GSLs by glycosidases must be facilitated by a different mechanism. It is now widely known that the in vivo hydrolyses of GSLs are facilitated by protein cofactors called sphingolipid activator proteins. In 1964, while studying the hydrolysis of sulfatide by arylsulfatase A isolated from porcine kidney, Mehl and Jatzkewitz (1964) separated the enzyme preparation into heat-labile and heat-stable protein fractions using carrier-free high voltage electrophoresis. The heat-labile fraction was found to be identical to arylsulfatase A that hydrolyzed synthetic arylsulfates. The heat-stable fraction, on the other hand, was enzymatically inactive and was able to stimulate the heat-labile arylsulfatase A to carry out hydrolysis of the sulfate from galactosylceramide-3-sulfate (Mehl and Jatzkewitz 1964).

Although arylsulfatase A is not a glycosidase, this was the first observation of the requirement for a protein cofactor (activator protein) for the enzymatic hydrolysis of a GSL. During the 1960s and 1970s, studies of the enzymatic hydrolysis of GlcCer (Ho et al. 1973), GM2 (Hechtman 1977; Li et al. 1973), and GM1 (Li and Li 1976; Li et al. 1979) established the requirement of activator proteins for glycosidases to hydrolyze sugar residues in GSLs. Among the known five sphingolipid activator proteins, four are proteolytic products of prosaposin (Kishimoto et al. 1992), whose gene is located on chromosome 10. The fifth activator protein, GM2 activator protein, is the product of a separate gene located on chromosome 5 (Sandhoff et al. 2013). Early historical reports about sphingolipid activator proteins have been reviewed (Li and Li 1984, 1999). In view of their biological importance, they were recently reviewed in detail (Sandhoff et al. 2013).

21.4 Conversion of Polysialogangliosides to GM1

Gangliosides are sialic acid-containing GSLs. It has been well recognized that sialic acids at the peripheral positions of the ganglio-series of gangliosides, such as the Neu5Ac in GD1a, GD1b, GT1b etc., are susceptible to sialidases without the assistance of a detergent. Even the sialic acids in gangliosides with shorter sugar chains, such as GM3 or GM4 (Neu5Acα2,3GalCer), are susceptible to sialidases; however, the sialic acid that is linked α-2,3 to the internal Gal of GM2 or GM1 is refractory to them. Thus, all mammalian brain polysialogangliosides with gangliotetraosyl chains are converted to GM1 as the final product upon digestion with brain sialidase as shown in Fig. 21.2 (Ohman et al. 1970). Among four types of mammalian

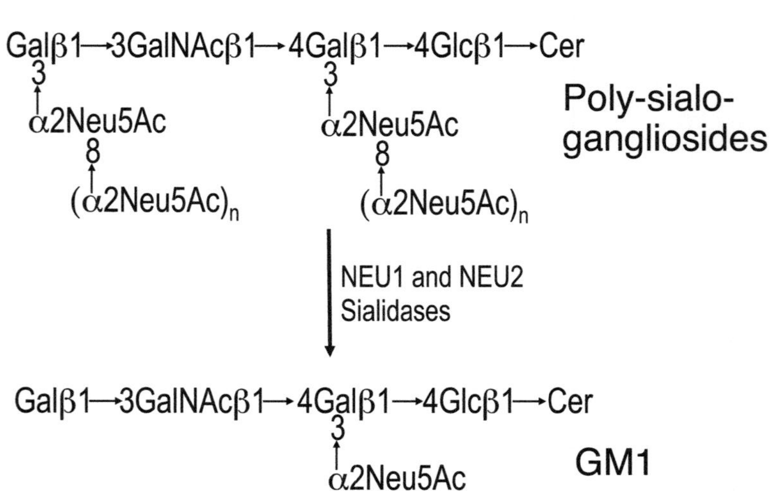

Fig. 21.2 Enzymatic conversion of brain poly-sialogangliosides to GM1. Shorthand representation is used for each monosaccharide

sialidases, NEU1 (lysosomes), NEU2 (cytosol), NEU3 (plasma membrane), and NEU4 (lysosomes, mitochondria, and ER), NEU3 and NEU4 have been shown to convert polysialogangliosides to GM1 (Miyagi et al. 1999; Miyagi and Yamaguchi 2012). The sialidase-resistant sialic acid linked to the internal Gal of GM1 or GM2 has been called the stable sialic acid. However, upon conversion of GM2 to GM3 by removal of GalNAc, the same sialic acid on GM3 becomes susceptible to sialidases without the assistance of a detergent. NMR analysis has revealed the specific through-space interactions between the GalNAc and Neu5Ac in GM1 or GM2 (Koerner et al. 1983) and these interactions may explain the resistance of the stable sialic acid to sialidases.

21.5 Catabolism of GM1

The structure of GM1 was established by Kuhn and Wiegandt in 1963 (Kuhn and Wiegandt 1963). Since the sugar chain of GM1 contains two non-reducing termini, Gal and Neu5Ac, it is obvious that GM1 can be catabolized via the GM2 pathway by the removal of Gal and then GalNAc (Fig. 21.3) or through the GA1 pathway by the removal of Neu5Ac before removing the Gal (Fig. 21.4). Through the intensive

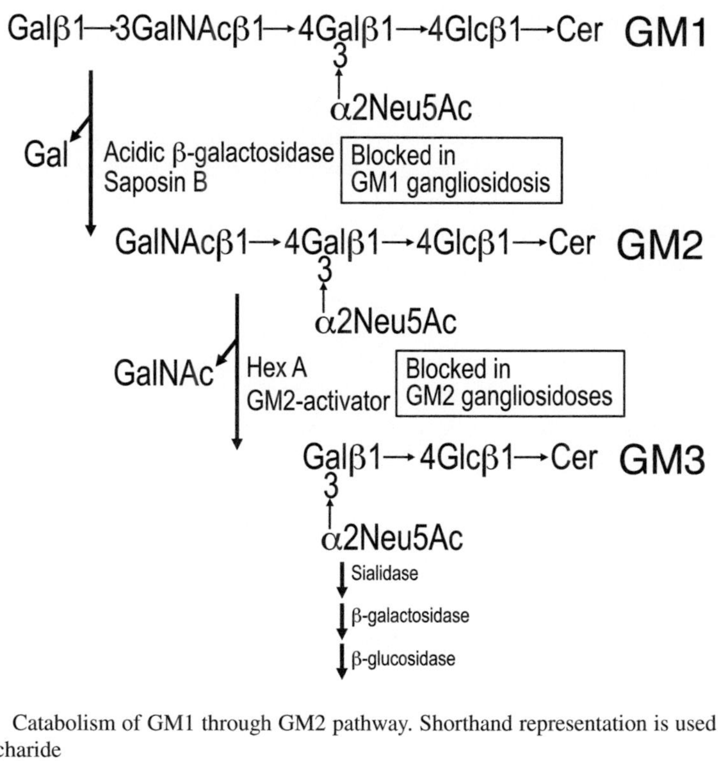

Fig. 21.3 Catabolism of GM1 through GM2 pathway. Shorthand representation is used for each monosaccharide

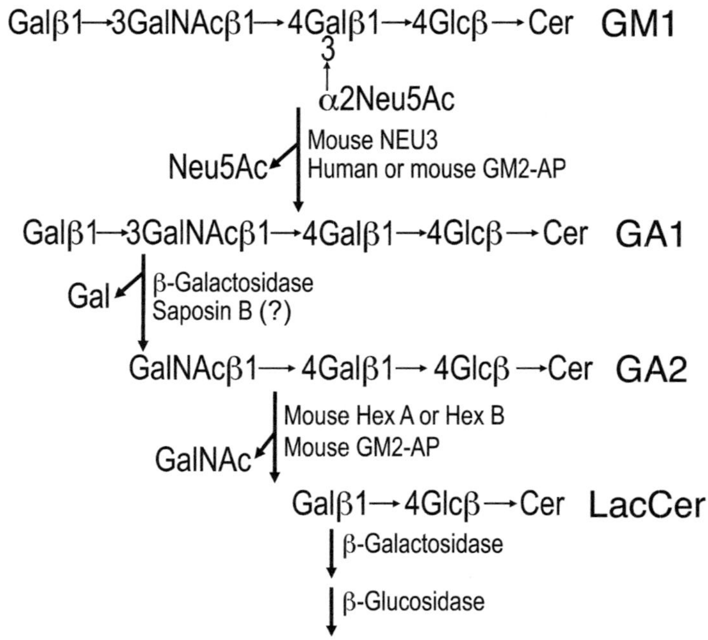

Fig. 21.4 Catabolism of GM1 through asialo-GM1 (GA1) pathway. Shorthand representation is used for each monosaccharide

studies of GM1 gangliosidosis and GM2 gangliosidosis (Tay–Sachs disease), the GM2 pathway as shown in Fig. 21.3 has been firmly established. Since neural tissue of patients with GM1 and GM2 gangliosidosis also respectively accumulates GA1 and GA2, there is no doubt that the GA1 pathway does exist. For the GA1 pathway, mouse NEU3 was found to convert GM1 to GA1 or GM2 to GA2 in the presence of human or mouse GM2-activator protein (GM2-AP) (Fig. 21.4) (Li et al. 2001). Conversion of GA1 to GA2 and subsequently to LacCer, as shown in Fig. 21.4, is still not well defined. It has been shown that mouse GM2-AP was able to stimulate mouse Hex B to convert GA2 to LacCer. Human GM2-AP, on the other hand, was unable to stimulate human Hex B to convert GA2 to LacCer. This provided a biochemical explanation as to why *Hexa* gene-disrupted mice, mouse model of variant B TSD, do not show excessive GM2-accumulation (Yuziuk et al. 1998). As discussed in the "Aglycone Specificities of Glycosidases" section, both acidic β-galactosidase and galactocerebrosidase are able to cleave LacCer.

21.6 Inborn Errors of Ganglioside Catabolism

The two well-established and best-studied inborn errors of ganglioside catabolism are GM1 gangliosidosis and GM2 gangliosidosis.

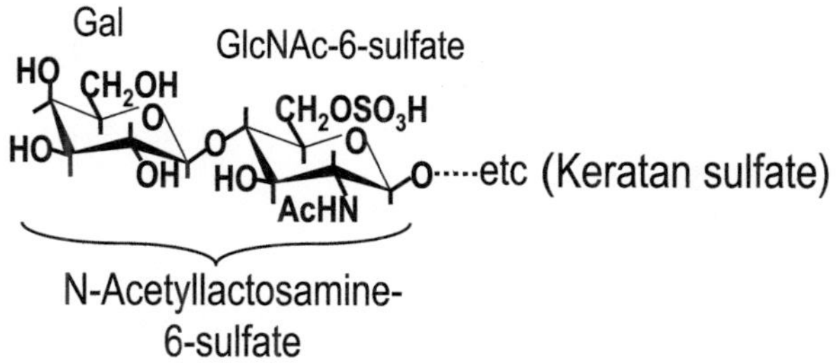

Fig. 21.5 The *N*-acetyllactosamine-6-sulfate disaccharide in keratan sulfate. Conformational representation is used for each monosaccharide

21.6.1 GM1 Gangliosidosis

As mentioned in Sect. 21.2.4, the acidic β-galactosidase converts GM1 to GM2 (Figs. 21.1 and 21.3). GM1 gangliosidosis or generalized gangliosidosis is caused by the deficiency of this β-galactosidase that leads to the cerebral accumulation of GM1 and GA1. The β-galactosidase deficiency in GM1 gangliosidosis was established by Okada and O'Brien (1968). The most interesting feature of GM1 gangliosidosis is that the same β-galactosidase is also responsible for degradation of the Galβ1,4-linked to GlcNAc-6-sulfate of the *N*-acetyllactosamine-6-sulfate unit in keratan sulfate (Fig. 21.5). Thus, the deficiency of acidic β-galactosidase leads to both GM1 gangliosidosis and Morquio type B disease. It should be noted that the glycone (Gal) of GM1 is linked to GalNAc through a β-1,3-linkage (Fig. 21.1) while Gal of keratan sulfate is linked to GlcNAc-6-sulfate through a β-1,4-linkage (Fig. 21.5). The fact that the aglycones in GM1 and keratan sulfate are quite different, but the same β-galactosidase cleaves the Gal from these two glycoconjugates underscores the complexity of glycone and aglycone specificities of glycosidases.

GM1 gangliosidosis (Morquio type B disease or mucopolysaccharidosis IVB) is an autosomal recessive disease caused by the deficiency of the acidic β-galactosidase, which is encoded by the gene *GLB1* located on chromosome 3. This β-galactosidase is a protein of 677 amino acid residues, including a putative signal sequence of 23 amino acids and 7 potential N-glycosylation sites. Upon maturation, the β-galactosidase forms an enzyme complex with NEU1 sialidase and protective protein/cathepsin A. This complex formation protects the enzyme from proteolysis in the lysosome. In addition to the accumulation of GM1, GA1 and keratan sulfate, high amounts of oligosaccharides with non-reducing terminal β-Gal derived from N- and O-glycans of glycoproteins have also been found in urine and visceral organs of patients. GM1 gangliosidosis has been extensively reviewed by Suzuki et al. (2013).

21.6.2 GM2 Gangliosidosis

GM2 gangliosidosis/TSD is one of the most intensively studied of the inborn errors of GSL catabolism. TSD is a lysosomal GM2 storage disease of the nervous system that becomes clinically evident by 5–6 months of age and usually fatal by the age of

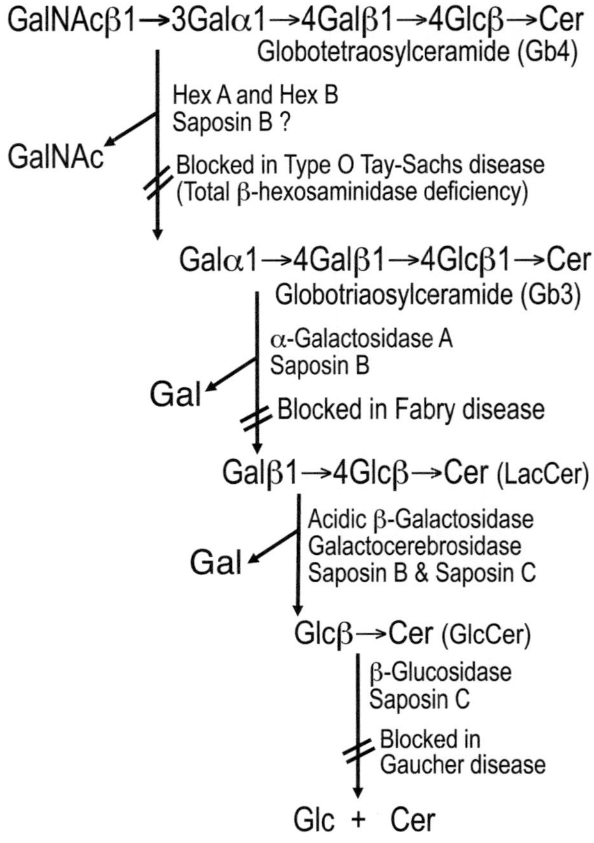

Fig. 21.6 Stepwise enzymatic hydrolysis of the sugar chain in Gb4. Shorthand representation is used for each monosaccharide

3–4 years. The disease is characterized by progressive retardation in development, motor neuron degeneration, dementia and blindness associated with a cherry red spot in the retina. TSD is caused by an impairment in the catabolism of GM2. As shown in Fig. 21.3, catabolism of GM2 in humans starts with the conversion of GM2 to GM3 by β-hexosaminidase A (Hex A) in the presence of GM2-AP. There are three β-hexosaminidase isozymes found in human tissues: Hex A consisting of one α- and one β-subunit; Hex B, consisting of two β-subunits; and minor Hex S, consisting of two α-subunits. Among the three isozymes, only Hex A is capable of hydrolyzing GalNAc from GM2 with the assistance of GM2-AP. Classical TSD (variant B TSD) and Sandhoff diseases (variant O TSD), respectively caused by mutations of *HEXA* (α-subunit) and *HEXB* (β-subunit) genes, show severe phenotypes due to the cerebral accumulation of GM2. Since the β-linked GlcNAc and GalNAc residues in oligosaccharides and glycopeptides can also be cleaved by Hex A or Hex B, significant tissue accumulation and urinary excretion of glycoprotein-derived oligosaccharides have also been found when both Hex A and Hex B are deficient (Sandhoff disease). Patients with Sandhoff disease also accumulate globotetraosylceramide (Gb4) that contains β-linked GalNAc at the non-reducing terminus (see Fig. 21.6). The third variant of TSD, variant AB TSD, is

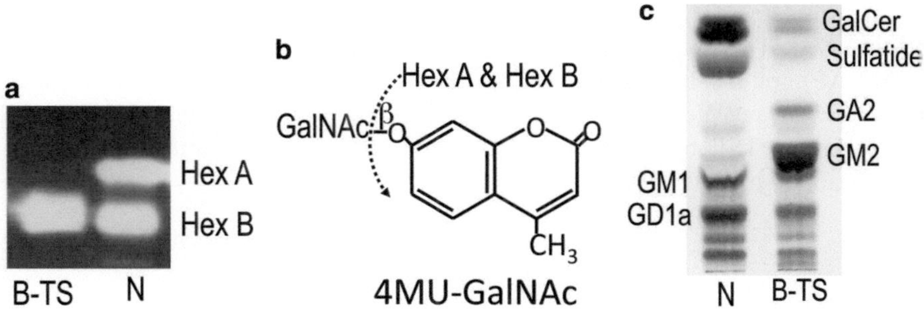

Fig. 21.7 (**a**) Cellulose acetate electrophoresis of Hex A and Hex B isozymes. (**b**) Hydrolysis of 4MU-GalNAc by both Hex A and Hex B. (**c**) Thin-layer chromatography of brain gangliosides extracted from a normal brain (N) and a variant B TS (B-TS) brain

caused by the deficiency of GM2-AP. Neither Hex A nor Hex B is deficient in the AB variant of TS patients. TSD has been extensively reviewed by Gravel et al. in OMMIB (Gravel et al. 2013).

GM2 Activator Protein (GM2-AP) Revisited

After the initial description of TSD in 1881 by Waren Tay (1881), the three most important advances in the understanding of TSD in the 1960s were: (1) structural elucidation of GM2 by Svennerholm (1962), Makita and Yamakawa (1963), and Ledeen and Salsman (1965); (2) revelation of the presence of two hexosaminidase isozymes, Hex A and Hex B, in human spleen by Robinson and Stirling (1968) and (3) detection of the absence of Hex A isozyme in classical TS patients by Okada and O'Brien (1969), Hultberg (1969), and Sandhoff (1969). It is remarkable that the three reports on the deficiency of Hex A in TSD appeared only 1 year after Robinson and Stirling's report (Robinson and Stirling 1968). As shown in Fig. 21.7a, the liver extract of a variant B TS patient was found to be devoid of Hex A isozyme when overlaying the cellulose acetate electrophoregram with 4MU-β-GalNAc (Fig. 21.7b), which is susceptible to both Hex A and Hex B without any cofactor. This result, together with the fact that GM2 accumulates in the brain of TS patients (Fig. 21.7c), was the basis for investigators in the 1960s to conclude that Hex A must be responsible for cleaving the β-linked GalNAc in GM2. However, there was no report on the hydrolysis of GalNAc in GM2 by Hex A until 1973 (Li et al. 1973). While studying the in vitro catabolism of GM2, Li et al. (1973) found that Hex A or Hex B extensively purified from human liver and urine using *p*-nitrophenyl-β-GlcNAc as substrate were not able to hydrolyze GalNAc in GM2, but the crude enzyme could effectively convert GM2 to GM3. These results suggested that a cofactor essential for the enzymatic hydrolysis of GM2 was removed from the Hex A during purification. They subsequently found that a heat-stable, non-dialyzable preparation obtained from the crude liver extract could stimulate the hydrolysis of GM2 by the purified Hex A but not Hex B (Li et al. 1973). This was the first report on the requirement of a protein cofactor (GM2-AP) for the hydrolysis of GM2 by Hex A. The clinical importance of GM2-AP was revealed by the findings that this activator protein was absent in the AB variant of TSD by Conzelman and Sandhoff (1978), Hechtman et al. (1982) and Hirabayashi et al. (1983).

Possible Mechanism of Action of GM2-AP

Although the crystal structure of GM2-AP has been solved (Wright et al. 2000), its mechanism of action is still not well understood. Two models for the possible mechanism of action of GM2-AP have been proposed by Furst and Sandhoff (1992): model 1 suggested that GM2-AP extracts a single GM2 molecule from a membrane or micelle to form a water-soluble 1:1 protein-GM2 complex to present the GM2 to the water-soluble Hex A; model 2 proposed the binding of GM2-AP to one molecule of membrane-bound GM2, lifting it out of the membrane, and the "activator GM2 complex" is the form recognized and cleaved by the water soluble Hex A. These two models, however, cannot explain why the water-soluble oligosaccharide derived from GM2, devoid of ceramide moiety, is also refractory to Hex A either in the presence or absence of GM2-AP (Li et al. 1999). These two models also cannot explain why Neu5Ac in GM2 is refractory to sialidase, but after removing the GalNAc from GM2, the GM3 becomes susceptible to sialidase. As stated in Sect. 21.4 of this chapter, NMR analysis has revealed the specific through-space interactions between the GalNAc and Neu5Ac in GM1 and GM2 (Koerner et al. 1983). Li et al. (1999, 2008) hypothesized that this interaction may be the reason why both GalNAc and Neu5Ac in GM2 are refractory to enzymatic hydrolysis and that GM2-AP may alleviate this interaction rendering GalNAc accessible to Hex A. Using a series of chemically synthesized and structurally modified GM2, they found that when GalNAcβ1,4Gal linkage in GM2 was converted to the GalNAcβ1,6Gal (6′GM2, Fig. 21.8), both GalNAc and Neu5Ac in 6′GM2 became susceptible to Hex A and sialidase in the absence of GM2-AP. Interestingly, the oligosaccharide derived from 6′GM2 was also susceptible to Hex A without GM2-AP (Li et al. 1999, 2008). Since GM2-AP cannot assist Hex A to hydrolyze the GalNAc from the water soluble-oligosaccharide derived from GM2, it appears that GM2-AP and Hex A require the ceramide moiety to carry out their action.

Revelation of B1 Variant of TSD, a New Variant of Variant AB GM2-Gangliosidosis

When examining brain samples from two variant AB TS patients reported by Goldman et al. (1980), Li et al. (1981) found that one of the brains had a deficiency of GM2-AP and the Hex A isolated from that brain was able to hydrolyze GM2 in the presence of exogenously added GM2-AP. Thus, this patient was a typical case

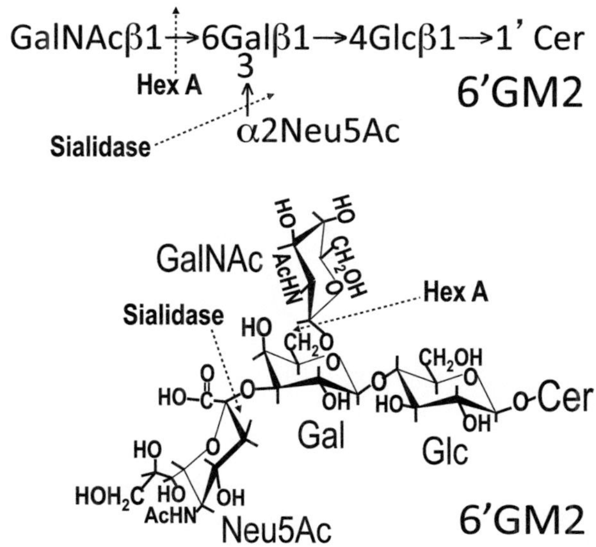

Fig. 21.8 Shorthand representation (*top*) and conformational representation (*bottom*) of the sugar chain in 6′GM2. *Dotted arrows* indicate enzymatic hydrolyses

of AB variant TSD. The other brain was found to have a normal GM2-AP, but had a defective Hex A that showed normal activity toward synthetic substrates, such as 4MU-β-GalNAc, but no activity toward GM2 in the presence of exogenously added GM2-AP. The level of GM2-AP in this sample was found to be three times higher than that found in two normal brain samples. This was the first revelation of a new variant of variant AB GM2-gangliosidosis. This new variant is now called the B1 variant (Gravel et al. 2013; Kytzia et al. 1983). Its Hex A was shown to be allelic with that seen in the α-subunit deficiency of Hex A (Sonderfeld et al. 1985), which is also unable to hydrolyze GM2 but retains activity toward synthetic substrates.

Unusual Taurine Conjugation of GM2 in TS Brain

Despite our clear understanding of the molecular and biochemical bases of TSD, very little is known about the effect of GM2 accumulation on disease progression and tissue dysfunction. To understand how the neural tissues of TS patients respond to and cope with the massive accumulation of GM2, Li et al. (2003) carried out a detailed analysis of GSLs in TS brains and revealed the presence of a novel GM2-derivative, a taurine-conjugated GM2 (TGM2), in TS brains. TGM2 was not found in normal brains. As shown in Fig. 21.9, TGM2 is a structurally modified GM2 formed by conjugating the carboxyl function of Neu5Ac in GM2 with taurine through an amide linkage. Since taurine-conjugation is a well-known mechanism for detoxification of xenobiotics by increasing their water solubility and facilitating their removal, the neural tissues of TS patients may regard the accumulated GM2 as a "quasi-xenobiotic" and use taurine-conjugation as a vehicle for its removal. Pathobiological significance of TGM2 remains to be elucidated.

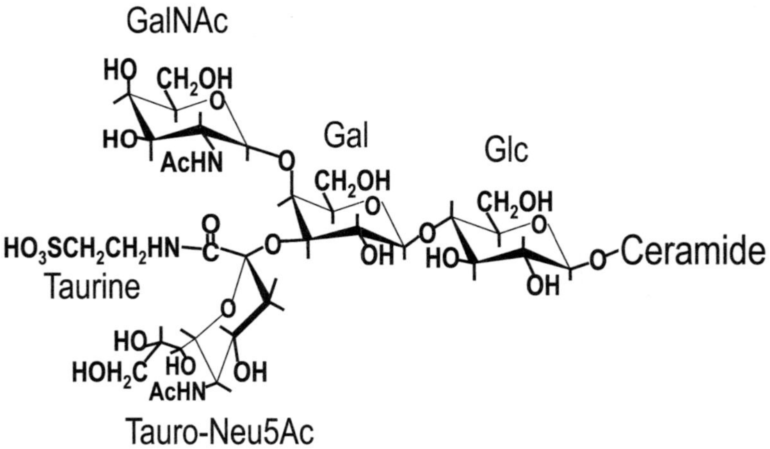

Fig. 21.9 Conformational representation of the sugar chain in taurine-conjugated GM2. *Tauro-Neu5Ac* taurine-conjugated Neu5Ac

21.7 Inborn Errors of Neutral GSL Catabolism

21.7.1 Catabolism of Globotetraosylceramide (Gb4)

Gb4 was first detected more than 60 years ago as a GalNAc-containing neutral GSL in human erythrocytes by Klenk and Lauenstein (1951). This GSL was subsequently named "globoside" by Yamakawa and Suzuki (1952), based on the observation that the glycolipid formed perfectly round globules when viewed under the microscope. The complete structure of Gb4 was established by Hakomori et al. (1971). As shown in Fig. 21.6, both Hex A and Hex B are able to convert Gb4 to Gb3 which is subsequently converted to LacCer, GlcCer and finally to Glc and Cer, respectively by the sequential action of α-galactosidase A, β-galactosidase and β-glucosidase. The fact that Hex B is unable to hydrolyze the GalNAc in GM2 but is capable of hydrolyzing it in Gb4, underscores once again the complexity of aglycone specificities of glycosidases. The activator protein responsible for the enzymatic conversion of Gb4 to Gb3 is still not well understood. Besides variant O TSD (Sandhoff disease), Fabry disease and Gaucher disease are the other two inborn errors associated with the catabolism of Gb4.

21.7.2 Fabry Disease

Fabry disease, caused by the deficiency of α-galactosidase A (Desnick et al. 2013), is a systemic disorder of Gb3 catabolism transmitted by an X-linked gene and resulting in progressive accumulation of mainly Gb3 in the kidney, heart, skin, and other visceral organs. The hemizygous male has atypical skin lesions, fever, burning pain in the extremities, and renal dysfunction. After the reports in 1898 by Anderson (1898) and Fabry (1898) describing patients with angiokeratoma corporis diffusion, over 60 years elapsed before Fabry disease was identified as a sphingolipidosis by Sweeley and Klionsky (1963) who reported the accumulation of Gal-Gal-Glc-Cer in the kidney of a Fabry patient. Gal-Gal-Glc-Cer was called ceramide trihexoside (CTH) in the 1960s. This GSL is now called globotriaosylceramide (Gb3). Since the biochemical basis of Fabry disease could be deduced from the anomeric configuration of the terminal Gal in Gb3, the determination of this anomeric configuration was the subject of intensive study in the 1960s. Based on NMR analysis Sweeley et al. (1970) erroneously reported in 1970 that the terminal Gal and the penultimate Gal in Gb3 were both β-linked (Galβ1,4Galβ1,4GlcCer). In 1971, five laboratories simultaneously reported the presence of α-linked terminal Gal in Gb3 (Bensaude et al. 1971; Clarke et al. 1971; Hakomori et al. 1971; Handa et al. 1971; Li and Li 1971). This reflected the immense interest in the structure of Gb3 at that time. The β-linked penultimate Gal in Gb3 was also determined in 1971 (Li and Li 1971) by using jack bean β-galactosidase. As shown in Fig. 21.6, conversion of Gb3 to LacCer is catalyzed by α-galactosidase A with the assistance of saposin B, which is a nonspecific activator protein (Li et al. 1988) that stimulates the enzymatic hydrolysis of a number of GSLs by interacting with the GSL-substrates (Kishimoto et al. 1992; Li et al. 1988; Sandhoff et al. 2013). α-Galactosidase A also cleaves the α-linked Gal from diGalCer (Galα1,4Galβ1,Cer) and blood group B glycotopes [Galα1,3(Fucα1,2)Galβ1,R]. Patients with saposin B deficiency show clinical phenotypes similar to that of metachromatic leukodystrophy with excessive urinary excretion of sulfatide, Gb3 and diGalCer (Sandhoff et al. 2013). The identity and the glycone specificity of α-galactosidase B as an α-*N*-acetylgalactosaminidase were revealed by Dean et al. (1977) and Schram et al. (1977) in 1977. Natural substrates of α-*N*-acetylgalactosaminidase are α-linked GalNAc in blood group A glycotopes [GalNAcα1,3(Fucα1,2)Galβ1,R] and the core structures of O-glycans (GalNAcα1,Ser/Thr). Kanzaki/Schindler Disease, an inborn error of glycoprotein catabolism, is caused by the deficiency of α-*N*-acetylgalactosaminidase (Desnick and Schindler 2013). The substrate specificity of α-*N*-acetylgalactosaminidase is another example indicating the complex nature of the glycone specificity of glycosidases.

21.7.3 Gaucher Disease

Gaucher disease is an inborn lysosomal glucocerebroside (Glcβ1,Cer) storage disorder caused by a deficiency in lysosomal glucocerebrosidase (β-glucosidase) activity resulting in the accumulation of GlcCer in tissue macrophages that become engorged Gaucher cells (Grabowski et al. 2013). Glucocerebrosidase is a specific type of β-glucosidase that hydrolyzes GlcCer to form Glc and ceramide in the presence of saposin C (Fig. 21.6). GlcCer can be derived from gangliosides and neutral GSLs.

Gaucher disease has been divided into three types: type 1, non-neuropathic adult form; type 2, acute neuropathic form; and type 3, subacute neuropathic form. After the first description in 1882 of the disease by Gaucher, a medical student, in his thesis (Gaucher 1882), the three key developments in our understanding of the biochemical basis of Gaucher disease have been: (1) structural elucidation of the storage GSL, GlcCer; (2) identification of the enzyme deficiency associated with the disease; and (3) detection of glucosylsphingosine (GlcSph) in tissues of patients.

The structure of stored GlcCer was not firmly established until the end of the 1970s. Between 1924 and 1939, several reports (Capper et al. 1934; Lieb 1924, 1927; Lieb and Mladenovic 1929; McConnell et al. 1939) suggested that the cerebroside accumulated in Gaucher spleen contained galactose as found in brain cerebroside. In 1940, Halliday et al. (1940) showed that the cerebroside isolated from a Gaucher spleen contained equimolar proportions of glucose, fatty acid and sphingosine and that the sugar moiety in Gaucher cerebroside, but not the normal brain cerebroside, was susceptible to emulsin β-glucosidase. In 1958, Rosenberg and Chargaff (1958), using infrared spectroscopy, conclusively showed that the crystalline cerebroside prepared from a Gaucher spleen contained β-linked glucose. In 1961, Shapiro and Flowers (1961) chemically synthesized galactocerebrosides (cerasine and phrenosine) and GlcCer. They found that the synthetic GlcCer was identical to the crystalline cerebroside prepared from a Gaucher spleen by Rosenberg and Chargaff (1958) and assigned the β-configuration for both cerebrosides. Finally the β-anomeric linkage of GlcCer was conclusively established by C-13 NMR spectroscopy in 1979 (Koerner et al. 1979).

For the enzyme deficiency in Gaucher disease, in 1965, Brady et al. (1965) using Glc-1-^{14}C-labeled-GlcCer as substrate showed a pronounced diminution of GlcCer-cleaving activity in Gaucher spleen samples. In the same year, Patrick (1965) showed the deficiency of glucocerebrosidase and β-glucosidase activities in Gaucher spleen specimens using intact GlcCer and *p*-nitrophenyl-β-glucoside as substrates. In 1968, Ockerman (1968) confirmed the β-glucosidase deficiency in Gaucher liver specimens using 4MU-β-glucoside as substrate. These three reports established the deficiency of glucocerebrosidase (β-glucosidase) in Gaucher disease. The most common mutation of glucocerebrosidase is N370S in type 1, and L444P in types 2 and 3 Gaucher disease.

Based on the seminal work by Miyatake and Suzuki in 1972 (Miyatake and Suzuki 1972), that tissues of Krabbe patients were not able to cleave both galactocerebroside (GalCer) and galactosphingosine (GalSph), Raghavan et al. (1973) found that Gaucher spleen was also deficient in GlcSph-cleaving activity and detected the presence of putative GlcSph in Gaucher spleen but not in normal spleen. This report may constitute the first report on the detection of GlcSph in Gaucher spleen. In 1974, Raghavan et al. (1974) isolated and characterized GlcSph from Gaucher's spleen. In 1982, Nilsson and Svennerholm (1982) reported that GlcSph accumulated in both cerebrum and cerebellum of brains from type 2 and type 3 Gaucher patients. GlcSph was shown to be toxic to neuronal cells (Schueler et al. 2003) and the accumulation of GlcSph in the brains of type 2 and type 3 Gaucher patients was found to correlate with CNS-involvement (Orvisky et al. 2002).

It is intriguing that chitotriosidase was found to be greatly elevated in the plasma of type 1 Gaucher patients and this activity has been used to monitor the efficacy of enzyme replacement therapy for type 1 Gaucher patients (Hollak et al. 1994). Recently, an association between Gaucher disease and parkinsonism has been reported (Sidransky et al. 2009). An in-depth coverage on various aspects of Gaucher disease can be found in the excellent book "Gaucher Disease" edited by Futerman and Zimran (2007).

21.7.4 Krabbe Disease (Globoid Cell Leukodystrophy)

Krabbe disease (globoid cell leukodystrophy) is a rare inherited degenerative disorder characterized by severe demyelination of the CNS and PNS, due to the deficiency of lysosomal galactocerebrosidase (GALC) responsible for the degradation of galactocerebroside (Galβ1,Cer). GalCer, initially called cerebroside, was the first GSL, described by Thudichum almost 140 years ago (Thudichum 1874). He divided cerebroside into cerasine (straight chain fatty acid containing GalCer) and phrenosine (hydroxy fatty acid containing GalCer). GalCer is one of the major constituents of myelin and this glycolipid is almost exclusively found in the CNS and PNS. Kidney is the only visceral organ found to contain GalCer. The β-galactosyl linkage in GalCer was established in 1961 through the chemical synthesis by Shapiro and Flowers (1961). Unlike GlcCer which can be elongated by adding more sugar residues to form a wide variety of GSLs, only diGalCer (Galα1,4GalCer), GalCer-3-sulfate (sulfatide) and GM4 (Neu5Acα2,3GalCer) have been found to contain GalCer. After the initial report of a new familial, infantile form of diffuse brain sclerosis by Krabbe (1916), the revelation of the biochemical basis of Krabbe disease as the deficiency of GALC/GalCer β-galactosidase was made 54 years later by Suzuki and Suzuki (1970). Unlike other lysosomal storage diseases, there is no apparent accumulation of GalCer, the substrate of the missing enzyme, GALC, in the brains of Krabbe patients. This is due to the fact that storage of GalCer is restricted to globoid cells. One of the striking features of Krabbe disease is a severe decrease of GalCer in the white matter. The work by Suzuki and Suzuki (1970) to establish the GALC deficiency in Krabbe disease was indeed remarkable. As discussed in the "Gaucher Disease" section, Miyatake and Suzuki (1972) reported that there was also a deficiency of GalSph (psychosine) β-galactosidase activity in Krabbe disease and put forth a psychosine hypothesis by proposing that the elevation in cytotoxic GalSph was the cause for the pathogenesis of Krabbe disease. After this seminal report, lyso-GSLs derived from storage GSLs have been detected in the tissues of various glycosphingolipid storage diseases: GlcSph in Gaucher disease (Nilsson and Svennerholm 1982; Orvisky et al. 2002; Raghavan et al. 1973, 1974; Schueler et al. 2003); lyso-GM1 and lyso-GA1 in GM1 gangliosidosis (Kobayashi et al. 1992) lyso-GM2 and lyso-GA2 in TSD (Kobayashi et al. 1992; Neuenhofer et al. 1986; Rosengren et al. 1987); and lyso-Gb3 in Fabry disease (Aerts et al. 2008). The detergent-like properties of lyso-sphingolipids may be detrimental to cell functions. It has also been shown that lyso-sphingolipids inhibit protein kinase C (Hannun and Bell 1987). The pathobiological significance of lyso-sphingolipids associated with sphingolipid storage diseases remains to be elucidated.

21.8 Conclusion

Recently, enzyme replacement (Brady 2006), substrate reduction (Radin 2000), and pharmacological chaperone (Boyd et al. 2013) therapies have become available for treating several sphingolipid storage diseases. Enzyme replacement therapy has been found to be effective for non-neuropathic type 1 Gaucher disease and Fabry disease. Gene therapy for lysosomal storage diseases (Sands and Davidson 2006) is also on the horizon. These therapies, extensively discussed in Chap. 22, are the fruits cumulated from the painstaking basic biochemical studies of these disorders.

Conflict of Interest The authors declare no conflicts of interest.

Acknowledgements The authors are grateful to Dr. Su-Chen Li for her invaluable input and suggestions. We would also like to thank Mr. Gilbert Estrada for his proofreading of the manuscript.

References

Aerts JM, Groener JE, Kuiper S, Donker-Koopman WE, Strijland A, Ottenhoff R, et al. Elevated globotriaosylsphingosine is a hallmark of Fabry disease. Proc Natl Acad Sci U S A. 2008;105:2812–7.

Anderson W. A case of angiokeratoma. Br J Dermatol. 1898;10:113–7.

Bensaude I, Callahan J, Philippart M. Fabry's disease as an α-galactosidosis: evidence for an α-configuration in trihexosyl ceramide. Biochem Biophys Res Commun. 1971;43:913–8.

Boyd RE, Lee G, Rybczynski P, Benjamin ER, Khanna R, Wustman BA, et al. Pharmacological chaperones as therapeutics for lysosomal storage diseases. J Med Chem. 2013;56:2705–25.

Brady RO. Enzyme replacement for lysosomal diseases. Annu Rev Med. 2006;57:283–96.

Brady RO, Kanfer JN, Shapiro D. Metabolism of glucocerebrosides. II. Evidence of an enzymatic deficiency in Gaucher's disease. Biochem Biophys Res Commun. 1965;18:221–5.

Capper A, Epstein H, Schless RA. Gaucher's disease. Report of a case with presentation of a table differentiating the lipoid disturbances. Am J Med Sci. 1934;188:84–93.

Chou MY, Li S-C, Kiso M, Hasegawa A, Li Y-T. Purification and characterization of sialidase L, a NeuAc α2→3Gal-specific sialidase. J Biol Chem. 1994;269:18821–6.

Clarke JT, Wolfe LS, Perlin AS. Evidence for a terminal α-D-galactopyranosyl residue in galactosylgalactosylglucosylceramide from human kidney. J Biol Chem. 1971;246:5563–9.

Conzelmann E, Sandhoff K. AB variant of infantile GM2 gangliosidosis: deficiency of a factor necessary for stimulation of hexosaminidase A-catalyzed degradation of ganglioside GM2 and glycolipid GA2. Proc Natl Acad Sci U S A. 1978;75:3979–83.

de Duve C. Lysosomes revisited. Eur J Biochem. 1983;137:391–7.

Dean KJ, Sung SS, Sweeley CC. The identification of α-galactosidase B from human liver as an α-*N*-acetylgalactosaminidase. Biochem Biophys Res Commun. 1977;77:1411–7.

Desnick RJ, Ioannou YA, Eng CM. Chapter 150: α-Galactosidase A deficiency: Fabry disease. In: Valle D, Vogelstein B, Kinzler KW, Antonarakis S, Ballabio A, editors. The online metabolic & molecular bases of inherited disease (Scriver's OMMBID). McGraw-Hill Global Education Holdings; 2013. Available from: http://www.ommbid.com/

Desnick RJ, Schindler D. Chapter 139: α-*N*-Acetylgalactosaminidase deficiency: Schindler disease. In: Valle D, Vogelstein B, Kinzler KW, Antonarakis S, Ballabio A, editors. The online metabolic & molecular bases of inherited disease (Scriver's OMMBID). McGraw-Hill Global Education Holdings; 2013. Available from: http://www.ommbid.com/

Dong DL, Hart GW. Purification and characterization of an *O*-GlcNAc selective *N-acetyl-β-D-glucosaminidase* from rat spleen cytosol. J Biol Chem. 1994;269:19321–30.

Fabry J. Ein Beitrag Zur Kenntnis der Purura haemorrhagica nodularis (Purpura papulosa hemorrhagica Habrae). Arch Dermatol Syph. 1898;43:187–200.

Furst W, Sandhoff K. Activator proteins and topology of lysosomal sphingolipid catabolism. Biochim Biophys Acta. 1992;1126:1–16.

Futerman AH, Zimran A, editors. Gaucher disease. Boca Raton, FL: CRC Press, Taylor and Francis Group; 2007.

Gaucher PCE. De l'epithelioma primitif de la rate, hypertrophie idiopathique de la rate sans leucemie. M.D. Thesis, Paris; 1882.

Goldman JE, Yamanaka T, Rapin I, Adachi M, Suzuki K, Suzuki K. The AB-variant of GM2-gangliosidosis. Clinical, biochemical, and pathological studies of two patients. Acta Neuropathol. 1980;52:189–202.

Grabowski GA, Petsko GA, Kolodny EH. Chapter 146: Gaucher disease. In: Valle D, Vogelstein B, Kinzler KW, Antonarakis S, Ballabio A, editors. The online metabolic & molecular bases of inherited disease (Scriver's OMMBID). McGraw-Hill Global Education Holdings; 2013. Available from: http://www.ommbid.com/

Gravel RA, Kaback MM, Proia RL, Sandhoff K, Suzuki K, Suzuki K. Chapter 153: The GM2. In: Valle D, Vogelstein B, Kinzler KW, Antonarakis S, Ballabio A, editors. The online metabolic & molecular bases of inherited disease (Scriver's OMMBID). McGraw-Hill Global Education Holdings; 2013. Available from: http://www.ommbid.com/

Hakomori S-I, Siddiqui B, Li Y-T, Li S-C, Hellerqvist CG. Anomeric structure of globoside and ceramide trihexoside of human erythrocytes and hamster fibroblasts. J Biol Chem. 1971;246: 2271–7.

Halliday N, Deuel Jr HJ, Tragerman LJ, Ward WE. On the isolation of a glucose-containing cerebroside from spleen in a case of Gaucher's disease. J Biol Chem. 1940;132:171–80.

Handa S, Ariga T, Miyatake T, Yamakawa T. Presence of α-anomeric glycosidic configuration in the glycolipids accumulated in kidney with Fabry's disease. J Biochem (Tokyo). 1971;69: 625–7.

Hannun YA, Bell RM. Lysosphingolipids inhibit protein kinase C: implications for the sphingolipidoses. Science. 1987;235:670–4.

Hechtman P. Characterization of an activating factor required for hydrolysis of GM2 ganglioside catalyzed by hexosaminidase A. Can J Biochem. 1977;55:315–24.

Hechtman P, Gordon BA, Ng Ying Kin NM. Deficiency of the hexosaminidase A activator protein in a case of GM2 gangliosidosis; variant AB. Pediatr Res. 1982;16:217–22.

Hirabayashi Y, Li Y-T, Li S-C. The protein activator specific for the enzymic hydrolysis of GM2 ganglioside in normal human brain and brains of three types of GM2 gangliosidosis. J Neurochem. 1983;40:168–75.

Ho MW, O'Brien JS, Radin NS, Erickson JS. Glucocerebrosidase: reconstitution of activity from macromolecular components. Biochem J. 1973;131:173–6.

Hollak CE, van Weely S, van Oers MHJ, Aerts JMFG. Marked elevation of plasma chitotriosidase activity. A novel hallmark of Gaucher disease. J Clin Invest. 1994;93:1288–92.

Hultberg B. *N*-Acetylhexosaminidase activities in Tay-Sachs disease. Lancet. 1969;2:1195.

Kishimoto Y, Hiraiwa M, O'Brien JS. Saposins: structure, function, distribution, and molecular genetics. J Lipid Res. 1992;33:1255–67.

Klenk E, Lauenstein K. Uber die zuckerhaltigen lipoide der formbestandteile des menschlichen blutes. Hoppe Seyler's Z Physiol Chem. 1951;288:220–8.

Klionsky DJ. Autophagy: from phenomenology to molecular understanding in less than a decade. Nat Rev Mol Cell Biol. 2007;8:931–7.

Kobayashi T, Goto I, Okada S, Orii T, Ohno K, Nakano T. Accumulation of lysosphingolipids in tissues from patients with GM1 and GM2 gangliosidoses. J Neurochem. 1992;59:1452–8.

Koerner Jr TAW, Cary LW, Li S-C, Li Y-T. Carbon 13 NMR spectroscopy of a cerebroside. Proof of the β-pyranosyl structure of D-glucosylceramide. J Biol Chem. 1979;254:2326–8.

Koerner Jr TA, Prestegard JH, Demou PC, Yu RK. High-resolution proton NMR studies of gangliosides. 1. Use of homonuclear two-dimensional spin-echo J-correlated spectroscopy for determination of residue composition and anomeric configurations. Biochemistry. 1983;22: 2676–87.

Krabbe K. A new familial, infantile form of diffuse brain sclerosis. Brain. 1916;39:74–114.

Kuhn R, Wiegandt H. Die konstitution der ganglio-*N*-tetraose und des gangliosids GI. Chem Ber. 1963;96:866–80.

Kytzia HJ, Hinrichs U, Maire I, Suzuki K, Sandhoff K. Variant of GM2-gangliosidosis with hexosaminidase A having a severely changed substrate specificity. EMBO J. 1983;2:1201–5.

Ledeen RW, Salsman K. Structure of the Tay-Sachs ganglioside I. Biochemistry. 1965;4: 2225–33.

Li Y-T, Li S-C. Anomeric configuration of galactose residues in ceramide trihexosides. J Biol Chem. 1971;246:3769–71.

Li S-C, Li Y-T. An activator stimulating the enzymic hydrolysis of sphingoglycolipids. J Biol Chem. 1976;251:1159–63.

Li Y-T, Li S-C. Activator proteins related to the hydrolysis of glycosphingolipids catalyzed by lysosomal glycosidases. In: Dingle JT, Dean RT, Sly W, editors. Lysosomes in biology and pathology, vol. 7. Amsterdam: Elsevier Science Publishers B.V.; 1984. p. 99–117.

Li Y-T, Li S-C. Enzymatic hydrolysis of glycosphingolipids. Anal Biochem. 1999;273:1–11.

Li Y-T, Mazzotta MY, Wan CC, Orth R, Li S-C. Hydrolysis of Tay-Sachs ganglioside by β-hexosaminidase A of human liver and urine. J Biol Chem. 1973;248:7512–5.

Li S-C, Nakamura T, Ogamo A, Li Y-T. Evidence for the presence of two separate protein activators for the enzymic hydrolysis of GM1 and GM2 gangliosides. J Biol Chem. 1979; 254:10592–5.

Li S-C, Hirabayashi Y, Li Y-T. A new variant of type-AB GM2-gangliosidosis. Biochem Biophys Res Commun. 1981;101:479–85.

Li S-C, Sonnino S, Tettamanti G, Li Y-T. Characterization of a nonspecific activator protein for the enzymatic hydrolysis of glycolipids. J Biol Chem. 1988;263:6588–91.

Li Y-T, Li S-C, Hasegawa A, Ishida H, Kiso M, Bernardi A, et al. Structural basis for the resistance of Tay-Sachs ganglioside GM2 to enzymatic degradation. J Biol Chem. 1999;274:10014–8.

Li S-C, Li Y-T, Moriya S, Miyagi T. Degradation of GM1 and GM2 by mammalian sialidases. Biochem J. 2001;360:233–7.

Li Y-T, Maskos K, Chou CW, Cole RB, Li S-C. Presence of an unusual GM2 derivative, taurine-conjugated GM2, in Tay-Sachs brain. J Biol Chem. 2003;278:35286–91.

Li Y-T, Li S-C, Kiso M, Ishida H, Mauri L, Raimondi L, et al. Effect of structural modifications of ganglioside GM2 on intra-molecular carbohydrate-to-carbohydrate interaction and enzymatic susceptibility. Biochim Biophys Acta. 2008;1780:353–61.

Lieb H. Cerebrosidspeicherung bei splenomegalie, typsu Gaucher. Hoppe Seyler's Z Physiol Chem. 1924;140:305–13.

Lieb H. Cerebrosidspeicherung bei morbus Gaucher III. Mitteilung. Hoppe Seyler's Z Physiol Chem. 1927;170:60–7.

Lieb H, Mladenovic M. Cerebrosidspeicherung bei morbus Gaucher II. Mitteilung. Hoppe Seyler's Z Physiol Chem. 1929;181:208–20.

Makita A, Yamakawa T. The glycolipids of the brain of Tay-Sachs' disease. The chemical structures of a globoside and main ganglioside. Jpn J Exp Med. 1963;33:361–8.

McConnell JS, Forbes JC, Apperly FL. Notes on chemical studies of a Gaucher spleen. Am J Med Sci. 1939;197:90–2.
Mehl E, Jatzkewitz H. Eine cerebrosidsulfatase aus schweineniere. Hoppe Seylers Z Physiol Chem. 1964;339:260–76.
Miyagi T, Yamaguchi K. Mammalian sialidases: physiological and pathological roles in cellular functions. Glycobiology. 2012;22:880–96.
Miyagi T, Wada T, Iwamatsu A, Hata K, Yoshikawa Y, Tokuyama S, et al. Molecular cloning and characterization of a plasma membrane-associated sialidase specific for gangliosides. J Biol Chem. 1999;274:5004–11.
Miyatake T, Suzuki K. Globoid cell leukodystrophy: additional deficiency of psychosine galactosidase. Biochem Biophys Res Commun. 1972;48:538–43.
Neuenhofer S, Conzelmann E, Schwarzmann G, Egge H, Sandhoff K. Occurrence of lysoganglioside lyso-GM2 (II^3-Neu5Ac-gangliotriaosylsphingosine) in GM2 gangliosidosis brain. Biol Chem Hoppe Seyler. 1986;367:241–4.
Nilsson O, Svennerholm L. Accumulation of glucosylceramide and glucosylsphingosine (psychosine) in cerebrum and cerebellum in infantile and juvenile Gaucher disease. J Neurochem. 1982;39:709–18.
Ockerman PA. Identity of β-glucosidase, β-xylosidase and one of the β-galactosidase activities in human liver when assayed with 4-methylumbelliferyl-β-D-glycosides studies in cases of Gaucher's disease. Biochim Biophys Acta. 1968;165:59–62.
Ohman R, Rosenberg A, Svennerholm L. Human brain sialidase. Biochemistry. 1970;9:3774–82.
Okada S, O'Brien JS. Tay-Sachs disease: generalized absence of a β-D-*N*-acetylhexosaminidase component. Science. 1969;165:698–700.
Okada S, O'Brien JS. Generalized gangliosidosis: β-galactosidase deficiency. Science. 1968;160:1002–4.
Orvisky E, Park JK, LaMarca ME, Ginns EI, Martin BM, Tayebi N, et al. Glucosylsphingosine accumulation in tissues from patients with Gaucher disease: correlation with phenotype and genotype. Mol Genet Metab. 2002;76:262–70.
Patrick AD. A deficiency of glucocerebrosidase in Gaucher's disease. Biochem J. 1965;97:17c–8.
Radin NS. Treating glucosphingolipid disorders by chemotherapy: use of approved drugs and over-the-counter remedies. J Inherit Metab Dis. 2000;23:767–77.
Raghavan SS, Mumford RA, Kanfer JN. Deficiency of glucosylsphingosine: β-glucosidase in Gaucher disease. Biochem Biophys Res Commun. 1973;54:256–63.
Raghavan SS, Mumford RA, Kanfer JN. Isolation and characterization of glucosylsphingosine from Gaucher's spleen. J Lipid Res. 1974;15:484–90.
Robinson D, Stirling JL. *N*-Acetyl-β-glucosaminidases in human spleen. Biochem J. 1968;107:321–7.
Rosenberg A, Chargaff E. A reinvestigation of the cerebroside deposited in Gaucher's disease. J Biol Chem. 1958;233:1323–6.
Rosengren B, Mansson JE, Svennerholm L. Composition of gangliosides and neutral glycosphingolipids of brain in classical Tay-Sachs and Sandhoff disease: more lyso-GM2 in Sandhoff disease? J Neurochem. 1987;49:834–40.
Sandhoff K. Variation of β-*N*-acetylhexosaminidase-pattern in Tay-Sachs disease. FEBS Lett. 1969;4:351–4.
Sandhoff K, Kolter T, Harzer K, Scheper U, Remmel N. Chapter 134: Sphingolipid activator proteins. In: Valle D, Vogelstein B, Kinzler KW, Antonarakis S, Ballabio A, editors. The online metabolic & molecular bases of inherited disease (Scriver's OMMBID). McGraw-Hill Global Education Holdings; 2013. Available from: http://www.ommbid.com/
Sands MS, Davidson BL. Gene therapy for lysosomal storage diseases. Mol Ther. 2006;13:839–49.
Schram AW, Hamers MN, Tager JM. The identity of α-galactosidase B from human liver. Biochim Biophys Acta. 1977;482:138–44.

Schueler UH, Kolter T, Kaneski CR, Blusztajn JK, Herkenham M, Sandhoff K, et al. Toxicity of glucosylsphingosine (glucopsychosine) to cultured neuronal cells: a model system for assessing neuronal damage in Gaucher disease type 2 and 3. Neurobiol Dis. 2003;14:595–601.

Shapiro D, Flowers HM. Synthetic studies on sphingolipids. VI. The total synthesis of cerasine and phrenosine. J Am Chem Soc. 1961;83:3327–32.

Sidransky E, Nalls MA, Aasly JO, Aharon-Peretz J, Annesi G, Barbosa ER, et al. Multicenter analysis of glucocerebrosidase mutations in Parkinson's disease. N Engl J Med. 2009;361: 1651–61.

Sonderfeld S, Brendler S, Sandhoff K, Galjaard H, Hoogeveen AT. Genetic complementation in somatic cell hybrids of four variants of infantile GM2 gangliosidosis. Hum Genet. 1985;71: 196–200.

Suzuki K, Suzuki Y. Globoid cell leucodystrophy (Krabbe's disease): deficiency of galactocerebroside β-galactosidase. Proc Natl Acad Sci U S A. 1970;66:302–9.

Suzuki Y, Nanba E, Matsuda J, Higaki K, Oshima A. Chapter 151: β-Galactosidase deficiency (β-galactosidosis): GM1 gangliosidosis and Morquio B disease. In: Valle D, Vogelstein B, Kinzler KW, Antonarakis S, Ballabio A, editors. The online metabolic & molecular bases of inherited disease (Scriver's OMMBID). McGraw-Hill Global Education Holdings; 2013. Available from: http://www.ommbid.com/

Svennerholm L. The chemical structure of normal human brain and Tay-Sachs gangliosides. Biochem Biophys Res Commun. 1962;9:436–41.

Sweeley CC, Klionsky B. Fabry's disease: classification as a sphingolipidosis and partial characterization of a novel glycolipid. J Biol Chem. 1963;238:3148–50.

Sweeley CC, Snyder Jr PD, Griffin CE. Chemistry of glycosphingolipids Fabry's disease. Chem Phys Lipids. 1970;4:393–408.

Tay W. Symmetrical changes in the region of the yellow spot in each eye of an infant. Trans Ophthalmol Soc UK. 1881;1:55–7.

Thudichum JLW. Researches on the chemical configuration of the brain. Report of the Medical Officer of the Privy Council and Local Government Board 3. No. 5; 1874; p. 113.

Tyagarajan K, Forte JG, Townsend RR. Exoglycosidase purity and linkage specificity: assessment using oligosaccharide substrates and high-pH anion-exchange chromatography with pulsed amperometric detection. Glycobiology. 1996;6:83–93.

Varki A, Esko JD, Colley KJ. Cellular organization of glycosylation. In: Varki A, Cummings RD, Esko JD, Freeze HH, Stanley P, Bertozzi CR, Hart GW, Etzler ME, editors. Essentials of glycobiology. 2nd ed. Cold Spring Harbor, NY: Cold Spring Harbor Laboratory Press; 2009. p. 37–46.

Wenger DA, Suzuki K, Suzuki Y, Suzuki K. Chapter 147: Galactosylceramide lipidosis: Globoid cell leukodystrophy (Krabbe disease). In: Valle D, Vogelstein B, Kinzler KW, Antonarakis S, Ballabio A, editors. The online metabolic & molecular bases of inherited disease (Scriver's OMMBID). McGraw-Hill Global Education Holdings; 2013. Available from: http://www.ommbid.com/

Wierzbicka-Wos A, Bartasun P, Cieslinski H, Kur J. Cloning and characterization of a novel cold-active glycoside hydrolase family 1 enzyme with β-glucosidase, β-fucosidase and β-galactosidase activities. BMC Biotechnol. 2013;13:22.

Woollen JW, Walker PG, Heyworth R. Studies on glucosaminidase. 6. *N*-Acetyl-β-glucosaminidase and *N*-acetyl-β-galactosaminidase activities of a variety of enzyme preparations. Biochem J. 1961;79:294–8.

Wright CS, Li S-C, Rastinejad F. Crystal structure of human GM2-activator protein with a novel β-cup topology. J Mol Biol. 2000;304:411–22.

Yamakawa T, Suzuki S. The chemistry of the lipids of posthemolytic residue or stroma of erythrocytes. III. Globoside, the sugar-containing lipid of human blood stroma. J Biochem (Tokyo). 1952;39:393–402.

Yuziuk JA, Bertoni C, Beccari T, Orlacchio A, Wu Y-Y, Li S-C, et al. Specificity of mouse GM2 activator protein and β-*N*-acetylhexosaminidases A and B. Similarities and differences with their human counterparts in the catabolism of GM2. J Biol Chem. 1998;273:66–72.

Chapter 22
Ganglioside Storage Diseases: On the Road to Management

Thomas N. Seyfried, Hannah E. Rockwell, Karie A. Heinecke, Douglas R. Martin, and Miguel Sena-Esteves

Abstract Although the biochemical and genetic basis for the GM1 and GM2 gangliosidoses has been known for decades, effective therapies for these diseases remain in early stages of development. The difficulty with many therapeutic strategies for treating the gangliosidoses comes largely from their inability to remove stored ganglioside once it accumulates in central nervous system (CNS) neurons and glia. This chapter highlights advances made using substrate reduction therapy and gene therapy in reducing CNS ganglioside storage. Information obtained from mouse and feline models provides insight on therapeutic strategies that could be effective in human clinical trials. In addition, information is presented showing how a calorie-restricted diet might facilitate therapeutic drug delivery to the CNS. The development of multiple new therapeutic approaches offers hope that longer-term management of these diseases can be achieved. It is also clear that multiple therapeutic strategies will likely be needed to provide the most complete management.

Keywords GM1 • GM2 • Sandhoff disease • Tay-Sachs disease • Imino sugar • Gene therapy • Adeno-associated virus (AAV) • Calorie restriction • Ketogenic diet

T.N. Seyfried (✉) • H.E. Rockwell • K.A. Heinecke
Biology Department, Boston College, Chestnut Hill, MA 02467, USA
e-mail: seyfridt@bc.edu

D.R. Martin
Scott-Ritchey Research Center and Department of Anatomy, Physiology & Pharmacology, Auburn University College of Veterinary Medicine, Auburn, AL 36849, USA

M. Sena-Esteves
Department of Neurology and Gene Therapy Center, University of Massachusetts Medical School, Worcester, MA 01605, USA

R.K. Yu and C.-L. Schengrund (eds.), *Glycobiology of the Nervous System*, Advances in Neurobiology 9, DOI 10.1007/978-1-4939-1154-7_22,

Abbreviations

6S-NBI-DGJ	Bicyclic 1-deoxygalactonojirimycin
AAV	Adeno-associated virus
AL	Ad libitum
CB	Cerebroside
CNS	Central nervous system
CR	Caloric restriction
GalCer	Galactosylceramide
GlcCer	Glucosylceramide
GlcT	Glucosyltransferase-1
GSL	Glycosphingolipid
KD-R	Restricted ketogenic diet
LacCer	Lactosylceramide
NB-DGJ	*N*-butyldeoxygalactonojirimycin
NB-DNJ	*N*-butyldeoxynojirimycin
NN-DGJ	*N*-nonyl-deoxygalactonojirimycin
PDMP	D-threo-1-phenyl-2-decanoylamino-3-morpholino-propanol
PNS	Peripheral nervous system
SD	Sandhoff disease
SRT	Substrate reduction therapy
TSD	Tay-Sachs disease

22.1 Introduction

The glycosphingolipids (GSLs) are components of the cell surface glycocalyx and include the gangliosides and the neutral GSLs. GSLs contain an oligosaccharide head group and a lipophilic ceramide tail that anchors the GSL in the outer surface of the plasma membrane. The gangliosides are distinguished from the neutral GSLs in having *N*-acetylneuraminic acid (sialic acid) as part of the oligosaccharide chain (Fig. 22.1). Mammalian gangliosides are synthesized by the stepwise addition of sugar residues to the oligosaccharide head group. This is accomplished through the action of a Golgi-bound multi-glycosyltransferase system, where the GSL product of one transferase serves as the substrate for another transferase (Giraudo and Maccioni 2003). Most mammalian gangliosides are synthesized through metabolic pathways that use GM3 as a precursor (Yu et al. 2011). Ganglioside catabolism involves the stepwise removal of the various sugar residues through the action of hydrolytic enzymes within the lysosomes (Gravel et al. 1995; Sango et al. 1996; Sekine et al. 1984). Ganglioside synthesis and turnover is most active during early stages of mammalian brain development but decreases with maturation (Baek et al. 2004; Brigande et al. 1998; Ebato et al. 1983; Hauser et al. 2004; Yu 1993). Although the precise function of GSLs is not known, they can modulate numerous membrane-related processes (Lee et al. 1998; Proia 2003; Schnaar 2004; Seyfried 1987; Varki 1993; Yu et al. 2011).

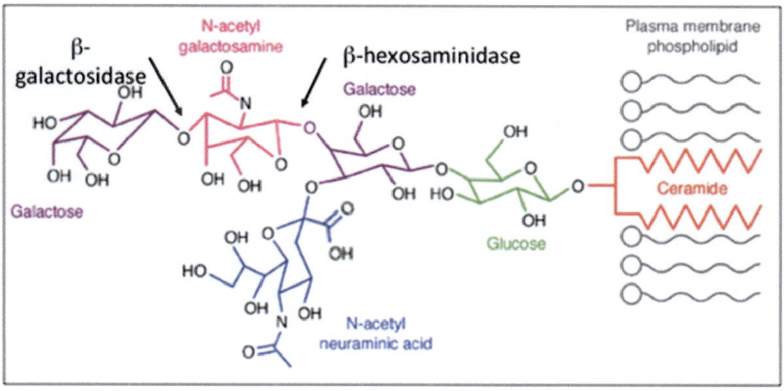

Fig. 22.1 The structure of ganglioside GM1 showing hydrolytic cleavage sites for β-galactosidase, which removes terminal galactose to form GM2 and β-hexosaminidase (Hex A) which removes the *N*-acetyl-galactosamine to form GM3. Inherited defects in the genes for β-galactosidase and β-hexosaminidase lead to the GM1 and the GM2 gangliosidoses, respectively

22.2 Ganglioside Storage Diseases and Pathogenesis

Most inherited mutations affecting ganglioside metabolism involve defects in catabolic pathways. These mutations cause the accumulation of either ganglioside GM1 and its asialo derivative, GA1 (GM1 gangliosidosis), or GM2 and its asialo derivative, GA2 (Tay-Sachs disease and Sandhoff disease), in the lysosomes throughout the CNS (Andersson et al. 2004; Gravel et al. 1995; Platt and Walkley 2004; Sandhoff and Harzer 2013; Sekine et al. 1984). The human GM2 gangliosidoses involve inherited defects in three genes that participate in the degradation of GM2 to GM3 and cause GM2 to accumulate in the CNS and viscera (Gravel et al. 1995) (Figs. 22.1 and 22.2). The *HEXA* and *HEXB* genes encode the α and β subunits of β-hexosaminidase, and defects in these genes cause Tay-Sachs disease (TSD) and Sandhoff disease (SD), respectively. GM2 accumulation can also arise from defects in a gene encoding a ganglioside activator protein (Gravel et al. 1995; Klima et al. 1991; Li et al. 1979). The severity of motor, mental, and visual dysfunction in the gangliosidoses is correlated with the degree of catabolic enzyme deficiency and with the level of ganglioside storage in the CNS leading to either early- or late-onset forms of the disorders (Denny et al. 2007; Gravel et al. 1995; O'Brien 1989; Sekine et al. 1984). Excessive ganglioside accumulation leads to progressive neurological deterioration (Platt and Walkley 2004). Inflammation and mitochondrial calcium imbalance can contribute to the pathophysiology (Jeyakumar et al. 2003; Myerowitz et al. 2002; Sano et al. 2009). In addition, myelin abnormalities (dysmyelination) occur in the human GM1 and GM2 gangliosidoses and in the canine and feline disease models (Folkerth et al. 2000; Kaye et al. 1992; Kroll et al. 1995; Van Der Voorn et al. 2004). Reductions in myelin-enriched glycolipids (cerebrosides and sulfatides) have been documented in adult brains from the GM1 and the GM2

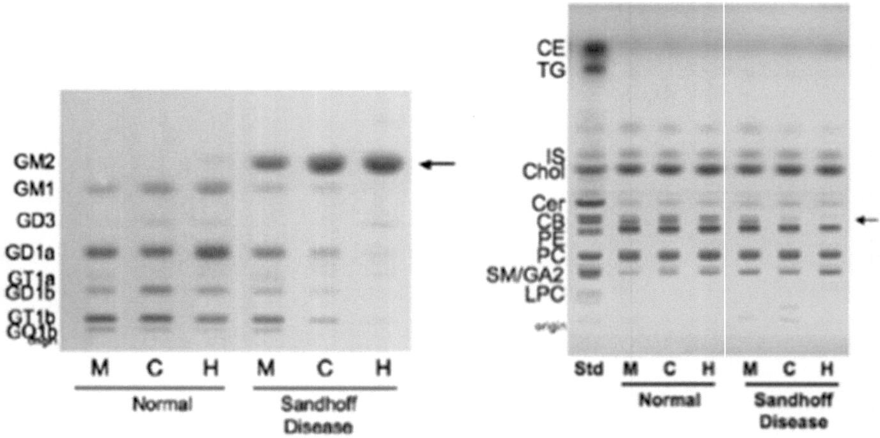

Fig. 22.2 High-performance thin-layer chromatography (HPTLC) analysis of cortical gangliosides (*left*) and neutral lipids (*right*) in normal and Sandhoff disease mice (M), cats (C), and humans (H). The results show massive GM2 storage (*arrow*) in each species with corresponding reduction of the myelin-enriched lipids, cerebrosides (CB) (*arrow*). The ganglioside plate was developed in one ascending run with chloroform-methanol-0.02 % aqueous calcium chloride (55:45:10 by volume). Gangliosides were visualized with resorcinol–HCl spray. The neutral lipid plate was developed to a height of 4.5 cm with chloroform-methanol-acetic acid-formic acid-water (35:15:6:2:1 by volume) and then developed to the top with hexanes-diisopropyl ether-acetic acid (65:35:2 by volume). The bands were visualized by charring with 3 % cupric acetate in 8 % phosphoric acid solution. The specific conditions for lipid analysis were as previously described (Baek et al. 2009)

gangliosidoses (Baek et al. 2009, 2010). Cerebroside reduction is proportional to GM2 accumulation across species in Sandhoff disease (Fig. 22.2).

Although the biochemical and genetic basis for the gangliosidoses has been known for decades, effective therapies for these diseases are in the early stages of development. The difficulty with many therapeutic strategies for treating the gangliosidoses comes largely from their inability to remove stored ganglioside once it accumulates in CNS neurons. Current therapeutic approaches focus on reducing ganglioside storage or managing CNS inflammation associated with storage. The association between ganglioside storage and inflammation is important in disease progression (Hayase et al. 2010; Jeyakumar et al. 2003; Myerowitz et al. 2002). Therapeutic strategies using bone marrow transplantation, enzyme replacement, or postnatal stem cell therapy have been labor intensive, costly, and only marginally effective to date (Andersson et al. 2004; Chavany and Jendoubi 1998; Ellinwood et al. 2004; Gravel et al. 1995; Lacorazza et al. 1996; Migita et al. 1995; Norflus et al. 1998; Paller et al. 1995; von Specht et al. 1979). Therapeutic synergy has been seen when an oral glycosphingolipid biosynthesis inhibitor imino sugar, *N*-butyldeoxynojirimycin (*N*B-DNJ), was combined with neural stem cell transplantation (Lee et al. 2007). Recent findings in mice showed that substrate reduction therapies (SRT) and adeno-associated viral (AAV)-mediated gene therapy were more effective than embryonic stem cell therapy in reducing CNS ganglioside storage (Arthur et al. 2012; Baek et al. 2010; Broekman et al. 2007; Cachon-Gonzalez et al. 2006, 2012). Further research will be needed on the therapeutic efficacy of substrate reduction therapy and gene therapy for managing ganglioside storage disease in animal models of the disease.

22.3 Mouse and Feline Models of the Gangliosidoses

The human GM1 gangliosidosis involves inherited mutations in the gene encoding lysosomal acid β-galactosidase (the *GLB1* gene), the enzyme that catabolizes GM1 to GM2 (O'Brien 1989). These mutations reduce or eliminate enzyme activity causing GM1 to accumulate in neural and nonneural tissues and produce motor and intellectual impairment (Gravel et al. 1995; O'Brien 1989). Histological evidence of ganglioside storage appears at about 3 weeks in the GM1 gangliosidosis mouse model, which lacks the functional gene encoding acid β-galactosidase (β-gal −/−) (Hahn et al. 1997). Extensive GM1 storage is observed in these mice by 5–6 weeks in neurons throughout the brain and spinal cord. Behavioral abnormalities are undetectable until about 4–6 months in the adult mouse mutants when CNS inflammation becomes apparent (Hauser et al. 2004; Jeyakumar et al. 2003). Mouse models for TSD and SD were produced through targeted disruption of the *HEXA* and *HEXB* genes, respectively (Sango et al. 1995). Behavioral abnormalities do not generally occur in the TSD mutant mouse until old age or after repeated pregnancies, and life span is not reduced (Jeyakumar et al. 2002). In contrast, the extensive accumulation of GM2 and asialo-GM2 (GA2) that occurs in SD mice causes severe behavioral abnormalities involving motor coordination and CNS inflammation. Life span is reduced with most SD mice dying between 130 and 140 days of age (Denny et al. 2006; Sango et al. 1995).

The feline GM1 and GM2 models were first reported in 1971 and 1977, respectively, and result from naturally occurring mutations that cause global storage in the CNS (Baker et al. 1971; Cork et al. 1977). GM1 cats synthesize normal amounts of a β-galactosidase enzyme that retains a low level of functionality, creating a moderate disease phenotype most representative of the juvenile-onset disease in humans (Martin et al. 2008). Symptom onset in GM1 cats occurs at 4.1 months of age, with the humane endpoint reached at 8 months. Feline GM2 gangliosidosis is more severe and is an effective model of infantile SD, with a disease course beginning at 1.7 months and ending at 4.5 months. Little residual enzymatic activity results from a β-hexosaminidase β subunit truncated at the carboxyl terminus by 8 amino acids, and protein levels are 10–20 % of normal (Martin et al. 2004). Though disease severity differs between GM1 and GM2 cats, symptoms are relatively consistent and begin with fine tremors of the head and tail that later involve the entire body. Also, mild hind limb weakness and ataxia in early stages progress to an inability to stand by the endpoint. Having considered the relatively consistent disease phenotype across species, our goal is to highlight some recent approaches for managing ganglioside storage diseases in mice and cats that may have future human application.

22.4 Substrate Reduction Therapy Using Imino Sugars and PDMP Analogues

Substrate reduction therapy (SRT) decreases the rate of GSL biosynthesis in order to counterbalance the impaired rate of catabolism thus reducing ganglioside accumulation and disease progression (Andersson et al. 2004; Butters et al. 2003; Larsen et al. 2012; Vunnam and Radin 1980). Platt and coworkers showed that *N*B-DNJ decreased CNS storage of GM2 ganglioside in adult TSD mice and SD mice and increased survival in SD mice (Jeyakumar et al. 1999; Neises et al. 1997). *N*B-DGJ as a galactose analogue could also reduce accumulation of GM1 and GM2 in early postnatal brains of mice with GM1 gangliosidosis and SD (Baek et al. 2008; Kasperzyk et al. 2005; Kasperzyk et al. 2004). The imino sugars, *N*B-DNJ and *N*B-DGJ, are competitive inhibitors of the ceramide-specific glucosyltransferase-1 (GlcT) that catalyzes the first step in GSL biosynthesis (Fischer et al. 1995) (Fig. 22.3). *N*B-DGJ is a more specific inhibitor of GlcT activity and GSL biosynthesis than is *N*B-DNJ because *N*B-DGJ does not inhibit *N*-glycan processing (Fischer et al. 1995). *N*B-DNJ (Zavesca®) is under clinical evaluation for the treatment of type 1 Gaucher's disease (glucosylceramide storage) and for late-onset

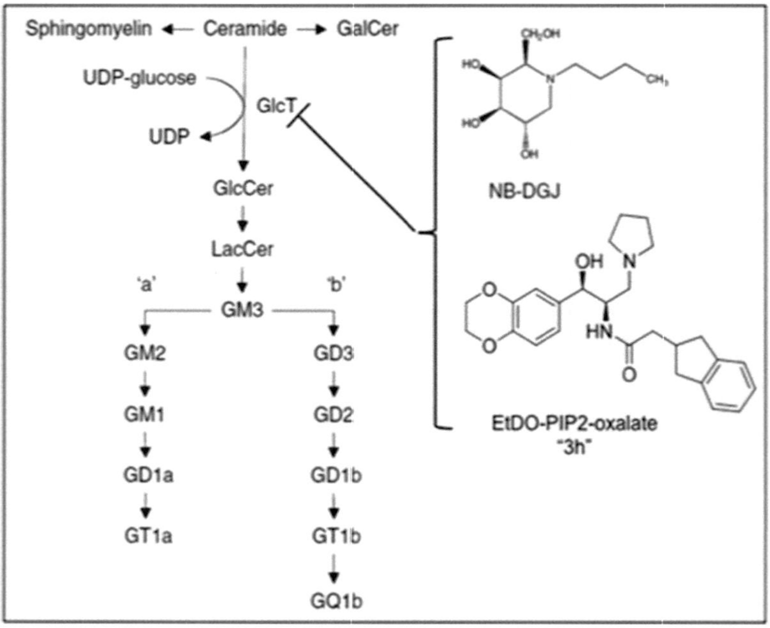

Fig. 22.3 Targeting glycosphingolipid synthesis in mammalian tissues with imino sugars (*NB-DGJ* & *N*B-DNJ) and **3 h**. Glucosyltransferase-1 (GlcT) catalyzes the first step in synthesis of gangliosides through the "a" and "b" pathways. Abbreviations: GalCer, galactosylceramide (cerebroside); GlcCer, glucosylceramide; LacCer, lactosylceramide. The *N*B-DGJ structure is from Andersson et al. (2004) and the **3 h** structure from Larsen et al. (2012)

Tay-Sachs disease (Badie et al. 2003; Kolodny et al. 2004; Lachmann 2003; Moyses 2003). Besides reducing ganglioside biosynthesis, some imino sugars can act as pharmacological chaperones to facilitate enzyme stabilization and transport to lysosomes, thereby enhancing the level of residual activity. In human and cat GM1 fibroblasts, residual activity of various β-gal mutants was increased up to 13.8-fold by *N*-nonyl-deoxygalactonojirimycin (NN-DGJ) (Rigat et al. 2012). Similarly, Takai and coworkers recently demonstrated reduced brain pathology through a chaperone mechanism using a bicyclic 1-deoxygalactonojirimycin (6S-NBI-DGJ) in GM1 transgenic mice expressing a R201C mutation in β-gal (Takai et al. 2013).

In addition to the imino sugars, analogues of D-threo-1-phenyl-2-decanoylamino-3-morpholino-propanol (PDMP) are also effective inhibitors of GLcT (Larsen et al. 2012; Radin 1982). Shayman and colleagues recently developed a new analogue inhibitor of GLcT, ethylenedioxy-PIP2 oxalate ("**3 h**"), that reduces GLcT activity at low nanomolar concentrations (Larsen et al. 2012) (Fig. 22.3). Since GlcCer is the common metabolic precursor required for the synthesis of most gangliosides and neutral glycolipids, the imino sugars and the "**3 h**" PDMP analogue will reduce the content in all tissues of GSLs containing the GlcCer core structure. It has recently been found that the "**3 h**" analogue reduces brain ganglioside storage to an extent similar to that seen with the imino sugars but at lower concentrations (Arthur et al. 2013). Further studies will be needed to assess the therapeutic efficacy of the "**3 h**" compound.

Earlier studies showed that the treatment of cultured whole mouse embryos with *N*B-DGJ inhibited the biosynthesis of glucosylceramide and gangliosides by over 90 %, indicating the drug's on-target effect (Brigande et al. 1998). Surprisingly, this GSL inhibition did not adversely affect embryo viability, growth, or morphogenesis, suggesting the feasibility of this drug as a nontoxic early intervention therapy. Moreover, *N*B-DGJ treatment significantly reduces brain ganglioside content in postnatal GM1 and GM2 gangliosidoses without altering brain growth, development, or behavior (Baek et al. 2008; Kasperzyk et al. 2004, 2005). Viewed collectively, these observations suggest *N*B-DGJ will have translational benefit to the clinic for treating early-onset ganglioside storage disease.

22.5 Gene Therapy

Gene therapy is emerging as a viable approach for targeting GSL storage in the GM1 and GM2 gangliosidoses (Baek et al. 2010; Broekman et al. 2007; Cachon-Gonzalez et al. 2006; Kyrkanides et al. 2005, 2009). It has been shown that infusion of adeno-associated virus (AAV) vectors encoding β-galactosidase or HEX A subunits in adults (bilateral thalamic infusion) or neonates (intracerebroventricular infusion) could effectively target CNS ganglioside storage in β-gal or Hexb knockout mice (Fig. 22.4 and Table 22.1) (Baek et al. 2010; Broekman et al. 2007). Mouse behavior and survival was also improved following gene therapy. Combined bilateral delivery of AAV vectors encoding HEX A to striatum and deep cerebellar nuclei

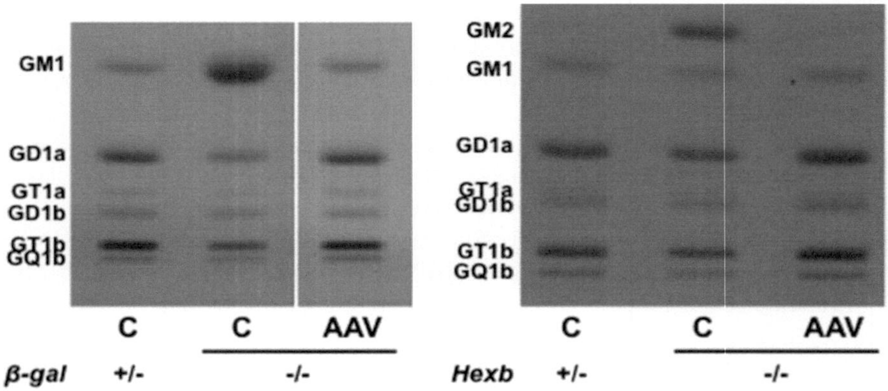

Fig. 22.4 Influence of AAV-gene therapy on the qualitative (HPTLC) and quantitative distribution of total cerebral cortex gangliosides in β-gal–/– and Hexb–/– mice. In these studies, AAV-gene therapy was administered (intracranially) to *adult* knockout mice that already expressed excessive elevations of GM1 and GM2. It is clear that AAV-gene therapy significantly reduced levels of stored GM1 and GM2 (Table 22.1). The reductions in total ganglioside were due largely to the reduction in either GM1 or GM2. In addition to cerebral cortex, reduced GM1 and GM2 storage in cerebellum and spinal cord of AAV-treated β-gal –/– and Hexb –/– mice was also found (not shown) (Baek et al. 2010). Reduced GM1 storage also restored normal cerebroside levels in the AAV-treated β-gal –/– mice suggesting a correction in myelin content (Baek et al. 2010). The data show the nearly complete removal of stored ganglioside in cerebral cortex of the AAV-treated β-gal –/– and Hexb –/– mice. The findings with the β-gal –/– mice were published (Baek et al. 2010). The findings with the Hexb –/– mice are unpublished. The conditions for AAV-gene therapy and ganglioside analysis were as previously described (Baek et al. 2010; Broekman et al. 2007). The HPTLC plate was developed and analyzed as shown in Fig. 22.2

Table 22.1 Cortical ganglioside concentration in adult Sandhoff and GM1 gangliosidosis mice[a]

Strain	Treatment	Age[b]	Total ganglioside sialic Acid content[c]	GM1[c]	GM2[c]
β-gal +/–	–	180	501.3 ± 9.0	79.3 ± 3.4	ND
β-gal –/–	–	180	1264.0 ± 70.8*	671.1 ± 11*	ND
β-gal –/–	AAV[d]	180	516.3 ± 7.7**	97.8 ± 9.3**	ND
Hexb +/–	–	90	465.1 ± 32.7	52.4 ± 5.8	3.4 ± 0.6
Hexb –/–	–	90	720.6 ± 29.5*	67.1 ± 2.5	179.1 ± 12.1*
Hexb –/–	AAV[d]	90	504.6 ± 10.5**	54.4 ± 2.7	15.2 ± 2**

ND, not detectable
[a]Values are expressed as mean ± standard error of the mean(SEM)
[b]Age in days
[c]Ganglioside values are expressed as μg sialic acid/100 mg dry weight
[d]Mice received thalamic injections of AAV-βgal or AAV-Hexα + β at 60 day of age
*Significant difference from the untreated +/– group at $p < 0.01$ by one-way ANOVA
**Significant difference from the untreated –/– group at $p < 0.01$ by one-way ANOVA
Baek et al. 2010

showed remarkable efficiency, increasing survival of SD mice from ~4 months to ~2 years of age and considerably reducing GM2 storage (Cachon-Gonzalez et al. 2006, 2012). These findings indicate that focal AAV-mediated gene delivery to specific structures in the brain is a highly effective approach for supplying the entire CNS with functional levels of lysosomal enzyme able to largely correct ganglioside storage throughout. Despite reduced ganglioside storage and increased survival, residual behavioral abnormalities persisted in the treated mice (Baek et al. 2010; Cachon-Gonzalez et al. 2006, 2012). The reason for residual behavioral abnormalities remains unknown but could result from either persistent storage in specific CNS regions, persistent inflammation, or substrate accumulation in the peripheral nervous system. PNS storage is probably not responsible for the behavioral abnormalities, as only traces of GM2 ganglioside storage in sciatic nerves of *Hexb–/–* mice were found (McNally et al. 2007). Further studies will be needed to determine whether gene therapy is also capable of targeting ganglioside storage in unique CNS areas including the retina, white matter, and optic nerve. The restoration of myelin-enriched cerebrosides in mice with GM1 gangliosidosis to levels seen in normal mice suggests that myelin function or pathology might be improved following removal of stored ganglioside (Baek et al. 2010; Broekman et al. 2007).

Initial gene therapy studies in cats are encouraging as well. In SD cats treated post-symptomatically (when pronounced storage was present), GM2 was reduced to 23 % of total ganglioside content at the thalamic injection site, compared to 72 % in untreated animals. Cats treated pre-symptomatically with bilateral thalamic injection of an AAV vector expressing feline hexosaminidase lived to ~10.5 months, compared to an untreated survival of 4.5 months (Bradbury et al. 2013). Further studies with multiple intracranial injection sites are ongoing and show further enhancement of survival benefit.

22.6 Caloric Restriction and the Restricted Ketogenic Diet (KD-R)

Caloric restriction (CR) is a natural dietary therapy that improves health, extends longevity, and reduces the effects of neuroinflammatory CNS diseases in rodents and humans (Chen et al. 2003; Duan et al. 2003; Greene et al. 2001; Mulrooney et al. 2011; Weindruch et al. 1988). CR is produced from a total dietary restriction and differs from acute fasting or starvation in that CR reduces total caloric energy intake without causing anorexia or deficiencies of any specific nutrients. A transition in brain energy substrate from predominantly glucose to predominantly ketone bodies is proposed to underlie the health benefits of CR (Chen et al. 2003; Greene et al. 2003). It has been found that rotarod performance (motor agility) and survival is significantly greater in adult SD mice maintained under CR instead of under ad libitum (AL) feeding (Denny et al. 2006). The improved motor performance and survival under CR was not due to reduced ganglioside storage, but was linked to reduced inflammation. Ketone body metabolism reduces inflammation possibly by

Table 22.2 Glycosphingolipid content in Hexb –/– mice[a]

Strain	N^b	Water content (%)	Ganglioside sialic acid (μg/100 mg dry weight)	GA2[c] (mg/100 mg dry weight)	NB-DNJ concentration[d] nmol/g tissue
SD	6	76.5±0.1	875±24	4.1±0.05	–
SD+NB-DNJ	5	76.2±0.6	795±18*	3.9±0.13	0.334±0.146
KD-R	4	75.1±1.1	821±27	4.0±0.18	–
KD-R+NB-DNJ	4	77.7±0.7	743±34*	4.0±0.04	1.163±0.038**

[a]Values represent the mean+SEM
[b]*N*, the number of independent samples analyzed
[c]Determined from densitometric scanning of HPTLC as shown in Fig 22.2
*Indicates that the value is significantly different from that of the SD mice at $p<0.05$ as determined from the two-tailed *t*-test
**Indicates that the value is significantly different from that of the SD+*N*B-DNJ mice at $p=0.0052$
Denny et al. 2010

suppressing generation of oxygen radicals (Veech 2004). As with CR, the calorie-restricted ketogenic diet (KD-R) also lowers blood glucose. The ketogenic diet is a low-carbohydrate-high-fat diet that is widely used to reduce refractory epileptic seizures in children but is also effective in managing symptoms of a wide range of neurological and neurodegenerative diseases (Freeman and Kossoff 2010; Kossoff and Hartman 2012). It is unclear if the therapeutic action of the KD results from reduced glucose, elevated ketone bodies, or some combination of these. It is interesting that the KD-R elevates blood ketones to a higher level than CR alone (Denny et al. 2010; Zhou et al. 2007). Research has found that KD-R resulted in a 3.5-fold increase of *N*B-DNJ in the CNS of adult SD mice, with greater inhibition of GM2 ganglioside storage (Denny et al. 2010) (Table 22.2). Moreover, the total forebrain ganglioside sialic acid and GM2 concentration was lower in the Hexb–/– mice when *N*B-DNJ was administered with the KD-R than when it was administered with standard lab chow. Further studies will be needed to evaluate the mechanism for this dietary effect on CNS drug delivery.

22.7 Summary

Despite knowledge of the genetic and biochemical defects involved, ganglioside storage diseases remain incurable (Sandhoff and Harzer 2013). The development of multiple new therapeutic approaches offers hope that longer-term management of these diseases can be achieved. It is also clear that multiple therapeutic strategies may be needed to provide the most complete management.

Conflict of Interest The authors declare that there are no conflicts of interest.

Acknowledgements This work was supported in part by National Institutes of Health Grants R01NS-055195 (TNS), R21NS053993 (MSE), and U01-NS064096 (TNS, DRM, MSE), the Boston College Research Expense Fund, the Scott-Ritchey Research Center, the Lysosomal Storage Disease Research Consortium, and the National Tay-Sachs and Allied Diseases Association, Inc.

References

Andersson U, Smith D, Jeyakumar M, Butters TD, Borja MC, Dwek RA, et al. Improved outcome of N-butyldeoxygalactonojirimycin-mediated substrate reduction therapy in a mouse model of Sandhoff disease. Neurobiol Dis. 2004;16(3):506–15.

Arthur JR, Lee JP, Snyder EY, Seyfried TN. Therapeutic effects of stem cells and substrate reduction in juvenile Sandhoff mice. Neurochem Res. 2012;37(6):1335–43.

Arthur JR, Wilson MW, Larsen SD, Rockwell HE, Shayman JA, Seyfried TN. Ethylenedioxy-PIP2 oxalate reduces ganglioside storage in juvenile Sandhoff disease mice. Neurochem Res. 2013;38(4):866–75. Research Support, N.I.H., Extramural Research Support, Non-U.S. Gov't.

Badie B, Schartner JM, Hagar AR, Prabakaran S, Peebles TR, Bartley B, et al. Microglia cyclooxygenase-2 activity in experimental gliomas: possible role in cerebral edema formation. Clin Cancer Res. 2003;9(2):872–7.

Baek RC, Kasperzyk JL, Platt FM, Seyfried TN. N-butyldeoxygalactonojirimycin reduces brain ganglioside and GM2 content in neonatal Sandhoff disease mice. J Neurochem. 2004;90 Suppl 1:89.

Baek RC, Kasperzyk JL, Platt FM, Seyfried TN. N-butyldeoxygalactonojirimycin reduces brain ganglioside and GM2 content in neonatal Sandhoff disease mice. Neurochem Int. 2008; 52(6):1125–33.

Baek RC, Martin DR, Cox NR, Seyfried TN. Comparative analysis of brain lipids in mice, cats, and humans with Sandhoff disease. Lipids. 2009;44(3):197–205.

Baek RC, Broekman ML, Leroy SG, Tierney LA, Sandberg MA, d'Azzo A, et al. AAV-mediated gene delivery in adult GM1-gangliosidosis mice corrects lysosomal storage in CNS and improves survival. PLoS One. 2010;5(10):e13468. Research Support, N.I.H., Extramural.

Baker Jr HJ, Lindsey JR, McKhann GM, Farrell DF. Neuronal GM1 gangliosidosis in a Siamese cat with beta-galactosidase deficiency. Science. 1971;174(4011):838–9. New York, NY.

Bradbury AM, Cochran JN, McCurdy VJ, Johnson AK, Brunson BL, Gray-Edwards H, et al. Therapeutic response in feline sandhoff disease despite immunity to intracranial gene therapy. Mol Ther. 2013;21(7):1306–15.

Brigande JV, Platt FM, Seyfried TN. Inhibition of glycosphingolipid biosynthesis does not impair growth or morphogenesis of the postimplantation mouse embryo. J Neurochem. 1998;70: 871–82.

Broekman ML, Baek RC, Comer LA, Fernandez JL, Seyfried TN, Sena-Esteves M. Complete correction of enzymatic deficiency and neurochemistry in the GM1-gangliosidosis mouse brain by Neonatal Adeno-associated virus-mediated gene delivery. Mol Ther. 2007;15(1):30–7.

Butters TD, Dwek RA, Platt FM. Therapeutic applications of imino sugars in lysosomal storage disorders. Curr Top Med Chem. 2003;3(5):561–74.

Cachon-Gonzalez MB, Wang SZ, Lynch A, Ziegler R, Cheng SH, Cox TM. Effective gene therapy in an authentic model of Tay-Sachs-related diseases. Proc Natl Acad Sci U S A. 2006; 103(27):10373–8.

Cachon-Gonzalez MB, Wang SZ, McNair R, Bradley J, Lunn D, Ziegler R, et al. Gene transfer corrects acute GM2 gangliosidosis–potential therapeutic contribution of perivascular enzyme flow. Mol Ther. 2012;20(8):1489–500. Research Support, Non-U.S. Gov't.

Chavany C, Jendoubi M. Biology and potential strategies for the treatment of GM2 gangliosidoses. Mol Med Today. 1998;4(4):158–65.

Chen JZ, Gokden N, Greene GF, Green B, Kadlubar FF. Simultaneous generation of multiple mitochondrial DNA mutations in human prostate tumors suggests mitochondrial hypermutagenesis. Carcinogenesis. 2003;24(9):1481–7.

Cork LC, Munnell JF, Lorenz MD, Murphy JV, Baker HJ, Rattazzi MC. GM2 ganglioside lysosomal storage disease in cats with beta-hexosaminidase deficiency. Science. 1977;196(4293):1014–7. New York, NY.

Denny CA, Kasperzyk JL, Gorham KN, Bronson RT, Seyfried TN. Influence of caloric restriction on motor behavior, longevity, and brain lipid composition in Sandhoff disease mice. J Neurosci Res. 2006;83(6):1028–38.

Denny CA, Alroy J, Pawlyk BS, Sandberg MA, d'Azzo A, Seyfried TN. Neurochemical, morphological, and neurophysiological abnormalities in retinas of Sandhoff and GM1 gangliosidosis mice. J Neurochem. 2007;101(5):1294–302.

Denny CA, Heinecke KA, Kim YP, Baek RC, Loh KS, Butters TD, et al. Restricted ketogenic diet enhances the therapeutic action of N-butyldeoxynojirimycin towards brain GM2 accumulation in adult Sandhoff disease mice. J Neurochem. 2010;113(6):1525–35.

Duan W, Guo Z, Jiang H, Ware M, Li XJ, Mattson MP. Dietary restriction normalizes glucose metabolism and BDNF levels, slows disease progression, and increases survival in huntingtin mutant mice. Proc Natl Acad Sci U S A. 2003;100(5):2911–6.

Ebato H, Seyfried TN, Yu RK. Biochemical study of heterosis for brain myelin content in mice. J Neurochem. 1983;40(2):440–6.

Ellinwood NM, Vite CH, Haskins ME. Gene therapy for lysosomal storage diseases: the lessons and promise of animal models. J Gene Med. 2004;6(5):481–506.

Fischer PB, Collin M, Karlsson GB, James W, Butters TD, Davis SJ, et al. The alpha-glucosidase inhibitor N-butyldeoxynojirimycin inhibits human immunodeficiency virus entry at the level of post-CD4 binding. J Virol. 1995;69(9):5791–7.

Folkerth RD, Alroy J, Bhan I, Kaye EM. Infantile G(M1) gangliosidosis: complete morphology and histochemistry of two autopsy cases, with particular reference to delayed central nervous system myelination. Pediatr Dev Pathol. 2000;3(1):73–86.

Freeman JM, Kossoff EH. Ketosis and the ketogenic diet, 2010: advances in treating epilepsy and other disorders. Adv Pediatr. 2010;57(1):315–29.

Giraudo CG, Maccioni HJ. Ganglioside glycosyltransferases organize in distinct multienzyme complexes in CHO-K1 cells. J Biol Chem. 2003;278(41):40262–71.

Gravel RA, Clarke JTR, Kaback MM, Mahuran D, Sandhoff K, Suzuki K. The GM2 gangliosidoses. In: Scriver CR, Beaudet al, Sly WS, Valle D, editors. The metabolic and molecular bases of inherited disease. 7th ed. New York: McGraw-Hill, Inc; 1995. p. 2839–79.

Greene AE, Todorova MT, McGowan R, Seyfried TN. Caloric restriction inhibits seizure susceptibility in epileptic EL mice by reducing blood glucose. Epilepsia. 2001;42(11):1371–8.

Greene AE, Todorova MT, Seyfried TN. Perspectives on the metabolic management of epilepsy through dietary reduction of glucose and elevation of ketone bodies. J Neurochem. 2003;86(3): 529–37.

Hahn CN, del Pilar MM, Schroder M, Vanier MT, Hara Y, Suzuki K, et al. Generalized CNS disease and massive GM1-ganglioside accumulation in mice defective in lysosomal acid beta-galactosidase. Hum Mol Genet. 1997;6(2):205–11.

Hauser EC, Kasperzyk JL, d'Azzo A, Seyfried TN. Inheritance of lysosomal acid beta-galactosidase activity and gangliosides in crosses of DBA/2J and knockout mice. Biochem Genet. 2004; 42(7–8):241–57.

Hayase T, Shimizu J, Goto T, Nozaki Y, Mori M, Takahashi N, et al. Unilaterally and rapidly progressing white matter lesion and elevated cytokines in a patient with Tay-Sachs disease. Brain Dev. 2010;32(3):244–7. Case Reports.

Jeyakumar M, Butters TD, Cortina-Borja M, Hunnam V, Proia RL, Perry VH, et al. Delayed symptom onset and increased life expectancy in Sandhoff disease mice treated with N-butyldeoxynojirimycin. Proc Natl Acad Sci U S A. 1999;96(11):6388–93.

Jeyakumar M, Smith D, Eliott-Smith E, Cortina-Borja M, Reinkensmeier G, Butters TD, et al. An inducible mouse model of late onset Tay-Sachs disease. Neurobiol Dis. 2002;10(3):201–10.

Jeyakumar M, Thomas R, Elliot-Smith E, Smith DA, van der Spoel AC, d'Azzo A, et al. Central nervous system inflammation is a hallmark of pathogenesis in mouse models of GM1 and GM2 gangliosidosis. Brain. 2003;126(Pt 4):974–87.

Kasperzyk JL, El-Abbadi MM, Hauser EC, D'Azzo A, Platt FM, Seyfried TN. N-butyldeoxygalactonojirimycin reduces neonatal brain ganglioside content in a mouse model of GM1 gangliosidosis. J Neurochem. 2004;89(3):645–53.

Kasperzyk JL, d'Azzo A, Platt FM, Alroy J, Seyfried TN. Substrate reduction reduces gangliosides in postnatal cerebrum-brainstem and cerebellum in GM1 gangliosidosis mice. J Lipid Res. 2005;46(4):744–51.

Kaye EM, Alroy J, Raghavan SS, Schwarting GA, Adelman LS, Runge V, et al. Dysmyelinogenesis in animal model of GM1 gangliosidosis. Pediatr Neurol. 1992;8(4):255–61.

Klima H, Tanaka A, Schnabel D, Nakano T, Schroder M, Suzuki K, et al. Characterization of full-length cDNAs and the gene coding for the human GM2 activator protein. FEBS Lett. 1991;289(2):260–4. Research Support, Non-U.S. Gov't Research Support, U.S. Gov't, P.H.S.

Kolodny EH, Neudorfer O, Gianutsos J, Zaroff C, Barnett N, Zeng B, et al. Late-onset Tay-Sachs disease: Natural history and treatment with OGT 918. J Neurochem 2004 (Suppl.) (in press).

Kossoff EH, Hartman AL. Ketogenic diets: new advances for metabolism-based therapies. Curr Opin Neurol. 2012;25(2):173–8.

Kroll RA, Pagel MA, Roman-Goldstein S, Barkovich AJ, D'Agostino AN, Neuwelt EA. White matter changes associated with feline GM2 gangliosidosis (Sandhoff disease): correlation of MR findings with pathologic and ultrastructural abnormalities. AJNR Am J Neuroradiol. 1995;16(6):1219–26.

Kyrkanides S, Miller JH, Brouxhon SM, Olschowka JA, Federoff HJ. Beta-hexosaminidase lentiviral vectors: transfer into the CNS via systemic administration. Brain Res Mol Brain Res. 2005;133(2):286–98.

Kyrkanides S, Yang M, Tallents RH, Miller JN, Brouxhon SM, Olschowka JA. The trigeminal retrograde transfer pathway in the treatment of neurodegeneration. J Neuroimmunol. 2009;209(1–2):139–42. Research Support, N.I.H., Extramural.

Lachmann RH. Miglustat. Oxford glycosciences/actelion. Curr Opin Investig Drugs. 2003;4(4): 472–9.

Lacorazza HD, Flax JD, Snyder EY, Jendoubi M. Expression of human beta-hexosaminidase alpha-subunit gene (the gene defect of Tay-Sachs disease) in mouse brains upon engraftment of transduced progenitor cells. Nat Med. 1996;2(4):424–9.

Larsen SD, Wilson MW, Abe A, Shu L, George CH, Kirchhoff P, et al. Property-based design of a glucosylceramide synthase inhibitor that reduces glucosylceramide in the brain. J Lipid Res. 2012;53(2):282–91.

Lee JP, Jeyakumar M, Gonzalez R, Takahashi H, Lee PJ, Baek RC, et al. Stem cells act through multiple mechanisms to benefit mice with neurodegenerative metabolic disease. Nat Med. 2007;13(4):439–47.

Lee MC, El-Abbadi M, Orosz CG, Yates AJ, Seyfried TN. A spontaneous metastatic brain tumor in the VM mouse: histological characteristics and ganglioside composition. 1998 (submitted).

Li SC, Nakamura T, Ogamo A, Li YT. Evidence for the presence of two separate protein activators for the enzymic hydrolysis of GM1 and GM2 gangliosides. J Biol Chem. 1979;254(21):10592–5. Research Support, U.S. Gov't, Non-P.H.S. Research Support, U.S. Gov't, P.H.S.

Martin DR, Krum BK, Varadarajan GS, Hathcock TL, Smith BF, Baker HJ. An inversion of 25 base pairs causes feline GM2 gangliosidosis variant. Exp Neurol. 2004;187(1):30–7. Research Support, Non-U.S. Gov't Research Support, U.S. Gov't, Non-P.H.S.

Martin DR, Rigat BA, Foureman P, Varadarajan GS, Hwang M, Krum BK, et al. Molecular consequences of the pathogenic mutation in feline GM1 gangliosidosis. Mol Genet Metab. 2008;94(2):212–21. Research Support, Non-U.S. Gov't.

McNally MA, Baek RC, Avila RL, Seyfried TN, Strichartz GR, Kirschner DA. Peripheral nervous system manifestations in a Sandhoff disease mouse model: nerve conduction, myelin structure, lipid analysis. J Negat Results Biomed. 2007;6:8. Research Support, N.I.H., Extramural Research Support, Non-U.S. Gov't.

Migita M, Medin JA, Pawliuk R, Jacobson S, Nagle JW, Anderson S, et al. Selection of transduced CD34+ progenitors and enzymatic correction of cells from Gaucher patients, with bicistronic vectors. Proc Natl Acad Sci U S A. 1995;92(26):12075–9.
Moyses C. Substrate reduction therapy: clinical evaluation in type 1 Gaucher disease. Philos Trans R Soc Lond B Biol Sci. 2003;358(1433):955–60.
Mulrooney TJ, Marsh J, Urits I, Seyfried TN, Mukherjee P. Influence of caloric restriction on constitutive expression of NF-kappaB in an experimental mouse astrocytoma. PLoS One. 2011;6(3):e18085.
Myerowitz R, Lawson D, Mizukami H, Mi Y, Tifft CJ, Proia RL. Molecular pathophysiology in Tay-Sachs and Sandhoff diseases as revealed by gene expression profiling. Hum Mol Genet. 2002;11(11):1343–50.
Neises GR, Woodman PG, Butters TD, Ornberg RL, Platt FM. Ultrastructural changes in the Golgi apparatus and secretory granules of HL-60 cells treated with the imino sugar N-butyldeoxynojirimycin. Biol Cell. 1997;89(2):123–31.
Norflus F, Tifft CJ, McDonald MP, Goldstein G, Crawley JN, Hoffmann A, et al. Bone marrow transplantation prolongs life span and ameliorates neurologic manifestations in Sandhoff disease mice. J Clin Invest. 1998;101(9):1881–8.
O'Brien JS. b-Galactosidase deficiency (GM1 gangliosidosis, galactosialidosis, and Morquio syndrome type B); ganglioside sialidase deficiency (Mucolipidosis IV). In: Scriver CR, Beaudet al, Sly WS, Valle D, editors. The metabolic basis of inherited disease. 6th ed. New York: McGraw-Hill Inc.; 1989. p. 1797–806.
Paller AS, Arnsmeier SL, Chen JD, Woodley DT. Ganglioside GT1b inhibits keratinocyte adhesion and migration on a fibronectin matrix. J Invest Dermatol. 1995;105(2):237–42.
Platt FM, Walkley SU. Lysosomal disorders of the brain. New York: Oxford University Press; 2004.
Proia RL. Glycosphingolipid functions: insights from engineered mouse models. Philos Trans R Soc Lond B Biol Sci. 2003;358(1433):879–83.
Radin NS. Inhibitors and stimulators of glucocerebroside metabolism. Prog Clin Biol Res. 1982;95:357–83.
Rigat BA, Tropak MB, Buttner J, Crushell E, Benedict D, Callahan JW, et al. Evaluation of N-nonyl-deoxygalactonojirimycin as a pharmacological chaperone for human GM1 gangliosidosis leads to identification of a feline model suitable for testing enzyme enhancement therapy. Mol Genet Metab. 2012;107(1–2):203–12. Research Support, Non-U.S. Gov't.
Sandhoff K, Harzer K. Gangliosides and gangliosidoses: principles of molecular and metabolic pathogenesis. J Neurosci. 2013;33(25):10195–208. Research Support, Non-U.S. Gov't.
Sango K, Johnson ON, Kozak CA, Proia RL. Beta-1,4-N-Acetylgalactosaminyltransferase involved in ganglioside synthesis: cDNA sequence, expression, and chromosome mapping of the mouse gene. Genomics. 1995;27(2):362–5.
Sango K, McDonald MP, Crawley JN, Mack ML, Tifft CJ, Skop E, et al. Mice lacking both subunits of lysosomal b-hexosaminidase display gangliosidosis and mucopolysaccharidosis. Nat Genet. 1996;14:348–52.
Sano R, Annunziata I, Patterson A, Moshiach S, Gomero E, Opferman J, et al. GM1-ganglioside accumulation at the mitochondria-associated ER membranes links ER stress to Ca(2+)-dependent mitochondrial apoptosis. Mol Cell. 2009;36(3):500–11. Research Support, N.I.H., Extramural Research Support, Non-U.S. Gov't.
Schnaar RL. Glycolipid-mediated cell-cell recognition in inflammation and nerve regeneration. Arch Biochem Biophys. 2004;426(2):163–72.
Sekine M, Ariga T, Miyatake T, Kase R, Suzuki H, Yamakawa T. Gangliosides and neutral glycolipids in guinea pig adrenal glands. J Biochem. 1984;96:237–44.
Seyfried TN. Ganglioside abnormalities associated with failed neural differentiation in a T-locus mutant mouse embryo. Dev Biol. 1987;123(1):286–91.
Takai T, Higaki K, Aguilar-Moncayo M, Mena-Barragan T, Hirano Y, Yura K, et al. A bicyclic 1-deoxygalactonojirimycin derivative as a novel pharmacological chaperone for GM1 gangliosidosis. Mol Ther. 2013;21(3):526–32. Research Support, Non-U.S. Gov't.

Van Der Voorn JP, Kamphorst W, Van Der Knaap MS, Powers JM. The leukoencephalopathy of infantile GM1 gangliosidosis: oligodendrocytic loss and axonal dysfunction. Acta Neuropathol (Berl). 2004;107(6):539–45.

Varki A. Biological roles of oligosaccharides: all of the theories are correct. Glycobiology. 1993;3(2):97–130.

Veech RL. The therapeutic implications of ketone bodies: the effects of ketone bodies in pathological conditions: ketosis, ketogenic diet, redox states, insulin resistance, and mitochondrial metabolism. Prostaglandins Leukot Essent Fatty Acids. 2004;70(3):309–19.

von Specht BU, Geiger B, Arnon R, Passwell J, Keren G, Goldman B, et al. Enzyme replacement in Tay-Sachs disease. Neurology. 1979;29:848–54.

Vunnam RR, Radin NS. Analogs of ceramide that inhibit glucocerebroside synthetase in mouse brain. Chem Phys Lipids. 1980;26(3):265–78.

Weindruch R, Kemnitz JW, Uno H. Interspecies variations in physiologic and antipathologic outcomes of dietary restriction (1988).

Yu RK. Development regulation of ganglioside metabolism. Prog Brain Res. 1993;101:31–44.

Yu RK, Tsai YT, Ariga T, Yanagisawa M. Structures, biosynthesis, and functions of gangliosides-an overview. J Oleo Sci. 2011;60(10):537–44. Research Support, N.I.H., Extramural.

Zhou W, Mukherjee P, Kiebish MA, Markis WT, Mantis JG, Seyfried TN. The calorically restricted ketogenic diet, an effective alternative therapy for malignant brain cancer. Nutr Metab (Lond). 2007;4:5.

Chapter 23
Dynamic Aspects of Neural Tumor Gangliosides

Stephan Ladisch and Yihui Liu

Abstract Gangliosides are important cell surface molecules in tumors of the nervous system. Initially thought of particularly as diagnostic disease markers, the biological properties and in vivo roles of tumor gangliosides have been recognized more recently. The latter affect the formation and progression of tumors of many types, including neural tumors. This chapter focuses on a prominent and almost universal dynamic property of tumor cells, the rapid synthesis and shedding of gangliosides, and consequences on the host response to tumors. In contemporary terms, these effects are understood to be functional properties of tumor cells affecting normal cells in the tumor microenvironment. Some of the most comprehensive studies have been possible in the human neuroectodermal tumors, neuroblastomas, including first observations of the shedding and activities of human tumor gangliosides. Citing other selected examples from tumors of different types, the impact and mechanisms of action of tumor gangliosides in facilitating tumor formation and progression will be delineated, leading to consideration of clinical relevance and therapeutic implications.

Keywords Gangliosides • Shedding • Neuroectodermal tumors • Angiogenesis • Immunosuppression • Tumor microenvironment • Tumor progression GD2 • Cerebrospinal fluid gangliosides • Growth factor signaling

S. Ladisch, M.D. (✉)
Center for Cancer and Immunology Research, Children's National Medical Center, 111 Michigan Ave. NW, Washington, DC 20010, USA

George Washington University School of Medicine, Washington, DC, USA
e-mail: sladisch@cnmc.org

Y. Liu, Ph.D.
Center for Cancer and Immunology Research, Children's National Medical Center, 111 Michigan Ave. NW, Washington, DC 20010, USA

R.K. Yu and C.-L. Schengrund (eds.), *Glycobiology of the Nervous System*, Advances in Neurobiology 9, DOI 10.1007/978-1-4939-1154-7_23,

23.1 Introduction

Cellular gangliosides are a component of the cell plasma membrane, present and concentrated on the extracellular surface. A glycosphingolipid, the ganglioside molecule consists of ceramide linked to a sialic acid-containing oligosaccharide. To place these amphiphatic molecules into the context of cancer, it is important to recognize that malignant transformation is a multistep process consisting of both genetic and epigenetic factors that collectively lead to increasing autonomy of malignant tumor cells (Foulds 1965). The tumor cell develops increasing independence from the multiple host control mechanisms for regulating cell growth (e.g., loss of normal cell cycle checkpoint controls, resistance to apoptosis, and increased expression of autocrine growth factors or receptors) (Hanahan and Weinberg 2000). Of particular importance, molecular interactions with the host (Hendrix et al. 2003), initiated by the tumor cell and causing alterations in the tumor microenvironment (TME), are also necessary to promote tumor progression. It is in this context that tumor gangliosides, which are among tumor-derived factors that may favor tumor cell survival, are increasingly being shown to be integral to the process of tumor formation and progression.

As cell surface molecules, initially tumor gangliosides were studied for their role in enabling tumor recognition and diagnosis, mainly by antibody or lectin detection. Their roles in cell-cell interactions by direct contact were also elucidated. These concepts and findings have been substantially developed and addressed elsewhere, e.g., by Hakomori (Hakomori 1985, 1996), and are outside the scope of this chapter. As cell surface molecules, acting as targets for antibody-mediated immunotherapy is also a characteristic of gangliosides on the tumor cell surface. This has been successfully applied in human neuroblastoma, in which therapy with a chimeric anti-GD2 ganglioside antibody in combination with GM-CSF and interleukin-2 caused significantly improved outcome as compared with standard therapy in patients with high-risk neuroblastoma (Yu et al. 2010).

This chapter addresses the observations that gangliosides are actively shed from the cell surface membranes of many tumor cells (Ladisch et al. 1983), but essentially not from normal nonproliferating cells. This dynamic process embodies these molecules with the ability to influence the functions of other cells in the tumor microenvironment and thereby to strongly influence tumor formation and progression. It should be noted that many of the studies defining the roles of gangliosides have been performed in systems other than tumors of the nervous system. However, since the multiple functional characteristics and activities of gangliosides are not dependent on tumor type (but rather on ganglioside structure, as will be discussed further), these diverse examples are included in this chapter with the intent of providing a global overview of dynamic properties of tumor gangliosides.

23.2 Tumor Cell Ganglioside Metabolism

The change from normal tissue to malignancy involves changes in ganglioside composition. These changes both mark the transition from normal to tumor tissue and may affect the outcome, i.e., tumor progression. Tumors may have highly variable composition of individual ganglioside species, evident both between tumor types and among individual tumors. Considering first the question of marking the transition, several examples, although not providing an explanation of cause, underline the likely significance of the details of altered characteristics of ganglioside metabolism resulting in qualitative changes in cell gangliosides. First, the ganglioside patterns of a series of gliomas demonstrated a greater proportion of structurally simpler gangliosides than those of the normal brain (in which GM1, GD1a, GD1b, and GT1b are the major gangliosides). The proportion of simpler gangliosides GM3 and GD3 was higher in tumors of higher grade (grade 3–4) than in those of lower grade (grade 1–2) (Nakamura et al. 1987). Analogous observations have been made in human neuroblastoma tumors. Specifically, higher proportions of complex gangliosides of the "b" pathway (GD1b, GT1b, and GQ1b) are associated with a less aggressive phenotype and a better clinical outcome (progression-free survival). In contrast, the simple "b" pathway gangliosides (GD3 and GD2) have been associated with tumors of poorer prognosis, including the aggressive forms of neuroblastoma (Hettmer et al. 2003; Kaucic et al. 2001; Schengrund et al. 1985).

23.2.1 Ganglioside Metabolism Is Upregulated in Tumor Cells During Tumor Progression

What could be the cause for overexpression of tumor gangliosides during tumor progression? Considering those genes that are responsible for ganglioside synthesis, genomic studies have revealed that certain genes are upregulated in tumors. This is particularly exemplified by recent findings regarding GM3 synthase (GM3S). Gene expression analysis of human melanoma showed GM3 synthase to be highly overexpressed in comparison to its expression in benign nevi or normal skin (Talantov et al. 2005). Since GM3 synthase itself is the key entry enzyme to the pathway for most human ganglioside synthesis, this may be the critical element in ganglioside expression in tumors. Conversely, other evidence comes from correlations between metastatic potential and the expression of sialidase, which results in the conversion of a ganglioside into a neutral glycosphingolipid; levels of sialidase are inversely correlated with the metastatic potential of mouse colon adenocarcinoma (Sawada et al. 2002). This kind of indirect correlative evidence from gene expression studies strongly suggests that the resulting ganglioside levels are associated with tumor growth and tumor progression. Together with direct ganglioside analyses, the findings point to control of ganglioside synthesis as important in tumor progression, making ganglioside synthesis also a potential target for therapeutic intervention in cancer.

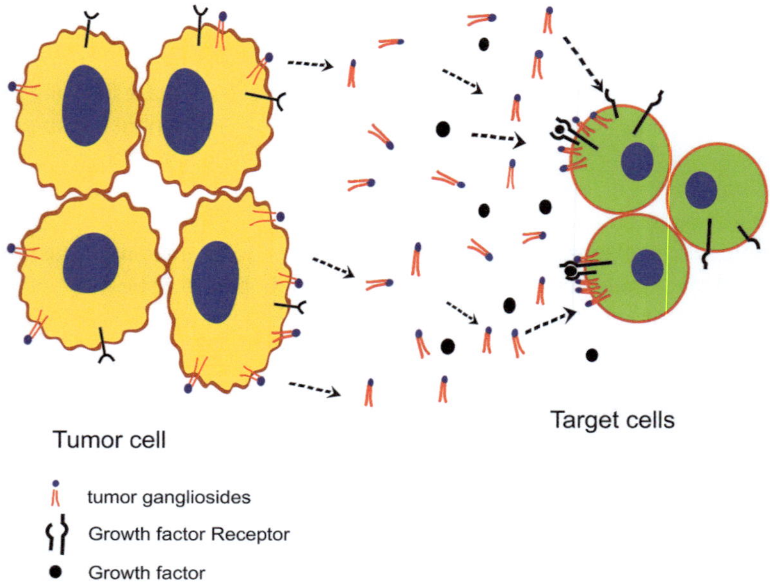

Fig. 23.1 Tumor ganglioside synthesis, shedding, and cellular interactions in the tumor microenvironment

23.2.2 Tumor Gangliosides Are Actively Shed into the Tumor Microenvironment

Over the years both by in vitro studies and in vivo, it has become clear that active synthesis and release of high levels of gangliosides into the tumor microenvironment are characteristic of many tumor cells. Our initial studies led us to propose the hypothesis (Ladisch et al. 1983), subsequently widely confirmed, that many tumors shed tumor gangliosides and that these molecules can then bind to other cells in this microenvironment (Fig. 23.1). There, they can affect the function of otherwise normal cells. This shedding and binding to normal host cells in the tumor microenvironment (TME) enhances tumor progression by modulating the function of these normal host cells critical in facilitating tumor progression enhances tumor progression. Characteristics of most types of tumors, i.e., not specifically and not only of neural tumors, since originally reported in 1983, ganglioside shedding by tumor cells has been widely documented (e.g., leukemias, lymphomas, neuroectodermal tumors such as neuroblastoma and melanoma, brain tumors, ovarian carcinoma, and renal cell carcinoma) (Biswas et al. 2006; Ladisch et al. 1983, 1997; Ladisch and Wu 1985; Merritt et al. 1994; Portoukalian et al. 1993; Ravindranath et al. 2007). And the rapid rate of ganglioside metabolism is associated with rapid shedding of these molecules, with rates of up to 0.5–1.0 %/24 h having been seen (Ladisch et al. 1983). This is of importance because of the multiple potential

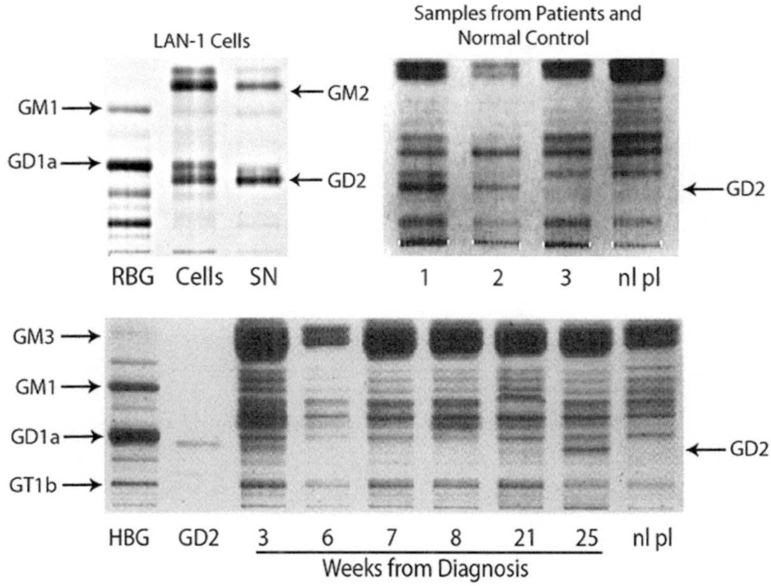

Fig. 23.2 Ganglioside shedding and clinical implications in human neuroblastoma. (**a**) Neuroblastoma tumor cell (LAN-1) shedding in vitro; (**b**) detection of GD2 in patient and normal plasma; (**c**) sequential analysis of circulating GD2 in a neuroblastoma patient, before and after surgical treatment (see text) (Adapted from Ladisch et al. 1987; Li and Ladisch 1991; Valentino et al. 1990)

functional roles of gangliosides acting upon normal cells, predominantly shown by in vitro studies in which normal cells found in the TME are enriched with exogenously added purified gangliosides.

23.2.3 Clinical Evidence for Shedding by Human Neural Tumors

Two specific examples of ganglioside shedding by neural tumors are highlighted here. The first is the human neuroectodermal tumor, neuroblastoma, which is characterized by poor prognosis and extensive ganglioside shedding. In fact, this tumor offers the most comprehensive example of the potential importance of tumor ganglioside in tumor progression and its treatment. This was the first human tumor in which the shedding of gangliosides into the peripheral circulation was clearly documented. This detection was possible because the characteristic ganglioside of neuroblastoma, GD2, is not detectable in normal human plasma, and yet ubiquitous in neuroblastoma tumors and shed by neuroblastoma cells (Fig. 23.2, top panels), making the ganglioside an excellent tumor marker (Ladisch and Wu 1985).

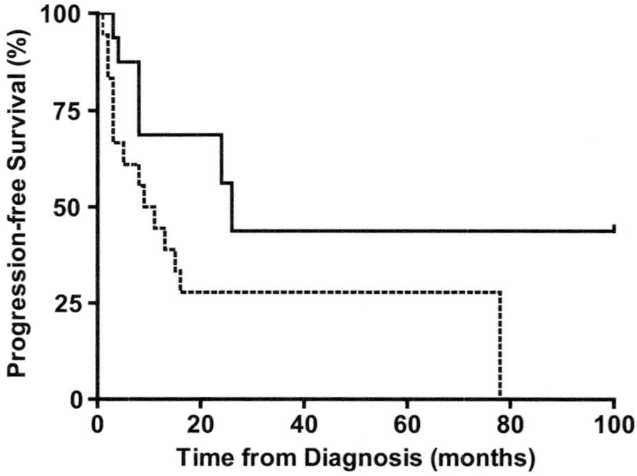

Fig. 23.3 Kaplan-Meier analysis of progression-free survival (PFS) of neuroblastoma patients. PFS of quartiles of high-risk neuroblastoma patients having the lowest (*dashed line*) and highest (*dotted line*) circulating GD2 concentrations at diagnosis (Adapted from Valentino et al. 1990)

Detection of circulating GD2 was further facilitated by the application of the new method of ganglioside isolation which overcame the problem of separating and purifying the gangliosides found in the presence of high concentrations of protein, such as in plasma (Ladisch and Gillard 1985). This approach complements antibody detection of tumor gangliosides, which is very useful in tumors but not so much in plasma.

By tracking circulating GD2 ganglioside levels, it was possible to follow (and even predict) the clinical course of a patient with neuroblastoma. In the specific case shown in Fig. 23.2, bottom panel, circulating GD2 was detectable in the patient at the time of diagnosis of neuroblastoma. Following surgical removal of the tumor, GD2 ganglioside disappeared from the circulation. Sequential studies were negative for GD2, until week 21, when traces of the gangliosides were again evident. This preceded a clinically detected relapse of neuroblastoma, evident finally also as a much higher circulating GD2 level in week 25 (Ladisch et al. 1987). Thus, circulating GD2 was a marker of tumor progression (prior to clinical detection), demonstrating the value of this marker and the significance of the shedding process.

23.2.4 Impact on Outcome

In fact, shed GD2 levels have correlated highly with patient clinical outcome. In an expanded study of a large group of patients, 74 patients with advanced stage (III and IV) neuroblastoma were found to have circulating tumor-derived GD2 ganglioside levels at the time of diagnosis that were inversely related to progression-free survival (PFS) ($p=0.018$) (Valentino et al. 1990). By Kaplan-Meier analysis (Fig. 23.3), the quartile of patients having the highest circulating GD2 levels (≥568 pmol/mL) had a strikingly worse outcome than the quartile of patients with the lowest GD2

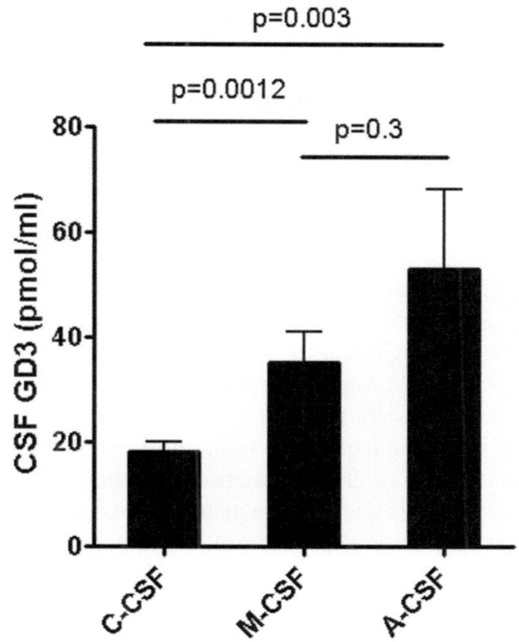

Fig. 23.4 Concentration of GD3 in cerebrospinal fluid (CSF) of controls (C-CSF), medulloblastoma patients (M-CSF), or astrocytoma patients (A-CSF) (Adapted from Ladisch et al. 1997)

levels (≤103 pmol/mL). Their median PFS was shorter (9 vs. 28 months), and the long-term survival rate lower (2-year PFS of 24 % vs. 70 %) ($p=0.013$). In conclusion, more rapid disease progression and lower survival rate are associated with high circulating GD2 levels at diagnosis. This leads to the hypothesis that shed neuroblastoma gangliosides were facilitating this process in the TME.

Another example addresses ganglioside shedding by a series of CNS tumors. In this case, detection of shed gangliosides is accomplished by examination of the cerebrospinal fluid draining the CNS tumors. CSF of patients with two types of pediatric brain tumors, medulloblastoma and astrocytoma, was evaluated. Using the same approach as for the plasma studies in neuroblastoma, we detected elevated levels of chemically quantified ganglioside GD3 ganglioside in the CSF of both the patients with medulloblastoma and those with astrocytoma (Fig. 23.4), reflecting shedding of the ganglioside by the tumors, themselves shown to be rich in this ganglioside (Ladisch et al. 1997).

The findings of these correlations and associations, as well as those of animal studies, clearly show the potential value of gangliosides as cell surface markers of neural tumors, with implications (a) for diagnosis and (b) as targets for therapeutic interventions (mainly antibody-mediated attacks on the tumor cell). The other important characteristic embodied in this shedding process is the dynamic property of these molecules, affecting normal cells in what might be considered a "paracrine" manner. The question is, what cellular functions do they affect?

23.3 Tumor Cell Ganglioside Function

Two important and now well-accepted biological properties of tumor cell gangliosides, potent immunosuppressive activity and proangiogenic activity, will be considered here. Both depend on shedding of tumor gangliosides, the uptake/binding of these gangliosides by other, normal, cells in the tumor microenvironment, and the modulation of the function of these "normal" cells. These cells include immunocytes (inhibition of antitumor immune responses) and vascular endothelial cells (enhancement of signaling and proliferation, leading to angiogenesis).

Historically, immunoinhibitory properties were the first functional characteristics of gangliosides that were associated with tumor cells. Parenthetically, these observations were being made at a time when the normal role of gangliosides in the nervous system was still a puzzle. Since that time, clear demonstrations of linkage of specific gangliosides to brain development have been made (see Chap. 9), and elucidation of distinct functions of endogenous cellular gangliosides continue. The focus here is on the consequences of exogenous ganglioside enrichment of a cell, by the pathophysiological process of tumor cell ganglioside shedding.

23.3.1 Immunosuppression

Immunoregulation by gangliosides has been extensively studied and will be briefly summarized here. It should be reiterated that most of these effects are characteristic of tumor cell gangliosides in general, even if not yet demonstrated in neural tumors in vivo. The initial findings of tumor ganglioside-induced immunosuppression were obtained in a murine lymphoma (Ladisch et al. 1983), in which the tumor gangliosides could be isolated, identified, and purified by a then new ganglioside purification method that permitted purification of tumor gangliosides from relatively small amounts of tissue, cells, or plasma (Ladisch and Gillard 1985). In this murine lymphoma model, very low (micromolar) concentrations of purified tumor gangliosides, isolated either from tumor cells in vitro or from the ascitic tumor in vivo, were strikingly inhibitory to lymphocyte proliferative responses to mitogens and antigens, the basis for cell-mediated immunity (Ladisch et al. 1983).

Cellular Mechanisms

Following this first demonstration of immunosuppressive activity of tumor gangliosides, subsequent studies identified the antigen-presenting cell as a primary target cell causing ganglioside-induced inhibition of the antigen-induced proliferative response (Ladisch et al. 1984). The affected antigen-presenting cells initially identified were the adherent monocytes (Ladisch et al. 1984) and, more recently, the dendritic cell (DC) (Shen and Ladisch 2002). Inhibition of DC maturation (Shurin et al. 2001) and induction of a state of tolerance of DC to TLR signaling (Shen et al. 2008) have been identified as mechanisms. In addition to these effects on myeloid cells involved in normal lymphoproliferative responses, exogenous gangliosides added in vitro inhibit IL2-dependent cell proliferation (Robb 1986), helper T cell proliferation (Chu and Sharom 1995), and natural killer cell activity (Bergelson et al. 1989; Grayson and Ladisch 1992), promote regulatory T cell development, inhibit T cell cytotoxicity (McKallip et al. 1999), and cause T cell apoptosis (Biswas et al. 2006; Chahlavi et al. 2005; Zhou et al. 1998). In vivo evidence corroborates these effects: in a murine model, tumor gangliosides inhibited immune responses, decreased tumor draining lymph node mass and cell numbers, and suppressed lymphoproliferative responses and the generation of tumor-specific cytotoxic T lymphocytes (Li et al. 1996; McKallip et al. 1999). Thus, a wide range of cellular immune functions are affected by exogenous (shed) gangliosides.

Molecular Mechanisms

The molecular mechanisms of ganglioside effects on immune cells, unlike the well-understood mechanisms of effects on fibroblasts or vascular endothelial cells (vide infra), remain to be fully elucidated. What is known is that exogenous tumor cell gangliosides inserted into the target cell (adherent monocyte or dendritic cell) membrane regulate membrane receptor activity and affect Toll-like receptor signaling, subsequently impeding nuclear localization and activation of NF-KB protein (Caldwell et al. 2003). Among mechanisms causing inhibition of T cell effector function (cytotoxicity) most recently, interference with granule exocytosis of CD8 T cells has recently been identified as a cause (Lee et al. 2012). Most likely, new model systems will provide additional insight into immunosuppressive mechanisms, both cellular and molecular, utilized by gangliosides to inhibit the host anti-tumor immune response.

Structure-Activity Relationships

With respect to relationship between immunoinhibitory activity and ganglioside structure, comprehensive studies were undertaken using normal brain gangliosides and human neuroblastoma tumor gangliosides, in the case of which HPLC purification of individual ganglioside ceramide species (Ladisch et al. 1989) allowed structure-activity relationships to be established. Antigen-induced lymphoproliferative responses were the indicator system. These studies revealed that with respect to carbohydrate structure, complexity and increasing number of sialic acid residues were associated with increased activity (Ladisch et al. 1992). With respect to ceramide structure, unsaturation and decreased fatty acyl group chain length of the ceramide portion were similarly associated with increased activity (Ladisch et al. 1994).

Brain Tumor Gangliosides

There have been few studies specifically addressing immunomodulation by brain tumor gangliosides. On the one hand, one report suggested that cytotoxic activity of NK cells may be affected by the immunoregulatory disturbances observed in patients with primary tumors in CNS (Peracoli et al. 1999). On the other hand, neurostatins (the O-acetylated forms of gangliosides GD1b and GT1b that are present in normal brain) have also been implicated, but as having an anti-tumorigenic effect (Nieto-Sampedro et al. 2011). Cytostatic for normal astroblasts, they were found to be cytotoxic for rat C6 glioma cells and cells of human astrocytoma grades III and IV, with ID_{50} values ranging from 200 to 450 nM. In contrast, they did not affect neurons or fibroblasts in concentrations of up to 4 μM or higher. The authors identified at least four different neurostatin-activated, cell-mediated antitumoral processes that they proposed to lead to tumor destruction: (1) inhibition of tumor neovascularization, (2) activation of microglia, (3) activation of natural killer cells, and (4) activation of cytotoxic lymphocytes (Nieto-Sampedro et al. 2011). Thus, there may be a balance between some anti- and pro-tumorigenic effects of specific gangliosides. While this remains to be elucidated, the neurostatin-activated processes clearly do not dominate in vivo in the case of human neural tumors.

23.3.2 Angiogenesis

The other functional property of gangliosides and particularly of tumor gangliosides released into the tumor microenvironment that has now been well characterized is the effect on angiogenesis. Here, the shedding of tumor gangliosides into TME may regulate the normal cells (vascular endothelial cells and fibroblasts) that are the structure for new blood vessels. This involves the processes of proliferation, migration, and tube formation in vivo. Initially, this effect on angiogenesis was observed in the laboratories of Gullino (Gullino 1995) and of Seyfried (Manfredi et al. 1999), who found a correlation between complex ganglioside expression and tumor proangiogenic activity, in the latter case in a murine brain tumor model in which a tumor rich in complex gangliosides had enhanced angiogenesis, compared to the parallel tumor containing only GM3 ganglioside.

In Vitro Findings

Direct observations of ganglioside promotion of tumor angiogenesis in vitro are using purified ganglioside, GD1a, pointed to a mechanism underlying the effect on responses of normal human umbilical vein endothelial cells (HUVEC) to VEGF (Lang et al. 2001). Preincubation of HUVEC with GD1a enhanced VEGF-induced cell proliferation; 10 μM GD1a caused a twofold increase in DNA synthesis. The migration of HUVEC across a VEGF gradient was also enhanced by 50 %, even with only a brief (1 h) preexposure of the cells to the same concentration of GD1a (Liu et al. 2006). These findings suggested that gangliosides shed by tumor cells can promote tumor angiogenesis by enhancing the VEGF response of endothelial cells in the tumor microenvironment.

In Vivo Observations

Further in vivo study has cemented this conclusion. Most recently, the development of a ganglioside knockout model allowed study of tumor formation in the complete absence of ganglioside in the tumor cell itself and without exogenous manipulation of the cells (e.g., pharmacologically). The model was developed by oncogene transformation of murine embryonic fibroblasts in which GM2 synthase and GM3 synthase had been knocked out (Liu et al. 2010). The resulting cells (termed DKO) were consequently constitutively, selectively, specifically, and completely devoid of gangliosides. Injecting these cells in vivo, tumor growth was strikingly affected. The DKO cells formed ganglioside-deficient tumors, much smaller than those formed by the corresponding litter-mate ganglioside-rich wild-type (WT) cells. Importantly, the DKO tumors had a striking paucity of tumor blood vessels (tenfold less than WT tumors), while the level of intrinsic tumor cell VEGF secretion was not reduced (Liu et al. 2013). This study definitively linked ganglioside depletion to impaired tumor angiogenesis and supports the interpretation of an effect of the gangliosides directly on the target (vascular endothelial) cells.

Molecular Mechanisms

A molecular mechanism by which gangliosides have this direct effect on angiogenesis has been elucidated. This relates to earlier studies of fibroblast proliferation in which we found that enrichment of the normal fibroblast cell membrane with exogenous gangliosides resulted in enhanced receptor binding affinities for their ligand (EGF and a series of other receptor tyrosine kinases) and enhanced the proliferation of these cells (Li et al. 2000, 2001). In this case of EGF-induced EGFR activation, in normal human dermal fibroblasts, membrane enrichment with GD1a enhanced the EGF-induced EGFR activity by increasing EGFR dimerization and the effective number of high-affinity EGFR, without increasing total receptor protein. Unexpectedly, GD1a enrichment also triggered increased EGFR dimerization in the absence of growth factor. This resulted in enhanced activation of the EGFR signal transduction cascade when EGF was added (Liu et al. 2004). Another recent example of regulating EGF-induced neural cell proliferation and differentiation by gangliosides (in this case, intrinsic cell membrane gangliosides, not shed molecules) is the regulation by ganglioside GD3 of neural stem cell survival. Using a GD3-synthase knockout mouse, the authors showed that GD3 upregulated EGF-EGFR signaling to sustain NSC self-renewal (Wang and Yu 2013).

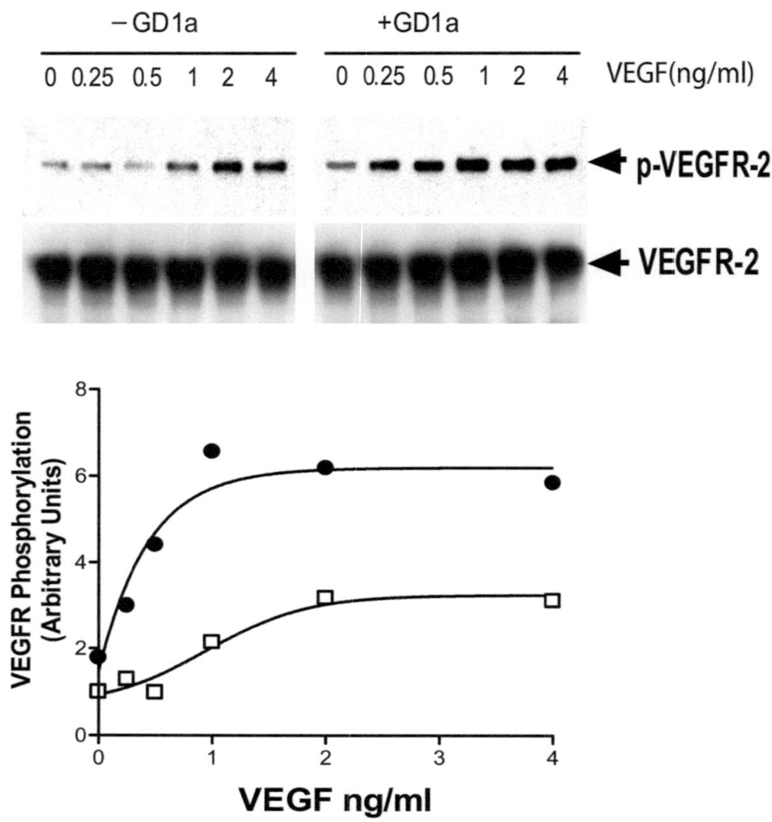

Fig. 23.5 Enhancement of VEGFR phosphorylation by membrane enrichment in GD1a ganglioside. Note particularly the amplification of receptor phosphorylation caused at low (<1 ng/ml) VEGF concentrations (Adapted from Liu et al. 2006)

In the case of angiogenesis, it is another receptor tyrosine kinase, VEGF, which was similarly affected as fibroblast signaling in our previous studies, resulting in enhanced proliferation, migration, and differentiation. In this system, gangliosides act to sensitize vascular endothelial cells to respond to subthreshold levels of VEGF (Liu et al. 2006). Ganglioside enrichment of human umbilical vein vascular endothelial cells (HUVEC) caused very low, normally barely stimulatory, VEGF concentrations to trigger robust VEGF receptor dimerization and autophosphorylation (Fig. 23.5), as well as activation of downstream signaling pathways, and cell proliferation and migration. Overall, by lowering the threshold for growth factor activation of these normal stromal cells, shed tumor gangliosides promote tumor progression, causing normal vascular endothelial cells to become increasingly autonomous from growth factor requirements by a process that we termed tumor-induced progression of the microenvironment (Liu et al. 2006).

23.4 Conclusion

With the identification of tumor cell gangliosides as potent immunosuppressive and proangiogenic molecules, the clinical relevance of tumor cell gangliosides, both extraneural and neural, is established. Their importance lies in the probability that the many properties elucidated through in vivo/in vitro experimental systems are operative or eventually will be proven to be so in human cancer. To facilitate further discovery, a *simple*, i.e., practical, sensitive, and specific approach to detecting shed tumor gangliosides in vivo would be valuable. This could provide more insight into prognosis, as the example of neuroblastoma demonstrates. Much remains to be discovered about antitumor immune responses in the central nervous system, and their modulation by tumor gangliosides, but the foundation for these studies has now been laid; it is likely that the mechanisms are analogous to those in other systems. In vivo tumor models in which ganglioside synthesis is constitutively and selective altered will undoubtedly also provide further insights. The murine DKO model may help because it is syngeneic and it allows study of tumor host interactions without exogenous manipulation, comparing ganglioside-poor to ganglioside-rich tumor cells that are genetically identical (except for knockout of the ganglioside synthesis enzymes).

The overall findings that gangliosides facilitate tumor progression suggest that strategies to interrupt tumor cell ganglioside metabolism should be pursued. Targeting synthesis (and elimination of shedding) of tumor gangliosides is likely to be a fruitful approach in neural tumors as in other tumors. This might be accomplished by metabolic or siRNA inhibition, or by otherwise suppressing the activity particularly of GM3 synthase, a key enzyme in human ganglioside synthesis. This approach could be operationally selective for the tumor cells and effective since ganglioside turnover in tumors is much more rapid than in most normal tissues.

Conflicts of Interest The authors declare no conflicts of interest.

References

Bergelson LD, Dyatlovitskaya EV, Klyuchareva TE, Kryukova EV, Lemenovskaya AF, Matveeva VA, et al. The role of glycosphingolipids in natural immunity. Gangliosides modulate the cytotoxicity of natural killer cells. Eur J Immunol. 1989;19(11):1979–83.

Biswas K, Richmond A, Rayman P, Biswas S, Thornton M, Sa G, et al. GM2 expression in renal cell carcinoma: potential role in tumor-induced T-cell dysfunction. Cancer Res. 2006;66(13): 6816–25.

Caldwell S, Heitger A, Shen W, Liu Y, Taylor B, Ladisch S. Mechanisms of ganglioside inhibition of APC function. J Immunol. 2003;171(4):1676–83. Epub 2003/08/07.

Chahlavi A, Rayman P, Richmond AL, Biswas K, Zhang R, Vogelbaum M, et al. Glioblastomas induce T-lymphocyte death by two distinct pathways involving gangliosides and CD70. Cancer Res. 2005;65(12):5428–38. Epub 2005/06/17.

Chu JW, Sharom FJ. Gangliosides interact with interleukin-4 and inhibit interleukin-4-stimulated helper T-cell proliferation. Immunology. 1995;84(3):396–403.

Foulds L. Multiple etiologic factors in neoplastic development. Cancer Res. 1965;25(8):1339–47. Epub 1965/09/01.

Grayson G, Ladisch S. Immunosuppression by human gangliosides. II. Carbohydrate structure and inhibition of human NK activity. Cell Immunol. 1992;139(1):18–29. Epub 1992/01/01.

Gullino PM. Prostaglandins and gangliosides of tumor microenvironment: their role in angiogenesis. Acta Oncol. 1995;34(3):439–41. Epub 1995/01/01.

Hakomori S. Aberrant glycosylation in cancer cell membranes as focused on glycolipids: overview and perspectives. Cancer Res. 1985;45(6):2405–14. Epub 1985/06/01.

Hakomori S. Tumor malignancy defined by aberrant glycosylation and sphingo(glyco)lipid metabolism. Cancer Res. 1996;56(23):5309–18. Epub 1996/12/01.

Hanahan D, Weinberg RA. The hallmarks of cancer. Cell. 2000;100(1):57–70. Epub 2000/01/27.

Hendrix MJ, Seftor EA, Kirschmann DA, Quaranta V, Seftor RE. Remodeling of the microenvironment by aggressive melanoma tumor cells. Ann N Y Acad Sci. 2003;995:151–61. Epub 2003/06/20.

Hettmer S, Malott C, Woods W, Ladisch S, Kaucic K. Biological stratification of human neuroblastoma by complex "B" pathway ganglioside expression. Cancer Res. 2003;63(21):7270–6. Epub 2003/11/13.

Kaucic K, Etue N, LaFleur B, Woods W, Ladisch S. Neuroblastomas of infancy exhibit a characteristic ganglioside pattern. Cancer. 2001;91(4):785–93. Epub 2001/03/10.

Ladisch S, Gillard B, Wong C, Ulsh L. Shedding and immunoregulatory activity of YAC-1 lymphoma cell gangliosides. Cancer Res. 1983;43(8):3808–13.

Ladisch S, Ulsh L, Gillard B, Wong C. Modulation of the immune response by gangliosides. Inhibition of adherent monocyte accessory function in vitro. J Clin Invest. 1984;74(6): 2074–81. Epub 1984/12/01.

Ladisch S, Gillard B. A solvent partition method for microscale ganglioside purification. Anal Biochem. 1985;146(1):220–31. Epub 1985/04/01.

Ladisch S, Wu ZL. Detection of a tumour-associated ganglioside in plasma of patients with neuroblastoma. Lancet. 1985;1(8421):136–8. Epub 1985/01/19.

Ladisch S, Wu ZL, Feig S, Ulsh L, Schwartz E, Floutsis G, et al. Shedding of GD2 ganglioside by human neuroblastoma. Int J Cancer. 1987;39(1):73–6. Epub 1987/01/15.

Ladisch S, Sweeley CC, Becker H, Gage D. Aberrant fatty acyl alpha-hydroxylation in human neuroblastoma tumor gangliosides. J Biol Chem. 1989;264(20):12097–105. Epub 1989/07/15.

Ladisch S, Becker H, Ulsh L. Immunosuppression by human gangliosides: I. Relationship of carbohydrate structure to the inhibition of T cell responses. Biochim Biophys Acta. 1992; 1125(2):180–8. Epub 1992/04/23.

Ladisch S, Li R, Olson E. Ceramide structure predicts tumor ganglioside immunosuppressive activity. Proc Natl Acad Sci U S A. 1994;91(5):1974–8. Epub 1994/03/01.

Ladisch S, Chang F, Li R, Cogen P, Johnson D. Detection of medulloblastoma and astrocytoma-associated ganglioside GD3 in cerebrospinal fluid. Cancer Lett. 1997;120(1): 71–8.

Lang Z, Guerrera M, Li R, Ladisch S. Ganglioside GD1a enhances VEGF-induced endothelial cell proliferation and migration. Biochem Biophys Res Commun. 2001;282(4):1031–7. Epub 2001/05/16.

Lee HC, Wondimu A, Liu Y, Ma JS, Radoja S, Ladisch S. Ganglioside inhibition of CD8+ T cell cytotoxicity: interference with lytic granule trafficking and exocytosis. J Immunol. 2012;189(7):3521–7. Epub 2012/09/08.

Li R, Gage D, McKallip R, Ladisch S. Structural characterization and in vivo immunosuppressive activity of neuroblastoma GD2. Glycoconj J. 1996;13(3):385–9.

Li R, Manela J, Kong Y, Ladisch S. Cellular gangliosides promote growth factor-induced proliferation of fibroblasts. J Biol Chem. 2000;275(44):34213–23. Epub 2000/06/22.

Li R, Liu Y, Ladisch S. Enhancement of epidermal growth factor signaling and activation of SRC kinase by gangliosides. J Biol Chem. 2001;276(46):42782–92. Epub 2001/09/06.

Li RX, Ladisch S. Shedding of human neuroblastoma gangliosides. Biochim Biophys Acta. 1991;1083(1):57–64. Epub 1991/04/24.

Liu Y, Li R, Ladisch S. Exogenous ganglioside GD1a enhances epidermal growth factor receptor binding and dimerization. J Biol Chem. 2004;279(35):36481–9. Epub 2004/06/25.

Liu Y, McCarthy J, Ladisch S. Membrane ganglioside enrichment lowers the threshold for vascular endothelial cell angiogenic signaling. Cancer Res. 2006;66(21):10408–14. Epub 2006/11/03.
Liu Y, Yan S, Wondimu A, Bob D, Weiss M, Sliwinski K, et al. Ganglioside synthase knockout in oncogene-transformed fibroblasts depletes gangliosides and impairs tumor growth. Oncogene. 2010;29(22):3297–306. Epub 2010/03/23.
Liu Y, Wondimu A, Yan S, Bobb D, Ladisch S. Tumor gangliosides accelerate murine tumor angiogenesis. Angiogenesis 2013. Epub 2013/10/30.
Manfredi MG, Lim S, Claffey KP, Seyfried TN. Gangliosides influence angiogenesis in an experimental mouse brain tumor. Cancer Res. 1999;59(20):5392–7. Epub 1999/10/28.
McKallip R, Li R, Ladisch S. Tumor gangliosides inhibit the tumor-specific immune response. J Immunol. 1999;163(7):3718–26. Epub 1999/09/22.
Merritt WD, Der-Minassian V, Reaman GH. Increased GD3 ganglioside in plasma of children with T-cell acute lymphoblastic leukemia. Leukemia. 1994;8(5):816–22.
Nakamura O, Ishihara E, Iwamori M, Nagai T, Matsutani M, Nomura K, et al. Lipid composition of human malignant brain tumors. No To shinkei. 1987;39(3):221–6. Epub 1987/03/01.
Nieto-Sampedro M, Valle-Argos B, Gomez-Nicola D, Fernandez-Mayoralas A, Nieto-Diaz M. Inhibitors of glioma growth that reveal the tumour to the immune system. Clin Med Insights Oncol. 2011;5:265–314. Epub 2011/11/16.
Peracoli MT, Montelli TC, Soares AM, Parise-Fortes MR, Alquati SA, Ueda A, et al. Immunological alterations in patients with primary tumors in central nervous system. Arq Neuropsiquiatr. 1999;57(3A):539–46. Epub 2000/02/10.
Portoukalian J, David MJ, Gain P, Richard M. Shedding of GD2 ganglioside in patients with retinoblastoma. Int J Cancer. 1993;53(6):948–51.
Ravindranath MH, Muthugounder S, Presser N, Selvan SR, Santin AD, Bellone S, et al. Immunogenic gangliosides in human ovarian carcinoma. Biochem Biophys Res Commun. 2007;353(2):251–8.
Robb RJ. The suppressive effect of gangliosides upon IL 2-dependent proliferation as a function of inhibition of IL 2-receptor association. J Immunol. 1986;136(3):971–6.
Sawada M, Moriya S, Saito S, Shineha R, Satomi S, Yamori T, et al. Reduced sialidase expression in highly metastatic variants of mouse colon adenocarcinoma 26 and retardation of their metastatic ability by sialidase overexpression. Int J Cancer. 2002;97(2):180–5. Epub 2002/01/05.
Schengrund CL, Repman MA, Shochat SJ. Ganglioside composition of human neuroblastomas. Correlation with prognosis. A Pediatric Oncology Group Study. Cancer. 1985;56(11):2640–6. Epub 1985/12/01.
Shen W, Ladisch S. Ganglioside GD1a impedes lipopolysaccharide-induced maturation of human dendritic cells. Cell Immunol. 2002;220(2):125–33. Epub 2003/03/27.
Shen W, Stone K, Jales A, Leitenberg D, Ladisch S. Inhibition of TLR activation and up-regulation of IL-1R-associated kinase-M expression by exogenous gangliosides. J Immunol. 2008; 180(7):4425–32. Epub 2008/03/21.
Shurin GV, Shurin MR, Bykovskaia S, Shogan J, Lotze MT, Barksdale Jr EM. Neuroblastoma-derived gangliosides inhibit dendritic cell generation and function. Cancer Res. 2001;61(1):363–9. Epub 2001/02/24.
Talantov D, Mazumder A, Yu JX, Briggs T, Jiang Y, Backus J, et al. Novel genes associated with malignant melanoma but not benign melanocytic lesions. Clin Cancer Res. 2005;11(20):7234–42. Epub 2005/10/26.
Valentino L, Moss T, Olson E, Wang HJ, Elashoff R, Ladisch S. Shed tumor gangliosides and progression of human neuroblastoma. Blood. 1990;75(7):1564–7. Epub 1990/04/01.
Wang J, Yu RK. Interaction of ganglioside GD3 with an EGF receptor sustains the self-renewal ability of mouse neural stem cells in vitro. Proc Natl Acad Sci U S A. 2013;110(47):19137–42. Epub 2013/11/08.
Yu AL, Gilman AL, Ozkaynak MF, London WB, Kreissman SG, Chen HX, et al. Anti-GD2 antibody with GM-CSF, interleukin-2, and isotretinoin for neuroblastoma. N Engl J Med. 2010;363(14):1324–34. Epub 2010/10/01.
Zhou J, Shao H, Cox NR, Baker HJ, Ewald SJ. Gangliosides enhance apoptosis of thymocytes. Cell Immunol. 1998;183(2):90–8.

Chapter 24
Galectins and Neuroinflammation

Hung-Lin Chen, Fang Liao, Teng-Nan Lin, and Fu-Tong Liu

Abstract Galectins, β-galactoside-binding lectins, play multiple roles in the regulation of immune and inflammatory responses. The major galectins expressed in the CNS are galectins 1, 3, 4, 8, and 9. Under normal physiological conditions, galectins maintain CNS homeostasis by participating in neuronal myelination, neuronal stem cell proliferation, and apical vesicle transport in neuronal cells. In neuronal diseases and different experimental neuroinflammatory disease models, galectins may serve as extracellular mediators or intracellular regulators in controlling the inflammatory response or conferring the remodeling capacity in damaged CNS tissues. In general, galectins 1 and 9 attenuate experimental autoimmune encephalomyelitis (a model of multiple sclerosis), while galectin-3 promotes inflammation in this model. In brain ischemic lesions, both galectins 1 and 3 are induced to help neuronal regeneration. The expression of galectin-1 is required for astrocyte-derived neurotrophic factor secretion, and recombinant galectin-1 promotes neuronal regeneration. Galectin-3 promotes microglial cell proliferation and attenuates ischemic damage and neuronal apoptosis after cerebral ischemia. In amyotrophic lateral sclerosis models, galectin-3 is deleterious to neuroregeneration, while intramuscular administration of oxidized galectin-1 can improve neuromuscular disorders. In axotomy and Wallerian degeneration, galectin-3 helps phagocytosis of macrophages to clear degenerate myelin in the injured PNS or CNS. Thus, galectins are important modulators participating in homeostasis of the CNS and neuroinflammation. Continued investigations of the roles of galectins in neuroinflammation promise to provide a better understanding of the mechanism of this process and lead to new therapeutic approaches.

Keywords Amyotrophic lateral sclerosis • Experimental autoimmune encephalomyelitis • Galectin • Multiple sclerosis • Wallerian degeneration

H.-L. Chen • F. Liao • T.-N. Lin • F.-T. Liu, M.D., Ph.D. (✉)
Institute of Biomedical Sciences, Academia Sinica,
128 Academia Road, Section 2, Nankang, Taipei, Taiwan
e-mail: ftliu@ibms.sinica.edu.tw

R.K. Yu and C.-L. Schengrund (eds.), *Glycobiology of the Nervous System*,
Advances in Neurobiology 9, DOI 10.1007/978-1-4939-1154-7_24,

Abbreviations

CNS	Central nervous system
BBB	Blood–brain barrier
PAMPs	Pathogen-associated molecular patterns
PRRs	Pattern recognition receptors
TCR	T-cell receptor
MHC	Major histocompatibility complex
APC	Antigen-presenting cells
Th	T helper cell
T_{FH}	Follicular helper T cell
Treg	Regulatory T cells
IFN-γ	Interferon-gamma
EAE	Experimental autoimmune encephalitis
MS	Multiple sclerosis
TLRs	Toll-like receptors
RLRs	Retinoic acid-inducible gene I-like receptors
NLR	Nucleotide-binding oligomerization domain-like receptor
BCR	B-cell receptor
TNF-α	Tumor necrosis factor-alpha
CCL20	CC chemokine ligand 20
ROS	Reactive oxygen species
AA	Arachidonic acid
5-LO	5-Lipoxygenase
CRDs	Carbohydrate-recognition domains
OLG	Oligodendrocyte
NSCs	Neural stem cells
MBP	Myelin basic protein
IL-1β	Interleukin-1β

24.1 Introduction to the Immune Responses in the CNS System

The central nervous system (CNS), composed of the brain and spinal cord, used to be considered as an "immune-privileged" organ due to the lack of an obvious lymphatic system, the limited numbers of professional antigen-presenting cells, and to the apparent absence of an immune response. The "immune privilege" of the CNS reflects its seclusion by the blood–brain barrier (BBB). The unique structure of the BBB limits entry of immune cells into the CNS (Hawkins and Davis 2005; Mrass and Weninger 2006). Additionally, the CNS is tolerant to immune responses due to CNS cells expressing regulatory molecules that suppress T-cell function. However, the concept of the "immune privilege" of the CNS has been reevaluated over the past two decades. Studies have demonstrated that the CNS does "talk" to the immune system and that the interactions between the CNS and immune system are critical for maintaining CNS homeostasis and neuronal function (Kipnis et al. 2002; Wolf et al. 2009a, b; Ziv et al. 2006). The important role of immune cells in CNS function is demonstrated by the fact that a deficiency of $CD4^+$ T cells in mouse models leads to impairment of cognitive function (Kipnis et al. 2004). Immune cells actively participate in the control of infection and injury in the CNS (Savarin and Bergmann 2008). The nerve system actively interacts with peripheral immune cells and is continuously under surveillance by resident microglia and immune cells such as dendritic cells, macrophages, and T cells that sense and eliminate insults that may damage it (Schwartz and Kipnis 2011).

24.1.1 Immune System

The immune system maintains tissue homeostasis and initiates host defense against pathogens. It is composed of two arms, innate (also referred to as nonspecific) immunity and adaptive (also referred to as specific) immunity. Innate immunity is the first line of host defense and is regulated by cells such as dendritic cells, macrophages, neutrophils, and natural killer cells. The innate immune system recognizes pathogen-associated molecular patterns (PAMPs), common structures present in pathogens, through pattern recognition receptors (PRRs). This recognition initiates innate immune responses. Adaptive immunity is mediated by T and B lymphocytes, which recognize specific antigens and facilitate cell-mediated immunity and humoral immunity, respectively. In cell-mediated immunity, T-cell receptors (TCR) of naïve $CD4^+$ T cells recognize peptides in complex with major histocompatibility complex (MHC) class II on antigen-presenting cells (APC, mainly dendritic cells) and subsequently activate naïve $CD4^+$ T cells to become effector $CD4^+$ T cells (T helper cells, Th). T helper cells can be grouped into several types of Th (Th1, Th2, Th17, T_{FH}, Treg) based on their signatures of cytokines and transcription factors. Effector $CD4^+$ T cells (Th1, Th2, Th17, T_{FH}, Treg) secrete distinct cytokines to regulate the functions of other immune cells. Th1 cells selectively secrete IFN-γ, Th2 cells selectively secrete IL-4, while Th17 and T_{FH} cells selectively secrete IL-17/IL-21 and IL-21, respectively (O'Shea and Paul 2010). Th1 cells mediate the immune response against intracellular pathogens, such as viruses, mycobacteria, and protozoa. Th2 cells mediate the host defense against helminthes and other parasites. Th2 cells also mediate allergic diseases. Th17 cells are important in host defense against Gram-negative extracellular bacteria, such as *Klebsiella pneumonia* and *Bacteroides fragilis* (Dubin and Kolls 2008), as well as fungi. The inflammatory nature of Th17 cells results in their being associated with a variety of autoimmune diseases, such as experimental autoimmune encephalomyelitis (EAE) in a mouse model equivalent to human multiple sclerosis (MS) (Cua et al. 2003; Langrish et al. 2005; Reboldi et al. 2009). T_{FH} cells regulate differentiation of naïve B cells into antibody-secreting cells (plasma cells) or memory B cells (Fazilleau et al. 2009).

Treg cells expressing $CD4^{+}CD25^{+}$ are important for the maintenance of peripheral tolerance and homeostasis (Sakaguchi et al. 2008; Shevach et al. 2006). T-cell receptors (TCR) on naïve $CD8^{+}$ T cells recognize peptides in complex with MHC class I on antigen-presenting cells (dendritic cells) in the presence of Th cells to form ternary clusters (APC, $CD8^{+}$, $CD4^{+}$) and subsequently activate naïve $CD8^{+}$ T cells to become effector $CD8^{+}$ T cells. Effector $CD8^{+}$ T cells recognize antigens on target cells and act to eliminate infected or damaged cells (Peters et al. 1991). In humoral immunity, naïve B cells, with help from T_{FH} cells, recognize antigens through B-cell receptors (BCR), are activated, and differentiate into memory B cells or plasma cells (antibody-secreting cells), which secrete antibodies that neutralize pathogens.

24.1.2 Innate Immunity

The dendritic cells and macrophages, sentinel antigen-presenting cells in the periphery tissues, are responsible for the uptake of foreign antigens, such as pathogens. PAMPs present in pathogens are recognized by dendritic cells or macrophages through PRRs. The PRRs are thus activated and trigger innate immune responses eliciting rapid and potent responses to limit pathogens before adaptive immunity is generated and to dictate the direction of the adaptive immune response. These PRRs include Toll-like receptors (TLRs), retinoic acid-inducible gene I (RIG-I)-like receptors (RLRs), nucleotide oligomerization domain (NOD)-like receptors (NLRs), and C-type lectin receptors (CLRs) (Beutler et al. 2007; Sancho and Reis e Sousa 2012; Takeuchi and Akira 2009; Ye and Ting 2008). PRRs can be expressed on the cell surface (e.g., most of TLRs, CLRs), in the endosomes (e.g., TLR3, TLR7-9), or in the cytoplasm (e.g., RLRs, NLRs). Different types of PRRs recognize different types of PAMPs. For example, bacterial pathogens are primarily recognized by cell surface membrane TLRs or NLRs (Beutler et al. 2007; Ye and Ting 2008); viral pathogens are commonly recognized by RLR and endosomal TLRs (Beutler et al. 2007; Takeuchi and Akira 2009). Activation of TLRs and RLRs usually leads to the activation of interferon response factors (IRFs) and NF-κB; IRFs induce transcription of type I interferons (type I IFNs), and NF-κB is a critical transcription factor for the induction of cytokines and chemokines (Beutler et al. 2007). Type I IFNs are the cardinal molecules that control viral spreading within the CNS during the innate immune response to viral infection. Some PRRs also sense endogenous damage-associated molecular patterns (DAMPs) generated by tissue injury or cell distress (Bianchi 2007). The DAMPs, such as high-mobility group box 1 (HMGB1), alert the immune system to sense cell damage, which is independent of controlling infection (Bianchi 2007). In the CNS, microglia and astrocytes that express several types of PRRs are the main resident cells that mediate innate immune responses (Lehnardt 2010; Olson and Miller 2004; Ransohoff and Brown 2012). Microglia sense the local milieu through PRRs and subsequently activate PRR downstream signaling pathways, leading to activation of NF-κB. This induces expression of proinflammatory cytokines, chemokines, and effector molecules that act to control the infection or injury (Jack et al. 2005). Similarly, astrocytes have recently been recognized as an important cell type in the CNS also participating in the innate immune response (Farina et al. 2007; Jack et al. 2005). Upon sensing PAMPs, astrocytes activate PRR signaling pathways and secret mediators to regulate both innate and adaptive immune responses. It has been reported that astrocytes have dual features; on the one hand, they promote inflammation through NF-κB-dependent pathways; on the other, proliferating reactive astrocytes confine lesions and restore brain homeostasis (Seil 2001; Seo et al. 2013).

24.1.3 Adaptive Immunity

Adaptive immunity is important for the clearance of pathogens at later stages of infection and generates memory lymphocytes that are rapidly activated and respond upon reencountering the same antigens. Adaptive immunity is mediated by T and B lymphocytes, which are developed in the thymus and bone marrow, respectively. T and B lymphocytes recognize antigens via the T-cell receptor (TCR) and B-cell receptor (BCR), respectively. TCRs and BCRs undergo gene rearrangement during T- and B-cell development, generating the great diversity in specificity of the TCRs and BCRs for recognition of a great number of antigens. Antigen-presenting cells (APC) in peripheral tissues take up antigens and migrate into peripheral lymphoid tissues where they encounter naïve $CD4^+$ T cells and activate naïve $CD4^+$ T cells, which undergo clonal expansion and become effector $CD4^+$ T cells. The effector $CD4^+$ T cells help the activation of B cells and $CD8^+$ T cells to become plasma cells and effector $CD8^+$ T cells, respectively. These effector lymphocytes then egress from the lymphoid tissues, enter the circulation, and are recruited into the inflamed or infected tissues to clear pathogens. The recruitment of effector lymphocytes expressing chemokine receptors into inflamed or infected tissues is orchestrated by chemokines that are secreted by those tissues. Because the CNS lacks lymphoid tissue, natural CNS infections are initiated at extra-neural sites. Upon neurotropic virus infection, virus-specific immune cells are activated in peripheral lymph nodes. The effector lymphocytes upregulate adhesion molecules and chemokine receptors on cell surfaces, which facilitates the ability of circulating effector lymphocytes to cross the BBB and enter into the CNS. These activated lymphocytes then mediate the elimination of viruses in the CNS. In general, $CD8^+$ T cells are believed to be the primary lymphocytes directly responsible for suppressing viral replication and viral clearance. Effector $CD8^+$ T cells mediate virus killing predominately by cytotoxic effects induced by secretion of perforin and granzymes (Peters et al. 1991). Perforin is a pore-forming protein required for delivery of granzymes into virus-infected cells (Bolitho et al. 2007), and granzymes are serine proteases that cleave caspases to induce apoptosis of virus-infected cells (Heusel et al. 1994). Although $CD8^+$ T cells have a protective effect during viral infection of the CNS, $CD8^+$ T cells may also cause pathological effects within the CNS during viral infection (Willing and Friese 2012). It is important to delicately control effector $CD8^+$ T cells in order to avoid their pathological effects during viral clearance.

Adaptive immunity is required for the destruction of pathogens by effector lymphocytes recognizing foreign antigens. However, recognition of self-antigens by lymphocytes can initiate autoimmunity and lead to tissue destruction. Normally, autoreactive lymphocytes are deleted during lymphocyte development, a process called central tolerance. If autoreactive lymphocytes escape clonal deletion, they may become anergic (tolerant) and fail to be activated, leading to peripheral tolerance. Patients with autoimmune diseases fail to develop immune tolerance and have autoreactive T cells able to specifically recognize self-peptides in the context of MHC, resulting in destruction of the tissue. For example, MS is a CNS autoimmune disease, in which patients have autoreactive $CD4^+$ T cells that specifically recognize myelin basic protein, which leads to destruction of the axon.

24.1.4 Neural Reflexes in Immunity

The immune system is known to functionally and anatomically connect to the nervous system, and recent studies have shown that the reflex neural circuits are important for maintaining immunological homeostasis and host health (Andersson and Tracey 2012a, b). Lymphocytes express various neurotransmitter receptors, and the receptor expression differs as a function of cell type and cell activation. In response to infectious agents, the efferent vagus nerve secretes norepinephrine, which culminates in the spleen in the vicinity of T cells that are capable of secreting acetylcholine. Acetylcholine interacts with the alpha 7 subunit of the nicotinic acetylcholine receptors (nAChR) expressed on cytokine-producing macrophages. Activation of the nAChR in macrophages leads to suppression of the synthesis and release of proinflammatory cytokines, such as tumor necrosis factor-alpha (TNF-α) (Rosas-Ballina et al. 2011). These observations support the conclusion that neurotransmitters stimulate immune cells that then transmit neural signals, generating a neuro-immune circuitry that controls the consequence of immune responses. Neural circuits have also been shown to regulate antibody production following B-cell exposure to blood-borne antigens (Mina-Osorio et al. 2012). Electrical stimulation of the vagus nerve or administration of nicotine significantly impairs the migration of B cells to the red pulp venous sinuses to generate plasma cells, resulting in a reduction of splenic antibody titers to blood-borne antigens (Mina-Osorio et al. 2012). The resultant efferent reflex signals regulate B-cell trafficking and influence the nature of the adaptive immune response.

Interestingly, a recent study using EAE, a mouse model for MS, demonstrated the existence of reflex neural circuits between immune cells and the CNS (Arima et al. 2012). During the earliest stages of EAE, autoreactive $CD4^+$ T cells, primed to attack myelinated neurons, access the CNS through a highly specific anatomical location – the fifth lumbar spinal cord (Arima et al. 2012). The fifth lumbar vertebra is near the dorsal root ganglia of sensory neurons that innervate the soleus muscles. The sensory nerve activation induces the expression of CCL20 in the dorsal blood vessels via activation of sympathetic neurons. CCL20 is a critical chemokine that attracts autoreactive $CD4^+$ T cells to the CNS. Decreasing the sensor input from the muscles by limiting their contraction reduces activity of sympathetic neurons, resulting in reduced CCL20 expression in endothelial cells and thus decreased recruitment of autoreactive T-cell entry to the CNS (Andersson and Tracey 2012a; Arima et al. 2012). This study demonstrates that neuro-immune interaction contributes to neuronal autoimmune disease. It is believed that the neuro-immune interactions contribute to not only the onset and progress of autoimmune diseases, neurodegenerative diseases, and traumatic brain injuries but also to the maintenance of neurogenesis, spatial learning memory, and mental/psychiatric disorders.

24.1.5 Neuroinflammation

CNS inflammation can be acute or chronic. Acute inflammation is defined as the immediate response to the initiation of injury, characterized by a short and effective process in the elimination of agents potentially harmful to the CNS and in the rapid resolution of inflammation and recurrence of homeostasis. If the host fails to recover from the acute inflammation, chronic inflammation then ensues, which is characterized by the dysregulation or overreaction of microglia and the decomposition or death of neurons and glia. Trauma and ischemic injury to CNS tissues usually exacerbate the initial damage and lead to development of chronic inflammation. Activated microglia play a critical role in CNS inflammation. They produce proinflammatory cytokines, TNF-α, IL-6, and IL-1β, which bind to their receptors on cells in the CNS and induce gene expression of a variety of cytokines and chemokines, subsequently recruiting leukocytes into the CNS and leading to inflammation. Activated microglia are also the source of reactive oxygen species (ROS) that induce inflammation and drive neurodegenerative diseases. The oxygenation pathway of arachidonic acid (AA) plays a critical role in CNS inflammation and resolution. The earlier inflammatory responses are associated with the phospholipase A2 (PLA_2) catalyzed release of AA from activated microglia and infiltrating leukocytes as well as activation of oxygenating enzymes, such as cyclooxygenase (COX-1/COX-2) and 5-lipoxygenase (5-LO). COX-1/COX-2 and 5-LO catalyze conversion of AA into prostaglandins (PGD_2, PGE_2) and leukotriene (LTB_4), respectively. Upon binding to their receptors, they initiate acute inflammation, which, as long as it is short lived, may be beneficial to the CNS. The progression of acute inflammation may lead to resolution of inflammation including tissue repair, clearance of pathogens, and resumption of homeostasis. The resolution of inflammation is mediated by anti-inflammatory cytokines (e.g., IL-10) and by pro-resolving lipid mediators (e.g., lipoxins, resolvins, and neuroprotectins). IL-10 is the most well-known anti-inflammatory cytokine, inhibiting both inflammatory and immune responses. IL-10 is reported to increase both cell survival and axonal regeneration after PNS injury, to reduce vulnerability of neurons to CNS ischemia and trauma, and to promote recovery after spinal cord injury (Vidal et al. 2013). The lipid mediators involved in inflammatory resolution include (1) cyclopentenones (15d-PGJ_2, PGA_2), produced by the nonenzymatic conversion of PGD_2 and PGE_2, which activate PPARγ and then attenuate cytokine production; (2) lipoxins (LXA_4), produced by enzymatic conversion of 12/15-LO that binds FPR2/ALX to stop cytokine production; (3) resolvins (RvD1), produced by the action of lipoxygenases on docosahexaenoic acid (DHA) that have potent pro-resolving properties and bind FPR2/ALX to terminate cytokine production; and (4) protectins (NPD1), also produced by enzymatic oxygenation of DHA that bind to a yet unknown receptor or PPARγ to attenuate cytokine production (Bazan et al. 2012; Carson 2012; Iadecola and Anrather 2011; Owens et al. 2001).

Inflammasomes are cytosolic multiprotein complexes that assemble in response to various insults (Martinon et al. 2002). Activation of inflammasomes leads to production of precursors of proinflammatory caspases, which subsequently cleave the precursor forms of IL-1β and IL-18. Recent studies have shown that IL-1β and IL-18 are implicated in the pathophysiology of numerous neurodegenerative diseases (Alzheimer's disease), autoimmune disease (MS), and CNS infection by pathogens (Heneka et al. 2013; Jha et al. 2010; Lippai et al. 2013; Ramos et al. 2012). The precise role of inflammasome signaling in CNS pathology warrants further investigation.

Galectins, β-galactoside-binding lectins, play important roles in the regulation of immune and inflammatory responses (Liu 2005; Liu and Rabinovich 2010; Rabinovich et al. 2002; Rabinovich and Toscano 2009; Thiemann and Baum 2011; Yang et al. 2008). Different members of the galectin family have been shown to play diverse roles in mediating pathological processes in various inflammatory diseases (Demetter et al. 2008; Dhirapong et al. 2009; Hokama et al. 2008; Liu and Rabinovich 2010). In neuronal diseases, galectins may serve as regulators of the inflammatory response and also confer remodeling capacity in damaged CNS tissues (Shin 2013). Here we review the roles of galectins in CNS pathology.

24.2 Expression and Functions (Nonimmune and Immune) of Galectins in the Nervous System

24.2.1 Introduction

The family of the β-galactoside-binding proteins galectins contains over 15 members across diverse species; they share homologous carbohydrate-recognition domains (CRDs) each with around 130 amino acids. Not all of the galectin members are expressed in humans. The variable expression of galectins in different species was reviewed in more detail by Cooper (2002) and Cummings and Liu (2009). Galectins are categorized into three types based on their protein structure. The prototype of galectins includes galectin-1, galectin-2, galectin-5, galectin-7, galectin-11, galectin-13, galectin-14, and galectin-15. They have only one CRD domain and can form homodimers or exist as monomers. The dimerization of prototype galectins was reported to facilitate binding to multivalent carbohydrates, which is critical for lectin function in cross-linking glycoconjugates. The tandem-repeat type of galectins (galectin-4, galectin-6, galectin-8, galectin-9, and galectin-12) have two non-identical but homologous CRDs connected by a variable intervening linker region. The variability of the CRDs confers different carbohydrate specificities, thus contributing to their ability to bind to different glycoconjugates. The chimeric type only consists of galectin-3. It has N-terminal repeats rich in proline, glycine, and tyrosine followed by a C-terminal CRD and has many unique properties compared to the other types.

Galectins have both extracellular and intracellular functions. Galectins do not have a classical signal sequence but may be secreted by a nonclassical secretory pathway. While a large variety of extracellular functions have been demonstrated in vitro, it is uncertain whether the effective extracellular concentrations of galectins required for these functions can be achieved physiologically or pathologically. Moreover, galectins reside predominantly in the cytoplasm of cells and have been shown to exert carbohydrate-independent functions by interacting with subcellular components. For example, galectin-3 was reported to interact with a number of different proteins, including β-catenin, nuclear thyroid-specific transcription factor (TTF-1), general transcription factor TFII-I, Gemin4, Bcl-2, and Alix. All of these appear to be independent of the protein's carbohydrate-binding function (Dagher et al. 1995; Liu et al. 2004, 2012; Park et al. 2001; Paron et al. 2003; Yu et al. 2002). In addition, the subcellular localization of galectin-3 was reported to be regulated by casein kinase, which can catalyze phosphorylation of the serine on the sixth amino acid position from the N-terminus, which facilitates nuclear export of the protein.

In this regard, it is important to clarify the roles of endogenous galectins in relation to their subcellular localization and distinguish them from those activities demonstrated by recombinant proteins.

24.2.2 Expression of Galectins in the CNS

The functions of galectins correlate with their expression pattern. Some galectins (galectin-1 and galectin-3) have wide tissue distribution, while others (galectin-4, galectin-7, galectin-8, galectin-9, and galectin-12) are distributed more specifically. The expression of galectins is also developmentally regulated and can be induced by physiological stimulation (Colnot et al. 1996). For example, expression of galectin-1 and galectin-3 can be induced in T lymphocytes, monocytes, macrophages, and Schwann cells. Expression correlates with the activation status of the cells (Blaser et al. 1998; Liu et al. 1995; Novak et al. 2012) and is governed by transcription factors (SP1, AP-1, and/or AP-2). Their expression is also affected by neoplastic transformation and infection by pathogens. P53 was reported to suppress expression of galectin-3 and to induce that of galectin-7 (Gaudin et al. 1997; Kopitz et al. 2003; Polyak et al. 1997). Some tissue-specific galectins often participate in the maintenance of homeostasis in different tissues, for example, galectin-7 in the epidermis, galectin-2 in the gastrointestinal tract, and galectin-12 in adipose tissue (Gendronneau et al. 2008; Thomsen et al. 2009; Yang et al. 2004).

The major galectins expressed in the CNS are galectin-1, galectin-3, galectin-4, galectin-8, and galectin-9, which were detected by immunohistochemical staining and real-time PCR analyses (Stancic et al. 2011). Galectin-1 was found to be expressed in olfactory receptor neurons (ORNs) at the basal layer of the olfactory epithelium and the subventricular zone (SVZ) of the lateral ventricle (Heilmann et al. 2000; Ishibashi et al. 2007). Galectin-3 can be found in ORN epithelium as well as in striatum, and its expression is induced in ischemic brain (Yan et al. 2009). Galectin-4 is expressed in immature neurons and decreased after oligodendrocyte (OLG) differentiation in the olfactory system (Stancic et al. 2012). The expression of galectin-8 and galectin-9 can be detected in astrocytes and astrocytic tumor cell lines by real-time PCR; however, their tissue distribution in the brain has not yet been characterized (Hadari et al. 1995; Yoshida et al. 2001). At the cellular level, galectin-1 is highly expressed in neuronal cells and neural stem cells (NSCs), while galectin-3 is primarily expressed by microglia (Lerman et al. 2012; Pasquini et al. 2011).

24.3 The Functions of Galectins in the CNS and PNS

Under normal physiological conditions, galectins have diverse functions in maintaining CNS homeostasis by participating in neuronal myelination, neuronal stem cell proliferation, and apical vesicle transportation. The physiological functions of galectins are summarized in this section. On the other hand, in CNS injury caused by autoinflammatory diseases or trauma, some galectins are induced in microglia cells and neurons and may contribute to either neuron regeneration or Wallerian degeneration.

24.3.1 Regulation of the CNS Functions by Galectins

The Roles of Galectin-1, Galectin-3, and Galectin-4 in Regulation of Axon Myelination by OLG

Axon myelination carried out by oligodendrocytes (OLGs) is important for proper saltatory nerve conduction. Galectin-1 and galectin-3 are both expressed in OLGs. However, only the expression of galectin-3 is correlated with OLG differentiation. In addition, galectin-3 knockout mice display loosely wrapped axons and a decrease in the frequency of myelinated axons (Pasquini et al. 2011). Sulfatides, sulfated galactosylceramides, are a major component of myelin and a ligand for galectin-4 (Stancic et al. 2012). Galectin-4 is therefore hypothesized to have an important role in regulating axon myelination by OLGs. Two different mechanisms have been proposed. First, immature neurons secrete galectin-4 that binds to oligodendrocytes suppressing differentiation and subsequently promoting their proliferation. Second, intracellular galectin-4 interacts with the cyclin-dependent kinase inhibitor p27 to enhance transcription of the myelin basic protein (MBP) gene in the OLGs (Wei et al. 2007).

The Role of Galectin-1 in Neural Stem Cell Development

Neural stem cells (NSCs) reside primarily in two areas of the brain: the dentate gyrus (DG) of the hippocampus and the SVZ near the olfactory bulb (OB). Galectin-1 can be detected in both areas. In vitro, galectin-1 was identified in conditioned medium of mouse bone marrow-derived stromal OP9 cells, which can promote neurosphere formation (Sakaguchi et al. 2006). Further, galectin-1 was shown to be expressed in a subset of NSCs that were positive for glial fibrillary acidic protein (GFAP). During adult neurogenesis in the DG of mice, deletion of galectin-1 resulted in an opposite effect on proliferation of progenitor cells in different strains, indicating that other genetic factors might contribute to its effect (Imaizumi et al. 2011; Kajitani et al. 2009). Although the function of endogenous galectin-1 remains to be investigated, studies using recombinant galectin-1, anti-galectin-1 antibody, and galectin-1 overexpression in NSCs demonstrated the therapeutic potential of galectin-1 in neuronal diseases, including brain ischemia and spinal cord injury (Ishibashi et al. 2007; Yamane et al. 2010).

Like other galectins, galectin-1 has both intracellular and extracellular functions; however, it is unique in that it can exist in both reduced and oxidized forms. The reduced form of galectin-1 promotes adult endogenous neural stem/progenitor cells (NSPCs) and neurosphere formation in vitro. The oxidized form of galectin-1 has no carbohydrate-binding activity, but can also promote regeneration of both CNS and PNS neurons and has a neural protective effect in a mouse model of neural disease (Chang-Hong et al. 2005). Although the mechanism of how galectin-1 regulates intracellular and extracellular pathways remains to be elucidated, the existing information suggests that galectin-1 may be employed as a therapeutic agent. The possible mechanism of how galectin-1 works in ischemia and spinal cord injury models is discussed in the following section.

The Role of Galectin-3, Galectin-4, and Galectin-9 in Regulation of Polarized Transport

Polarized transport is critical in the functions of neurons. Galectin-3, galectin-4, and galectin-9 were reported to regulate epithelial polarity by escorting surface receptors and sorting membrane constituents, which is important for maintaining epithelium morphogenesis (Koch et al. 2010; Mishra et al. 2010; Velasco et al. 2013). A potential mechanism controlling the apical sorting of neurotrophin is based on the assumption that secreted galectin-3 is internalized with the apical membrane through non-clathrin-mediated endocytosis, and it then escorts newly synthesized receptors (i.e., neurotrophin/p75NTR) to the apical surface (Straube et al. 2013).

Galectin-4, which is expressed by hippocampal and cortical neurons, controls sorting of neural cell adhesion molecules (NCAMs) also known as L1CAM, CD171, and L1, which are transmembrane proteins. The L1 clustering and polarization was reported to be mediated by extracellular galectin-4 via carbohydrate binding. In addition, trafficking of L1 by galectin-4 to the axon membrane is mediated through its binding to sulfatide-containing microtubule-associated carriers. Knockdown of galectin-4 attenuated axon growth by perturbing L1 clustering and the effect rescued by addition of recombinant galectin-4. Immunostaining indicated that in contrast to its restricted localization on the membrane when associated with L1, intracellular galectin-4 is homogeneously distributed in the cytoplasm. Therefore, it is suggested that the partition of L1 is mediated by secreted galectin-4 that can interact with L1 extracellularly. The mechanism of how galectin-4 is secreted remains unknown. However, its interaction with L1 has been hypothesized to mediate its partition in the extracellular membrane through a recycling process in which receptor (L1) proteins interact with extracellular galectin-4 prior to being endocytosed to perinuclear endosomes (Rab11-positive). They are then trafficked back to restricted regions of the membrane in a step that precedes axon growth (Velasco et al. 2013).

24.3.2 *The Functions of Individual Galectins in CNS*

Galectin-1 is the most studied of the galectins in the nervous system. The expression pattern of galectin-1 in the CNS was reviewed by Sakaguchi et al. (2007). In adult tissue, galectin-1 can be detected in the spinal cord, cerebellum, forebrain, and olfactory bulb. In addition, it is present in sensory neurons, motor neurons, Schwann cells, glial cells, and astrocytes. Galectin-1 was reported to be able to regulate NSC and astrocyte differentiation (Uehara et al. 2001; Akazawa et al. 2004; Endo 2005), and glial cell line-derived neurotrophic factor (GDNF)-induced neurite outgrowth from cultured adult rat DRG neurons (Takaku et al. 2013). It can also participate in sensory axon pathfinding and axonal regeneration through both carbohydrate-dependent and carbohydrate-independent mechanisms. As mentioned above, galectin-1 can exist in both oxidized and reduced forms and utilize different mechanisms to regulate cell functions. The oxidized form of galectin-1 does not have carbohydrate-binding activity but can function at a substantially lower concentration than the reduced form. Table 24.1 summarizes the functions of galectin-1 in relation to its subcellular localization and whether the function was demonstrated in response— to administration of recombinant galectin-1.

Galectin-3 interacts with several intracellular molecules and participates in regulating macrophage activation and cell apoptosis. In the nervous system, galectin-3 is expressed in microglia, astrocytes, oligodendrocytes, and Schwann cells and regulates their function during nerve injury. In addition, the expression of galectin-3 is upregulated during oligodendrocyte differentiation (Pasquini et al. 2011) and induced in many injury and disease models disease models, such as ALS, ischemia, EAE, and axotomy (Jiang et al. 2009; Lalancette-Hébert et al. 2012; Lerman et al. 2012).

Table 24.1 The function of galectin-1 in neuron

Expression	Intra/extra function	CRD dependent	Function	References
Peripheral neurons; sensory and motor neurons	Extra	No	rhGal-1 promotes axonal growth by stimulating neurofilaments and L1 expression. In addition, rhGal-1 helps to recruit Schwann cells	Horie et al. (1999), Horie and Kadoya (2000)
Dorsal root ganglia (DRG)	Extra	No	rhGAL-1/Ox may be a factor for functional restoration of injured peripheral nerves	Kadoya et al. (2005)
Primary afferent neurons	Intra	N/A	Galectin-1 is transported centrally by dorsal root axons. The neuronal and glial expression of galectin-1 is tightly correlated with regenerative success	McGraw et al. (2005)
Nonneuronal cells in the rat olfactory nerve	Extra	Yes (to glycolipid and laminin)	In vitro, recombinant galectin-1 promotes primary olfactory neuron adhesion to laminin (in both a cell-to-cell and cell-to- ECM manner)	Mahanthappa et al. (1994)
GDNF family dependent subgroup	Extra	N/A	Galectin-1 can contribute to the establishment of neuropathic pain after peripheral nerve injury	Imbe et al. (2003)
Rat embryonic cortical cells	Intra	N/A	ΔFoxB transactivates galectin-1 expression after ischemia	Kurushima et al. (2005)
Adult neural stem cells	Extra	Yes	Infusion of rGalectin-1 into lateral ventricle increases the neurosphere formation from SVZ	Ishibashi et al. (2007)

N/A: no answer

Galectin-3 expressed in different cells has distinct functions in the regulation of neural inflammation, neuron repair, and myelination. The endogenous functions of galectin-3 are best revealed by the aberrant phenotypes displayed by knockout mice, which are summarized in Table 24.2.

Table 24.2 The endogenous function of galectins in neuron

Galectin	Phenotype in KO mice	Proposed mechanism	Exogenous methods	Reference
Galectin-1	Pilocarpine-induced neuronal death is abolished in the forebrain of galectin-1 KO mice	Galectin-1 is proapoptotic in a subpopulation of GABAergic interneuron	N/A	Bischoff et al. (2012)
	Increased numbers of type 1 cells, doublecortin-positive immature progenitors, and neuronal nuclei-positive newborn neurons in the SVZ of galectin-1 KO mice	N/A	N/A	Imaizumi et al. (2011)
	Galectin-1 KO mice have reduced neural progenitor cell proliferation	N/A	N/A	Kajitani et al. (2009)
	Galectin-1 KO mice have an attenuated rate of functional recovery of whisking movement after a facial nerve crush	N/A	N/A	McGraw et al. (2004)
	Galectin-1 KO mice have a delay in elimination of nerve ending after section of the sciatic nerve	Section of sciatic nerve upregulates p75NTR and galectin-1 in terminal Schwann cells, which helps degeneration of processes	Functional blocking antibody against galectin-1 prevents degeneration processes, and recombinant galectin-1 causes WT neurons to degenerate	Plachta et al. (2007)
	Galectin-1 KO mice exhibit impaired spatial learning in the water maze and contextual fear learning	Hippocampal-expressed galectin-1 contributes to learning and memory	N/A	Sakaguchi et al. (2011)

Galectin-3	Galectin-3 KO mice have disrupted ependymal cilia and SVZ astrocytes	Galectin-3 regulates SVZ neuroblast migration to the olfactory bulb	Anti-galectin-3 antibody administration reproduces the phenotype of galectin-3 KO mice Recombinant galectin-3 inhibits SVZ neuroblast migration	Comte et al. (2011)
	Galectin-3 KO mice have defective injury-induced microglia activation and proliferation and increase ischemic lesions (associated with apoptotic neurons and IGF-1 expression)	Galectin-3 interacts with IGF-1 receptor N-glycans	IGF-1-mediated mitogenic microglia response is reduced by the N-glycan inhibitor tunicamycin	Lalancette-Hébert et al. (2012)
	$SOD1^{G93A}$/Gal-3 KO mice display more rapid progression of ALS disease stage compared with SOD1G93A/Gal-3 WT mice	N/A	Deletion of galectin-3 increases microglia activation, TNF-α level, and oxidative injury in $SOD1^{G93A/G}$	Lerman et al. (2012)
	Galectin-3 KO mice exhibit earlier functional recovery and faster neuron regeneration after sciatic neuron crush	Galectin-3 inhibits Schwann cell and macrophage functions in myelination and clearance of axon debris	N/A	Narciso et al. (2009)
	Galectin-3 KO mice have a decrease in the frequency of myelinated axons, myelin turns (lamellae), and γ-ratio in the corpus callosum and striatum	Conditioned medium from galectin-3 KO microglia cells promotes oligodendrocyte differentiation. Altered myelin structure in galectin-3 KO mice leads to behavioral abnormalities	Exogenous galectin-3 promotes OLG differentiation but does not affect astrocytes	Pasquini et al. (2011)

N/A: no answer

Recently, galectin-4 was shown to regulate axon growth and apical protein sorting in the nervous system (Velasco et al. 2013). The expression of galectin-4 is downregulated just before the onset of myelination. When immature OLGs were treated with galectin-4, OLG maturation was retarded, while a subset of the immature OLGs reverted to a morphologically less complex progenitor stage, while displaying a concomitant increase in proliferation (Stancic et al. 2012). Therefore, galectin-4 is also hypothesized to regulate the differentiation of OLGs.

Only a few studies deal with the functions of galectin-9 in the nervous system. It was found to be aberrantly expressed in biopsy samples from patients with multiple sclerosis (MS). Its subcellular localization was changed from the cytoplasm where it was found in inactive MS lesions to the nucleus in active ones (Stancic et al. 2011). The expression of galectin-9 is induced by IL-1β in human primary astrocytes, and the induced protein is located in the perinuclear region and membrane fraction (Yoshida et al. 2001). This indicates that galectin-9 has important intracellular functions in addition to extracellular ones.

24.3.3 Endogenous Functions of Galectins Revealed by Studies of Knockout Mice

Endogenous functions of galectins have been investigated by using genetic knockout mice. In general, mice deficient in galectins are viable, fertile, and without major overt phenotypes (Poirier and Robertson 1993). Earlier studies therefore postulated that the expression of complementary galectins may be altered in knockout mice. However, only a few studies support this possibility. Further investigations revealed their functions in different disease models and in mice with different genetic backgrounds. Results of studies done using knockout mice are summarized in Table 24.2.

24.4 Galectins and Neuronal Inflammation

24.4.1 Multiple Sclerosis (EAE Model)

Galectins have been implicated in the regulation of several chronic inflammatory diseases, including rheumatoid arthritis and diabetes (Mensah-Brown et al. 2009; Ohshima et al. 2003; Radosavljevic et al. 2012). In the chronic progressive inflammatory demyelination disease, MS, expression of galectin-1, galectin-3, galectin-8, and galectin-9 are elevated (Stancic et al. 2011). In general, endogenous galectin-1 and galectin-9 prevent auto-inflammation, while galectin-3 is proinflammatory (Jiang et al. 2009; Toscano et al. 2007; Zhu et al. 2005).

Application of recombinant galectin-1 prevented induction of EAE by injection of guinea pig myelin basic protein (GP-BP) in complete Freud adjuvant (CFA) and also inhibited delayed-type hypersensitivity. It was proposed that administration of recombinant galectin-1 prevents sensitization of encephalitogenic GP-BP-specific T cells, possibly as the result of an increase in suppressor T cells (Offner et al. 1990).

Galectin-1 knockout mice develop more severe EAE than WT mice (Toscano et al. 2007). The former also develop higher amounts of autoreactive Th1 and Th17 cells, which may be due to reduced apoptosis of Th1 and Th17 cells (Toscano et al. 2007). The increase in autoreactive Th1 and Th17 cells in galectin-1 knockout mice correlates with the severity of EAE. The molecular mechanism proposed for the effect of galectin-1 on apoptosis is that extracellular galectin-1 interacts with Th1 and Th17 cell surface glycans to induce apoptosis (Toscano et al. 2007).

Similarly, galectin-9 was reported to interact with Tim-3 (a cell surface marker of Th1 cells), to promote Th1 cell apoptosis (Zhu et al. 2005). However, how galectins are secreted in vivo is not clear, and whether they can be present at levels matching the effective concentrations achieved by using recombinant galectins in vitro remains unclear. The cells targeted by recombinant galectins in vivo also remain to be investigated. Regardless of these issues, administration of recombinant proteins in EAE may be of therapeutic value, as might silencing of endogenous galectins, even though they act through different mechanisms. Administration of recombinant galectin-9 in mice was shown to selectively reduce IFN-secreting Th1 cells and to ameliorate EAE (Zhu et al. 2005).

In addition to CNS injury, the phagocytosis of myelin by microglia and macrophages in the EAE model is important for neuron degeneration and regeneration. In CNS injury in vivo, most microglia and macrophages do not phagocytose myelin. However, galectin-3 is upregulated along with MAC-1 and F4/80 in microglia and macrophages, and it actively participates in myelin degeneration in the spinal cord and optic nerve areas in EAE mice (Reichert and Rotshenker 1999).

Galectin-3 knockout mice develop milder EAE than WT mice when it is induced by myelin oligodendrocyte glycoprotein (MOG) (Jiang et al. 2009). This phenotype is accompanied by lower levels of lymphocyte infiltration, a decrease in Th-17 and IFN-γ-producing cells, and an elevation in the number of Treg cells. Moreover, dendritic cells isolated from galectin-3 knockout mice produce more IL-10 and are prone to drive Th2 differentiation and produce less IL-17 and IFN-r(Jiang et al. 2009). In summary, the endogenous functions of galectins in different cell types may be carried out via diverse mechanisms that together contribute to regulation of neural inflammatory diseases.

24.4.2 Ischemic Lesion Model

Brain ischemia is followed by synchronized inflammatory processes. Proinflammatory cytokines, including tumor necrosis factor-α (TNF-α) and IL-1β, are elevated 24 h after ischemic injury. Subsequent activation of resident rather than infiltrating microglia and astrocytes is responsible for the clearance of damaged myelin and cell debris. The activated microglia also participates in production of proinflammatory cytokines. In addition to inflammatory cytokines, activated innate immune cells also serve to generate neurotropic and neuroprotective substances that facilitate neuron regeneration after ischemia. Both galectin-1 and galectin-3 are induced and were reported to help neuron regeneration after ischemic injury. However, their modes of action are different.

Galectin-1 is primarily expressed in astrocytes and neurons, while galectin-3 which is very low in resting cells is overexpressed in activated microglia and macrophages. The expression of galectin-1 is required for astrocyte-derived neurotrophic factor (BDNF) secretion, and recombinant galectin-1 promotes neuron regeneration (Ishibashi et al. 2007; Qu et al. 2011). Although recombinant galectin-1 works differently from the endogenous protein, systemic administration of galectin-1 was reported to regulate inflammation and stimulate neuronal regeneration.

Galectin-3 is required for microglia activation after ischemic injury and therefore is responsible for several of their immune functions, including expression of neuroprotective IGF-1, TLR2, and CD68. Galectin-3 KO mice have defective microglial cell proliferation and exhibit exacerbating ischemic damage and neuronal apoptosis after cerebral ischemia. A deficiency in galectin-3 is also responsible for aberrant responses to IGF-1-induced mitogenic signaling and the overexpression of IL-6 and suppressor of cytokine signaling 3 (SOCS3). Selective ablation of galectin-3-positive cells resulted in increased neuronal apoptosis, which is associated with increased expression of proinflammatory cytokines after transient middle cerebral artery occlusion (Lalancette-Hébert et al. 2012).

24.4.3 Amyotrophic Lateral Sclerosis (ALS) and the Mouse Model SOD1^{G93A}

ALS displays an overt neuroinflammatory response, with motor neuron degeneration. Increased mRNA and protein expression of galectin-1, galectin-3, and galectin-9 were observed in the ALS mouse model SOD1^{G93A}. Sporadic ALS patients also have increased expression of galectin-3 and galectin-9. This indicates that galectins may have a protective role in ALS pathogenesis. In hSOD1G93A transgenic mice, astrocyte and microglia produce an excess of reactive oxygen species and proinflammatory cytokines. This creates a neurotoxic environment for disease development. Because galectin-3 can be detected in the presymptomatic stage and increases to the end stage of ALS, it is thought to participate in the pathogenesis of ALS. In contrast, expression of galectin-1 only increases at the end stage of the disease while that of galectin-9 increases after symptoms appear. Galectin-9 is expressed by cultured astrocytes upon stimulation by IL-1β, and it can be inhibited by dexamethasone. Although this suggests galectin-9 may participate in neuroinflammation, its function in ALS has not yet been investigated.

In B6SJL SOD1^{G93A} mice, immunofluorescence staining of the lumbar spinal cord indicated that galectin-3 is expressed in microglia cells and a small percentage of astrocytes. C57BL6 SOD1^{G93A}/Gal-$^{3-/-}$ mice underwent a rapid disease progression and severity as evaluated by functionally defined disease stages and the neurological symptoms correlated with microglia function. Deletion of galectin-3 resulted in increased neurodegeneration that was associated with elevated TNF-α and protein carbonyls, representative of oxidative injury (Lerman et al. 2012). Although increased microglia infiltration is associated with neuroprotection in the pathogenesis of ALS, SOD1^{G93A}/Gal-3$^{-/-}$ mice have both increased microglia infiltration and neurodegeneration. This phenotype is also inconsistent with the neuroprotective function of microglial galectin-3 found in ischemia and EAE studies. These findings support the as yet untested hypothesis that in the ALS mouse model SOD1^{G93A}, galectin-3 in microglia may be instrumental to the alternative M2 phenotype with anti-inflammatory functions. Exogenous galectins may also alter the inflammatory environment to ameliorate disease progression; for example, intramuscular administration of oxidized galectin-1 improved SOD1 neurological function, although the expression of galectin-1 is not altered in ALS patient samples (Chang-Hong et al. 2005).

24.4.4 Axotomy and Wallerian Degeneration

Upon traumatic injury, neurons go through an ordered process of degeneration and regeneration. This process is orchestrated by an interplay of several types of cells including neurons, macrophages, microglia, Schwann cells, oligodendrocytes, and fibroblasts and is characterized as Wallerian degeneration. The cellular and molecular events of Wallerian degeneration were reviewed by Rotshenker (2011). Major events in this process include inactivation of innate immunity for efficient clearance of damaged myelin, where galectin-3 plays an important role. Neurotoxic glycoproteins of myelin released by damaged axons perturb neuron regeneration, which requires active clearance of the toxic debris by macrophages and Schwann cells in order for recovery to occur. Galectin-3 has an important function in regulating phagocytosis by macrophages. The expression of galectin-3 is elevated during macrophage activation, and this correlates with the ability of the macrophages to phagocytose degenerate myelin in the injured PNS or CNS (Reichert et al. 1994). The proposed mechanism by which galectin-3 activates phagocytosis by macrophages and Schwann cells is through its interaction with K-Ras-GTP. The K-Ras-GTP signaling then activates PI3K, PLC, and cPKC to promote phagocytosis. Although galectin-3 is required for the phagocytic activity of macrophages, knockout mice were reported to display faster regeneration of neurons after surgical compression of sciatic nerve fibers. This was accompanied by increased proliferation of Schwann cells and activated macrophages. In summary, galectins are induced during neuron damage and participate in the process of neuron regeneration, activation of astrocytes, microglia phagocytosis, and Schwann cell proliferation.

In conclusion, galectins are of important immune modulatory functions in both physiological and pathological conditions. They have pleiotropic functions when expressed in different types of cells and subcellular localizations. Their expression is often correlated with physiological functions (galectin-1 in neurogenesis, galectin-3 in phagocytosis) and pathological conditions (galectin-1 and galectin-3 with EAE severity). Within the cells, endogenous galectins regulate gene expression and signaling pathways to modulate cytokine production. Exogenous galectins can induce immune cells apoptosis and activation to modulate the inflammatory response. Exogenous galectins may or may not have the same targets or utilize the same mechanisms (CRD dependent or independent) as endogenous ones in modulating the cellular functions; however, they are of great therapeutic potential in inflammatory diseases.

Conflict of Interest The authors declare that they have no conflict of interest.

References

Andersson, U., and Tracey, K.J. (2012a). Neural reflexes in inflammation and immunity. The Journal of experimental medicine *209*, 1057–1068.

Andersson, U., and Tracey, K.J. (2012b). Reflex principles of immunological homeostasis. Annual review of immunology *30*, 313–335.

Arima, Y., Harada, M., Kamimura, D., Park, J.H., Kawano, F., Yull, F.E., Kawamoto, T., Iwakura, Y., Betz, U.A., Marquez, G., *et al.* (2012). Regional neural activation defines a gateway for autoreactive T cells to cross the blood-brain barrier. Cell *148*, 447–457.

Bazan, N.G., Halabi, A., Ertel, M., and Petasis, N.A. (2012). Chapter 34 - Neuroinflammation. In Basic Neurochemistry (Eighth Edition), T.B. Scott, J.S. George, R.W. Albers, G.J.S.R.W.A. Donald L. PriceA2 - Scott T. Brady, and L.P. Donald, eds. (New York, Academic Press), pp. 610–620.

Beutler, B., Eidenschenk, C., Crozat, K., Imler, J.L., Takeuchi, O., Hoffmann, J.A., and Akira, S. (2007). Genetic analysis of resistance to viral infection. Nature reviews Immunology *7*, 753–766.

Bianchi, M.E. (2007). DAMPs, PAMPs and alarmins: all we need to know about danger. Journal of leukocyte biology *81*, 1–5.

Bischoff, V., Deogracias, R., Poirier, F., and Barde, Y.-A. (2012). Seizure-induced neuronal death is suppressed in the absence of the endogenous lectin Galectin-1. J Neurosci *32*, 15590–15600.

Blaser, C., Kaufmann, M., Müller, C., Zimmermann, C., Wells, V., Mallucci, L., and Pircher, H. (1998). Beta-galactoside-binding protein secreted by activated T cells inhibits antigen-induced proliferation of T cells. Eur J Immunol *28*, 2311–2319.

Bolitho, P., Voskoboinik, I., Trapani, J.A., and Smyth, M.J. (2007). Apoptosis induced by the lymphocyte effector molecule perforin. Current opinion in immunology *19*, 339–347.

Carson, M.J. (2012). Chapter 33 - Molecular Mechanisms and Consequences of Immune and Nervous System Interactions. In Basic Neurochemistry (Eighth Edition), T.B. Scott, J.S. George, R.W. Albers, G.J.S.R.W.A. Donald L. PriceA2 - Scott T. Brady, and L.P. Donald, eds. (New York, Academic Press), pp. 597–609.

Chang-Hong, R., Wada, M., Koyama, S., Kimura, H., Arawaka, S., Kawanami, T., Kurita, K., Kadoya, T., Aoki, M., Itoyama, Y., *et al.* (2005). Neuroprotective effect of oxidized galectin-1 in a transgenic mouse model of amyotrophic lateral sclerosis. Exp Neurol *194*, 203–211.

Colnot, C., Ripoche, M.A., Scaerou, F., Foulis, D., and Poirier, F. (1996). Galectins in mouse embryogenesis. Biochem Soc Trans *24*, 141–146.

Comte, I., Kim, Y., Young, C.C., van der Harg, J.M., Hockberger, P., Bolam, P.J., Poirier, F., and Szele, F.G. (2011). Galectin-3 maintains cell motility from the subventricular zone to the olfactory bulb. J Cell Sci *124*, 2438–2447.

Cooper, D.N.W. (2002). Galectinomics: finding themes in complexity. Biochim Biophys Acta *1572*, 209–231.
Cua, D.J., Sherlock, J., Chen, Y., Murphy, C.A., Joyce, B., Seymour, B., Lucian, L., To, W., Kwan, S., Churakova, T., *et al.* (2003). Interleukin-23 rather than interleukin-12 is the critical cytokine for autoimmune inflammation of the brain. Nature *421*, 744–748.
Cummings, R.D., and Liu, F.T. (2009). Galectins. In Essentials of Glycobiology, A. Varki, R.D. Cummings, J.D. Esko, H.H. Freeze, P. Stanley, C.R. Bertozzi, G.W. Hart, and M.E. Etzler, eds. (Cold Spring Harbor NY, The Consortium of Glycobiology Editors, La Jolla, California).
Dagher, S.F., Wang, J.L., and Patterson, R.J. (1995). Identification of galectin-3 as a factor in pre-mRNA splicing. Proc Natl Acad Sci U S A *92*, 1213–1217.
Dubin, P.J., and Kolls, J.K. (2008). Th17 cytokines and mucosal immunity. Immunological reviews *226*, 160–171.
Farina, C., Aloisi, F., and Meinl, E. (2007). Astrocytes are active players in cerebral innate immunity. Trends in immunology *28*, 138–145.
Fazilleau, N., Mark, L., McHeyzer-Williams, L.J., and McHeyzer-Williams, M.G. (2009) Follicular helper T cells: lineage and location. Immunity *30*, 324–335.
Gaudin, J.C., Arar, C., Monsigny, M., and Legrand, A. (1997). Modulation of the expression of the rabbit galectin-3 gene by p53 and c-Ha-ras proteins and PMA. Glycobiology *7*, 1089–1098.
Gendronneau, G., Sidhu, S.S., Delacour, D., Dang, T., Calonne, C., Houzelstein, D., Magnaldo, T., and Poirier, F. (2008). Galectin-7 in the control of epidermal homeostasis after injury. Mol Biol Cell *19*, 5541–5549.
Hadari, Y.R., Paz, K., Dekel, R., Mestrovic, T., Accili, D., and Zick, Y. (1995). Galectin-8. A new rat lectin, related to galectin-4. J Biol Chem *270*, 3447–3453.
Hawkins, B.T., and Davis, T.P. (2005). The blood-brain barrier/neurovascular unit in health and disease. Pharmacological reviews *57*, 173–185.
Heilmann, S., Hummel, T., Margolis, F.L., Kasper, M., and Witt, M. (2000). Immunohistochemical distribution of galectin-1, galectin-3, and olfactory marker protein in human olfactory epithelium. Histochem Cell Biol *113*, 241–245.
Heneka, M.T., Kummer, M.P., Stutz, A., Delekate, A., Schwartz, S., Vieira-Saecker, A., Griep, A., Axt, D., Remus, A., Tzeng, T.C., *et al.* (2013). NLRP3 is activated in Alzheimer's disease and contributes to pathology in APP/PS1 mice. Nature *493*, 674–678.
Heusel, J.W., Wesselschmidt, R.L., Shresta, S., Russell, J.H., and Ley, T.J. (1994). Cytotoxic lymphocytes require granzyme B for the rapid induction of DNA fragmentation and apoptosis in allogeneic target cells. Cell *76*, 977–987.
Horie, H., Inagaki, Y., Sohma, Y., Nozawa, R., Okawa, K., Hasegawa, M., Muramatsu, N., Kawano, H., Horie, M., Koyama, H., *et al.* (1999). Galectin-1 regulates initial axonal growth in peripheral nerves after axotomy. J Neurosci *19*, 9964–9974.
Horie, H., and Kadoya, T. (2000). Identification of oxidized galectin-1 as an initial repair regulatory factor after axotomy in peripheral nerves. Neurosci Res *38*, 131–137.
Iadecola, C., and Anrather, J. (2011). The immunology of stroke: from mechanisms to translation. Nat Med *17*, 796–808.
Imaizumi, Y., Sakaguchi, M., Morishita, T., Ito, M., Poirier, F., Sawamoto, K., and Okano, H. (2011). Galectin-1 is expressed in early-type neural progenitor cells and down-regulates neurogenesis in the adult hippocampus. Mol Brain *4*, 7.
Imbe, H., Okamoto, K., Kadoya, T., Horie, H., and Senba, E. (2003). Galectin-1 is involved in the potentiation of neuropathic pain in the dorsal horn. Brain Res *993*, 72–83.
Ishibashi, S., Kuroiwa, T., Sakaguchi, M., Sun, L., Kadoya, T., Okano, H., and Mizusawa, H. (2007). Galectin-1 regulates neurogenesis in the subventricular zone and promotes functional recovery after stroke. Exp Neurol *207*, 302–313.
Jack, C.S., Arbour, N., Manusow, J., Montgrain, V., Blain, M., McCrea, E., Shapiro, A., and Antel, J.P. (2005). TLR signaling tailors innate immune responses in human microglia and astrocytes. J Immunol *175*, 4320–4330.
Jha, S., Srivastava, S.Y., Brickey, W.J., Iocca, H., Toews, A., Morrison, J.P., Chen, V.S., Gris, D., Matsushima, G.K., and Ting, J.P. (2010). The inflammasome sensor, NLRP3, regulates CNS

inflammation and demyelination via caspase-1 and interleukin-18. The Journal of neuroscience : the official journal of the Society for Neuroscience *30*, 15811–15820.
Jiang, H.R., Al Rasebi, Z., Mensah-Brown, E., Shahin, A., Xu, D., Goodyear, C.S., Fukada, S.Y., Liu, F.T., Liew, F.Y., and Lukic, M.L. (2009). Galectin-3 deficiency reduces the severity of experimental autoimmune encephalomyelitis. J Immunol *182*, 1167–1173.
Kadoya, T., Oyanagi, K., Kawakami, E., Hasegawa, M., Inagaki, Y., Sohma, Y., and Horie, H. (2005). Oxidized galectin-1 advances the functional recovery after peripheral nerve injury. Neurosci Lett *380*, 284–288.
Kajitani, K., Nomaru, H., Ifuku, M., Yutsudo, N., Dan, Y., Miura, T., Tsuchimoto, D., Sakumi, K., Kadoya, T., Horie, H., *et al.* (2009). Galectin-1 promotes basal and kainate-induced proliferation of neural progenitors in the dentate gyrus of adult mouse hippocampus. Cell Death Differ *16*, 417–427.
Kipnis, J., Cohen, H., Cardon, M., Ziv, Y., and Schwartz, M. (2004). T cell deficiency leads to cognitive dysfunction: implications for therapeutic vaccination for schizophrenia and other psychiatric conditions. Proceedings of the National Academy of Sciences of the United States of America *101*, 8180–8185.
Kipnis, J., Mizrahi, T., Hauben, E., Shaked, I., Shevach, E., and Schwartz, M. (2002). Neuroprotective autoimmunity: naturally occurring CD4+CD25+ regulatory T cells suppress the ability to withstand injury to the central nervous system. Proceedings of the National Academy of Sciences of the line United States of America *99*, 15620–15625.
Koch, A., Poirier, F., Jacob, R., and Delacour, D. (2010). Galectin-3, a novel centrosome-associated protein, required for epithelial morphogenesis. Mol Biol Cell *21*, 219–231.
Kopitz, J., André, S., von Reitzenstein, C., Versluis, K., Kaltner, H., Pieters, R.J., Wasano, K., Kuwabara, I., Liu, F.-T., Cantz, M., *et al.* (2003). Homodimeric galectin-7 (p53-induced gene 1) is a negative growth regulator for human neuroblastoma cells. Oncogene *22*, 6277–6288.
Kurushima, H., Ohno, M., Miura, T., Nakamura, T.Y., Horie, H., Kadoya, T., Ooboshi, H., Kitazono, T., Ibayashi, S., Iida, M., *et al.* (2005). Selective induction of DeltaFosB in the brain after transient forebrain ischemia accompanied by an increased expression of galectin-1, and the implication of DeltaFosB and galectin-1 in neuroprotection and neurogenesis. Cell Death Differ *12*, 1078–1096.
Lalancette-Hébert, M., Swarup, V., Beaulieu, J.M., Bohacek, I., Abdelhamid, E., Weng, Y.C., Sato, S., and Kriz, J. (2012). Galectin-3 is required for resident microglia activation and proliferation in response to ischemic injury. J Neurosci *32*, 10383–10395.
Langrish, C.L., Chen, Y., Blumenschein, W.M., Mattson, J., Basham, B., Sedgwick, J.D., McClanahan, T., Kastelein, R.A., and Cua, D.J. (2005). IL-23 drives a pathogenic T cell population that induces autoimmune inflammation. The Journal of experimental medicine *201*, 233–240.
Lehnardt, S. (2010). Innate immunity and neuroinflammation in the CNS: the role of microglia in Toll-like receptor-mediated neuronal injury. Glia *58*, 253–263.
Lerman, B.J., Hoffman, E.P., Sutherland, M.L., Bouri, K., Hsu, D.K., Liu, F.-T., Rothstein, J.D., and Knoblach, S.M. (2012). Deletion of galectin-3 exacerbates microglial activation and accelerates disease progression and demise in a SOD1(G93A) mouse model of amyotrophic lateral sclerosis. Brain Behav *2*, 563–575.
Lippai, D., Bala, S., Petrasek, J., Csak, T., Levin, I., Kurt-Jones, E.A., and Szabo, G. (2013). Alcohol-induced IL-1beta in the brain is mediated by NLRP3/ASC inflammasome activation that amplifies neuroinflammation. Journal of leukocyte biology *94*, 171–182.
Liu, F.T., Hsu, D.K., Zuberi, R.I., Kuwabara, I., Chi, E.Y., and Henderson, W.R. (1995). Expression and function of galectin-3, a beta-galactoside-binding lectin, in human monocytes and macrophages. Am J Pathol *147*, 1016–1028.
Liu, L., Sakai, T., Sano, N., and Fukui, K. (2004). Nucling mediates apoptosis by inhibiting expression of galectin-3 through interference with nuclear factor kappaB signalling. Biochem J *380*, 31–41.
Liu, W., Hsu, D.K., Chen, H.-Y., Yang, R.-Y., Carraway, K.L., 3rd, Isseroff, R.R., and Liu, F.-T. (2012). Galectin-3 Regulates Intracellular Trafficking of EGFR through Alix and Promotes Keratinocyte Migration. J Invest Dermatol *132*, 2828–2837.

Mahanthappa, N.K., Cooper, D.N., Barondes, S.H., and Schwarting, G.A. (1994). Rat olfactory neurons can utilize the endogenous lectin, L-14, in a novel adhesion mechanism. Development *120*, 1373–1384.

Martinon, F., Burns, K., and Tschopp, J. (2002). The inflammasome: a molecular platform triggering activation of inflammatory caspases and processing of proIL-beta. Molecular cell *10*, 417–426.

McGraw, J., Gaudet, A.D., Oschipok, L.W., Kadoya, T., Horie, H., Steeves, J.D., Tetzlaff, W., and Ramer, M.S. (2005). Regulation of neuronal and glial galectin-1 expression by peripheral and central axotomy of rat primary afferent neurons. Exp Neurol *195*, 103–114.

McGraw, J., McPhail, L.T., Oschipok, L.W., Horie, H., Poirier, F., Steeves, J.D., Ramer, M.S., and Tetzlaff, W. (2004). Galectin-1 in regenerating motoneurons. Eur J Neurosci *20*, 2872–2880.

Mensah-Brown, E.P.K., Al Rabesi, Z., Shahin, A., Al Shamsi, M., Arsenijevic, N., Hsu, D.K., Liu, F.-T., and Lukic, M.L. (2009). Targeted disruption of the galectin-3 gene results in decreased susceptibility to multiple low dose streptozotocin-induced diabetes in mice. Clin Immunol *130*, 83–88.

Mina-Osorio, P., Rosas-Ballina, M., Valdes-Ferrer, S.I., Al-Abed, Y., Tracey, K.J., and Diamond, B. (2012). Neural signaling in the spleen controls B-cell responses to blood-borne antigen. Mol Med *18*, 618–627.

Mishra, R., Grzybek, M., Niki, T., Hirashima, M., and Simons, K. (2010). Galectin-9 trafficking regulates apical-basal polarity in Madin-Darby canine kidney epithelial cells. Proc Natl Acad Sci U S A *107*, 17633–17638.

Mrass, P., and Weninger, W. (2006). Immune cell migration as a means to control immune privilege: lessons from the CNS and tumors. Immunological reviews *213*, 195–212.

Narciso, M.S., Mietto, B.d.S., Marques, S.A., Soares, C.P., Mermelstein, C.d.S., El-Cheikh, M.C., and Martinez, A.M.B. (2009). Sciatic nerve regeneration is accelerated in galectin-3 knockout mice. Exp Neurol *217*, 7–15.

Novak, R., Dabelic, S., and Dumic, J. (2012). Galectin-1 and galectin-3 expression profiles in classically and alternatively activated human macrophages. Biochim Biophys Acta *1820*, 1383–1390.

O'Shea, J.J., and Paul, W.E. (2010). Mechanisms underlying lineage commitment and plasticity of helper CD4+ T cells. Science *327*, 1098–1102.

Offner, H., Celnik, B., Bringman, T.S., Casentini-Borocz, D., Nedwin, G.E., and Vandenbark, A.A. (1990). Recombinant human beta-galactoside binding lectin suppresses clinical and histological signs of experimental autoimmune encephalomyelitis. J Neuroimmunol *28*, 177–184.

Ohshima, S., Kuchen, S., Seemayer, C.A., Kyburz, D., Hirt, A., Klinzing, S., Michel, B.A., Gay, R.E., Liu, F.-T., Gay, S., *et al.* (2003). Galectin 3 and its binding protein in rheumatoid arthritis. Arthritis Rheum *48*, 2788–2795.

Olson, J.K., and Miller, S.D. (2004). Microglia initiate central nervous system innate and adaptive immune responses through multiple TLRs. J Immunol *173*, 3916–3924.

Owens, T., Wekerle, H., and Antel, J. (2001). Genetic models for CNS inflammation. Nat Med *7*, 161–166.

Park, J.W., Voss, P.G., Grabski, S., Wang, J.L., and Patterson, R.J. (2001). Association of galectin-1 and galectin-3 with Gemin4 in complexes containing the SMN protein. Nucleic Acids Res *29*, 3595–3602.

Paron, I., Scaloni, A., Pines, A., Bachi, A., Liu, F.T., Puppin, C., Pandolfi, M., Ledda, L., Di Loreto, C., Damante, G., *et al.* (2003). Nuclear localization of Galectin-3 in transformed thyroid cells: a role in transcriptional regulation. Biochem Biophys Res Commun *302*, 545–553.

Pasquini, L.A., Millet, V., Hoyos, H.C., Giannoni, J.P., Croci, D.O., Marder, M., Liu, F.T., Rabinovich, G.A., and Pasquini, J.M. (2011). Galectin-3 drives oligodendrocyte differentiation to control myelin integrity and function. Cell Death Differ *18*, 1746–1756.

Peters, P.J., Borst, J., Oorschot, V., Fukuda, M., Krahenbuhl, O., Tschopp, J., Slot, J.W., and Geuze, H.J. (1991). Cytotoxic T lymphocyte granules are secretory lysosomes, containing both perforin and granzymes. The Journal of experimental medicine *173*, 1099–1109.

Plachta, N., Annaheim, C., Bissière, S., Lin, S., Rüegg, M., Hoving, S., Müller, D., Poirier, F., Bibel, M., and Barde, Y.-A. (2007). Identification of a lectin causing the degeneration of neuronal processes using engineered embryonic stem cells. Nat Neurosci *10*, 712–719.

Poirier, F., and Robertson, E.J. (1993). Normal development of mice carrying a null mutation in the gene encoding the L14 S-type lectin. Development *119*, 1229–1236.

Polyak, K., Xia, Y., Zweier, J.L., Kinzler, K.W., and Vogelstein, B. (1997). A model for p53-induced apoptosis. Nature *389*, 300–305.

Qu, W.-S., Wang, Y.-H., Ma, J.-F., Tian, D.-S., Zhang, Q., Pan, D.-J., Yu, Z.-Y., Xie, M.-J., Wang, J.-P., and Wang, W. (2011). Galectin-1 attenuates astrogliosis-associated injuries and improves recovery of rats following focal cerebral ischemia. J Neurochem *116*, 217–226.

Radosavljevic, G., Volarevic, V., Jovanovic, I., Milovanovic, M., Pejnovic, N., Arsenijevic, N., Hsu, D.K., and Lukic, M.L. (2012). The roles of Galectin-3 in autoimmunity and tumor progression. Immunol Res *52*, 100–110.

Ramos, H.J., Lanteri, M.C., Blahnik, G., Negash, A., Suthar, M.S., Brassil, M.M., Sodhi, K., Treuting, P.M., Busch, M.P., Norris, P.J., *et al.* (2012). IL-1beta signaling promotes CNS-intrinsic immune control of West Nile virus infection. PLoS pathogens *8*, e1003039.

Ransohoff, R.M., and Brown, M.A. (2012). Innate immunity in the central nervous system. The Journal of clinical investigation *122*, 1164–1171.

Reboldi, A., Coisne, C., Baumjohann, D., Benvenuto, F., Bottinelli, D., Lira, S., Uccelli, A., Lanzavecchia, A., Engelhardt, B., and Sallusto, F. (2009). C-C chemokine receptor 6-regulated entry of TH-17 cells into the CNS through the choroid plexus is required for the initiation of EAE. Nature immunology *10*, 514–523.

Reichert, F., and Rotshenker, S. (1999). Galectin-3/MAC-2 in experimental allergic encephalomyelitis. Exp Neurol *160*, 508–514.

Reichert, F., Saada, A., and Rotshenker, S. (1994). Peripheral nerve injury induces Schwann cells to express two macrophage phenotypes: phagocytosis and the galactose-specific lectin MAC-2. J Neurosci *14*, 3231–3245.

Rosas-Ballina, M., Olofsson, P.S., Ochani, M., Valdes-Ferrer, S.I., Levine, Y.A., Reardon, C., Tusche, M.W., Pavlov, V.A., Andersson, U., Chavan, S., *et al.* (2011). Acetylcholine-synthesizing T cells relay neural signals in a vagus nerve circuit. Science *334*, 98–101.

Rotshenker, S. (2011). Wallerian degeneration: the innate-immune response to traumatic nerve injury. J Neuroinflammation *8*, 109.

Sakaguchi, M., Arruda-Carvalho, M., Kang, N.H., Imaizumi, Y., Poirier, F., Okano, H., and Frankland, P.W. (2011). Impaired spatial and contextual memory formation in galectin-1 deficient mice. Mol Brain *4*, 33.

Sakaguchi, M., Imaizumi, Y., and Okano, H. (2007). Expression and function of galectin-1 in adult neural stem cells. Cell Mol Life Sci *64*, 1254–1258.

Sakaguchi, M., Shingo, T., Shimazaki, T., Okano, H.J., Shiwa, M., Ishibashi, S., Oguro, H., Ninomiya, M., Kadoya, T., Horie, H., *et al.* (2006). A carbohydrate-binding protein, Galectin-1, promotes proliferation of adult neural stem cells. Proc Natl Acad Sci U S A *103*, 7112–7117.

Sakaguchi, S., Yamaguchi, T., Nomura, T., and Ono, M. (2008). Regulatory T cells and immune tolerance. Cell *133*, 775–787.

Sancho, D., and Reis e Sousa, C. (2012). Signaling by myeloid C-type lectin receptors in immunity and homeostasis. Annual review of immunology *30*, 491–529.

Savarin, C., and Bergmann, C.C. (2008). Neuroimmunology of central nervous system viral infections: the cells, molecules and mechanisms involved. Current opinion in pharmacology *8*, 472–479.

Schwartz, M., and Kipnis, J. (2011). A conceptual revolution in the relationships between the brain and immunity. Brain, behavior, and immunity *25*, 817–819.

Seil, F.J. (2001). Interactions between cerebellar Purkinje cells and their associated astrocytes. Histology and histopathology *16*, 955–968.

Seo, T.B., Chang, I.A., Lee, J.H., and Namgung, U. (2013). Beneficial function of cell division cycle 2 activity in astrocytes on axonal regeneration after spinal cord injury. Journal of neurotrauma *30*, 1053–1061.

Shevach, E.M., DiPaolo, R.A., Andersson, J., Zhao, D.M., Stephens, G.L., and Thornton, A.M. (2006). The lifestyle of naturally occurring CD4+ CD25+ Foxp3+ regulatory T cells. Immunological reviews *212*, 60–73.

Stancic, M., Slijepcevic, D., Nomden, A., Vos, M.J., de Jonge, J.C., Sikkema, A.H., Gabius, H.-J., Hoekstra, D., and Baron, W. (2012). Galectin-4, a novel neuronal regulator of myelination. Glia *60*, 919–935.

Stancic, M., van Horssen, J., Thijssen, V.L., Gabius, H.-J., van der Valk, P., Hoekstra, D., and Baron, W. (2011). Increased expression of distinct galectins in multiple sclerosis lesions. Neuropathol Appl Neurobiol *37*, 654–671.

Straube, T., von Mach, T., Hönig, E., Greb, C., Schneider, D., and Jacob, R. (2013). PH-dependent recycling of galectin-3 at the apical membrane of epithelial cells. Traffic.

Takaku, S., Yanagisawa, H., Watabe, K., Horie, H., Kadoya, T., Sakumi, K., Nakabeppu, Y., Poirier, F., and Sango, K. (2013). GDNF promotes neurite outgrowth and upregulates galectin-1 through the RET/PI3K signaling in cultured adult rat dorsal root ganglion neurons. Neurochem Int *62*, 330–339.

Takeuchi, O., and Akira, S. (2009). Innate immunity to virus infection. Immunological reviews *227*, 75–86.

Thomsen, M.K., Hansen, G.H., and Danielsen, E.M. (2009). Galectin-2 at the enterocyte brush border of the small intestine. Mol Membr Biol *26*, 347–355.

Toscano, M.A., Bianco, G.A., Ilarregui, J.M., Croci, D.O., Correale, J., Hernandez, J.D., Zwirner, N.W., Poirier, F., Riley, E.M., Baum, L.G., *et al.* (2007). Differential glycosylation of TH1, TH2 and TH-17 effector cells selectively regulates susceptibility to cell death. Nat Immunol *8*, 825–834.

Velasco, S., Díez-Revuelta, N., Hernández-Iglesias, T., Kaltner, H., André, S., Gabius, H.-J., and Abad-Rodríguez, J. (2013). Neuronal Galectin-4 is required for axon growth and for the organization of axonal membrane L1 delivery and clustering. J Neurochem *125*, 49–62.

Vidal, P.M., Lemmens, E., Dooley, D., and Hendrix, S. (2013). The role of "anti-inflammatory" cytokines in axon regeneration. Cytokine & growth factor reviews *24*, 1–12.

Wei, Q., Eviatar-Ribak, T., Miskimins, W.K., and Miskimins, R. (2007). Galectin-4 is involved in p27-mediated activation of the myelin basic protein promoter. J Neurochem *101*, 1214–1223.

Willing, A., and Friese, M.A. (2012). CD8-mediated inflammatory central nervous system disorders. Current opinion in neurology *25*, 316–321.

Wolf, S.A., Steiner, B., Akpinarli, A., Kammertoens, T., Nassenstein, C., Braun, A., Blankenstein, T., and Kempermann, G. (2009a). CD4-positive T lymphocytes provide a neuroimmunological link in the control of adult hippocampal neurogenesis. J Immunol *182*, 3979–3984.

Wolf, S.A., Steiner, B., Wengner, A., Lipp, M., Kammertoens, T., and Kempermann, G. (2009b). Adaptive peripheral immune response increases proliferation of neural precursor cells in the adult hippocampus. FASEB journal : official publication of the Federation of American Societies for Experimental Biology *23*, 3121–3128.

Yamane, J., Nakamura, M., Iwanami, A., Sakaguchi, M., Katoh, H., Yamada, M., Momoshima, S., Miyao, S., Ishii, K., Tamaoki, N., *et al.* (2010). Transplantation of galectin-1-expressing human neural stem cells into the injured spinal cord of adult common marmosets. J Neurosci Res *88*, 1394–1405.

Yan, Y.-P., Lang, B.T., Vemuganti, R., and Dempsey, R.J. (2009). Galectin-3 mediates post-ischemic tissue remodeling. Brain Res *1288*, 116–124.

Yang, R.-Y., Hsu, D.K., Yu, L., Chen, H.-Y., and Liu, F.-T. (2004). Galectin-12 is required for adipogenic signaling and adipocyte differentiation. J Biol Chem *279*, 29761–29766.

Ye, Z., and Ting, J.P. (2008). NLR, the nucleotide-binding domain leucine-rich repeat containing gene family. Current opinion in immunology *20*, 3–9.

Yoshida, H., Imaizumi, T., Kumagai, M., Kimura, K., Satoh, C., Hanada, N., Fujimoto, K., Nishi, N., Tanji, K., Matsumiya, T., *et al.* (2001). Interleukin-1beta stimulates galectin-9 expression in human astrocytes. Neuroreport *12*, 3755–3758.

Yu, F., Finley, R.L., Jr., Raz, A., and Kim, H.R. (2002). Galectin-3 translocates to the perinuclear membranes and inhibits cytochrome c release from the mitochondria. A role for synexin in galectin-3 translocation. J Biol Chem *277*, 15819–15827.

Zhu, C., Anderson, A.C., Schubart, A., Xiong, H., Imitola, J., Khoury, S.J., Zheng, X.X., Strom, T.B., and Kuchroo, V.K. (2005). The Tim-3 ligand galectin-9 negatively regulates T helper type 1 immunity. Nat Immunol *6*, 1245–1252.

Ziv, Y., Ron, N., Butovsky, O., Landa, G., Sudai, E., Greenberg, N., Cohen, H., Kipnis, J., and Schwartz, M. (2006). Immune cells contribute to the maintenance of neurogenesis and spatial learning abilities in adulthood. Nature neuroscience *9*, 268–275.

Chapter 25
Glycoconjugates and Neuroimmunological Diseases

Hugh J. Willison

Abstract A wide range of neuroimmunological diseases affect the central and peripheral nervous systems. These disorders are caused by autoimmune attack directed against structurally and functionally diverse nervous system antigens. One such category comprises peripheral nervous system (PNS) diseases, termed peripheral neuropathies, in which the target antigens for autoantibody-directed nerve injury are glycan structures borne by glycoproteins and glycolipids, particularly gangliosides that are concentrated in peripheral nerve. The archetypal PNS disorder is the acute paralytic disease, Guillain–Barré syndrome (GBS) in which autoantibodies against glycolipids arise in the context of acute infections that precede the clinical onset, notably *Campylobacter jejuni* enteritis. In addition, several chronic autoimmune neuropathies are associated with IgM antibodies directed against nerve glycans including sulphated glucuronic acid epitopes present on myelin-associated glycoprotein and sulphated glucuronyl paragloboside, a range of disialylated gangliosides including GD1b and GD3, and GM1 ganglioside. This chapter describes the immunological, pathological and clinical features of these disorders in the context of our broader knowledge of the glycobiology underpinning this neuroimmunological field.

Keywords Gangliosides • Glycolipids • Autoantibody • Peripheral nerve • Neuropathy

H.J. Willison (✉)
Institute of Infection, Immunity and Inflammation, College of Medical, Veterinary and Life Sciences, University of Glasgow, Glasgow, UK

Glasgow Biomedical Research Centre, University of Glasgow,
Room B330, 120 University Place, Glasgow G12 8TA, UK
e-mail: hugh.willison@glasgow.ac.uk

R.K. Yu and C.-L. Schengrund (eds.), *Glycobiology of the Nervous System*, Advances in Neurobiology 9, DOI 10.1007/978-1-4939-1154-7_25,

25.1 Introduction

A wide range of neuroimmunological diseases affect the central and peripheral nervous systems. These disorders are caused by autoimmune attack directed against structurally and functionally diverse nervous system antigens. One such category comprises peripheral nervous system (PNS) diseases, termed peripheral neuropathies, in which the target antigens for autoantibody-directed nerve injury are glycan structures borne by glycoproteins and glycolipids, particularly gangliosides that are concentrated in peripheral nerve. The archetypal PNS disorder is the acute paralytic disease, Guillain–Barré syndrome (GBS) in which autoantibodies against glycolipids arise in the context of acute infections that precede the clinical onset, notably *Campylobacter jejuni* (*C. jejuni*) enteritis. In addition, several chronic autoimmune neuropathies are associated with IgM antibodies directed against nerve glycans including sulphated glucuronic acid epitopes present on myelin-associated glycoprotein and sulphated glucuronyl paragloboside, a range of disialylated gangliosides including GD1b and GD3, and GM1 ganglioside. This chapter describes the immunological, pathological and clinical features of these disorders in the context of our broader knowledge of the glycobiology underpinning this neuroimmunological field.

25.2 The Structural and Molecular Composition of the Peripheral Nervous System in Relation to Carbohydrate Autoantigens

The peripheral nervous system (PNS) is formed by many different neuronal cell types within various specialised regions, supported by myelin-forming and non-myelinating Schwann cells, and composed of a very wide range of molecular components. In functional terms the PNS comprises motor, sensory and autonomic neurons. In the case of motor neurons, cell bodies within the spinal cord extend axonal projections via the nerve trunks to form motor nerve terminals at neuromuscular synapses in muscles up to one metre away. Electrical impulse propagation over these distances is facilitated by myelin, and saltatory conduction. In the sensory arm of the PNS, the diversity of sensory fibre types and sizes, the different types of sensory ending, and the presence of the dorsal root ganglion with peripheral axons and central projections into the dorsal horn of the spinal cord contribute to an even greater level of complexity. Autonomic nerves are also regarded as PNS components. The endoneurial compartment, containing intrafascicular nerve fibres, is comprised of connective tissue rich in collagen and extracellular matrix material, and is surrounded by the perineurium. Within the endoneurial compartment lie the nerve axons enveloped by Schwann cells and their basal laminae, resident endoneurial macrophages and fibroblasts, and a vascular supply. The molecular diversity of the basal lamina includes many glycoyslated and/or glycan-binding molecules such as laminins, dystroglycans, glycosaminoglycans and integrins, some of which are autoantigens in neuropathy and other autoimmune diseases. The PNS is an immunologically complex structure with well developed innate and adaptive immune function, and many sites within the PNS are subject to a wide range of inflammatory and autoimmune responses (Kieseier et al. 2006; Meyer Zu et al. 2008). Much of the peripheral nervous system resides within an immunologically privileged site within the blood–nerve barrier (the PNS equivalent of the blood–brain barrier), with the exception of the dorsal and ventral spinal roots and the sensory and motor nerve terminals that lie in sites more freely exposed to circulating factors (Kanda 2013).

The PNS is affected by an extensive range of inherited, metabolic, degenerative and autoimmune disorders. In the context of autoimmunity to glycans, the PNS contains many glycosylated molecules potentially capable of acting as antigens in autoimmune peripheral neuropathies, notably myelin and axonal glycolipids, gangliosides and the myelin glycoproteins. P0, peripheral myelin protein 22 (PMP-22) and myelin associated glycoprotein (MAG). The structural complexity of the PNS, combined with the limited access that autoantibodies might have to specialised compartments such as compacted myelin, makes considerations about the relevance of such antigens as autoantibody targets a highly complex one. Schwann cells synthesise and maintain the compacted myelin membranes that enwrap large calibre peripheral axons where some neuropathy specific glycoprotein antigens such as PMP22 and PO are localised (Scherer 1997). It is believed that the molecular composition of compact myelin is common to all fibre types although it remains possible that there could be subtle differences in the content of particular antigens, including gangliosides that could affect the clinical manifestations of particular autoimmune neuropathies (Ogawa-Goto et al. 1992). Other antigens, such as MAG, are found in specialised uncompacted Schwann cell membranes including the adaxonal membrane that is separated from the axonal membrane by the periaxonal space (Yinchang et al. 1998). Peripheral nerve myelin contains many glycolipids and gangliosides that are important antigens for autoantibody responses. The myelin sheath also develops specialisations associated with non-compact myelin at paranodal regions where certain glycolipids may be more readily exposed to autoantibodies than they are in compact myelin membranes (Salzer et al. 2008). Despite much research over many decades, the nature and pathological role of immune responses to glycolipids in central nervous system disorders such as multiple sclerosis has not been clarified, and is not considered further in this chapter.

25.3 Glycolipid Antigens in Autoimmune Neuropathy

Galactocerebroside (GalC) is one of the simplest glycosphingolipids, comprising a single galactose residue linked to ceramide (monogalactosylceramide). Historically, it was the first glycolipid to be identified as a key antigen in experimental autoimmune neuritis (EAN), a rodent model of autoimmune neuropathy, as well as in its central nervous system (CNS) counterpart, experimental autoimmune encephalomyelitis

(EAE) (Saida et al. 1979). Sulphatide is synthesised from GalC following sulphation of the 3rd carbon of galactose, and is also a prominent lipid antigen in PNS and CNS myelin. GalC and sulphatide reside within the exofacial leaflet of the lipid bilayer, where the oligosaccharide is exposed to the extracellular environment and thus available to autoantibody binding. Functionally GalC and sulphatide play important roles in the development and maintenance of myelin, as demonstrated by deficiency states created by targeted disruption of the UDP-galactose ceramide galactosyl-transferase (CGT), the key enzyme in galactolipid biosynthesis (Popko et al. 1999). These mice have a phenotype characterised by slow nerve conduction velocity and myelin breakdown. Deletion of the galactosyl ceramide sulphotransferase (CST) gene that specifically inhibits the synthesis of sulphatide, whilst maintaining GalC results in a less severe phenotype than CGT deficiency, principally comprising ultrastructural abnormalities in the paranodal regions, indicating a critical role for sulphatide in the localization and maintenance of glial–axonal interactions in the nodal complex (Honke 2013).

Peripheral nerve also has an abundance of neolacto-series gangliosides that are localised mainly in myelin. LM1 is also termed sialosylneolactotetraosylceramide, or sialosylparagloboside, and its higher homologue sialosyllactosaminyl-paragloboside is referred to as Hex-LM1 (Svennerholm et al. 1994). Although their function has not been fully elucidated, both LM1 and Hex-LM1 have been identified as autoantigens in both acute and chronic autoimmune neuropathies (Gu et al. 2012). Substitution of sialic acid on the terminal saccharide chain of LM1 and Hex-LM1 for a 3 sulphated glucuronic acid generates sulphated glucuronyl paragloboside (SGPG) and its higher lactosaminyl homolog, sulphated glucuronyl lactosaminyl paragloboside (SGLPG). Both were discovered directly as a result of studying autoimmune neuropathy patients with IgM paraproteins reactive to MAG, with which they share the immunologically cross-reactive sulphated glucuronic acid structure. These sulphated glucuronyl glycolipids have been identified in myelin- and axon-enriched fractions of motor and sensory nerves from human subjects and have a slightly greater abundance in sensory nerve fractions compared with motor nerve, which may account for the relative predominance of sensory features in patients with anti-SGPG antibodies (Ilyas et al. 1986). The structures of the two glycolipid antigens, SGPG and SGLPG were characterised independently and revealed that the minimal carbohydrate requirement for the HNK-1 epitope resides in the terminal disaccharide residue, SO_4-3GlcAβ1-3Galβ1-. (Ariga 1987; Chou 1986).

Gangliosides are principally enriched in neuronal membranes but are also minor constituents of myelin (Ledeen and Yu 1982). Structural diagrams of the major gangliosides of relevance to autoimmune neuropathy are shown in Fig. 25.1. These notably include GM1, GD1b, GD1a and GQ1b, but many other gangliosides have also been identified as antibody targets. The term ganglioside refers to glycosphingolipids that contain one or more sialic acid residues linked to the oligosaccharide core. In man, most ganglioside sialic acid is in the N-acetyl form, as opposed to the N-glycolyl form, that is common to many other species. These sialic acid residues are a critical part of the epitope involved in autoantibody binding. The amphipathic

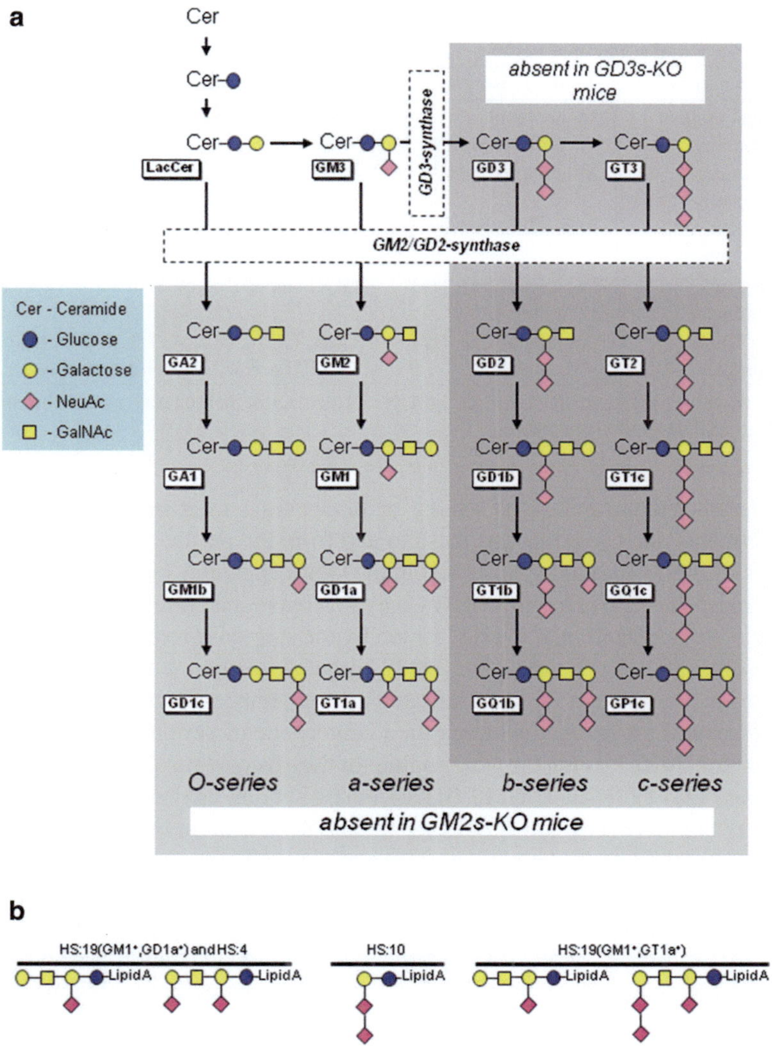

Fig. 25.1 Glycan structures. (**a**) A schematic representation of the key structures and enzymes in ganglioside biosynthesis. Note that GalNAc transferase (GalNAcT) and GD3-synthase (GD3s) deficiency result in the loss of all complex and b series gangliosides respectively. (**b**) GBS associated *C. jejuni* LOS structures, although linked to Lipid A rather than ceramide, are structural mimics of gangliosides. Strains of *C. jejuni* can express multiple structural variants of LOS. Examples shown are HS19 and HS:4 strains that bear oligosaccharides of GM1, GD1a and GT1a, and HS:10 strain that bears GD3

nature of gangliosides, comprising a hydrophobic ceramide moiety that is embedded in the lipid membrane and a hydrophilic oligosaccharide moiety which is exposed to the extracellular space makes the oligosaccharide ideally situated as a recognition platform for lectins, including bacterial toxins, and for autoantibody binding.

Table 25.1 Examples of peripheral neuropathy syndromes associated with anti-glycolipid antibodies

Clinical syndrome	Antibody isotype	Antibody specificity
Multifocal motor neuropathy	Monoclonal or polyclonal IgM	GM1, GM2, GM1:GalC complex
Chronic ataxic neuropathy	Monoclonal IgM	Disialosyl epitopes (CANOMAD) (on GD1b, GT1b, GQ1b, GD3)
"MAG negative" neuropathy[a]	Monoclonal IgM	GM2, LM1, GalNAc-GD1b, GalNAc-GD1a and unidentified glycolipids/gangliosides
Miller Fisher syndrome	Polyclonal IgG, IgM and IgA	GQ1b and GT1a, other disialosyl epitopes
Guillain–Barré syndrome	Polyclonal IgG, IgM and IgA	GM1, GD1a, LM1, GT1b, GD1b, GalNAc-GD1a, GalC, glycolipid complexes

[a]"MAG negative" refers to the 50 % of IgM paraproteinaemic neuropathy cases who are anti-MAG/anti-SGPG antibody negative

They are also found in a wide variety of intracellular membrane and endosomal compartments and cycle through these to and from the plasma membrane as small vesicles. Gangliosides are synthesised in the Golgi apparatus by the stepwise addition of saccharides in reactions catalysed by glycosyltransferases and sialyltransferases, and are degraded in a reverse fashion by the step-wise removal of saccharides in lysosomes (Sandhoff and Kolter 2003). Targeted disruption of genes involved in ganglioside biosynthesis has demonstrated the importance of gangliosides in a wide range of neural processes, including the maintenance of peripheral nerve integrity and function. Mice lacking the gene coding for the glycosyltransferase, 1,4 GalNAc-transferase (1,4 GalNAc-T; GM2/GD2-synthase) which lack all complex gangliosides (see Fig. 25.1) develop age-related axonal degeneration in the CNS and PNS, and also develop demyelination in peripheral nerves with pathological features resembling those seen in MAG knock-out mice, a known ligand for complex gangliosides. In double knock-out mice that lack both GM2/GD2- and GD3-synthases, therefore expressing only GM3, peripheral nerve degeneration has also been observed, combined with mutilating skin lesions and nerve fibre proliferation in the skin, suggesting that reduced sensory function is present (Inoue et al. 2002).

When considered as antigen targets, ganglioside localization and distribution in the plasma membrane is crucially important. Gangliosides are concentrated in association with cholesterol in clusters referred to as lipid rafts where their close lateral interactions in the plane of the plasma membrane may enhance their ability to bind antibody with high avidity or affinity, and also affect antibody clearance by endocytosis (Fewou et al. 2012). Additionally, activated complement components and complement regulators such as CD59 that reside in rafts may in turn affect the normal functioning of gangliosides and the pathological effects of anti-ganglioside antibodies (Willison et al. 2008). The relative contribution of these factors may vary from site to site within peripheral nerve membranes, and between anti-ganglioside antibodies of differing immunoglobulin class and reactivity. The distribution of gangliosides in different membrane and raft domains in the PNS and the relevance that this may have to their role as antigens in autoimmune neuropathy is a subject of ongoing study. However, when considering the pathogenic relationship between the presence of an antibody and neuropathy, it is clearly important to have detailed knowledge of the glycosphingolipid composition and distribution within the PNS in both humans and in experimental animals used for modelling these disorders, and furthermore in species from which gangliosides are purified for experimental and diagnostic use. These issues are not straightforward, and a complete map of the ganglioside composition of human nerves, and a comparison between species used for experimental modelling and between individuals, would be a valuable resource (Chiba et al. 1997).

25.4 Clinical Syndromes Associated with Specific Anti-glycolipid Antibody Profiles

25.4.1 Paraproteins and IgM Antibodies

The discovery in the 1980s of the first human autoimmune neuropathy in which the antigen specificity was identified to be a peripheral nerve glycoconjugate paved the way for subsequent decades of discovery of new anti-glycan antibodies linked to clinical neuropathy syndromes (Table 25.1). In this first syndrome, patients with IgM paraproteins and neuropathy were found to have IgM reactivity with a carbohydrate determinant present on nerve glycoproteins and glycolipids bearing the carbohydrate epitope HNK-1 (Leu-7, CD57, for more detail see Chap. 6). These molecules were identified as MAG, sulphoglucuronosyl paragloboside (SGPG) and sulphoglucuronosyl lactosaminyl paragloboside (SGLPG) and subsequently human NCAM, J1 glycoprotein, P0 and PMP22 were also found to bear this epitope (Burger et al. 1990). Structural analysis revealed that the critical antigenic epitope was a 3-sulphate-glucuronic acid residue present on both glycoproteins and glycolipids, thereby accounting for the antibody reactivity with multiple molecules, all bearing the same antigenic determinant. Some anti-MAG antibodies additionally cross-react with the glycolipid sulphatide. Antibodies to sulphatide (that do not react with sulphated glucuronic acid) are in turn associated with predominantly sensory neuropathies and also form part of the natural autoantibody repertoire (Lopate et al. 1997). The clinical phenotype of anti-MAG IgM paraproteinaemic neuropathy is usually that of a late onset, chronic and slowly progressive sensory dominant demyelinating neuropathy (Nobile-Orazio et al. 1994). Tremor is a common feature. A characteristic clinical electrophysiological pattern is present in affected patients. On sural nerve biopsy, chronic segmental demyelination and axonal degeneration are seen, along with extensive IgM deposits in myelin that correlate with a "widely spaced" myelin appearance on electron microscopy. Animal modelling has unequivocally shown that the IgM antibody is directly pathogenic {Quarles, 1990 86/id}. Most patients have a good prognosis and the modest clinical benefit of prolonged treatment with toxic immunotherapy is usually outweighed by the high rate of adverse events. Subsequent to the discovery of "anti-MAG" antibodies that account for ~50 % of patients with IgM paraproteinaemic neuropathy, a further group of patients were found to have IgM paraproteins that react with the disialosyl epitope (NeuAcα2-8NeuAcα2-3 Gal) common to all b-series gangliosides including GD1b, GD3, GT1b and GQ1b. The clinical phenotype of this paraproteinaemic neuropathy syndrome is a highly characteristic chronic sensory ataxic neuropathy, and often referred to under the acronym CANOMAD: *c*hronic *at*axic *n*europathy, *o*phthalmoplegia, Ig*M* paraprotein, cold *a*gglutinins and *d*isialosyl antibodies (Willison et al. 2001).

Another syndrome subject to much attention is a form of motor neuropathy associated with anti-GM1 ganglioside antibodies. These antibodies are also of the IgM class but do not normally occur as IgM paraproteins, despite the early reports of this association in which isolated patients were originally described with unusual forms of motor neuron disease (Kornberg and Pestronk 1995). A multifocal motor neuropathy (MMN) with conduction block was identified by clinical and electrophysiological parameters at that time and it soon became apparent that this was strongly associated with the presence of anti-GM1 IgM antibodies in over 50 % of affected patients. Three main patterns of anti-GM1 antibody specificity were recognised: those that react with the terminal Gal beta1-3 GalNAc structure common to GM1, GD1b and asialo-GM1; those that react with the internal sialylated epitope common to GM1 and GM2, and those that react with GM1 monospecifically. The clinical picture of multifocal motor neuropathy is usually of slowly progressive, asymmetrical limb weakness with minimal or no sensory impairment, often with distal onset in an upper limb. It is characterised electrophysiologically by partial and focal motor conduction block across a segment of nerve with sparing of motor conduction in adjacent sensory fibres. Over time, the clinical pattern may become confluent, with significant motor axonal degeneration. Multifocal motor neuropathy usually responds well to intravenous immunoglobulin which needs to be given regularly. Other agents, including cyclophosphamide have not been shown in systematically controlled trials to improve outcome although case series have been reported.

25.4.2 Guillain–Barré Syndrome

An important group of acute peripheral neuropathies associated with anti-glycolipid antibodies are the Guillain–Barré syndromes (GBS), now the foremost cause of post-infectious neuromuscular paralysis worldwide following the eradication of poliomyelitis (Hughes and Cornblath 2005). The clinical onset of GBS is rapid, and in around 20 % of cases leads to total paralysis requiring mechanical ventilation. Whilst recovery is usual, around 20 % of affected individuals are still unable to walk 1 year after onset. The cause of GBS is complex; however, the acute, monophasic, post-infectious nature of GBS and its response to treatment with plasma exchange or intravenous immunoglobulin strongly suggest a contributory role for autoantibodies. The commonest form of GBS arises from segmental demyelination of peripheral nerve (acute inflammatory demyelinating polyneuropathy, AIDP), mediated by antibody and complement deposition on Schwann cell and myelin membranes that results in macrophage-mediated destruction of the myelin sheath (Hafer-Macko et al. 1996b). Although the precise targets for immune attack in AIDP remain unknown, accumulating evidence suggest that the antibodies may bind to glycolipids and gangliosides enriched in Schwann cell membranes. These include LM1, SGPG and GM1 ganglioside, identified in a small proportion of cases (Kusunoki 2003).

Two GBS variants, acute motor axonal neuropathy (AMAN) and Miller Fisher syndrome (MFS) provide evidence to support the view that anti-glycolipid autoantibodies are important in GBS. In AMAN, rather than the myelin sheath, the primary target is the axolemmal membrane either in the motor nerve roots or distal nerve terminals (Hafer-Macko et al. 1996a; Ho et al. 1997). Immune attack leads to axonal conduction block due to axonal injury which can either be temporary and reversible, or more prolonged if axonal transection with Wallerian degeneration occurs. AMAN is very strongly associated with anti-ganglioside antibodies directed to sialylated epitopes on GM1a, GM1b, GD1a and GalNAc-GD1a (Willison and Yuki 2002). The second widely studied variant is MFS. In contrast to the generalised limb and respiratory weakness that occurs in GBS, symptoms in MFS are restricted to limb ataxia, tendon reflex loss and extraocular nerve paralysis (causing paralysis of eye movements, ophthalmoplegia). This disorder is very strongly associated with antibodies to GQ1b and GT1a gangliosides in the majority of cases (Chiba et al. 1993).

25.5 The Origin and Nature of Anti-ganglioside Antibodies

Carbohydrates differ fundamentally from protein antigens in that the former cannot be presented by MHC Class II molecules to receptors borne by T cells (T cell receptor, TCR) whose function is to help the immune system generate adaptive immune responses: Thus they are referred to as T cell independent (TI) antigens (Defrance et al. 2011). This long standing immunological dogma has been partially deconstructed over recent years through deeper understanding of both MHC and TCR biology and the CD1 system of antigen presentation in which TCRs can recognise glycans when they are presented by CD1 molecules (Mori and De 2012). Notwithstanding the detailed immunological pathways by which they arise, humoral responses to TI antigens provide important components to both innate and adaptive anti-microbial immunity, including protection against lipopolysaccharide (LPS) encapsulated bacteria. Within the B cell pool, both innate $CD5^+$ B1 cells and marginal zone (MZ) B cells account for the early antibody response to a microbial glycan challenge, whereas follicular dendritic (FO) B2 cells participate later in the antibody response by providing classical responses to T dependent protein antigens generated within germinal centres in secondary lymphoid tissues. Since many microbial oligosaccharide structures are also expressed in man, uncontrolled

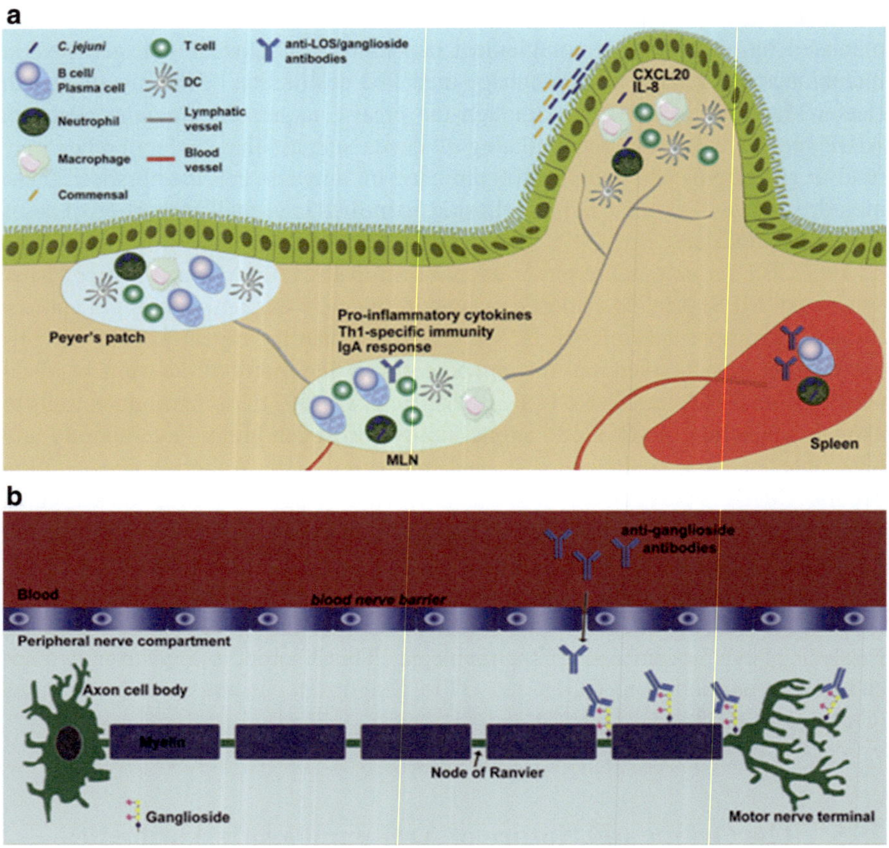

Fig. 25.2 Immune response to *C. jejuni* LOS and the resulting anti-ganglioside antibody-mediated pathology. (**a**) A gastrointestinal infection with *C. jejuni* will result in the activation of intestinal epithelial cells, with subsequent recruitment of DCs, macrophage and neutrophils. The transition of this immune response to the mesenteric lymph nodes (MLN) or the generation of a response in the Peyer's patches will result in a predominately Th1 response for bacterial clearance. Importantly, a robust IgA response is part of this response. It is hypothesised that in individuals who go on to develop GBS the gut microbiota is altered and that the immune response that is generated to LOS results in an anti-ganglioside antibody response, which could be due the contribution of inducible B cell-helper neutrophils (iN_{BH}) in either the Peyer's patches, MLN or spleen. (**b**) The resulting anti-ganglioside antibody response crosses the blood–nerve barrier and interacts with gangliosides that are expressed at the motor nerve terminal axonal membrane, node of Ranvier and/or the myelin sheath. Distal nerve injury and injury to the nerve roots may be favoured as these sites lack an effective blood–nerve barrier. Complement is activated, macrophages are recruited, and demyelination and/or axon injury ensues. Reproduced from (Willison and Goodyear 2013)

immune responses directed to these structures may play a role in autoimmune diseases and the balance between providing anti-microbial protection whilst avoiding autoimmunity is therefore critically controlled (Fig. 25.2). In GBS, this immunological control over B cell responses to microbial glycans is dysregulated (Willison and Goodyear 2013).

The commonest and most widely studied bacterial infection that triggers GBS is the food and water borne pathogen *C. jejuni*, responsible for acute diarrhoeal disease throughout the world (Yuki 2001). Campylobacter species synthesise a truncated form of LPS that lacks the branching O-side chain, and more simply comprises a core oligosaccharide attached to the lipid A molecule, hence the term lipo-oligosaccharide (LOS). Remarkably, the glycans present on the LOS of GBS-associated *C. jejuni* isolates contain ganglioside-mimicking structures including GM1, GM2, GD1a, GT1a and GD3. These are immunologically recognisable by the anti-ganglioside antibodies in the serum of GBS patients triggered by *C. jejuni* infection. Thus, anti-ganglioside antibodies most likely arise through an immune response against *C. jejuni* LOS and subsequently cause neuropathy by cross-reacting with peripheral nerve gangliosides of the same structural type (Fig. 25.1). This process for the induction of autoimmune disease through shared antigens between microbes and host tissue is referred to as the molecular mimicry hypothesis.

Only a small proportion (~1 %) of Campylobacter serotypes carry these ganglioside mimics, and an equally small proportion of infected individuals raise an immune response against these shared epitopes when infected; hence the overall risk of developing GBS following Campylobacter is low. What determines the host factors that lead to this break in immunological tolerance when infected with a Campylobacter species bearing a ganglioside-mimicking LOS remains unknown. However, the biosynthetic machinery by which Campylobacter species synthesise ganglioside-mimicking glycans has been studied in detail. Evidence indicates that variability in the activity of the *C. jejuni* sialyltransferase enzyme (cstII) between bacterial strains due to a coding sequence polymorphism directs the synthesis of particular glycans and hence dictates the antibody activity and neuropathy subtype (Yuki 2007). Thus, cstII with α2,3-activity alone results in LOS with GM1 and GD1a like mimics, inducing the acute motor axonal neuropathy (AMAN) variant of GBS, whereas bifunctional α2,3- and α2,8-cstII activity generates disialosyl mimics and leads to Miller Fisher syndrome (MFS) (Yuki 2007). Other microorganisms that trigger GBS, including Haemophilus influenza and Mycoplasma pneumonia, also bear ganglioside and glycolipid mimics, indicating that this mechanism might be a more general mechanism of the immunological induction pathway for GBS, rather than being restricted to Campylobacter infection. In this regard, analysis of a strain of *Haemophilus influenza* from a patient with MFS also revealed a homologous bifunctional sialyltransferase (Lic3B) producing a disialoysl group linked to the terminal galactose of the LOS (Houliston et al. 2007). One mechanism by which cross reactivity can occur between gangliosides which share such groups, such as bacterial GD1c and host GQ1b has been demonstrated by epitope mapping using saturation transfer difference NMR spectroscopy. For one mAb, only a tightly confined area on the terminal sialic acid, common to the above molecules was required for binding, indicating that molecular identity between discretely specific regions with the bacterial core OS and ganglioside may be sufficient to induce antibody cross-reactivity (Houliston et al. 2009).

25.6 The Pathological Effects of Anti-glycolipid Antibodies in Man and Experimental Neuropathy Models

The demonstration that anti-glycolipid antibodies directly contribute to neuropathy pathogenesis has been difficult to demonstrate but nevertheless is an important step in understanding these disorders. Clearly, detailed pathogenesis studies requiring nerve tissue are impractical to study in man and as such proof of causality, along with immunopathological features of anti-glycolipid antibody-mediated neuropathy have been mainly studied in experimental animals. An important prerequisite for establishing a direct pathological role for an autoantibody is that it is capable of binding its ligand in physiologically and anatomically intact tissue. Gangliosides are primarily localised to raft domains of the extracellular leaflet of plasma membranes, especially at synapses, where they are available for anti-ganglioside antibody binding. Anti-ganglioside antibodies would be expected to bind any ganglioside-containing glial, neuronal or axonal membranes provided that they can gain access through the blood–nerve barrier, and that binding is not subject to steric inhibition locally within the membrane (Greenshields et al. 2009; Kusunoki et al. 1999). Additionally, since gangliosides traffic to and from the plasma membrane through endosomal sorting pathways, any bound antibody has to remain on the membrane surface for sufficient time to activate the local pro-inflammatory pathway in order to exert pathological effects through inflammatory cell recruitment or complement activation (Fewou et al. 2012). Potentially important axonal and glial sites of injury exist, including proximal and distal peripheral nerve membranes where the blood–nerve barrier is relatively deficient or absent. The axolemmal membrane is encased by myelin in the internodes, but at the nodes of Ranvier and motor nerve terminals it is exposed and available for antibody binding (McGonigal et al. 2010). In the dorsal root ganglion, the plasma membrane of the neuronal cell body is also accessible to autoantibodies. Paranodal, non-compacted myelin membranes, as well as abaxonal Schwann cell plasma membranes will be similarly exposed. One important explanation for the lack of CNS involvement in anti-glycolipid antibody associated neuropathies, despite the wide distribution of gangliosides in the CNS, would thus be the prevention from significant antibody entry provided by the blood–brain barrier.

Having determined that a target membrane is accessible, it is then necessary to establish whether the density of the glycolipid within the membrane is sufficient to mediate a pathological injury upon antibody binding (Goodfellow et al. 2005). Biochemical and immunohistological approaches to establishing this have merits and limitations. Whereas biochemical analysis can usefully show the overall glycolipid composition of different nerves, it provides limited information about microanatomical distribution that could be crucial to disease pathogenesis. Immunohistology or other in situ ligand binding studies (e.g. using ganglioside binding bacterial toxins such as cholera toxin B subunit, a high affinity ligand for GM1) can reveal the much-needed fine structural detail about ganglioside distribution at the cellular and sub-cellular level. For these studies, high quality reagents such as affinity purified antisera or monoclonal antibodies are essential. However, many anti-ganglioside antibodies may cross-react with structurally similar gangliosides and other glycoconjugate antigens, making extrapolation of results to ganglioside localization difficult.

Active and passive immunisation have both been widely used to demonstrate causality and investigate effects of anti-glycolipid antibodies in animal models. In active immunisation, the antigen is administered by injection in an immunologically active form such that the recipient generates an antibody response to the antigen, which then proceeds to damage the target tissue. In passive immunisation, the recipient animal is simply injected with the relevant antibody directed towards the particular target antigen under study.

The first studies looking at pathological effects of anti-glycolipid antibodies were conducted over 30 years ago when an animal model of demyelinating neuropathy was induced by active immunisation of rabbits with GalC (Saida et al. 1979). Rabbits developed flaccid weakness, sensory loss and respiratory paralysis. At autopsy animals had prominent demyelinating lesions in the spinal roots with splitting and vesiculation of the myelin sheaths. Lesions were infiltrated with macrophages which phagocytozed myelin. The distribution of demyelinating lesions corresponded with areas known to have a defective blood–nerve barrier. Subsequently, intraneural injection of rabbit anti-galactocerebroside serum produced focal demyelinating lesions in rat sciatic nerves with similar pathology. Local application of anti-galactocerebroside antibody to rat ventral roots was also shown to induce acute conduction block, and activity was lost after complement inactivation, suggesting that the pathological activity of anti-galactocerebroside serum is complement-mediated (Sumner 1982). Since these seminal studies were conducted, this overall approach combining both active and passive immunisation has been applied to a wide range of other anti-glycolipid antibodies.

Local microinjection of anti-MAG IgM antibodies into feline nerves has been shown to cause focal demyelination at the injection site, also characterised by macrophage mediated myelin stripping (Hays et al. 1987). Again, these effects only occurred when serum was added fresh, or supplemented with a fresh source of complement, suggesting that antibodies caused demyelination by complement fixation, as is the case with galactocerebroside. In a series of interesting passive immunisation experiments in which purified anti-MAG IgM was transfused into young chickens, demyelination of nerve with prominent widening of myelin lamellae was observed, closely resembling the "widely spaced myelin" characteristic of human biopsies (Tatum 1993). These abnormalities in chickens were obtained in the absence of an external source of complement, suggesting that multiple pathomechanisms may be at play. Active immunisation experiments in rabbits have also been performed with SGPG, resulting in an immune response with anti-SGPG antibody induction, hindlimb weakness and slowed nerve conduction velocity in the sciatic nerve (Yu et al. 1990). Studies have thus demonstrated that myelin membranes can be readily injured by different categories of anti-glycolipid antibodies.

An interesting model of sensory neuropathy that correlates with the human disease counterpart was induced in rabbits by active immunisation with GD1b ganglioside (Kusunoki et al. 1996). IgM paraproteins and IgG antibodies that bind to GD1b and related disialylated gangliosides are strongly associated with the development of sensory ataxic neuropathy. Since immunohistochemical studies showed that GD1b is localised to large DRG neurons in human and rabbit, the rabbit presented a good candidate for an animal model of dorsal root ganglionopathy. About 50 % of rabbits immunised with GD1b developed ataxia with splayed limbs, but retained normal muscle power. IgG and IgM antibody titres specific for GD1b were more prominently raised in affected than unaffected rabbits, indicating the importance of the fine specificity of the immune response for GD1b in particular. Pathological examination of affected rabbits showed axonal degeneration in the dorsal roots and some neuronal cell bodies in the DRG had either degenerated or disappeared. The ventral root was entirely spared. Lymphocytic infiltration was not observed in the affected regions, indicating an entirely antibody-mediated injury had taken place. This was further supported by demonstrating that degeneration of rabbit sensory neurons could be induced by passive transfer of anti-GD1b antiserum. The loss of primary sensory neurons that mediate proprioceptive sensation prompted this group to investigate the expression of trkC in DRG in their rabbit model, as these neurons are dependent upon neurotrophin-3-mediated trkC signalling (Hitoshi et al. 1999). TrkC was reduced in the DRG of diseased rabbits suggesting that the anti-GD1b antibody induced downregulation of trkC expression. This may be one of the pathogenic mechanisms by which anti-GD1b antibodies mediate both human and experimental sensory ataxic neuropathy.

Motor axons in peripheral nerve are the major target for anti-GM1 and -GD1a ganglioside antibodies in AMAN, and it would therefore be expected that a similar model could be generated in experimental animals. In human pathological studies on spinal roots and nerves from AMAN autopsy cases, motor axonal degeneration is seen in the ventral spinal roots, accompanied by Wallerian degeneration. Some motor fibres have lengthening of the node of Ranvier in the absence of florid demyelination, accompanied by condensation of the axonal cytoplasm. Infiltrating macrophages with extensive processes invade the periaxonal space abutting the nodal and internodal axolemma and displace the adaxonal Schwann cell membrane and myelin sheath (Hafer-Macko et al. 1996a). By immunohistochemistry, intense IgG and complement deposits bound to the nodal and internodal axolemma in the periaxonal space are seen. In addition to AMAN pathology in the ventral roots, it has also been demonstrated that very distal motor axonal degeneration might take place in AMAN, and that this could also be an explanation for the rapid recovery that sometimes occurs (Ho et al. 1997). Using motor point biopsy from an AMAN case, denervated neuromuscular junctions and reduced intramuscular nerve fibre densities were seen, indicating that if confined to this site, these changes could be rapidly reversible through axonal regeneration over relatively short distances. Based on these human data a variety of animal models of AMAN have been generated.

In a rabbit model of AMAN induced by active immunisation with gangliosides and resulting in induction of anti-GM1 antibodies, animals become severely paralysed (Yuki et al. 2001). Pathological features indicative of motor axonal degeneration with nodal injury in the spinal roots are present, including lengthening of nodes in ventral root with complement deposition. A progressive disruption of the nodal architecture was observed, with the clusters of voltage gated sodium channels being dispersed and cell adhesion molecules critical to the axoglial junction being disrupted as the inflammatory lesion extended towards the juxtaparanodal region (Susuki et al. 2003). In murine studies, AMAN-associated antisera and anti-GM1 and -GD1a monoclonal antibodies derived from both humans and mice immunised with *C. jejuni* LOS have been used to induce disease (Goodfellow et al. 2005; McGonigal et al. 2010). Axons within the motor nerve terminal and distal motor nerve were severely affected, and antibody deposition and complement fixation were florid. Deposition of terminal complement pores allowed uncontrolled calcium ingress triggering a sequence of destructive events, including calpain activation and disruption of nodal and nerve terminal architecture, with concomitant paralysis (O'Hanlon et al. 2003). One crucial experimental modification used in mice was to study animals expressing abnormally large amounts of GD1a, made possible through the targeted deletion of the biosynthetic enzyme GD3-synthase (Fig. 25.3). This resulted in an absence of b series gangliosides and a compensatory increase in the levels of a series gangliosides, including GM1 and GD1a. These mice developed a highly destructive and paralytic motor nerve terminal lesion, which from a functional perspective disguised more proximal injury to the node of Ranvier. Ingress of calcium through complement pores results in calpain activation as a major pathological event. Inhibition of calpain has some partially protective effects. However, an inhibitor of complement component C5 (that results in an inability to form the terminal membrane attack complex [MAC]) had a major neuroprotective effect, completely preventing any structural and functional changes at the node of Ranvier and the motor nerve terminal (Halstead et al. 2008). Thus a major pathway of acute pathological injury mediated by anti-glycolipid antibodies in GBS and variants converges at MAC.

Finally, a wide range of experimental studies have been conducted on the effects of anti-GQ1b antibodies that are found in MFS. At the pre-synaptic neuromuscular junction (NMJ), complex gangliosides including GQ1b are known to act as receptors for botulinum toxins (Bullens et al. 2002). It therefore seems logical that anti-GQ1b antibodies might also bind at the NMJ where they could induce nerve terminal injury. A wide variety of toxin, lectin and antibody based immunohistological studies have shown that the pre-synaptic NMJ is rich in gangliosides (Martin 2003). Using in vitro mouse hemidiaphragm preparations, it was first demonstrated that anti-GQ1b antibodies associated with MFS bind the motor nerve terminal where they locally activate complement (Plomp and Willison 2009). Antibody binding to gangliosides in pre-synaptic membranes leads to (a) local complement fixation and MAC pore formation, (b) uncontrolled calcium influx into the nerve terminal through MAC pores, thereby bypassing voltage gated calcium channels, that provokes, (c) spontaneous exocytosis followed by conduction block, (d) calpain-mediated structural degradation of the terminal axonal cytoskeleton assessed by measuring neurofilament (NF) levels and (e) calcium-mediated mitochondrial injury, with (f) resultant paralysis. MAC formation is essential, since C6 deficient conditions abolish the effect. The term "acute synaptic necrosis" has been used to

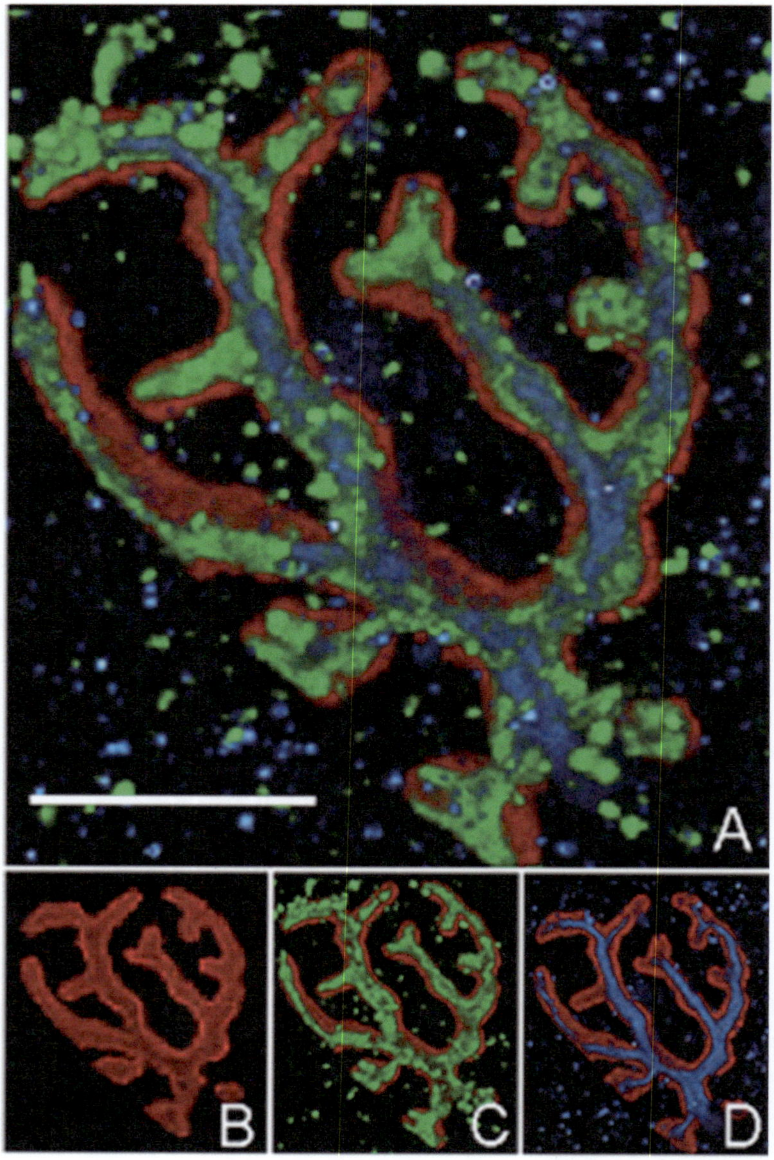

Fig. 25.3 Anti-GD1a antibody-induced autoimmune nerve terminal injury. Immunofluorescent localisation of antibody deposits at GD3s$^{-/-}$ NMJs following exposure of living triangularis sterni muscles ex vivo to an anti-GD1a antibody raised in GalNAcT$^{-/-}$ mice immunised with *C. jejuni* LOS. Reconstructed confocal images (3 colour composite image in *panel a*) show the localisation of post-synaptic AChRs (Texas red-BTx staining, *red*, *panel b*), anti-GD1a antibody (*green*, *panel c*) and neurofilament in the terminal axon arborisations (anti-NF antibody, *blue*, *panel d*). Anti-GD1a antibody is localised directly over the over end plate gutters, ensheathing the terminal axon. Wild-type mice show faint anti-GD1a antibody staining in the same distribution and no staining is seen in the GD1a deficient GalNAcT$^{-/-}$ mice. Scale bar = 10 μm. Modified from (Goodfellow et al. 2005)

describe these events in distinction from other forms of synapse elimination including synaptosis (Gillingwater and Ribchester 2003). In addition to inducing injury to the axon terminal, anti-ganglioside antibodies with different specificities for GQ1b, GT1a and GD3 also target perisynaptic Schwann cells (pSCs) for immune attack (Halstead et al. 2005). An interesting study that has widespread implications indicates that antibody uptake at the motor nerve terminal through endocytosis might limit the extent of injury through masking effects (Fig. 25.4). The extent to which this regulates antibody effects at other sites is unknown.

Clinical clues suggest that the concentration of particular gangliosides in specific neuronal membranes might influence the extent to which antibodies are able to induce injury. The most obvious example of this is MFS, in which anti-GQ1b antibodies appear to target the oculomotor and bulbar nerves because GQ1b and GT1a are enriched in those sites (Chiba et al. 1997). These findings support the view that the regional distribution of gangliosides in different areas of the peripheral nervous system might account for the localisation of the clinical features and pathology, even though in the mouse nerve membranes in other sites, including the diaphragm as described above, may be affected.

Overall the above animal studies, supported by human autopsy and biopsy studies, clearly indicate that anti-ganglioside antibodies have major pathological effects in animal models and point towards some of the pathological pathways that might be involved, and amenable to therapeutic intervention. A major pathological driver appears to be at least in part mediated by complement activation and the recent development of novel complement therapeutics that prevent MAC formation thus represent a clear direction for future trials in this clinical area (Willison et al. 2008).

25.7 The Emerging Role of Glycolipid Complexes as Antigens in Inflammatory Neuropathy

There is a long-standing body of evidence that *cis* interactions (lateral interactions in the plane of the plasma membrane) between heterogeneous membrane glycolipids can substantially enhance or inhibit their ability to bind to antibodies and other lectins (Regina and Hakomori 2008; Rinaldi et al. 2010). The biological significance of these interactions in living nerve membranes is of major interest to the anti-glycolipid antibody and neuropathy field (Fig. 25.5). The presence of "cryptic" glycolipid antigens in plasma membranes and the importance of accessory lipids and glycolipids in modulating antibody binding is long recognised, especially when studying antibody binding in immunohistological analyses (Schwarz and Futerman 1996). One study that very clearly illustrated this point demonstrated that an anti-GM3 monoclonal antibody that bound well to melanoma cell lines expressing GM3 alone was no longer able to bind when GM2, GD2 or GD3 gangliosides were co-expressed with GM3, even when the latter formed 50 % of the total ganglioside composition of the cells (Lloyd et al. 1992). This indicates that the antibody-binding GM3 epitope is capable of being masked by adjacent gangliosides. A more recent

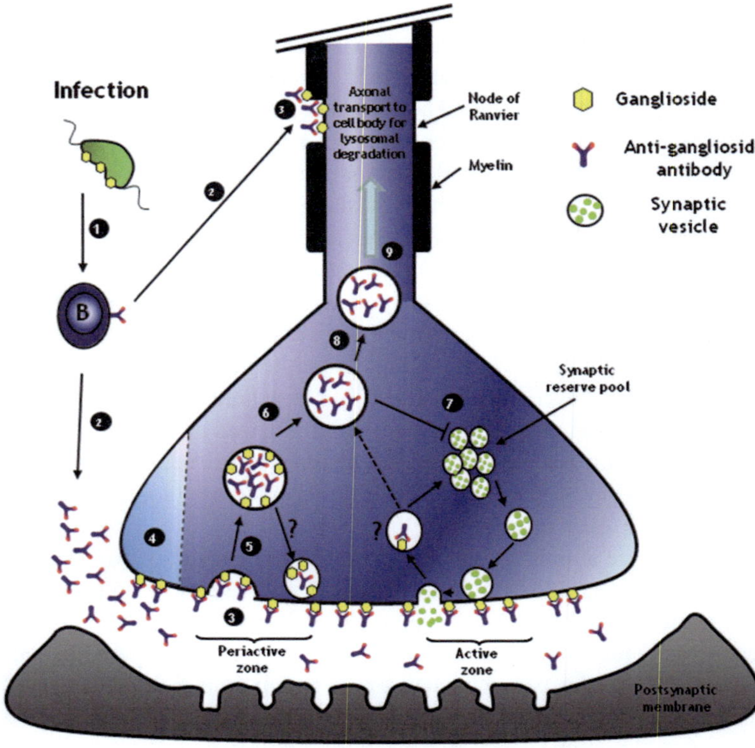

Fig. 25.4 An antibody uptake mechanism that underlies the attenuation of pre-synaptic nerve terminal injury in anti-ganglioside antibody-mediated motor axonal neuropathy. (*1*) Following an infection (e.g. Campylobacter enteritis), B lymphocytes are stimulated to produce anti-microbial antibodies which through molecular mimicry also bind to self gangliosides. (*2*) Circulating anti-ganglioside antibodies bind ganglioside-rich neural membranes in accessible sites lying outside the blood–nerve barrier, including the pre-synaptic nerve terminal membrane and nodal axolemma in distal motor nerves. (*3*) At the pre-synaptic membrane, where vesicular uptake pathways are highly active, bound antibodies are internalised by clathrin-mediated endocytosis. The extent to which antibody uptake may also occur during synaptic vesicle recycling is unknown (?). In contrast to the pre-synaptic membrane, at the nodal axolemma antibodies remain on the surface. (*4*) At 4 °C, endocytosis is inhibited and antibody uptake does not take place. Antibody remaining on the plasma membrane is available for complement fixation and subsequent lytic injury, whereas endocytosed antibody becomes cryptic to complement fixation. (*5*) Following uptake, endocytic vesicles are depleted from the clathrin coating and sorted to the early endosome. (*6*) Endocytic vesicles at the early endosome are not targeted to the synaptic recycling pool (*7*), rather are sorted to the retrograde transport pathway (*8*). The extent of local recycling of antibody to the plasma membrane is unknown (?) (*9*) Vesicles containing antibodies are transported to the neuronal cell body and targeted to the lysosome for degradation. Reproduced from (Fewou et al. 2012)

study has shown masking of GM1 by GD1a on the axolemmal membranes of motor axons that is of relevance in anti-GM1 antibody-associated motor neuropathies (Greenshields et al. 2009). It is equally recognised that combinations of oligosaccharide groups from different glycans can form a distinct selectin epitope termed a

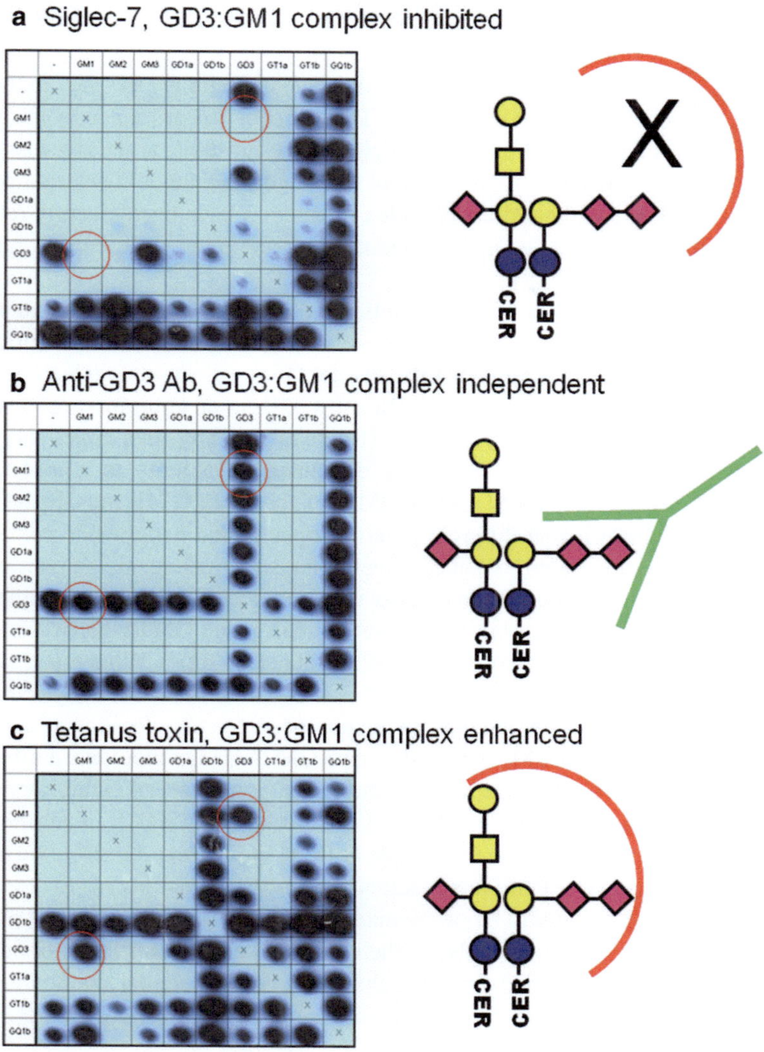

Fig. 25.5 Complex enhanced, complex independent and complex attenuated binding as demonstrated for GD3:GM1 heteromeric complexes. Glycoarrays are printed using a thin layer chromatography autospotter with 9 single gangliosides and all their 1:1 heteromeric combinations (complexes). All complexes are spotted in duplicate symmetrically across the diagonal (x). The *red circles* indicate the GD3:GM1 heteromeric complexes duplicated on each grid. After printing, arrays are overlaid with glycan binding ligands to detect patterns of binding, and then developed using peroxidise-labelled probes and chemiluminescence. The binding pattern of tetanus toxin (TeNT) HC fragment, a monoclonal antibody reactive with the terminal disialosyl epitope common to GD3 and GQ1b, and siglec-7-Fc is markedly different with respect to GD3 series complexes. Binding to other single and complexed gangliosides is also seen, but is not discussed herein. TeNT HC-HRP only binds GD3:GM1, whilst failing to bind to either component ganglioside in isolation ("complex enhanced"), the mAb CGM3 binds all GD3 complexes equally ("complex independent"), whereas the siglec-7-Fc fusion protein binds GD3 in isolation but is completely inhibited form binding in the presence of specific second gangliosides, including GM1, GM2, independent of the presence of GM3 (complex independent) and partially inhibited by GD1a and GT1a (complex attenuated). In addition, TeNT HC is inhibited from binding GQ1b, and to a lesser extent GT1b, by pre-complexing with GM2. Reproduced from (Rinaldi et al. 2009)

"clustered saccharide patch", without detectable binding to the individual oligosaccharides themselves (Varki 1994). This concept was prompted by the observation that the selectins had unusually high affinity for their target glycans which was unexpected for apparently monovalent carbohydrate ligands. These patterns of lectin- and antibody-carbohydrate complex interaction have been recently referred to as complex enhanced, complex attenuated or complex independent binding (Rinaldi and Willison 2008; Varki 1994).

A series of studies on sera from patients with GBS and chronic motor neuropathies has rekindled interest in this area. A strikingly unusual observation provided evidence that serum IgG antibodies from certain GBS patients showed strong reactivity to the heteromeric mixture of GD1a and GD1b in an equimolar ratio, whilst failing to bind to either ganglioside alone (Kaida et al. 2004). In a larger study that comprised screening of GBS sera, they demonstrated activity to complexes of GD1a:GD1b, GM1:GD1a, GM1:GD1b, GM1:GT1b, GD1a:GT1b and GD1b:GT1b, indicating that clustering of different glycoepitopes within ganglioside complexes in the plasma membrane might act as novel ligands for pathogenic autoantibodies (Kaida et al. 2004). This phenomenon also appear to occur with IgM antibodies: for example, anti-GM1 antibodies in MMN preferentially bind GM1 when it is in complex with GalC (Galban-Horcajo et al. 2012; Kaida et al. 2004). What remains to be established is how these findings observed in solid-phase immunoassays correlate with glycolipid behaviour in the plasma membrane. One study looking at the interaction between GM2 and GM3 has used an anti-GM2:GM3 monoclonal antibody to identify the complex on cell surfaces and shown that it mediates an inhibitory effect on cell motility via a CD82/cMe pathway (Todeschini et al. 2008). In autoimmune neuropathy, serum antibodies to the complex of GM1:GD1a exert neurotoxic effects on motor axon terminals. Until larger series of monoclonal antibodies to a range of complexes are isolated, it will be problematic to investigate this in depth at either a biophysical or mechanistic level. Nevertheless it is likely that further antibodies to complexes will be indentified provided that analytical screening methodology is designed with this concept in mind.

25.8 Conclusions

Recent years have witnessed abundant progress in our understanding of the glycobiology and glycoimmunology of inflammatory neuropathies, and in doing so has both informed basic and clinical science. Clinical and serological data clearly shows a disease-specific correlation between peripheral neuropathies and particular anti-glycolipid antibodies. Many interesting patterns have emerged that support the view that anti-glycolipid antibodies play an active role in pathogenesis. Experimental evidence obtained from human and animal studies continues to support the model of post-infectious neuropathy as a disease involving molecular mimicry between bacterial and neural oligosaccharides. One notable emerging area concerns the interacting relationship between glycolipids and other lipids in the plane of the plasma membrane and how this might either positively or negatively influence auto-antibody binding.

Conflict of Interest The author declares no conflict of interest.

References

Bullens RW, O'Hanlon GM, Wagner E, Molenaar PC, Furukawa K, Furukawa K, Plomp JJ, Willison HJ. Complex gangliosides at the neuromuscular junction are membrane receptors for autoantibodies and botulinum neurotoxin but redundant for normal synaptic function. J Neurosci. 2002;22:6876–84.

Burger D, Simon M, Peruisseau G, Steck AJ. The epitope(s) recognised by HNK-1 antibody and IgM paraprotein in neuropathy is present on several N-linked oligosaccharide structures on human P0 and myelin-associated glycoprotein. J Neurochem. 1990;54:1569–75.

Chiba A, Kusunoki S, Obata H, Machinami R, Kanazawa I. Serum anti-GQ1b IgG antibody is associated with ophthalmoplegia in Miller Fisher syndrome and Guillain-Barre syndrome: clinical and immunohistochemical studies. Neurology. 1993;43:1911–7.

Chiba A, Kusunoki S, Obata H, Machinami R, Kanazawa I. Ganglioside composition of the human cranial nerves, with special reference to pathophysiology of Miller Fisher syndrome. Brain Res. 1997;745:32–6.

Defrance T, Taillardet M, Genestier L. T cell-independent B cell memory. Curr Opin Immunol. 2011;23:330–6.

Fewou SN, Rupp A, Nickolay LE, Carrick K, Greenshields KN, Pediani J, Plomp JJ, Willison HJ. Anti-ganglioside antibody internalization attenuates motor nerve terminal injury in a mouse model of acute motor axonal neuropathy. J Clin Invest. 2012;122:1037–51.

Galban-Horcajo F, Fitzpatrick AM, Hutton AJ, Dunn SM, Kalna G, Brennan KM, Rinaldi S, Yu RK, Goodyear CS, Willison HJ. Antibodies to heteromeric glycolipid complexes in multifocal motor neuropathy. Eur J Neurol. 2012;20:62–70.

Gillingwater TH, Ribchester RR. The relationship of neuromuscular synapse elimination to synaptic degeneration and pathology: insights from WldS and other mutant mice. J Neurocytol. 2003;32:863–81.

Goodfellow JA, Bowes T, Sheikh K, Odaka M, Halstead SK, Humphreys PD, Wagner ER, Yuki N, Furukawa K, Furukawa K, Plomp JJ, Willison HJ. Overexpression of GD1a ganglioside sensitizes motor nerve terminals to anti-GD1a antibody-mediated injury in a model of acute motor axonal neuropathy. J Neurosci. 2005;25:1620–8.

Greenshields KN, Halstead SK, Zitman FM, Rinaldi S, Brennan KM, O'Leary C, Chamberlain LH, Easton A, Roxburgh J, Pediani J, Furukawa K, Furukawa K, Goodyear CS, Plomp JJ, Willison HJ. The neuropathic potential of anti-GM1 autoantibodies is regulated by the local glycolipid environment in mice. J Clin Invest. 2009;119:595–610.

Gu Y, Chen ZW, Siegel A, Koshy R, Ramirez C, Raabe TD, Devries GH, Ilyas AA. Analysis of humoral immune responses to LM1 ganglioside in guinea pigs. J Neuroimmunol. 2012; 246:58–64.

Hafer-Macko C, Hsieh ST, Li CY, Ho TW, Sheikh K, Cornblath DR, McKhann GM, Asbury AK, Griffin JW. Acute motor axonal neuropathy: an antibody-mediated attack on axolemma. Ann Neurol. 1996a;40:635–44.

Hafer-Macko CE, Sheikh KA, Li CY, Ho TW, Cornblath DR, McKhann GM, Asbury AK, Griffin JW. Immune attack on the Schwann cell surface in acute inflammatory demyelinating polyneuropathy. Ann Neurol. 1996b;39:625–35.

Halstead SK, Morrison I, O'Hanlon GM, Humphreys PD, Goodfellow JA, Plomp JJ, Willison HJ. Anti-disialosyl antibodies mediate selective neuronal or Schwann cell injury at mouse neuromuscular junctions. Glia. 2005;52:177–89.

Halstead SK, Zitman FM, Humphreys PD, Greenshields K, Verschuuren JJ, Jacobs BC, Rother RP, Plomp JJ, Willison HJ. Eculizumab prevents anti-ganglioside antibody-mediated neuropathy in a murine model. Brain. 2008;131:1197–208.

Hays AP, Latov N, Takatsu M, Sherman WH. Experimental demyelination of nerve induced by serum of patients with neuropathy and an anti MAG IgM M protein. Neurology. 1987;37: 242–56.

Hitoshi S, Kusunoki S, Murayama S, Tsuji S, Kanazawa I. Rabbit experimental sensory ataxic neuropathy: anti-GD1b antibody-mediated trkC downregulation of dorsal root ganglia neurons. Neurosci Lett. 1999;260:157–60.

Ho TW, Hsieh ST, Nachamkin I, Willison HJ, Sheikh K, Kiehlbauch J, Flanigan K, McArthur JC, Cornblath DR, McKhann GM, Griffin JW. Motor nerve terminal degeneration provides a potential mechanism for rapid recovery in acute motor axonal neuropathy after Campylobacter infection. Neurology. 1997;48:717–24.

Honke K. Biosynthesis and biological function of sulfoglycolipids. Proc Jpn Acad Ser B Phys Biol Sci. 2013;89:129–38.

Houliston RS, Jacobs BC, Tio-Gillen AP, Verschuuren JJ, Khieu NH, Gilbert M, Jarrell HC. STD-NMR used to elucidate the fine binding specificity of pathogenic anti-ganglioside antibodies directly in patient serum. Biochemistry. 2009;48:220–2.

Houliston RS, Koga M, Li J, Jarrell HC, Richards JC, Vitiazeva V, Schweda EK, Yuki N, Gilbert M. A Haemophilus influenzae strain associated with Fisher syndrome expresses a novel disialylated ganglioside mimic. Biochemistry. 2007;46:8164–71.

Hughes RA, Cornblath DR. Guillain-Barre syndrome. Lancet. 2005;366:1653–66.

Ilyas AA, Dalakas MC, Brady RO, Quarles RH. Sulfated glucuronyl glycolipids reacting with anti-myelin-associated glycoprotein monoclonal antibodies including IgM paraproteins in neuropathy: species distribution and partial characterization of epitopes. Brain Res. 1986;385:1–9.

Inoue M, Fujii Y, Furukawa K, Okada M, Okumura K, Hayakawa T, Furukawa K, Sugiura Y. Refractory skin injury in complex knock-out mice expressing only the GM3 ganglioside. J Biol Chem. 2002;277:29881–8.

Kaida K, Morita D, Kanzaki M, Kamakura K, Motoyoshi K, Hirakawa M, Kusunoki S. Ganglioside complexes as new target antigens in Guillain-Barre syndrome. Ann Neurol. 2004;56:567–71.

Kanda T. Biology of the blood-nerve barrier and its alteration in immune mediated neuropathies. J Neurol Neurosurg Psychiatry. 2013;84:208–12.

Kieseier BC, Hartung HP, Wiendl H. Immune circuitry in the peripheral nervous system. Curr Opin Neurol. 2006;19:437–45.

Kornberg AJ, Pestronk A. Chronic motor neuropathies: diagnosis, therapy, and pathogenesis. Ann Neurol. 1995;37:S43–50.

Kusunoki S. Anti-ganglioside antibodies in Guillain-Barre syndrome; useful diagnostic markers as well as possible pathogenetic factors. Intern Med. 2003;42:457–8.

Kusunoki S, Hitoshi S, Kaida K, Arita M, Kanazawa I. Monospecific anti-GD1b IgG is required to induce rabbit ataxic neuropathy. Ann Neurol. 1999;45:400–3.

Kusunoki S, Shimizu J, Chiba R, Ugawa Y, Hitoshi S, Kanazawa I. Experimental sensory neuropathy induced by sensitisation with ganglioside GD1b. Ann Neurol. 1996;39:324–31.

Ledeen RW, Yu RK. Gangliosides: structure, isolation, and analysis. Methods Enzymol. 1982;83:139–91.

Lloyd KO, Gordon CM, Thampoe IJ, DiBenedetto C. Cell surface accessibility of individual gangliosides in malignant melanoma cells to antibodies is influenced by the total ganglioside composition of the cells. Cancer Res. 1992;52:4948–53.

Lopate G, Parks BJ, Goldstein JM, Yee WC, Friesenhahn GM, Pestronk A. Polyneuropathies associated with high titre antisulphatide antibodies: characteristics of patients with and without serum monoclonal proteins. J Neurol Neurosurg Psychiatry. 1997;62:581–5.

Martin PT. Glycobiology of the neuromuscular junction. J Neurocytol. 2003;32:915–29.

McGonigal R, Rowan EG, Greenshields KN, Halstead SK, Humphreys PD, Rother RP, Furukawa K, Willison HJ. Anti-GD1a antibodies activate complement and calpain to injure distal motor nodes of Ranvier in mice. Brain. 2010;133:1944–60.

Meyer Zu HG, Hu W, Hartung HP, Lehmann HC, Kieseier BC. The immunocompetence of Schwann cells. Muscle Nerve. 2008;37:3–13.

Mori L, De LG. T cells specific for lipid antigens. Immunol Res. 2012;53:191–9.

Nobile-Orazio E, Manfredini E, Carpo M, Meucci N, Monaco S, Ferrari S, Bonetti B, Cavaletti G, Gemignani F, Durelli L, Barbieri S, Allaria S, Sgarzi M, Scarlato G. Frequency and clinical correlates of anti-neural IgM antibodies in neuropathy associated with IgM monoclonal gammopathy. Ann Neurol. 1994;36:416–24.

O'Hanlon GM, Humphreys PD, Goldman RS, Halstead SK, Bullens RW, Plomp JJ, Ushkaryov Y, Willison HJ. Calpain inhibitors protect against axonal degeneration in a model of anti-ganglioside antibody-mediated motor nerve terminal injury. Brain. 2003;126:2497–509.

Ogawa-Goto K, Funamoto N, Ohta Y, Abe T, Nagashima K. Myelin gangliosides of human peripheral nervous system: an enrichment of GM1 in the motor nerve myelin isolated from cauda equina. J Neurochem. 1992;59:1844–9.

Plomp JJ, Willison HJ. Pathophysiological actions of neuropathy-related anti-ganglioside antibodies at the neuromuscular junction. J Physiol. 2009;587:3979–99.

Popko B, Dupree JL, Coetzee T, Suzuki K, Suzuki K. Genetic analysis of myelin galactolipid function. Adv Exp Med Biol. 1999;468:237–44.

Regina TA, Hakomori SI. Functional role of glycosphingolipids and gangliosides in control of cell adhesion, motility, and growth, through glycosynaptic microdomains. Biochim Biophys Acta 2008;1780:421–33.

Rinaldi S, Brennan KM, Goodyear CS, O'Leary C, Schiavo G, Crocker PR, Willison HJ. Analysis of lectin binding to glycolipid complexes using combinatorial glycoarrays. Glycobiology. 2009;19:789–96.

Rinaldi S, Brennan KM, Willison HJ. Heteromeric glycolipid complexes as modulators of autoantibody and lectin binding. Prog Lipid Res. 2010;49:87–95.

Rinaldi S, Willison HJ. Ganglioside antibodies and neuropathies. Curr Opin Neurol. 2008;21:540–6.

Saida T, Saida K, Dorfman SH, Silberberg DH, Sumner AJ, Manning MC, Lisak RP, Brown MJ. Experimental allergic neuritis induced by sensitization with galactocerebroside. Science. 1979;204:1103–6.

Salzer JL, Brophy PJ, Peles E. Molecular domains of myelinated axons in the peripheral nervous system. Glia. 2008;56:1532–40.

Sandhoff K, Kolter T. Biosynthesis and degradation of mammalian glycosphingolipids. Philos Trans R Soc Lond B Biol Sci. 2003;358:847–61.

Scherer SS. Molecular genetics of demyelination: new wrinkles on an old membrane. Neuron. 1997;18:13–6.

Schwarz A, Futerman AH. The localization of gangliosides in neurons of the central nervous system: the use of anti-ganglioside antibodies. Biochim Biophys Acta. 1996;1286:247–67.

Sumner A. Electrophysiological and morphological effects of the injection of Guillain-Barré sera in the sciatic nerve of the rat. Rev Neurol (Paris). 1982;138:17–24.

Susuki K, Nishimoto Y, Yamada M, Baba M, Ueda S, Hirata K, Yuki N. Acute motor axonal neuropathy rabbit model: immune attack on nerve root axons. Ann Neurol. 2003;54:383–8.

Svennerholm L, Bostrom K, Fredman P, Jungbjer B, Lekman A, Mansson JE, Rynmark BM. Gangliosides and allied glycosphingolipids in human peripheral nerve and spinal cord. Biochim Biophys Acta. 1994;1214:115–23.

Tatum AH. Experimental paraprotein neuropathy, demyelination by passive transfer of human IgM anti-myelin-associated glycoprotein. Ann Neurol. 1993;33:502–6.

Todeschini AR, Dos Santos JN, Handa K, Hakomori SI. Ganglioside GM2/GM3 complex affixed on silica nanospheres strongly inhibits cell motility through CD82/cMet-mediated pathway. Proc Natl Acad Sci U S A. 2008;105:1925–30.

Varki A. Selectin ligands. Proc Natl Acad Sci USA. 1994;91:7390–7.

Willison HJ, Goodyear CS. Glycolipid antigens and autoantibodies in autoimmune neuropathies. Trends Immunol. 2013;34:453–9.

Willison HJ, Halstead SK, Beveridge E, Zitman FM, Greenshields KN, Morgan BP, Plomp JJ. The role of complement and complement regulators in mediating motor nerve terminal injury in murine models of Guillain-Barre syndrome. J Neuroimmunol. 2008;201–202:172–82.

Willison HJ, O'Leary CP, Veitch J, Blumhardt LD, Busby M, Donaghy M, Fuhr P, Ford H, Hahn A, Renaud S, Katifi HA, Ponsford S, Reuber M, Steck A, Sutton I, Schady W, Thomas PK, Thompson AJ, Vallat JM, Winer J. The clinical and laboratory features of chronic sensory ataxic neuropathy with anti-disialosyl IgM antibodies. Brain. 2001;124:1968–77.

Willison HJ, Yuki N. Peripheral neuropathies and anti-glycolipid antibodies. Brain. 2002;125: 2591–625.

Yinchang Y, Crawford TO, Griffin JW, Tu P, Lee VMY, Li C, Roder J, Trapp BD. Myelin-associated glycoprotein is a myelin signal that modulates the caliber of myelinated axons. J Neurosci. 1998;18:1953–62.

Yu RK, Ariga T, Kohriyama T, Kusunoki S, Maeda Y, Miyatani N. Autoimmune mechanisms in peripheral neuropathies. Ann Neurol. 1990;27:S30–5.

Yuki N. Infectious origins of, and molecular mimicry in, Guillain-Barre and Fisher syndromes. Lancet Infect Dis. 2001;1:29–37.

Yuki N. Campylobacter sialyltransferase gene polymorphism directs clinical features of Guillain-Barre syndrome. J Neurochem. 2007;103:S150–8.

Yuki N, Yamada M, Koga M, Odaka M, Susuki K, Tagawa Y, Ueda S, Kasama T, Ohnishi A, Hayashi S, Takahashi H, Kamijo M, Hirata K. Animal model of axonal Guillain-Barre syndrome induced by sensitization with GM1 ganglioside. Ann Neurol. 2001;49:712–20.

Index

R.K. Yu and C.-L. Schengrund (eds.), *Glycobiology of the Nervous System*, Advances in Neurobiology 9, DOI 10.1007/978-1-4939-1154-7,

P

Q

R

S

T

MIX
Papier aus verantwortungsvollen Quellen
Paper from responsible sources
FSC® C105338

If you have any concerns about our products,
you can contact us on
ProductSafety@springernature.com

In case Publisher is established outside the EU,
the EU authorized representative is:
Springer Nature Customer Service Center GmbH
Europaplatz 3, 69115 Heidelberg, Germany

Printed by Libri Plureos GmbH
in Hamburg, Germany